AF789925

Models for Planning Wildlife Conservation in Large Landscapes

Models for Planning Wildlife Conservation in Large Landscapes

Joshua J. Millspaugh
Department of Fisheries and Wildlife Sciences,
University of Missouri, Columbia, Missouri

Frank R. Thompson, III
U.S. Forest Service Northern Research Station,
Columbia, Missouri

ELSEVIER

AMSTERDAM • BOSTON • HEIDELBERG • LONDON
NEW YORK • OXFORD • PARIS • SAN DIEGO
SAN FRANCISCO • SINGAPORE • SYDNEY • TOKYO
Academic Press is an imprint of Elsevier

Academic Press is an imprint of Elsevier
30 Corporate Drive, Suite 400, Burlington, MA 01803, USA
525 B Street, Suite 1900, San Diego, California 92101-4495, USA
84 Theobald's Road, London WC1X 8RR, UK

This book is printed on acid-free paper. ∞

Copyright © 2009, Elsevier Inc. All rights reserved.

Cover photo credit for Greater sage-grouse: Gary Kramer.
Cover photo credit for sagebrush landscape: USDI Bureau of Land Management.

No part of this publication may be reproduced or transmitted in any form or by any means, electronic or mechanical, including photocopy, recording, or any information storage and retrieval system, without permission in writing from the publisher.

Permissions may be sought directly from Elsevier's Science & Technology Rights Department in Oxford, UK: phone: (+44) 1865 843830, fax: (+44) 1865 853333, E-mail: permissions@elsevier.com. You may also complete your request on-line via the Elsevier homepage (*http://elsevier.com*), by selecting "Support & Contact" then "Copyright and Permission" and then "Obtaining Permissions."

Library of Congress Cataloging-in-Publication Data
Millspaugh, Joshua J.
Models for planning wildlife conservation in large landscapes / Joshua Millspaugh, Frank R. Thompson.
p. cm.
Includes bibliographical references and index.
ISBN 978-0-12-373631-4 (hard cover : alk. paper) 1. Wildlife conservation–Simulation methods. 2. Habitat conservation–Simulation methods. 3. Landscape ecology–Simulation methods. I. Thompson, Frank R. (Frank Richard) II. Title.
QL82.M55 2009
639.901′1–dc22
2008018400

British Library Cataloguing-in-Publication Data
A catalogue record for this book is available from the British Library.

ISBN: 978-0-12-373631-4

For information on all Academic Press publications visit our Web site at *www.books.elsevier.com*

Printed in the United States of America
09 10 11 12 9 8 7 6 5 4 3 2 1

Working together to grow libraries in developing countries
www.elsevier.com | www.bookaid.org | www.sabre.org
ELSEVIER BOOK AID International Sabre Foundation

Contents

Preface ix
List of Contributors xiii
List of Reviewers xix

CHAPTER 1 General Principles for Developing Landscape Models for Wildlife Conservation 1
Joshua J. Millspaugh, Robert A. Gitzen, David R. Larsen, Michael A. Larson, and Frank R. Thompson, III

CHAPTER 2 Application of Population Viability Analysis to Landscape Conservation Planning 33
Steven R. Beissinger, Emily Nicholson, and Hugh P. Possingham

CHAPTER 3 Multispecies Conservation Planning on U.S. Federal Lands 51
Barry R. Noon, Kevin S. McKelvey, and Brett G. Dickson

CHAPTER 4 Geographic Approaches to Biodiversity Conservation: Implications of Scale and Error to Landscape Planning 85
Curtis H. Flather, Kenneth R. Wilson, and Susan A. Shriner

CHAPTER 5 Social and Economic Considerations for Planning Wildlife Conservation in Large Landscapes 123
Robert G. Haight and Paul H. Gobster

CHAPTER 6 Landscape Considerations for Conservation Planning on Private Lands 153
Jonathan B. Haufler and Brian J. Kernohan

CHAPTER 7 A Multiscale Stepwise Approximation Approach for Wildlife Conservation Assessment and Planning 177
John R. Probst and Eric J. Gustafson

CHAPTER 8 An Emerging Agency-Based Approach to Conserving Populations Through Strategic Habitat Conservation 201
Rex R. Johnson, Charles K. Baxter, and Michael E. Estey

CHAPTER 9 A Review of Methods for Quantifying Wildlife Habitat in Large Landscapes 225
Michael A. Larson, Joshua J. Millspaugh, and Frank R. Thompson, III

CHAPTER 10 Wildlife Habitat-Relationships Models: Description and Evaluation of Existing Frameworks 251
Jeffrey L. Beck and Lowell H. Suring

CHAPTER 11 Lessons Learned from Using GIS to Model Landscape-Level Wildlife Habitat 287
Gary J. Roloff, Michael L. Donovan, Daniel W. Linden, and Marshall L. Strong

CHAPTER 12 A Review of LANDIS and Other Forest Landscape Models for Integration with Wildlife Models 321
Hong S. He

CHAPTER 13 Simulating Landscape Change Using the Landscape Management System 339
Chadwick D. Oliver, James B. McCarter, Jeffrey M. Comnick, Kevin Ceder, and Christopher S. Nelson

CHAPTER 14 Development and Application of Habitat Suitability Models to Large Landscapes 367
William D. Dijak and Chadwick D. Rittenhouse

CHAPTER 15 Modeling Understory Vegetation and Its Response to Fire 391
Donald McKenzie, Crystal L. Raymond, and Samuel A. Cushman

CHAPTER 16 Validation of Landscape-Scale Decision Support Models That Predict Vegetation and Wildlife Dynamics 415
Stephen R. Shifley, Chadwick D. Rittenhouse, and Joshua J. Millspaugh

CHAPTER 17 Methods for Determining Viability of Wildlife Populations in Large Landscapes 449
H. Resit Akçakaya and Barry W. Brook

CHAPTER 18 Dynamic Landscape Metapopulation Models and Sustainable Forest Management 473
Sarah Bekessy, Brendan Wintle, Ascelin Gordon, Ryan Chisholm, Lisa Venier, and Jennie Pearce

CHAPTER 19 Habitat Networks for Terrestrial Wildlife: Concepts and Case Studies 501
Mary M. Rowland and Michael J. Wisdom

CHAPTER 20 Landscape-Level Planning for Conservation of Wetland Birds in the U.S. Prairie Pothole Region 533
Neal D. Niemuth, Ronald E. Reynolds, Diane A. Granfors, Rex R. Johnson, Brian Wangler, and Michael E. Estey

CHAPTER 21 Plum Creek's Central Cascades Habitat Conservation Plan and Modeling for the Northern Spotted Owl 561
Lorin L. Hicks, William O. Vogel, Dale R. Herter, Richard J. Early, and Henning C. Stabins

CHAPTER 22 Application of Models to Conservation Planning for Terrestrial Birds in North America 593
Jane Fitzgerald, Wayne E. Thogmartin, Randy Dettmers, Tim Jones, Christopher Rustay, Janet M. Ruth, Frank R. Thompson, III, and Tom Will

CHAPTER 23 Modeling Bird Responses to Predicted Changes in Land Cover in an Urbanizing Region 625
Jeffrey A. Hepinstall, John M. Marzluff, and Marina Alberti

CHAPTER 24 A Decision Framework for Choosing Models in Large-Scale Wildlife Conservation Planning 661
Frank R. Thompson, III and Joshua J. Millspaugh

Preface

Including wildlife priorities in land management decisions is a common goal for state and federal agencies, nongovernment organizations, and industrial and nonindustrial private landowners. Conservation efforts increasingly occur at larger spatial scales with greater consideration of multispecies strategies, collaboration across agency and ownership boundaries, and with consideration of population performance. These strategies and approaches are driven by increased recognition that successful wildlife conservation and natural resources planning at large scales must consider more than just site-level habitat management. We must consider the use of models that predict vegetation and corresponding wildlife impacts, ecological theory, and social and economic factors.

Although theoretical advancements in landscape ecology have driven us to large-scale conservation activities, several technological advancements have also contributed to our ability to model and study wildlife dynamics at large spatial scales. Spatially explicit information about land use and vegetation composition and structure is now available for landscapes around the world. Advancements in remote sensing and related technologies have increased the resolution and quantity of landscape data, and efforts have been made to increase the availability of landscape data. Advancements have also been made in desktop computers and software to model vegetation dynamics and quantify spatial patterns in large landscapes. Geographic information systems (GIS) allow managers and researchers to study the spatially explicit effects of management decisions and other disturbances on vegetation at large spatial scales and allow for inclusion of complex spatial processes in models of wildlife-habitat relationships. These developments have changed the way wildlife habitat is modeled and how wildlife priorities are considered in land management planning. As we considered the latest ecological theory, analytical techniques, and technological advancements in our own studies, we sensed the need for a book that consolidates the conceptual basis and practical approaches to modeling and conserving wildlife in large landscapes.

This book focuses on practical approaches, concepts, and tools to model and conserve wildlife in large landscapes. The book is intended for conservation and wildlife biologists, managers, resource planners, and students interested in wildlife habitat models and conservation planning. We secured manuscripts from a diverse group of scientists from university, state and federal agencies, industrial landowners, private consulting firms, and nongovernmental organizations. By default, this diversity of personnel also ensured a mix of expertise related to landscape simulation, wildlife population viability, ecological modeling, wildlife management, and natural resources planning. We felt such diversity was important to demonstrate the variety of philosophies, approaches, and strategies being implemented. We also invited international participants to further broaden the perspective of available ideas and methods.

The book is roughly divided into three components. The first third of the book largely addresses critical concepts that should be considered in large-scale conservation activities. In addition to consideration of social and economic issues, chapters address the conceptual basis of multispecies assessment approaches; population viability analysis; reserve design; issues of scale and error in landscape planning; and practical approaches used by federal, state, and private landowners. The second third of the book is primarily methods based. Here, the reader will find reviews of available methods and software for modeling vegetation and wildlife dynamics. Chapters in this section discuss approaches for modeling overstory and understory vegetation, methods to validate vegetation and wildlife models, ideas for integrating vegetation and wildlife population viability, and issues with using spatially explicit data from within large landscapes. Although there is a strong focus on methods, several case studies highlight application of these methods. The last third of the book integrates theory and methods by presenting detailed, practical case studies from a diversity of ecosystems and for a diversity of species. These chapters illustrate approaches to modeling and conserving wildlife, including both game and nongame species, in forested and nonforested environments, and in urban environments and industrial landscapes to broad-scale and comprehensive approaches that cover a diversity of ownerships. However, even the concept chapters contain case studies and the application chapters also contain important theory. For this reason, we have no formal section designators in the book.

As with any edited book, much of the focus and applications relate directly to the editors' research. Thus, given much of our large-scale modeling work has related to bird conservation over the past 10 years, the primary focus of the applications contained within the book use bird examples. Additionally, many large-scale conservation programs have involved birds (e.g., North American Waterfowl Management Plan, the Partners in Flight North American Landbird Conservation Plan, and the U.S. Shorebird Conservation Plan). Concepts and applications involving reserve design, coarse filters, and community or species richness approaches are discussed, but we emphasized multiple-use landscapes and approaches that ultimately step down to modeling species habitat, abundance, or viability because even most large-scale approaches usually eventually address focal or high-concern species. Although we attempted to provide a broad suite of applications, there are admittedly outstanding examples of wildlife conservation at large spatial scales that were not adequately discussed in this book.

We greatly appreciate the willingness of chapter authors to participate in this project. They were remarkably patient, timely, and responsive to our requests. Clearly, peer-review is critical to an edited volume such as this one. We are grateful to those who reviewed papers. A complete "List of Reviewers" is included in the book. Several people graciously reviewed two manuscripts. We thank several agencies and personnel for their support, assistance, and many contributions. The School of Natural Resources, Department of Fisheries and Wildlife Sciences, University of Missouri, and the U.S. Forest Service, Northern Research Station supported production of this book. Our thanks are extended to Jen Reidy for her constructive and meticulous reviews of the literature cited sections. Dick DeGraaf and John Marzluff provided constructive and valuable comments on the production of this book. Bill Dijak produced the GIS image for the back cover. We thank Gary Kramer for the Greater sage-grouse photo and USDI Bureau of Land Management for the sagebrush landscape photo used on the front cover. We gratefully acknowledge our spouses, Rami Woods and Fran Thompson, for their patience during the production of this book (and other work). We thank the U.S. Forest Service; U.S. Fish and Wildlife Service; National Park Service; National Science Foundation; Missouri Department of Conservation; South Dakota Department of Game, Fish and Parks; University of Missouri; and Boise Cascade Corporation for support of research that has been the impetus for our interest in the use of habitat models and in wildlife studies. We are sincerely grateful to all these people and organizations.

Joshua J. Millspaugh

Frank R. Thompson, III

List of Contributors

H. Resit Akçakaya
Department of Ecology and Evolution, Stony Brook University, Stony Brook, New York, U.S.A.

Marina Alberti
College of Architecture and Urban Planning, University of Washington, Seattle, Washington, U.S.A.

Charles K. Baxter
Lower Mississippi Valley Joint Venture, U.S. Fish and Wildlife Service, Vicksburg, Mississippi, U.S.A.

Jeffrey L. Beck
Department of Renewable Resources, University of Wyoming, Laramie, Wyoming, U.S.A.

Steven R. Beissinger
Department of Environmental Science, Policy & Management, University of California, Berkeley, California, U.S.A.

Sarah Bekessy
School of Global Studies, Social Science and Planning, RMIT University, University of Melbourne, Melbourne 3001, Victoria, Australia

Barry W. Brook
Research Institute for Climate Change and Sustainability, University of Adelaide, SA 5005, Australia

Kevin Ceder
College of Forest Resources, University of Washington, Seattle, Washington, U.S.A.

Ryan Chisholm
Department of Ecology and Evolutionary Biology, Princeton University, Princeton, New Jersey, U.S.A.

Jeffrey M. Comnick
Olympic Natural Resources Center, College of Forest Resources, University of Washington, Forks, Washington, U.S.A.

Samuel A. Cushman
Rocky Mountain Research Station, U.S. Forest Service, Missoula, Montana, U.S.A.

Randy Dettmers
U.S. Fish and Wildlife Service, Hadley, Massachusetts, U.S.A.

Brett G. Dickson
Center for Environmental Sciences, Northern Arizona University, Flagstaff, Arizona, U.S.A.

William D. Dijak
Northern Research Station, U.S. Forest Service, Columbia, Missouri, U.S.A.

Michael L. Donovan
Michigan Department of Natural Resources, Wildlife Division, Lansing, Michigan, U.S.A.

Richard J. Early
Plum Creek Timber Company, Columbia Falls, Montana, U.S.A.

Michael E. Estey
Habitat and Population Evaluation Team, U.S. Fish and Wildlife Service, Bismarck, North Dakota, U.S.A.

Jane Fitzgerald
American Bird Conservancy, Brentwood, Missouri, U.S.A.

Curtis H. Flather
Rocky Mountain Research Station, U.S. Forest Service, Fort Collins, Colorado, U.S.A.

Robert A. Gitzen
Department of Fisheries and Wildlife Sciences, University of Missouri, Columbia, Missouri, U.S.A.

Paul H. Gobster
Northern Research Station, U.S. Forest Service, Evanston, Illinois, U.S.A.

Ascelin Gordon
School of Social Science and Planning, RMIT University, University of Melbourne, Melbourne 3001, Victoria, Australia

Diane A. Granfors
Habitat and Population Evaluation Team, U.S. Fish and Wildlife Service, Fergus Falls, Minnesota, U.S.A.

Eric J. Gustafson
Northern Research Station, U.S. Forest Service, Rhinelander, Wisconsin, U.S.A.

Robert G. Haight
Northern Research Station, U.S. Forest Service, St. Paul, Minnesota, U.S.A.

Jonathan B. Haufler
Ecosystem Management Research Institute, Seeley Lake, Montana, U.S.A.

Hong S. He
School of Natural Resources, University of Missouri, Columbia, Missouri, U.S.A.

Jeffrey A. Hepinstall
Warnell School of Forestry and Natural Resources, University of Georgia, Athens, Georgia, U.S.A.

Dale R. Herter
Raedeke Associates, Inc., Seattle, Washington, U.S.A.

Lorin L. Hicks
Plum Creek Timber Company, Columbia Falls, Montana, U.S.A.

Rex R. Johnson
Division of Bird Habitat Conservation, U.S. Fish and Wildlife Service, Fergus Falls, Minnesota, U.S.A.

Tim Jones
U.S. Fish and Wildlife Service, Laurel, Maryland, U.S.A.

Brian J. Kernohan
Forest Capital Partners, LLC, Portland, Oregon, U.S.A.

David R. Larsen
Department of Forestry, University of Missouri, Columbia, Missouri, U.S.A.

Michael A. Larson
Forest Wildlife Populations and Research Group, Minnesota Department of Natural Resources, Grand Rapids, Minnesota, U.S.A.

Daniel W. Linden
Department of Fisheries and Wildlife, Michigan State University, East Lansing, Michigan, U.S.A.

John M. Marzluff
College of Forest Resources, University of Washington, Seattle, Washington, U.S.A.

James B. McCarter
College of Forest Resources, University of Washington, Seattle, Washington, U.S.A.

Kevin S. McKelvey
Rocky Mountain Research Station, U.S. Forest Service, Missoula, Montana, U.S.A.

Donald McKenzie
Pacific Wildland Fire Sciences Lab, U.S. Forest Service, Seattle, Washington, U.S.A.

Joshua J. Millspaugh
Department of Fisheries and Wildlife Sciences, University of Missouri, Columbia, Missouri, U.S.A.

Christopher S. Nelson
College of Forest Resources, University of Washington, Seattle, Washington, U.S.A.

Emily Nicholson
Division of Biology, Imperial College London, Silwood Park Campus, Berkshire, United Kingdom

Neal D. Niemuth
Habitat and Population Evaluation Team, U.S. Fish and Wildlife Service, Bismarck, North Dakota, U.S.A.

Barry R. Noon
Department of Fish, Wildlife, and Conservation Biology, Graduate Degree Program in Ecology, Colorado State University, Fort Collins, Colorado, U.S.A.

Chadwick D. Oliver
School of Forestry and Environmental Studies, Yale University, New Haven, Connecticut, U.S.A.

Jennie Pearce
Pearce & Associates Ecological Research, Sault Ste. Marie, Ontario, Canada

Hugh P. Possingham
The Ecology Centre, School of Integrative Biology, University of Queensland, St. Lucia, Queensland, Australia

John R. Probst
Northern Research Station, U.S. Forest Service, Rhinelander, Wisconsin, U.S.A.

Crystal L. Raymond
College of Forest Resources, University of Washington, Seattle, Washington, U.S.A.

Ronald E. Reynolds
Habitat and Population Evaluation Team, U.S. Fish and Wildlife Service, Bismarck, North Dakota, U.S.A.

Chadwick D. Rittenhouse
Department of Fisheries and Wildlife Sciences, University of Missouri, Columbia, Missouri, U.S.A.

Gary J. Roloff
Department of Fisheries and Wildlife, Michigan State University, East Lansing, Michigan, U.S.A.

Mary M. Rowland
Pacific Northwest Research Station, U.S. Forest Service, La Grande, Oregon, U.S.A.

Christopher Rustay
Playa Lakes Joint Venture, Albuquerque, New Mexico, U.S.A.

Janet M. Ruth
U.S. Geological Survey, Fort Collins Science Center, Arid Lands Field Station, Albuquerque, New Mexico, U.S.A.

Stephen R. Shifley
Northern Research Station, U.S. Forest Service, Columbia, Missouri, U.S.A.

Susan A. Shriner
National Wildlife Research Center, U.S.D.A. Wildlife Services, Fort Collins, Colorado, U.S.A.

Henning C. Stabins
Plum Creek Timber Company, Columbia Falls, Montana, U.S.A.

Marshall L. Strong
Wildlife Division, Michigan Department of Natural Resources, Lansing, Michigan, U.S.A.

Lowell H. Suring
Terrestrial Wildlife Ecology Unit, U.S. Forest Service, Boise, Idaho, U.S.A.

Wayne E. Thogmartin
U.S. Geological Survey, Upper Midwest Environmental Sciences Center, La Crosse, Wisconsin, U.S.A.

Frank R. Thompson, III
Northern Research Station, U.S. Forest Service, Columbia, Missouri, U.S.A.

Lisa Venier
Great Lakes Forestry Centre, Canadian Forest Service, Sault Ste Marie, Ontario, Canada

William O. Vogel
U.S. Fish and Wildlife Service, Lacey, Washington, U.S.A.

Brian Wangler
Habitat and Population Evaluation Team, U.S. Fish and Wildlife Service, Bismarck, North Dakota, U.S.A.

Tom C. Will
Division of Migratory Birds, U.S. Fish and Wildlife Service, Fort Snelling, Minnesota, U.S.A.

Kenneth R. Wilson
Department of Fish, Wildlife, and Conservation Biology, Colorado State University, Fort Collins, Colorado, U.S.A.

Brendan Wintle
School of Botany, University of Melbourne, Melbourne 3010, Victoria, Australia

Michael J. Wisdom
Pacific Northwest Research Station, U.S. Forest Service, La Grande, Oregon, U.S.A.

List of Reviewers

H. Resit Akçakaya
Department of Ecology and Evolution, Stony Brook University, Stony Brook, New York, U.S.A.

Richard K. Baydack
Department of Environment and Geography, University of Manitoba, Winnipeg, Manitoba, Canada

William M. Block
U.S. Forest Service, Rocky Mountain Research Station, Flagstaff, Arizona, U.S.A.

Mark Burgman
Australian Centre of Excellence for Risk Analysis, School of Botany, University of Melbourne, Parkville 3010, Australia

Brian S. Cade
U.S. Geological Survey, Fort Collins Science Center, Fort Collins, Colorado, U.S.A.

James H. Devries
Ducks Unlimited Canada-IWWR, Stonewall, Manitoba, Canada

William D. Dijak
Northern Research Station, U.S. Forest Service, Columbia, Missouri, U.S.A.

Robert A. Gitzen
Department of Fisheries and Wildlife Sciences, University of Missouri, Columbia, Missouri, U.S.A.

Eric J. Gustafson
Northern Research Station, U.S. Forest Service, Rhinelander, Wisconsin, U.S.A.

Jonathan B. Haufler
Ecosystem Management Research Institute, Seeley Lake, Montana, U.S.A.

Jeffrey A. Hepinstall
Warnell School of Forestry and Natural Resources, University of Georgia, Athens, Georgia, U.S.A.

Lorin L. Hicks
Plum Creek Timber Company, Columbia Falls, Montana, U.S.A.

Mevin B. Hooten
Department of Mathematics and Statistics, Utah State University, Logan, Utah, U.S.A.

Brian J. Kernohan
Forest Capital Partners, LLC, Portland, Oregon, U.S.A.

J.P. (Hamish) Kimmins
Department of Forest Sciences, University of British Columbia, Vancouver, British Columbia, Canada

Bruce G. Marcot
U.S. Forest Service, Pacific Northwest Research Station, Portland, Oregon, U.S.A.

John M. Marzluff
College of Forest Resources, University of Washington, Seattle, Washington, U.S.A.

Neal D. Niemuth
Habitat and Population Evaluation Team, U.S. Fish and Wildlife Service, Bismarck, North Dakota, U.S.A.

Emily Nicholson
Division of Biology, Imperial College London, Silwood Park Campus, Berkshire, United Kingdom

Charles H. Nilon
Department of Fisheries and Wildlife Sciences, University of Missouri, Columbia, Missouri, U.S.A.

Barry R. Noon
Department of Fish, Wildlife, and Conservation Biology, Graduate Degree Program in Ecology, Colorado State University, Fort Collins, Colorado, U.S.A.

Reed F. Noss
Department of Biology, University of Central Florida, Orlando, Florida, U.S.A.

Volker Radeloff
Department of Forest and Wildlife Ecology, University of Wisconsin, Madison, Wisconsin, U.S.A.

Terrell D. Rich
Partners in Flight, U.S. Fish and Wildlife Service, Boise, Idaho, U.S.A.

Chadwick D. Rittenhouse
Department of Fisheries and Wildlife Sciences, University of Missouri, Columbia, Missouri, U.S.A.

Gary J. Roloff
Department of Fisheries and Wildlife, Michigan State University, East Lansing, Michigan, U.S.A.

Brent A. Rudolph
Michigan Department of Natural Resources, Rose Lake Wildlife Research Center, East Lansing, Michigan, U.S.A.

Winston P. Smith
Pacific Northwest Research Station, U.S. Forest Service, La Grande, Oregon, U.S.A.

Stephanie Snyder
U.S. Forest Service, Northern Research Station, St. Paul, Minnesota, U.S.A.

Scott Stephens
Ducks Unlimited, Inc., Bismarck, North Dakota, U.S.A.

Alexandra D. Syphard
Conservation Biology Institute, San Diego, California, U.S.A.

Wayne E. Thogmartin
U.S. Geological Survey, Upper Midwest Environmental Sciences Center, La Crosse, Wisconsin, U.S.A.

Denis White
U.S. Environmental Protection Agency, Corvallis, Oregon, U.S.A.

Michael J. Wisdom
Pacific Northwest Research Station, U.S. Forest Service, La Grande, Oregon, U.S.A.

CHAPTER

General Principles for Developing Landscape Models for Wildlife Conservation

1

Joshua J. Millspaugh, Robert A. Gitzen, David R. Larsen, Michael A. Larson, and Frank R. Thompson, III

Models are abstract descriptions of systems or processes (Starfield and Bleloch 1991, Haefner 1996). In other words, a model is a formal framework for organizing and synthesizing existing knowledge about an ecological system. Models have become pervasive tools in natural resources management, large-scale planning, and landscape ecology (Shenk and Franklin 2001, Scott et al. 2002). Models help address fundamental questions about wildlife habitat relationships and habitat management. For example, models are useful for evaluating the potential impacts of management alternatives (Morrison et al. 1998, Larson et al. 2004, Shifley et al. 2006), predicting species occurrence (Scott et al. 2002), and assessing economic implications of management decisions (Haight and Gobster, this volume).

Landscape models take many forms, including statistical models that quantify relationships and patterns among variables (e.g., Niemuth et al., this volume; Hepinstall et al., this volume), conceptual models that offer a qualitative construct of a system, and simulation models that project landscape features into the future (e.g., He, this volume; Oliver et al., this volume). Landscape models can produce output that is as difficult to analyze and understand as data from the original system. For examining and presenting the results from landscape simulation models, ecologists need tools that facilitate interpretation of complex multivariate patterns (Shifley et al., this volume). For this reason, visualization tools are often used with landscape models because they make complex data easier to understand (McGaughey 1997, 1999).

Because of the usefulness and widespread application of models, researchers and decision makers should be well informed about potential strengths and limitations of these models. Here, we review principles underlying the construction and use of models, with an emphasis on their application to large-scale wildlife conservation planning. In addition to outlining general principles of modeling, we offer advice about using models in an adaptive management framework, addressing uncertainty, and making models useful and transparent. We also encourage a focus on viability and population objectives (Johnson et al., this volume) in modeling and we present a broadened concept of viability for species of conservation concern and game species as an important measure in understanding wildlife response in large landscapes. To communicate results from landscape models, we need tools for visualizing these results. Therefore, we end the chapter by briefly discussing some basic theory, dangers, and utility of visualization software. We refer readers to other relevant papers and books, such as Box (1979), Starfield and Bleloch (1991), Hilborn and Mangel (1997), Starfield (1997), Williams et al. (2002), Shenk and Franklin (2001), and Scott et al. (2002), that further discuss philosophical considerations of modeling in natural resources.

USES OF MODELS

Modeling has become widespread in natural resources management because models can be incredibly useful and practical tools. Johnson (2001) defined three categories of purposes for models: explanation, prediction, and decision making.

1. *Explanatory models* are used to describe or decipher the workings of systems. Such models attempt to identify the mechanisms involved in the system.

2. *Predictive models* are used to forecast future states of systems or results of management actions. Prediction is a common use of landscape models and allows the user to determine the potential impacts of various proposed management actions (e.g., Shifley et al. 2006). The opportunity to ask "what if?" questions is especially attractive to natural resource managers.

3. *Decision-support models* are used to identify management strategies that will produce desired results. Optimization techniques are one useful example of decision-support models used in planning resource management (Moore et al. 2000).

A given model may be used for more than one purpose. For example, habitat suitability models may be used to investigate the relative importance of key habitat characteristics and simultaneously predict future habitat suitability. Many of the habitat suitability and population models discussed in this book and elsewhere are decision-support models that allow managers to assess the relative trade-offs of management actions.

PHILOSOPHY OF MODELING

In this section, we summarize general principles that modelers and end users should consider when working with models, regardless of the model purpose. We re-emphasize points frequently made in introductions to modeling, especially Starfield (1997).

Every Biologist Constructs Models

Some biologists view modeling as a mathematical art of little relevance to real-world management problems. However, every biologist constructs models. Every scientist and manager has an intellectual framework of hypotheses about how his or her focal system is organized, what factors drive changes in key resources, how the system will respond to management actions, and what the major uncertainties and holes are in this framework. Whether these scientists and managers admit it, this framework is the basis for a conceptual model that can be translated easily into narratives, diagrams, pictures, equations, and even computer programs (i.e., into quantitative models).

There are multiple potential purposes for formalizing one's intellectual framework into a model, whether conceptual or quantitative. Regardless of whether one constructs a landscape simulation model or draws a diagram on the back of a napkin, constructing a model forces biologists to confront their assumptions about the system and the support for these assumptions. It prompts them to consider the most critical uncertainties inhibiting scientists and managers from better understanding the system. It can act as a framework for integrating new information and is a tool for more rigorous thought about the system (White 2001). Finally, it forces the biologists to expose hypotheses and assumptions to critiques from others. In the case of complex, high-profile management decisions, a manager may be unable to recommend and defend (perhaps in court) a course of action without well-developed quantitative models (Swartzman 1996, Starfield 1997, Walters and Martell 2004:3–4).

Models Are Useful Despite a Lack of Data or Understanding

As frameworks for the organization and synthesis of existing information, "all models are wrong, but some are useful" (Box 1979). Ultimately, we seek a sophisticated, accurate understanding of natural systems, precise estimates of important parameters and their dynamics, and good knowledge about the specific effects of various management alternatives. In such an optimal situation, we might have at least moderate confidence in model predictions, even though there is still significant uncertainty. For example, even biologists who are skeptical about most models are comfortable using predictive results in this situation

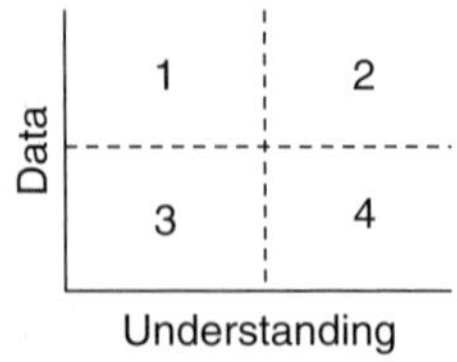

FIG. 1-1

A classification of modeling from Holling (1978). The x-axis represents understanding of a system (from limited to complete), and the y-axis represents the quality and quantity of data (from incomplete to adequate) that are available for use in model-building. Ecological models typically are based on limited data and incomplete understanding of systems, and thus fall in region 3 (Starfield and Bleloch 1991).

(e.g., daily weather forecasts produced from atmospheric models) despite knowing that such forecasts are often inaccurate.

However, in wildlife habitat modeling, we usually possess limited data and an incomplete understanding of the system (Holling 1978; Fig. 1-1). Models can be especially useful tools for decision making and for prioritizing efforts to address these gaps in our understanding. The argument that modeling should not be used unless data are adequate is just as misguided as arguing that no new management actions should be tried unless we completely understand the system and can predict the specific effects with high certainty. Managers have to act in the face of uncertainty; models help them make as defensible a choice as is currently feasible. Similarly, researchers have to justify why they are proposing studies of a particular aspect of the resource. Model building helps us evaluate the relative importance of various influences on a system and identify data that should be collected (Starfield 1997; Shifley et al., this volume).

Models Should Be Constructed for Specific Purposes

A model can be seen as a structural framework for our current knowledge and as a tool for exploring uncertainties in our knowledge. To create a useful framework or tool, we need clear, specific objectives for the modeling effort. The purpose of the model should determine its structure; scope, resolution, and complexity; its user interface and output; and how it is evaluated (Starfield 1997, Nichols 2001, Kettenring et al. 2006).

In defining the purpose for the model, we should address multiple issues:

1. Who are the intended end users of the model? What are the technical skill levels of these end users?

2. How will the model be used: for evaluating management alternatives, determining high priorities for future research, communicating what we know to other stakeholders, or simply clarifying for our own benefit what we know and need to learn about the system?

3. What spatial and temporal context do we want to explore? For example, do we care about breeding season patterns only, modeling short-term forecasts or long-term dynamics, a specific management area or an ecological province?

4. How will the model be evaluated?

5. Are we building the model for long-term use? How will it be updated as our understanding of the system improves?

Predicting the Future Is a Lofty Goal

Ecological systems are driven by factors with high variability and unpredictability, and observed ecological patterns are shaped partially by random processes (e.g., Hubbell 2001, Fuentes et al. 2006). Modeling experts understand that even the best model rarely can accurately forecast the future condition of natural systems (e.g., Boyce 2001, White 2001, Walters and Martell 2004:10–11), except sometimes over short time spans. In the face of this variability, the predictive value of models usually comes not in forecasting the expected future condition of a resource, but in projecting a range of potential conditions given the likelihood of different stochastic events (Clark and Schmitz 2001). However, with increasing time, previously undocumented events or misunderstood processes are likely to move the system beyond a range of variability predictable from our current knowledge (but see Brook et al. 2000).

Therefore, why bother with predictive, quantitative modeling at all? First, short-term predictive accuracy sometimes is an important goal for management. For example, when using spring-collected data to set autumn hunting regulations each year, we want highly accurate model-based predictions about what fall abundance will be that year. Second, we learn a great deal by making specific but inaccurate predictions and assessing why our underlying hypothesis was inadequate. Biologists should fear acting on ill-defined assumptions and weak logic far more than acting on inaccurate predictions and incomplete models. Certainly, we should not be over-confident in our predictions; we have to weigh the risks of acting on a wrong prediction versus the risks of inaction and other alternatives. For this reason, having a model and its output account for known sources and magnitudes of uncertainty is important. Third, we can make defensible management decisions by comparing the predicted results of various management alternatives. Moreover, we may be able to predict the relative benefits of these alternatives far more accurately than we can predict specific outcomes of each (Beissinger and Westphal 1998, Walters and Martell 2004:5).

Useful Versus Truthful Models

The process of model evaluation and validation is a critical step in modeling (Johnson 2001; Shifley et al., this volume). However, this evaluation should focus not on how well the model captures "truth" (verification), but how well the model performs for its intended purpose (Oreskes et al. 1994; Starfield 1997; Shifley et al., this volume). Even when quantitative prediction is the primary objective of modeling, accuracy can be judged only in terms of the desired use of the model. For example, when one is predicting occurrence of below-freezing nightly temperatures, specificity (correctly predicting frost episodes) of 70% might be fine for southern Canada in October but terrible for Florida fruit-growing areas in February. Our primary purposes for modeling usually are to quantify major uncertainties about the system and to provide a framework for improving our understanding of the system. In the context of resource management, a "good" model is one that promotes a better decision than could be made without it (Starfield 1997, Johnson 2001).

Model Complexity and Credibility

Because of uncertainty surrounding our knowledge of the system and limited data, the use of complex models may not improve one's understanding of a system. Occam's razor is a logical guiding principle in habitat suitability modeling: the simplest model that is consistent with existing knowledge is likely to be most appropriate and is most likely to produce reliable insights. Models should be no more complex than necessary to capture the key relevant features of the system. That is, one should construct the simplest model that fulfills one's purpose adequately (Starfield 1997, Nichols 2001; for a contrasting view see Walters and Hilborn 1978:168).

When a more complex model produces different results than a simpler model, it is easy to assume that this means the complex model is more realistic (more "truthful"). This is a dangerous assumption. Increased complexity does not guarantee increased accuracy. As model complexity increases, the necessary assumptions multiply, and the body of data needed to parameterize the model is stretched thinner. Small errors that would have minor effects in a simple model may propagate; the uncertainty in the model output may be far greater than the sum of the uncertainties in the input parameters (Haefner 1996:186–187). With careful simulations and sensitivity analyses, the modeler can examine how dependent model results are on the additional assumptions and parameter values needed for the complex model (Bekessy et al., this volume). However, this requires an honest assessment of the degree of uncertainty about values for which data are extremely limited. This uncertainty may be far greater than the range of variability from a few published studies.

Often, model development should proceed incrementally starting with a general, very simple model. For example, a landscape habitat model could consist simply of a cover-type map and expert rankings of suitability for each cover type. If possible, each step up in complexity could be added as a new module that can be turned on or off easily; sometimes separate incremental models will be required (e.g., adding a vegetation simulator; adding a subroutine to account for spatial arrangement and size of patches when estimating suitability, incorporating demographic simulation to model density rather than qualitative suitability). At each step, the model can be explored, tested, and compared with simpler models before adding additional complexity. Logistically, this approach often helps point out errors in the structure or calculation of the more complex model, particularly if the two models can be used to run nearly identical scenarios. Scientifically, this approach helps the modeler be explicit about the purpose of the additional complexity, additional assumptions needed, and data limitations.

Consider Alternative Models

When only one model is built, modelers and users risk getting too attached to the model and its assumptions, and too forgiving of its faults. Instead, biologists should simultaneously construct and compare two or more competing models, each based on alternative hypotheses about the system being modeled (Holling 1978:100–101, Nichols et al. 1995, Haefner 1996:22, Mangel et al. 2001, Conroy and Moore 2001, Hill et al. 2007). Such models may differ in their underlying structure or may form a nested set of models (e.g., models with and without density-dependent effects on population growth). The multimodel approach can be seen as a direct extension of having multiple working hypotheses (Platt 1964, Ford 2000:290).

Comparing multiple models helps users to clearly examine effects of alternative assumptions about the system, check for structural and computation errors, and compare projected effects of alternative management policies. It may have intuitive benefits when there are multiple stakeholders with differing views about the ecological dynamics of a resource and likely effects of management actions (Williams et al. 2002:663–684, Conroy and Moore 2001).

Models Should Be Transparent

Models should be completely transparent in their objectives, assumptions, model structure, data used, mathematical details, and limitations. Unless users of a model are examining a test scenario in which they know what the outcome should be, these users can tell little about the quality of a model from its output. This is particularly true in landscape habitat models, where colorful maps created using geographic information systems (GIS) may give us undue confidence in the underlying model. Both modelers and model users share blame for the frequent situation in which a model is treated as a black box and the user is encouraged to worry only about running the model and interpreting the output, not about the model itself. Just as we can fully assess the results and discussion of a scientific paper only after critically examining the methods, we can assess the value of a model only by carefully considering the model formulation—structure, assumptions, calculations, and data sources.

When a model is constructed for other users, it should be documented well enough to (1) provide a sufficient recipe for the model such that another modeler could rebuild the model from scratch and replicate its output (or at least reproduce similar average values and variability in results for stochastic models); and (2) provide a clear description of the purpose and hypotheses guiding the model, the assumptions involved, the ecological spatial and temporal domain for which the model was constructed and validated, a description of how the model was evaluated and tested, and a discussion of its potential limitations (Benz et al. 2001, Kettenring et al. 2006).

Modelers need to document explicitly what data were used for constructing and parameterizing the model, and model users need to critically assess limitations in these available data. Citing the data sources is an obvious step, but a citation title tells the user nothing about the sample size, sampling limitations, or temporal duration of the study. Usually, one can get only a vague idea of the study location and timing (e.g., spring versus fall) from the citation title. Therefore, adequate documentation for a model usually should provide more details about studies critical to the model, including the study area, habitat types or ecological province, sample size, season, temporal duration, sex and age classes addressed, and any strong criticisms of the study (e.g., occupancy estimated naively rather than with models accounting for detectability). Last, transparency regarding the visualization process and procedures is important in understanding whether images accurately project model results. This discussion suggests that modelers must spend significant time documenting their models. However, model users share equal responsibility for demanding adequate documentation. These users need to assess carefully the model's assumptions, data support, and limitations before focusing on interpreting model output.

Use of Models from Other Regions Should Be Done Judiciously

Biologists need to use great caution when using generic models or models developed for other regions or purposes. Models are developed to meet specific objectives and are influenced by available data, knowledge of the system, and assumptions, which makes them difficult to apply universally. Although many general models are structurally similar (e.g., matrix models for demographic analyses; Caswell 2001), specific models are uniquely suited for specific regions and applications. When adapting existing models to a new situation, biologists need to assess the compatibility of that model's purpose, assumptions, and data requirements with the details of their own problem (Kettenring et al. 2006; Probst and Gustafson, this volume). In some cases (e.g., Landscape Management System; Oliver et al., this volume), one may refine a general model for local applications by inputting site-specific data. However, the user still must be cautious that the underlying assumptions and relationships in the model are appropriate for the situation at hand.

Even if the objectives of the existing model are fairly close to those of the current situation, the biologist must evaluate whether specified relationships are appropriate and relevant to the new system, and that parameters in the model can be estimated precisely. For example, Mladenoff and Sickely (1998) applied a resource selection function model developed for wolves (*Canis lupus*) in the upper Midwest United States to the Northeast United States to project the suitability of wolf habitat. In doing so, there is an implicit assumption that the important features (e.g., whether or not variables such as road density or prey type) and their form and strength of the relationships that affect the suitability of wolf habitat in Wisconsin translate to potential wolf habitat in New England.

Managers Should Be Involved in Model Construction

A management-oriented model is built to help managers determine the course of action to take, the risks associated with alternative actions, and the uncertainties that must be addressed to make better decisions. Increasingly, managers should see modeling as an invaluable tool for the scientific component of management, and model building should be an art they practice regularly on their own computer or notepad (Starfield 1997). We believe that managers should be involved heavily when management-oriented models are constructed.

Constructing large-scale, complex models usually requires a modeling team, including a manager. To construct useful management-oriented models, the modelers need clear management objectives or a range of objectives, clearly defined management options, detailed understanding of the populations and landscapes being modeled, and clear ideas about how the models can be used to improve subsequent management decisions. Usually, the manager is best able to supply much of this expertise, and educates the rest of the modeling team (Kendall 2001). In some cases, the manager may not be accustomed to heavily quantitative, partially data-driven, structured approaches for decision making. Therefore, part of the modeling process will be spent educating the manager to consider how the model can be most useful. In either case, the resulting models will be useful only if the manager is an integral part of the modeling group throughout all stages of model development (Clark and Schmitz 2001).

The format and output of models must carefully consider the targeted end user. For management-oriented tools, there often is a trade-off between user-friendliness and technical sophistication. One might program cutting-edge spatially explicit, data-driven landscape habitat models using completely open-source, free GIS and statistical software with state-of-the-art numerical tools. For some managers who are technically proficient, this might be ideal. Other managers might have no interest in installing and learning to use new programs. A more useful first step for them might be a spatially implicit spreadsheet-based model (with careful assessment of whether the simple model is adequate for their purposes) or a relatively sophisticated model in their preferred GIS package. Similarly, some users may require readily available help files to explain concepts that might be elementary to other audiences; they may pay more attention to simple graphical displays of output than complex graphs or numerical summaries. In any case, modelers and managers should collaborate carefully as models are developed to produce a user interface that will facilitate their application of the model.

AVOIDING UNRELIABLE MODELS

A key to successful modeling is the avoidance of common missteps that make models unreliable. Ineffective or unreliable models maintain the following characteristics (Starfield 1997):

1. *Explicit accounting for processes that are not relevant or well understood.* In an attempt to increase the utility of a model, there is sometimes a tendency to incorporate all processes, including those that are not necessary (in the context of model objectives) or well understood. One should only account for processes that are relevant and understood well enough to be included in the model. For example, in some landscapes, disturbances from insects might be of minor importance. In an attempt to be all inclusive, the modeler might be tempted to include an insect disturbance module when projecting habitat conditions, although it is not particularly relevant or important to the outcome.

2. *Dependence on parameters that cannot be estimated precisely.* Many parameters might be considered important, but some might be difficult to estimate with a meaningful degree of precision. Such imprecision compounds uncertainty and propagates error. With greater uncertainty, it becomes even more important to limit the number of input parameters (Mangel et al. 2001). Input parameters need to be estimated with enough confidence that they are helpful in the modeling process. It is also possible to consider other ways to structure the model that relies on parameters about which more is known (Nichols 2001). On the other hand, if the purpose of a model is to assess our understanding of a system, one needs to account for the full uncertainty in critical parameters, regardless of how great the uncertainty is.

3. *Dependence on too many parameters.* Much has been written recently in the natural resources literature about model parsimony (Hilborn and Mangel 1997, Burnham and Anderson 2002) and the dangers of over-fitting models. There is a direct trade-off between bias and variance and although richer models are preferred when the purpose is prediction, there are also practical considerations. When one is modeling large landscapes, additional parameters require additional data collection and synthesis, which can take considerable time and effort (Roloff et al., this volume; Shifley et al., this volume). Regardless of whether one is developing statistical models or models based on expert opinion, it is advisable to make models parsimonious.

4. *Uncritical application of pre-existing models.* Statistical models, such as resource selection functions, are specific to the data used to construct them (Hicks et al., this volume), and inferences made from them are limited by the sampling design used to collect the data (Mangel et al. 2001). In many cases, subtle differences in model structure, assumptions, or other issues affect whether a model is suitable for a new application in a new environment. For these reasons, applying models from one system to another should be done judiciously (Probst and Gustafson, this volume).

MODELS AND ADAPTIVE RESOURCE MANAGEMENT (ARM)

Managers and scientists should continually and systematically try to improve management by formally examining the outcomes of management actions and policies, and assessing how well they meet well-defined management objectives (Nichols et al. 1995, Taylor et al. 1997). Landscape models, like other models, are often developed with an incomplete understanding of system properties (Starfield 1997). Landscape models are best viewed within an adaptive management framework. Such an approach inherently acknowledges that (1) there is uncertainty in our knowledge of the system to be managed; (2) improving management decisions would be facilitated by a reduction in uncertainty; and (3) management decisions must be made and revisited periodically. Through adaptive management, alternative models are considered working hypotheses to be developed, evaluated, and refined as new data become available. Management prescriptions are treated as experiments and opportunities for learning by confronting model predictions with data from a purposefully designed monitoring program. Therefore, models serve as a mechanism to evaluate uncertainty and our understanding of the system; they also facilitate decisions based on the best available data. Last, models facilitate refinement of management actions as additional data become available. However, because of the complexity and variability inherent in ecological systems, appropriate model refinements are difficult to identify and incorporate in short time periods. Therefore, the Adaptive Resource Management (ARM) philosophy is an essential framework for scientifically defensible and effective management (Walters and Hilborn 1978). Several principles provide the background context for ARM:

1. Current management actions are based on hypotheses about what processes control the system being managed. Every manager and advising scientist is operating from models, based on these hypotheses, which predict how the system will respond to specific management actions and policies. These models may be conceptual or quantitative, formal or informal.

2. There is uncertainty about the current usefulness of these hypotheses (i.e., whether predictions arising from these hypotheses will be accurate).
3. Management effects and general changes in the system need to be assessed so that this uncertainty about current management actions can be reduced. This monitoring and assessment also ensures that hypotheses can be refined and expanded as management goals or ecological conditions (e.g., climate, presence of disease) change (Williams et al. 2002:231).
4. To justify and defend specific management actions and additional research, the manager needs to make explicit these hypotheses, models, and uncertainties.

In this context, quantitative management-focused models are an element of the ARM process. At the least, these models are a vital tool for predicting comparative outcomes of alternative management options. When used at a deeper level, they provide the central skeleton for an ARM program. The model set encompasses what we think we know about the system, helps us assess the risks of incorrect assumptions, and can help prioritize (e.g., through sensitivity analyses) which of these gaps in our knowledge are most limiting in predicting system behavior. Whenever there are multiple alternative proposed management actions and quantifiable management objectives for which we want to select an optimal management strategy, quantitative models facilitate structured decision making based on the predicted effectiveness and risks of each strategy (Nichols 2001).

Such models can be updated and improved continuously as new data accumulate (i.e., from monitoring or separate research) and as hypotheses are refined. Moreover, these models help us focus monitoring on variables and areas that are most information-rich in evaluating management effectiveness and for improving our models (Nichols and Williams 2006). Models are most useful when they focus on specific questions and do not incorporate unnecessary complexity. However, as information increases, ARM models may be expanded to incorporate all major ecological factors driving viability of the species in question, allowing useful predictions even if there are major changes in management objectives and the ecological environment beyond what was envisioned when the models were originally constructed (Holling 1978:66).

ADDRESSING UNCERTAINTY IN MODELS

In nearly all resource management situations, we have an incomplete or flawed understanding, and snapshots of data from systems characterized by high variation and unpredictability. These facts are not arguments against modeling. Instead, this uncertainty often is the primary rationale *for* constructing models.

Models are useful because they help examine this uncertainty and its potential causes. However, this requires that modelers and users think thoroughly about several major forms of uncertainty affecting model development.

Uncertainty in Our Underlying Hypotheses and in Model Structure

Regardless of model complexity, there will always be more than one plausible hypothesis about critical components and processes in the ecological system. Moreover, we may be able to represent a single hypothesis with several alternate models or equations (Ford 2000). We can address this uncertainty by having multiple working hypotheses (Chamberlin 1890), by specifying the contrasting assumptions and predictions of each hypothesis, and by representing each hypothesis with one or more models. Such models may be nested, differing only in which parameters are turned on or off (i.e., are allowed to affect the calculations) in each model. For example, a simple landscape suitability model focusing on patch type may be expanded by adding additional parameters to incorporate effects of patch size and inter-patch distance. Alternative models may be non-nested, with large differences in their underlying structure (e.g., comparing an expert ranking of suitability versus a metapopulation demographic model).

For some purposes, qualitative comparison of these alternate models may be sufficient. However, strategies for comparing and ranking multiple models, and making decisions in a multimodel framework, have been among the most important tools developed by quantitative ecologists over the last few decades (e.g., Holling 1978, Walters and Hilborn 1978, Burnham and Anderson 2002, Williams et al. 2002:643–864). Previously, biologists often focused on ranking the "best" model and then drawing inference only from that model. Variance estimates from this final top-ranked model are underestimates of the true uncertainty in parameter estimates (Harrell 2001). Rather than ignoring model-selection uncertainty, biologists can incorporate it to produce inference unconditional on any single model. For example, with statistical modeling, biologists can weight models by Akaike's Information Criterion (AIC; e.g., Burnham and Anderson 2002) and use these weights to produce unconditional estimates of parameter values and confidence intervals about these estimates. Whether by such objective criteria or by subjective expert rankings, prior model weights can be incorporated into Bayesian data analysis (e.g., Link and Barker 2006) or into a simulation approach to integrate output from multiple models producing compatible output.

It may be tempting to see the multimodel approach as a problem that ultimately will be cured when we have collected sufficient data. Instead, we should recognize that most natural resource dynamics are the result of numerous varying influences. In a multimodel framework, each model may capture a particular element of "truth"—therefore, examining the composite picture of multiple models may be preferred to acting on a single "best" model (Hobbs and Hilborn 2006). Monitoring outcomes and using these data to continuously update and reweight our alternate models can facilitate decision making that incorporates what we know about system dynamics, our management objectives, and estimates of the current state of the system (e.g., Kendall 2001, Williams et al. 2002, Hill et al. 2007).

Uncertainty in Parameter Estimates

The purposes of a model should determine whether a deterministic or stochastic approach is appropriate (e.g., Kettenring et al. 2006). Stochastic models are intuitively attractive because even in complex situations, we can address uncertainty about which parameter values we should use, and we can account for natural variability in these parameters. In each model run we draw a random sample from the assumed probability distribution for each parameter. However, this requires that we have suitable (given the purpose of the model) input values for this distribution. Often, published estimates for some parameters may help us pick an average value for the parameter with some confidence. However, determining how much uncertainty we have about the parameter, or how much it varies spatially and temporally, is even more challenging.

In some cases, there is a tendency for modelers to be overly optimistic about the adequacy of the underlying data for estimating parameter uncertainty and natural variability. Parameter distributions (e.g., mean and variance) for input into stochastic models should be defined after an honest and careful consideration of limitations in available data. Modelers and model users need to carefully consider potential biases in reported parameter estimates and whether reported variances are adequate for capturing uncertainty in the current model. For example, failure to account for incomplete detectability will produce biased estimates of abundance, survival, and occupancy (e.g., MacKenzie, et al., 2006). Even in a modest local study, published estimates of a parameter and its uncertainty may be biased if the landscape was sampled nonprobabilistically (Anderson 2001) or if the study did not cover enough years to cover the temporal domain we are modeling.

Occasionally, published studies may overestimate process variability (e.g., yearly variation in juvenile survival). For example, in individual studies both process (e.g., temporal and spatial) and sampling (measurement) variability are often lumped into a single variance estimate. More frequently, estimates from a few studies may not capture the full range of temporal and spatial variation in the system being studied.

Even estimates from high-quality studies from other locations and times need to be used with caution. For complex and even some simple landscape models, local data for the focal species are usually insufficient for adequately parameterizing the model or even for guiding the structure of the model. For a wide-ranging species, the only suitable estimates of many parameters may come from other regions. For example, for all but the most heavily studied species, there are few studies which use rigorous telemetry or capture-recapture modeling to estimate dispersal rates through various landscape matrix conditions. Even the basic assumptions underlying a viability model (e.g., habitat variables included in estimating occurrence) may vary greatly among regions. Species with broad geographic ranges typically show surprising spatial and temporal flexibility and variability in habitat associations, demographic parameters, and responses to management actions (Wolff 1995, Converse et al. 2006, Murphy and Lovett-Doust 2007, Whittingham et al. 2007). Yet, even relatively simple habitat suitability models may rely heavily on relationships documented only from other regions.

Modelers should be explicit about sources of input-parameter values and assumptions made in converting published estimates into input values. When models must rely heavily on studies from other regions, modelers need to be very explicit about this, and end users need to be particularly cautious in evaluation and testing. If parameters are used from studies in other regions, the estimates of uncertainty about these parameters in these studies may be gross underestimates, given the additional uncertainty induced by transferring estimates among regions. In some cases, one may be able to better place bounds on the assumed parameter distribution by integrating comparable estimates from multiple systems or closely related species with similar life histories. In other cases, the modeler may need to use very conservative bounds on the parameter. If a conservative distribution would be ridiculously wide, the modeler may need to reduce model complexity to eliminate parameters about which we know too little, or the modeler may rerun the model under several carefully defined, plausible scenarios for the distributions of these parameters.

In most cases, formal sensitivity analyses—systematically examining how changes in input values affect model output—can be invaluable in assessing effects of parameter uncertainty and assumptions (e.g., Johnson 2001). For example, if model results are highly sensitive to parameters relying heavily on data from other regions, obtaining better within-region estimates of their values may be a high priority for additional research even if very precise estimates were available from other regions.

Uncertainty in Whether the Model Works the Way It Is Intended

Regardless of the purpose of the model or the accuracy of model predictions, we need to carefully check that the model has adequately captured our underlying hypothesis, that the structure is logical and correct, that there are no mathematical mistakes, and that the model has been programmed correctly. Frequently, this is referred to as "verification" (e.g., Haefner 1996; Johnson 2001; Shifley et al., this volume). For example, our programming code may make simple or major mistakes (e.g., missing parenthesis, reading the wrong location for an input parameter). We may have incorrectly derived a mathematical equation. We may forget to initialize the random number generator for each run, producing the same set of random numbers for each simulation, or we may be using a generator which produces some duplicate streams of numbers because its period is too small for the number of simulations we are running.

Verification is an art rather than a rote process. The following are a small number of the many strategies the biologist/modeler can use for verification.

1. As the model is built, output the result of each calculation and any random number streams used for these calculations. Check that these make sense before proceeding.
2. Work through your code and assumptions with a colleague who has equal or higher experience with your programming language.
3. Compare a deterministic model to results generated from a stochastic version in which variability is turned off or set to nearly zero.
4. Replicate the model or submodels in both a user-friendly and more mathematically robust and efficient program (e.g., a desktop spreadsheet package versus programs such as R or MATLAB) and compare results.
5. Have target users explore model output, and evaluate whether it seems plausible, over a broad range of conditions within its intended domain of use.

Uncertainty and Biases Caused by the Model User

Regardless of how well a model has been constructed and evaluated, the intended users may use poor input data, fail to select correct program options, and misinterpret output. The modeler can take specific steps to reduce such problems, such as trying to minimize input errors with traps for inadmissible parameters. The modeler and intended users can work closely together to develop and evaluate the model. Most importantly, users must take responsibility for how they apply the model, rather than assuming their responsibility is simply to induce the model to produce output as quickly as possible. Visualization of model output is another potential source of uncertainty and bias in landscape planning models (see below).

VIABILITY AS A GENERAL MANAGEMENT METRIC

Landscape models used in conservation planning output different metrics of species performance that include habitat suitability (Dijak and Rittenhouse, this volume), the probability of occurrence (Hicks et al., this volume), relative or absolute population size (Niemuth et al., this volume; Johnson et al., this volume), and population trajectories or viability (Akçakaya and Brook, this volume; Bekessy et al., this volume). Along this continuum of performance metrics (see Larson et al., this volume), population trajectories and viability are the most ambitious goals.

Viability is a useful and effective concept in wildlife conservation (Beissinger and Westphal 1998; Beissinger et al., this volume). It developed out of concern for populations in danger of extirpation or extinction. The concept of viability is effective because consideration of values and risk, which are implied in nearly all definitions of viability, are fundamental to making sound management decisions (Akçakaya and Brook, this volume; Bekessy et al., this volume). We argue that viability, and its associated risk-assessment framework, is a useful concept in any population management situation. There is still value in indices of habitat suitability, but associating some measure of risk with those indices might aid planning efforts.

The concept of viability, due to its origin in conservation biology, has retained connotations relevant mostly to small population sizes, endangerment, and extinction. Land management agencies in the United States and elsewhere, however, have begun perpetuating a mandate of population viability for all desirable species. We propose expanding the concept of viability to be applicable to conservation planning for all species by including criteria in addition to population persistence so that it applies to all species and to populations of all sizes, integrates biological and human dimensions considerations, and is quantitatively explicit. There are three hierarchical levels of viability in our definition (Table 1-1). If a population is deemed viable at one level, it also satisfies criteria for viability at higher levels. The hierarchical structure of the definition allows explicit incorporation of multiple, potentially overlapping concepts of viability that are implicit in the variety of definitions currently available.

Table 1-1 Hierarchical Structure of a Comprehensive Definition of Population Viability

Level	Abbreviated Definition	Quantitative Assessment Criteria
Primary viability	Continued existence of a population	Probability of remaining extant (or going extinct) during a specified time interval
Secondary viability	Resilience of a population[a]	Probability of returning to a desired abundance or remaining extant given a specified reduction in population size or landscape condition
Tertiary viability	Ability of a population to provide desired benefits[b]	Change in the level of service provided, target population size

[a]*Includes the concepts of genetic diversity and adaptability inherent in previous definitions of viability.*
[b]*Benefits may be a harvestable surplus, recreation, aesthetics, an ecosystem function, or maintenance of secondary viability without direct human intervention.*

Primary viability.— Primary viability is the ability of a population to remain extant. It is nearly synonymous with the classical and most widely understood definition of viability. That is, the abundance of a viable population will remain above zero some higher quasiextinction threshold (Ginzburg et al. 1982). Primary viability refers to the persistence of a population (Connell and Sousa 1983), so it requires explicit specification of (1) an abundance threshold below which the population will be considered extinct or extirpated; (2) a time interval, usually in years from present; and (3) a probability of remaining extant or, conversely, going extinct during the interval (Table 1-1).

Secondary viability.— Secondary viability, in addition to primary viability, is the resilience of a population (i.e., the probability of returning to a desired level after a change). The idea that populations may or may not return to a potential equilibrium state after a perturbation of a given magnitude has a long history of its own (Holling 1973). Dodson et al. (1998) listed resilience as one of many factors by which to assess population viability. In pragmatic terms, viable populations are resilient to fluctuations in abundance, where fluctuations may be due to deterministic threats or stochasticity (e.g., catastrophes). We propose quantifying secondary viability using a more conservative application of criteria for primary viability. Therefore, we must explicitly specify (1) the expected reduction in abundance or extension of the time interval over which viability is evaluated, and (2) an acceptable reduction in probability of remaining extant (or, conversely, increase in probability of extinction). For example, if a population of 10,000 individuals has an 80% chance of persistence for 100 years, it might satisfy criteria for secondary viability if its predicted probability of persistence is reduced by <50% for a period of 200 years or for an initial population size of only 5,000. Dennis et al. (1991) suggested that the conservation status of a population would change if its abundance declined by an order of magnitude, which is a possible rule of thumb for evaluating secondary viability. Our definition of secondary viability includes the ideas of genetic diversity and adaptability mentioned in traditional, generic definitions of viability. For example, a population that satisfies criteria for primary viability (i.e., genetic stochasticity is not a threat over a short time period) may not satisfy criteria for secondary viability because global climate change is shifting the distribution of appropriate environmental conditions faster than the population can adapt or individuals can disperse.

Tertiary viability.— Tertiary viability is the ability of a population to provide desired benefits, in addition to remaining extant and resilient. This third facet is the catch-all that makes our definition comprehensive and equally applicable to rare and common species and declining, stable, and increasing populations. Examples of desired benefits are a harvestable surplus, recreation, aesthetics, or an ecosystem function. All these benefits can be quantified and evaluated, and might be addressed by planning for a target population size that is in excess of what is required for primary or secondary viability. In general, (1) the relationship between direct human uses and their effects on the wildlife population must be formalized in a model, and (2) the maximum level of human use that still maintains tertiary viability must be quantified. The concept of maximum sustained yield (MSY, Hjort et al. 1933) is one example of a quantitative criterion for establishing a level of harvest by humans that maintains tertiary viability. Certainly, a population sustaining exploitation by humans (i.e., satisfies criteria for tertiary viability) should satisfy criteria for primary and secondary viability. We propose that tertiary viability also include the maintenance of primary and secondary viability without direct human intervention. Decisions regarding the degree of human intervention will necessarily be based on subjective, normative judgments, but they may still be quantitatively explicit.

Suggestions for Modeling Viability in Large Landscapes

The benefits of using "viability" as a general framework management is that it helps us focus on defining quantitative management objectives and evaluation criteria for comparing models and management scenarios. This is a necessary step in using models in an adaptive management framework. In focusing on viability, managers and modelers should consider the following suggestions:

1. Organize viability problems in a risk assessment framework (Harwood 2000). We propose that acknowledgment and rigorous, quantitative comparisons of risks associated with management alternatives should not be restricted to conservation of endangered species. Given a pragmatic concept of viability that includes exploited populations, managers should consider a decision analysis and risk assessment framework for all their decisions (Maguire 1986, 1991). For example, Nicholls et al. (1996) applied risk assessment to populations facing mostly deterministic rather than stochastic threats. Hatter (1998) and Tyutyunov et al. (2002) applied risk assessment to decisions about harvested populations.

2. State explicitly what is and is not known (i.e., based on sound empirical data) about the population and factors limiting or regulating it.

3. When assessing population viability at the landscape level, a metapopulation structure may not be appropriate. Although the theory and applications related to metapopulation dynamics are well developed, managers must realize that most populations are not structured as true metapopulations (Harrison and Taylor 1997, Elmhagen and Angerbjörn 2001). Nonetheless, it remains useful to recognize interactions between somewhat local populations of management interest and populations of the same species in a much larger landscape.

4. Several methods exist for evaluating viability; population viability analysis (PVA; Reed et al. 2002) is only one. In many situations PVA is not appropriate because the methods were developed for specific purposes related

to small, declining, relatively well-studied populations. Several habitat-based approaches were outlined by Andelman et al. (2001).

5. An ideal performance measure for assessing viability should integrate all aspects of viability. It must balance realism with generality and understandability. Generally, managers will have to select several measures (e.g., population size, growth rate, age structure, genetic diversity) to fully assess viability.

6. Viability assessment should be spatially explicit when possible (c.f., Hof and Raphael 1993). In fact, managers can learn more about the system and be more confident in subsequent management actions if viability is assessed at multiple scales (Marcot et al. 2001, Mitchell et al. 2001). The extent and resolution of spatial and temporal dimensions can be varied (Probst and Gustafson, this volume), as can levels of biological organization (e.g., population versus multispecies versus community).

7. Focus on relative differences in viability instead of trying to establish a single quantitative measure of viability for a population (Reed et al. 2002). Pragmatically, many management decisions lend themselves to relative comparisons between populations, land units, or management alternatives anyway. Furthermore, estimates of viability are based on many simplifying assumptions and, when considered in isolation, cannot usually be trusted to provide robust recommendations (c.f., Brook et al. 2000). Qualitative comparisons of viability estimates over space or time, however, are considered more informative because biases due to the model are consistent for all scenarios under consideration.

VISUALIZATION

Communicating the modeling process and results to stakeholders is a critical step in conservation planning, and visualization is an important tool for this purpose. Visualization is an important and often misunderstood part of analyzing data and assessing model output. Visualization and visualization tools are analogous to well formulated figures (Day and Gastel 2006). In the case of landscape modeling, the purpose of visualization is not simply to produce a photorealistic picture of the simulation. The role of visualization in science and resource management is to effectively convey complex information and relationships about the state and change of the resource, in a clear and unambiguous way. Whether or not a realistic picture is necessary depends on the objectives and use of the output.

Several methods of visualizing data in natural resources contexts have been used, and each has strengths and weaknesses because of the way an image is perceived by the human brain (e.g., Orland 1992; McGaughey 1997, 1999, 2000; Brabyn 2003; Lang and Langanke 2005). The human brain processes visual data quickly. While generally quite accurate, the observer makes a number of assumptions about what is expected in a particular environment. However, if the environment is unfamiliar or assumptions change, observer perception might be incorrect. The effect of this biological legacy is that when observing data, humans often perceive patterns that are difficult to confirm statistically and that are highly dependent on the context and assumptions of the observer. This inherent human tendency toward subjective visual interpretation plays a role in how people interpret visual data.

To illustrate this point, consider three classic optical illusions (Figs. 1-2, 1-3, and 1-4). In Fig. 1-2, the observer often has difficulty determining the correct extension of the line that passes underneath the rectangle (Poggendorff 1863). In Fig. 1-3, the parallel lines appear to be curved because of the visual distraction of the background lines (Hering 1861). Last, Fig. 1-4 illustrates the effect of local context on the perception of size, where the center dots are the same size, but do not appear so at first glance. These figures highlight the point that while people trust visual images as a realistic portrayal of a situation, a person's assumptions and the context of the image affect perception. When presenting the results of landscape models, we must be careful to provide an appropriate context for accurate transmission of results and implications.

Most people are familiar with traditional data visualization techniques such as graphs, maps, and photographs. These tools can be of limited value in

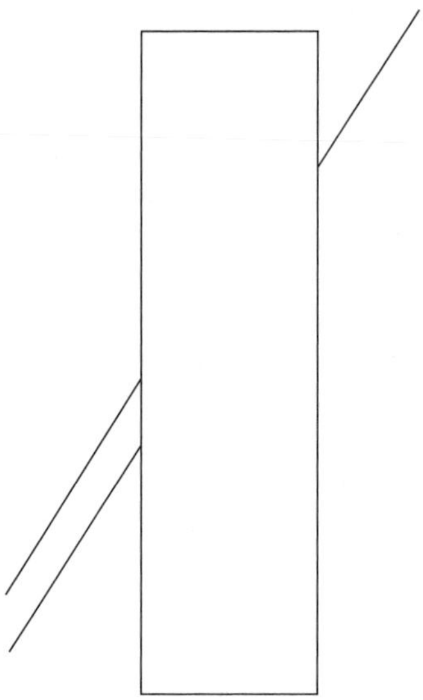

FIG. 1-2

The Poggendorff illusion (1863) in which one can easily misinterpret the association of lines passing underneath the rectangle.

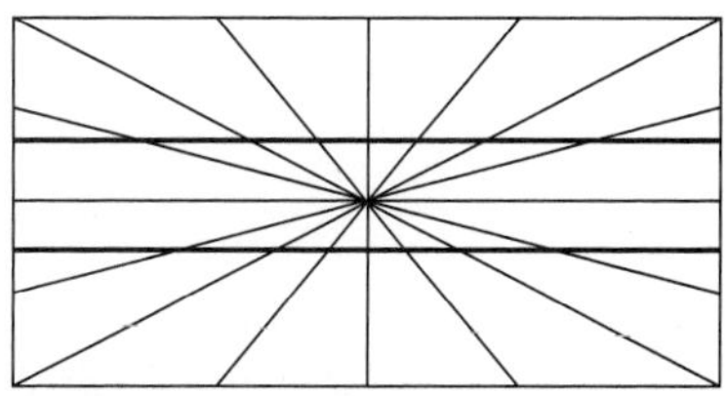

FIG. 1-3

The Hering illusion (1861) illustrates parallel straight lines that are perceived as curved.

FIG. 1-4

In this illusion the center dots are the same size but can be easily interpreted as differing because of the context of the neighboring dots.

relaying complex situations because they allow only two or sometimes three dimensions of data to be displayed in a two-dimensional context. In complex situations, these tools can easily oversimplify the relationships of the data. However, a tool that helps the user understand complex relationships is more useful than one that creates a pleasing image (Tufte 1986, 1990). Sometimes, realistic detail that does not inherently convey important information can simply be distracting. For example, in some large-scale modeling activities, we can produce images of woody vegetation on a landscape surface (Oliver et al., this volume). In this context, the adage "one cannot see the forest through the trees" is true. Trees that make up the forest image can distract from our understanding of the relationship of the trees that create the forest structure. Given these issues, how do we create visualization output that facilitates our understanding of important principles?

Issues, Solutions, and Examples

Motion.— One helpful tool in visualization that is not available in traditional printed images is motion. In some cases an interactive two-dimensional image might allow for better interpretation of the relationships than a rotated three-dimensional view. The reason is that many of the objects of interest are hidden in the three-dimensional view but can be easily singled out in the two-dimensional view. In this context, we use motion for interpretation the same way we use motion in a real landscape. Moving objects (e.g., wildlife) are much easier to see than the same objects remaining motionless in the same landscape. Additionally, for objects that do not move, such as trees, we often change our point of view to more fully comprehend the shape and size of the object. Motion allows users to move among the visualized objects, which enables a better understanding of their relationship in the landscape.

When one is using motion to visualize output, it is important that interactive opportunities take place at a reasonable speed so the user can adequately experience the visualized objects. An older tool in this area of visualization is called a "Flyby" and is available in many GIS software packages. These packages allow one to create a landscape with a digital elevation model (DEM) and other coded data and pass over the landscape on a predefined path, creating a feeling of flying over the landscape. These techniques have been used to effectively convey aspects of a landscape through the use of motion.

The software in Google Earth™ (<http://earth.google.com/>) is another example of observer perspective motion that is widely available. Using this software, the observer can alter his or her perspective from a vertical map (i.e., looking straight down on the earth) to a low oblique (i.e., looking at the earth at an angle with no horizon) to a high oblique (i.e., looking at the earth at an angle with the horizon visible) or even a sky view from the specified location. With these various perspectives, the observer can move around the space. At the present time, man-made objects such as buildings are the only available objects that can stand above the landscape surface. However, the Google Earth™ KML Gallery contains many useful examples relevant to conservation at large scales (e.g., forest logging in southeastern Australia; nature preserves at the Cornwall Wildlife Trust in England; conservation work by the African Wildlife Foundation; the migratory patterns of different bird species across Europe and Asia). This tool is relatively simple to use and very accessible to a wide variety of people over the Internet.

At a tree stand level, software often illustrates a representative plot (e.g., 1 ha) of the forest that displays the trees of interest (McGaughey 1997, 1999; Davidson 1995). In many currently available software packages for visualizing trees, objects of interest are presented in a three-dimensional cube that can be rotated. Again, motion aids in understanding the relationships between trees displayed in the image. However, this view might not be helpful in understanding the objects. New software developments (Scott 2006) allow not only map and profile views, but also allow interactive selection of the objects of interest. This allows one to observe the object in context and quickly move to the neighbors to understand the relationships between the objects (see Fig. 1-5).

Larger Context.— An important aspect of visualization software is the ability to place objects in a larger context. With vegetation, this allows one to quickly access several aesthetic qualities affected by proposed resource management. Envision, a product of the U.S. Forest Service, was one of the first software

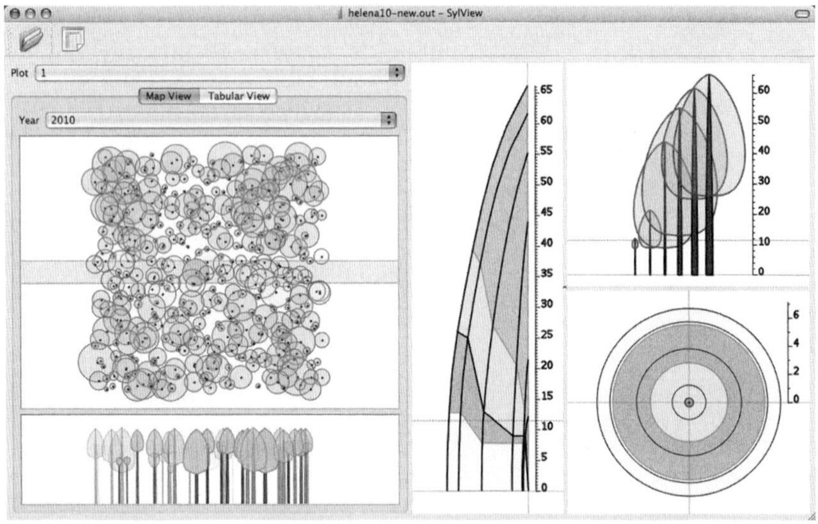

FIG. 1-5

Example of plot and tree level visualization illustrating a plot in the forest and the wood character produced in one of the trees in the plot.

packages to allow such actions (McGaughey 2000). Using Envision, a collection of forest stands is processed to determine stand average characteristics. When one combines the model with a stand map and a digital terrain model, it is possible to display the size and status of many stands on a forested landscape. DeGraaf et al. (2005) offers an alternative to this same problem; their approach models the landscape using traditional forest growth models to determine stand characteristics.

Issues of Human Perceptions.— Forest scientists have studied human perceptions of visualized forests versus photographs (Stewart et al. 1984, Kellomäki and Pukkala 1989, Pukkala 1998, Tyrväinen and Tahvanainen 1999, Lange 2001, Karjalainen and Tyrväinen 2002). Stewart et al. (1984) compared field observations with photographs taken at field sites (Fig. 1-6). They found that photographs provided a good analog to the field visit. In the other cited studies, people's perception of the aesthetic beauty of forests was compared between visualizations prepared from data collected from a forest landscape and photographs taken in the same landscape. These studies collectively demonstrate that in terms of aesthetic perceptions, visualizations are a good analog for photographs.

Issues of Misrepresentation.— With the development of new technology, software exists that can create visually believable landscapes that are not physically or biologically possible (Fig. 1-7). Sheppard (2001) argued for a code of ethics for those developing resource visualizations to avoid the creation of unrealistic but believable landscapes. Visualization should reflect as honestly as

FIG. 1-6

The image on the left is a photorealistic image created from data from 1998. The image on the right is a photograph taken in 1998 (Wilson and McGaughey 2000).

FIG. 1-7

The image is from a New England hardwood stand generated using NE-twigs for the stand conditions and Visual Nature Studio 2 (<http://3dnature.com/>) for the image. This image is the work of Anna Lester (DeGraaf et al. 2005). (See Color Insert.)

possible the data or analyses. However, how are users of visualization to know if the images portray a viable future reality or a fiction? These constraints are not that different from those placed on other forms of data. Sheppard (1989) suggested that all visualizations adhere to the following principles: accuracy, representativeness, visual clarity, interest, legitimacy, and access. Additionally the person preparing the visualizations should provide the source data in other formats to the readers when requested. Wilson and McGaughey (2000) also explored the issue of representing a landscape in a way that presents useful information that accurately summarizes the model and its output. Thus, visualization tools can be helpful and powerful to illustrating alternatives; however, they can also be misleading and confusing in presenting conditions that cannot possibly exist or by making certain conditions appear different than the data would indicate.

SUMMARY

Models have become a necessary tool for the land manager, particularly when large landscapes are considered. Models are formal frameworks for organizing and synthesizing existing knowledge of an ecological system. We reviewed principles underlying the construction and use of models, with an emphasis on habitat suitability modeling. We believe most landscape models are best viewed within an adaptive management framework because such an approach inherently acknowledges uncertainty in our knowledge of the system to be managed while recognizing that management decisions must be made. We discussed the attributes of effective models and believe that "good" models are those that promote a better decision than could be made without them (see Starfield 1997, Johnson 2001). Ineffective or unreliable models account for processes that are not relevant or well understood, depend on parameters that cannot be estimated precisely, or are dependent on too many parameters. We discussed the basic issues and approaches for addressing uncertainty in landscape models.

Population viability is an increasingly important concept in wildlife conservation, and it is useful because it focuses attention on values and risk. Several definitions of population viability are commonly used in biology; most relate to small or declining populations. We argued that the concept of viability and the quantitative risk assessment tools associated with it should be applied to the management of all wildlife populations, including potentially stable populations of abundant species, many of which are exploited by humans. We provided a definition of viability that incorporates the wide variety of current uses of the term and described a broader concept of viability that places the term in the pragmatic context of resource management. Our comprehensive, hierarchical, quantitatively explicit definition consists of persistence (i.e., primary viability), resilience (i.e., secondary viability), and the ability to provide desired services (i.e., tertiary viability).

Visualization technology has progressed with the rapid advancement of computer technology. We presented the topic of data visualization in a number of forms to help the reader understand the differing motives for creating visualization and the role of the various techniques in producing these images. Visualization software can be a powerful tool to illustrate alternative management options presented by landscape models. A major issue in the current field is the need to use advanced visualization technology in a way that accurately represents conditions being portrayed, including uncertainty in the model structure or output. When one supplies the underlying information in other forms, visualization can be evaluated for the adherence to Sheppard's (2001) principles.

LITERATURE CITED

Andelman, S. J., S. Beissinger, J. F. Cochrane, L. Gerber, P. Gomez-Priego, C. Groves, J. Haufler, R. Holthausen, D. Lee, L. Maguire, B. Noon, K. Ralls, and H. Regan. 2001. Scientific standards for conducting viability assessments under the National Forest Management Act: Report and recommendations of the NCEAS working group. National Center for Ecological Analysis and Synthesis, University of California, Santa Barbara, California, USA.

Anderson, D. R. 2001. The need to get the basics right in wildlife field studies. *Wildlife Society Bulletin* 29:1294–1297.

Beissinger, S. R., and M. I. Westphal. 1998. On the use of demographic models of population viability in endangered species management. *Journal of Wildlife Management* 62:821–841.

Benz, J., R. Hoch, and T. Legović. 2001. ECOBAS—modeling and documentation. *Ecological Modeling* 138:3–15.

Box, G. E. P. 1979. Robustness in scientific model building. Pages 201–236 *in* R. L. Launer and G. N. Wilkinson, editors. *Robustness in statistics.* Academic Press, New York, New York, USA.

Boyce, M. S. 2001. Population viability analysis: Development, interpretation, and application. Pages 123–136 *in* T. M. Shenk and A. B. Franklin, editors. *Modeling in natural resource management: Development, interpretation, and application.* Island Press, Washington, D.C., USA.

Brabyn, L. 2003. The use of geographical information systems for analysing and visualising biodiversity data. *International Journal of Environmental Technology and Management* 3:157–165.

Brook, B. W., J. J. O'Grady, A. P. Chapman, M. A. Burgman, H. R. Akçakaya, and R. Frankham. 2000. Predictive accuracy of population viability analysis in conservation biology. *Nature* 404:385–387.

Burnham, K. P., and D. R. Anderson. 2002. *Model selection and multimodel inference: A practical information-theoretic approach.* Second edition. Springer-Verlag, New York, New York, USA.

Caswell, H. 2001. *Matrix population models: Construction, analysis, and interpretation.* Second edition. Sinauer Associates, Sunderland, Massachusetts, USA.

Chamberlin, T. C. 1890. The method of multiple working hypotheses. *Science* (old series) 15:92; reprinted in *Science* 148:754–759.

Clark, W. R., and R. A. Schmitz. 2001. When modelers and field biologists interact: Progress in resource science. Pages 197–208 *in* T. M. Shenk and A. B. Franklin, editors. *Modeling in natural resource management: Development, interpretation, and application.* Island Press, Washington, D.C., USA.

Connell, J. H., and W. P. Sousa. 1983. On the evidence needed to judge ecological stability or persistence. *American Naturalist* 121:789–824.

Conroy, M. J., and C. T. Moore. 2001. Simulation models and optimal decision making in natural resource management. Pages 91–104 *in* T. M. Shenk and A. B. Franklin, editors. *Modeling in natural resource management: Development, interpretation, and application.* Island Press, Washington, D.C., USA.

Converse, S. J., G. C. White, K. L. Farris, and S. Zack. 2006. Small mammals and forest fuel reduction: national-scale responses to fire and fire surrogates. *Ecological Applications* 16:1717–1729.

Davidson, M. 1995. *The sylvan display program.* Thesis. University of Missouri, Columbia, Missouri, USA.

Day, R. A., and B. Gastel. 2006. *How to write and publish a scientific paper.* Sixth edition. Greenwood Press, Connecticut, USA.

DeGraaf, R. M., M. Yamasaki, W. B. Leak, and A. M. Lester. 2005. *Landowner's guide to wildlife habitat: Forest management for the New England region.* University of Vermont, Burlington, USA.

Dennis, B., P. L. Munholland, and J. M. Scott. 1991. Estimation of growth and extinction parameters for endangered species. *Ecological Monographs* 61:115–144.

Dodson, J. J., R. J. Gibson, R. A. Cunjak, K. D. Friedland, C. G. de Leaniz, M. R. Gross, R. Newbury, J. L. Nielsen, M. E. Power, and S. Roy. 1998. Elements in the development of conservation plans for Atlantic salmon (*Salmo salar*). *Canadian Journal of Fisheries and Aquatic Sciences* 55(Supplement 1): 312–323.

Elmhagen, B., and A. Angerbjörn. 2001. The applicability of metapopulation theory to large mammals. *Oikos* 94:89–100.

Ford, E. D. 2000. *Scientific method for ecological research*. Cambridge University Press, Cambridge, United Kingdom.

Fuentes, M., T. G. F. Kittel, and D. Nychka. 2006. Sensitivity of ecological models to their climate drivers: Statistical ensembles for forcing. *Ecological Applications* 16:99–116.

Ginzburg, L. R., L. B. Slobodkin, K. Johnson, and A. G. Bindman. 1982. Quasiextinction probabilities as a measure of impact on population growth. *Risk Analysis* 2:171–181.

Haefner, J. W. 1996. *Modeling biological systems: Principles and applications*. Chapman and Hall, New York, New York, USA.

Harrell, F. E., Jr. 2001. *Regression modeling strategies: With applications to linear models, logistic regression, and survival analysis*. Springer-Verlag, New York, New York, USA.

Harrison, S., and A. D. Taylor. 1997. Empirical evidence for metapopulation dynamics. Pages 27–42 *in* I. Hanski and M. Gilpin, editors. *Metapopulation biology: Ecology, genetics, and evolution.* Academic Press, San Diego, California, USA.

Harwood, J. 2000. Risk assessment and decision analysis in conservation. *Biological Conservation* 95:219–226.

Hatter, I. W. 1998. A Bayesian approach to moose population assessment and harvest decisions. *Alces* 34:47–58.

Hering, C. 1861. "Wo ist der Beweis für diese Symptome?" *Homoeopathische Vierteljahreschrift* 12:236–289.

Hilborn, R., and M. Mangel. 1997. *The ecological detective: Confronting models with data*. Princeton University Press, Princeton, New Jersey, USA.

Hill, S. L., G. M. Watters, A. E. Punt, M. K. McAllister, C. Le Quére, and J. Turner. 2007. Model uncertainty in the ecosystem approach to fisheries. *Fish and Fisheries* 8:315–336.

Hjort, J., G. Jahn, and P. Ottestad. 1933. The optimum catch. *Hvalrådets skrifter* 7:92–127.

Hobbs, N. T., and R. Hilborn. 2006. Alternatives to statistical hypothesis testing in ecology: A guide to self teaching. *Ecological Applications* 16:5–19.

Hof, J. G., and M. G. Raphael. 1993. Some mathematical programming approaches for optimizing timber age-class distributions to meet multispecies wildlife population objectives. *Canadian Journal of Forestry Research* 23:828–834.

Holling, C. S. 1973. Resilience and stability of ecological systems. *Annual Review of Ecology and Systematics* 4:1–23.

Holling, C. S. 1978. *Adaptive environmental assessment and management*. Wiley and Sons, New York, New York, USA.

Hubbell, S. P. 2001. *The unified neutral theory of biodiversity and biogeography.* Princeton University Press, Princeton, New Jersey, USA.

Johnson, D. H. 2001. Validating and evaluating models. Pages 105–119 *in* T. M. Shenk and A. B. Franklin, editors. *Modeling in natural resource management: Development, interpretation, and application.* Island Press, Washington, D.C., USA.

Karjalainen, E., and L. Tyrväinen. 2002. Visualization in landscape preference research: A review of Finnish forest visualization systems. *Landscape and Urban Planning* 59:13–28.

Kellomäki, S., and T. Pukkala. 1989. Forest landscape: A method of amenity evaluation based on computer simulation. *Landscape and Urban Planning* 18:117–125.

Kendall, W. L. 2001. Using models to facilitate complex decisions. Pages 147–170 *in* T. M. Shenk and A. B. Franklin, editors. *Modeling in natural resource management: Development, interpretation, and application.* Island Press, Washington, D.C., USA.

Kettenring, K. M., B. T. Martinez, A. M. Starfield, and W. M. Getz. 2006. Good practices for sharing ecological models. *BioScience* 56:59–64.

Lang, S., and T. Langanke. 2005. Multiscale GIS tools for site management. *Journal of Nature Conservation* 13:185–196.

Lange, E. 2001. The limits of realism: Perceptions of visual landscapes. *Landscape and Urban Planning* 54:163–182.

Larson, M. A., F. R. Thompson, III, J. J. Millspaugh, W. D. Dijak, and S. R. Shifley. 2004. Linking population viability, habitat suitability, and landscape simulation models for conservation planning. *Ecological Modelling* 180:103–118.

Link, W. A., and R. J. Barker. 2006. Model weights and the foundations of multimodel inference. *Ecology* 87:2626–2635.

MacKenzie, D. I., J. D. Nichols, J. A. Royle, K. H. Pollock, L. L. Bailey, and J. E. Hines. 2006. *Occupancy estimation and modeling: Inferring patterns and dynamics of species occurrence*. Elsevier, San Diego, California, USA.

Maguire, L. A. 1986. Using decision analysis to manage endangered species populations. *Journal of Environmental Management* 22:345–360.

Maguire, L. A. 1991. Risk analysis for conservation biologists. *Conservation Biology* 5:123–125.

Mangel, M., Ø. Fiksen, and J. Giske. 2001. Theoretical and statistical models in natural resource management and research. Pages 57–72 *in* T. M. Shenk and A. B. Franklin, editors. *Modeling in natural resource management: Development, interpretation, and application.* Island Press, Washington, D.C., USA.

Marcot, B. G., R. S. Holthausen, M. G. Raphael, M. M. Rowland, and M. J. Wisdom. 2001. Using Bayesian belief networks to evaluate fish and wildlife population viability under land management alternatives from an environmental impact statement. *Forest Ecology and Management* 153:29–42.

McGaughey, R. 1999. Stand Visualization System. U.S. Forest Service, Pacific Northwest Research Station, Seattle, Washington, USA. <http://forsys.cfr.washington.edu/svs.html>. Accessed 27 December 2007.

McGaughey, R. 2000. EnVision—Environmental Visualization System. U.S. Forest Service, Pacific Northwest Research Station, Seattle, Washington, USA. <http://forsys.cfr.washington.edu/envision.html>. Accessed 27 December 2007.

McGaughey, R. J. 1997. Visualizing forest stand dynamics using the stand visualization system. *ACSM/ASPRS Annual Convention and Exposition/Proceedings* 4:248–257.

Meffe, G. K., and C. R. Carroll. 1997. What is conservation biology? Pages 3–27 *in* G. K. Meffe, C. R. Carroll, and contributors. *Principles of conservation biology.* Second edition. Sinauer Associates, Sunderland, Massachusetts, USA.

Mitchell, M. S., R. A. Lancia, and J. A. Gerwin. 2001. Using landscape-level data to predict the distribution of birds on a managed forest: Effects of scale. *Ecological Applications* 11:1692–1708.

Mladenoff, D. J., and T. A. Sickley. 1998. Assessing potential gray wolf restoration in the Northeastern United States: A spatial prediction of favorable habitat and potential population levels. *Journal of Wildlife Management* 62:1–10.

Moore, C. T., M. J. Conroy, and K. Boston. 2000. Forest management decisions for wildlife objectives: System resolution and optimality. *Computers and Electronics in Agriculture* 27:25–39.

Morrison, M. L., B. G. Marcot, and R. W. Mannan. 1998. *Wildlife-habitat relationships: Concepts and applications*. Second edition. University of Wisconsin Press, Madison, Wisconsin, USA.

Murphy, H. T., and J. Lovett-Doust. 2007. Accounting for regional niche variation in habitat suitability models. *Oikos* 116:99–110.

Nichols, J. D. 2001. Using models in the conduct of science and management of natural resources. Pages 11–34 *in* T. M. Shenk and A. B. Franklin, editors. *Modeling in natural resource management: Development, interpretation, and application.* Island Press, Washington, D.C., USA.

Nichols, J. D., F. A. Johnson, and B. K. Williams. 1995. Managing North American waterfowl in the face of uncertainty. *Annual Review of Ecology and Systematics* 26:177–199.

Nichols, J. D., and B. K. Williams. 2006. Monitoring for conservation. *Trends in Ecology and Evolution* 21:668–673.

Nicholls, A. O., P. C. Viljoen, M. H. Knight, and A. S. van Jaarsveld. 1996. Evaluating population persistence of censused and unmanaged herbivore populations from the Kruger National Park, South Africa. *Biological Conservation* 76:57–67.

Oreskes, N., K. Shrader-Frechette, and K. Belitz. 1994. Verification, validation, and confirmation of numerical models in the earth sciences. *Science* 263:641–646.

Orland, B. 1992. Papers presented at a workshop on data visualization techniques in environmental management. *Landscape and Urban Planning* 21:237–345.

Platt, J. R. 1964. Strong inference. *Science* 146:347–353.

Poggendorff, J. C. 1863. Biographisch-literarisches Handwörterbuch zur Geschichte der exacten Wissenschaften enthaltend Nachweisungen über Lebensverhältnisse und Leistungen von Mathematikern, Astronomen, Physikern, Chemikern, Mineralogen, Geologen usw. aller Völker und Zeiten. 2 volumes, Leipzig, Germany.

Pukkala, J. C. 1998. MONSU metsäsuunnitteluohjelma. Ohjelmiston toiminta ja käyttö (MONSU forest planning system. Function and use of the Software) . 61 pp. (In Finnish).

Reed, J. M., L. S. Mills, J. B. Dunning, Jr., E. S. Menges, K. S. McKelvey, R. Frye, S. R. Beissinger, M.-C. Anstett, and P. Miller. 2002. Emerging issues in population viability analysis. *Conservation Biology* 16:7–19.

Scott, I. 2006. *Sylview: A visualization system for forest management*. Thesis. University of Missouri, Columbia, Missouri, USA.

Scott, J. M., P. J. Heglund, M. L. Morrison, J. B. Haufler, M. G. Raphael, W. A. Wall, and F. B. Samson, editors. 2002. *Predicting species occurrences: Issues of accuracy and scale*. Island Press, Washington D.C., USA.

Shenk, T. M., and A. B. Franklin, editors. 2001. *Modeling in natural resource management: Development, interpretation, and application.* Island Press, Washington, D.C., USA.

Sheppard, S. R. J. 1989. *Visual simulation: A user's guide for architects, engineers, and planners.* Van Nostrand Reinhold, New York, New York, USA.

Sheppard, S. R. J. 2001. Guidance for crystal ball gazers: Developing a code of ethics for landscape visualization. *Landscape and Urban Planning* 54:183–199.

Shifley, S. R., F. R. Thompson, III, W. D. Dijak, M. A. Larson, and J. J. Millspaugh. 2006. Simulated effects of forest management alternatives on landscape structure and habitat suitability in the Midwestern United States. *Forest Ecology and Management* 229:361–377.

Starfield, A. M. 1997. A pragmatic approach to modeling for wildlife management. *Journal of Wildlife Management* 61:261–270.

Starfield, A. M., and A. L. Bleloch. 1991. *Building models for conservation and wildlife management.* Second edition. Interaction Book Company, Edina, Minnesota, USA.

Stewart, T. R., P. Middleton, M. Downtown, and D. Ely. 1984. Judgments of photographs vs. field observations in studies of perceptions and judgment of the visual environment. *Journal of Environmental Psychology* 4:283–302.

Swartzman, G. 1996. Resource modeling moves into the courtroom. *Ecological Modelling* 92:277–288.

Taylor, B., L. Kremsater, and R. Ellis. 1997. *Adaptive management of forests in British Columbia*, B.C. Ministry of Forests, Victoria, British Columbia, Canada.

Tufte, E. R. 1986. *The visual display of quantitative information*. Graphics Press, Cheshire, Connecticut, USA.

Tufte, E. R. 1990. *Envisioning information*. Graphics Press, Cheshire, Connecticut, USA.

Tyrväinen, L., and L. Tahvanainen. 1999. Using computer graphics for assessing the scenic value of large-scale rural landscapes. *Scandinavian Journal of Forest Research* 13:282–288.

Tyutyunov, Y., I. Senina, C. Jost, and R. Arditi. 2002. Risk assessment of the harvested pike-perch population of the Azov Sea. *Ecological Modelling* 149:297–311.

Walters, C. J., and R. Hilborn. 1978. Ecological optimization and adaptive management. *Annual Review of Ecology and Systematics* 9:157–188.

Walters, C. J., and S. J. D. Martell. 2004. *Fisheries ecology and management*. Princeton University Press, Princeton, New Jersey, USA.

White, G.C 2001. Why take calculus? Rigor in wildlife management. *Wildlife Society Bulletin* 29:380–386.

Whittingham, M. J., J. R. Krebs, R. D. Swetnam, J. A. Vickery, J. D. Wilson, and R. P. Freckleton. 2007. Should conservation strategies consider spatial generality? Farmland birds show regional not national patterns of habitat association. *Ecology Letters* 10:25–35.

Williams, B. K., J. D. Nichols, and M. J. Conroy. 2002. *Analysis and management of animal populations: Modeling, estimation, and decision making*. Academic Press, San Diego, California, USA.

Wilson, J. S., and R. J. McGaughey. 2000. Presenting landscape-scale forest information: What is sufficient and what is appropriate? *Journal of Forestry* 98:21–28.

Wolff, J.O 1995. On the limitations of species-habitat association studies. *Northwest Science* 69:72–76.

CHAPTER

Application of Population Viability Analysis to Landscape Conservation Planning

2

Steven R. Beissinger, Emily Nicholson, and Hugh P. Possingham

The concept of population viability is a central tenet of landscape planning and management. Maintaining biological diversity and therefore the viability of all species presents a significant constraint in designing working landscapes that can produce the goods and services society desires. Large-scale landscape planning to maintain biodiversity usually integrates coarse-scale assessments of land use change with systematic evaluations of its effects on the likelihood of species becoming extinct years in the future, or their population viability (Andelman et al. 2001; Noon et al. 2003; Noon et al., this volume).

Population viability is challenging, however, to define and to evaluate. There is no standard definition of a viable population in terms of the time horizon or level of acceptable risk of extinction, and there is no standard definition of what comprises a population viability analysis (PVA). The first PVAs were quantitative modeling exercises that estimated the risk of extinction within a specified time period (Shaffer 1981, Gilpin and Soulé 1986). The output of a probabilistic estimate of extinction from a stochastic model differentiated PVA from earlier deterministic models used to guide the recovery of endangered species (Beissinger 2002), but PVA has been used to describe analysis outcomes ranging from workshops to the results of expert panels. There are recent suggestions, however, that the term *PVA* should be restricted to analytical or simulation models that use data to estimate the risk of extinction or a closely related measure of population health, such as the proportion of simulated populations that end below some size after a specified period of time or quasiextinction (Ralls et al. 2002, Reed et al. 2002). A viable population is one that does not exceed this measure of risk. An acceptable risk of extinction, and the time horizons over which they should be assessed, should be set at levels that scientists, decision makers, and society are willing to accept.

Extinction risk is difficult to assess in part because it is a stochastic process affected by many factors. Estimates of long-term means and variances for birth and death rates, population growth rate, immigration and emigration rates, and density-dependent relationships are needed (Beissinger and Westphal 1998). Genetic and catastrophic processes should be incorporated because they can play critically important roles in population viability when a long time period is considered—but unfortunately they are often ignored (Mangel and Tier 1994, Allendorf and Ryman 2002). Assessing extinction risk must project these processes into the future in relation to potential scenarios of environmental change. Models have become a useful tool for integrating the effects of these multiple forces on the risk of extinction.

In this chapter, we briefly review available methods for assessing population viability and discuss which models are useful under different circumstances. We then discuss how they may be used in making decisions about landscape design and conclude with a discussion of the uncertainties inherent in population viability analysis.

METHODS OF MODELING POPULATION VIABILITY

There are many methods for modeling population viability (Table 2-1). They include demographic models that assess the impact of management on the rate of population growth or risk of extinction, analyses of occupancy using presence-

Table 2-1 Types of Models and Analyses that have been Used to Assess Viability and the Kinds of Data they Require (Modified From Andelman et al. 2001)

		General Data Needs				
Type	**Analysis**	**Habitat Maps**	**Occurrence**	**Demography**	**Dispersal**	**Pop. Estimate**
Demographic	Matrix	No	No	Population	No	No
	Stochastic PVA	No	No	Population	No	Population
	Metapopulation	Current, Future	Population, Patches	Patch	Population	Patch
	Spatially Explicit	Current, Future	Individuals, Patches	Individual	Individual	Patch
Occupancy	Incidence Function	Current	Population, Patches	No	No	No
	Patch Turnover	Current	Population, Patches	No	No	No
Population	Trend Analysis	No	No	No	No	Population, Index
Genetic	Effective Size	No	No	Population	No	Population

absence data, population trend analysis, and genetic models that assess the loss of genetic diversity. We briefly review these methods below, but see Beissinger and Westphal (1998), Andelman et al. (2001), Possingham et al. (2001), Morris and Doak (2002), and Beissinger et al. (2006) for more in-depth discussions.

Demographic models explicitly incorporate birth and death rates, and to varying degrees the processes that affect them, and are often used to evaluate population viability (Beissinger and McCullough 2002). Demographic models vary in complexity from deterministic matrix models of a single population to stochastic, spatially explicit individual-based models (IBMs) that keep track of each individual on specific landscapes (Table 2-1).

Deterministic single-population matrix models (hereafter "*matrix models*") use a set of equations (one for each age or stage class), which are often formulated into a matrix, to predict population size at time $t+1$ from information on the survival, growth, and reproduction of individuals at time t. Matrix models can be used to make estimates of the geometric rate of population growth (λ), to indicate population characteristics such as the distribution of individuals among age classes (stable age distribution) or the reproductive value of age classes, and to evaluate the relative influence of demographic rates on population change (sensitivity or elasticity analysis; Caswell 2001). Matrix population models have limited use in assessing viability because they only indicate whether and how fast a population is increasing or decreasing, and do not give an estimate of the risk of extinction. Nevertheless, a sensitivity analysis of a matrix provides useful insights into population viability because it can indicate how changes in specific demographic rates impact rates of population change (e.g., reproductive success versus adult survival). Recently, this sort of analysis has been adapted to include the costs of conservation actions to assist decision makers choosing among management options (Baxter et al. 2006).

Stochastic single population PVA models (hereafter *stochastic PVA models*) project populations for 50, 100, or more years into the future by allowing demographic rates to change using the Monte Carlo method, which samples rates randomly from predetermined distributions. Each run of a stochastic model follows a unique trajectory and yields a different ending population size, so models must be run 500 to 1000 times to explore the full range of parameter values and portray the distribution of possible ending population sizes (Harris et al. 1987, Burgman et al. 1993). In addition to estimates of mean age- or stage-specific survival and fecundity, stochastic models require estimates of variance in fecundity and survival for each age or stage class to model the effects of demographic and environmental stochasticity. Stochastic PVA models should also include carrying capacity and its variance (Ginzburg et al. 1990), and the frequency and effects of catastrophes. They can also incorporate processes that may occur when populations become small, such as Allee effects or inbreeding depression. Stochastic PVA models yield the proportion of runs that end at population size zero ("extinction" rate) or at a small size such as <25 individuals ("quasiextinction" rate), or the mean or median year of extinction for populations that went extinct ("time to extinction"). Perhaps the most complete descriptor of model results is the cumulative probability function for ending population size, or the "quasiextinction function" (Ginzburg et al. 1982, Burgman et al. 1993). All these outputs can be expressed in relation to changes in landscape management or planning scenarios.

Demographic models of metapopulations can examine relationships between landscape structure and population dynamics by incorporating site-specific effects on demography. Compared to stochastic PVA models, they incorporate additional details of patch-specific demography, dispersal, and landscape dynamics to project population dynamics into the future (Hanski 2002; Harrison and Ray 2002; Akçakaya and Brook, this volume). These so-called *patch models* track individuals in populations as they reside only within these habitat patches and their movement among them. Population dynamics within each patch are modeled as in stochastic single-population PVA models, with the additional steps of determining the number of individuals that will disperse from and migrate to each patch. Patch metapopulation models provide output on population trajectories, risk of population extinction or decline, and related measures of population size.

Spatially explicit population models (SEPMs) are complex simulation models that may be built for single populations or a metapopulation to model movements of individuals across a diverse landscape, or the response of populations to changing landscape structure. They incorporate exact spatial locations of individuals, habitats, barriers to dispersal, and other landscape characteristics. SEPMs are often individual-based models (IBMs), where individuals are placed in known locations and assigned demographic traits based on the habitat where they are located. Two common forms of spatially explicit population models are as follows:

1. *Grid or cell-based models* track population sizes in equal-sized cells, which are typically the building blocks of larger habitat patches and are influenced by the inputs and outputs of neighboring cells. They are often used for abundant organisms, where modeling the movement or fate of each individual may be intractable, such as plants, insects, or rodents (Bradstock et al. 1996, Price and Gilpin 1996); and
2. *Individual-based models* track the location and behavior of every individual. They have been applied to determine the distribution of suitable habitat at multiple scales (Hatten and Paradzick 2003), simulate responses to regional management practices (Boone and Hunter 1996), predict regions of future human-wildlife conflicts (Treves et al. 2004), and simulate the effects of forest management policies (McKelvey et al. 1993, Lamberson et al. 1994, Liu et al. 1995, Walters et al. 2002).

Output from SEPMS is composed of averages over individuals to yield population statistics, such as population size at a specific time, population trajectories, or time to extinction.

Stochastic patch occupancy models (SPOMs) rely on simple presence-absence data obtained from single or multiple surveys to estimate the probability that a patch is occupied and to project metapopulation dynamics into the future (Day and Possingham 1995, Hanski 1998, Thomas and Hanski 2004). Etieene et al. (2004) reviewed the two types of SPOMs: (1) The incidence function approach combines the pattern of patch occupancy with relationships derived from biogeography to estimate colonization and extinction rates (Hanski 1999); and (2) the patch-turnover approach uses observed patch colonizations and extinctions (Sjogren-Gulve and Ray 1996, Kindvall 2000, Sjogren-Gulve and Hanski 2000). An important recent advance for parameterizing occupancy models accounts for imperfect detection to estimate patch occupancy (MacKenzie et al. 2006). Occupancy models can be applied to simulate metapopulation dynamics by substituting new patch-area, patch-isolation, or other landscape characteristics into a simulation model (Hanski 1999, Hanski and Gaggiotti 2004).

Population trend analysis creates models of population fluctuations from count data, such as the number of individuals in a population surveyed over multiple years, to project populations into the future and estimate extinction risk (Morris and Doak 2002). Count data may be relatively easy and cheap to collect compared to detailed demographic data. The simplest conceptual model of population growth is $N_{t+1} = \lambda N_t$ where N_t is the number of individuals in the population in year t, and λ is the population growth rate. λ and variation in λ may be dependent or independent of density, and are functions of many of the same factors that influence demographic rates discussed previously, as well as immigration and emigration (Peery et al. 2006). Dennis et al. (1991) developed an approach that extracts a maximum likelihood estimator of growth rate and its confidence interval, which has become widely used in analyses of extinction risk. It assumes that population changes can be approximated by a simple diffusion process with drift, which may work as long as density dependence is not strong, data gaps in the time series are few, and sampling error is not extensive (Dennis et al. 1991, Holmes and Fagan 2002, Morris and Doak 2002, Holmes 2004, Staples et al. 2004). Applications have been diverse and growing (Buenau and Gerber 2004, Gonzalez-Suarez et al. 2006, Thomson and Schwartz 2006). Accurate prediction of extinction risk, however, often requires decades-long time series of data (Fieberg and Ellner 2000, Lotts et al. 2004).

The concept of *effective population size* or N_e has been used to evaluate genetic considerations into PVA (Haig and Ballou 2002). N_e, defined as the number of individuals in a population that contribute genes to the next generation, measures the effect of genetic drift on a natural population. It is a useful way to evaluate the magnitude of genetic loss over time in a small population by comparing how it deviates from an "ideal" population, or by examining the relationship between N_e and the actual or census population size (N) (Waples 2002). Characteristics of an ideal population are both genetic and demographic, and include large and stable size, equal genetic contribution by both sexes, no inbreeding, equal family size, and nonoverlapping generations. Most organisms violate some or all of these characteristics. N_e is often substantially less than N, usually only 10–25% of total population size (Frankham 1995, Waples 2002). N_e was one of the first measures of population viability. Frankel and Soulé (1981) suggested that small populations should try to attain an N_e of 50 individuals over the short term and N_e of 500 individuals over the long term (the "50/500 rule"). More recently, Franklin and Frankham (1998) argued that an N_e of 500–1000 would be appropriate, and Lande (1995) suggested a minimum effective size of 5000 was more appropriate. Debate continues and no clear consensus has yet emerged on what level of N_e is needed to maintain viability (Waples 2002, Beissinger et al. 2006). In viability assessments, it seems most satisfactory to consider how various landscape scenarios affect the factors that contribute to N_e, but to avoid comparing N_e estimates to specific numeric goals (Beissinger et al. 2006).

WHEN SHOULD PVA MODELS BE USED AND WHICH ONE(S)?

All methods of assessing viability have some utility, and they all explicitly or implicitly use a model—even expert opinion PVAs (Burgman 2005). Methods that use formal mathematical models for analyzing viability are usually preferable to less quantitative methods, such as indices of rarity and expert opinion, when there is some knowledge of demography, dispersal, habitat use, and threats. The advantages of a formal model often lie mostly in its transparency and repeatability (Burgman and Possingham 2000). Creating a formal model often involves a necessary trade-off between generality, precision, and realism (Levins 1968, Beissinger et al. 2006). The range of possible methods in Table 2-1 offers a trade-off between complexity of analysis and generality of results. None might provide perfect analysis alone, but all can provide important insights when used together as one of several analyses that compose a viability assessment. Statistical models (e.g., incidence function models) may yield precise model outputs for some situations, but their generality may extend only to a specific set of data and they may have limited applicability due to assumption of a stationary landscape configuration required by the model (Hanski 2002). The most realistic models for landscape planning are often stochastic PVA, patch occupancy, metapopulation, or spatially explicit models that can be developed for specific species, management scenarios, or landscapes. These models, however, often yield less precise or more uncertain results because they are stochastic and because all model parameters can rarely be estimated from data specific to the system of interest.

The choice of a PVA model will often be determined by the objectives of the analysis, the data that are available, and the assumptions that are realistic to make (Ralls et al. 2002). The typical objective of a viability assessment in large-scale landscape planning is to determine which land management actions would be most beneficial or least injurious to a set of species. This concerns comparing risk to a species posed by different management options relative to one another (i.e., *relative risk*), rather than estimating an accurate risk or impact from a particular management action (i.e., *absolute risk*). A model that attempts to estimate absolute risk should strive to develop an accurate, comprehensive representation of reality and would include all factors influencing the probability of extinction (Ralls et al. 2002). Building models that can accurately estimate absolute extinction risk and the data required for these models are, for most species, still beyond our capabilities (Ludwig 1996, 1999; Fieberg and Ellner 2000). Models designed to compare differences among management options (i.e., estimate relative risks of extinction), however, may be less comprehensive than a model intended to estimate absolute risk of extinction and can be useful for assisting in decision making. The general consensus in the literature is that because of uncertainties in estimates of extinction risk, PVA is probably most useful in ranking alternative management policies' or scenarios' uncertainty (Possingham et al. 1993, Lindenmayer and Possingham 1996, Beissinger and Westphal 1998, Reed et al. 2002, Drechsler et al. 2003, Wintle et al. 2003).

Analyzing viability for large landscapes requires many species to be considered and increases the demand for data, affecting which analyses are appropriate. More complex models require greater technical skills and more time to create. Viability assessments for most species should be conducted from the scale of one or more National Forests to the scale of bioregions by comparing relative differences of the impacts of different planning scenarios. Thus, the potential for building PVA models is likely to be limited to a few or at most a dozen of the 50–300 species that are likely to need assessment when managing large-scale landscapes. Expert opinion in some form, elicited in a structured fashion and clearly documented, may be used to assess viability for the majority of affected species (Andelman et al. 2001, Burgman 2005).

USING PVA TO INFORM LARGE-SCALE LANDSCAPE PLANNING, MANAGEMENT, AND POLICY DECISIONS

We now consider how PVAs can be used to aid landscape planning and policy decisions, and issues surrounding the use of PVA in this context. What sort of large-scale landscape management decisions might a PVA help inform? Examples include (1) deciding which areas are priorities for protection or land acquisition, such as Strategic Habitat Conservation Areas identified by the Florida Fisheries and Wildlife Commission (Kautz and Cox 2001); (2) assessing and designing alternative management regimes, such as logging, to minimize the impact on biodiversity (Montgomery et al. 1994, Fox et al. 2004, Nalle et al. 2004);

(3) setting take levels for harvested species, such as fisheries or game (Milner-Gulland et al. 2001, Sethi et al. 2005); and (4) testing theory for biodiversity conservation and landscape design to be applied through rules of thumb or other ways to real-world decisions (McCarthy et al. 2001, Carroll et al. 2003, Lindenmayer et al. 2003, Nicholson et al. 2006).

PVAs have also been used to provide information that can, in turn, be used in either systematic conservation planning or to design landscapes. Examples of such use include (1) setting minimum population or patch sizes for single species or multispecies systematic conservation planning (e.g., Strategic Habitat Conservation Area in Florida [Burgman et al. 2001, Kautz and Cox 2001, Cowling et al. 2003]); (2) ascribing the value of a given area to viability, such as using the intrinsic growth rate λ of an area or the contribution of an area to the overall growth rate (Calkin et al. 2002, Noss et al. 2002, Carroll et al. 2003); and (3) parameterizing statistical approximation models (Calkin et al. 2002, Haight et al. 2002, Nalle et al. 2004). They may also be used to rank alternative management decisions in a decision analysis context, as we discuss next.

Decision Theory and PVA

To get the most out of a PVA when making management decisions, one should invoke a decision analysis framework (Maguire 1986), where the goals and constraints of the problem are explicitly defined (Shea et al. 1998, Possingham et al. 2002, Drechsler and Burgman 2004). In a single species example, the problem formulation may be as simple as comparing the viability of the species under a business-as-usual scenario versus an active management plan (e.g., captive breeding or preservation of critical habitat). It may be framed as a cost-benefit analysis: how much benefit (e.g., reduction in extinction risk) is gained as the amount of area protected increases or timber harvest is forgone (Montgomery et al. 1994)?

When we have the luxury of constructing PVA models for more than one species, we have the capacity to consider a much wider range of biodiversity needs, but it also entails a more complex decision model (Nicholson and Possingham 2006). How do we balance the competing needs of several species when each has different management preferences (Noon et al., this volume)? For example, a species with large area requirements may have higher viability when one or two large areas of habitat are protected, whereas a dispersal-limited species would benefit more from good corridors connecting areas of suitable habitat, while another species needs more frequent burns or predator removal. When one makes decisions that take into account the needs of different species, explicit goals and a structure for making decisions (i.e., decision theory or decision analysis) become even more important (Drechsler and Burgman 2004, Nicholson and Possingham 2006).

There are many ways in which assessment of viability for many species can inform management decisions. Multicriteria decision analysis can be used to make a decision based on the ranking of the management alternatives for each of the species (e.g., Drechsler and Burgman 2004, Nicholson and Possingham 2007). Similar to a voting system, an overall ranking can be derived by finding the management alternative that ranks highest for the most species. Species may be weighted to emphasize keystone species or species with higher risks of extinction. Alternatively, the probabilities of extinction can be combined to form a benefit or utility function. The simplest of these functions might be to add the extinction risks together to yield a cost, the expected number of extinctions, which we attempt to minimize. However, there are several ways to derive an overall benefit function, including minimizing the chance of any species becoming extinct or of all becoming extinct (Hof and Raphael 1993, Nicholson and Possingham 2006). An index can also be developed to combine assessments of viability for multiple species across a landscape (Akçakaya 2000, Carroll et al. 2003, Root et al. 2003). For example, Root et al. (2003) generated a multiple-species index of conservation value, combining species-specific habitat suitability, extinction risk, and contribution to population viability estimated using a stochastic metapopulation model.

Using PVA in an Optimization Framework

Population models may also be used in an optimization framework, where an algorithm is used to find the landscape configuration that maximizes the viability of one or more species. There are few examples of population models being used directly in designing optimal landscapes, primarily because the simulations needed to assess the likelihood of extinction in a stochastic model constrain the size and complexity of the problems that can be included in an optimization framework. Some methods exist for finding the optimal landscape design for a single species (e.g., Hof and Raphael 1997, Hof et al. 1999, Moilanen and Cabeza 2002), but they tend to be based on simpler types of population models. Stochastic individual-based models or stage-based matrix models could not be used in such an optimization framework because computing power limits the number of simulations required (at least for the time being). Because there are so many spatial options to explore in optimal landscape design, finding the best option can only be achieved if the performance of each option can be evaluated extremely quickly, i.e., in fractions of a second. Simulation-based PVA does not run this quickly.

The computational constraints of single species problems become even more exacerbated when trying to consider multiple species within a landscape optimization framework (Noon et al., this volume; Haight and Gobster, this volume). The few examples of multiple-species optimization based on population models have been simple deterministic or approximation models. Nicholson et al. (2006) used a spatially realistic metapopulation model, which approximated the results of a stochastic approximation model (Frank and Wissel 2002), to maximize the expected number of persistent species in a reserve system for a case study of 10 species. In a case study of 92 species, Hof and Bevers

(Hof and Raphael 1993, Hof et al. 1994, Bevers et al. 1995) used simple area-based viability functions to maximize the persistence across multiple species using a variety of utility functions, including maximizing the expected number of viable species and maximizing the chance that all species persist. Rothley (2002) used a deterministic Lotka-Volterra model for two-species predator-prey conservation. Araújo and Williams (Araújo and Williams 2000, Williams and Araújo 2000) used the probability of occurrence to approximate probability of persistence when maximizing expected number of persistent species for 174 tree species. McCarthy et al. (2006) developed theory, rather than found optimal solutions, for a general case where dispersal between populations was ignored.

Each of the models used in the preceding examples relied on simplifications and assumptions, but they provide frameworks that can be built upon as computational power increases to permit more complex models. Their consideration of landscape processes, such as local extinction dynamics and dispersal among populations, provide an improvement on more static methods for landscape design that only consider the presence of species or landscape indices, rather than species persistence (Williams et al. 2005, Nicholson et al. 2006). There is no doubt that we will need to further develop these ideas as we strive to include the idea of conservation adequacy into landscape planning.

The Certainty of Needing to Deal with Uncertainty in PVAs

The only certainty in PVA is that there is uncertainty, and a lot of it. Uncertainties can exist in the estimates for the model parameters, the structure of the model, GIS habitat layers, and other model components. Dealing with uncertainty forms an important component of a decision theory approach to management (Shea et al. 1998). Therefore, any discussion of population modeling cannot be complete without at least a brief consideration of the impact of uncertainty on management decisions, and an overview of some of the tools that are available for considering and including uncertainty when making management decisions using PVAs.

Regan et al. (2002) reviewed types of uncertainty in ecology and conservation biology, and what they mean for decision making. Epistemic uncertainty is uncertainty associated with knowledge of the state of a system. The parameters the model misestimated may be due to measurement error, systematic error (bias), natural variation, and inherent randomness (stochasticity). Also, results of the model may be uncertain due to the way in which the model structure is represented and due to subjective judgment in the interpretation of the results. Linguistic uncertainty arises because language can be ambiguous and vague (Regan et al. 2002).

Uncertainty also results from the inherent trade-off in the duration of the time span analyzed in a PVA and the accuracy of its predictions (Beissinger and Westphal 1998). Viability assessments should ensure that activities to be conducted over the planning period do not compromise the long-term persistence of a species. Assessments over long time spans are especially important for long-lived organisms with long generation times (Lande 2002). Yet, errors (i.e., uncertainties) are propagated with each time step (often one year) that the model or assessment evaluates into the future.

Since uncertainty in model outcomes will occur regardless of the measures, time span, and model used in the analysis, good PVAs will analyze and present this uncertainty. Analysis of uncertainty in model performance should include a sensitivity analysis to understand the influence of individual parameters on model outcome (Mills and Lindberg 2002, Burgman et al. 2005, Wintle et al. 2005). An increasingly large array of methods is available for making decisions under uncertainty (Burgman 2005). Cross and Beissinger (2001) and Cariboni et al. (2007) reviewed sensitivity methods for PVA models and ecological models, respectively. Perhaps the most comprehensive way to deal with uncertainty is to include it directly in the model using a Bayesian PVA (Goodman 2002, Wade 2002). Another method for explicitly considering uncertainty in the decision process is information gap decision theory (Ben-Haim 2006), which provides a framework for making decisions where severe parameter uncertainty exists. Several examples exist of the application of information gap theory to conservation problems that use population models (Burgman 2005, Regan et al. 2005, Halpern et al. 2006, Nicholson and Possingham 2007). Attention should also be paid to interactions between parameters and the influence of functions or relationships built into the model, as well as basic model structures (e.g., density dependence).

SUMMARY

Population viability analysis has an important role to play in ecosystem management and conservation planning for large landscapes. The concept of population viability is central to landscape planning because the modern theory of conservation planning demands that we consider the notion of adequacy (Possingham et al. 2006). Population viability analysis represents a "fine filter" approach to conservation planning compared to the coarse filter approach, which posits that the majority of species can be protected by conserving examples of natural vegetation communities (Noon et al. 2003). Coarse filter approaches to landscape planning use distribution maps of land cover to inform the conservation of entire species assemblages (Groves et al. 2002), but have had equivocal success (Scott et al. 2002). A credible assessment approach to maintain a diversity of plant and animal communities in large landscapes is one that combines both coarse and fine filters using PVA for rare and at-risk species, for indicator species that provide information on the state of a given ecosystem, and for focal species that play significant functional roles in ecosystems (Noon et al. 2003).

LITERATURE CITED

Akçakaya, H. R. 2000. Conservation and management for multiple species: Integrating field research and modeling into management decisions. *Environmental Management* 26:S75-S83.

Allendorf, F. W., and N. Ryman. 2002. The role of genetics in population viability analysis. Pages 50-85 *in* S. R. Beissinger and D. R. McCullough, editors. *Population viability analysis.* University of Chicago Press, Chicago, Illinois, USA.

Andelman, S. J., S. Beissinger, J. F. Cochrane, L. Gerber, P. Gomez-Priego, C. Groves, J. Haufler, R. Holthausen, D. Lee, L. Maguire, B. Noon, K. Ralls, and H. Regan. 2001. Scientific standards for conducting viability assessments under the National Forest Management Act: Report and recommendations of the NCEAS Working Group. National Center for Ecological Analysis and Synthesis, University of California, USA. <http://www.fs.fed.us/psw/rsl/projects/wild/lee/lee3.pdf>.

Araújo, M. B., and P. H. Williams. 2000. Selecting areas for species persistence using occurrence data. *Biological Conservation* 96:331-345.

Baxter, P. W. J., M. A. McCarthy, H. P. Possingham, P. Menkhorst, and N. McLean. 2006. Accounting for management costs in sensitivity analyses of matrix population models. *Conservation Biology* 20:893-905.

Beissinger, S. R. 2002. Population viability analysis: Past, present, future. Pages 5-17 *in* S. R. Beissinger and D. R. McCullough, editors. *Population viability analysis.* University of Chicago Press, Chicago, Illinois, USA.

Beissinger, S. R., and D. R. McCullough editors. 2002. *Population viability analysis.* University of Chicago Press, Chicago, Illinois, USA.

Beissinger, S. R., J. R. Walters, D. G. Catanzaro, K. G. Smith, J. B. Dunning, Jr., S. M. Haig, B. R. Noon, and B. Stith. 2006. Modeling approaches in avian conservation and the role of field biologists. *Ornithological Monographs* 59.

Beissinger, S. R., and M. I. Westphal. 1998. On the use of demographic models of population viability in endangered species management. *Journal of Wildlife Management* 62:821-841.

Ben-Haim, Y. 2006. *Information-gap decision theory: Decisions under severe uncertainty.* Second edition. Academic Press, London, United Kingdom.

Bevers, M., J. Hof, B. Kent, and M. G. Raphael. 1995. Sustainable forest management for optimizing multispecies wildlife habitat: A coastal Douglas-fir example. *Natural Resource Modeling* 9:1-23.

Boone, R. B., and M. L. Hunter. 1996. Using diffusion models to simulate the effects of land use on grizzly bear dispersal in the Rocky Mountains. *Landscape Ecology* 11:51-64.

Bradstock, R. A., M. Bedward, J. Scott, and D. A. Keith. 1996. Simulation of the effect of spatial and temporal variation in fire regimes on the population viability of *Banksia* species. *Conservation Biology* 10:776-784.

Buenau, K. E., and L. R. Gerber. 2004. Developing recovery and monitoring strategies for the endemic Mount Graham red squirrels (*Tamiasciurus hudsonicus grahamensis*) in Arizona. *Animal Conservation* 7:17-22.

Burgman, M. A. 2005. *Risks and decisions for conservation and environmental management.* Cambridge University Press, Cambridge, United Kingdom.

Burgman, M. A., S. Ferson, and H. R. Akçakaya. 1993. *Risk assessment in conservation biology.* Chapman and Hall, London, United Kingdom.

Burgman, M. A., D. B. Lindenmayer, and J. Elith. 2005. Managing landscapes for conservation under uncertainty. *Ecology* 86:2007-2017.

Burgman, M., and H. P. Possingham. 2000. Population viability analysis for conservation: The good, the bad and the undescribed. Pages 97-112 *in* A. G. Young and G. M. Clarke, editors. *Genetics, demography and viability of fragmented populations.* Cambridge University Press, London, United Kingdom.

Burgman, M. A., H. P. Possingham, A. J. J. Lynch, D. A. Keith, M. A. McCarthy, S. D. Hopper, W. L. Drury, J. A. Passioura, and R. J. Devries. 2001. A method for setting the size of plant conservation target areas. *Conservation Biology* 15:603-616.

Calkin, D. E., C. A. Montgomery, N. H. Schumaker, S. Polasky, J. L. Arthur, and D. J. Nalle. 2002. Developing a production possibility set of wildlife species persistence and timber harvest value. *Canadian Journal of Forest Research* 32:1329–1342.

Cariboni, J., D. Gatelli, R. Liska, and A. Saltelli. 2007. The role of sensitivity analysis in ecological modelling. *Ecological Modelling* 203:167–182.

Carroll, C., R. F. Noss, P. C. Paquet, and N. H. Schumaker. 2003. Use of population viability analysis and reserve selection algorithms in regional conservation plans. *Ecological Applications* 13:1773–1789.

Caswell, H. 2001. *Matrix population models: Construction, analysis, and interpretation*. Second edition. Sinauer Associates, Sunderland, Massachusetts, USA.

Cowling, R. M., R. L. Pressey, M. Rouget, and A. T. Lombard. 2003. A conservation plan for a global biodiversity hotspot—the Cape Floristic Region, South Africa. *Biological Conservation* 112:191–216.

Cross, P. C., and S. R. Beissinger. 2001. Using logistic regression to analyze the sensitivity of PVA models: A comparison of methods based on African wild dog models. *Conservation Biology* 15:1335–1346.

Day, J. R., and H. P. Possingham. 1995. A stochastic metapopulation model with variability in patch size and position. *Theoretical Population Biology* 48:333–360.

Dennis, B., P. L. Munholland, and J. M. Scott. 1991. Estimation of growth and extinction parameters for endangered species. *Ecological Monographs* 61:115–143.

Drechsler, M., and M. A. Burgman. 2004. Combining population viability analysis with decision analysis. *Biodiversity and Conservation* 13:115–139.

Drechsler, M., K. Frank, I. Hanski, R. B. O'Hara, and C. Wissel. 2003. Ranking metapopulation extinction risk: From patterns in data to conservation management decisions. *Ecological Applications* 13:990–998.

Etienne, R. S., C. J. F. ter Braak, and C. C. Vos. 2004. Application of stochastic patch occupancy models to real metapopulations. Pages 105–132 *in* I. Hanski and O. Gaggiotti, editors. *Ecology, genetics, and evolution of metapopulations.* Elsevier Academic Press, New York, New York, USA.

Fieberg, J., and S. P. Ellner. 2000. When is it meaningful to estimate an extinction probability? *Ecology* 81:2040–2047.

Fox, J. C., T. J. Regan, S. A. Bekessy, B. A. Wintle, M. J. Brown, J. M. Meggs, K. Bonham, R. Mesibov, M. A. McCarthy, S. A. Munks, P. Wells, R. Brereton, K. Graham, J. Hickey, P. Turner, M. Jones, W. E. Brown, N. Mooney, S. Grove, K. Yamada, and M. A. Burgman. 2004. Linking landscape ecology and management to population viability analysis: Report 2: Population viability analysis for eleven forest dependent species. University of Melbourne report prepared for Forestry Tasmania.

Frank, K., and C. Wissel. 2002. A formula for the mean lifetime of metapopulations in heterogeneous landscapes. *American Naturalist* 159:530–552.

Frankel, O. H., and M. E. Soulé. 1981. *Conservation and evolution*. First edition. Cambridge University Press, London, United Kingdom.

Frankham, R. 1995. Effective population size-adult population size ratios in wildlife: A review. *Genetical Research* 66:95–107.

Franklin, I. R., and R. Frankham. 1998. How large must populations be to retain evolutionary potential? *Animal Conservation* 1:69–73.

Gilpin, M. E., and M. E. Soulé. 1986. Minimum viable populations: Processes of species extinction. Pages 19–34 *in* M. E. Soulé, editor. *Conservation biology: The science of scarcity and diversity.* Sinauer Associates, Sunderland, Massachusetts, USA.

Ginzburg, L. R., S. Ferson, and H. R. Akçakaya. 1990. Reconstructibility of density dependence and the conservative assessment of extinction risks. *Conservation Biology* 4:63–70.

Ginzburg, L. R., L. B. Slobodkin, K. Johnson, and A. G. Bindman. 1982. Quasiextinction probabilities as a measure of impact on population growth. *Risk Analysis* 2:171–181.

Gonzalez-Suarez, M., K. E. McCluney, D. Aurioles, and L. R. Gerber. 2006. Incorporating uncertainty in spatial structure for viability predictions: A case study of California sea lions (*Zalophus californianus californianus*). *Animal Conservation* 9:219–227.

Goodman, D. 2002. Predictive Bayesian population analysis: A logic for listing criteria, delisting criteria and recovery plans. Pages 447–469 *in* S. R. Beissinger and D. R. McCullough, editors. *Population viability analysis.* University of Chicago Press, Chicago, Illinois, USA.

Groves, C. R., D. B. Jensen, L. L. Valutis, K. H. Redford, M. L. Shaffer, J. M. Scott, J. V. Baumgartner, J. V. Higgins, M. W. Beck, and M. G. Anderson. 2002. Planning for biodiversity conservation: Putting conservation science into practice. *BioScience* 52:499–512.

Haig, S. M., and J. D. Ballou. 2002. Pedigree models for managing small populations in the wild. Pages 388–405 *in* S. R. Beissinger and D. R. McCullough, editors. *Population viability analysis.* University of Chicago Press, Chicago, Illinois, USA.

Haight, R. G., B. Cypher, P. A. Kelly, S. Phillips, H. P. Possingham, K. Ralls, A. M. Starfield, P. J. White, and D. Williams. 2002. Optimizing habitat protection using demographic models of population viability. *Conservation Biology* 16:1386–1397.

Halpern, B. S., H. M. Regan, H. P. Possingham, and M. A. McCarthy. 2006. Accounting for uncertainty in marine reserve design. *Ecology Letters* 9:2–11.

Hanski, I. 1998. Metapopulation dynamics. *Nature* 396:41–49.

Hanski, I. 1999. *Metapopulation ecology.* Oxford University Press, New York, New York, USA.

Hanski, I. 2002. Animal metapopulations in highly fragmented landscapes and PVA. Pages 86–108 *in* S. R. Beissinger and D. R. McCullough, editors. *Population viability analysis.* University of Chicago Press, Chicago, Illinois, USA.

Hanski, I., and O. Gaggiotti. 2004. Metapopulation biology: Past, present, and future. Pages 3–22 *in* I. Hanski and O. Gaggiotti, editors. *Ecology, genetics, and evolution of metapopulations.* Elsevier Academic Press, New York, New York, USA.

Harris, R. B., L. A. Maguire, and M. L. Shaffer. 1987. Sample sizes for minimum viable population estimation. *Conservation Biology* 1:72–76.

Harrison, S., and C. Ray. 2002. Plant population viability and metapopulation-level processes. Pages 109–122 *in* S. R. Beissinger and D. R. McCullough, editors. *Population viability analysis.* University of Chicago Press, Chicago, Illinois, USA.

Hatten, J. R., and C. E. Paradzick. 2003. A multiscaled model of southwestern willow flycatcher breeding habitat. *Journal of Wildlife Management* 67:774–788.

Hof, J., M. Bevers, L. Joyce, and B. Kent. 1994. An integer programming approach for spatially and temporally optimizing wildlife populations. *Forest Science* 40:177–191.

Hof, J., and M. G. Raphael. 1997. Optimization of habitat placement: A case study of the northern spotted owl in the Olympic Peninsula. *Ecological Applications* 7:1160–1169.

Hof, J., C. H. Sieg, and M. Bevers. 1999. Spatial and temporal optimization in habitat placement for a threatened plant: The case of the western prairie fringed orchid. *Ecological Modelling* 115:61–75.

Hof, J. G., and M. G. Raphael. 1993. Some mathematical-programming approaches for optimizing timber age-class distributions to meet multispecies wildlife population objectives. *Canadian Journal of Forest Research* 23:828–834.

Holmes, E. E. 2004. Beyond theory to application and evaluation: Diffusion approximations for population viability analysis. *Ecological Applications* 14:1272–1293.

Holmes, E. E., and W. F. Fagan. 2002. Validating population viability analysis for corrupted data sets. *Ecology* 83:2379–2386.

Kautz, R. S., and J. A. Cox. 2001. Strategic habitats for biodiversity conservation in Florida. *Conservation Biology* 15:55–77.

Kindvall, O. 2000. Comparative precision of three spatially realistic simulation models of metapopulation dynamics. *Ecological Bulletin* 48:101–110.

Lamberson, R. H., B. R. Noon, C. Voss, and K. S. McKelvey. 1994. Reserve design for territorial species: The effects of patch size and spacing on the viability of the northern spotted owl. *Conservation Biology* 8:185–195.

Lande, R. 1995. Mutation and conservation. *Conservation Biology* 9:782–791.

Lande, R. 2002. Incorporating stochasticity in population viability analysis. Pages 18-40 *in* S. R. Beissinger and D. R. McCullough, editors. *Population viability analysis.* University of Chicago Press, Chicago, Illinois, USA.

Levins, R. 1968. *Strategies of adaptation*. Pages 10-38 *in Evolution in changing environments*. Princeton University Press, Princeton, New Jersey, USA.

Lindenmayer, D. B., and H. P. Possingham. 1996. Ranking conservation and timber management options for leadbeater's possum in southeastern Australia using population viability analysis. *Conservation Biology* 10:235-251.

Lindenmayer, D. B., H. P. Possingham, R. C. Lacy, M. A. McCarthy, and M. L. Pope. 2003. How accurate are population models? Lessons from landscape-scale tests in a fragmented system. *Ecology Letters* 6:41-47.

Liu, J., J. B. Dunning, Jr., and H. R. Pulliam. 1995. Potential effects of a forest management plan on Bachman's sparrows (*Aimophila aestivalis*): Linking a spatially explicit model with GIS. *Conservation Biology* 9:62-75.

Lotts, K. C., T. A. Waite, and J. A. Vucetich. 2004. Reliability of absolute and relative predictions of population persistence based on time series. *Conservation Biology* 18:1224-1232.

Ludwig, D. 1996. Uncertainty and the assessment of extinction probabilities. *Ecological Applications* 6:1067-1076.

Ludwig, D. 1999. Is it meaningful to estimate a probability of extinction? *Ecology* 80:298-310.

MacKenzie, D. I., J. D. Nichols, J. A. Royle, K. H. Pollock, L. L. Bailey, and J. E. Hines. 2006. *Occupancy estimation and modeling*. Academic Press, San Diego, California, USA.

Maguire, L. A. 1986. Using decision-analysis to manage endangered species populations. *Journal of Environmental Management* 22:345-360.

Mangel, M., and C. Tier. 1994. Four facts every conservation biologist should know about persistence. *Ecology* 75:607-614.

McCarthy, M. A., H. P. Possingham, J. R. Day, and A. J. Tyre. 2001. Testing the accuracy of population viability analysis. *Conservation Biology* 15:1030-1038.

McCarthy, M. A., C. J. Thompson, and N. S. G. Williams. 2006. Logic for designing nature reserves for multiple species. *American Naturalist* 167:717-727.

McKelvey, D., B. R. Noon, and R. H. Lamberson. 1993. Conservation planning for species occupying fragmented landscapes: The case of the northern spotted owl. Pages 424-452 *in* P. M. Kareiva, J. G. Kingslover, and R. B. Huey, editors. *Biotic interactions and global change.* Sinauer Associates, Sunderland, Massachusetts, USA.

Mills, L. S., and M. S. Lindberg. 2002. Sensitivity analysis to evaluate the consequences of conservation actions. Pages 338-366 *in* S. R. Beissinger and D. R. McCullough, editors. *Population viability analysis.* University of Chicago Press, Chicago, Illinois, USA.

Milner-Gulland, E. J., K. Shea, H. Possingham, T. Coulson, and C. Wilcox. 2001. Competing harvesting strategies in a simulated population under uncertainty. *Animal Conservation* 4:157-167.

Moilanen, A., and M. Cabeza. 2002. Single-species dynamic site selection. *Ecological Applications* 12:913-926.

Montgomery, C. A., G. M. Brown, Jr., and D. M. Adams. 1994. The marginal cost of species preservation: The northern spotted owl. *Journal of Environmental Economics and Management* 26:111-128.

Morris, W. F., and D. F. Doak. 2002. *Quantitative conservation biology: Theory and practice of population viability analysis*. Sinauer Associates, Sunderland, Massachusetts, USA.

Nalle, D. J., C. A. Montgomery, J. L. Arthur, S. Polasky, and N. H. Schumaker. 2004. Modeling joint production of wildlife and timber. *Journal of Environmental Economics and Management* 48:997-1017.

Nicholson, E., and H. P. Possingham. 2006. Objectives for multiple-species conservation planning. *Conservation Biology* 20:871-881.

Nicholson, E., and H. P. Possingham. 2007. Making conservation decisions under uncertainty for the persistence of multiple species. *Ecological Applications* 17:251–265.

Nicholson, E., M. I. Westphal, K. Frank, W. A. Rochester, R. L. Pressey, D. B. Lindenmayer, and H. P. Possingham. 2006. A new method for conservation planning for the persistence of multiple species. *Ecology Letters* 9:1049–1060.

Noon, B. R., D. D. Murphy, S. R. Beissinger, M. L. Shaffer, and D. Dellasala. 2003. Conservation planning for U.S. National Forests: Conducting comprehensive biodiversity assessments. *BioScience* 53:1217–1220.

Noss, R. F., C. Carroll, K. Vance-Borland, and G. Wuerthner. 2002. A multicriteria assessment of the irreplaceability and vulnerability of sites in the Greater Yellowstone Ecosystem. *Conservation Biology* 16:895–908.

Peery, M. Z., B. H. Becker, and S. R. Beissinger. 2006. Combining demographic and count-based approaches to identify source-sink dynamics of a threatened seabird. *Ecological Applications* 16:1516–1528.

Possingham, H. P., D. B. Lindenmayer, and M. A. McCarthy. 2001. Population viability analysis. Pages 831–844 *in* S. Levin, editor. *The encyclopedia of biodiversity.* Volume 4, Academic Press, San Diego, California, USA.

Possingham, H. P., D. B. Lindenmayer, and T. W. Norton. 1993. A framework for the improved management of threatened species based on PVA. *Pacific Conservation Biology* 1:39–45.

Possingham, H. P., D. B. Lindenmayer, and G. N. Tuck. 2002. Decision theory for population viability analysis. Pages 470–489 *in* S. R. Beissinger and D. R. McCullough, editors. *Population viability analysis.* University of Chicago Press, Chicago, Illinois, USA.

Possingham, H. P., K. A. Wilson, S. Andelman, and C. H. Vynne. 2006. Protected areas: Goals, limitations, and design. Pages 509–533 *in* M. J. Groom, G. K. Meffe, and C. R. Carroll, editors. *Principles of conservation biology.* Third edition. Sinauer Associates, Sunderland, Massachusetts, USA.

Price, M. V., and M. Gilpin. 1996. Modelers, mammalogists, and metapopulations: Designing Stephens' kangaroo rat reserves. Pages 217–240 *in* D. R. McCullough, editor. *Metapopulations and wildlife conservation.* Island Press, Washington, D.C., USA.

Ralls, K., S. R. Beissinger, and J. F. Cochrane. 2002. Guidelines for using PVA in endangered species management. Pages 521–550 *in* S. R. Beissinger and D. R. McCullough, editors. *Population viability analysis.* University of Chicago Press, Chicago, Illinois, USA.

Reed, J. M., L. S. Mills, J. B. Dunning, Jr., E. S. Menges, K. S. McKelvey, R. Frye, S. R. Beissinger, M.-C. Anstett, and P. Miller. 2002. Emerging issues in population viability analysis. *Conservation Biology* 16:7–19.

Regan, H. M., Y. Ben-Haim, B. Langford, W. G. Wilson, P. Lundberg, S. J. Andelman, and M. A. Burgman. 2005. Robust decision making under severe uncertainty for conservation management. *Ecological Applications* 15:1471–1477.

Regan, H. M., M. Colyvan, and M. A. Burgman. 2002. A taxonomy and treatment of uncertainty for ecology and conservation biology. *Ecological Applications* 12:618–628.

Root, K. V., H. R. Akçakaya, and L. Ginzburg. 2003. A multispecies approach to ecological valuation and conservation. *Conservation Biology* 17:196–206.

Rothley, K. D. 2002. Dynamically-based criteria for the identification of optimal bioreserve networks. *Environmental Modeling and Assessment* 7:123–128.

Scott, J. M., P. J. Heglund, M. L. Morrison, J. B. Haufler, M. G. Raphael, W. A. Wall, and F. B. Samson. 2002. *Predicting species occurrences: Issues of accuracy and scale.* Island Press, Washington, D.C., USA.

Sethi, G., C. Costello, A. Fisher, M. Hanemann, and L. Karp. 2005. Fishery management under multiple uncertainty. *Journal of Environmental Economics and Management* 50:300–318.

Shaffer, M. L. 1981. Minimum population sizes for species conservation. *BioScience* 31:131–134.

Shea, K., P. Amarasekare, P. Karelva, M. Mangel, J. Moore, W. W. Murdoch, E. Noonburg, A. M. Parma, M. A. Pascual, H. P. Possingham, C. Wilcox, and D. Yu. 1998. Management of populations in conservation, harvesting and control. *Trends in Ecology and Evolution* 13:371–374.

Sjogren-Gulve, P., and I. Hanski. 2000. Metapopulation viability analysis using occupancy models. *Ecological Bulletin* 48:53–71.

Sjogren-Gulve, P., and C. Ray. 1996. Using logistic regression to model metapopulation dynamics: Large-scale forestry extirpates the pool frog. Pages 111–137 *in* D. R. McCullough, editor. *Metapopulations and wildlife conservation.* Island Press, Washington, D.C., USA.

Staples, D. F., M. L. Taper, and B. Dennis. 2004. Estimating population trend and process variation for PVA in the presence of sampling error. *Ecology* 85:923–929.

Thomas, C. D., and I. Hanski. 2004. Metapopulation dynamics in changing environments: Butterfly responses to habitat and climate change. Pages 489–514 *in* I. Hanski and O. Gaggiotti, editors. *Ecology, genetics, and evolution of metapopulations.* Elsevier Academic Press, New York, New York, USA.

Thomson, D. M., and M. W. Schwartz. 2006. Using population count data to assess the effects of changing river flow on an endangered riparian plant. *Conservation Biology* 20:1132–1142.

Treves, A., L. Naughton-Treves, E. K. Harper, D. J. Mladenoff, R. A. Rose, T. A. Sickley, and A. P. Wydeven. 2004. Predicting human-carnivore conflict: A spatial model derived from 25 years of data on wolf predation on livestock. *Conservation Biology* 18:114–125.

Wade, P. R. 2002. Bayesian population viability analysis. Pages 213–238 *in* S. R. Beissinger and D. R. McCullough, editors. *Population viability analysis.* University of Chicago Press, Chicago, Illinois, USA.

Walters, J. R., L. B. Crowder, and J. A. Priddy. 2002. Population viability analysis for red-cockaded woodpeckers using an individual-based model. *Ecological Applications* 12:249–260.

Waples, R. S. 2002. Definition and estimation of effective population size in the conservation of endangered species. Pages 147–168 *in* S. R. Beissinger and D. R. McCullough, editors. *Population viability analysis.* University of Chicago Press, Chicago, Illinois, USA.

Williams, J. C., C. S. ReVelle, and S. A. Levin. 2005. Spatial attributes and reserve design models: A review. *Environmental Modeling and Assessment* 10:163–181.

Williams, P. H., and M. B. Araújo. 2000. Using probability of persistence to identify important areas for biodiversity conservation. *Proceedings of the Royal Society of London Series B* 267:1959–1966.

Wintle, B. A., R. P. Kavanagh, M. A. McCarthy, and M. A. Burgman. 2005. Estimating and dealing with detectability in occupancy surveys for forest owls and arboreal marsupials. *Journal of Wildlife Management* 69:905–917.

Wintle, B. A., M. A. McCarthy, C. T. Volinsky, and R. P. Kavanagh. 2003. The use of Bayesian model averaging to better represent uncertainty in ecological models. *Conservation Biology* 17:1579–1590.

CHAPTER

Multispecies Conservation Planning on U.S. Federal Lands

3

Barry R. Noon, Kevin S. McKelvey, and Brett G. Dickson

Numerous laws directly, or indirectly, mandate the conservation of all species on public lands in the United States (Goble and Freyfogle 2002, Nagle and Ruhl 2002). Key laws relevant to the management of biodiversity on federal lands include the National Forest Management Act (NFMA; Noon et al. 2003) and the Federal Endangered Species Act (ESA; 1973). However, there are also many companion laws that govern the use of these same lands that are in conflict with a goal of maximizing the conservation of species and their habitats (e.g., Multiple Use Sustained Yield Act, 1960). Consequently, land managers must balance the trade-offs between the conservation of species and the exploitation of resources for short-term human use. It is commonplace for species and people to compete for the same set of limiting resources.

Most prominent among the laws protecting species and their habitats is the ESA, which stipulates that no federal action, or federally sanctioned action, on public or private lands shall jeopardize the continued existence of any species. Given the importance of maintaining biodiversity for both ethical and practical reasons—for example, to sustain environmental goods and services critical to human welfare (Hooper et al. 2005)—it is imperative that the scientific community provide land managers with the knowledge and tools needed to meet their conservation mandate.

Despite the importance of multispecies conservation planning from both a legal and practical perspective, we believe that current scientific understandings and methods provide only limited guidance to land managers. Studying multiple species and the range of spatial and temporal scales that they span has been identified as one of the key challenges in conservation biology (e.g., MacNally et al. 2002, Fischer et al. 2004*a*). However, as we discuss in the following text, progress is slow in part due to the complexity of the problem and to the lack of sufficient information on the abundance, distribution, life histories, and ecological relations of most species. In addition, much conservation science has been crisis driven, responding to the conservation needs of individual species experiencing threats to their persistence. A clear expression of the crisis-driven nature of conservation biology is that the vast majority of scientific publications in this discipline focus on individual species at risk or small groups of imperiled species sharing similar life histories or habitats. The difficulties associated with directly evaluating multiple species coupled with this single-species emphasis in the primary literature has led to the adoption of surrogate measures; all studies that we are aware of that propose guidelines for multispecies conservation planning ultimately default to surrogate-based approaches. These surrogates tend to be based on small sets of species with presumed importance to general diversity and/or vegetation-based proxies. Both of these approaches, however, rely on largely untested premises.

In this chapter, we review past approaches and some noteworthy recent advances in multispecies conservation planning. We are encouraged by recent attempts to extend conservation planning from single to multiple species but find that even the most innovative new methods fall far short of addressing the hundreds to thousands of species found in most management areas. As discussed in the following sections, the challenges to multispecies conservation planning are great but not insurmountable if rigorous surrogate-based approaches are adopted.

THE CHALLENGE OF MULTISPECIES CONSERVATION PLANNING

In the following section we briefly review theories and general principles from ecology that suggest, at some level of resolution, that all species are fundamentally distinct. The purpose of this section is not to challenge these concepts but rather to evaluate whether they present insurmountable challenges to multispecies conservation planning on federal lands. Our discussion is mostly general and applies broadly to plant and animal species. However, the context of our discussion throughout this chapter is mostly on the conservation of vertebrate communities.

Niche Theory

We need look no further than ecological theory to understand why multispecies conservation planning is so difficult and why it seems impractical in most circumstances. Based on niche theory (e.g., Hutchinson 1957, Pulliam 2000), no two species can long occupy the same niche. Thus, all coexisting, sympatric species must differ along at least one critical niche dimension. Persistence for a species whose niche is included as a subset of another species is possible only until a critical, shared resource becomes limiting. Adjunct to this theory is the understanding that there must be some limit to the similarity of coexisting species (MacArthur 1967). The implication of niche theory to multispecies conservation planning is that it will be impossible to manage for all species unless all relevant resources (i.e., niche axes) are sustained at sufficient levels to support viable populations.

Trophic Structure

Most species can be unambiguously positioned within a food web in terms of what they eat and who eats them. The foundational species for all food webs are the autotrophic species which fix carbon via the process of photosynthesis and create food for herbivorous species. Herbivores, in turn, support one or more trophic levels of predatory species. As a result, it is impossible to consider the conservation of any species without also considering its contribution to a community food web and its reliance on other species. In most ecological communities, many species occupy a given trophic level, and functional redundancies within any given food web are expected (Pimm 2002). However, due to niche-based differences, each species contributes in some unique way to food web dynamics and the stability of trophic hierarchies.

Allometric Scaling Relationships

The view of the niche as a static multidimensional resource space (Hutchinson 1957) does not explicitly incorporate concepts of time and space that are essential considerations for multispecies conservation planning. Scaling relationships are important topics in ecology—for example, in metapopulation biology (e.g., With 2004) and landscape ecology (e.g., Wiens 1989), where key areas of research focus on dispersal abilities and movement behaviors in heterogeneous landscapes. Many studies have demonstrated that species uniquely scale their spatial environment and that patterns of space-use (area requirements) are strongly correlated with differences in body mass (Cotgreave and Harvey 1992, Silva and Downing 1994). These allometric relations presumably reflect underlying differences among species in terms of their movement behaviors and food resource requirements. For example, space-use relationships differ predictably between birds and mammals and probably reflect the much greater movement ability provided by flight.

Differences in temporal scaling are also related to differences in body mass and taxonomy (McArdle et al. 1990). Such differences often lead to differences in demography. For example, age at first reproduction and life span have clear relationships to reproductive potential and survival rates, respectively. These aspects of life history, as well as growth potential, have long been recognized as being related to body mass (Peters 1983) and have led to a somewhat simplistic partitioning of species into r- or K-selected life histories. Even though allometric scaling relationships demonstrate important differences among species, they offer one possible way to group species according to similarities in space use and temporal scaling of the environment (Theobald and Hobbs 2002). We revisit this point below.

Demography

At some level of resolution, all species differ in the fine points of their demography. For example, species differ in terms of their life history schedules—age- or stage-specific birth and survival rates—and in differences in sensitivity of population growth to changes in various vital rates (e.g., Wisdom et al. 2000). Such differences can lead to differential sensitivities to various environmental threats (e.g., habitat loss and fragmentation) and thus to different extinction likelihoods in the face of environmental change. In addition, some species persist as metapopulations, where persistence is determined not just by local population dynamics but by the relative extinction and colonization rates among local populations. These rates, in turn, are affected by distance among local populations, characteristics of the landscape matrix that separates these populations, and on the ability of individuals to move among local populations. The likelihood of successful movement among spatially disjunct populations will vary greatly by species due to both their mobility and fecundity (Noon and McKelvey 1996). Further, habitat requirements among species can be widely divergent, and the population-level effects associated with a particular landscape configuration will also vary between species. This variability, coupled with ongoing habitat loss and fragmentation, suggests that each species may have unique conservation requirements expressed as the amount and spatial arrangement of its habitat on the landscape.

Movement and Landscape Connectivity

Animal movement is motivated by a complex suite of environmental cues and ecological processes, including the location of conspecifics, prey, and competitors in time and space. However, the movement patterns of species cannot be separated from the habitat and landscape features that connect individuals or populations. Taylor et al. (2006:29) characterize landscape connectivity as a dynamic emergent property resulting from "the interaction between a behavioral process (movement) and the physical structure of the landscape." For multiple species on a heterogeneous landscape, connectedness depends on the organisms under investigation and how relevant landscape attributes are distributed. Due to differences in life and ecological histories (e.g., body size, mobility, migration rates), co-occurring species can simultaneously perceive a landscape as both connected and disconnected (Pearson et al. 1996, With et al. 1997, Taylor et al. 2006). Thus, the methods used to quantify connectivity require careful consideration of how organisms differentially interact with the landscape during movement events (Tischendorf and Fahrig 2000) and how these events are influenced by management or land-use activities. With increasing levels of habitat loss and fragmentation, for example, dispersal events can be disrupted (King and With 2002), forcing species to navigate novel environments with features that may threaten their distribution or persistence (Gardner and Gustafson 2004, Aurambout et al. 2005).

BROAD STRATEGIES FOR MULTISPECIES CONSERVATION PLANNING

Given the preceding discussion, it is clear that all species are fundamentally distinct from one another in at least one significant dimension. Taken at face value, this recognition makes the challenge of managing for the conservation of all or most species appear insurmountable. The task seems all the more daunting in the context of an ongoing and accelerating human-caused transformation of the landscape. However, the necessity of striving to conserve as much biodiversity as possible remains a scientific, legal, and ethical imperative. As a result, the challenge of multispecies conservation planning becomes redefined in terms of identifying and implementing management strategies that will achieve some constrained maximal conservation outcome. The solution will be unavoidably constrained for at least three reasons. First, only some fraction of the available landscape will be devoted to a primary objective of species conservation. Second, beyond the land itself, the amount of resources allocated to conservation will compete with other societal objectives. Third, the earth's resources are finite, and all users are participants in a zero-sum game; resources exploited for direct human uses are generally not available to support the rest of the natural world (Vitousek et al. 1997).

Most comprehensive conservation strategies are similar in that they invoke a set of key conservation planning principles. For example, the selection of reserve sites is based on characteristics such as their representation, resilience, and redundancy (Shaffer and Stein 2000) or complementarity, irreplaceability, and vulnerability (Sarkar et al. 2006). A conservation strategy has representation and complementarity if it provides for habitat for each species at one or more locations on the landscape. Resilience means that reserve areas are sufficiently large so as to incorporate normal disturbance regimes without compromising the persistence of any species. Redundancy implies that the spatial distribution of species and their habitats is sufficiently broad that the dynamics of local populations of a species are spatially decoupled. Irreplaceability puts an emphasis on sites that support unique species found nowhere else, and vulnerability ranks sites by their degree of threat. Collectively, these characteristics act as constraints on site selection.

The design of most comprehensive multispecies conservation planning efforts invoke some form of a "coarse filter" and/or "fine filter" approach (e.g., Hunter et al. 1988, Hunter 1991, Noss and Cooperrider 1994, Noss 1996, Haufler et al. 1996, Cushman et al. 2008). The coarse filter is usually considered to function at broad spatial scales (100s to 1000s of km^2) and to reflect underlying ecological processes that are operative over long temporal scales (decades to centuries). The fine filter is most often used in reference to individual species or groups of functionally related species (e.g., species guilds, Block et al. 1987). In general, coarse filters are composed of fairly broad ecological, often vegetative, classifications and do not utilize direct measurements of vertebrate species.

Some authors have referred to a collection of surrogate species as a coarse filter of sorts (e.g., Tognelli 2005). However, in our discussion, we consider all direct measures of species to represent a fine filter approach.

Filtering approaches are not necessarily confined to this coarse/fine dichotomy. Decisions concerning reserve design and location can also be cast in terms of biotic and abiotic "filters" that account for both coarse- and fine-scale processes important to biodiversity. The multiple filter analogy invokes hierarchical levels of biological organizational levels, which are strongly correlated with hierarchies of spatial scales (e.g., Poff 1997; Probst and Gustafson, this volume).

Filter-based approaches to multispecies conservation generally seek to characterize ecological systems in terms of indicators of function, structure, and composition (Lindenmayer et al. 2000). Noss (1990), for example, proposed an exhaustive set of candidate indicators based on four scales of ecological organization—landscape, community-ecosystem, population-species, and genetic—and three primary attributes of biological diversity—function, structure, and composition. Function-based indicators include direct measures of processes and their rates. Examples include primary productivity, rates of nutrient cycling, and water flows. Structure-based indicators, measured at local and landscape scales, include elements such as vegetation structural complexity, among-patch vegetation heterogeneity, landscape connectivity, and landscape pattern (i.e., the distribution and abundance of different patch types). While these metrics are often assumed to constitute a "coarse filter" because of their ability to predict broad-scale patterns of biological diversity (Hunter 1991; Haufler et al. 1996, 1999), both function- and structure-based indicators can be measured at multiple spatial scales ranging from local to regional. In addition, there are composition-based indicators that entail direct measurement at the species level and include information on each species' distribution, life history, demography, or behavior. These are often referred to as "fine filter" assessments because they evaluate the effects of management practices on individual species (Haufler et al. 1996).

Coarse Filter Approaches

Coarse filter approaches are often based on broad-scale criteria such as climate, soils, geology, and vegetation cover types (Kintsch and Urban 2002). For example, the conservation planning framework of The Nature Conservancy uses ecoregional classifications (e.g., Bailey 1995) as an initial way of partitioning the landscape into a set of conservation areas that span the native species and ecosystems of a region (Groves et al. 2002). Ecosystems are often proposed as the appropriate level for conservation planning at broad spatial scales (Noss 1996). However, ecosystems can seldom be portrayed as discrete geographical units because they do not have tangible boundaries that allow them to be mapped (Pickett and Cadenasso 2002). Often, the boundaries established for ecosystems are artificial and vary according to one's goals. Thus, the spatial and temporal dimensions of an ecosystem are usually user-defined, rendering them a dimensional conceptual unit (Fauth 1997). Moreover, the communities and habitat features that define ecosystems are not always neatly contained within a single jurisdictional boundary. As a result, ecosystems have limited utility as a map-based unit for conservation planning. For the purposes of conservation planning at broad spatial scales, ecosystems are often referred to as a target conservation unit, but in practice, ecosystems are usually equated with dominant vegetation communities or physical landscape features.

For example, on public lands in the Pacific Northwest, reference is made to managing the late-seral Douglas-fir ecosystem to sustain native biodiversity (Forest Ecosystem Management Assessment Team 1993). In the southeastern United States, management may be targeted to sustain key processes characteristic of bottomland hardwood forest communities. In the Rocky Mountains, we might be interested in sustaining the species and processes characteristic of alpine lake ecosystems. In order to define ecosystems and their boundaries in this way, it is common practice to partition the landscape into patch types. The default coarse filter for most public land management decisions defines patches on the basis of dominant vegetation communities and their successional stages.

If coarse filter attributes are defined as the characteristics of a vegetative patch mosaic, the relationship between these metrics and the dynamics of any animal species or group of species is at least two steps removed from reality. The first assumption is that vegetation composition and structure per se can be used as a surrogate for a species' habitat. The second is that various successional stages or structural configurations of vegetation types can be related to habitat quality and thus species abundance or probability of occurrence. Given the ubiquity of both of these assumptions, it is surprising how few formal tests have been conducted. Most of the widely used habitat relationship models, for example, are heuristic rather than quantitative (Schulte et al. 2006). The relationships between these heuristic understandings and broad landscape characteristics are often based in expert opinion (Marcot et al. 2001, Raphael et al. 2001). In many cases, consistency in expert opinion is equated to expected model performance, a tenuous assumption at best. As such, the efficacy of these approaches is entirely unknown.

Statistical explorations of multiscale habitat relationships suggest that evaluations based on broadly defined cover-types may not provide robust surrogates for multispecies distribution or abundance patterns (MacNally et al. 2002). For example, Cushman and McGarigal (2004) tested the correlations between multiscale vegetative characteristics and breeding bird abundance in three watersheds in Oregon. Data included detailed plot level information, stand type and seral stage, and compositional characteristics of the surrounding landscape. For all species, total variance explained was about 60%, with the plot variables having the greatest explanatory power. Landscape composition variables only explained 4% of the variance in each of the three drainages. Patch characteristics alone explained 5–6% of variance in bird abundance (also see Cushman et al. 2008).

While limited in area and only pertaining to breeding birds, Cushman and McGarigal's (2004) results suggest that the efficacy of coarse filter approaches should not be assumed. It is, in fact, unlikely that proxy-on-proxy relationships will have high explanatory power, since power declines multiplicatively with each proxy step. Thus, it is unlikely that coarse filter approaches alone will be sufficient for effective multispecies conservation.

Mesofilter Approaches

Hunter (2005) recently proposed the concept of the mesofilter to bridge the gap between more traditional coarse and fine filter approaches. The mesofilter concentrates on habitat elements that are too small to be the focus of reserve design strategies, but that often may be limiting the populations of some species. Sample elements retained by the mesofilter include large logs and snags, riparian zones, seeps and springs, and rock outcrops (Schulte et al. 2006). For many species, these fine-grained habitat features may be more important than vegetation community type; however, collection of these data is limited to ground-based plots. As such, spatial mapping of these attributes and extrapolation to unsampled areas must be accomplished through statistical imputation methods (Ohmann and Gregory 2002) and the precision of model estimates will vary greatly between habitat elements.

Fine Filter Approaches

Fine filter approaches involve the direct measurement of species' spatial distribution patterns (presence/absence or abundance) rather than application of a surrogate or index metrics. However, the number of species that constitute an ecosystem is enormous. It is impossible to track them all and, even if possible, the relationships between multiple population trajectories and ecosystem function remains obscure. For these reasons, fine filter approaches address some subset of species that may go unprotected by the coarse filter or, alternatively, act as presumed surrogates for other species or the ecosystem as a whole.

Threatened, at-risk, and rare species.— A logical approach to conservation planning is to focus on those species most at risk of local or global extinction. This is the approach taken for conservation of North American bird species under The Partners in Flight program (Panjabi et al. 2005). This has also been the approach adopted for recovery planning for species listed under the ESA. For obvious reasons, the vast majority of recovery efforts for listed species have focused on single, and not multiple, species. An exception to single species planning for imperiled species has been the application of multispecies Habitat Conservation Plans to private lands affected by the ESA (Noss et al. 1997).

Perhaps the highest profile example of the imperiled species approach is the forest planning exercise that led to large changes in federal land management in the Pacific Northwest (Forest Ecology Management Assessment Team 2003). The Forest Ecology Management Assessment Team started with a reserve structure carefully designed to provide for a well-distributed population of northern spotted owls (*Strix occidentalis*; Thomas et al. 1990). This plan was then examined for adequacy in protecting a suite of additional species representing a wide variety of animals, plants, and fungi (Raphael and Molina 2007). Because of the necessary niche separation between species, the spatial extent of the reserve design expanded greatly in this process, and protections were added to areas between reserves.

Focal Species.— In the 2000 NFMA regulations applicable to Forest Service lands, focal species were defined as surrogate measures used in the evaluation of ecological sustainability, including species and ecosystem diversity (Noon and Dale 2003). The key characteristic of a focal species is that its status and trend provide insights to the integrity of the larger ecological system to which it belongs. Focal species serve an umbrella function in terms of encompassing habitats needed for many other species, play a key role in maintaining community structure or processes, are sensitive to changes likely to occur in the area, or otherwise serve as an indicator of ecological sustainability (Committee of Scientists 1999:38–39). The focal species concept described in the 2000 regulations differed subtly, but significantly, from the previous "management indicator species" concept used by the Forest Service. Rather than acting as an indicator of the outcome of a specific management prescription, the status and trend of focal species are to allow induction upward to the conservation status of entire suites of species. As such, the concept is inclusive of the variety of surrogate species concepts such as keystone, umbrella, indicator, and engineering species.

Potential Focal Species Categories.— The following is a list of species "types" that may serve as conservation targets, or surrogates for unmeasured species, for the purpose of multispecies conservation planning (see Lambeck 1997, Caro and O'Doherty 1999, Favreau et al. 2006). The unifying principle across all these categories is that that the status and trend of these species types should provide information beyond their own measurement.

1. *Indicator species*: "An organism whose characteristics (presence or absence, population density, dispersion, reproductive success) are used as an index of attributes too difficult, inconvenient, or expensive to measure for other species or environmental conditions of interest" (Landres et al. 1988:317, Landres 1992). In addition, Patton (1987) described an indicator as an organism so intimately associated with particular environmental conditions that its presence indicates the existence of those conditions. Indicator species can be further broken down into three categories (Caro and O'Doherty 1999):

- *Early warning indicator*: Provides an early warning of a stressor acting on a key ecosystem process (traditional interpretation of an indicator species from ecotoxicology). For example, changes in lichen communities in forest ecosystems may act as indicators of stress arising from atmospheric pollution.
- *Population surrogate indicator*: Species whose status and trend are indicative of the status and trends of other species. This species type is related to the guild indicator concept of Block et al. (1987).
- *Biodiversity indicator*: A species, or more commonly a taxonomic group, that acts as a surrogate for a number of poorly known taxonomic groups.

2. *Umbrella species*: A species that needs such large areas of habitat that managing for its viability addresses the viability concerns of numerous other species with similar habitat but smaller area requirements (after Wilcox 1984). The principal requirement of an umbrella species is that its range is large compared to sympatric species. Many large-bodied, wide-ranging animals are candidate umbrella species—for example, bison in prairie ecosystems and carnivores in forested ecosystems. Additional discussions of the umbrella species concept are found in Lambeck (1997), Andelman and Fagan (2000), Fleishman et al. (2000, 2001), and Roberge and Angelstam (2004).

3. *Keystone species*: Species which significantly affect one or more key ecological processes or elements to an extent that greatly exceeds what would be predicted from their abundance or biomass (Mills et al. 1993, Power et al. 1996). For example, sea otters prey strongly on the herbivores that consume kelp forests. Diverse kelp forests, in turn, support diverse communities of vertebrates and invertebrates.

4. *Strong interactors*: A species whose dynamics are affected by, or greatly affect, the dynamics of other species in the community (Christianou and Ebenman 2005). Energy-maximizing predators in freshwater aquatic ecosystems often exert strong top-down control on the structure and composition of their prey communities.

5. *Link species*: A species that occupies a key position in a food web and efficiently transfers energy and matter between trophic levels. For example, prairie dogs in grassland ecosystems efficiently convert primary plant productivity into animal biomass. Prairie dog biomass, in turn, supports a diverse predator community (Smith and Lomolino 2004). These are similar to the bottom-up members of food webs discussed by Jordán et al. (1999).

6. *Ecological engineers*: Species that directly or indirectly control the availability of resources to other organisms by causing physical state changes in biotic or abiotic materials (Jones et al. 1994, 1997). For example, beavers alter and restructure riparian habitats and greatly increase the diversity of both plant and animal communities as a consequence.

As mentioned previously, fine filter refers to the direct measurement of individual species or guilds. While fine filter measurements, being direct, would seem to be free from the proxy-on-proxy issues that plague the coarse filter, this is not the case. For the measured species, presence/absence or relative abundance measures are, in fact, proxy statistics for density or population size. They have the advantage that they are mechanistically linked to the desired measurements, but the inevitable loss of explanatory power associated with working through an index of population size rather than size itself should not be ignored. Far more critical for the purpose of multispecies monitoring, however, is the issue associated with using population trend data for one species as a proxy for the trends of many other species. The potential problems with this approach were elucidated by Landres et al. (1988) and Landres (1992) but, to our knowledge, there have been too few tests of the surrogate species approaches to fully evaluate its efficacy (Favreau et al. 2006).

Mixing Coarse and Fine Filter Approaches

Root et al. (2003) developed a multispecies conservation algorithm that assessed habitat importance for 40 species in California based on an integration of vegetation maps (the coarse filter) and wildlife habitat relationships models (the surrogate fine filter). They used data from the California GAP Analysis Project to create a vegetation polygon map where each polygon on the map was designated a habitat suitability score separately for each species. A composite map, based on the individual species maps, combined information across all species into a single value for each polygon based on the sum of the habitat suitability scores. In addition, Root et al. (2003) developed a weighted multispecies conservation score for each polygon using a weighting factor based on a formal population viability assessment or different indices of threat faced by each of the 40 species. The final map, at the scale of the entire state of California, provided a multispecies conservation score for each vegetation polygon that reflected both occurrence and degree of threat. Use of this approach requires information on the habitat relationships for each species expressed relative to the vegetation classes in the GAP database, and knowledge of how suitability varies by class. This information is available for many species from habitat relationships models but suitability ratings are based on assumed abundance relationships, which are generally unknown.

Comprehensive conservation planning is not restricted to coarse and fine filter approaches; other conceptual tools are often used to complement the filter approach. For example, Schulte et al. (2006) reviewed conceptual tools used for multispecies conservation planning in large-scale forest management plans. They found that the concept of filters (coarse, meso, and fine) are most widely used but that other principles drawn from landscape ecology including reserve design principles, matrix management, and connectivity via networks or corridors are used to refine the filter-based approaches. However, they point out that currently there is too little empirical support at the temporal and spatial scales required for comprehensive conservation planning to support the use of any of these conceptual tools without close monitoring of the desired outcomes.

RECENT ADVANCES IN MULTISPECIES CONSERVATION PLANNING

Because multispecies conservation planning is now recognized as one of the major challenges to biodiversity conservation, there has been a significant increase in the number of researchers addressing this problem. Also, with advances in spatial modeling techniques (e.g., Guisan and Zimmerman 2000) and the increased availability of spatial data, moving from single to multiple species evaluations is increasingly possible. Because of data limitations and computing requirements, these approaches are currently limited in the number of species that can be simultaneously evaluated. However, we believe the new methods discussed in this chapter are significant for having made the important first step of moving from single to multiple species conservation planning while largely avoiding the use of surrogate-based filters.

To facilitate discussion of recent advances in multispecies conservation planning, we have partitioned the topic into various categories. These categories, however, are not distinct, but rather they overlap somewhat in data requirements, methods, and objectives. For example, habitat suitability models (e.g., Root et al. 2003, Zielinski et al. 2006*a*), probability of occurrence models (Moilanen et al. 2005), and population viability models (Haight et al. 2002, Carroll et al. 2003) can all serve as input to reserve selection algorithms.

Probability Surfaces for Species Occurrences

A wide variety of methods exist for predicting the occurrence of species at local to broad spatial scales (see Guisan and Zimmermann 2000, Scott et al. 2002, Elith et al. 2006). The recent coupling of statistical techniques with geographic information systems (GIS) has increased capabilities for modeling and visualizing the occurrence of individual species in time and space, typically as raster-based probabilistic surfaces. In addition, a useful conceptual model to map the spatial distribution of multiple species at several spatial scales has recently been developed (Fischer et al. 2004*b*, Fischer and Lindenmayer 2006). However, we are aware of very few empirical examples of models developed to simultaneously predict the probability of occurrence for multiple species. Logistic regression is perhaps the most common statistical approach to modeling the probability of species occurrence. In these models, binary information on the presence-absence of a single species (dependent variable) is used to estimate a conditional probability of occurrence as a linear function of multiple independent predictor variables (e.g., Pereira and Itami 1991, Carroll et al. 1999, Klute et al. 2002; see Hosmer and Lemeshow 2000). Often, values for the predictor variables are extracted directly from digital data layers in a GIS (e.g., landcover type and elevation). Once the relevant parameters have been estimated, these models can be implemented within a GIS to produce a probabilistic, cartographic surface encompassing the inferential or spatial extent of the independent variables (Niemuth et al., this volume). Numerous tools, including cross-validation techniques and receiver operating characteristic curves, for example, can be used to evaluate model performance and prediction errors (see Fielding and Bell 1997; Shifley et al., this volume).

One common extension of multiple logistic regression is the probabilistic predictions of habitat selection by individual species (e.g., resource selection functions; Manly et al. 2002). Similarly, new methods for estimating probability of site occupancy by a species use multiple logistic regression as their statistical foundation, and permit the adjustment of estimates using models of detection probability (see MacKenzie et al. 2006). Other methods for predicting probability of occurrence that are easily implemented in a GIS include Mahalanobis distance (Clark et al. 1993, Knick and Dyer 1997), Bayesian weights-of-evidence (Bonham-Carter et al. 1989), and kernel density estimation (Worton 1989, Bailey and Gatrell 1995). Segurado and Araújo (2004) and Elith et al. (2006) provide a current and comprehensive review of other methods used to predict the occurrence and distribution of species and that can be used to generate probabilistic surfaces.

When reliable information on the spatial location of multiple species is available, and when a suite of meaningful covariates can be specified, probabilistic models of the occurrence of single species can be used to inform multispecies conservation planning efforts. It is possible to combine multiple probabilistic surfaces derived from one of the statistical models described earlier to create a single, synthetic surface predicting the occurrence of >1 species of interest. For example, Moilanen et al. (2005) described an approach to quantifying priority areas for multispecies conservation planning using a connectivity algorithm and probability of occurrence surfaces developed for seven indicator species in eastern Australia (after Wintle et al. 2005). Using the computer program ZONATION (Moilanen et al. 2005), they hierarchically aggregated, or "zoned," priority areas based on concentrations of high probability values, which were then assumed to simultaneously support high connectivity and persistence for multiple species. Recently, the ZONATION algorithm has been used to identify a reserve design for 23 species of butterflies in the United Kingdom (Early and Thomas 2007).

Significant advances have been made recently in the integration of environmental data with presence-only location data (e.g., Tsoar et al. 2007). For example, program MAXENT (Phillips et al. 2006) creates a probability of occurrence grid for multiple species based on the spatial intersection of environmental and species presence data via a GIS interface. Predicted probabilities of habitat suitability can then be thresholded to identify candidate reserve boundaries.

Improved and reliable models of species occurrence will be necessary to assess the impacts of management and environmental changes on the collective properties of biodiversity (Guisan et al. 2006). Probabilistic surfaces and associated visualization techniques (Millspaugh et al., this volume), derived from any number of statistical models, can convey complex information that is relatively easy to interpret and communicate. However, because no single best model can be applied to all species in all environments, emphasis should be placed on identifying the model that best matches the ecology of the species, community, or system under investigation (Barry and Elith 2006), and on the data, assumptions, and goals of the analysis (Segurado and Araújo 2004).

Generation of empirically based probability surfaces, whether for single or multiple species, requires sampling animal occurrence with sufficient intensity to assign probabilities of detection to covariate GIS elements. Further, it requires that GIS variables provide high explanatory power. These data requirements currently make these approaches most useful when applied to a small group of focal species that collectively act as comprehensive biodiversity surrogates.

Addressing Landscape Connectivity for Multiple Species

For all but rare endemic species, conservation objectives will not be achieved with a single reserve or a single population. Rather, local populations widely scattered across the landscape, but connected by movement, will be necessary. Few of these populations will be large enough to avoid problems faced by small populations, such as extirpation due to stochastic factors and inbreeding depression. Connectivity maintenance is therefore one of the most critical aspects of multispecies conservation. Connectivity, however, is notoriously difficult to directly measure. Not only are observed dispersal events rare, but the results of observed dispersals are seldom known. The probability of locating an organism in a distant location is low, and the probability of subsequently tracking an individual to determine mating behavior and reproductive success is very small. Past dispersal events that did affect the population through breeding do, however, leave characteristic genetic patterns. Areas of high mixing are relatively genetically similar, and isolated populations are more divergent. Historically, genetic samples were collected from populations on either side of a putative barrier, and the null hypothesis that both samples came from the same population is tested. When rejected, the putative barrier was assumed to be real.

While not without merit, this approach assumes *a priori* that we can identify important barriers, an assumption that is significantly weakened if population substructure exists on either side of the barrier. Recently, a growing number of techniques seek to characterize a surface of genetic differences based on evaluation of individuals, and relate these surfaces to mapped landscape characteristics such as topography. Dubbed "landscape genetics" (Manel et al. 2003), this approach is new, and its possibilities largely unexplored. Cushman et al. (2006) used landscape genetic methods to correlate 108 different landscape resistance patterns to patterns of genetic differences in black bears (*Ursus americana*). Bear genetic patterns best correlated with low resistance movement through mid-elevation forests and avoidance of both high and low elevation zones and open areas. Cushman et al.'s (2006) methods are broadly applicable to other species—requiring a spatially distributed genetic sample of individuals from a species of interest and mapped landscape attributes to define a plausible suite of potential resistance surfaces.

The reserve design algorithm ZONATION directly addresses connectivity by creating a grid-based connectivity surface (Moilanen et al. 2005). For each species considered, the likelihood of successful dispersal from location *i* to location *j* is estimated from Hanski's incidence function equation (Hanski 1994) based on the occupancy status of cells and their distance apart. Every cell with a connectivity value above a threshold is selected to produce clusters of cells and incipient reserves.

McRae (2006) used concepts from electronic circuit theory to predict gene flow and genetic structuring in heterogeneous landscapes. Briefly, the isolation-by-resistance (IBR) model uses circuits as models of population or landscape networks. The model represents populations or raster habitat cells connected by dispersal as nodes connected by resistors. Effective conductance calculated among nodes can be used to predict gene flow, whereas effective resistance is used to predict genetic differentiation. Greater connectivity among populations or habitat patches is predicted when more connected pathways are available. The IBR model can accommodate a broad range of available habitat data, from simple range maps to quantitative habitat models. The model could easily be extended to predictions for multiple species by assigning conductance values to cells that characterize the minimum habitat requisites of a set of species of interest, or by combining output values from individual IBR analyses into a single conservation solution.

Reserve Selection Algorithms

In the Northwest Forest Plan (Forest Ecosystem Management Assessment Team 1993), areas were added to the northern spotted owl reserve structure based on expert opinion. Because so many species were evaluated, and each species added conservation requirements to the plan, reserves were large, and between-reserve constraints were significant. The result for public lands in the

Cascades and Coast Ranges of Washington, Oregon, and northern California was that timber harvest was reduced to 5% of preplan levels. While this approach favors species conservation, in many areas large reductions in resource extraction are not socially acceptable. This understanding has led to the idea of developing more efficient reserve systems by either seeking maximal coverage (greatest number of species protected) for a fixed area or expenditure, or a minimum reserve set (smallest expenditure) for a target set of species using the tools of linear and nonlinear optimization (e.g., Pressey et al. 1993, Bevers et al. 1995, Possingham et al. 2000, Cabeza and Moilanen 2003).

There is an extensive literature on methods to optimally select areas for inclusion into a biodiversity reserve design (e.g., Williams et al. 2005, Sarkar et al. 2006). Many methods focus on selecting the smallest possible number (area) of reserves that will achieve the objective of including all species in at least one reserve (Pressey et al. 1993, Possingham et al. 2000 [program SITES]). This has been called the minimum representation problem (Possingham et al. 2000). There are many variants on these methods, but most cast the problem as an optimization exercise where the goal is to minimize social and economic costs constrained by biodiversity objectives (McDonnell et al. 2002). The optimization algorithms are generally based on the principle of comprehensive species representation at minimal economic cost, a goal achieved by optimally selecting some minimal set of complementary reserve sites. Data requirements for these methods include information on the presence-absence of each species on each candidate reserve site. Since the number of possible solutions is equal to 2^n, where n is the number of candidate sites, finding optimal solutions often requires nonlinear optimization software, and the methods can be very computer intensive.

Other objectives for reserve selection can be addressed in the optimization problem. Most common are a site's irreplaceability and its vulnerability (Lawler et al. 2003). Irreplaceable sites contain species that are found nowhere else among the candidate sites and thus must be included in any solution in order to obtain complete representation (Pressey et al. 1994). To sustain the objective of full representation requires identification of sites that are vulnerable to loss and contain species that are poorly represented elsewhere among candidate sites (Wilson et al. 2005). Sites can also be prioritized for inclusion if they contain species that are threatened with extinction or generally at risk because of their rarity (e.g., Pressey et al. 2003).

Most reserve selection algorithms are based on presence-absence distribution data and cannot directly address issues of persistence. However, a key assumption of reserve selection algorithms is that the selected sites and their configuration allow for species persistence across the reserve system. Even though this goal is an obvious one, it is much more difficult to implement because it requires far more information. To design reserve systems with persistence explicitly evaluated as a conservation objective requires additional information on each species' ecology and life history, including information on dispersal abilities and how persistence likelihood scales with local population size (Carroll et al. 2003, Nicholson and Possingham 2006, Nicholson et al. 2006). Multispecies optimization approaches that include persistence objectives as design criteria are currently limited to a relatively small number of well-studied species (e.g., Zielinski et al. 2006*a*).

Maximizing the Persistence Likelihood of Multiple Species

Most reserve selection algorithms do not directly address the persistence likelihoods of species within the reserved areas. Rather, the algorithms represent a static spatial analysis of biodiversity patterns and do not generally include information on demographic processes. To incorporate population dynamics, for example, reserve algorithms can include objective functions that minimize the extinction risk of a set of focal or umbrella species (reviewed in Nicholson and Possingham 2006). These algorithms are similar to those used in static reserve selection in that they invoke realistic constraints such as available area or level of funding that can be dedicated for conservation. Recent research in this area has built on and extended earlier work aimed at optimizing the spatial pattern of multispecies wildlife habitat in the Pacific Northwest (e.g., Hof and Raphael 1993, Bevers et al. 1995).

Nicholson et al. (2006) used a spatially realistic metapopulation model to estimate extinction risk for a set of 10 well-studied species. Their optimization objective was to maximize persistence likelihood simultaneously across the 10 species, and their model incorporated spatially explicit information on the location, quality, and configuration of habitat, as well as detailed information on the ecology and demography of the focal species. Their model invoked a patch-based view of the landscape, where patches were allowed to vary in quality by species, but matrix properties were not explicitly incorporated. In addition to distributional data for each species, their models required difficult-to-acquire life history data including home range size, number of female dispersers per home range, mean dispersal distance, how extinction risk scales with patch area, and habitat quality values for the different patch types. The authors found that optimizing an objective function that minimized extinction risk outperformed other objective functions based on maximizing the area of habitat given a financial constraint.

Holzkämper et al. (2006) also used a spatial optimization approach to simultaneously address the conservation of three bird species. The species chosen had differing habitat requirements, and this introduced the important reality that multispecies conservation planning involves inescapable trade-offs among the habitat requirements of the species considered. Their model does not view the landscape in terms of a competing set of reserve patches as in Nicholson et al. (2006), but rather as composed of contiguous polygons (patches) of adjacent land use types. Given information on the species habitat requirements and assuming similar area requirements for each species, their optimization model aims to maximize habitat suitability for the three species by identifying the optimal configuration and composition of the landscape. This model also has strict data requirements for parameterization, including knowledge of the species distributions, territory sizes, and habitat suitability values in terms of a set of environmental covariates and landscape metrics.

Use of spatial optimization methods for multispecies planning, in the future, is likely to be an important advance. However, these methods are unlikely to be widely available for conservation planning in the near term, particularly for land managers whose responsibilities often include oversight for diverse taxa and hundreds of species. The requirements for information on distribution, habitat relationships, and, for the extinction models, demographic information may preclude their practical use for most large-scale conservation planning efforts. The exception may be the case where viability-based algorithms focus on a small number of well-studied species that collectively span a range of life histories, habitat, and area requirements that they serve as comprehensive surrogates for the unmeasured species (e.g., Carroll 2003).

Genetic Monitoring

Recently, abundance estimation for many species has undergone a transformation from live trapping and tagging to noninvasive sampling and the use of DNA "tags" for Capture/Mark/Recapture (CMR) or other analyses. Even poor quality DNA can be used reliably to determine species presence-absence and distribution patterns (Zielinski et al. 2007), information that is fundamental to all conservation strategies. These methods, while transformative for many species, underutilize genetic data—genetic differences between individuals devolve into counts. While most widely applied in aquatic systems, genetic data can be directly used to monitor population performance (Schwartz et al. 2007). The advantages of this approach lie in the wealth of information associated with each genetic sample. For example, estimation of population connectivity and effective population size are much more sensitive to the number of loci and alleles per locus than they are sample size. At sample sizes >100, power is insensitive to sample size and is entirely dependent on the genes analyzed. Additionally, because genes integrate across a local population, rigorously representative sampling is much less important. Because it is possible to make strong inferences from samples from a small proportion of the population and from imperfectly collected data, genetic monitoring provides promising and cost-effective approaches for population analyses for multiple species at large spatial scales.

Given that few species have any history of data collection to evaluate trends, genetic methods, being less sensitive to data quality, offer the possibility of retrospective studies, where samples from the past such as museum specimens can be genetically compared to current samples to infer trend

(Schwartz et al. 2007). In an excellent example of a retrospective study, Østergaard et al. (2003) analyzed genetic heterozygosity in brown trout (*Salmo trutta*) in Demark from 1944 to 1997. Older samples were obtained from scale collections. Østergaard et al. (2003) concluded that the population was stable, but maintenance of genetic diversity was dependent on gene flow between small local populations.

A SAMPLE STRATEGY FOR MULTISPECIES CONSERVATION

The Nature Conservancy (TNC) has been a leader in the development of broad-scale conservation strategies that have multispecies conservation as a priority objective. For example, Poiani et al. (2000) proposed that biodiversity conservation on TNC lands focus on ecosystem- and landscape-level concepts. Specifically, they proposed a strategy based on three types of functional conservation areas—sites, landscapes, and networks—defined by ecosystems and species targeted for conservation. Sites aim to conserve species or ecosystems that exist at local scales (meters to thousands of hectares); landscapes seek to conserve many ecosystems and multiple species at scales ranging up to millions of hectares; and networks are integrated sites or landscapes that provide for movement and connectivity. A distinguishing aspect of the Poiani et al. (2000) approach is the consideration of ecological processes (e.g., fire, flood, hydrology) as part of their evaluation of functionality.

Four spatial scales of assessment are used in their conservation framework—local, intermediate, coarse, and regional (Poiani et al. 2000). Each of these scales can be associated with plant or animal species or ecosystems. Ecosystems are broadly defined to include plant and animal communities and ecological process but, in practice, often default to boundaries defined by vegetation community types similar to traditional coarse filter approaches. Ecosystem scale can thus vary as a function of the degree of resolution applied to the plant community data. Animal species with diverse ecologies are also used as surrogates to represent a range of scales based on differences in the species' mobilities, home range sizes, and migration patterns.

Though not explicitly cast in terms of a coarse filter–fine filter framework, the four scales of analysis used by Poiani et al. (2000) have elements that can be aligned with traditional coarse and fine filter approaches. A distinguishing characteristic of their approach, however, is that species are used as coarse as well as fine filter elements for conservation planning. For example, similar to the umbrella species concept (Fleishman et al. 2000), wide-ranging animals such as migrating ungulates and top-level predators are specified as conservation targets for regional scale assessments.

Currently, TNC uses a comprehensive conservation planning process that extends the Poiani et al. (2000) approach and draws on elements of both coarse and fine filter strategies (Groves et al. 2002). Five criteria are used to identify priority conservation areas on the landscape: degree of existing protection, conservation value, threat, feasibility, and leverage (Groves et al. 2000). The process of identifying conservation value begins by identifying conservation targets that may be defined on the basis of spatial scale and biological organization. Levels in these two hierarchies include broad-scale physical features of the landscape (e.g., soils, geology, topography, and climate), ecoregions (combinations of physical features and dominant vegetation community types [Bailey 1995]), biological communities, and individual species. A key objective of the TNC approach is to conserve ecosystems, but the reality is that ecosystems are usually defined on the basis of dominant vegetation types (the coarse filter).

In the TNC biological hierarchy, the finest scale includes information on the genetic diversity of target species to address viability concerns of small, isolated populations. Second, the population abundance of individual species—particularly rare, keystone, or umbrella species—is evaluated. Third, representative biological communities for a given planning area provide a coarse filter target, whereby maintenance of community dominant species indicates whether critical interactions and processes are also being maintained. Fourth, ecosystems, when they can be delineated on the basis of disturbance regimes or geological or topographic features, are included as conservation targets. Finally, portions of entire landscapes including multiple ecosystems are evaluated for inclusion in the final conservation strategy. Full implementation of the TNC conservation planning algorithm is data intensive, requiring information on local population sizes, genetic diversity, and spatial distribution patterns. In practice, however, surrogate measures are used for many of these data elements, and recourse is made to the coarse filter components as needed.

More recently, practical approaches to land use and management planning have begun to emerge that link comprehensive biodiversity mapping efforts with science-based public processes. The Forest Ecosystem Restoration Analysis (ForestERA) Project (Sisk et al. 2006) is an interdisciplinary approach to landscape-scale (e.g., 10^4–10^6 ha) resource management and planning that integrates spatially explicit data and models within a collaborative decision-making framework. Using GIS-based tools, the ForestERA process couples fine and coarse filter information to map and prioritize large areas for restoration and conservation on lands that are typically under the jurisdiction of the Forest Service. For example, empirically derived models of indicator species occurrence and demographic rates of focal species are commonly selected and combined to represent "high value" landscape or biodiversity features. Using a series of ranking and weighting exercises, this multispecies information is, in turn, confronted by "risk" models, including fire hazard and post-fire erosion, and alternative management actions, including prescribed fire and mechanical thinning. By manipulating the original model parameters, stakeholders can collectively predict, evaluate, and compare the effects of various risk factors and management scenarios on the conservation of multiple species or their habitats.

RECOMMENDATIONS

We are encouraged by recent advances in quantitative methods to inform multispecies conservation planning—diverse life histories and ecological requirements can now be simultaneously addressed for multiple species. However, we perceive a significant gap between the research and practitioner communities in the arena of public lands conservation. As a result, the innovative methods described in recent scientific literature may not be accessible to most land managers. This suggests a need for increased collaboration between the research and management communities.

Given the sheer number of species that must be managed and conserved on most national forests, national parks, wildlife refuges, or other public land areas in the United States, the use of biodiversity surrogates is inescapable (see Margules and Pressey 2000). Despite significant increases in data availability and modeling methods in recent years, we believe the most feasible approach to multispecies conservation planning on public lands at this time remains a combined coarse filter–fine filter approach applied at broad spatial scales (see Haufler et al. 1999). Substantial increases in the quality and quantity of spatial data from remote sensing have increased the precision and resolution of coarse filter data. These data may allow the identification and mapping of broad vegetation types and stages that serve as useful habitat surrogates for many species. In addition, for the fine filter we advocate for a judicious selection of focal species based on the principles of complementarity and comprehensiveness similar in concept to their use in reserve selection algorithms. A comprehensive species set has the property of spanning, to the extent possible, the range of niche types, trophic positions, allometrically based spatial and temporal scales, demographics, and movement behaviors that characterize the entire species pool. Given that the size of the focal species set will be constrained by budget and data limits, the minimal species set will have the property that member species maximally complement each other's ecologies and life histories.

The knowledge needed to select an (constrained) optimal set of focal species is incomplete at this time (Favreau et al. 2006), and there is debate on the utility of the surrogate species approach (e.g., Lambeck 2002, Lindenmayer et al. 2002). However, we believe categories of focal species and criteria for their selection may be sufficiently advanced that the goals of complementarity and comprehensiveness can be approximated. In addition to focal species selection methods based on complementary life histories and ecologies, existing species categories long recognized by ecologists as useful concepts should be considered.

Identifying Focal Species

To begin the process of filtering the species pool in search of focal species, one may find it useful to move sequentially through the following steps that include both ecological and social criteria (cf. with Lambeck 1997, Hilty and Merenlender 2000, Beazley and Cardinal 2004):

1. Determine legal responsibilities of the public land managers relative to sustaining biological diversity and the viability of individual species. That is, clarify the requirements for species protection within the larger context of land-use planning, multiple use, and management.
2. Make a list of all the species in the affected area for which the land manager has a legal responsibility. This list could be prioritized, for example, in terms of risk.
3. Apply an initial filter based on the criteria listed in Table 3-1.
4. From this list, attempt to identify species that fall into one of the focal species categories identified (see earlier list).
5. Cross-classify the species list from (4) according to various ecological attributes, including body size (as a surrogate for many allometric relationships), home range size, demographic characteristics (e.g., life span, generation length, reproductive potential), trophic position, and habitat specialization (and other niche dimensions). This step will allow complementary species sets to be identified within each focal species category.
6. Remove species whose dynamics are largely independent of management activities occurring on public lands.

Table 3-1 Desired Attributes of Focal Species as Surrogate Measures for Multispecies Conservation Planning

1	Taxonomic status is well established
2	Ecology, life history, and demography are sufficiently "known" to allow direct or indirect estimates of relative abundance and spatial distribution
3	Relatively high detectability allowing for precise estimates of population status (presence-absence or abundance)
4	Low sampling variability (consistent and high detectability across time and space)
5	Low process variation in demographic rates, allowing more reliable inferences to causal factors affecting population status
6	Attributes (4) and (5) allow for high statistical power to detect trends in presence-absence or abundance
7	Known relationships between environmental stressors and population status

7. Further filter species by retaining only those that can be monitored with available resources, but retaining several species within each focal category.

8. Select a final set of focal species that spans the range of focal species categories, encompasses a wide range of ecological attributes, and, if possible, represents different taxonomic groups. This list should be critically evaluated to see whether it meets the criteria of comprehensiveness and complementarity.

After one follows these steps, the hope is that a near optimal set of fine filter species will have been identified. The set of focal species should complement and provide an opportunity to evaluate the utility of the vegetation-based coarse filter approach. In addition, if the set of focal species is sufficiently small with well-known distributions, habitat relationships, and demographic characteristics, many of the new methods for multiple species assessment discussed previously may be feasible.

APPLICATION TO FEDERAL PUBLIC LANDS

Conservation practice on public land areas in the United States is constrained by a plethora of often conflicting legal statutes and by profound limits on data availability. These points can be illustrated by considering the management of Forest Service lands in the United States. Probably the most important statute to consider is the NFMA, which requires the Forest Service to maintain biological diversity in perpetuity. Thus, the Forest Service has a formal legal mandate to engage in multispecies conservation. Between 1982 and 2005, the biodiversity requirements in NFMA were interpreted to mean maintaining the viability of native vertebrate species across their ranges. The approach was to choose "management indicator species," similar in intent to focal species, and to manage for and monitor their populations. Although the efficacy of this approach has been questioned (Landres et al. 1988, Landres 1992), there is no way to evaluate its effectiveness because the requirements were never fully implemented. To the best of our knowledge, the Forest Service has never monitored the populations of the indicator species designated in their management plans in a manner that would allow a rigorous and comprehensive assessment of their response to management and the degree to which unmeasured species were being conserved.

In 2005, the Forest Service issued new regulations to implement the NFMA, dropped their requirements for viability assessments of selected vertebrate species, and shifted over to a coarse filter model for conservation (Noon et al. 2005). The agency asserts that maintaining a diversity of natural vegetation community types will maintain ecological systems and their component species (Federal Register, 70:3:1023). As a result, the agency is no longer required to implement mesofilter or fine filter approaches to conservation planning, or to directly monitor species responses to management actions.

While maintaining a diversity of vegetation types is far easier to implement and monitor than is the maintenance of viable populations of multiple species, there is little guidance concerning how vegetation diversity alone can be used to infer the conservation status of unmeasured species. The reality is that the composition and configuration of vegetation types and successional stages needed to sustain multiple species over the long term is unknown (Cushman et al. 2008). The challenge of managing multiple animal species through vegetation surrogates is further complicated by current data limitations. Without additional funding, vegetation will be mapped based almost exclusively on spectral imagery with little ground-truthing. Forest vegetation data, such as species composition and structure, collected over large spatial extents are generally limited to Forest Inventory and Analysis (FIA; U.S. Forest Service 2003) data collected at plots located on a 4.8 km grid (see www.fia.fs.fed.us). These data are designed to provide broad-scale assessments of forest resources and have only secondarily been modified to include additional measurements believed relevant to wildlife populations. While more than adequate for their designed purpose, FIA plots carry little spatial information, do not extend to nonforested areas, and primarily measure trees instead of providing comprehensive floristic information. Further, statistical models relating the coarse-grained habitat data available from satellite imagery or FIA data and fauna presence and/or abundance have not been generated (however, see Fitzgerald et al., this volume). Thus, land management agencies have little quantitative information to provide even an indirect check on whether the populations of specific species are likely being maintained by use of vegetation surrogates as a coarse filter.

In recent years, public land management agencies have seen reduced budgets while fixed costs have steadily risen. This trend is not likely to reverse in the near future, and therefore proposals that require large infusions of money are likely dead on arrival. Thus, for the near term, public land management agencies in the United States will need primarily to work within the confines of existing data. This is unfortunate, as virtually all the promising new approaches to multiple species conservation cannot be implemented given these data. Although we believe that a mix of coarse and fine filter approaches likely has the greatest merit, we acknowledge that, given the existing data and knowledge base, federal land management agencies may be able to implement only coarse filter approaches. Under this scenario, the emphasis should be on validation of the coarse filter approach (see Schulte et al. 2006), identification of mesofilter elements, and selection of fine filter species based on the criteria for focal species selection discussed previously. With these constraints in mind, we propose the following steps to improve multispecies conservation on federal public lands:

1. Make sure that existing data are fully utilized. For many species, mesofilter elements will likely have the greatest relationship to habitat quality; therefore, plot inventory data should be expanded such that these elements can be mapped and spatially evaluated. Currently, a variety of approaches that would serve as a coarse filter are under development, including gradient modeling and machine-learning or expert-system methods (Frescino et al. 2001, Moisen and Frescino 2002, Ohmann and Gregory 2002, Olden and Jackson 2002, Stockwell 2006). Although the general efficacy of these methods is unknown, priority should be given to this area of research.

2. Recognize that certain habitat elements and landcover classes have been greatly reduced from their historical levels (e.g., large diameter trees and old-growth forest). Identify, and prioritize for conservation, species associated with these elements and classes. Where large, relatively pristine areas of rare habitats exist, manage to ensure their persistence.

3. Recognize that it will be some time before more complete data on species distributions and habitat relationships become available. In the interim, give conservation priority to habitat elements and landscape classes, such as older forests or large diameter trees, which cannot be rapidly replaced.

4. Given uncertainties about the relationships between vegetation mosaics and fauna populations, any multispecies conservation strategy must be considered an untested hypothesis. Testing the hypothesis is a formidable task, but it may be possible to use currently collected fauna information as a check on land management. For instance, at the landscape scale, FIA plot composition has been correlated with breeding bird survey data (Fearer 2006) and with the distribution of forest carnivores (Zielinski et al. 2006*b*). Similarly, population information on game species collected by state Fish and Game agencies can be correlated to broad landscape vegetation characteristics.

5. Take advantage of new monitoring methodologies. Genetic monitoring methods (Schwartz et al. 2007, Zielinski et al. 2007) allow the population attributes of many species to be inferred at greatly reduced costs, potentially providing direct feedback. Agencies therefore should prototype genetic collection and analysis approaches to accelerate adoption of this technology.

6. Based on the methods discussed previously, choose to monitor a small group of focal species with complementary and comprehensive ecologies. Develop detailed habitat relationships for these species, manage for their viability, and track their populations. If the set of focal species is small enough, and sufficient data exist on their ecologies and life history, optimization methods based on maximizing their combined persistence

likelihoods may be possible (Nicholson and Possingham 2006, Nicholson et al. 2006). Because of the limited nature of this fine filter, and because of the requirement of additional funding for this work, for pragmatic reasons we suggest that the set of focal species include some of public interest, such as game species or those listed species under the ESA.

SUMMARY

Federal land managers in the United States are responsible for sustaining native biodiversity and obligated not to put any species at risk of extinction. In other words, they must manage in a way that benefits and sustains multiple species. However, even assuming well-intentioned and adequately funded managers, difficulties arise because most species are poorly known, and the quantitative methods and technologies that allow the concurrent evaluation of ≥ 2 species are just now being developed. These novel methods are significant, but their data requirements and technical challenges currently limit their utility to the conservation manager. As a result, conservation planning now, and in the foreseeable future, will depend on the continued use of broad-scale environmental proxies and species-based surrogates. These surrogate measures, encompassing both coarse filter and fine filter elements, must be judiciously selected so as to allow inference to the viability of all the unmeasured species.

LITERATURE CITED

Andelman, S. J., and W. F. Fagan. 2000. Umbrellas and flagships: Efficient conservation surrogates or expensive mistakes? *Proceedings of the National Academy of Sciences* 97:5954–5959.

Aurambout, J. P., A. G. Endress, and B. M. Deal. 2005. A spatial model to estimate habitat fragmentation and its consequences on long-term persistence of animal populations. *Environmental Monitoring and Assessment* 109:199–225.

Bailey, R. G. 1995. Descriptions of the Ecoregions of the United States. U.S. Forest Service, Miscellaneous Publication number 1391, Washington, D.C., USA.

Bailey, T. C., and A. C. Gatrell. 1995. *Interactive spatial data analysis*. Addison-Wesley Publishers, Reading, Massachusetts, USA.

Barry, S., and J. Elith. 2006. Error and uncertainty in habitat models. *Journal of Applied Ecology* 43:413–423.

Beazley, K., and N. Cardinal. 2004. A systematic approach for selecting focal species for conservation in forests of Nova Scotia and Maine. *Environmental Conservation* 31:91–101.

Bevers, M., J. Hof, B. Kent, and M. G. Raphael. 1995. Sustainable forest management for optimizing multispecies wildlife habitat: A coastal Douglas-fir example. *Natural Resources Modeling* 9:1–23.

Block, W. M., L. A. Brennan, and R. J. Gutiérrez. 1987. Evaluation of guild-indicator species for use in resource management. *Environmental Management* 11:265–269.

Bonham-Carter, G. F., F. P. Agterberg, and D. F. Wright. 1989. Weights of evidence modeling: A new approach to mapping mineral potential. Pages 171–183 *in* F. P. Agterberg and G. F. Bonham-Carter, editors. *Statistical applications in the earth sciences.* Geological Survey of Canada paper 89-9, Ottawa, Ontario, Canada.

Cabeza, M., and A. Moilanen. 2003. Site selection algorithms and habitat loss. *Conservation Biology* 17:1402–1413.

Caro, T. M., and G. O'Doherty. 1999. On the use of surrogate species in conservation biology. *Conservation Biology* 13:805–814.

Carroll, C., R. F. Noss, P. C. Paquet, and N. H. Schumaker. 2003. Use of population viability analysis and reserve selection algorithms in regional conservation plans. *Ecological Applications* 13:1773–1789.

Carroll, C., W. J. Zielinski, and R. F. Noss. 1999. Using presence-absence data to build and test special habitat models for the fisher in the Klamath Region, USA. *Conservation Biology* 13:1344–1359.

Christianou, M., and B. Ebenman. 2005. Keystone species and vulnerable species in ecological communities: Strong or weak interactors? *Journal of Theoretical Biology* 235:95–103.

Clark, J. D., J. E. Dunn, and K. G. Smith. 1993. A multivariate model of female black bear habitat use for a geographic information system. *Journal of Wildlife Management* 57:519–526.

Committee of Scientists (COS). 1999. Sustaining the people's lands: Recommendations for stewardship of the national forests and grasslands into the next century. U.S. Forest Service, COS report. <http://www.fs.fed.us/forum/nepa/rule/cosreport.html>.

Cotgreave, P., and P. H. Harvey. 1992. Relationships between body size, abundance, and phylogeny in bird communities. *Functional Ecology* 6:248–256.

Cushman, S. A., and K. McGarigal. 2004. Hierarchical analysis of forest bird species-environment relationships in the Oregon Coast Range. *Ecological Applications* 14:1090–1105.

Cushman, S. A., K. S. McKelvey, C. H. Flather, and K. McGarigal. 2008. Do forest community types provide a sufficient basis to evaluate biological diversity? *Frontiers in Ecology and the Environment* 6:13–17.

Cushman, S. A., K. S. McKelvey, J. Hayden, and M. K. Schwartz. 2006. Gene flow in complex landscapes: Testing multiple hypotheses with causal modeling. *American Naturalist* 168:486–499.

Early, R., and C. D. Thomas. 2007. Multispecies conservation planning: Identifying landscapes for the conservation of viable populations using local and continental species priorities. *Journal of Applied Ecology* 44:253–262.

Elith, J., C. H. Graham, R. P. Anderson, M. Dudik, S. Ferrier, A. Guisan, R. J. Hijmans, F. Huettmann, J. R. Leathwick, A. Lehmann, J. Li, L. G. Lohmann, B. A. Loiselle, G. Manion, C. Moritz, M. Nakamura, Y. Nakazawa, J. M. C. Overton, A. T. Peterson, S. J. Phillips, K. Richardson, R. Scachetti-Pereira, R. E. Schapire, J. Soberón, S. Williams, M. S. Wisz, and N. E. Zimmermann. 2006. Novel methods improve prediction of species' distributions from occurrence data. *Ecography* 29:129–151.

Fauth, J. E. 1997. Identifying potential keystone species from field data—An example from temporary ponds. *Ecology Letters* 2:36–43.

Favreau, J. M., C. A. Drew, G. R. Hess, M. J. Rubino, F. H. Koch, and K. A. Eschelbach. 2006. Recommendations for assessing the effectiveness of surrogate species approaches. *Biodiversity and Conservation* 15:3949–3969.

Fearer, T. M. 2006. *Evaluating population-habitat relationships of forest breeding birds at multiple spatial and temporal scales using forest inventory and analysis data*. Dissertation, Virginia Polytechnic Institute and State University, Blacksburg, Virginia, USA.

Fielding, A. H., and J. F. Bell. 1997. A review of methods for the assessment of prediction errors in conservation presence/absence models. *Environmental Conservation* 24:38–49.

Fischer, J., and D. B. Lindenmayer. 2006. Beyond fragmentation: The continuum model for faunal research and conservation in human-modified landscapes. *Oikos* 112:473–480.

Fischer, J., D. B. Lindenmayer, and A. Cowling. 2004*a*. The challenge of managing multiple species at multiple scales: Reptiles in an Australian gazing landscape. *Journal Applied Ecology* 41:32–44.

Fischer, J., D. B. Lindenmayer, and I. Fazey. 2004*b*. Appreciating ecological complexity: Habitat contours as a conceptual landscape model. *Conservation Biology* 18:1245–1253.

Fleishman, E., R. B. Blair, and D. D. Murphy. 2001. Empirical validation of a method for umbrella species selection. *Ecological Applications* 11:1489–1501.

Fleishman, E., D. D. Murphy, and P. F. Brussard. 2000. A new method for selection of umbrella species for conservation planning. *Ecological Applications* 10:569–579.

Forest Ecosystem Management Assessment Team. 1993. Forest ecosystem management: An ecological, economic, and social assessment. U.S. Government Printing Office, Report number 1993-793-071, Washington, D.C., USA.

Frescino, T. S., T. C. Edwards, Jr., and G. G. Moisen. 2001. Modeling spatially explicit forest structural attributes using generalized additive models. *Journal of Vegetation Science* 12:15–26.

Gardner, R. H., and E. J. Gustafson. 2004. Simulating dispersal of reintroduced species within heterogeneous landscapes. *Ecological Modelling* 171:339–358.

Goble, D. D., and E. T. Freyfogle. 2002. *Wildlife law.* University Case Book Series, Foundation Press, New York, New York, USA.

Groves, C. R., D. B. Jensen, L. L. Valutis, K. H. Redford, M. L. Shaffer, J. M. Scott, J. V. Baumgartner, J. V. Higgins, M. W. Beck, and M. G. Anderson. 2002. Planning for biodiversity conservation: Putting conservation science into practice. *Bioscience* 52:499–512.

Groves, C., L. Valutis, D. Vosick, B. Neely, K. Wheaton, J. Touval, and B. Runnels. 2000. *Designing a geography of hope: A practitioner's handbook for ecoregional conservation planning.* Second edition. The Nature Conservancy. <http://conserveonline.org/docs/2000/11/GOH2-v2.pdf>.

Guisan, A., A. Lehmann, S. Ferrier, M. Austin, J. M. C. Overton, R. Aspinall, and T. Hastie. 2006. Making better biogeographical predictions of species' distributions. *Journal of Applied Ecology* 43:386–392.

Guisan, A., and N. E. Zimmermann. 2000. Predictive habitat distribution models in ecology. *Ecological Modelling* 135:147–186.

Haight, R. G., B. Cypher, P. A. Kelly, S. Philips, H. P. Possingham, K. Ralls, A. M. Starfield, P. J. White, and D. Williams. 2002. Optimizing habitat protection using demographic models of population viability. *Conservation Biology* 16:1386–1397.

Hanski, I. 1994. A practical model of metapopulation dynamics. *Journal of Animal Ecology* 63:151–162.

Haufler, J. B., C. A. Mehl, and G. J. Roloff. 1996. Using a coarse-filter approach with species assessment for ecosystem management. *Wildlife Society Bulletin* 24:200–208.

Haufler, J. B., C. A. Mehl, and G. J. Roloff. 1999. Conserving biological diversity using a coarse-filter approach with a species assessment. Pages 107–125 *in* R. K. Baydack, H. Campa, III, and J. B. Haufler, editors. *Practical approaches to the conservation of biological diversity.* Island Press, Washington, D.C., USA.

Hilty, J., and A. Merenlender. 2000. Faunal indicator taxa selection for monitoring ecosystem health. *Biological Conservation* 92.185–197.

Hof, J. G., and M. G. Raphael. 1993. Some mathematical programming approaches for optimizing timber age class distributions to meet multispecies wildlife population objectives. *Canadian Journal of Forest Research* 23:828–834.

Holzkämper, A., A. Lausch, and R. Seppelt. 2006. Optimizing landscape configuration to enhance habitat suitability for species with contrasting habitat requirements. *Ecological Modelling* 198:277–292.

Hooper, D. U., F. S. Chapin, III, J. J. Ewel, A. Hector, P. Inchausti, S. Lavorel, J. H. Lawton, D. M. Lodge, M. Loreau, S. Naeem, B. Schmid, H. Setälä, A. J. Symstad, J. Vandermeer, and D. A. Wardle. 2005. Effects of biodiversity on ecosystem functioning: A consensus of current knowledge. *Ecological Monographs* 75:3–35.

Hosmer, D. W., and S. Lemeshow. 2000. *Applied logistic regression.* Second edition. John Wiley and Sons, New York, New York, USA.

Hunter, M. L., Jr. 1991. Coping with ignorance: The coarse filter strategy for maintaining biodiversity. Pages 266–281 *in* K. A. Kohm, editor. *Balancing on the brink of extinction: The Endangered Species Act and lessons for the future.* Island Press, Washington, D.C., USA.

Hunter, M. L., Jr. 2005. A mesofilter conservation strategy to complement fine and coarse filters. *Conservation Biology* 19:1025–1029.

Hunter, M. L., Jr., G. L. Jacobson, Jr., and T. Webb, III. 1988. Paleoecology and the coarse-filter approach to maintaining biological diversity. *Conservation Biology* 2:375–385.

Hutchinson, G. E. 1957. Concluding remarks. Cold Springs Harbor Symposium. *Quantitative Biology* 22:415–427.

Jones, C. G., J. H. Lawton, and M. Shachak. 1994. Organisms as ecosystem engineers. *Oikos* 69:373–386.

Jones, C. G., J. H. Lawton, and M. Shachak. 1997. Positive and negative effects of organisms as physical ecosystem engineers. *Ecology* 78:1946–1957.

Jordán, F., A. Takács-Sánta, and I. Molnár. 1999. A reliability theoretical quest for keystones. *Oikos* 86:453–462.

King, A. W., and K. A. With. 2002. Dispersal success on spatially structured landscapes: When do spatial pattern and dispersal behavior really matter? *Ecological Modelling* 147:23–39.

Kintsch, J. A., and D. L. Urban. 2002. Focal species, community representation, and physical proxies as conservation strategies: A case study in the Amphibolite Mountains, North Carolina, U.S.A. *Conservation Biology* 16:936–947.

Klute, D. S., M. J. Lovallo, and W. M. Tzilkowski. 2002. Autologistic regression modeling of American woodcock habitat use with spatially dependent data. Pages 335–344 *in* J. M. Scott, P. J. Heglund, M. L. Morrison, J. B. Haufler, M. G. Raphael, W. A. Wall, and F. B. Samson, editors. *Predicting species occurrences: Issues of accuracy and scale.* Island Press, Washington, D.C., USA.

Knick, S. T., and D. L. Dyer. 1997. Distribution of black-tailed jackrabbit habitat determined by GIS in southwestern Idaho. *Journal of Wildlife Management* 61:75–85.

Lambeck, R. J. 1997. Focal species: A multi-species umbrella for nature conservation. *Conservation Biology* 11:549–551.

Lambeck, R. J. 2002. Focal species and restoration ecology: Response to Lindenmayer et al. *Conservation Biology* 16:849–856.

Landres, P. B. 1992. Ecological indicators: Panacea or liability? Pages 1295–1318 *in* D. H. McKenzie, D. E. Hyatt, and V. J. McDonald, editors. *Ecological indicators*. Volume 2. Elsevier Applied Science, New York, New York, USA.

Landres, P. B., J. Verner, and J. W. Thomas. 1988. Ecological uses of vertebrate indicator species: A critique. *Conservation Biology* 2:316–328.

Lawler, J. J., D. White, and L. L. Master. 2003. Integrating representation and vulnerability: Two approaches for prioritizing areas for conservation. *Ecological Applications* 13:1762–1772.

Lindenmayer, D. B., A. D. Manning, P. L. Smith, H. P. Possingham, J. Fischer, I. Oliver, and M. A. McCarthy. 2002. The focal-species approach and landscape restoration: A critique. *Conservation Biology* 16:338–345.

Lindenmayer, D. B., C. R. Margules, and D. B. Botkin. 2000. Indicators of biodiversity for ecologically sustainable forest management. *Conservation Biology* 14:941–950.

MacArthur, R. H., and R. Levins. 1967. The limiting similarity, convergence and divergence of coexisting species. *American Naturalist* 101:377–385.

MacKenzie, D. I., J. D. Nichols, J. A. Royle, K. H. Pollock, L. L. Bailey, and J. E. Hines. 2006. *Occupancy estimation and modeling: Inferring patterns and dynamics of species occurrence*. Elsevier Academic Press, Amsterdam, The Netherlands.

MacNally, R., A. F. Bennett, G. W. Brown, L. F. Lumsden, A. Yen, S. Hinkley, P. Lillywhite, and D. A. Ward. 2002. How well do ecosystem-based planning units represent different components of biodiversity? *Ecological Applications* 12:900–912.

Manel, S., M. K. Schwartz, G. Luikart, and P. Taberlet. 2003. Landscape genetics: Combining landscape ecology and population genetics. *Trends in Ecology and Evolution* 18:189–197.

Manly, B. F. J., L. L. McDonald, D. A. Thomas, T. L. McDonald, and W. P. Erickson. 2002. *Resource selection by animals: Statistical design and analysis for field studies*. Second edition. Kluwer Academic, Dordrecht, The Netherlands.

Marcot, B. G., R. S. Holthausen, M. G. Raphael, M. M. Rowland, and M. J. Wisdom. 2001. Using Bayesian belief networks to evaluate fish and wildlife population viability under land management alternatives from an environmental impact statement. *Forest Ecology and Management* 153:29–42.

Margules, C. R., and R. L. Pressey. 2000. Systematic conservation planning. *Nature* 405:243–253.

McArdle, B. H., K. J. Gaston, and J. H. Lawton. 1990. Variation in the size of animal populations: Patterns, problems, and artifacts. *Journal of Animal Ecology* 59:439–454.

McDonnell, M. D., H. P. Possingham, I. R. Ball, and E. A. Cousins. 2002. Mathematical methods for spatially cohesive reserve design. *Environmental Modeling and Assessment* 7:107–114.

McRae, B. H. 2006. Isolation by resistance. *Evolution* 60:1551–1561.

Mills, L. S., M. E. Soulé, and D. F. Doak. 1993. The keystone-species concept in ecology and conservation. *Bioscience* 43:219–224.

Moilanen, A., A. M. A. Franco, R. I. Early, R. Fox, B. Wintle, and C. D. Thomas. 2005. Prioritizing multiple-use landscapes for conservation: Methods for large multi-species planning problems. *Proceeding of the Royal Society of London, Series B* 272:1885–1891.

Moisen, G. G., and T. S. Frescino. 2002. Comparing five modelling techniques for predicting forest characteristics. *Ecological Modelling* 157:209–225.

Nagle, J. C., and J. B. Ruhl. 2002. *The law of biodiversity and ecosystem management.* University Case Book Series, Foundation Press, New York, New York, USA.

Nicholson, E., and H. P. Possingham. 2006. Objectives for multi-species conservation planning. *Conservation Biology* 20:871–881.

Nicholson, E., M. I. Westphal, K. Frank, W. A. Rochester, R. L. Pressey, D. B. Lindenmayer, and H. P. Possingham. 2006. A new method for conservation planning for the persistence of multiple species. *Ecology Letters* 9:1049–1060.

Noon, B. R., and V. H. Dale. 2003. Broad-scale ecological science and its application. Pages 34–51 *in* K. J. Gutzwiller, editor. *Applying landscape ecology in biological conservation.* Springer, New York, New York, USA.

Noon, B. R., and K. S. McKelvey. 1996. A common framework for conservation planning: Linking individual and metapopulation models. Pages 139–166 *in* D. R. McCullough, editor. *Metapopulations and wildlife conservation.* Island Press, Washington, D.C., USA.

Noon, B. R., D. D. Murphy, S. R. Beissinger, M. L. Shaffer, and D. DellaSala. 2003. Conservation planning for U.S. National Forests: Conducting comprehensive biodiversity assessments. *Bioscience* 53:1217–1220.

Noss, R. F. 1996. Ecosystems as conservation targets. *Trends in Ecology and Evolution* 11:351.

Noss, R. F., and A. Y. Cooperrider. 1994. *Saving nature's legacy.* Island Press, Washington, D.C., USA.

Noss, R. F., M. A. O'Connell, and D. D. Murphy. 1997. *The science of conservation planning: Habitat conservation under the Endangered Species Act.* Island Press, Washington, D.C., USA.

Ohmann, J. L., and M. J. Gregory. 2002. Predictive mapping of forest composition and structure with direct gradient analysis and nearest-neighbor imputation in coastal Oregon, USA. *Canadian Journal of Forest Resources* 32:725–741.

Olden, J. D., and D. A. Jackson. 2002. A comparison of statistical approaches for modelling fish species distributions. *Freshwater Biology* 47:1976–1995.

Østergaard, S., M. M. Hansen, V. Loeschcke, and E. E. Nielson. 2003. Long-term temporal changes of genetic composition in brown trout (*Salmo trutta L.*) populations inhabiting an unstable environment. *Molecular Ecology* 12:3123–3135.

Panjabi, A. O., E. H. Dunn, P. J. Blancher, W. C. Hunter, B. Altman, J. Bart, C. J. Beardmore, H. Berlanga, G. S. Butcher, S. K. Davis, D. W. Demarest, R. Dettmers, W. Easton, H. G. de Silva Garza, E. E. Iñigo-Elias, D. N. Pashley, C. J. Ralph, T. D. Rich, K. V. Rosenberg, C. M. Rustay, J. M. Ruth, J. S. Wendt, and T. C. Will. 2005. Partners in Flight handbook on species assessment. Version 2005. Partners in Flight Technical Series number 3, Rocky Mountain Bird Observatory website: <http//:www.rmbo.org/pubs/downloads/Handbook2005.pdf>.

Patton, D. R. 1987. Is the use of "management indicator species" feasible? *Western Journal of Applied Forestry* 2:33–34.

Pearson, S. M., M. G. Turner, R. H. Gardner, and R. V. O'Neill. 1996. An organism-based perspective of habitat fragmentation. Pages 77–95 *in* R. C. Szaro and D. W. Johnston, editors. *Biodiversity in managed landscapes: Theory and practice.* Oxford University Press, New York, New York, USA.

Pereira, J. M., and R. M. Itami. 1991. GIS-based habitat modeling using logistic multiple regression: A study of the Mount Graham red squirrel. *Photogrammetric Engineering and Remote Sensing* 57:1475–1486.

Peters, R. H. 1983. *The ecological implications of body size.* Cambridge University Press, Cambridge, United Kingdom.

Phillips, S. J., R. P. Anderson, and R. E. Schapire. 2006. Maximum entropy modeling of species geographic distributions. *Ecological Modelling* 190:231–259.

Pickett, S. T. A., and M. L. Cadenasso. 2002. The ecosystem as a multidimensional concept: Meaning, model, and metaphor. *Ecosystems* 5:1–10.

Pimm, S. L. 2002. *Food webs.* University of Chicago Press, Chicago, Illinois, USA.

Poff, N. L. 1997. Landscape filters and species traits: Towards mechanistic understanding and prediction in stream ecology. *Journal of the North American Benthological Society* 16:391–409.

Poiani, K. A., B. D. Richter, M. G. Anderson, and H. E. Richter. 2000. Biodiversity conservation at multiple scales: Functional sites, landscapes, and networks. *Bioscience* 50:133–146.

Possingham, H., I. Ball, and S. Andelman. 2000. Mathematical methods for identifying representative reserve networks. Pages 291–306 *in* S. Ferson and M. Burgman, editors. *Quantitative methods for conservation biology.* Springer, New York, New York, USA.

Power, M. E., D. Tilman, J. A. Estes, B. A. Menge, W. J. Bond, L. S. Mills, G. Daily, J. C. Castilla, J. Lubchenco, and R. T. Paine. 1996. Challenges in the quest for keystones. *Bioscience* 46:609–620.

Pressey, R. L., R. M. Cowling, and M. Rouget. 2003. Formulating conservation targets for biodiversity pattern and process in the Cape Floristic Region, South Africa. *Biological Conservation* 112:99–127.

Pressey, R. L., C. J. Humphries, C. R. Margules, R. I. Vane-Wright, and P. H. Williams. 1993. Beyond opportunism: Key principles for systematic reserve selection. *Trends in Ecology and Evolution* 8:124–128.

Pressey, R. L., I. R. Johnson, and P. D. Wilson. 1994. Shades of irreplaceability: Towards a measure of the contribution of sites to a reservation goal. *Biodiversity and Conservation* 3:242–262.

Pulliam, H. R. 2000. On the relationship between niche and distribution. *Ecology Letters* 3:349–361.

Raphael, M. G., and R. Molina. 2007. *Conservation of rare and little-known species: Biological, social, and economic considerations.* Island Press, Washington, D.C., USA.

Raphael, M. G., M. J. Wisdom, M. M. Rowland, R. S. Holthausen, B. G. Wales, B. G. Marcot, and T. D. Rich. 2001. Status and trends of habitats of terrestrial vertebrates in relation to land management in the interior Columbia River basin. *Forest Ecology and Management* 153:63–88.

Roberge, J.-M., and P. Angelstam. 2004. Usefulness of the umbrella species concept as a conservation tool. *Conservation Biology* 18:76–85.

Root, K. V., H. R. Akçakaya, and L. Ginzburg. 2003. A multispecies approach to ecological valuation and conservation. *Conservation Biology* 17:196–206.

Sarkar, S., R. L. Pressey, D. P. Faith, C. R. Margules, T. Fuller, D. M. Stoms, A. Moffett, K. A. Wilson, K. J. Williams, P. H. Williams, and S. Andelman. 2006. Biodiversity conservation planning tools: Present status and challenges for the future. *Annual Review of Environment and Resources* 31:123–159.

Schulte, L. A., R. J. Mitchell, M. L. Hunter, Jr., J. F. Franklin, R. K. McIntyre, and B. J. Palik. 2006. Evaluating the conceptual tools for forest biodiversity conservation and their implementation in the U.S. *Forest Ecology and Management* 232:1–11.

Schwartz, M. K., G. Luikart, and R. S. Waples. 2007. Genetic monitoring as a promising tool for conservation and management. *Trends in Ecology and Evolution* 22:25–33.

Scott, J. M., P. J. Heglund, M. L. Morrison, J. B. Haufler, M. G. Rafael, W. A. Wall, and F. B. Samson, editors. 2002. *Predicting species occurrences: Issues of accuracy and scale.* Island Press, Washington, D.C., USA.

Segurado, P., and M. B. Araújo. 2004. An evaluation of methods for modelling species distributions. *Journal of Biogeography* 31:1555-1568.

Shaffer, M. L., and B. A. Stein. 2000. Safeguarding our precious heritage. Pages 301-321 *in* B. A. Stein, L. S. Kutner, and J. S. Adams, editors. *Precious heritage: The status of biodiversity in the United States.* Oxford University Press, Oxford, United Kingdom.

Silva, M., and J. A. Downing. 1994. Allometric scaling of minimal mammal densities. *Conservation Biology* 8:732-743.

Sisk, T. D., J. W. Prather, H. M. Hampton, E. N. Aumack, Y. Xu, and B. G. Dickson. 2006. Participatory landscape analysis to guide restoration of ponderosa pine ecosystems in the American Southwest. *Landscape and Urban Planning* 78:300-310.

Smith, G. A., and M. V. Lomolino. 2004. Black-tailed prairie dogs and the structure of avian communities on the shortgrass plains. *Oecologia* 138:592-602.

Stockwell, D. R. B. 2006. Improving ecological niche models by data mining large environmental datasets for surrogate models. *Ecological Modelling* 192:188-196.

Taylor, P. D., L. Fahrig, and K. A. With. 2006. Landscape connectivity: A return to the basics. Pages 29-43 *in* K. R. Crooks and M. J. Sanjayan, editors. *Connectivity conservation.* Cambridge University Press, Cambridge, United Kingdom.

Theobald, D. M., and N. T. Hobbs. 2002. Functional definition of landscape structure using a gradient-based approach. Pages 667-672 *in* J. M. Scott, P. J. Heglund, M. L. Morrison, J. B. Haufler, M. G. Raphael, W. A. Wall, and F. B. Samson, editors. *Predicting species occurrences: Issues of accuracy and scale.* Island Press, Washington, D.C., USA.

Thomas, J. W., E. D. Forsman, J. B. Lint, E. C. Meslow, B. R. Noon, and J. Verner. 1990. *A conservation strategy for the northern spotted owl: Interagency Scientific Committee to address the conservation of the northern spotted owl.* U.S. Forest Service, Pacific Northwest Research Station, Portland, OR.

Tischendorf, L., and L. Fahrig. 2000. How should we measure landscape connectivity? *Landscape Ecology* 15:633-641.

Tognelli, M. F. 2005. Assessing the utility of indicator groups for the conservation of South American terrestrial mammals. *Biological Conservation* 121:409-417.

Tsoar, A., O. Allouche, O. Steinitz, D. Rotem, and R. Kadmon. 2007. A comparative evaluation of presence-only methods for modelling species distributions. *Diversity and Distributions* 13:397-405.

U.S. Forest Service. 2003. *Forest inventory and analysis.* Pacific Northwest Research Station, Portland, Oregon, USA.

Vitousek, P. M., H. A. Mooney, J. Lubchenco, and J. M. Melillo. 1997. Human domination of the Earth's ecosystems. *Science* 277:494-499.

Wiens, J. A. 1989. Spatial scaling in ecology. *Functional Ecology* 3:385-397.

Wilcox, B. A. 1984. *In situ* conservation of genetic resources: Determinants of minimum area requirements. Pages 639-647 *in* J. A. McNeely and K. R. Miller, editors. *National Parks: Conservation and development.* Smithsonian Institution Press, Washington, D.C., USA.

Williams, J. C., C. C. ReVelle, and S. A. Levin. 2005. Spatial attributes and reserve design models: A review. *Environmental Modeling and Assessment* 10:163-181.

Wilson, K., R. L. Pressey, A. Newton, M. Burgman, H. Possingham, and C. Weston. 2005. Measuring and incorporating vulnerability into conservation planning. *Environmental Management* 35:527-543.

Wintle, B. A., J. Elith, and J. M. Potts. 2005. Fauna habitat modelling and mapping: A review and case study in the Lower Hunter Central Coast region of NSW. *Austral Ecology* 30:719-738.

Wisdom, M. J., L. S. Mills, and D. F. Doak. 2000. Life stage simulation analysis: Estimating vital rate effects on population growth for conservation. *Ecology* 81:628-641.

With, K. A. 2004. Metapopulation dynamics: Perspectives from landscape ecology. Pages 23-44 *in* I. Hanski and O. E. Gaggiotti, editors. *Ecology, genetics, and evolution of metapopulations.* Elsevier Academic Press, Amsterdam, The Netherlands.

With, K. A., R. H. Gardner, and M. G. Turner. 1997. Landscape connectivity and population distributions in heterogeneous environments. *Oikos* 78:151-169.

Worton, B. J. 1989. Kernel methods for estimating the utilization distribution in home-range studies. *Ecology* 70:164-168.

Zielinski, W. J., C. Carroll, and J. R. Dunk. 2006*a*. Using landscape suitability models to reconcile conservation planning for two key forest predators. *Biological Conservation* 133:409-430.

Zielinski, W. J., R. L. Truex, J. R. Dunk, and T. Gaman. 2006*b*. Using forest inventory data to assess fisher resting habitat suitability in California. *Ecological Applications* 16:1010-1025.

Zielinski, W. J., M. J. Mazurek, and J. Zinck. 2007. Identifying the species of bats roosting in redwood basal hollows using genetic methods. *Northwest Science* 81:155-162.

CHAPTER

Geographic Approaches to Biodiversity Conservation: Implications of Scale and Error to Landscape Planning

4

Curtis H. Flather, Kenneth R. Wilson, and Susan A. Shriner

Conservation science is concerned with understanding why distribution and abundance patterns of species vary in time and space. Although these patterns have strong signatures tied to the availability of energy and nutrients, variation in climate, physiographic heterogeneity, and differences in the structural complexity of natural vegetation, it is becoming more difficult to ignore the role that humans play in shaping the composition of species assemblages across landscapes (Gaston 2006). The amount of net primary productivity that goes directly to support humans has become a common, if not contentious, measure of the human footprint on ecosystems. Mean estimates of the proportion of total terrestrial net primary productivity that is appropriated by humans range from 25–40% (Vitousek et al. 1986, Rojstaczer et al. 2001, Imhoff et al. 2004), and the more that is co-opted by humans, the less there is available to support other species (Haberl et al. 2002, Gaston et al. 2003). Although these estimates have low precision (Haberl et al. 2002), there is widespread agreement that human impacts on ecosystems are substantial and growing (Laurance 2001, Wackernagel et al. 2002, Palmer et al. 2004).

The impacts of human activities on biodiversity are projected to have broad global ramifications (Sala et al. 2000), but the spatial extent of those impacts will be uneven (Cincotta et al. 2000, Imhoff et al. 2004, Evans et al. 2006). Similarly, an uneven spatial distribution of biodiversity is among the most conspicuous of patterns in macroecology (Gaston 2000, Myers et al. 2000). It is this dual pattern of regional concentrations of biodiversity and regional concentrations of human impact—areas of concentration that are often spatially correlated (Balmford et al. 2001, Luck et al. 2004, Gaston 2006)—that has encouraged the development of geographic approaches to conservation and fostered an underlying "hope" that a significant portion of species diversity could be conserved in a relatively small fraction of the landscape (Reid 1998).

A geographic perspective has also been triggered by the rapidly growing availability of spatially explicit data on the occurrence and abundance of species (Blackburn and Gaston 1998, Pärtel 2006). In the >15 years since urgent calls for spatially explicit data to assist land managers and policy makers with broad-scale environmental problems (Brown and Roughgarden 1990, Lubchenco et al. 1991), there has been substantial progress on making ecological data sets more readily available in digital format (Graham et al. 2004). The global coverage of these data is far from complete (Hortal et al. 2007, Soberón et al. 2007), but where they do exist the potential uses of geographic data for testing ecological hypotheses and describing nature are numerous (Guisan and Thuiller 2005). A common practical application of these data has been the development of conservation prioritization schemes that ultimately lead to geographically explicit conservation designs.

Establishment of conservation areas, whether they are focused strictly on biodiversity conservation or on conservation that allows for some degree of multiple-use resource management, has become a vital component of most regional strategies to mitigate the oft cited erosion of contemporary biodiversity resources. However, conservation of all resources is impossible due to limited financial resources. Thus, managers are frequently confronted with the problem of "... where should scarce conservation resources be spent?" On the surface this seems a simple question, but unequivocal answers have eluded conservation scientists for a number of reasons. Two important confounding factors in conservation design are scale and error.

Because scale affects the detection of biodiversity patterns across broad geographic areas (Willis and Whittaker 2002), it is also expected to affect our choice of where to focus conservation activities (Shriner et al. 2006). On the other hand, error in our understanding of species occupancy across the landscape, error in our understanding of the environmental attributes that are important to species habitat selection, and error in our measurement of species occurrence or habitat, all contribute to uncertainty in our characterization of biodiversity patterns and the conservation strategies derived from those patterns.

In this chapter we review the implications of scale and error effects on conservation design. Then we provide an overview of geographic-based conservation approaches before examining scale and error effects in detail using data from our work on biodiversity patterns in the southwestern United States. Finally, we present our thoughts on the implications of scale and error effects to conservation planning and some suggestions for future research needs.

GEOGRAPHIC APPROACHES TO BIODIVERSITY CONSERVATION

Setting Context

A geographical perspective in ecology and applied conservation has a long history. Early 19th and 20th century phyto- and zoogeographers focused on documenting the distributional patterns of flora and fauna for the purpose, among others, of delineating realms of biotic similarity according to climatic, physiographic, and evolutionary criteria (Hooker 1859, Wallace 1876, Shelford 1913). It is tempting to classify this early work as merely descriptive—a view promoted by some contemporaries who have characterized this perspective as "the search for patterns of animal and plant life that can be put on a map" (MacArthur 1972*b*:1). However, it is clearly more than simple description. There are numerous examples where explanation for the observed patterns in species occurrence is tied to land mass proximity, dispersal capacity, dispersal agents, allometry, energetics, and evolutionary principles (Spellerberg and Sawyer 1999, Gaston and Blackburn 2000).

How humans may have influenced these geographic patterns has, at times, been ignored (Stott 1984, Spellerberg and Sawyer 1999). For some ecological questions, this is legitimate because an understanding of the factors and processes that affect the natural geography of biodiversity is important (Gaston and Blackburn 2000:295–300). However, accounting for how humans alter this natural geography is equally important if we are to counteract the erosion of biodiversity that is attributed to human activities (Rapport et al. 1985, Cox and Moore 1993, Balmford et al. 1998). So while we do not deny that interesting ecological patterns can be studied in the absence of invoking any human causation, the focus of this chapter is on using geographic-based conservation to stem biodiversity losses attributable to human influences. Thus, we will not discuss floral or faunal realms, range-abundance relationships, latitudinal gradients of species diversity, or species-area relationships, all of which are topics in geographical ecology (MacArthur 1972*b*, Gaston and Blackburn 2000). Rather, we focus on conservation planning issues that are motivated explicitly by human actions, are relevant over broad landscapes, and have a spatially explicit, and therefore geographic, component.

Another context setting issue for our chapter concerns the word "biodiversity." A question that is basic to any conservation plan is: "What exactly are we proposing to conserve?" Answering this question requires the definition of conservation targets—those biodiversity features that we wish to ensure long-term persistence of through conservation plan implementation (Groves 2003). Historically, conservation targets have focused on species, populations, ecosystems, scenery, landscapes, and perhaps the most inclusive target, biodiversity

(Bakker et al. 2000, Redford et al. 2003). In this chapter we restrict our discussion to cases in which species are the primary focus of biodiversity conservation. However, our discussion and examples are not unique to a species focus. Ecosystems (Hoekstra et al. 2005) or genes (Neel and Cummings 2003)—the other mentioned elements that round out the triad of biodiversity features (Dirzo and Raven 2003)—could be subject to the same geographic perspective as reviewed in this chapter.

A final context setting issue concerns the notion of "reserves" in geographic conservation planning. Reserves are often defined as lands strictly managed for the conservation of biological resources and permanently protected from human development (Noss et al. 1999). Although reserves are most certainly a critical component of large-scale conservation planning efforts (Pimm and Lawton 1998, Noss et al. 1999), sole reliance on reserves will be insufficient to meet conservation objectives because of biological, economic, social, and political constraints (Langholz and Lassoie 2001, Liu et al. 2001). Consequently, it is important to accommodate a continuum of human uses within units of land comprising the conservation plan. Thus, we use the term "conservation area" (*sensu* Groves 2003) to define geographic units that are to be managed in a way that maintains the biodiversity features (in our case, species) associated with those units. We will restrict our usage of the term "reserve" when referring to that subset of conservation areas where land, by some formal designation, is managed strictly for biodiversity conservation.

Species Criteria Used in Biodiversity Conservation Planning

One of the early attempts to relate species geography to applied conservation biology is attributed to Wilson and Willis (1975). Focusing on the number of species that could be conserved in a network of habitat patches, they proposed a set of geographic rules for conservation design (e.g., contiguous habitat patches will conserve more species than fragmented habitat patches of the same area; when fragmentation is unavoidable, minimize separation distances). Although these rules quickly gained the status of conventional wisdom, subsequent work exposed their failure to apply generally (Simberloff 1988, Hof and Flather 1996). Our intent here is not to rehash past criticisms of these geographic conservation rules in particular, but to use this work as a stepping-off point to review two basic species criteria used in geographic-based biodiversity conservation planning.

One is based on a simple *species count* where the composition of the species pool being counted is immaterial except to the extent that the pool is often constrained by some common taxonomic (e.g., birds, arthropods) or ecological (e.g., habitat specialist, endemic, rare) attribute. Species count, also called *species richness* (we use count and richness interchangeably), is certainly the simplest and most easily understood criterion (Purvis and Hector 2000). For this reason, focusing on areas where species richness is concentrated has received extensive consideration in conservation planning (Brooks et al. 2006, Ceballos and Ehrlich 2006). There are important ecological reasons why a particular area supports more species than surrounding areas and species richness should be the criterion of choice when mean site diversity is an important conservation goal (Williams et al. 1996, Shriner et al. 2006).

However, species counts are nothing more than an integer representation of a nameless list of species. This failure to consider species identity forms the basis for this criterion's main criticism. In particular, within biogeographic regions, areas of high species counts have been found to have many species in common (Lennon et al. 2001). Moreover, there is evidence to suggest that spatial variation in species counts appears to be driven largely by the landscape occupancy pattern among common and widespread species—species that are unlikely to be most deserving of our conservation efforts (Brooks et al. 2006). In the absence of composition information, the degree to which comprehensive conservation of the species pool is being achieved is impossible to evaluate (Flather et al. 1997).

To address this weakness, systematic conservation strategies began to rely on a second criterion in conservation planning—*species representation*. The criterion is focused on ensuring some target set of species pool members are adequately represented in the conservation plan. The use of species representation in conservation planning traces back to Australian ecologists and geographers of the early 1980s (see Margules and Usher 1981, Kirkpatrick 1983) who noted that if the goal of conservation is to conserve biodiversity broadly, then we should be focusing on adding units of land to conservation networks that contribute the greatest marginal increase in species coverage (Sarkar et al. 2006). Conservation planning under this criterion becomes a search for units of land whose occupying species complement those, as opposed to being redundant with, species covered either in extant conservation areas or among a set of potential units of land that are being considered for conservation status (Vane-Wright et al. 1991). As such, species representation as a conservation planning criterion is linked inextricably to the concept of *complementarity*.

Because species are not distributed randomly, but occur on the landscape with varying degrees of spatial structure, species composition tends to be more similar among sites that are near to one another (Nekola and White 1999). This underlying structure in compositional similarity results in a well-known problem with conservation designs based on the species representation criterion—namely, sites selected to maximize complementarity are often well dispersed throughout a region, making management of these areas more costly, logistically difficult, and subject to elevated edge effects and dispersal constraints (Bedward et al. 1992, Possingham et al. 2000, Williams et al. 2005). Furthermore, there is evidence that strict application of a species representation criterion could result in the selection of sites that disproportionately represent areas that are at the periphery of species' geographic ranges, which may predispose these so-called marginal populations to future extinction events (Araújo and Williams 2001). These weaknesses can be addressed explicitly by including a variety of spatial constraints (e.g., adjacency requirements, minimization of boundary lengths, focus on core distributions) that effectively broaden the species representation criterion to include additional ecological considerations (Sarkar and Margules 2002).

A commonly expressed weakness of geographically explicit conservation planning, regardless of the species criterion used, is that it often fails to address the persistence of species (Lambeck and Hobbs 2002, Wiersma and Urban 2005). In their simplest forms, richness and representation criteria are based on the presence-absence pattern of species across the landscape. However, the likelihood of species persistence increases as the population size increases. Unfortunately, abundance data are not generally available for many species, making the oft noted data constraints associated with conservation planning (Lamoreux et al. 2006) even more severe.

Addressing the persistence issue has resulted in the rapid expansion of species criteria that get used in conservation planning. A few of these criteria are *redundancy*—a measure of species incidence across conservation networks such that representation occurs at least *k* times (ReVelle et al. 2002); *irreplaceability*—a measure that reflects the importance of a potential unit of land to the overall conservation design and is sensitive to unique or rare targets like local endemics (Pressey et al. 1994, Cabeza and Moilanen 2001); *vulnerability*—a measure of threat, either to species or habitat persistence or habitat conversion (Redford et al. 2003, Ricketts et al. 2005); and *robustness*—a measure that merges notions of redundancy and vulnerability by quantifying the degree to which conservation goals are maintained in the face of anthropogenic or natural disturbance (O'Hanley et al. 2007). Although these additional criteria complicate the subject of conservation planning, we find it reassuring that, for the most part, these emerging criteria can be thought of as variations on the fundamental criteria that we began this section with—namely, species counts or representation. For example, many of these new criteria simply invoke weighting schemes that permit the conservation practitioner to emphasize species or areas differentially.

The emergence of these additional criteria offers extreme flexibility in tailoring a conservation plan to the idiosyncrasies of a locale or species. With this flexibility comes the burden of choice—and the set from which to make that choice is growing. Since a choice must be made, it is tempting to evaluate the effectiveness of this growing list of species criteria in terms of which is "the best." MacArthur (1972*a*:259) once made the observation that "[a]nyone familiar with the history of science knows it [science] is done in the most astonishing ways...", as a commentary on ecology's search for *the* scientific method. The same observation holds for geographic approaches to conservation. The land management landscape is too complicated by land ownership patterns and conflicting resource values to allow a single "best" approach to geographically based conservation planning to emerge. Therefore, the preference for a particular species criterion, or set of criteria, will be dictated more by underlying value structures and circumstances specific to the conservation problem being addressed rather than any simple prescription of how geographically based conservation planning should be done (Redford et al. 2003).

Data Types for Conservation Planning

The fundamental datum for any geographic approach is a spatial location for a given species. Although there may be other attributes of species or areas brought along in the datum record, "knowing" where each species occurs on the landscape is the minimum prerequisite needed to characterize biodiversity across some area of interest (Ferrier 2002). How does the conservation practitioner go about obtaining information on species identity and location? There are four broad classes of data types or sources for such information.

First, and most obvious, are survey-based data that provide spatially explicit *empirical observations* of species occupancy. Much of these data take the form of simple locational records where a species was observed or collected. Data of this form that are collected on a probabilistic sample, where both presence and absence can be discerned, are extremely valuable to landscape-level conservation planning. However, such data tend to be restricted taxonomically and geographically—being available for only certain species in a particular locale. For this reason, when the planning area is large, conservation planners must make use of existing data from museum, herbarium, academic, or private collections (Bender et al. 2005). These data often represent an accumulated set of observations because they have been collected by a number of individuals, over varying periods of time, using a number of field collection techniques, and visiting sites in an opportunistic fashion (Funk and Richardson 2002). Consequently, the data are characterized by a number of inherent biases (Williams et al. 2002). Furthermore, the data often only record the presence of an individual, making it difficult to distinguish true absence from areas that have not been surveyed (Ferrier 2002).

Short of designing new surveys to eliminate the noted constraints of using existing data—which under most circumstances is infeasible given the time and financial constraints associated with most planning efforts—the conservationist must look for supplemental data. A second data type that is often used is that generated by *expert judgment* (Groves 2003). When data gaps are severe, expert panels may be the only source for species occurrence information. However, expert-generated data are difficult to evaluate, difficult to replicate, and often highly variable among comparable experts (Ferrier 2002, Neel and Cummings 2003). These well-known shortcomings of expert-generated data have motivated some to call for a shift away from expert judgments to evidence-based conservation (Sutherland et al. 2004) for biodiversity planning efforts.

This brings us to our third data type—*predicted occurrence* of species. Spatial interpolation that is based only on the geometry of location records is perhaps the simplest form of filling in species occupancy gaps (see Rapoport 1982). Such approaches are ecologically neutral and have given way to methodologies that quantify the covariation between species locations and some set of environmental attributes. Data of this type start with empirical observations of species that are then linked geographically with environmental predictors that allow the planner to extrapolate occupancy across a region of interest. Such approaches are often discussed under the rubric of species distribution modeling and have become commonplace in ecology over the past decade (Fortin et al. 2005). In addition to standard linear modeling approaches such as logistic and autologistic models (Augustin et al. 1996), sophisticated methods such as maximum entropy (Phillips et al. 2006), ecological niche factor analysis (Hirzel et al. 2002), Bayesian modeling (Gelfand et al. 2003), and genetic algorithm for rule-set production (Stockwell and Noble 1991) have emerged as tools for predicting species distributions. These models can be classified according to the type of data required. Some require only those locations where a species is known to occur and are classified as presence-only models (e.g., Zaniewski et al. 2002). Others also require information about where a species is known not to occur and are classified as presence-absence models (e.g., Engler et al. 2004). Although the rapidly growing number of approaches to predict species occupancy across a landscape (see Elith et al. [2006] for a review) has the potential to overwhelm conservation planners, these approaches do provide repeatable results that can be evaluated quantitatively.

In the best of all worlds planners would have comprehensive biodiversity data across all taxonomic groups with adequate spatial coverage. This is far from the case, and many have commented that our understanding of biodiversity is woefully incomplete (Brown and Roughgarden 1990, Pimm and Gittleman 1992, Flather and Sieg 2000). One approach for overcoming this data constraint is to assume that the biodiversity pattern of well-studied taxa can be used as a surrogate for other, less well-known taxa (Caro and O'Doherty 1999, Marcot and Flather 2007). In this sense, *surrogates* represent a fourth data type. Like predicted occurrence, this data type is inferential rather than measured *per se*. Although there has been recent evidence in support of surrogacy (Lennon et al. 2004, Lamoreux et al. 2006), the support is certainly not general (Ceballos and Ehrlich 2006). This latter finding is consistent with a growing number of papers that have cautioned conservation planners against blindly using surrogacy in geographic conservation approaches (Flather et al. 1997, Ricketts et al. 1999, Hess et al. 2006*b*).

Overview of Geographic Approaches

Once species criteria are selected and the data are in hand, the conservation planner must decide where, geographically, biodiversity conservation efforts will be put into practice. This involves formalizing the conservation problem and implementing some algorithm to solve it (Sarkar et al. 2006). There are two broad classes of approaches for solving the conservation network delineation problem: (1) those that focus on some quantile of a frequency distribution, and (2) those that focus on efficiently meeting an explicitly stated conservation objective.

The former are commonly discussed as a *hotspot* analysis, and this approach is linked directly with the species count criterion discussed earlier. Reducing the conflict between human land use intensification and areas possessing globally significant counts of species remains an important challenge to biodiversity conservation (Burgess et al. 2007). As noted by Ceballos and Ehrlich (2006), few topics in conservation planning have received more attention than species diversity hotspots. Originally, this approach involved the enumeration of species that were most threatened or vulnerable to human activities within some geographic unit (Myers 1988). Those units could be countries (e.g., Sisk et al. 1994), a systematic grid (e.g., Balmford et al. 2001), or a habitat patch (e.g., Wilson and Willis 1975). The patch-based definition is appropriate for local conservation problems, but is intractable at regional, continental, or global scales. Geographic units that are defined by administrative or political boundaries lend themselves to macroecologic investigation but suffer from unequal areas with little ecological basis that can skew evaluations of conservation importance. For this reason, species occupancy patterns across some systematic grid have become the more common empirical basis for hotspot analysis. The flexibility of this approach has resulted in usage that extends well beyond its species richness roots with the term "hotspot" being invoked any time the analysis seeks to identify geographic areas that rank particularly high (i.e., some upper quantile) on one or more axes of species (genus, family) richness, levels of endemism, numbers of rare or threatened species, intensity of threat, or indicator of ecosystem degradation (Prendergast et al. 1993, Flather et al. 1998, Reid 1998, Hof et al. 1999, Margules and Pressey 2000). The designation of the upper quantile (i.e., the frequency distribution threshold that identifies those geographic units as "hot") varies in the literature but is usually $\leq$10%. However, because there is no ecological justification for the choice of quantile, the conservation planner is left to make this subjective decision.

Although the hotspot approach has played a central role in conservation planning, it has been criticized for the same reason that the species count criterion has been criticized; namely, it ignores species composition (Possingham and Wilson 2005, Fleishman et al. 2006). At the heart of this criticism is the principle of efficiency. Given limited conservation resources, an efficient strategy is one that concentrates on the fewest high-quality sites that meet the conservation objective (Redford et al. 2003). Because hotspot approaches ignore composition, they are generally thought to be inefficient unless the number of conservation units is constrained to be very small (Reid 1998, but see Shriner et al. 2006).

Efforts to develop algorithms that identify those conservation areas that give the biggest biodiversity bang for the conservation buck have resulted in an extremely rich set of tools that have their origins in the operations research literature—namely, *optimization* analysis (Haight and Gobster, this volume). The use of optimization models in biological conservation is increasing as evidenced by Rodrigues and Gaston's (2002) list of >30 optimal conservation design studies in the previous decade. Although there is an impressive variety of algorithms, they tend to fall into two broad classes: (1) those that define conservation networks based on iterative or stepwise algorithms, and (2) those that seek exact optimal solutions. Although they share the goal of designing efficient conservation strategies, they differ in that iterative algorithms are often referred to as inexact heuristics, since they can only approximate an efficient design (Cabeza and Moilanen 2001); while those based on a closed-form optimization formulation offer a globally optimal prescription (Hof and Flather 2007). So, why would conservation planners choose to use an inexact heuristic? There are a number of reasons that procedures not offering true optimal solutions get used. Many realistic conservation problems are unsolvable in closed form, while heuristic approaches tend to be intuitive, simple, and appear to provide reasonably good solutions when compared to exact solutions (Pressey et al. 1997).

Factors Affecting Our Ability to Describe Biodiversity

To this point we have implicitly ignored a number of factors that are known to affect our ability to characterize biodiversity and to develop tenable conservation plans. In this chapter we wish to highlight two: scale and error.

There is a substantial literature on the subject of scale and its effects on ecological study and the conservation recommendations derived from that research (Wiens 1989, Hoekstra et al. 1991, Schneider 2001, Willis and Whittaker 2002). There is growing evidence that patterns of covariation can shift when the analysis scale is changed (Lennon et al. 2001, Hess et al. 2006*a*, Pautasso 2007), and this explains, to a large degree, why unequivocal conservation recommendations have been so difficult to make.

Similarly, geographic conservation planning requires accurate data on the identity and location of species and the adequacy of biodiversity surveys. Data limitations have long been recognized as an important constraint associated with geographic-based conservation efforts (Prendergast et al. 1999), and there is a concern that data quality is not keeping pace with the growth in algorithmic sophistication (Possingham et al. 2000). Moreover, the ease with which large quantities of biodiversity data are now being made available raises additional doubts concerning their reliability (Cherrill and McClean 1995). Like scale, the impact of error in geographic-based conservation plans is rarely considered outside scientific journals, and even then, the sensitivity of geographic-based analyses to varying levels of error is seldom explored.

How do scale and error potentially affect geographically based conservation planning efforts? In order to explore some of the answers to this question, we now turn to a set of case studies that are motivated by our biodiversity assessment work in the southwestern United States. We present these case studies not as definitive works on the issues of scale or error, but rather offer them as examples of the kinds of issues that can emerge when practitioners consider their potential impacts on biodiversity conservation planning.

CONSIDERATION OF SCALE IN CONSERVATION PLANNING: DO BROAD BRUSHES COVER A GNAT'S ASSETS?

Before reviewing the specifics of our work on scale effects in conservation planning, we first need to define what we mean by scale. Probably the most frequent definition of scale in an ecological context (as opposed to a cartographic context) refers to the relationship between grain and extent of a particular investigation (Wiens 1989, Schneider 2001). *Grain* refers to the physical size or time period of the observation unit, whereas *extent* refers to the overall area or time period of the study or the geographic or temporal dimension to which inferences are drawn. It is the combined characterization of grain and extent that defines the scale of any investigation or conservation planning effort.

Given the biodiversity data deficiencies discussed previously (see "Data Types for Conservation Planning"), the input data into either hotspot or optimization analyses often stem from relatively coarse-grained observation units (Shriner et al. 2006); e.g., commonly analysis units are $\geq$10,000 km^2 (see Andelman and Willig 2003, Larsen and Rahbek 2003, Orme et al. 2005) for global, continental, and regional extents. How does this analysis scale compare with conservation implementation scale? We examined data from the World Database on Protected Areas (WDPA; WDPA Consortium 2004) to characterize the size distribution of currently implemented conservation areas. We restricted our examination of these data to those conservation areas that are terrestrial and classified by the World Conservation Union (IUCN) to category I and II (i.e., chief purpose is biodiversity conservation), qualifying them as biodiversity reserves. Furthermore, we purged all reserves with areas $\leq$1 ha because a large proportion of these very small reserves protect historic monuments or unique geologic formations (see Shriner et al. 2006). The median was ~5 km^2 (n = 8,967) and nearly 75% of the reserves were <62 km^2 (Fig. 4-1).

This high variability in conservation planning scales does raise a question of whether there is a "right" scale to analyze biodiversity patterns. Although Wiens (1989) made a plea, >15 years ago, for objective approaches by which ecologists can define appropriate investigative scales, we still tend to treat scale with an exploration of system behavior resulting from varying grain or extents in an arbitrary manner. The danger with such an approach is that it becomes difficult

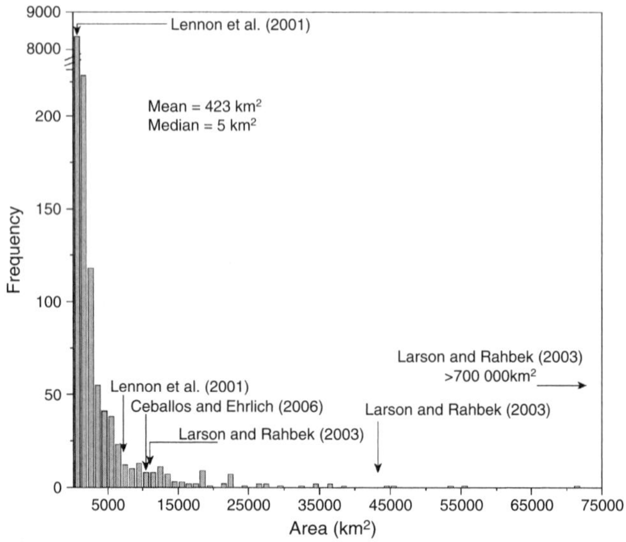

FIG. 4-1

Frequency distribution of the size of terrestrial conservation reserves dedicated to strict biodiversity conservation (IUCN categories I and II) from the World Database on Protected Areas.

to distinguish scale artifacts from ecologically meaningful patterns (Wiens 1989). This concern is particularly relevant to geographically based conservation efforts given the discrepancy between conservation analysis scales and the scales of conservation implementation (Fig. 4-1).

A common argument to address the scale mismatch between conservation data/analyses and implementation is to couch conservation planning as a hierarchical process whereby coarse-grained analyses provide a broad brush depiction of biodiversity that serves to focus conservation attention on regions that warrant attention (Ferrier 2002; Probst and Gustafson, this volume). The actual identification of land units that will comprise a particular conservation network can then be identified with finer scale data within these priority regions (Pressey et al. 1993, Harris et al. 2005, Fjeldså 2007). This approach implicitly assumes that conservation priorities are nested geographically, which is to say that conservation analyses at coarse (regional) scales are consistent with conservation actions that are implemented at fine (local) scales (Larsen and Rahbek 2003). Do we have evidence that conservation designs based on fine-scale data are generally nested within coarse-scaled designs?

The Approach and Database

We examined conservation designs with species count and species representation criteria using richness hotspot and optimization approaches. We based our analyses on predicted range maps for mammals and birds developed by the Arizona and New Mexico Gap Analysis Programs (Thompson et al. 1996, Halvorson et al. 2001) available at 90 m and 100 m grid resolution, respectively. For the purposes of this analysis we assumed that these range maps reflected the "true" occupancy pattern of species across these two states. We used these "known" distributions to derive 1, 100, 625, 2500, and 10,000 km^2 grid cell representations of each species' distribution based on a binary rule that classified a cell as occupied if any portion of the species' true geographic range intersected a particular grid cell. We had suitable range map data for four state-species groups: Arizona birds (279 species), New Mexico birds (324 species), Arizona mammals (129 species), and New Mexico mammals (138 species).

For the richness hotspot analysis we first generated the total species count within each grid cell, at each of the grain sizes, by simply summing occupancies across all species. We defined hotspots as grid cells exceeding the 95th quantile as in Prendergast et al. (1993); that is, we identified those 5% of grid cells with the highest richness estimate. For the optimization analysis we used the occurrence data for each grid cell, at each grain size, to select that set of cells such that each species was represented at least once in the set. We used the MARXAN conservation design software, in particular simulated annealing (Ball and Possingham 2000), to identify the most efficient (minimum area) set of cells meeting the representation objective. Simulated annealing is an inexact heuristic that has been shown to perform well in conservation design applications (Possingham et al. 2000).

Total area of the conservation network designed under our representative criteria varied greatly with grain size (Shriner et al. 2006:1665); networks based on small grain units reached solutions after affecting <1% of the total extent, whereas large grain units affected nearly 20% of the extent. These size differences confound interpretation of scale effects because networks based on smaller grain units are more likely to overlap networks based on large grain units given the larger total area included in the large-grain solution. For this reason, we further constrained the conservation network solutions from MARXAN based on the notion of irreplaceability. An irreplaceability score was generated by MARXAN that reflected the number of times any one grid cell was selected as a member of the "best" network in 1000 realizations of the design solution with the score ranging from 0 (never selected) to 1000 (always selected). We rank-ordered grid cells based on these irreplaceability scores and selected those cells that exceeded the 95th quantile as in the hotspot analysis. Because of the numerical intensity of this analysis, we only explored scale effects for optimally designed conservation areas for Arizona birds. Furthermore, we did not develop a conservation network based on an irreplaceability ranking at the 1 km^2 grain because too few cells were selected to meet the 5% area goal (i.e., we met our representativeness criterion with <5% of the extent's area).

Results

The amount of overlap observed for species hotspots varied from 0.0% to 63.1% with an overall mean of 28.1% (Table 4-1). The degree of overlap in conservation networks was different between taxa (grand mean across birds in both states = 25.6%; grand mean across mammals in both states = 31.6%), but was very similar between states (28.7% for Arizona and 28.6% for New Mexico).

Conservation areas designed under a species representation criterion showed similarly low degrees of overlap across grain sizes. Percentage overlap in pair-wise grain comparisons for Arizona birds varied from a low of 15.3% to a high of 44.2% (mean = 23.9%). Although network overlap was generally low, it is noteworthy that the selected conservation units tended to cluster in certain geographic locales (Fig. 4-2). So, while conservation areas showed low overlap, there was a high degree of adjacency, suggesting that there may be some underlying ecological mechanism causing the proximity of network units at different scales. This pattern of proximity notwithstanding, it is also notable that there are some portions of the state that were selected uniquely for conservation focus at individual scales (e.g., the south central portion of Arizona at the 100 km^2 grain size; Fig. 4-2).

Table 4-1 Pairwise Comparisons of Percent Overlap for Richness Hotspot Reserves Developed at Five Grain Sizes: 1 km^2, 100 km^2, 625 km^2, 2500 km^2 and 10,000 km^2. Percent Overlap is Calculated by Dividing the Area of Overlap by the Area of the Smaller Reserve

	Arizona	New Mexico	Arizona	New Mexico
Map Comparison (km^2)	**Birds**		**Mammals**	
10,000, 2500	16.7	37.5	17.8	50.0
10,000, 625	21.9	18.8	0.0	25.3
10,000, 100	26.5	11.2	0.0	12.4
10,000, 1	33.4	13.0	5.1	8.7
2500, 625	17.6	34.1	54.5	56.9
2500, 100	27.8	21.4	43.4	36.1
2500, 1	14.6	19.6	27.7	21.8
625, 100	47.1	42.7	63.1	50.2
625, 1	25.1	18.4	40.5	23.9
100, 1	32.7	29.4	57.8	36.2
Mean	26.3	24.6	31.0	32.1

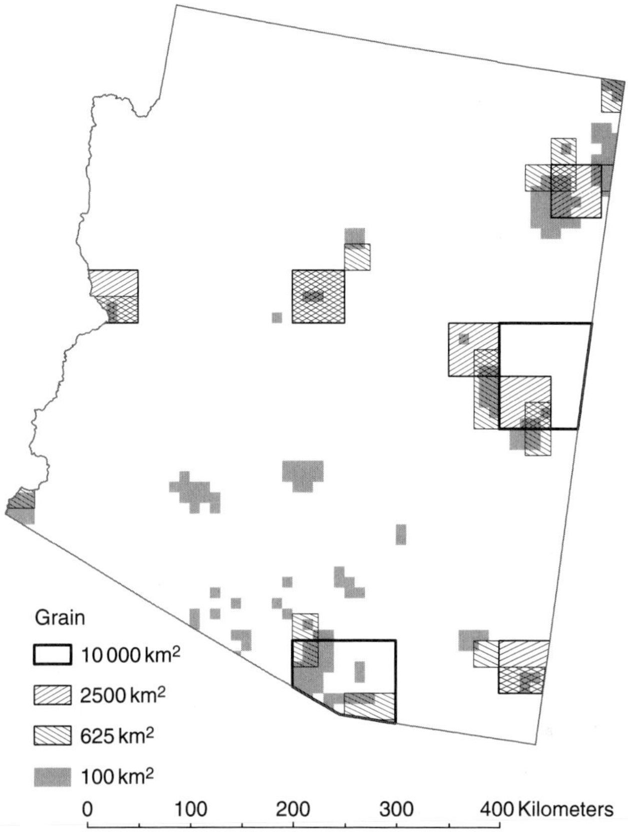

FIG. 4-2

The spatial overlap of conservation areas selected at four conservation unit grain sizes (100 km^2, 625 km^2, 2500 km^2, and 10,000 km^2). The conservation networks selected at each grain size were based on an irreplaceability criterion and an inexact heuristic optimization (i.e., simulated annealing using MARXAN).

Although tangential to an evaluation of scale effects, a comparison of conservation networks selected under species counts and representation criteria is also interesting. Like the other overlap comparisons, richness hotspots and representation reserves showed low overlap across all grain sizes (Fig. 4-3). In fact, the two criteria appear to be focusing on very different aspects of the state's geography. Overlap was minimal at the coarsest grain (0.0% at 10,000 km^2 grain) and reached a maximum (17.5%) at the 2500 km^2 grain. Reserves based on the representation criterion also had higher degrees of species coverage than reserves based on species counts (as expected), but it was surprising that hotspot reserves at the finest scale (1 km^2) did cover most species used in the analysis.

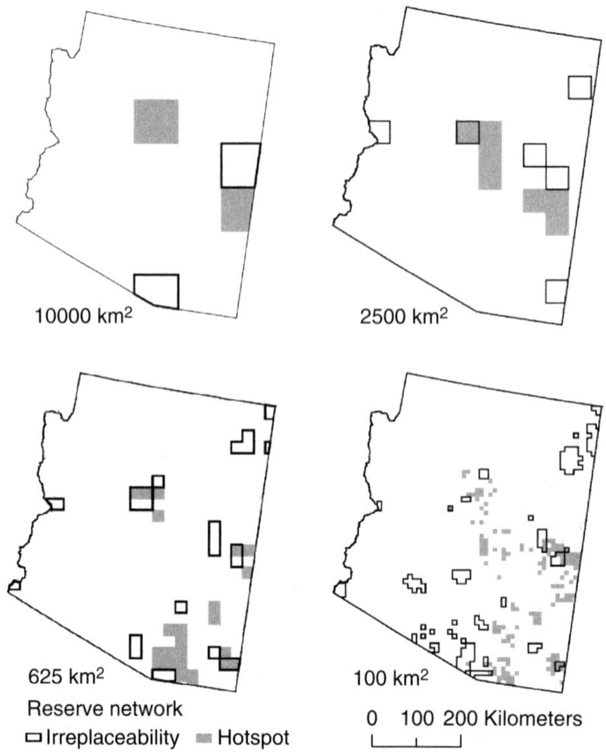

FIG. 4-3

Spatial congruity between reserve networks selected using hotspot and optimization analysis at four reserve unit grain sizes (100 km^2, 625 km^2, 2500 km^2, and 10,000 km^2).

Implications

Because it is rarely feasible politically to establish large conservation networks (Margules and Usher 1981:99), it is legitimate to ask: "If the observation units in a geographically based conservation analysis can't be implemented (i.e., too large), then what is the point of coarse grain assessments for biodiversity conservation?" Justifications for retaining coarse grain biodiversity assessments tend to proceed according to the following logic: (1) Conservation practitioners do not hold detailed knowledge of species occurrence because comprehensive biodiversity data are limited spatially; (2) coarse-grain assessments allow some of those spatial limitations to be relaxed because the heterogeneity in sampling effort becomes less detectable at large grain sizes; and (3) therefore, coarse grain assessments provide a means of identifying broad areas that should be the focus of more detailed conservation study. This logic is based on the assumption that conservation designs form a spatial hierarchy where coarse-grain designs subsume fine-grain designs. That is, fine-grain designs would be nested within coarse-grain designs.

The results from our case study do not support the assumption of nested designs for conservation analyses across spatial scales. Our results clearly indicate that conservation planning outcomes can be scale dependent (but see Fjeldså 2007) and that conservation planners should proceed cautiously when conservation priorities are based on analysis scales that are disparate from implementation scales. The low spatial overlap of conservation networks based on different scales (observation grain in our case) suggests that a simple comparison of biodiversity analysis scales with conservation implementation scales would be a useful attribute for conservation planners to consider in judging whether their plan is likely to be an efficient mechanism for conserving biodiversity.

This does present a quandary for conservation planners. In the absence of fine-scale biodiversity data, there may be no opportunity for avoiding inefficient conservation plans. Obviously, a solution to this predicament is to develop fine-scale species occupancy data for biodiversity conservation planning. Alternatively, conservation practitioners could implement conservation plans at the coarser scales that correspond to current knowledge. Ultimately, the best approaches will likely blend management at coarse scales for large spatial extents while simultaneously integrating fine-scale management, potentially resulting in more effective conservation of the species in question.

Given the prohibitive costs associated with collecting primary species occupancy data, it would seem important to also invest in detailed distributional modeling efforts that can accurately predict species occupancy. Efforts to assess our ability to predict species distribution are often overlooked (Wilson et al. 2005), even though the uncertainty in our predictions can be substantial (Flather et al. 1997, Elith et al. 2002). Moreover, it is important to understand how robust our conservation plans may be to species distribution errors—which is the subject of the next case study.

ERROR AND UNCERTAINTY: THE DIVINER'S LAMENT

As in life, sins of commission and omission are often fatal. I can write with feeling on this subject, having dowsed. . .things that I ought not to have dowsed and left undowsed those things I ought to have dowsed. (Terry Ross, founder and president of the American Society of Dowsers, Danville Vermont. First published in Rod & Pendulum, *No. 68, September 1992).*

The objective of the dowser is not unlike that of the modeler who is attempting to predict the occupancy pattern of species across the landscape. Whether using the L-Rod, Y-Rod, or the pendulum to predict the location of water, or GLMs, GAMs, or maximum entropy to predict species occurrence, the underlying

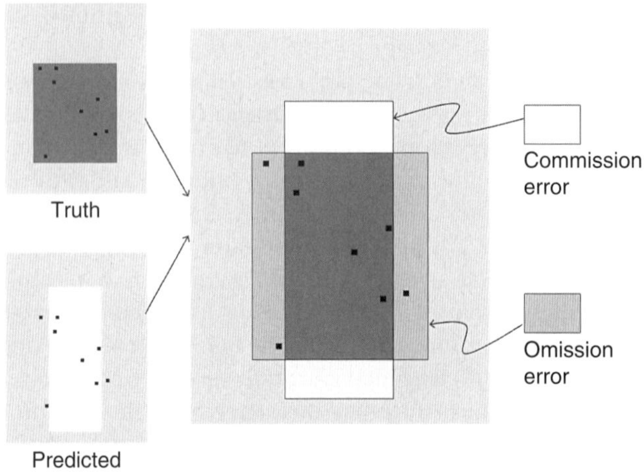

FIG. 4-4

Conceptual model illustrating commission (false presence) and omission (false absence) errors associated with predicting species distributions.

goal is the same—assign presence or absence in a way that minimizes error and is subject to validation by the well driller or species survey.

The representation of landscape occupancy by a species can be wrong in two ways (Fig. 4-4). A species can mistakenly be predicted to occur in an area (false presence or commission error), or a species can mistakenly be predicted to be absent in an area (false absence or omission error). The sources of uncertainty that lead to such errors are many. Sampling errors occur with species surveys due to the sample units selected, identification errors, incomplete or biased sampling, and imperfect detectability; models used to predict occupancy may introduce error because of imperfect habitat relationships or model misspecification; and cartographic errors can manifest due to inaccuracies (thematic or locational) in species point observations or map layers used in predicting occupancy—not to mention data transcription errors or errors associated with summarizing information across multiple observation grains. As is the case with all geographically based conservation planning endeavors, the outcome is only as reliable as the underlying data (Burgess et al. 2007:174). There are numerous examples in the literature demonstrating the sensitivity of geographically based conservation plans to errors (Dean et al. 1997, Gaston and Rodrigues 2003, Wilson et al. 2005), yet explicit consideration of error in conservation prioritization schemes is still wanting (Rondinini et al. 2006).

In this case study, we review the findings of our efforts in the southwestern United States to examine the effects of omission and commission errors on analyses to support systematic conservation designs. In particular, we were interested in the following questions: (1) How robust are spatially explicit conservation recommendations to errors in the underlying biodiversity data?; (2) does the type of error differentially affect design sensitivity?; (3) does the manner that errors manifest on the landscape (random versus spatially constrained) affect conservation plans?; and (4) does the number of species of conservation interest affect the conservation plan's sensitivity to error?

The Approach and Database

Our general approach and data are similar to the previous case study, and we refer the reader to that section for the details. The underlying data were again based on the Arizona and New Mexico Gap Analysis Programs, and the distribution maps available were treated as the "known" landscape occupancy pattern for each species. As before, *true* hotspots were generated for birds and mammals based on the known landscape occupancy maps. Species richness hotspots were again defined as the set of map cells that exceeded the 95th quantile for species richness.

Errors of omission or commission were imposed on the known occupancy maps for each species in two fundamentally different ways: (1) *spatially random* in which error occurred in a simple, spatially random fashion, and (2) *boundary correlated* in which errors were spatially correlated and more likely to occur at the boundary of the known species distribution. In the spatially random error case, for omission, all cells where the species was known to be present were equally likely to be selected and changed to an absence; and for commission, all cells where the species was known to be absent were equally likely to be selected and changed to a presence. We simulated the boundary correlated error case by weighting the probability that a given cell would be selected as an error cell based on its proximity to a range boundary using the following distance decay function:

$$\text{Prob}(i) = 1 - (1 - \theta^{D_i})^{\beta},$$

where Prob(i) is the probability that cell i is selected for error imposition, D_i is the distance that cell i is from a range boundary, and θ and β are parameters that affect the maximum distance from the range boundary and the rate of decay that a cell could be considered for misclassification, respectively. We used a combination of θ and β such that little error occurred beyond 30 km of a known range boundary with most errors occurring within 7–10 km.

A Monte Carlo simulation experiment was developed that compared species richness hotspots that included overlays of $n = 10, 25, 50, 75, 100$, or 125 species, and again the overlay of these species distributions without error was considered truth. To examine error, as each species was overlaid, either spatially random or boundary correlated error was generated with each species having either (1) all omission error, (2) all commission error, (3) balanced omission and commission error, or (4) a random combination of omission and commission error. The resulting landscape occupancy maps with error were used to develop new species richness hotspots. Our response variable in this analysis was the percent overlap between hotspots defined using the known (truth) occupancy maps and the error imposed maps. Because the rules used to generate the original landscape occupancy maps varied by state and by taxon (birds versus mammals), we also considered state and taxa as factors that may explain observed variations in hotspot overlap across the simulation experiment. A total of 30 replicates for each combination of error factors was run for the experiment, each with a fixed error rate (20%). The numerical intensity of the MARXAN optimization analyses prevented us from examining how error would affect conservation designs based on a species representation criterion.

Results

Mean percent overlap between error and true hotspots varied from 62–93%. Spatially random error (mean overlap = 86%) had less of an impact on hotspot coincidence than did boundary error (mean overlap = 77%) (Fig. 4-5). Commission error (mean overlap = 85%) had less of an impact than omission error (mean overlap 75%). In general, the percent overlap increased as the number of species increased when the error was random, whereas it remained essentially stable when the error was spatially constrained to occur near range boundaries. A case that deviates notably from this pattern is boundary omission errors where we observed a monotonic decline in error as the number of species increased (Fig. 4-5b).

Looking at all possible factors that could affect the degree of overlap between true and error imposed hotspots revealed that taxa (bird versus mammal) had the

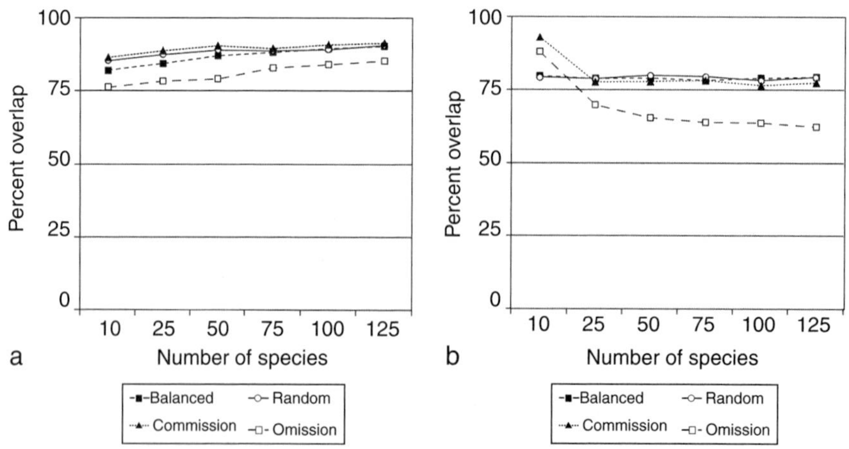

FIG. 4-5

Percent overlap between species richness hotspots for birds and mammals in Arizona and New Mexico for (a) spatially random error and (b) boundary error. Each species had a fixed error rate (20%) that was imposed as all omission, all commission, balanced omission and commission, and random mix of omission and commission.

Table 4-2 Analysis of Variance Results for 5 Factors That May Affect the Degree of Overlap Between True Species Richness Hotspots and Error Imposed Hotspots in Arizona and New Mexico. The Factors are Taxa (Birds, Mammals), Error Type (All Omission, All Commission, Balanced Omission and Commission, Random Omission and Commission), Number of Species (10, 25, 50, 100, 125), Location (Spatially Random, Constrained to Range Boundary), and State (Arizona, New Mexico).

Source	DF	Type III SS	MS
Taxa	1	22 863	22 863
Error Type	3	25 412	8471
No. of Species	5	29 006	5801
Location	1	1617	1617
State	1	1527	1527

Note: Statistical significance is not reported, since rejection of the null can be guaranteed by simply increasing the number of realizations in the simulation experiment.

greatest impact on observed variation in hotspot overlap (Table 4-2). This was followed by error type (all omission, all commission, balance, and random) and the total number of species in the analysis. The spatial location of the error (i.e., whether it was distributed randomly or associated with range boundaries) had relative low explanatory power relative to the other factors.

Implications

The results from this simulation experiment were surprising for a number of reasons. First, the dominating effect of taxa was not anticipated *a priori*. In hindsight, this factor is likely related to the differences in mean landscape occupancy between birds and mammals. Birds, being a more vagile taxon, are able to occupy more of the landscape than mammals. This part of the United States is characterized by isolated montane habitats set in a hostile (at least to endothermic vertebrates) arid matrix. This particular abiotic context likely resulted in a higher proportion of mammalian species with smaller, and therefore restricted, distributions when compared to birds. This fundamental difference in range size and shape is a likely explanation for the pattern we observed. Such speculation could be refuted or confirmed by repeating our analysis on species with widespread versus restricted landscape occupancy patterns. We suspect that this is not a taxon *per se* effect, but rather points to landscape occupancy as a key attribute affecting the robustness of conservation designs based on hotspot criteria.

Another surprising finding was the moderate effect attributed to the number of species overlaid in the analysis. Our expectation was linked to work that has quantified error propagation and the observed decline in overall map accuracy as the number of overlay operations increased (see Veregin 1989). We actually observed an increase in the robustness of hotspot selection under random error as we increased the number of species, and essentially no decline in robustness for error that was spatially constrained to range boundaries as we increased the number of species of conservation concern from 10 to 125. This is likely related to our use of the hotspot criterion in this case study. It is probable that the actual estimate of species richness in any one cell showed much greater degrees of error than the geographic location of the 95th quantile. So while the magnitude of the richness estimate may be sensitive to the number of species overlays, the relative ranking of cells for hotspot identification appears to be somewhat immune to both omission and commission errors.

Finally, the generally high degree of overlap between true hotspots and error imposed hotspots was not anticipated. We intentionally chose a high fixed error rate (20% of the true distribution of the species) to ensure that an effect would be observed. We were surprised that the degree of overlap exceeded 75% in most of the error simulations conducted (Fig. 4-5). This pattern is likely related to the fact that richness is compositionally neutral. We suspect that if optimization based on a species representation criterion had been used to prioritize conservation units, the design solutions would have shown more sensitivity, and therefore be less robust to the kinds of error we imposed in our simulation experiment. One exception to the generally high degree of overlap we observed involved omission errors. In both random and boundary error cases, omission error resulted in the lowest overlap with the true hotspot maps, and in the case of boundary error overlap actually declined with the number of species considered. A potential explanation for this result is that commission error tends to expand the general range of a species such that areas of overlap between species distributions are broadened with minimal impact on the underlying pattern of high species counts. On the other hand, omission error is more likely to erode areas of overlap between species distributions such that true areas of high species counts may look more like their neighbors, lessening the signal of the underlying pattern of species richness. This finding suggests that models that favor commission over omission error may lead to more robust hotspot identification.

An important limitation of this particular case study is that our findings may be conditioned on the ecological circumstances that define this region of the United States. Distinguishing whether the results we observed are general, or are simply specific to this particular geography, will require repeating our analysis in other geographic locales. Furthermore, repeating this analysis using other geographic approaches (e.g., optimization) and other species criteria (representation) would more completely inform the conservation planner about the likely impacts of error. Although we end this case study with a call for additional research on the impacts of error, these are not the only avenues for future work that will extend our conservation planning capability.

FUTURE RESEARCH NEEDS

Geographically based conservation planning to conserve biodiversity is deceptively simple: Information on the location and identity of species can be used to prioritize where limited conservation resources should be focused. However, designing effective biodiversity conservation strategies across extensive landscapes is remarkably complex. This complexity derives from a number of sources, and reductions in that complexity can be realized by extending research efforts into a number of areas that include improving data availability and quality, improving the inferential basis for spatially explicit representation of biodiversity, extending geographic approaches to planning, incorporating consideration of ecological processes, and improving the accessibility of geographically based conservation analyses to practitioners.

Data Availability and Quality

The species has been regarded by many as the fundamental unit of biodiversity (Huston 1993). Although species surveys are central to any geographically based assessment of biodiversity, important barriers to comprehensive biodiversity inventories remain. Perhaps the most obvious need for extending future work concerns the development of monitoring protocols that are economically feasible and ecologically tenable (see Haight and Gobster, this volume).

Part of the difficulty with the availability and quality of primary biodiversity data relates to substantial knowledge gaps in the systematics of some taxa and the fluid nature of taxonomic classifications over time. The emerging discipline of biodiversity informatics (see Bisby 2000), which focuses on the development of a comprehensive taxonomic accounting of all species, would help further efforts to monitor biodiversity patterns. However, even among taxa with relatively well-described taxonomies, most have no data from which to describe species occupancy pattern over the geographic scales necessary to support spatially explicit, landscape-wide analyses for conservation planning. We have the technical wherewithal to design and implement species monitoring programs, but we lack the financial resources to make comprehensive monitoring programs a reality in the near future. Even the simpler task of a taxonomically comprehensive accounting of species is decades, not years, away (Lawler 2001). Furthermore, there is a need to move from biodiversity data derived from accumulated records with an unknown statistical foundation, to data that have their basis in a probabilistic sample of both presence (recorded) and absence (not recorded) (e.g., Pollock et al. 2002). Failure to do so will continue to limit our use of formal inferential procedures (Anderson 2001) to estimate and predict important attributes of biodiversity.

Inferential Basis for Conservation Planning

Although taxonomically comprehensive and spatially extensive species inventories are not going to be available in the foreseeable future, conservation planners cannot sit idle waiting for ideal biodiversity data to appear while land-use decisions are made. For this reason, research that explores how to make the best use of extant data needs to continue. Of particular importance are efforts that extend our capacity to infer species occupancy across space (species distribution models) and to infer overall biodiversity patterns from a few taxa (surrogacy).

Given the incomplete spatial coverage of species surveys, distribution models permit planners to extend survey data to infer occupancy across large geographic areas—portions of which have not been surveyed (Guisan and Thuiller 2005). Although there has been an explosion of species distribution modeling approaches (see Scott et al. 2002), there are a number of modeling challenges that remain, including better representation of species movement to capture source-sink or metapopulation dynamics; determining if interspecific interactions need to be incorporated into species occupancy models; and the extension of species models to those that treat multispecies attributes as the response variable (see Noon et al., this volume). Certainly, more evaluative efforts like that of Elith et al. (2006) are needed to quantify the performance of these modeling approaches and to better inform planners about which approaches are appropriate given the situation specific to the planning context.

The assumption that biodiversity patterns from a few well-studied taxa can represent the pattern among the throng of little-known taxa is necessary given that most biodiversity remains nameless to science. This assumption has been the focus of widespread empirical testing with equivocal outcomes. Although a number of studies have found little evidence supporting the surrogacy assumption (Flather et al. 1997, Ricketts et al. 1999, Ceballos and Ehrlich 2006, Noon et al., this volume), others observed sufficient positive co-occurrence patterns among taxa (Lamoreux et al. 2006) to provide hope that we can simplify the biodiversity conservation challenge by focusing on a small subset of species to derive tenable conservation plans. In particular, Lennon et al. (2004) recently found evidence that variation in species richness across the landscape is associated with relatively few, more common, species. Such contradictory findings point to an important research need—namely, to identify those ecological circumstances when it is tenable to use occurrence patterns of a few taxa to represent the pattern for other taxa (Marcot and Flather 2007).

Extending Geographic Approaches

Improvements in data, whether empirical or model-based, will make substantial improvements in the accuracy of conservation plans. However, data by themselves will not address all limitations associated with geographically based conservation planning; we also need research to extend the capability of the analytical approaches themselves. An admittedly partial list includes explicit incorporation of scale effects, extending the capability of optimization, and consideration of landscape context effects.

Much of what is done to conserve biodiversity takes place at very local scales (Oldfield et al. 2004, Turner et al. 2006). This contrasts considerably with many recent analyses of global conservation priority, which are often several orders of magnitude larger in their analysis grain. Although this disparity is acknowledged (Harris et al. 2005, Burgess et al. 2007), the potential implications of the disparity between coarse-grain prioritization to fine-grained implementation has not been fully evaluated. Our analysis of biodiversity patterns in the southwestern United States suggests that conservation priorities can be scale dependent and that large-grain analyses may not efficiently identify where fine-grained implementation should occur (but see Probst and Gustafson, this volume). However, there is a need to repeat our multiscale comparisons across a broad range of ecosystems to determine if the lack of spatial coincidence in multiscale priority setting that we observed is a general pattern.

Another area of future research concerns the need for continued improvements in applying optimization analyses to conservation planning problems. The literature points to an inherent trade-off in optimization approaches, namely that there is a choice between obtaining an exact optimal solution to simple conservation problems or an approximate optimum to ecologically complex conservation problems (Fig. 4-6). Incorporation of mechanistic ecological detail is a strength of simulation modeling where system responses are explored numerically. Simulation models, however, are hard pressed to prescribe how lands should be managed to obtain efficient conservation objectives. The latter is a strength of optimization, but optimization suffers from constraints associated with obtaining analytical solutions to complex ecological circumstances. For this reason there is potential for fruitful research that explores the joint use of these strategies. A combination of approaches may offer planners the ecological detail from simulation approaches and the analytical power of optimization to prescribe the best solution (Williams et al. 2005, Hof and Flather 2007).

Traditionally, geographically based conservation planning has represented landscapes in a binary fashion with places of biodiversity conservation focus being embedded in a benign matrix of concealed heterogeneity. After plan implementation, these conserved lands are cordoned off in an attempt to isolate them from the threats attributable to "human enterprises" (*sensu* Vitousek et al. 1997). However, research is showing that this matrix is far from benign. The matrix of semi-natural and intensely managed lands can have significant impacts on designed conservation networks (Williams et al. 2006). On the one hand, human uses within and outside the conservation area boundary can erode the biodiversity elements featured in the conservation areas (Liu et al. 2001). On

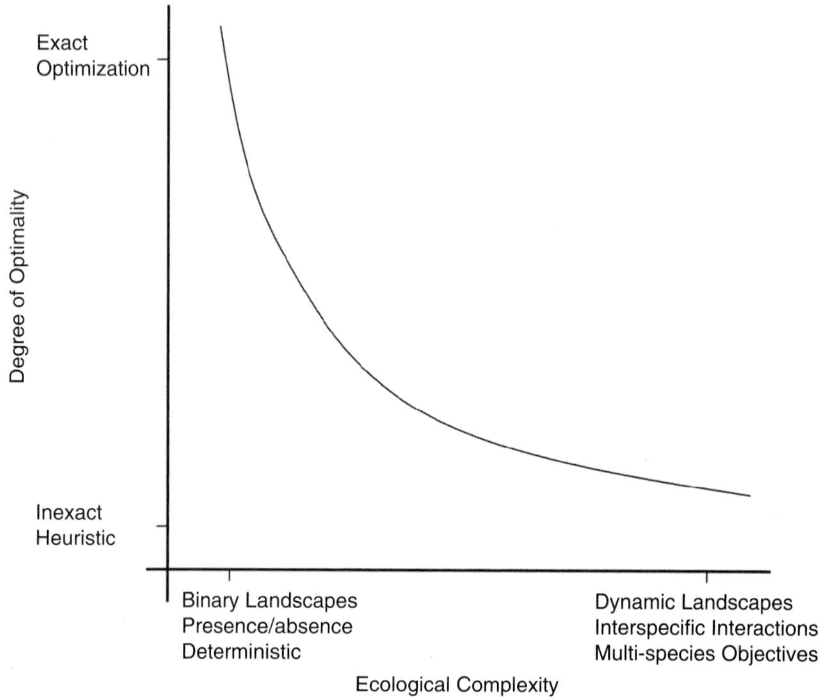

FIG. 4-6

The trade-off between degree of optimality and ecological complexity that can be addressed with exact and inexact optimization analyses for conservation planning (adapted from Hof and Flather 2007).

the other hand, the so-called matrix is not devoid of biodiversity and therefore makes a contribution to the overall mix of species that inhabit a conservation network (Ricketts 2001). Treating the matrix as biologically depauperate may be justified from the standpoint of accounting for the worst-case scenario. However, such a strategy may in fact be setting the conservation bar too high and limiting our flexibility in considering reasonably good solutions in the real world where strong economic pressures make implementation of "the best" strategy politically infeasible (Possingham et al. 2000). For this reason, there is a need for geographic approaches that take a broader perspective—one that considers the landscape whole as well as the conserved parts. This is easier said than done, but it is a perspective that is extending landscape ecology from its traditional patch-based focus to one that treats the landscape as a continuum (Fischer and Lindenmayer 2006)—a shift that is also seen as facilitating the incorporation of ecological process into conservation planning and land management.

Incorporating Ecological Process

Geographic approaches to conservation planning have been dominated by the examination of patterns of noteworthy biodiversity features across the landscape (Pressey 2004). These biodiversity patterns manifest from a complex interaction of ecological processes (e.g., species dispersal, spatially explicit demographics, disturbance, succession, interspecific interactions, primary productivity) with human uses of the landscape (e.g., subsistence, land use conversion, resource extraction). These interactions are the drivers of ecosystem dynamics and the transitory nature of species occupancy patterns across the landscape (Sarkar et al. 2006). The observed temporal turnover in species begs the question: "Can static conservation designs protect the full complement of species suggested by the planning analyses?" (Cabeza and Moilanen 2001).

Unfortunately, conservation science has yet to determine the best approaches for including the underlying ecological and socioeconomic processes into geographical approaches to conservation planning (Williams et al. 2005). For this reason, there is a growing recognition that the biodiversity patterns used to justify a particular conservation design may not be conserved over time. This realization was the motivation behind the efforts of Leroux et al. (2007) to examine the effects of disturbance on the effectiveness and efficiency of static conservation designs. The risks associated with ignoring dynamics varied across conservation targets with some targets (population levels of a focal species) failing to be maintained in a dynamic world, whereas other targets (vegetation representation) were conserved with a high probability under most circumstances (Leroux et al. 2007:1963).

Demonstrating the potential consequences for failing to consider ecosystem dynamics in conservation designs is important, but the planning tools to incorporate such effects analytically are still in their infancy. Given the impending changes to ecosystems in response to climate change (Thuiller 2007), this perhaps represents one of the most important gaps in our ability to effectively plan for the long-term persistence of species across broad landscapes (Groves 2003). Closing that gap will require continuing efforts to incorporate both pattern and process in conservation planning (Ferrier 2002), including an ability to explicitly consider the dynamic nature of ecosystems and the scheduling of conservation strategies over time (Meir et al. 2004).

Accessibility to Practitioners

Geographic-based approaches have a long conservation history, and recent developments in computer technology and analytical capability have allowed impressive advances to the science of biodiversity conservation. These advances notwithstanding, several investigators have made a somewhat disturbing observation—namely, that much of this advance has remained in the ivory towers of academia with little impact on applied conservation (Prendergast et al. 1999).

Although there are clearly exceptions to this observation—most notably, the success that Australian ecologists have had in infusing spatially explicit conservation science into the planning process (see Margules and Pressey 2000)—it is cause for concern that the interaction between research and management on conservation planning remains limited (Cabeza and Moilanen 2001, Flather et al. 2002). Making these geographically based conservation approaches more accessible to conservation practitioners will fail if researchers rely solely on published journal articles to communicate advances. Rather, it will require concerted efforts directed toward meaningful and long-term collaboration on real applied problems and will also require research that demonstrates what is gained by using these planning tools. Two areas that we see as particularly important are evaluative monitoring and characterizing uncertainty.

We are not implying a complete absence of research that has demonstrated the value of conservation planning approaches. However, much of this demonstration has been done with simulated data and evaluation of hypothetical scenarios (Cabeza and Moilanen 2001). Clearly, there are several key logistical constraints. Implementation of conservation plans can take many years, and monitoring to detect species responses, particularly as it relates to detecting extinctions, can take generations (Reed et al. 2003). However, unless we attempt long-term monitoring of implemented conservation plans, we will remain exposed to an apathy of inaction (Prendergast et al. 1999) or the relegation of conservation areas to those of low economic value (Prance 2000). Fortunately, real-world examples that are attempting to use many of the methodologies outlined here are beginning to appear. For example, in the U.S., The Nature Conservancy has seemingly taken the lead in implementing research in large-scale conservation planning and prioritization (Valutis and Mullen 2000, Groves et al. 2002), and has completed a number of conservation assessments in various regions, e.g., the Pacific Northwest coast (Vander Schaaf et al. 2006). Still other conservation planning efforts such as the Northwest Forest Plan (Forest Ecosystem Management Assessment Team 1993), Greater Yellowstone Ecosystem, Wyoming (Noss et al. 2002), Florida (Oetting et al. 2006), South Africa (Smith et al. 2006), and Canada (Beazley et al. 2005) are utilizing these methods to evaluate existing conservation networks and proposed land acquisition for biodiversity conservation.

A related research need is a more comprehensive effort toward the characterization of uncertainty. As noted by Margules and Pressey (2000:251), conservation planning is "...riddled with uncertainty," and our case study has shown how this uncertainty can affect conservation planning. Geographical displays of species distributions or priority conservation areas are often presented as if they are known, and this tendency impedes consideration of uncertainty in the planning process. Given the burgeoning number of modeling approaches that are now available to either predict species occurrence or to prescribe efficient conservation designs, comparative evaluations (like that of Elith et al. 2006) of what works and under which set of circumstances remain an important research opportunity.

Commensurate with research efforts to quantify our uncertainty, we also need research that asks: "How accurate do we need to be?" We know that the propagation of errors from a multitude of sources can be substantial (Elith et al. 2002). What we know less about is how robust our conservation plans are to these various sources of uncertainty and whether robust conservation designs can be derived from approaches that explicitly incorporate uncertainty in the underlying data (e.g., Bini et al. 2006, Moilanen et al. 2006). Ultimately, this issue comes down to determining when moderately precise models will be adequate and when increased accuracy will be necessary (Wiens 2002).

Until we make these analyses more accessible to those that engage in conservation planning by demonstrating their utility and quantifying their uncertainties in ways that go beyond measures of deviance, kappa statistics, or area under receiver operating curves, these approaches will be less accessible to practitioners than they might otherwise be. And ultimately, this rapid development of sophisticated geographically based planning tools will do little to advance conservation decisions if these tools are not understood or used appropriately by those burdened with the responsibility for recommending what, where, and how biodiversity conservation occurs on the landscape.

SUMMARY

Establishment of conservation areas, whether focused strictly on biodiversity conservation or on conservation allowing some degree of multiple-use resource management, continues to be an important regional strategy in the conservation of contemporary biodiversity resources. Given that financial resources for species conservation are limited, responsible stewards must decide where on the landscape management actions should be implemented to maximize conservation benefits. On the surface this seems a simple objective, but unequivocal approaches to attain that objective have eluded conservation scientists for a number of reasons. Two important factors that affect landscape planning for biodiversity conservation are scale and error. Because scale affects our detection and description of biological diversity across the landscape, it also affects our choice of where to focus conservation efforts. Similarly, error in measurement and prediction of species occupancy across the landscape contributes to uncertainty in biodiversity patterns and to the conservation designs derived from those patterns. This chapter focused on the implications of scale and error effects to geographically based conservation planning. We provided an overview of geographic conservation approaches before examining scale and error effects in detail using data from a case study in the southwestern United States. Finally, we provided suggestions on how conservation practitioners can address scale and error in conservation plans and offered our thoughts on future research needs.

LITERATURE CITED

Andelman, S. J., and M. R. Willig. 2003. Present patterns and future prospects for biodiversity in the Western Hemisphere. *Ecology Letters* 6:818–824.

Anderson, D. R. 2001. The need to get the basics right in wildlife field studies. *Wildlife Society Bulletin* 29:1294–1297.

Araújo, M. B., and P. H. Williams. 2001. The bias of complementarity hotspots toward marginal populations. *Conservation Biology* 15:1710–1720.

Augustin, N. H., M. A. Mugglestone, and S. T. Buckland. 1996. An autologistic model for the spatial distribution of wildlife. *Journal of Applied Ecology* 33:339–347.

Bakker, J. P., A. P. Grootjans, M. Hermy, and P. Poschod. 2000. How to define targets for ecological restoration?—Introduction. *Applied Vegetation Science* 3:3–6.

Ball, I. R, and H. P. Possingham. 2000. MARXAN (v 1.8.6): Marine reserve design using spatially explicit annealing: A manual.

Balmford, A., G. M. Mace, and J. R. Ginsberg. 1998. The challenges to conservation in a changing world: Putting processes on the map. Pages 1–28 *in* G. M. Mace, A. Balmford, and J. R. Ginsberg, editors. *Conservation in a changing world.* Cambridge University Press, Cambridge, United Kingdom.

Balmford, A., J. L. Moore, T. Brooks, N. Burgess, L. A. Hansen, P. Williams, and C. Rahbek. 2001. Conservation conflicts across Africa. *Science* 291:2616–2619.

Beazley, K., L. Smandych, T. Snaith, F. MacKinnon, P. Austen-Smith, Jr., and P. Duinker. 2005. Biodiversity considerations in conservation system planning: Map-based approach for Nova Scotia, Canada. *Ecological Applications* 15:2192–2208.

Bedward, M., R. L. Pressey, and D. A. Keith. 1992. A new approach for selecting fully representative reserve networks: Addressing efficiency, reserve design and land suitability with an iterative analysis. *Biological Conservation* 62:115–125.

Bender, D. J., C. H. Flather, K. R. Wilson, and G. C. Reese. 2005. Regional biodiversity data to support biodiversity assessments: Terrestrial vertebrate and butterfly data from the Southwest. U.S. Forest Service, Rocky Mountain Research Station, General Technical Report RMRS-GTR-152, Fort Collins, Colorado, USA.

Bini, L. M., J. A. F. Diniz-Filho, T. F. L. V. B. Rangel, R. P. Bastos, and M. P. Pinto. 2006. Challenging Wallacean and Linnean shortfalls: Knowledge gradients and conservation planning in a biodiversity hotspot. *Diversity and Distributions* 12:475–482.

Bisby, F. A. 2000. The quiet revolution: Biodiversity informatics and the Internet. *Science* 289:2309–2312.

Blackburn, T. M., and K. J. Gaston. 1998. Some methodological issues in macroecology. *American Naturalist* 151:68–83.

Brooks, T. M., R. A. Mittermeier, G. A. B. da Fonseca, J. Gerlach, M. Hoffmann, J. F. Lamoreux, C. G. Mittermeier, J. D. Pilgrim, and A. S. L. Rodrigues. 2006. Global biodiversity conservation priorities. *Science* 313:58–61.

Brown, J. H., and J. Roughgarden. 1990. Ecology for a changing Earth. *Bulletin of the Ecological Society of America* 71:173–188.

Burgess, N. D., A. Balmford, N. J. Cordeiro, J. Fjeldså, W. Küper, C. Rahbek, E. W. Sanderson, J. P. W. Scharlemann, J. H. Sommer, and P. H. Williams. 2007. Correlations among species distributions, human density and human infrastructure across the high biodiversity tropical mountains of Africa. *Biological Conservation* 134:164–177.

Cabeza, M., and A. Moilanen. 2001. Design of reserve networks and the persistence of biodiversity. *Trends in Ecology and Evolution* 16:242–248.

Caro, T. M., and G. O'Doherty. 1999. On the use of surrogate species in conservation biology. *Conservation Biology* 13:805–814.

Ceballos, G., and P. R. Ehrlich. 2006. Global mammal distributions, biodiversity hotspots, and conservation. *Proceedings of the National Academy of Sciences* 103:19374–19379.

Cherrill, A., and C. McClean. 1995. An investigation of uncertainty in field habitat mapping and the implications for detecting land cover change. *Landscape Ecology* 10:5–21.

Cincotta, R. P., J. Wisnewski, and R. Engelman. 2000. Human population in the biodiversity hotspots. *Nature* 404:990–992.

Cox, C. B., and P. D. Moore. 1993. *Biogeography: An ecological and evolutionary approach*. Fifth edition. Blackwell Science, Oxford, United Kingdom.

Dean, D. J., K. R. Wilson, and C. H. Flather. 1997. Spatial error analysis of species richness for a gap analysis map. *Photogrammetric Engineering and Remote Sensing* 63:1211–1217.

Dirzo, R., and P. H. Raven. 2003. Global state of biodiversity and loss. *Annual Review of Environment and Resources* 28:137–167.

Elith, J., M. A. Burgman, and H. M. Regan. 2002. Mapping epistemic uncertainties and vague concepts in predictions of species distributions. *Ecological Modelling* 157:313–329.

Elith, J., C. H. Graham, R. P. Anderson, M. Dudík, S. Ferrier, A. Guisan, R. J. Jijmans, F. Heuttmann, J. R. Leathwick, A. Lehmann, L. Jin, L. G. Lohmann, B. A. Loiselle, G. Manion, C. Moritz, M. Nakamura, Y. Nakazawa, J. M. C. Overton, A. T. Peterson, and S. J. Phillips. 2006. Novel methods improve prediction of species' distributions from occurrence data. *Ecography* 29:129–151.

Engler, R., A. Guisan, and L. Rechsteiner. 2004. An improved approach for predicting the distribution of rare and endangered species from occurrence and pseudo-absence data. *Journal of Applied Ecology* 41:263–274.

Evans, K. L., B. J. van Rensburg, K. J. Gaston, and S. L. Chown. 2006. People, species richness and human population growth. *Global Ecology and Biogeography* 15:625–636.

Ferrier, S. 2002. Mapping spatial pattern in biodiversity for regional conservation planning: Where to from here? *Systematic Biology* 51:331–363.

Fischer, J., and D. B. Lindenmayer. 2006. Beyond fragmentation: The continuum model for fauna research and conservation in human-modified landscapes. *Oikos* 112:473–480.

Fjeldså, J. 2007. How broad-scale studies of patterns and processes can serve to guide conservation planning in Africa. *Conservation Biology* 21:659–667.

Flather, C. H., M. Bevers, and J. Hof. 2002. Prescribing habitat layouts: Analysis of optimal placement for landscape planning. Pages 428–453 *in* K. J. Gutzwiller, editor. *Concepts and applications of landscape ecology in biological conservation*. Springer-Verlag, New York, New York, USA.

Flather, C. H., M. S. Knowles, and I. A. Kendall. 1998. Threatened and endangered species geography: Characteristics of hot spots in the conterminous United States. *BioScience* 48:365–376.

Flather, C. H., and C. H. Sieg. 2000. Applicability of Montreal Process criterion 1—Conservation of biological diversity—to rangeland sustainability. *International Journal of Sustainable Development and World Ecology* 7:81–96.

Flather, C. H., K. R. Wilson, D. J. Dean, and W. C. McComb. 1997. Identifying gaps in conservation networks: Of indicators and uncertainty in geographic-based analyses. *Ecological Applications* 7:531–542.

Fleishman, E., R. F. Noss, and B. R. Noon. 2006. Utility and limitations of species richness metrics for conservation planning. *Ecological Indicators* 6:543–553.

Forest Ecosystem Management Assessment Team. 1993. *Forest ecosystem management: An ecological, economic, and social assessment*. Bureau of Land Management, Report number 1993-793-071, Portland, Oregon, USA.

Fortin, M.-J., T. H. Keitt, B. A. Maurer, M. L. Taper, D. M. Kaufman, and T. M. Blackburn. 2005. Species' geographic ranges and distributional limits: Pattern analysis and statistical issues. *Oikos* 108:7–17.

Funk, V. A., and K. S. Richardson. 2002. Systematic data in biodiversity studies: Use it or lose it. *Systematic Biology* 51:303–316.

Gaston, K. J. 2000. Global patterns in biodiversity. *Nature* 405:220–227.

Gaston, K. J. 2006. Biodiversity and extinction: Macroecological patterns and people. *Progress in Physical Geography* 30:258–269.

Gaston, K. J., and T. M. Blackburn. 2000. *Pattern and process in macroecology.* Blackwell Science, Oxford, United Kingdom.

Gaston, K. J., T. M. Blackburn, and K. K. Goldewijk. 2003. Habitat conversion and global avian biodiversity loss. *Proceedings of the Royal Society of London, Series B* 270:1293–1300.

Gaston, K. J., and A. S. L. Rodrigues. 2003. Reserve selection in regions with poor biological data. *Conservation Biology* 17:188–195.

Gelfand, A. E., J. A. Silander, S. Wu, A. Latimer, P. O. Lewis, A. G. Rebelo, and M. Holder. 2003. Explaining species distribution patterns through hierarchical modeling. *Bayesian Analysis* 1:41-92.

Graham, C. H., S. Ferrier, F. Huettman, C. Moritz, and A. T. Peterson. 2004. New developments in museum-based informatics and applications in biodiversity analysis. *Trends in Ecology and Evolution* 19:497–503.

Groves, C. R. 2003. *Drafting a conservation blueprint: A practitioner's guide to planning for biodiversity.* Island Press, Washington, D.C., USA.

Groves, C. R., D. B. Jensen, L. L. Valutis, K. H. Redford, M. L. Shaffer, J. M. Scott, J. V. Baumgartner, J. V. Higgins, M. W. Beck, and M. G. Anderson. 2002. Planning for biodiversity conservation: Putting conservation science into practice. *BioScience* 52:499–512.

Guisan, A., and W. Thuiller. 2005. Predicting species distribution: Offering more than simple habitat models. *Ecology Letters* 8:993–1009.

Haberl, H., F. Krausmann, K.-H. Erb, and N. B. Schulz. 2002. Human appropriation of net primary production. *Science* 296:1968–1969.

Halvorson, W., K. A. Thomas, and L. Graham. 2001. *Arizona gap project final report.* U.S. Geological Survey, Sonoran Desert Field Station, Special Technical Report, University of Arizona, Tucson, Arizona, USA.

Harris, G. M., C. N. Jenkins, and S. L. Pimm. 2005. Refining biodiversity conservation priorities. *Conservation Biology* 19:1957–1968.

Hess, G. R., R. A. Bartel, A. K. Leidner, K. M. Rosenfeld, M. J. Rubino, S. B. Snider, and T. H. Ricketts. 2006*a*. Effectiveness of biodiversity indicators varies with extent, grain, and region. *Biological Conservation* 132:448–457.

Hess, G. R., F. H. Kock, M. J. Rubino, K. A. Eschelbach, C. A. Drew, and J. M. Favreau. 2006*b*. Comparing the potential effectiveness of conservation planning approaches in central North Carolina, USA. *Biological Conservation* 128:358–368.

Hirzel, A. H., J. Hausser, D. Chessel, and N. Perrin. 2002. Ecological-niche factor analysis: How to compute habitat-suitability maps without absence data? *Ecology* 83:2027–2036.

Hoekstra, J. M., T. M. Boucher, T. H. Ricketts, and C. Roberts. 2005. Confronting a biome crisis: Global disparities of habitat loss and protection. *Ecology Letters* 8:23–29.

Hoekstra, T. W., T. F. H. Allen, and C. H. Flather. 1991. Implicit scaling in ecological research: On when to make studies of mice and men. *BioScience* 41:148–154.

Hof, J., and C. H. Flather. 1996. Accounting for connectivity and spatial correlation in the optimal placement of wildlife habitat. *Ecological Modelling* 88:143–155.

Hof, J., and C. H. Flather. 2007. Optimization of landscape pattern. Pages 143–160 *in* J. Wu and R. Hobbs, editors. *Key topics and perspectives in landscape ecology.* Cambridge University Press, Cambridge, United Kingdom.

Hof, J., C. Flather, T. Baltic, and S. Davies. 1999. Projections of forest and rangeland condition indicators for a national assessment. *Environmental Management* 24:383–398.

Hooker, J. D. 1859. *On the flora of Australia: Its origins, affinities, and distribution, being an introductory essay to the flora of Tasmania.* Lovell Reeve, London, United Kingdom.

Hortal, J., J. M. Lobo, and A. Jiménez-Valverde. 2007. Limitations of biodiversity databases: Case study on seed-plant diversity in Tenerife, Canary Islands. *Conservation Biology* 21:853–863.

Huston, M. 1993. Biological diversity, soils, and economics. *Science* 262:1676–1680.

Imhoff, M. L., L. Bounoua, T. Ricketts, C. Loucks, R. Harriss, and W. T. Lawrence. 2004. Global patterns in human consumption of net primary production. *Nature* 429:870–873.

Kirkpatrick, J. B. 1983. An iterative method for establishing priorities for the selection of nature reserves: An example from Tasmania. *Biological Conservation* 25:127–134.

Lambeck, R. J., and R. J. Hobbs. 2002. Landscape and regional planning for conservation: Issues and practicalities. Pages 360–380 *in* K. J. Gutzwiller, editor. *Applying landscape ecology in biological conservation.* Springer-Verlag, New York, New York, USA.

Lamoreux, J. F., J. C. Morrison, T. H. Ricketts, D. M. Olson, E. Dinerstein, M. W. McNight, and H. H. Shugart. 2006. Global tests of biodiversity concordance and the importance of endemism. *Nature* 440:212–214.

Langholz, J. A., and J. P. Lassoie. 2001. Perils and promise of privately owned protected areas. *BioScience* 51:1079–1085.

Larsen, F. W., and C. Rahbek. 2003. Influence of scale on conservation priority setting—A test on African mammals. *Biodiversity and Conservation* 12:599–614.

Laurance, W. F. 2001. Future shock: Forecasting a grim fate for the Earth. *Trends in Ecology and Evolution* 16:531–533.

Lawler, A. 2001. Up for the count? *Science* 294:769–770.

Lennon, J. J., P. Koleff, J. J. D. Greenwood, and K. J. Gaston. 2001. The geographical structure of British bird distributions: Diversity, spatial turnover and scale. *Journal of Animal Ecology* 70:966–979.

Lennon, J. J., P. Koleff, J. J. D. Greenwood, and K. J. Gaston. 2004. Contribution of rarity and commonness to patterns of species richness. *Ecology Letters* 7:81–87.

Leroux, S. J., F. K. A. Schmiegelow, S. G. Cumming, R. B. Lessard, and J. Nagy. 2007. Accounting for system dynamics in reserve design. *Ecological Applications* 17:1954–1966.

Liu, J., M. Linderman, Z. Ouyang, L. An, J. Yang, and H. Zhang. 2001. Ecological degradation in protected areas: The case of Wolong Nature Reserve for giant pandas. *Science* 292:98–101.

Lubchenco, J., A. M. Olson, L. B. Brubaker, S. R. Carpenter, M. M. Holland, S. P. Hubbell, S. A. Levin, J. A. MacMahon, P. A. Matson, J. M. Melillo, H. A. Mooney, C. H. Peterson, H. R. Pulliam, L. A. Real, P. J. Regal, and P. G. Risser. 1991. The sustainable biosphere initiative: An ecological research agenda. *Ecology* 72:371–412.

Luck, G. W., T. H. Ricketts, G. C. Daily, and M. Imhoff. 2004. Alleviating spatial conflict between people and biodiversity. *Proceedings of the National Academy of Sciences* 101:182–186.

MacArthur, R. H. 1972*a*. Coexistence of species. Pages 253–259 *in* J. Behnke, editor. *Challenging biological problems.* Oxford University Press, Oxford, United Kingdom.

MacArthur, R. H. 1972*b*. *Geographical ecology: Patterns in the distribution of species.* Harper and Row, New York, New York, USA.

Marcot, B. G., and C. H. Flather. 2007. Species-level strategies for conserving rare or little-known species. Pages 125–164 *in* M. G. Raphael and R. Molina, editors. *Conservation of rare or little-known species: Biological, social, and economic considerations.* Island Press, Washington, D.C., USA.

Margules, C. R., and R. L. Pressey. 2000. Systematic conservation planning. *Nature* 405:243–253.

Margules, C., and M. B. Usher. 1981. Criteria used in assessing wildlife conservation potential: A review. *Biological Conservation* 21:79–109.

Meir, E., S. Andelman, and H. P. Possingham. 2004. Does conservation planning matter in a dynamic and uncertain world? *Ecology Letters* 7:615–622.

Moilanen, A., B. A. Wintle, J. Elith, and M. Burgman. 2006. Uncertainty analysis for regional-scale reserve selection. *Conservation Biology* 20:1688–1697.

Myers, N. 1988. Threatened biotas: 'Hotspots' in tropical forests. *Environmentalist* 8:187–208.

Myers, N., R. A. Mittermeier, C. G. Mittermeier, G. A. B. da Fonseca, and J. Kent. 2000. Biodiversity hotspots for conservation priorities. *Nature* 403:853–858.

Neel, M. C., and M. P. Cummings. 2003. Genetic consequences of ecological reserve design guidelines: An empirical investigation. *Conservation Genetics* 4:427–439.

Nekola, J. C., and P. S. White. 1999. The distance decay of similarity in biogeography and ecology. *Journal of Biogeography* 26:867–878.

Noss, R. F., C. Carroll, K. Vance-Borland, and G. Wuerthner. 2002. A multicriteria assessment of the irreplaceability and vulnerability of sites in the Greater Yellowstone Ecosystem. *Conservation Biology* 16:895-908.

Noss, R. F., E. Dinerstein, B. Gilbert, M. Gilpin, B. J. Miller, J. Terborgh, and S. Trombulak. 1999. Core areas: Where nature reigns. Pages 99-128 *in* M. E. Soulé and J. Terborgh, editors. *Continental conservation: Scientific foundations of regional reserve networks.* Island Press, Washington, D.C., USA.

Oetting, J. B., A. L. Knight, and G. R. Knight. 2006. Systematic reserve design as a dynamic process: F-TRAC and the Florida Forever program. *Biological Conservation* 128:37-46.

O'Hanley, J. R., R. L. Church, and J. K. Gilless. 2007. The importance of *in situ* site loss in nature reserve selection: Balancing notions of complementarity and robustness. *Biological Conservation* 135:170-180.

Oldfield, T. E. E., R. J. Smith, S. R. Harrop, and N. Leader-Williams. 2004. A gap analysis of terrestrial protected areas in England and its implications for conservation policy. *Biological Conservation* 120:303-309.

Orme, C. D. L., R. G. Davies, M. Burgess, F. Eigenbrod, N. Pickup, V. A. Olson, A. J. Webster, T. S. Ding, P. C. Rasmussen, R. S. Ridgely, A. J. Stattersfield, P. M. Bennett, T. M. Blackburn, K. J. Gaston, and I. P. F. Owens. 2005. Global hotspots of species richness are not congruent with endemism or threat. *Nature* 436:1016-1019.

Palmer, M., E. Bernhardt, E. Chornesky, S. Collins, A. Dobson, C. Duke, B. Gold, R. Jacobson, S. Kingsland, R. Kranz, M. Mappin, M. L. Martinez, F. Micheli, J. Morse, M. Pace, M. Pascual, S. Palumbi, O. J. Reichman, A. Simons, A. Townsend, and M. Turner. 2004. Ecology for a crowded planet. *Science* 304:1251-1252.

Pärtel, M. 2006. Data availability for macroecology: How to get more out of regular ecological papers. *Acta Oecologica* 30:97-99.

Pautasso, M. 2007. Scale dependence of the correlation between human population presence and vertebrate and plant species richness. *Ecology Letters* 10:16-24.

Phillips, S. J., R. P. Anderson, and R. E. Schapire. 2006. Maximum entropy modeling of species geographic distributions. *Ecological Modelling* 190:231-259.

Pimm, S. L., and J. L. Gittleman. 1992. Biological diversity: Where is it? *Science* 255:940.

Pimm, S. L., and J. H. Lawton. 1998. Planning for biodiversity. *Science* 279:2068-2069.

Pollock, K. H., J. D. Nichols, T. R. Simons, G. L. Farnsworth, L. L. Bailey, and J. R. Sauer. 2002. Large scale wildlife monitoring studies: Statistical methods for design and analysis. *Environmetrics* 13:105-119.

Possingham, H., I. Ball, and S. Andelman. 2000. Mathematical methods for identifying representative reserve networks. Pages 291-306 *in* S. Ferson and M. Burgman, editors. *Quantitative methods for conservation biology.* Springer-Verlag, New York, New York, USA.

Possingham, H., and K. A. Wilson. 2005. Turning up the heat on hotspots. *Nature* 436:919-920.

Prance, G. 2000. The failure of biogeographers to convey the conservation message. *Journal of Biogeography* 27:51-53.

Prendergast, J. R., R. M. Quinn, and J. H. Lawton. 1999. The gaps between theory and practice in selecting nature reserves. *Conservation Biology* 13:484-492.

Prendergast, J. R., R. M. Quinn, J. H. Lawton, B. C. Eversham, and D. W. Gibbons. 1993. Rare species, the coincidence of diversity hotspots and conservation strategies. *Nature* 365:335-337.

Pressey, R. L. 2004. Conservation planning and biodiversity: Assembling the best data for the job. *Conservation Biology* 18:1677-1681.

Pressey, R. L., C. J. Humphries, C. R. Margules, R. I. Vane-Wright, and P. H. Williams. 1993. Beyond opportunism: Key principles for systematic reserve selection. *Trends in Ecology and Evolution* 8:124-128.

Pressey, R. L., I. R. Johnson, and P. D. Wilson. 1994. Shades of irreplaceability: Towards a measure of the contribution of sites to a reservation goal. *Biodiversity and Conservation* 3:242-262.

Pressey, R. L., H. P. Possingham, and J. R. Day. 1997. Effectiveness of alternative heuristic algorithms for identifying indicative minimum requirements for conservation reserves. *Biological Conservation* 80:207–219.

Purvis, A., and A. Hector. 2000. Getting the measure of biodiversity. *Nature* 405:212–219.

Rapoport, E. H. 1982. *Areography: Geographical strategies of species*. Pergamon Press, Oxford, United Kingdom.

Rapport, D. J., H. A. Regier, and T. C. Hutchinson. 1985. Ecosystem behavior under stress. *American Naturalist* 125:617–640.

Redford, K. H., P. Coppolillo, E. W. Sanderson, G. A. B. da Fonseca, E. Dinerstein, C. Groves, G. Mace, S. Maginnis, R. A. Mittermeier, R. Noss, D. Olson, J. G. Robinson, A. Vedder, and M. Wright. 2003. Mapping the conservation landscape. *Conservation Biology* 17:116–131.

Reed, D. H., J. J. O'Grady, B. W. Brook, J. D. Ballou, and R. Frankham. 2003. Estimates of minimum viable population sizes for vertebrates and factors influencing those estimates. *Biological Conservation* 113:23–34.

Reid, W. V. 1998. Biodiversity hotspots. *Trends in Ecology and Evolution* 13:275–280.

ReVelle, C. S., J. C. Williams, and J. J. Boland. 2002. Counterpart models in facility location science and reserve selection science. *Environmental Modeling and Assessment* 7:71–80.

Ricketts, T. H. 2001. The matrix matters: Effective isolation in fragmented landscapes. *American Naturalist* 158:87–99.

Ricketts, T. H., E. Dinerstein, T. Boucher, T. M. Brooks, S. H. M. Butchart, M. Hoffman, J. F. Lamoreux, J. Morrison, M. Parr, J. D. Pilgrim, A. S. L. Rodrigues, W. Sechrest, G. E. Wallace, K. Berlin, J. Bielby, N. D. Burgess, D. R. Church, N. Cox, D. Knox, C. Loucks, G. W. Luck, L. L. Master, R. Moore, R. Naidoo, R. Ridgely, G. E. Schatz, G. Shire, H. Strand, W. Wettengel, and E. Wikramanakaye. 2005. Pinpointing and preventing imminent extinctions. *Proceedings of the National Academy of Sciences* 102:18497–18501.

Ricketts, T. H., E. Dinerstein, D. M. Olson, and C. Loucks. 1999. Who's where in North America: Patterns of species richness and the utility of indicator taxa for conservation. *BioScience* 49:369–381.

Rodrigues, A. S. L., and K. J. Gaston. 2002. Optimisation in reserve selection procedures—Why not? *Biological Conservation* 107:123–129.

Rojstaczer, S., S. M. Sterling, and N. J. Moore. 2001. Human appropriation of photosynthesis products. *Science* 294:2549–2552.

Rondinini, C., K. A. Wilson, L. Boitani, H. Grantham, and H. P. Possingham. 2006. Tradeoffs of different types of species occurrence data for use in systematic conservation planning. *Ecology Letters* 9:1136–1145.

Sala, O. E., F. S. Chapin, III, J. J. Armesto, E. Berlow, J. Bloomfield, R. Dirzo, E. Huber-Sanwald, L. F. Huenneke, R. B. Jackson, A. Kinzig, R. Leemans, D. M. Lodge, H. A. Mooney, M. Oesterheld, N. L. Poff, M. T. Sykes, B. H. Walker, M. Walker, and D. H. Wall. 2000. Global biodiversity scenarios for the year 2100. *Science* 287:1770–1774.

Sarkar, S., and C. R. Margules. 2002. Operationalizing biodiversity for conservation planning. *Journal of Biosciences* 27:299–308.

Sarkar, S., R. L. Pressey, D. P. Faith, C. R. Margules, T. Fuller, D. M. Stoms, A. Moffett, K. A. Wilson, K. J. Williams, P. H. Williams, and S. Andelman. 2006. Biodiversity conservation planning tools: Present status and challenges for the future. *Annual Review of Environment and Resources* 31:123–159.

Schneider, D. C. 2001. The rise of the concept of scale in ecology. *BioScience* 51:545–553.

Scott, J. M., P. J. Heglund, M. L. Morrison, J. B. Haufler, M. G. Raphael, W. A. Wall, and F. B. Samson. 2002. *Predicting species occurrences: Issues of accuracy and scale*. Island Press, Washington, D.C., USA.

Shelford, V. E. 1913. *Animal communities in temperate America*. The University of Chicago Press, Chicago, Illinois, USA.

Shriner, S. A., K. R. Wilson, and C. H. Flather. 2006. Reserve networks based on richness hotspots and representation vary with scale. *Ecological Applications* 16:1660–1673.

Simberloff, D. 1988. The contribution of population and community biology to conservation science. *Annual Review of Ecology and Systematics* 19:473-511.

Sisk, T. D., A. E. Launer, K. R. Switky, and P. R. Ehrlich. 1994. Identifying extinction threats: Global analyses of the distribution of biodiversity and the expansion of the human enterprise. *BioScience* 44:592-604.

Smith, R. J., P. S. Goodman, and W. S. Matthews. 2006. Systematic conservation planning: A review of perceived limitations and an illustration of the benefits, using a case study from Maputaland, South Africa. *Oryx* 40:400-410.

Soberón, J., R. Jiménez, J. Golubov, and P. Koleff. 2007. Assessing completeness of biodiversity databases at different spatial scales. *Ecography* 30:152-160.

Spellerberg, I. F., and J. W. D. Sawyer. 1999. *An introduction to applied biogeography*. Cambridge University Press, Cambridge, United Kingdom.

Stockwell, D. R. B., and I. R. Noble. 1991. Induction of sets of rules from animal distribution data—A robust and informative method of data-analysis. *Mathematics and Computers in Simulation* 33:385-390.

Stott, P. 1984. History of biogeography. Pages 1-24 *in* J. A. Taylor, editor. *Biogeography: Recent advances and future directions.* Barnes and Noble Books, Totowa, New Jersey, USA.

Sutherland, W. J., A. S. Pullin, P. M. Dolman, and T. M. Knight. 2004. The need for evidence-based conservation. *Trends in Ecology and Evolution* 19:305-308.

Thompson, B. C., P. J. Cist, J. S. Prior-Magee, R. A. Deitner, D. L. Garber, and M. A. Hughes. 1996. Gap analysis of biological diversity conservation in New Mexico using geographic information systems U.S. Geological Survey, New Mexico Cooperative Fish and Wildlife Research Unit, Research Completion Report, Las Cruces, New Mexico, USA.

Thuiller, W. 2007. Climate change and the ecologist. *Nature* 448:550-552.

Turner, W. R., D. S. Wilcove, and H. M. Swaine. 2006. Assessing the effectiveness of reserve acquisition programs in protecting rare and threatened species. *Conservation Biology* 20:1657-1669.

Valutis, L., and R. Mullen. 2000. The Nature Conservancy's approach to prioritizing conservation action. *Environmental Science and Policy* 3:341-346.

Vander Schaaf, D., G. Wilhere, Z. Ferdaña, K. Popper, M. Schindel, P. Skidmore, D. Rolph, P. Iachetti, G. Kittel, R. Crawford, D. Pickering, and J. Christy. 2006. *Pacific Northwest Coast Ecoregional assessment.* The Nature Conservancy, Portland, Oregon, USA.

Vane-Wright, R. I., C. J. Humphries, and P. H. Williams. 1991. What to protect?—Systematics and the agony of choice. *Biological Conservation* 55:235-254.

Veregin, H. 1989. Error modeling for the map overlay operation. Pages 3-18 *in* M. Goodchild and S. Gopal, editors. *The accuracy of spatial databases.* Taylor and Francis, New York, New York, USA.

Vitousek, P. M., P. R. Ehrlich, A. H. Ehrlich, and P. A. Matson. 1986. Human appropriation of the products of photosynthesis. *BioScience* 36:368-373.

Vitousek, P. M., H. A. Mooney, J. Lubchenco, and J. M. Melillo. 1997. Human domination of Earth's ecosystems. *Science* 277:494-499.

Wackernagel, M., N. B. Schulz, D. Deumling, A. C. Linares, M. Jenkins, V. Kapos, C. Monfreda, J. Loh, N. Myers, R. Norgaard, and J. Randers. 2002. Tracking the ecological overshoot of the human economy. *Proceedings of the National Academy of Sciences* 99:9266-9271.

Wallace, A. R. 1876. *The geographical distribution of animals. With a study of the relations of living and extinct faunas as elucidating the past changes of the Earth's surface.* Volumes 1 and 2. Macmillan, London, United Kingdom.

Wiens, J. A. 1989. Spatial scaling in ecology. *Functional Ecology* 3:385-397.

Wiens, J. A. 2002. Predicting species occurrences: Progress, problems, and prospects. Pages 739-749 *in* J. M. Scott, P. J. Heglund, M. L. Morrison, J. B. Haufler, M. G. Raphael, W. A. Wall, and F. B. Samson, editors. *Predicting species occurrences: Issues of accuracy and scale.* Island Press, Washington, D.C., USA.

Wiersma, Y. F., and D. L. Urban. 2005. Beta diversity and nature reserve system design in the Yukon, Canada. *Conservation Biology* 19:1262-1272.

Williams, J. C., C. S. ReVelle, and S. A. Levin. 2005. Spatial attributes and reserve design models: A review. *Environmental Monitoring and Assessment* 10:163-181.

Williams, N. S. G., J. W. Morgan, M. A. McCarthy, and M. J. McDonnell. 2006. Local extinction of grassland plants: The landscape matrix is more important than patch attributes. *Ecology* 87:3000-3006.

Williams, P., D. Gibbons, C. Margules, A. Rebelo, C. Humphries, and R. Pressey. 1996. A comparison of richness hotspots, rarity hotspots, and complementarity areas for conserving diversity of British birds. *Conservation Biology* 10:155-174.

Williams, P. H., C. R. Margules, and D. W. Hilbert. 2002. Data requirements and data sources for biodiversity priority area selection. *Journal of Biosciences* 27:327-338.

Willis, K. J., and R. J. Whittaker. 2002. Species diversity—scale matters. *Science* 295:1245-1248.

Wilson, E. O., and E. O. Willis. 1975. Applied biogeography. Pages 523-534 *in* M. L. Cody and J. M. Diamond, editors. *Ecology and evolution of communities.* Harvard University Press, Cambridge, Massachusetts, USA.

Wilson, K. A., M. I. Westphal, H. P. Possingham, and J. Elith. 2005. Sensitivity of conservation planning to different approaches to using predicted species distribution data. *Biological Conservation* 122:99-112.

World Database on Protected Areas Consortium. 2004. World database on protected areas. World Conservation Union (IUCN) and UNEP-World Conservation Monitoring Centre (UNEP-WCMC). CD-ROM. Center for Applied Biodiversity Science, Conservation International, Washington, D.C., USA.

Zaniewski, A. E., A. Lehmann, and J. M. C. Overton. 2002. Predicting species spatial distributions using presence-only data: A case study of native New Zealand ferns. *Ecological Modelling* 157:261-280.

CHAPTER

Social and Economic Considerations for Planning Wildlife Conservation in Large Landscapes

5

Robert G. Haight and Paul H. Gobster

People conserve wildlife for a variety of reasons. People conserve wildlife because they enjoy wildlife-related activities such as recreational hunting, wildlife viewing, or ecotourism that satisfy many personal and social values associated with people's desire to connect with each other and with nature (Decker et al. 2001). People conserve wildlife because it provides tangible benefits such as food, clothing, and other products. People conserve wildlife because they recognize that species are integral parts of larger ecosystems that perform a number of valuable services including nutrient cycling, water purification, and climate regulation (Daily 1997). People also conserve wildlife for its option value or potential to produce future benefits, such as new pharmaceuticals (Fisher and Hanneman 1986). Finally, people conserve wildlife for its existence value even if they will never see or use it (Bishop and Welsh 1992).

Because wildlife provides benefits to the public at large, government agencies and private organizations take responsibility for wildlife conservation. Programs for wildlife conservation typically protect species and habitat from human activities such as hunting, timber harvesting, or housing. As a result, conservation programs may impose substantial costs on other parts of society. Although it seems reasonable to evaluate conservation programs with an assessment of their benefits and costs, in practice, quantifying benefits is difficult, if not impossible. We are far from being able to obtain definitive estimates of wildlife benefits associated with nonconsumptive recreation activities, option values, existence values, and ecosystem services.

An alternative approach to evaluating conservation programs involves efficiency and trade-off analysis. Because public and private groups involved in wildlife conservation often have multiple objectives and limited resources to carry out their programs, efficiency analysis plays an important role in the evaluation of alternative programs. Efficiency analysis involves determining the strategy that maximizes a conservation objective given limited resources. Trade-off analysis involves the analysis of competing conservation goals in terms of how much of one goal must be given up to achieve another goal. Both types of analyses focus on the cost-effectiveness of alternative strategies and sidestep the difficult problem of estimating the total benefits of conservation.

In this chapter we describe key reasons why people conserve wildlife. We first examine contemporary attitudes and values associated with activities such as recreation, landscape restoration, and amenity migration. We then discuss ways to determine cost-effective habitat protection strategies and to identify the trade-offs among various conservation goals in case studies of habitat protection. We conclude with directions for future research. By "wildlife conservation," we mean a wide range of activities to protect and restore individual species and assemblages, from hands-on management of animals to land acquisition for habitat protection and restoration. Our definition of "large landscape" is similarly wide ranging and refers to conservation programs such as protecting old growth forest for northern spotted owl (*Strix occidentalis*) habitat on thousands of hectares in the western United States as well as programs to protect small habitat remnants from encroaching urban development in the Chicago, Illinois, USA metropolitan area. In these ways the largeness of landscapes is a social construction that depends on particular conservation goals. Finally, the term "human dimensions" describes the range of perceptions, attitudes, values, uses, and other interactions that people have with respect to natural resources such as wildlife (e.g., Decker et al. 2001).

PEOPLE-WILDLIFE INTERACTIONS AND TRENDS

We see three important trends in people-wildlife interactions: (1) direct interactions with wildlife through consumptive and nonconsumptive uses that are largely recreational in nature, (2) the restoration of landscapes and the wildlife that depends on them, and (3) indirect impacts on wildlife caused by "amenity migration" where people are increasingly purchasing and building seasonal or permanent homes on forested and other natural lands because of their amenity values. While these trends are occurring to varying degrees across the United States, we focus our discussion on data and examples within the Midwest.

Recreation

The transition from unregulated market and subsistence hunting to regulated recreational hunting at the turn of the 20th century helped many wildlife species to successfully rebound. But while an important part of contemporary recreational hunting is consumptive in nature and aimed at harvesting game, social scientists have come to understand how hunting also satisfies a wide range of human values. These include aesthetic values in viewing wildlife, personal values in the development and testing of self-reliance skills, social values such as camaraderie and the passing down of traditions across generations, and ecological values such as understanding ecological principles and developing an ethical relationship with wildlife (e.g., Dizard 2003). The balance of these values, however, can shift across time and location, affecting how hunting as a wildlife-related activity is engaged and perceived. For instance, urbanization and the severance of rural ties to the land can disrupt long-held social values and uses and is thought to be partly responsible for declines in hunting participation (Heberlein and Ericsson 2005). Income, education, and race/ethnicity are additional forces that are affecting a shift in expressed values, suggesting a continued decline in hunting participation in future years (Manfredo et al. 2003, Lopez et al. 2005).

Longitudinal statistics from the National Survey of Fishing, Hunting, and Wildlife-Associated Recreation (U.S. Fish and Wildlife Service 2002*a*) reinforce these conceptual studies, and document that nationally the percentage of the U.S. population that hunts dropped from 10% in 1955 to 6% in 2001. Variations within these broad numbers are illustrated by statistics in Wisconsin, where in 2001, 9% of urban residents hunted compared to 24% of rural residents. Furthermore, the Wisconsin resident hunting population was 98% white non-Hispanic and 87% male versus 86% and 50%, respectively, for the entire state population; and half as many college graduates hunted (21%) compared to those with only high school diplomas (45%). The drop in participation over time should not discount the importance of hunting because in many states and localities hunting continues to be a major driver of social and economic activity. In Wisconsin, resident and nonresident hunters spent more than $800 million on licenses, equipment, trips, and other items in 2001 (U.S. Fish and Wildlife Service 2002*b*).

Manfredo and his colleagues suggest that this shift in the balance of values has drawn people away from recreational hunting and made them "protectionists" in their views of wildlife (Manfredo and Zinn 1996, Manfredo et al. 2003). But has this shift resulted in increased nonconsumptive recreation? The National Survey refers to observing, photographing, and feeding fish and wildlife as wildlife-watching activities, and distinguishes "residential" activities close to home from "nonresidential" activities more than a kilometer away from home. Here data for wildlife watching over the period 1980–2001 also show a significant drop in activity nationally, with an 18% decrease in the number of people that fed wildlife close to their home and a 19% decrease in those who took wildlife-watching trips away from home. Despite this drop, there are still many more who engage in nonconsumptive versus consumptive recreation, with 30% of U.S. residents engaging in some form of wildlife viewing. The large majority of this activity involves birds, especially people feeding and observing them in residential areas. The population of wildlife watchers is also much more broad-based than hunters. Again looking at statistics from Wisconsin, 47% of urban residents and 63% of rural residents participated in wildlife watching in 2001, and watchers were well distributed across gender, age, income, and education categories. While there may be some overlap in economic impact by those who watch wildlife and also hunt or fish during the same trip, the contribution of wildlife watchers is nonetheless considerable and in Wisconsin amounted to $1.3 billion in 2001 including more than $137 million for wild bird food alone (U.S. Fish and Wildlife Service 2002*b*).

These recreation data do not speak directly to planning wildlife conservation in large landscapes, but some key characteristics can be inferred. For Midwestern hunting, most of it focuses on species that favor early successional and mixed woodland-agricultural habitat rather than large, undisturbed landscapes. Bear hunting is one exception that has a small but dedicated cadre of participants and is concentrated in the large blocks of Northwoods forestland. Another exception might be waterfowl hunting, which often takes place on farmland but depends on significant wetland resources nearby. With residential bird watching and feeding as top activities, wildlife watching is also predominantly focused on fragmented habitat, but again there are important exceptions. Although places such as Yellowstone National Park may be better known destinations for watching charismatic megafauna such as grizzly bear (*Ursus arctos horribilis*) and timber wolves (*Canis lupus lycaon*) (e.g., Montag et al. 2005), the Midwest is also becoming known for this type of ecotourism. In Ely, Minnesota, the International Wolf Center has established itself as a center for "wolf country learning vacations" (International Wolf Center 2006).

Restoration

Landscape restoration is becoming a major means of land management as people increasingly value the existence of native species and an understanding of ecological principles (Gobster and Hull 2000). Landscape restoration involves the re-establishment of vegetation structure, native plant species, and natural disturbance processes such as fire that maintain plant communities; and the removal of roads, invasive species, and human activities such as cattle grazing or off-road recreation that are incompatible with the native ecosystem. Efforts to restore landscapes also involve the reintroduction of native wildlife species, which may play important roles in maintaining ecosystem structure and function. Consequently, large landscape restoration efforts can be controversial, as they involve a range of potentially conflicting management goals and human values (Gobster and Hull 2000).

Efforts to "re-wild" North America (Foreman 2004) include long-range visions for huge proposals such as The Wildlands Project for the Florida Everglades (Noss and Cooperrider 1994) as well smaller scale efforts that have been accomplished or are now underway. Often inherent in these proposals is the re-establishment of viable populations of large mammals (Maehr et al. 2001), but restoration can also focus on smaller birds, mammals, insects, and fish that are rare, threatened, or endangered.

Recent examples of habitat restoration in the Midwest illustrate the range of goals associated with large landscape planning for wildlife conservation. In 1980, the State of Missouri acquired the 1600 ha Prairie State Park for prairie restoration and bison reintroduction (Boyd 2003). The small herd of 78 bison (*Bison bison*) had high educational and symbolic value, but when the herd contracted brucellosis in 1990, the bison were removed because of the economic risk they posed to local livestock operations. A disease-free herd was reinstated into a fenced-in park, but the issue of disease transmission remains a key problem in reintroducing wild, free-ranging herds in large, unfenced landscapes such as Yellowstone National Park (Animal and Plant Health Inspection Service 2006).

The U.S. Forest Service manages jack pine (*Pinus banksiana*) for the federally endangered Kirtland's warbler (*Dendroica kirtlandii*) on the Huron-Manistee National Forests in central Michigan. The warbler depends on large, dense stands of young jack pine, which in turn depend on fire for regeneration. The low, sandy plains provide ideal ecological conditions for warbler restoration efforts, and while many people value the idea of restoring endangered species, they may not be supportive of jack pine management, as the monotypic stands have low scenic value (Schroeder et al. 1993). Additionally, concerns about using fire as a management tool stem back to the 1980 Mack Lake fire, a prescribed fire that escaped and killed 1 person, destroyed 44 homes, and burned more than 8000 ha of forestland before it was brought under control (Simard et al. 1983).

In 1996, the U.S. Forest Service acquired 7700 ha of the former Joliet Arsenal in Will County, Illinois, and established the Midewin National Tallgrass Prairie with a goal of restoring the tallgrass prairie and other native plant communities. There was early public interest in reintroducing bison and elk (*Cervus elaphus*) to the site, but the 2002 Prairie Plan recommended this be deferred to a future date. The site is on the Chicago metropolitan fringe and nearby residential growth and expected recreational demand increase the complexity of reintroduction issues, and fencing, removal of toxics from the former arsenal, and prairie plant re-establishment are needed before reintroduction can be considered (U.S. Forest Service 2002).

In the 1970s, the gray wolf (*Canis lupus*) was listed as an endangered species in the eastern United States, and its recovery plan prohibited hunting and facilitated natural recolonization in parts of Minnesota, Wisconsin, and Michigan. While the wolf became a cherished symbol of the wilderness forests, its movement into agricultural areas was greeted with much less enthusiasm. Today, many rural residents view the wolf as a threat to livestock, poultry, and pets (Chavez et al. 2005).

Amenity Migration

Landscape fragmentation can seriously impact the ability of wildlife managers to sustain species that require large blocks of undisturbed habitat. Land ownership parcelization and development can have significant impacts on landscapes (Sampson and DeCoster 2000). This trend is occurring nationwide but is especially acute near regions of the country with substantial surface water resources, public lands, and other amenity resources. For example, recreation has long been an important use of the Lake States Northwoods, and access to lakes for summer fishing and forests for fall hunting has been a major driving force behind private land acquisition. Recently, human demographic change has fueled an increased demand for owning a piece of the Northwoods. This amenity migration is resulting in further subdivision of private lands and development in the form of seasonal and permanent homes (Gobster and Schmidt 2000, Hammer et al. 2004). In a study of stakeholder perceptions of parcelization and development in the Wisconsin Northwoods, Gobster and Rickenbach (2004) identified four areas of interest and concern that highlight many of the social, environmental, and economic impacts: patterns, drivers, effects, and response strategies.

Stakeholders identified a number of trends in parcelization and development patterns. These included new development and land subdivision along small lakes and rivers and in forest areas that had formerly not been considered amenity attractions. They also spoke of a number of places in the Northwoods where private lands were being advertised for sale bordering national forest and state wild river properties. These patterns of parcelization and development could compromise critical habitat areas needed for wildlife as well as constrict the effectiveness of large blocks of public land by eroding the buffer of undeveloped private forestland that now surrounds them. Finally, stakeholders were concerned that the size of private forestland parcels considered "big" is steadily eroding. In northern Wisconsin where once 30 or 40 ha was thought to be a sizeable piece of land to own, 15 ha is now considered large by many.

Human demographic change is a major driving force behind amenity migration, and as more of the baby boom generation retires, more of them are purchasing and developing seasonal and retirement homes in amenity areas such as northern Wisconsin. Another driver is globalization. Many stakeholders noted the substantial transfer of locally owned industrial forests to multinational corporations, and they feared this transfer would "cream off" attractive vacation properties and fragment these large blocks of private forestland. Last, stakeholders discussed changes in technology such as the mound septic system that has led to increased home building in wet and rocky areas formerly unsuited to development (see also LaGro 1996).

The effects of parcelization and development on wildlife were well summarized by one stakeholder: "If you come to it from the aspect of wildlife, period, it's probably not a bad thing because fragmented property can support all kinds of wildlife. But if you come to it from the position of the diversity of wildlife, or wildlife that was historically present in Wisconsin, then it's probably a growing problem and it'd be a bad thing." In this respect, other stakeholders mentioned direct impacts to species including wolves, bears, lynxes, goshawks, and woodland and grassland songbirds. They also talked about indirect effects including loss of habitat because of invasive plants and loss of songbirds because of cowbird parasitism. One participant mentioned that changing landowner values are leading to a decline in timber harvesting and a subsequent "mapleization of the north," where the loss of earlier successional trees such as oaks and hickories will affect important food sources to many animals.

A final area of interest involved land use strategies to minimize or mitigate the negative effects of parcelization and development. Those strategies include conservancy zoning, where individual landowners cluster development and leave the larger proportion of their land in relatively undisturbed forest cover, and incentive programs such as the Wisconsin Managed Forest Law, which provides a tax break to landowners who develop a conservation plan for their property that may include managing their land for wildlife values (Gobster and Rickenbach 2004). Cross boundary management among private and public landowners is a growing method in which large landscapes can be more effectively managed to meet wildlife goals (e.g., Harper and Crow 2006). Another type of government payment program involves land acquisition for reserves to protect wildlife habitat and provide open space for recreation activities. Reserve-based modeling approaches to large-scale conservation planning are discussed in detail in following sections.

COST-EFFECTIVE WILDLIFE CONSERVATION

A cornerstone of wildlife conservation planning is establishing and expanding habitat reserves (Noss and Cooperrider 1994). Reserves are typically public lands protected from development and managed in part with wildlife objectives. Reserves have a variety of forms including public parks dedicated to nonconsumptive wildlife viewing, wilderness areas in national forests, or multiple-use lands managed for key species.

As we discussed in the previous section, residents of small towns and large cities alike are concerned about the environmental impacts of rapid growth and large-scale conversion of undeveloped to developed land. One result is that local governments and private land trusts have instituted policies to acquire land or conservation easements to preserve undeveloped land within or on the fringe of towns and cities. From 1996 through 2004, voters approved 1062 of 1373 referenda for open space and parks and authorized the use of $26.4 billion (2000 constant dollars) to acquire open space or development rights (Nelson et al. 2007). Agency planners have a variety of objectives for open space acquisition, including habitat protection for wildlife as well as economic efficiency (Ruliffson et al. 2002). In response, biologists and economists have developed reserve selection and design models, which suggest cost-effective ways to protect open space to attain wildlife objectives.

Reserved-based modeling approaches to large-scale conservation planning have been around since the 1980s and are the subject of a rich and growing literature (Kingsland 2002; Flather et al., this volume; Noon et al., this volume). We discuss three broad categories of models: reserve selection models, reserve design models, and reserve design models with population dynamics (Table 5-1). Following Williams et al. (2005), we distinguish the terms "site," "reserve," and "reserve system." A site is a selection unit—a piece of land that

Table 5-1 Reserve-Based Modeling Approaches to Large-Scale Conservation Planning: A. Reserve Selection Models, B. Reserve Design Models, and C. Reserve Design Models With Population Dynamics

Problem	Objective	Reference
A. Reserve selection models		
Maximum species covering	Maximize number of species protected for a given budget	Church et al. 1996
Bi-criteria species covering	Maximize number of species protected and some other conservation objective	Church et al. 2000, Ruliffson et al. 2003
Maximum expected species covering	Maximize expected number of species protected for a given budget	Camm et al. 2002, Arthur et al. 2004
Dynamic species covering	Maximize expected number of species protected at end of horizon	Costello and Polasky 2004, Haight et al. 2005, Turner and Wilcove 2006
B. Reserve design models		
Reserve proximity	Minimize sum of pairwise distances between reserves	Önal and Briers 2002
Reserve connectivity	Maximize number of adjacent reserves	Nalle et al. 2002
Reserve compactness	Minimize boundary length of reserves	Fischer and Church 2003
C. Reserve design models with population dynamics		
Metapopulation size	Maximize metapopulation size	Hof et al. 2001
Safe minimum standard	Maximize probability of metapopulation persistence	Montgomery et al. 1994, Moilanen and Cabeza 2002, Haight and Travis 2008
Surviving populations	Maximize expected number of surviving populations	Haight et al. 2004*a*

may be selected for protection. A site is undeveloped open space belonging to one or more cover types, including forest, grassland, pasture, or farm, that provide habitat for wildlife. A reserve is a single site or a contiguous cluster of sites that has been selected for protection. A reserve system is a set of multiple, spatially separated reserves. Reserve selection models identify sites to protect to maximize some measure of biological diversity (e.g., species richness). Reserve design models incorporate spatial attributes of the selected sites (e.g., connectivity) as conservation objectives. Reserve design models with population dynamics identify sites to protect to achieve objectives related to population size or persistence. We begin by discussing reserve selection models with an objective of maximizing species richness in the selected sites subject to a budget constraint. The problem is used to explain basic economic principles of cost-effectiveness, marginal cost, and trade-off analysis.

Reserve Selection Models

Reserve selection models are based on information about the distribution of species or other conservation features (e.g., habitat types) among sites and targets for protecting those features. For convenience, we will use species as the feature of interest. Each site is described by a list of species that it contains, and a species is covered or represented if at least one site that contains the species is selected for protection. Early models selected the minimum number of sites that represented all species from a list of target species (e.g., Margules et al. 1988). Selecting sites to minimize the cost of protecting all species is called the species set covering problem, an analogue of the location set covering problem from facility location science (ReVelle et al. 2002). Recognizing that resources may limit the number of sites selected for protection, later models maximized the number of species or conservation features that could be represented within a given number of sites (e.g., Church et al. 1996). This latter type of model is called the maximal species covering problem (ReVelle et al. 2002), and it provides case-specific policy guidance on sets of sites that efficiently achieve conservation goals and trade-offs between conservation goals. Cabeza and Moilanen (2001), ReVelle et al. (2002), and Rodrigues and Gaston (2002) summarize applications of reserve selection models.

Maximum Species Covering Problem.— Here, we describe an application of the maximal species covering problem in a case study in the Midwestern United States. The application is in the Lake County portion of the Fox River watershed northwest of the city of Chicago (Fig. 5-1). In response to rapid population growth and conversion of open space to housing and commercial development, Lake County planners are interested in acquiring land to protect rare animals and plants and provide equitable access to recreation. It is important to note that the focus was on rare animals and plants rather than all animals and plants. To help planners identify cost-effective sets of sites, we formulate a maximal species covering problem and analyze the cost of increasing the number of species represented in the selected sites.

The analysis is conducted using data for 31 privately owned open-space sites in the Lake County portion of the Fox River watershed (see Haight et al. 2005 for details). The sites vary in size from 1 to 313 ha, with a median of 29 ha (Table 5-2). Each site is described by a list of rare plants and animals present. Collectively, 27 rare species occur in the 31 sites, and species richness of individual sites varies from 1 to 9 species. Because the budget constraint places an upper bound on total area of sites selected, we expect that smaller sites with

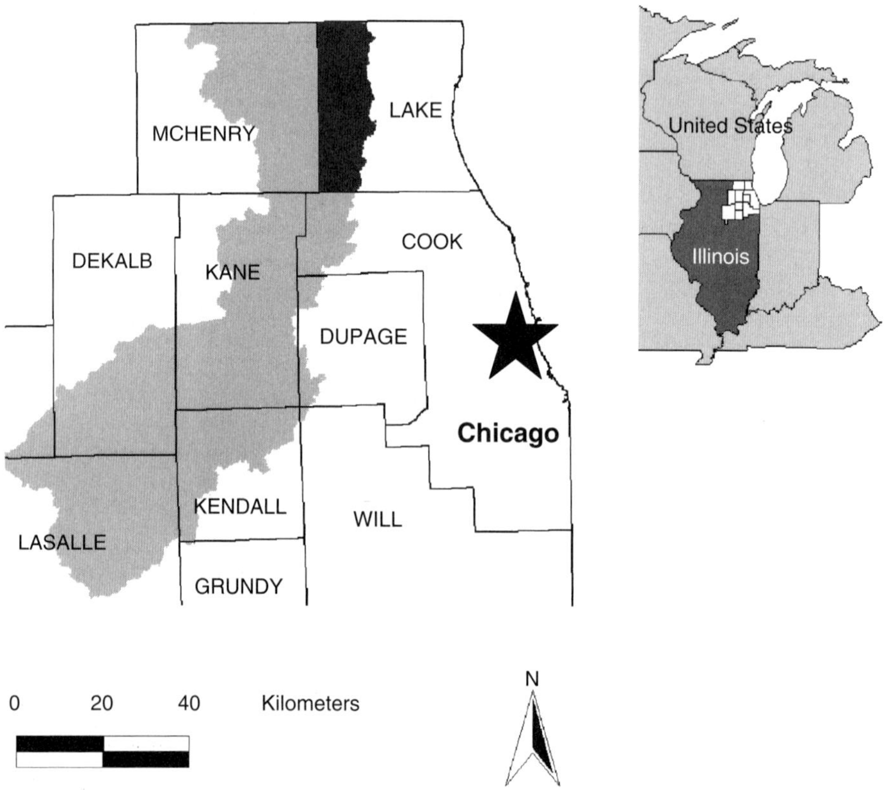

FIG. 5-1

Fox River watershed (shaded gray) in counties of northeastern Illinois, USA. The study area (shaded black) is the northeastern portion of the watershed located in Lake County.

more species may be preferable, and we list the number of species per unit area in Table 5-2.

The maximal species covering problem is a linear-integer programming problem with a cost constraint that limits resources spent for site protection. The problem is solved in seconds using commercial software on a laptop computer. The model has the following notation:

i, I = index and set of species in need of protection,
j, J = index and set of potential reserve sites,
B = upper bound on budget,
c_j = cost of protecting site j,
M_i = set of sites that contain species i,
x_j = 0–1 variable: 1 if site j is selected for protection, 0 otherwise,
y_i = 0–1 variable: 1 if species i is represented in at least one protected site, 0 otherwise.

Table 5-2 Attributes of Open-Space Sites in the Fox River Watershed of Lake County, Illinois, USA

Site	Area (ha)	Number of Species	People With Access (1000s)	Species per ha	Rank	People With Access per ha (1000s)	Rank
1	37	1	0.0	0.03	26	0.00	30
2	40	2	8.0	0.05	19	0.20	18
3	65	2	2.7	0.03	25	0.04	24
4	24	2	9.3	0.08	13	0.39	13
5	9	1	2.9	0.11	10	0.31	15
6	47	3	3.3	0.06	16	0.07	23
7	1	5	1.8	5.00	1	1.80	1
8	16	1	17.6	0.06	17	1.09	5
9	39	4	36.1	0.10	11	0.93	6
10	121	5	9.3	0.04	23	0.08	22
11	141	2	3.3	0.01	28	0.02	27
12	29	2	0.0	0.07	14	0.00	31
13	22	1	33.8	0.05	21	1.55	3
14	9	5	2.7	0.56	3	0.30	16
15	84	7	21.4	0.08	12	0.26	17
16	23	1	9.1	0.04	22	0.39	12
17	5	4	3.1	0.82	2	0.64	8
18	14	3	32.9	0.21	6	2.33	2
19	13	2	6.0	0.16	8	0.48	10
20	30	2	26.7	0.07	15	0.88	7
21	7	1	10.5	0.14	9	1.44	4
22	189	9	2.7	0.05	20	0.01	28
23	313	2	32.5	0.01	31	0.10	20
24	80	1	2.4	0.01	29	0.03	26
25	10	2	5.8	0.20	7	0.57	9
26	142	1	5.8	0.01	30	0.04	25
27	92	2	35.2	0.02	27	0.38	14
28	17	4	0.2	0.23	5	0.01	29
29	24	1	2.7	0.04	24	0.11	19
30	7	2	2.9	0.29	4	0.42	11
31	37	2	3.1	0.05	18	0.08	21

The model is formulated as follows:

$$\text{Maximize} : \sum_{i \in I} y_i \qquad (1)$$

$$\sum_{j \in M_i} x_j \geq y_i \qquad \text{for all } i \in I \qquad (2)$$

$$\sum_{j \in J} c_j x_j \leq B \qquad (3)$$

$$x_j, y_i \in \{0, 1\} \qquad \text{for all } i \in I \text{ and } j \in J \qquad (4)$$

The objective (Eq. 1) is to maximize the number of species that are represented or covered in the set of selected sites. Eq. 2 enforces the logic of covering: a species is considered covered ($y_i = 1$) if at least one site that contains the species is selected for protection. Eq. 3 is the budget constraint that limits how much can be spent on site protection. Eq. 4 describes the integer restrictions on the decision variables.

The cost constraint (Eq. 3) is a key part of the maximal species covering problem because it represents the decision maker's goal of staying within a budget. In our application, we use areas of sites as proxies for site costs because we do not know the dollar value of every site. We therefore assume that the decision maker has an overall area budget for selecting sites. Solving the problem for a given value of the budget level B allows the determination of an efficient set of sites, where efficiency means that there are no other sets of sites that provide a higher level of species coverage and stay within the budget. Solving the problem with increasing levels of B allows construction of a cost curve showing the cost of increasing the number of species covered.

We determine the optimal sets of sites to protect for budgets ranging from 1 to 618 ha and plot the cost curve in Fig. 5-2. The slope of the cost curve is the marginal cost of species protection, which is the area required to protect an additional species. Marginal cost is small (4 ha per species) as coverage increases from 5 to 20 species, moderate (34 ha per species) in the range of 20 to 25 species, and large (195 ha per species) for levels of species coverage greater than 25.

As the budget increases, the optimal set of sites is not always a matter of adding another site to the previously selected set. For example, to increase species coverage from 20 to 22 species, one site can be added to the list of protected sites (Table 5-3). However, increasing species coverage above 22 species involves dropping one site and adding up to four others. Nevertheless, there is consistency in sites selected for protection. Six sites (7, 8, 17, 18, 21, and 30) are selected whenever the budget is greater than 50 ha. These sites are small (<16 ha), rank in the top 10 in terms of species per ha, and contain endemics (Table 5-2).

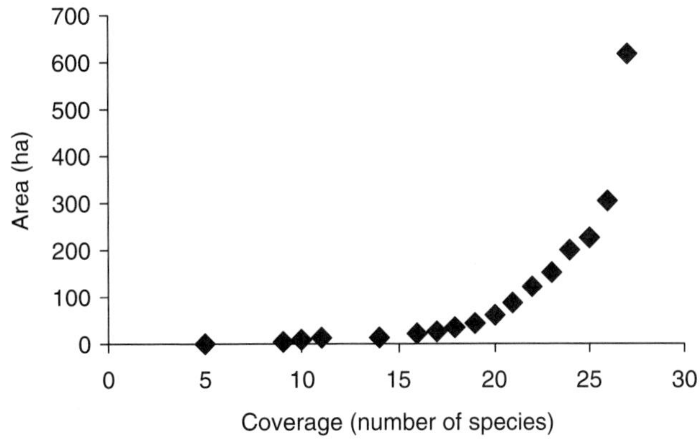

FIG. 5-2

Cost curve showing area protected versus number of species covered for the site selection options in the Fox River watershed of Lake County, Illinois, USA.

Table 5-3 Optimal Sets of Sites Selected for Increasing Area Budgets in the Fox River Watershed of Lake County, Illinois, USA

Objective Values		Sites Protected											
Species	**Area (ha)**	**3**	**7**	**8**	**14**	**15**	**17**	**18**	**20**	**21**	**22**	**23**	**30**
20	59		X	X	X		X	X		X			X
22	123	X	X	X	X		X	X		X			X
25	228	X	X	X		X	X	X	X	X			X
27	618	X	X	X			X	X		X	X	X	X

Bi-Criteria Covering Problem.—Metropolitan planners may have a variety of objectives for land acquisition including habitat protection for rare species, public accessibility, and economic efficiency (Ruliffson et al. 2002). In this section, we extend the maximum species covering problem to handle a second objective of maximizing the accessibility of open space sites to urban populations in the county (Ruliffson et al. 2003, Haight et al. 2005).

Multiobjective site selection models are useful tools for investigating the opportunities for simultaneously meeting multiple conservation objectives (Rothley 1999, Church et al. 2000, Marianov et al. 2004). Analyses typically determine the trade-off between objectives—the pareto-optimal curve that displays the best value of one objective given a required achievement of the other. In addition, important information can be obtained by analyzing the site selection decisions associated with alternative solutions along the trade-off curve, including identification of sites that should be selected no matter what the decision maker's position on the relative importance of the two objectives (Schilling et al. 1982). We show how to display model solutions in terms of both the objectives and decisions in our multiobjective analysis.

There are 34 towns in western Lake County. Based on the 2000 U.S. Census, the towns collectively held 222,000 people, and individual towns were home to 1,000 to 30,000 people. We assume that people in a town have access to a site if the site is within 3.2 km (2.0 miles) of the town. Based on the average distance between each town and each site, we know towns that are within the required distance of each site, and based on the population of each town, we list the total population with access to each site (Table 5-2). Almost all sites have at least 2,000 people within 3.2 km, and five sites have more than 30,000 people within 3.2 km. We also compute the number of people with access per unit area as an approximate index of site desirability.

In addition to the notation listed previously, the bi-criteria site selection model has the following:

k, K = index and set of towns,
Q_1 = number of species represented in the protected sites,
Q_2 = number of people with access to the protected sites,
r_k = number of people in town k,
N_k = set of sites that are within 3.2 km of town k,
w = objective weight: $0 \leq w \leq 1$,
z_k = 0–1 variable: 1 if town k has at least one protected site within 3.2 km, 0 otherwise.

The model is formulated as follows

$$\text{Maximize} : wQ_1 + (1-w)Q_2 \tag{5}$$

$$Q_1 = \sum_{i \in I} y_i \tag{6}$$

$$Q_2 = \sum_{k \in K} r_k z_k \tag{7}$$

$$\sum_{j \in N_k} x_j \geq z_k \quad \text{for all k} \in K \tag{8}$$

$$\sum_{j \in M_i} x_j \geq y_i \quad \text{for all } i \in I \tag{9}$$

$$\sum_{j \in J} c_j x_j \leq B \tag{10}$$

$$x_j, y_i, z_k \in \{0, 1\} \quad \text{for all } i \in I, j \in J, k \in K \tag{11}$$

The objective (Eq. 5) is to maximize the weighted sum of the two objective functions: the number of species represented in protected sites (Eq. 6) and the number of people with access to protected sites (Eq. 7). Public access is the number of towns with access weighted by population size (r_k). The weight w represents the decision maker's position on the relative importance of the two objectives. When w is closer to one, more weight is placed on maximizing the number of species covered. When w is closer to zero, more weight is put on maximizing the number of people with access to protected sites. Eq. 8 is the condition under which town k has access to protected sites (i.e., $z_k = 1$): at least one site that is within 3.2 km of town k must be selected for protection. Eqs. 9–11 are the species coverage definition, the budget constraint, and the integer restrictions on the decision variables, respectively.

The analysis focuses on how the optimal set of protected sites varies as we trade off species representation and public access under different budget levels. We compute optimal site selections for problems in which the objective function weight is decreased from 1.0 to 0.0 in increments of 0.05 subject to area constraints of 81 ha and 200 ha. The curves showing the trade-offs between species representation and public access have concave shapes in which species representation drops as public access increases (Fig. 5-3). The points on each curve represent nondominated sets of sites and their relative performance with respect to the two objectives under a given level of budget. For each nondominated set of sites, improvement in one objective cannot be achieved without simultaneously causing degradation in the value of the other objective. As a result, the points on each trade-off curve represent a frontier beyond which no better solutions can be found.

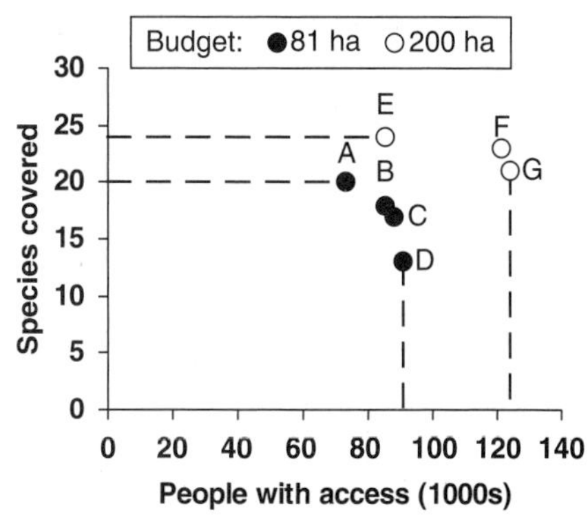

FIG. 5-3

Trade-offs between open-space protection objectives of maximizing species coverage and maximizing public access under different area budgets in the Fox River watershed of Lake County, Illinois, USA.

Among the nondominated solutions for a given budget, the best depends on the decision maker's preference for the two objectives. If species representation is most important and the budget is 81 ha, the choice is alternative A, in which species representation is 20 (74% of the maximum representation without a budget constraint) and public access is 73,000 people (33% of the maximum accessibility). The dashed horizontal line between the y-axis and point A indicates that we found a number of solutions that had the same species representation as alternative A but with less public access. A move from alternative A to alternative B or C increases public access more than 20% and reduces species representation 15%. The highest level of public access under this budget (point D, 91,000 people) is obtained with a 35% reduction in species representation. The dashed vertical line from point D to the x-axis indicates that there are a number of solutions with the same level of public access as alternative D but with less species representation. Increasing the budget from 81 ha to 200 ha shifts the trade-off curve up and to the right while reducing the trade-off between the objectives (Fig. 5-3).

To complement the trade-off curves, it is important to look at the site selection decisions and identify core sites, which are sites selected for protection regardless of the weights given to the objective functions. With a budget of 81 ha, three core sites (7, 18, and 20) are protected in all four solutions (Table 5-4). With a budget of 200 ha, there are four additional core sites (8, 15, 17 and 21). The core sites are typically small (<30 ha) and have relatively large numbers of species and people with access (Table 5-2). As a result, the

Table 5-4 Objective Function Values and Sites Selected for Protection for Nondominated Solutions with Area Budgets of 81 ha (Solutions A,B,C,D) and 200 ha (Solutions E,F,G) in the Fox River Watershed of Lake County, Illinois, USA

	Objective Value		Sites Protected												
Solution	**Species**	**People (1000s)**	**3**	**7**	**8**	**14**	**15**	**16**	**17**	**18**	**19**	**20**	**21**	**30**	**Area Protected (ha)**
A	20	73		X		X			X	X		X	X	X	73
B	18	85		X	X	X			X	X		X			75
C	17	88		X	X				X	X		X	X	X	80
D	13	91		X	X					X	X	X	X		81
E	24	85	X	X	X		X		X	X			X	X	198
F	23	121		X	X	X	X	X	X	X		X	X	X	196
G	21	124		X	X		X	X	X	X	X	X	X		193

core sites rank in the top half in terms of species per hectare and people with access per hectare, two indices of desirability (Table 5-2).

There is a lot of overlap in the sets of sites selected for protection in the alternative solutions under each budget (Table 5-4). As a result of the overlap in composition, the choice between alternative solutions on a trade-off curve involves shifting a small portion of the area budget between a few sites. For example, with a budget of 81 ha, moving from alternative A to alternative C shifts about 11% of the budget from protecting site 14 to protecting site 8. Moving from alternative C to alternative D involves a shift of about 15% of the budget from protecting sites 17 and 30 to protecting site 19. Because adjacent solutions on the trade-off curve often differ in only a few sites, decisions about which alternative to select can focus on the strengths and weaknesses of those few sites.

Maximum Expected Species Covering Problem.—In many cases information about the presence and absence of species in sites is uncertain, and presence is expressed as a probability of occurrence. The species covering problem can be extended to handle this information and maximize the expected number of species covered subject to a budget constraint. Let p_{ij} be the probability that species i exists at site j where the probabilities are independent across sites. Defining v_i as the probability that species i is not covered in the sites selected for protection, we can write

$$v_i = \prod_{j \in J} (1 - p_{ij})^{x_j} \qquad \text{for all } i \in I \qquad (12)$$

where x_j is the 0–1 decision variable for whether site j is selected for protection. Eq. 12 follows from the fact that a selected set of sites fails to cover a given species i if that species is absent from all the selected sites. The independence assumption allows us to write v_i as a product of absence probabilities over all sites. The problem is to determine the values of the site selection variables to maximize the expected number of species covered subject to a budget constraint:

$$\text{Maximize}: \sum_{i \in I} (1 - v_i) \qquad (13)$$

$$\sum_{j \in J} c_j x_j \leq B \qquad (14)$$

$$x_j \in \{0, 1\} \qquad \text{for all } j \in J \qquad (15)$$

The problem in Eqs. 12–15 is nonlinear and cannot be converted to an equivalent linear integer program because the objective function is the sum of terms that involve the products of the decision variables x_j. Nevertheless, a linear approximation of the nonlinear problem can be formulated and solved using commercial software (Camm et al. 2002), and the model has been illustrated using probabilistic occurrence data for 403 terrestrial vertebrates in 147 candidate sites in western Oregon, USA (Arthur et al. 2004).

Dynamic Species Covering Problem.—The maximum species covering problems described so far have taken a static approach to conservation planning. Static models are designed to select cost-effective sets of sites to protect biodiversity given current information about species occurrence and site availability. The models assume that decisions are made all at once and protection takes place rapidly before site degradation or loss. This may be a reasonable first-pass approach to the immediate problem of slowing biodiversity loss; however, planning is a dynamic process that incorporates new information as it unfolds. Researchers have begun to develop methods to address sequential site selection problems with budget restrictions and uncertainties about site degradation and loss with the objective of maximizing the expected number of species covered in protected sites at the end of a planning horizon.

One approach to sequential site selection is building a stochastic dynamic programming model that includes periodic budget constraints and uncertainty about future site availability (Costello and Polasky 2004). The optimal solution includes the set of sites to protect now along with a policy or rule that describes the sites to protect in the future depending on species already protected and sites that are available. Unfortunately, dynamic programming is computationally intensive and has been used to solve problems with fewer than about 10 sites, far less than can be handled with heuristic algorithms. Simple rules for site selection based on current gaps in species coverage and current threats to habitat loss perform reasonably well on small problems in comparison with optimal policies obtained from dynamic programming (Costello and Polasky 2004), and similar heuristics have been applied to large, practical problems (Meir et al. 2004, Turner and Wilcove 2006).

Another approach to sequential site selection involves a two-period linear-integer model in which uncertainty about future site availability is represented with a set of probabilistic scenarios (Snyder et al. 2004, Haight et al. 2005). The decision variables include the set of sites to protect now and sites to protect in the second period, depending on availability. The linear-integer formulation allows solution of realistic-sized problems with commercial software on personal computers. Furthermore, the formulation can be expanded to model multiple objectives and constraints that allow for budget allocation between periods.

Reserve Design Models

A weakness of site selection models is their ignorance of the effects of size and spatial arrangement of reserves on species dynamics, and as a result, there is no guarantee that species represented will persist (Cabeza and Moilanen 2001). One way to improve the models is to include spatial objectives for reserve design that are related to species persistence. For example, some species require large areas of compact and contiguous habitat for survival, whereas other species can survive in disjunct habitat patches as long as they are relatively close together. Reserve proximity, contiguity, and compactness can be formulated as spatial objectives in linear-integer programming models (see Williams et al. 2005 for a review), and we give examples of each type of model in the following sections. For ease of presentation, each model includes a spatial objective combined with a species coverage constraint. By varying the level of the constraint, trade-offs between the spatial and coverage objectives can be obtained. The models can be easily expanded with budget constraints and other conservation objectives.

Reserve Proximity Problem.—A reserve system in which the reserves are close together may be preferred to facilitate movement of individuals between reserves. Shorter migration distances facilitate recolonization of areas where a species has become locally extinct and help prevent the loss of genetic diversity because of inbreeding. One way to reduce the distances between reserves is to minimize the sum of distances between all pairs of selected sites. Letting d_{jk} be the distance between sites j and k and u_{jk} be a 0–1 variable for whether both sites j and k are selected, the problem can be written as follows:

$$\text{Minimize}: \sum_{j\in J}\sum_{k>j} d_{jk}u_{jk} \tag{16}$$

$$U_{jk} \geq x_j + x_k - 1 \qquad \text{for all } j,k \in J, k > j \tag{17}$$

$$\sum_{j\in M_i} x_j \geq R \qquad \text{for all } i \in I \tag{18}$$

$$x_j \in \{0,1\}, u_{jk} \in \{0,1\} \qquad \text{for all } j,k \in J \tag{19}$$

The objective (Eq. 16) minimizes the sum of the pairwise distances between selected sites subject to a species coverage constraint (Eq. 18) that requires each species i to be represented in at least R selected sites. Eq. 17 enforces the definition of u_{jk} by requiring both $x_j = 1$ and $x_k = 1$ for $u_{jk} = 1$. Önal and Briers (2002) apply this formulation to the problem of selecting a subset of 131 pond sites in Oxfordshire, United Kingdom, to protect 256 invertebrate species.

Reserve Connectivity Problem.—Another objective of reserve design is to maximize the structural connectivity of the selected sites. Structural connectivity refers to the physical contiguity of sites and is desirable to create larger reserves or corridors between reserves. In situations in which the landscape is subdivided into contiguous polygons representing candidate sites, structural connectivity can be promoted by selecting sites for protection that are adjacent to each other. The objective is to maximize the number of adjacent pairs of selected sites:

$$\text{Minimize}: \sum_{j\in J}\sum_{k\in A_j, k>j} u_{jk} \tag{20}$$

$$u_{jk} \geq x_j + x_k - 1 \qquad \text{for all } j \in J, k \in A_j, k > j \tag{21}$$

$$\sum_{j\in M_i} x_j \geq R \qquad \text{for all } i \in I \tag{22}$$

$$x_j \in \{0,1\}, u_{jk} \in \{0,1\} \qquad \text{for all } j, k \in J \tag{23}$$

where the set A_j represents all sites that are adjacent to (share a boundary with) site j. Nalle et al. (2002) employed a similar formulation to the problem of selecting a subset of 4181 sites in Josephine County, Oregon, to protect examples of 13 habitat types.

Reserve Compactness Problem.—The shape of reserves in a reserve system may be important for species survival, and many authors advocate creating compact reserves that are nearly circular and have low edge-to-area ratios. Compact reserves are better for edge-intolerant species such as tropical songbirds that prefer large areas of interior habitat for nesting. In reserve design models, compactness is measured by the total length of the boundaries (perimeters) of all the reserves. Total boundary length is the difference between the length of the boundaries of all the selected sites and two times the length of the shared boundaries between the selected sites. Letting b_j be the length of the boundary of site j and sb_{jk} be the length of the shared boundary between sites j and k, the problem of minimizing total boundary length can be written as follows:

$$\text{Minimize}: \sum_{j\in J} b_j x_j - 2\sum_{j\in J}\sum_{k\in A_j, k>j} sb_{jk} u_{jk} \tag{24}$$

$$u_{jk} \geq x_j + x_k - 1 \qquad \text{for all } j \in J, k \in A_j, k > j \tag{25}$$

$$\sum_{j\in M_i} x_j \geq R \qquad \text{for all } i \in I \tag{26}$$

$$x_j \in \{0,1\}, u_{jk} \in \{0,1\} \qquad \text{for all } j, k \in J \tag{27}$$

In the objective function (Eq. 24), the boundary length of the reserve system is calculated by adding the boundary lengths of the selected sites and then subtracting twice the length of the boundaries shared by selected sites that are adjacent. Fischer and Church (2003) utilized this model to analyze trade-offs between total area and compactness of reserve systems to protect examples of 55 plant community types in northern California forests.

Reserve Design Models with Population Dynamics

While the reserve design models discussed in the preceding sections include spatial objectives, they do not model species' population dynamics. In this section, we discuss reserve design models that explicitly incorporate population dynamics. We begin with a discussion of deterministic reserve design models that aim to maximize the size of the metapopulation (collection of subpopulations residing in separate sites) based on estimates of population growth and dispersal (Hof and Bevers 2002). Then, we describe two types of reserve design problems that incorporate stochastic models of population dynamics. These are important because the fields of wildlife management and conservation biology have a long history of developing stochastic models of population viability, which help managers predict the likelihood that wildlife populations survive under various levels of habitat protection (Boyce 1992; Beissinger and Westphal 1998; Beissinger et al., this volume). There are two broad types of viability models: Demographic models predict the birth, death, and migration of individuals in one or more localized populations (e.g., Liu et al. 1995); and incidence function models predict the extinction of local populations and colonization of empty habitat patches (Hanski 1994). Both types of models incorporate uncertainty in one or more demographic parameters, and Monte Carlo methods are used to sample from the underlying distributions and simulate populations many times for different combinations of parameter values. Thus, stochastic population models yield probabilistic results, which are typically summarized by performance measures such as the probability that the ending metapopulation size exceeds a threshold or the expected number of surviving populations.

Metapopulation Size Problem.—A simple way to model change in the size of a metapopulation is to estimate the growth rate (per capita reproduction minus net mortality) of each subpopulation and a matrix of dispersal parameters that govern movement of individuals between subpopulations. Given these parameters, Bevers and Flather (1999) formulated a system of linear difference equations for metapopulation dynamics and explored the effects of patch size, number, and spatial arrangement on the size of hypothetical metapopulations. Because the model is a system of linear equations, it can be put into a linear programming model for site selection to maximize metapopulation size subject to budget constraints. Hof et al. (2001) described an application to black-tailed prairie dog (*Cynomys ludovicianus*) conservation in the Buffalo Gap National Grassland in South Dakota, USA. First, they identified 601 patches of prairie dog habitat covering approximately 20,000 ha and defined choice variables for the amount of each patch that is zoned for prairie dog colonies. Then, they developed a model of the prairie dog population in which each subpopulation grows exponentially until patch carrying capacity is reached, emigration is limited to subpopulations that exceed patch carrying capacity, and the number of dispersers that reach each patch depends on inter-patch distances. Finally, they explored the effects of budget constraints on total population size over an 8-year horizon and suggested priority locations for habitat expansion. While this reserve design model contains some basic elements of population dynamics, it ignores features of population models (e.g., age-dependent birth and mortality rates, density-dependent emigration rates, and parameter uncertainty) that are difficult to formulate in linear programs. Later we discuss ways to incorporate stochastic demographic models of population viability into reserve design problems.

Safe-Minimum-Standard Problem.—In the United States, the Endangered Species Act requires the U.S. Fish and Wildlife Service and other agencies to prepare recovery plans for threatened and endangered species. Recovery plans usually include population size goals assuming that species are viable when those goals are attained. Recognizing that population dynamics and species survival are uncertain, scientists have defined population viability in probabilistic terms as the likelihood of survival over some time period (e.g., Boyce 1992, Beissinger and Westphal 1998). We define population viability as a safe minimum standard—the likelihood that population size exceeds a minimum size target at the end of the planning horizon—and we assume that site selection affects the probability of exceeding the target (also see Millspaugh et al., this volume). Then, we can estimate the trade-off between higher probabilities and the costs of attaining them. These cost curves, first developed by Montgomery et al. (1994) and Haight (1995), quantify important components of the social costs and benefits of species protection.

Suppose we have a set of disjunct sites that can support subpopulations of an endangered species and a limited budget for habitat protection. By "disjunct," we mean that sites are physically separated from each other; however, individuals can move between sites. The objective is to determine the sites to protect to maximize the viability of the metapopulation. A metapopulation is considered viable if its size is greater than a predefined minimum population size (the safety standard). Because of uncertainty in population dynamics, population size at the end of the horizon is uncertain and the viability objective is probabilistic. As before, we define 0–1 decision variables x_j for all $j \in J$ for site protection. In addition, we define random variable $N(x)$ as the size of the metapopulation in ending period T as a function of the decision variables and parameter n as the target population in period T. The safe-minimum standard problem is

$$\text{Maximize}: \text{prob}[N(x) \geq n] \tag{28}$$

$$\sum_{j \in J} c_j x_j \leq B \tag{29}$$

$$x_j \in \{0, 1\} \qquad \text{for all } j \in J \tag{30}$$

The objective (Eq. 28) is to maximize the probability that the metapopulation exceeds a predetermined size target at the end of the management horizon subject to a budget constraint (Eq. 29) and binary restrictions on the decision variables (Eq. 30). A solution is a cost-effective set of sites to protect to maximize the likelihood of exceeding the population size target. By increasing the budget B and re-solving the problem, we can estimate the cost of attaining higher levels of certainty of attaining the target. The model explicitly recognizes that species survival is not certain and that the decision to save a species is not an all or nothing choice. Rather, the model measures the performance of a conservation plan in terms of the probability of attaining a population size target and allows determination of the cost of attaining higher probabilities.

This is a difficult optimization problem because the objective function is estimated using a stochastic population model, which typically has nonlinear relationships and random variables that cannot readily be put into classical integer and mixed-integer programming formulations. Instead, tools are needed to join simulation and optimization to find good approximations of optimal reserve design. One approach is simulation optimization in which the probability of metapopulation persistence is estimated via stochastic simulation until a suitable approximation of the optimal reserve design is found. A disadvantage of simulation optimization is computational intensity: Multiple replications of the stochastic population model may be required to obtain a useful estimate of the probability of persistence for each set of sites evaluated. Simulation optimization strategies are beginning to be developed and tested with incidence function models (Moilanen and Cabeza 2002) and demographic models (Haight and Travis 2008) of population dynamics. An excellent application of the safe-minimum-standard problem is a study of the cost of protecting old growth forest for northern spotted owl habitat in the Pacific Northwest (Montgomery et al. 1994).

Surviving Populations Problem.—In some cases, populations of an endangered species exist in disjunct sites that are isolated enough that migration between sites is inconsequential. If we have information for each site about the relationship between risk of population extinction and the amount of habitat in the site, we can formulate a model for determining the amount of habitat to add to each site to maximize the expected number of populations that survive over the management horizon. Here, the decision variable x_j is the amount of habitat to add to site j and the parameter a_j is the amount of already-protected habitat. In addition, we define $N_j(a_j + x_j)$ as a random variable for the population size in site j in ending period T as a function of the total amount of habitat in the site, and n as the minimum viable population size. Then, prob $[N_j(a_j + x_j) < n]$ is defined as extinction risk and the optimization problem

$$\text{Maximize}: \sum_{j \in j} 1 - \text{prob}\left[N_j(a_j + x_j) \leq n\right] \tag{31}$$

$$\sum_{j \in J} c_j x_j \leq B \tag{32}$$

$$x_j \geq 0 \qquad \text{for all } j \in J \tag{33}$$

is to maximize the expected number of populations that survive over the management horizon (Eq. 31) subject to a budget constraint on the total cost of added habitat (Eq. 32). The probability of extinction of each population depends on the amount of habitat, which is the sum of the already-protected habitat and the newly added habitat.

Haight et al. (2004*a*) used this formulation to address a problem of allocating a fixed budget for habitat protection among disjunct populations of the endangered San Joaquin kit fox (*Vulpes macrotis mutica*) in California to maximize the expected number of surviving populations. A key part of the problem is estimating prob$[N_j\,(a_j{+}x_j) \leq n]$ as a function of the amount of habitat in each site. They used response surface analysis in the following way. First, a stochastic demographic model of a disjunct kit fox population was used to predict extinction risk in 100 years in habitat patches of increasing size. For each patch area, the estimator of extinction risk was the percentage of 1000 independent simulations in which population size was less than 10 individuals in 100 years. Then, the predictions were used to estimate a relationship between extinction risk and patch area. The risk-area relationship was a logistic function estimated using a form of logistic regression called the minimum logit chi-squared method (Maddala 1983). Logistic regression describes a binary response as a function of one or more explanatory variables. In this case, the binary response was extinction or persistence of a population in a habitat patch, and the explanatory variable was patch area. The minimum logit chi-squared method of estimation is appropriate when there are multiple observations of the binary response for each level of the explanatory variable. Risk-area curves were estimated for each of eight populations and then incorporated into the optimization model (Eqs. 31–33). The results included priorities for reserve expansion under increasing budgets and a cost curve showing funding required for incremental increases in the number of surviving populations.

Discussion of Modeling Approaches

Reserve selection and design models provide guidance to planners about cost-effective ways to achieve wildlife objectives and trade-offs. The type of model to use depends on the scope of the problem, the management objective, and the information available. With an objective of maximizing the number of species within protected sites, reserve selection models provide information to decision makers about sets of sites that protect the most species within the budget for acquiring land, and the models provide the marginal cost of increasing the number of species protected. Sometimes the marginal cost of protecting the last species within the scope of the problem is very high (e.g., Fig. 5-2), which suggests that funding could be invested in other conservation projects with greater benefits. While reserve selection models provide a first-pass solution, they ignore reserve design features such as proximity, connectivity, and shape that may affect species dynamics and persistence. These design features can be included as objectives and analyzed in terms of their trade-offs with species representation under a given budget. A limitation of reserve selection and design models is their ignorance of species dynamics, and there is no guarantee that species will persist in the resulting reserve system. Reserve design models can be formulated with species dynamics, but they are complicated by the difficulties of adequately representing birth, death, and migration as functions of available habitat and by the computational intensity of finding optimal or near-optimal reserve designs. Nevertheless, applications have addressed reserve design problems for single species for which there is considerable knowledge of population dynamics.

A big limitation of reserve-based models is their assumption of a static time horizon: Decisions are made all at once, and habitat protection takes place rapidly before site degradation or loss. Researchers are beginning to address sequential site selection problems to optimize conservation objectives subject to budget constraints and uncertainties about site degradation and loss (Costello and Polasky 2004, Meir et al. 2004, Snyder et al. 2004, Turner and Wilcove 2006). The idea is to develop adaptive decision rules for selecting sites to protect depending on sites already protected, those currently available, and available funding. Decision rules like these can be compared with rules used in practice to see if efficiency gains can be obtained.

FUTURE DIRECTIONS

Government agencies and private organizations design and evaluate wildlife conservation programs based in part on their benefits and costs to society. While it is relatively easy to quantify the costs of conservation programs in terms of foregone economic activity, we are far from able to obtain definitive estimates of wildlife benefits. One place where we can make progress is gathering and analyzing data on participation in recreational activities related to wildlife in large landscapes. This is especially true for watching large mammals and bird species that form the basis of eco-tourism and eco-learning programs. Information on economic expenditures as well as the nonmarket values of such experiences could go far to help quantify the benefits of wildlife conservation efforts. Regional and national level analyses in the United States would be particularly helpful; information tends to be very general or anecdotal in nature. For wildlife-based ecotourism, most detailed studies focus on national parks or adventures in Africa and Latin America than on opportunities in places like the Midwestern U.S. (Krüger 2005).

Recognizing that estimates of benefits of wildlife conservation are not available, planners with well-defined conservation objectives evaluate the cost-effectiveness of alternative conservation plans and the trade-offs among their objectives using a variety of analytical models, including reserve selection and design models discussed previously. While reserve selection and design models focus on one important element of conservation planning, they ignore activities such as fire management, invasive species detection and eradication, and vegetation management that restore and enhance habitat for targeted species. In some cases habitat restoration is the only available management option because creating and expanding reserves is neither feasible nor affordable. Investment models with a wider range of conservation actions are needed to evaluate and prioritize reserve expansion versus other habitat restoration activities that are beneficial and possibly more cost effective. While such investment models are beginning to appear (e.g., Wilson et al. 2007), their success will depend on the participation of stakeholders and experts who help identify conservation objectives, threats to achieving those objectives, mitigation activities, economic costs, and local constraints on implementation.

In terms of the human dimensions of restoration, more research is needed on how to anticipate and work through conflicts that juxtapose restoration with other human values. This is especially true for restoration projects in urbanized landscapes. In places like Lake County, Illinois, discussed previously, trade-off modeling and conflict resolution and negotiation might help stakeholder groups better understand how options such as consolidation of acquisitions or restoration practices such as prescribed burning might be balanced with issues such as spatial equity in the distribution of open space or aesthetic considerations.

Finally, we need a better understanding of the patterns, drivers, and impacts of amenity migration as they pertain to wildlife conservation in large landscapes. Recent work examining the influence of housing density on landscape fragmentation (Hawbaker et al. 2006), bird populations (Lepczyk et al., 2008), and fire (Haight et al. 2004) is a good first step. Further interdisciplinary collaboration can merge this work with wildlife modeling efforts to look at potential impacts of housing and urban development on different types of species that depend on large landscapes.

SUMMARY

We described social and economic considerations for wildlife conservation planning in large landscapes. The social value of wildlife for recreational hunting provided an important justification for early landscape conservation efforts, but in more recent decades there has been a shift toward an appreciation of the value of wildlife for recreational viewing. However, human-wildlife conflicts have increased, and parcelization and development of open space provided by private forests, grasslands, pastures, and farms have inhibited wildlife conservation planning. Because people are concerned about the loss of open space, local governments and private land trusts have instituted policies to acquire land or conservation easements to preserve undeveloped land within or on the fringe of towns and cities. Planners have a variety of objectives for land acquisition, including wildlife habitat protection and restoration as well as economic efficiency. In response, biologists and economists have developed reserve selection and design models, which suggest cost-effective ways to protect open space to attain wildlife objectives. We describe three reserve-based modeling approaches to large-scale conservation planning: reserve selection models, reserve design models, and reserve design models with wildlife population dynamics. Models are presented with real-life applications and used to explain basic economic principles of cost-effectiveness, marginal cost, and trade-off analysis.

LITERATURE CITED

Animal and Plant Health Inspection Service. 2006. Brucellosis—Cattle and Bison. <http://www.aphis.usda.gov/vs/nahps/brucellosis/cattle.htm> accessed 29 December 2006.

Arthur, J. L., J. D. Camm, R. G. Haight, C. A. Montgomery, and S. Polasky. 2004. Weighing conservation objectives: Maximum expected coverage versus endangered species protection. *Ecological Applications* 14:1936–1945.

Beissinger, S. R., and M. I. Westphal. 1998. On the use of demographic models of population viability in endangered species management. *Journal of Wildlife Management* 62:821–841.

Bevers, M., and C. H. Flather. 1999. Numerically exploring habitat fragmentation effects on populations using cell-based coupled map lattices. *Theoretical Population Biology* 55:61–76.

Bishop, R. C., and M. P. Welsh. 1992. Existence values in benefit-cost analysis and damage assessment. *Land Economics* 68:405–417.

Boyce, M. S. 1992. Population viability analysis. *Annual Review of Ecology and Systematics* 23:481–506.

Boyd, D. P. 2003. *Conservation of North American bison: Status and recommendations*. Thesis, University of Calgary, Alberta, Canada.

Cabeza, M., and A. Moilanen. 2001. Design of reserve networks and the persistence of biodiversity. *Trends in Ecology and Evolution* 16:242–248.

Camm, J. D., S. K. Norman, S. Polasky, and A. Solow. 2002. Nature reserve site selection to maximize expected species covered. *Operations Research* 50:946–955.

Chavez, A. S., E. M. Gese, and R. S. Krannich. 2005. Attitudes of rural landowners toward wolves in northwestern Minnesota. *Wildlife Society Bulletin* 33:517–527.

Church, R. L., R. A. Gerrard, A. D. Hollander, and D. M. Stoms. 2000. Understanding the tradeoffs between site quality and species presence in reserve site selection. *Forest Science* 46:157–167.

Church, R. L., D. M. Stoms, and F. W. Davis. 1996. Reserve selection as a maximal coverage location problem. *Biological Conservation* 76:105–112.

Costello, C., and S. Polasky. 2004. Dynamic reserve site selection. *Resource and Energy Economics* 26:157–174.

Daily, G. C., editor. 1997. *Nature's services: Societal dependence on natural ecosystems*. Island Press, Washington, D.C., USA.

Decker, D. J., T. L. Brown, and W. F. Siemer, editors. 2001. *Human dimensions of wildlife management in North America.* The Wildlife Society, Bethesda, Maryland, USA.

Dizard, J. 2003. *Mortal stakes: Hunters and hunting in contemporary America*. University of Massachusetts Press, Amherst, Massachusetts, USA.

Fischer, D. T., and R. L. Church. 2003. Clustering and compactness in reserve site selection: An extension of the biodiversity management area selection model. *Forest Science* 49:555–565.

Fisher, A. C., and W. M. Hanneman. 1986. Option values and the extinction of species. *Advances in Applied Microeconomics* 4:169–190.

Foreman, D. 2004. *Rewilding North America: A vision for conservation in the 21st century.* Island Press, Washington, D.C., USA.

Gobster P. H., and R. B. Hull, editors. 2000. *Restoring nature: Perspectives from the social science and humanities*. Island Press, Washington, D.C., USA.

Gobster, P. H., and M. G. Rickenbach. 2004. Private forestland parcelization and development in Wisconsin's Northwoods: Perceptions of resource-oriented stakeholders. *Landscape and Urban Planning* 69:165–182.

Gobster, P. H., and T. L. Schmidt. 2000. The use of amenity indicators in anticipating private forest-land parcelization: A look at the Lake States' Northwoods. Pages 171-180 *in* L. A. DeCoster and R. E. Sampson, editors. *Fragmentation 2000: A conference on sustaining private forests in the 21st century.* The Sampson Group, Annapolis, Maryland, USA.

Haight, R. G. 1995. Comparing extinction risk and economic cost in wildlife conservation planning. *Ecological Applications* 5:767-775.

Haight, R. G., B. Cypher, P. A. Kelly, S. Phillips, K. Ralls, and H. P. Possingham. 2004. Optimizing reserve expansion for disjunct populations of San Joaquin kit fox. *Biological Conservation* 117:61-72.

Haight, R. G., S. A. Snyder, and C. S. ReVelle. 2005. Metropolitan open-space protection with uncertain site availability. *Conservation Biology* 19:327-337.

Haight, R. G., and L. E. Travis. 2008. Reserve design to maximize species persistence. *Environmental Modeling and Assessment* 13:243-253.

Hammer, R. B., S. I. Stewart, R. L. Winkler, V. C. Radeloff, and P. R. Voss. 2004. Characterizing dynamic spatial and temporal residential density patterns from 1940-1990 across the North Central United States. *Landscape and Urban Planning* 69:183-199.

Hanski, I. 1994. A practical model of metapopulation dynamics. *Journal of Animal Ecology* 63:151-162.

Harper, C., and T. Crow. 2006. *Cooperating across boundaries: Partnerships to conserve open space in rural America.* U.S. Forest Service, Report number FS-861, Washington, D.C., USA.

Hawbaker, T. J., V. C. Radeloff, M. K. Clayton, R. B. Hammer, and C. E. Gonzalez-Abraham. 2006. Road development, housing growth, and landscape fragmentation in Northern Wisconsin: 1937-1999. *Ecological Applications* 16:1222-1237.

Heberlein, T. A., and G. Ericsson. 2005. Ties to the countryside: Accounting for urbanites attitudes toward hunting, wolves, and wildlife. *Human Dimensions of Wildlife* 10:213-227.

Hof, J., and M. Bevers. 2002. *Spatial optimization in ecological applications.* Columbia University Press, New York, New York, USA.

Hof, J., M. Bevers, D. W. Uresk, and G. L. Schenbeck. 2001. Optimizing habitat location for black-tailed prairie dogs in southwestern South Dakota. *Ecological Modelling* 147:11-21.

International Wolf Center. 2006. IWC homepage. <http://www.wolf.org/wolves/index.asp>. Accessed 29 December 2006.

Kingsland, S. E. 2002. Creating a science of nature reserve design: Perspectives from history. *Environmental Modeling and Assessment* 7:61-69.

Krüger, O. 2005. The role of ecotourism in conservation: Panacea or Pandora's box? *Biodiversity and Conservation* 14:579-600.

LaGro, J. A. 1996. Designing without nature: Unsewered residential development in rural Wisconsin. *Landscape and Urban Planning* 35:1-9.

Lepczyk, C. A., C. H. Flather, V. C. Radeloff, A. M. Pidgeon, R. D. Hammer, and J. Liu. 2008. Human impacts on regional avian diversity and abundance. *Conservation Biology* 22: 405-416.

Liu, J., J. B. Dunning, and H. R. Pulliam. 1995. Potential effects of a forest management plan on Bachman's sparrow (*Aimophila aestivalis*): Linking a spatially explicit model with GIS. *Conservation Biology* 9:62-75.

Lopez, R. R., A. Lopez, R. N. Wilkins, C. C. Torres, R. Valdez, J. G. Teer, and G. Bowser. 2005. Changing Hispanic demographics: Challenges in natural resource management. *Wildlife Society Bulletin* 2005 33:553-564.

Maddala, G. S. 1983. *Limited-dependent and qualitative variables in econometrics.* Cambridge University Press, New York, New York, USA.

Maehr, D., R. F. Noss, and J. L. Larkin, editors. 2001. *Large mammal restoration: Ecological and sociological challenges in the 21st century.* Island Press, Washington, D.C., USA.

Manfredo, M. M., T. L. Teel, and A. D. Bright. 2003. Why are public values toward wildlife changing? *Human Dimensions of Wildlife* 8:287-306.

Manfredo, M. J., and H. C. Zinn. 1996. Population change and its implications for wildlife management in the new west: A case study of Colorado. *Human Dimensions of Wildlife* 1:62–74.

Margules, C. R., A. O. Nicholls, and R. L. Pressey. 1988. Selecting networks of reserves to maximize biological diversity. *Biological Conservation* 43:63–76.

Marianov, V., S. A. Snyder, and C. S. ReVelle. 2004. Trading off species protection and timber production in forests managed for multiple objectives. *Environment and Planning* 31:847–862.

Meir, E., S. Andelman, and H. P. Possingham. 2004. Does conservation planning matter in a dynamic and uncertain world? *Ecology Letters* 7:615–622.

Moilanen, A., and M. Cabeza. 2002. Single-species dynamic site selection. *Ecological Applications* 12:913–926.

Montag, J. M., M. E. Patterson, and W. A. Freimund. 2005. The wolf viewing experience in the Lamar Valley of Yellowstone National Park. *Human Dimensions of Wildlife* 10:273–284.

Montgomery, C. A., G. M. Brown, Jr., and D. M. Adams. 1994. The marginal cost of species preservation: The northern spotted owl. *Journal of Environmental Economics and Management* 26:111–128.

Nalle, D. J., J. L. Arthur, and J. Sessions. 2002. Designing compact and contiguous reserve networks with a hybrid heuristic algorithm. *Forest Science* 48:59–68.

Nelson, E. J., M. Uwasu, and S. Polasky. 2007. Voting on open space: What explains the appearance and support of municipal-level open space conservation referenda in the United States? *Ecological Economics* 62:580–593.

Noss, R. F., and A. Y. Cooperrider. 1994. *Saving nature's legacy: Protecting and restoring biodiversity.* Island Press, Washington, D.C., USA.

Önal, H., and R. A. Briers. 2002. Incorporating spatial criteria in optimum reserve network selection. *Proceedings of the Royal Society of London Series B* 269:2437–2441.

ReVelle, C. S., J. C. Williams, and J. J. Boland. 2002. Counterpart models in facility location science and reserve selection science. *Environmental Modeling and Assessment* 7:71–80.

Rodrigues, A. S. L., and K. J. Gaston. 2002. Optimisation in reserve selection procedures—why not? *Biological Conservation* 107:123–129.

Rothley, K. D. 1999. Designing bioreserve networks to satisfy multiple, conflicting demands. *Ecological Applications* 9:741–750.

Ruliffson, J. A., P. H. Gobster, R. G. Haight, and F. R. Homans. 2002. Niches in the urban forest: Organizations and their role in acquiring metropolitan open space. *Journal of Forestry* 100:16–23.

Ruliffson, J. A., R. G. Haight, P. H. Gobster, and F. R. Homans. 2003. Metropolitan natural area protection to maximize public access and species representation. *Environmental Science and Policy* 6:291–299.

Sampson, N., and L. DeCoster. 2000. Forest fragmentation: Implications for sustainable private forests. *Journal of Forestry* 98:4–8.

Schilling, D. A., A. McGarity, and C. ReVelle. 1982. Hidden attributes and the display of information in multiobjective analysis. *Management Science* 28:236–242.

Schroeder, H. W., P. H. Gobster, and R. Frid. 1993. *Visual quality of human-made clearings in Central Michigan conifers.* U.S. Forest Service, North Central Forest Experiment Station, Research paper NC-313, St. Paul, Minnesota, USA.

Simard, A. J., D. A. Haines, R. W. Blank, and J. S. Frost. 1983. *The Mack Lake Fire.* U.S. Forest Service, North Central Forest Experiment Station, General Technical Report NC-83, St. Paul, Minnesota, USA.

Snyder, S. A., R. G. Haight, and C. S. ReVelle. 2004. A scenario optimization model for dynamic reserve site selection. *Environmental Modeling and Assessment* 9:179–187.

Turner, W. R., and D. S. Wilcove. 2006. Adaptive decision rules for the acquisition of nature reserves. *Conservation Biology* 20:527–537.

U.S. Fish and Wildlife Service. 2002*a*. *2001 National survey of fishing, hunting, and wildlife-associated recreation.* U.S. Fish and Wildlife Service, Report FHW/01-NAT, Washington, D.C., USA.

U.S. Fish and Wildlife Service. 2002*b*. *2001 National survey of fishing, hunting, and wildlife-associated recreation—Wisconsin*. U.S. Fish and Wildlife Service, Report FHW/01-WI Rev, Washington, D.C., USA.

U.S. Forest Service. 2002. *The Prairie Plan (Record of Decision, Final Environmental Impact Statement, and Land and Resource Management Plan)*, Midewin National Tallgrass Prairie. U.S. Forest Service, Wilmington, Illinois, USA.

Williams, J. C., C. S. ReVelle, and S. A. Levin. 2005. Spatial attributes and reserve design models: A review. *Environmental Modeling and Assessment* 10:163–181.

Wilson, K. A., E. C. Underwood, S. A. Morrison, K. R. Klausmeyer, W. W. Murdoch, B. Reyers, G. Wardell-Johnson, P. A. Marquet, P. W. Rundel, M. F. McBride, R. L. Pressey, M. Bode, J. M. Hoekstra, S. Andelman, M. Looker, C. Rondinini, P. Kareiva, M. R. Shaw, and H. P. Possingham. 2007. Conserving biodiversity efficiently: What to do, where and when. *PLoS Biology* 5:1850–1861.

CHAPTER

Landscape Considerations for Conservation Planning on Private Lands 6

Jonathan B. Haufler and Brian J. Kernohan

To many, the image of conservation brings to mind visions of wild lands managed as national parks and forests for purposes of maintaining our natural heritage. Certainly, public lands play a significant conservation role, as these lands remain the primary habitat for a number of species such as grizzly bears (*Ursus arctos horribilis*) and wolverines (*Gulo gulo*). However, the reality is that private lands offer some of the greatest opportunities for conservation planning and management. Private lands comprise 70% of the land ownership in the United States excluding Alaska (U.S. Department of Agriculture 1997), and this percentage is >90% in many states in the eastern United States. Analyses of species listed under the Endangered Species Act (ESA; Natural Heritage Data Network 1993; Groves et al. 2000) or identified as imperiled by the Natural Heritage Data Base (Groves et al. 2000) reveal that private lands are critical to the maintenance of these species, with approximately half of all listed species not occurring on federal lands. Private lands were reported by Groves et al. (2000) to have at least one occurrence of over half of the imperiled species and two-thirds of the listed species in the United States.

Very few private lands are wild lands. However, many acres exist as working lands, with private forestlands and ranchlands providing large areas where conditions can still resemble native ecosystems in the area and provide many conservation benefits. However, knowing what is on these lands, modeling their future conditions, and predicting their conservation contributions are problematic for a number of reasons. Haufler and Kernohan (2001) discussed various considerations for management of private lands for ecological purposes. In this chapter, we expand on these considerations with particular emphasis on conservation planning and modeling at landscape scales. Our experience with private land conservation is from the United States; therefore, the discussion presented here does not include considerations from other countries.

CHALLENGES TO PRIVATE LAND PLANNING FOR CONSERVATION

Despite the large amount of private land across the United States, conservation planning on these lands presents significant challenges. One significant challenge is the lack of information on what is present in terms of conservation elements on private lands, a need exacerbated by concerns for private property rights and proprietary information. Another significant challenge is that private landowners frequently have a distrust of planning conducted by and promoted to them by agencies, both state and federal. However, few private landowners have the technical skills to conduct conservation planning on their own, and funding mechanisms from state or federal sources to assist private landowners in conducting their own planning have been largely nonexistent.

Conservation planning for wildlife has largely focused on the needs of individual species of concern or special interest. Landowners have been encouraged to address the needs of a number of species, often with differing requirements, all on their individual properties. This has led to confusion by private landowners, causing them to question incentive programs offered to conduct habitat improvements. Fear of regulatory constraints if listed species or species of concern move onto private lands has limited involvement by private landowners for conservation planning. Complexities in legal agreements designed to help balance conservation objectives with economic and other objectives of private lands often prove problematic and push landowners toward avoidance of conservation initiatives rather than participation in these efforts.

Current recognition by ecologists for the need for conservation planning and management at landscape scales to address complex conservation issues is a relevant concern to managers. However, to private landowners these concerns may not be a significant inducement, particularly as they focus on their lands and economic needs, often with limited involvement or even competition with their neighbors and with distrust of agencies and their agendas. Collaborative initiatives that address these landscape complexities must overcome these landowner concerns if effective large-scale conservation planning is to be supported.

Each of these challenges can be addressed for private landowners, but only if each is recognized as a legitimate concern and if landowners are considered equal partners in conservation planning initiatives. We discuss each of the challenges in more detail and explore possible solutions in the following sections.

Private Property Rights

All private landowners have an economic investment in their land, and most expect to maintain an economic value or return from their land (Haufler and Kernohan 2001). To realize an economic return, private landowners are often strident when it comes to private property rights. However, private landowners typically value a healthy environment, wish to contribute to the maintenance of biological diversity, and understand that their ownership is a part of a larger landscape. Because of the perceived conflict between economic return and conservation, private landowners are wary of becoming involved in conservation planning across landscapes, particularly when government agencies are involved. Economic investment, along with diverse views of the extent of reasonable government intervention in land-management decisions, has led to recent debates on property rights (Haufler and Kernohan 2001).

It should be recognized that private landowners have little obligation to engage in conservation efforts other than limited regulatory constraints. Excluding land-use conversions to development or other similar changes, the constraints currently imposed on private landowners are species or site specific. For example, the ESA requires landowners to prevent "take" of any threatened or endangered species, and forest practices acts regulate site-specific forest management activities. Although these types of constraints regulate management on private property, neither form of constraint removes the rights of private property owners or requires a landowner to engage in landscape-level conservation planning. In contrast, public land management agencies have guidelines, regulations, and legislation to guide conservation planning on public lands. Recognizing these differences, the effectiveness of conservation planning across large, mixed-ownership landscapes will require all participants to recognize and respect private landowner objectives and property rights.

Funding of Privately Led Conservation Planning

The prevailing view of conservation planning on private lands in the United States is that such planning will be conducted by federal or state agencies with funding provided to these agencies for that purpose. The results of planning are then provided to landowners so that they can put management practices in place on their lands that meet the identified planning objectives. A number of funding programs exist to support on-the-ground application of conservation practices on private lands. For example, Farm Bill practices included in the Wildlife Habitat Incentive Program, Wetland Reserve Program, and Environmental Quality Incentive Program can produce conservation benefits for wildlife (Natural Resource Conservation Service 2006). However, with very few exceptions, no funding programs exist at the federal level to pay private landowners or groups to conduct their own planning. The assumption is that landowners do not have the knowledge or abilities to conduct planning. Therefore, it is assumed that planning funds should go to the federal agencies or in some cases state agencies, where the planning expertise is perceived to exist.

This view has some merits, in that it encourages federal agencies to provide technical assistance to private landowners and consider the broader mixed-ownership landscapes that occur in many areas. However, it overlooks several key concerns. First, many private landowners, particularly those owning sizable working ranchlands and forestlands are often distrustful of government agencies. They may strongly resist any efforts from agencies to direct what happens on their lands. They often resent the assumption that they do not know what is best for their lands and do not believe that agency personnel will provide information that will be of value to them.

A recent government focus has been the promotion of collaborative conservation initiated from the grassroots level. However, the allocation of federal funding has not shifted to support this approach. A few programs, such as the Conservation Innovation Grants of the Farm Bill, have provided some funds for nongovernment-led planning of private lands, but these are relatively small amounts (Natural Resource Conservation Service 2006). While a detailed review of all 50 state programs was not conducted, generally, the state view is the same or even more biased toward agency-conducted planning rather than landowner-led planning. For example, a majority of the recently completed State Comprehensive Wildlife Conservation Plans (required for each state to continue to receive state wildlife grant funding from the federal government) included private lands in the planning process in only a token manner through invitations to Farm Bureau representatives or other landowner-focused organizations. An exception to this has been the sage grouse (*Centrocercus urophasianus*) working groups formed under the direction of the Western Association of Fish and Wildlife Agencies that have been established within regions of each of the 11 states involved in addressing the needs of this species. This program has included private landowners as part of each working group. However, even this program is top-down driven by the state agencies and designed to conform to their approach to addressing the needs of this species.

Until the ability and value of having landowners lead conservation planning initiatives is recognized and direct funding to assist such initiatives is available, many landowners will be resistant to government-led conservation programs. While many landowners will need government assistance to complete planning projects, the prevailing attitude that landowners cannot lead planning efforts is a major impediment to acceptance of government conservation programs.

Engagement of Private Landowners

Another challenge of conservation planning across large landscapes is engaging private landowners in conservation planning. Recognizing that private landowners own a significant amount of land important for biodiversity conservation throughout the United States, the need to engage them in conservation planning efforts is not a trivial matter. Reasons why private landowners might engage in conservation planning or implementation include (1) the desire to be good environmental stewards, (2) regulatory relief, and (3) economic benefits. As mentioned previously, private landowners typically value a healthy environment and therefore are often inclined to help create such an environment by being good stewards of their land. For private landowners that own large acreage (e.g., large industrial or private forest management companies), there is often a recognition that they have a responsibility to manage their lands in a manner that is beneficial to the communities in which they operate, much like a public land management agency would view management of public lands held within their trust.

In contrast to "doing the right thing" by being good environmental stewards, private landowners have several admittedly self-serving reasons to engage in conservation planning including regulatory relief and economic benefit. When engagement in conservation efforts can clearly result in relief from regulatory pressure (e.g., ESA), then private landowners are more inclined to participate. An example of this is enrollment of private land into a statewide safe harbor agreement for red-cockaded woodpeckers (*Picoides borealis*) in Louisiana (Louisiana Department Wildlife and Fisheries 2006) that offers assurances that landowners who provide habitat for this species would not then be restricted from normal economic activities should the species occupy their land. Similarly, the ability to realize economic return from involvement in a conservation planning effort could provide ample reason for private landowners to engage. Economic benefit might come by way of a receipt of a "social license" to operate because of their participation, or the effort might actually identify management activities that benefit the environment as well as provide a return on the investment for private landowners. Current discussions of the Cooperative Sagebrush Initiative (www.sandcounty.net) are investigating mechanisms to pay private landowners for improvements to sagebrush (*Artemisia* spp.) ecosystems that could provide off-site mitigation benefits.

Possibly the greatest challenge to engaging private landowners in conservation across large landscapes is the need to balance conservation and production through working landscapes. Private landowners might engage for the reasons stated previously, but planning for working landscapes whereby a balance is achieved between conservation and production is critical. Consider the United States as a mosaic of wild lands, urban lands, and multipurpose lands. Wild lands (e.g., wilderness, national parks) are effectively being "managed," through legislation and regulation that typically limits the management that occurs. Similarly, urban land serves a purpose, but this purpose is not to contribute to broad biodiversity conservation objectives. Consequently, what is left is the space in between that consists largely of private land intermixed with public land with a multipurpose objective.

Proprietary Information

Top-down regulatory approaches continue to mandate specific responses from private landowners in land-management practices, but typically result in increased animosity toward regulatory agencies as well as toward the environmental feature that is the focus of the regulation (Haufler and Kernohan 2001).

A secondary fallout of this model is the lack of trust in public agencies when it comes to sharing information. For example, the ESA creates a disincentive for private landowners to divulge the location of a threatened or endangered species because of fear that regulatory pressure might result. Consequently, when a private landowner becomes aware of the presence of a sensitive plant or animal species, the current regulatory environment suggests that it is best to keep that information out of the public sector for fear that the information might be used to limit economic return. This attitude extends to the point of private landowners being unwilling to allow government agencies to have access to their lands for surveys or inventory work. While advances in remote sensing are providing new sources of information with increased accuracy that include information on private lands, major gaps in information exist. Current remote sensing cannot gather information such as the composition of grasslands on private ranchlands or understory characteristics of forests on private industrial forestlands.

Data on flora are not as sensitive as data on fauna because vegetation does not carry the same level of regulation on private lands through the ESA. However, the potential loss of competitive advantage/market share is a concern, particularly within the forest industry. Industrial companies do not want competitors to know what types and amounts of forest inventory exist on their lands, information that might lead to competitive strategies concerning supply and production of certain types of forest products. Similarly, information such as inventories of commercial trees and commercial production information is subject to antitrust laws designed to prevent collusion in the market place (Thompson et al. 2004). Such laws and regulations restrict the amount of proprietary information that can be shared among competitors. Consequently, modeling landscapes and landscape management that includes private landowners are challenged by reluctance to share proprietary data and proprietary harvest schedules in the case of the forest industry. The challenge is to create mechanisms whereby information from private landowners can be shared for purposes of planning and management that does not expose them to management risks, jeopardize their rights, or subject them to legal penalties.

Focus on Species Management

The vast majority of conservation planning has focused on the needs of single species or various groupings of species. The concept of coarse filters (ecosystem focus) and fine filters (species focus) as two ways of addressing conservation planning was introduced over 20 years ago (The Nature Conservancy 1982), but effective use of coarse filters has not been widely used to date. This is due to several reasons. First, agencies have interpreted the ESA to focus on species. Their resulting emphasis on management of listed species or species of concern has overridden other conservation planning efforts. Second, most biologists have been trained with a strong focus on the species level of biological organization, with little emphasis on community and landscape levels of organization. Consequently, biologists have the greatest comfort level working at the species level and have generated the greatest amount of information on the needs and management of species as opposed to information on ecosystems and landscapes. In contrast, foresters and range conservationists have had more focus on community and landscape dynamics, but typically for production of desired forest and range products, not as it relates to providing for the needs of species or for biodiversity conservation.

Private landowners are typically more in tune with the grass, shrub, and forest ecosystems that comprise their working landscapes than they are with the fish and wildlife using these lands. They hear about concerns for various species, but often do not have any idea what the species even looks like or why it might be important to maintain. They may be approached by private conservation organizations, suggesting that they manage their lands for a number of species of concern. Many times, these species may have habitat requirements that conflict, leaving landowners wondering what is being asked of them. The range conservationists or foresters that a landowner may be working with for economic objectives may also be unclear about what conditions are being sought by the biologists. An often heard statement by these land managers is "Just tell me what conditions you want, and I can produce them," to which the answer from a biologist may be "it's not that simple; we need lots of different conditions." Finding ways to simplify conservation planning needs and putting them in terms that engage foresters, range conservationists, and landowners is a challenge that could yield increased conservation dividends. The role of landscape models discussed in this book is crucial in this process.

Mixed Ownerships

The preceding challenges, while applicable on mixed-ownership landscapes, are primarily concerns for private lands. Mixed-ownership landscapes add further complications. Where checkerboard ownership patterns occur, what happens on one parcel of land will have an influence on adjoining parcels of lands, at various scales of influence. For example, at broad scales, Stribley and Haufler (1999) found that the conditions in the surrounding landscape determined the probability of the presence of cowbirds (*Molothrus ater*), with site conditions secondary to the landscape influences. At mid-scales, where sections of land (e.g., 250 ha) occur in a checkerboard ownership pattern, habitat requirements for many species are dependent on or influenced by the availability of habitat variables across multiple ownerships, meaning that aggregates of land parcels will be required to provide the composite needs of many species (Freemark et al. 2002, Wiens 2002). In addition, for adjoining parcels, concerns may exist that what is present or occurs on one parcel can move to or influence the adjacent parcel. Prairie dogs (*Cynomys* spp.) occurring on federal lands may spread onto adjacent private lands, causing landowners to seek their removal from the federal lands. Conversely, noxious weeds occurring on some private lands may be a source for movement onto adjacent federal lands, stretching the ability of federal funding to control this problem. Finally, especially for rangelands, federal lands may be an important component of working landscapes, with grazing permittees using the federal lands in pastures that are of mixed ownership. In these areas, balancing economic needs with conservation objectives will require broader landscape plans that integrate the two objectives in workable solutions.

An additional challenge of mixed-ownership landscapes is providing appropriate agreements and mutual assurances across ownerships. Federal lands are managed under regulations that require specified processes, public input, interagency reviews, and similar bureaucracy. Private lands seeking to be involved in mixed-ownership conservation planning operate under very different requirements and utilize different types of agreements. Finding ways for private and federal lands to effectively coordinate and enter into parallel agreements is a challenge. The time lags and complexities of federal processes that involve extensive public input and review, while important for management of these lands, can strain cooperative relationships with private landowners who can make management decisions more quickly.

SOLUTIONS TO INCREASE EFFECTIVENESS OF CONSERVATION PLANNING

Our discussion of solutions to increase effectiveness of conservation planning on private lands focuses on a preferred approach and mechanisms to implement the approach. An ecosystem diversity approach is described that provides efficiencies in planning while effectively dealing with proprietary information concerns and concerns with single species management. Mechanisms explored to facilitate an ecosystem diversity approach include increasing incentives while recognizing regulatory constraints.

Use of an Ecosystem Diversity Approach

Haufler and Kernohan (2001) identified several principles for land management in mixed-ownership landscapes that are applicable to private lands. Two of these they termed the "coarse filter principle" and the "ecological site and historical disturbance principle." The coarse filter principle emphasized the need to use a planning approach that focused primarily on ecosystem diversity rather than on individual species. The ecological site and historical disturbance principle expanded on the coarse filter principle to provide some direction for its implementation.

There are various reasons why these principles are applicable to conservation planning for private lands. First, there are far too many species to plan for individually, and the complexity of species complicates conservation planning, especially on private lands (Haufler and Kernohan 2001). Private landowners find management difficult when presented with a list of species for which management actions are desired, particularly when these species present different and even conflicting habitat needs. For example, ranchers willing to consider conservation planning of their grasslands might be provided habitat information about long-billed curlew (*Numenius americanus*), upland sandpiper (*Bartramia longicauda*), chestnut-collared longspur (*Calcarius ornatus*), lark bunting (*Calamospiza melanocorys*), McCown's longspur (*Calcarius mccownii*), mountain plover (*Charadrius montanus*), and short-eared owl (*Asio flammeus*). Each of these species has different habitat requirements, and some have dramatically opposing habitat requirements, such as mountain plovers and chestnut-collared longspurs. Mountain plovers occupy very open prairie sites with little vegetation, while chestnut-collared longspurs prefer sites with varying densities and heights of grasses. Producing these conditions would require substantially different types of management. The rancher may be interested in being a good environmental steward, but may be confused and frustrated by the uncertainty of what management is desired. Consequently, the lands that might have been included in a conservation initiative are not.

To address the number of species potentially required under fine filter approaches, researchers and managers have proposed a number of fine filter substitutes or surrogates to simplify the conservation planning process (Noon et al., this volume). These surrogate methods use selected species to address the needs of a broader group of species. For example, sage grouse have been suggested as an "umbrella" species for conservation planning for sagebrush ecosystems. Lambeck (1997), Noon and Dale (2002), and Groves (2003) described different types of species proposed for surrogate conservation planning. Groves (2003) listed the following categories of surrogates: declining or at-risk species (threatened, endangered, and imperiled), endemic species, flagship species, umbrella species, focal species, keystone species, and indicator species. Noon and Dale (2002) listed the additional categories of ecological engineers, link species, and phylogenetically distinct species. Carignan and Villard (2002) listed dispersal-limited species, resource-limited species, process-limited species, and species linked with specific habitat features. Recent research has investigated how well these substitutes or surrogates address the objectives of providing for all species. Numerous difficulties and limitations of using species groupings or surrogates for conservation planning have been reported (Flather et al. 1997; Niemi et al. 1997; Pearson and Carroll 1998; van Jaarsveld et al. 1998; Carroll et al. 2001; Fleishman et al. 2001, 2002; Lawler et al. 2003; Su et al. 2004). For example, Fleishman et al. (2001) evaluated the use of a set of umbrella species for selection of conservation areas and found that these were no more effective than random species when used as surrogates for cross-taxon representation.

All the preceding reasons make fine filter approaches to conservation planning problematic, both for the scientific community and for private landowners. As an alternative, coarse filter, or ecosystem diversity approaches offer many advantages (Haufler et al. 1999). While it is not possible to describe, characterize, and track the habitat needs of all species in an area, it is feasible to classify, describe, and track ecosystem diversity in a planning landscape. Hughes et al. (2000) and Vos et al. (2002) discussed how use of ecosystem approaches is the direction in which conservation planning is generally heading. Various tests of coarse filter strategies have shown that they can be effective for biodiversity conservation planning (Nichols et al. 1998, Wessels et al. 1999, Ben Wu and Smeins 2000, Kintsch and Urban 2002, Oliver et al. 2004). However, ecosystem diversity approaches are not without their critics, as both Lindenmayer et al. (2002) and Noon et al. (2003; this volume) have questioned the use of landscape surrogates for addressing species needs and distributions.

While the scientific evaluation of coarse filter and fine filter approaches continues, coarse filter approaches offer advantages for use on private lands. One of these advantages is that landowners can often relate better to objectives described in ecosystem diversity terms than in terms of species habitat. For example, stating that a grassland be composed of a certain mix of grass and forb species and that within a specified area at least some percentage of the grassland has at least a minimum grass height maintained throughout the growing season is a description that a rancher can understand and produce. The economics of producing this condition can be identified and appropriate incentives provided to the landowner for producing these conditions. Similarly, a forest landowner can understand a prescription to apply on a certain type of site to maintain specific tree species in specified size classes and densities, and with certain understory characteristics. Further, focusing on providing appropriate ecosystem diversity opens up additional funding opportunities for on-the-ground treatments, including various practices within Farm Bill programs including prescribed burning, herbicide control of invasive species, and seeding with native species.

Another advantage to an ecosystem diversity focus is that in its basic form, it only requires mapping of private lands to an appropriate classification of ecosystem diversity. While this can be a complex task, it can preclude the need for specific information on occurrences of species of concern or detailed information such as forest inventory data. Landowners might be much more amenable to allowing this level of information to be provided or collected for their lands, allowing for more effective conservation planning and protection of proprietary information.

The examples discussed in following sections both utilize the same ecosystem diversity approach, one that has the goal of providing adequate representation of all native ecosystems in an area (Haufler et al. 1996, 1999). It identifies ecosystem diversity as a component of the differences in ecological sites in a planning area (abiotic factors), and the role that historical disturbance regimes played in maintaining various plant communities and associated animal use of these communities over time. As such, it relies on development of a historical reference of ecosystem diversity, and compares current conditions to this reference. While the goal is not a return to historical conditions, the base assumption is that historical conditions were what the biodiversity of the area evolved with and adapted to. This ecosystem diversity approach can provide a scientifically based method of addressing biodiversity conservation in an effective and efficient manner. While the authors advocate use of this approach, they acknowledge that other ecosystem diversity approaches may also be effective in addressing many of the challenges for private lands discussed previously.

Increased Incentives Within a Regulatory Environment

Haufler and Kernohan (2001) suggested that regulation has a role in management of private lands for conservation, and possibly a central role when it comes to providing for the good of society (e.g., quality of waters downstream from private property). The debate is not whether private lands conservation can fit within a regulatory environment, but rather over how much the rights and needs of society should be dominant or subservient to the rights of property owners (Haufler and Kernohan 2001). Increased incentive programs and voluntary action are likely to produce greater conservation gains in the long term than trying to force solutions through regulation. Increased incentives partially solve private property rights, funding, and engagement challenges.

Various incentive programs exist to reduce the fear of regulatory restrictions, particularly under the ESA. Programs such as Habitat Conservation Plans, Safe Harbor Agreements, and Candidate Conservation Agreements with Assurances are designed to provide assurances to private landowners that their involvement in conservation initiatives either will reduce regulatory constraints or will not result in increased regulatory restrictions. However, even with efforts at streamlining these programs, they remain complex, time consuming, costly, and difficult to complete for all but the biggest and/or wealthiest landowners. Additional simplification of these programs and development of new programs that make involvement of landowners even easier would go a long way toward reducing the fear of increased regulatory constraints. Additionally, conservation planning that focuses on ecosystem diversity needs to be able to fit more easily into assurance programs. At present, species must be identified, and their individual conservation status assessed in order for assurance programs to be applicable. Incorporating ecosystem diversity approaches into assurance programs would encourage involvement of additional private lands.

Two levels of conservation planning need to be recognized for private lands. One level involves maintaining lands in uses that allow them to provide conservation benefits as specific conservation lands, or as lands maintained in working landscapes. The primary focus is to keep lands from being converted to development. Conservation easements are a powerful tool that can be used to achieve this objective. The second level of conservation planning addresses the question of whether or not the right conditions exist on conservation or working lands to meet the conservation objectives. This is where the ecosystem diversity approach, discussed previously, comes into play. Where inadequacies in conditions are identified (e.g., adequate representation of specific ecosystems is lacking), incentives can be utilized to produce or maintain the desired conditions within the planning area.

Conservation easements are a flexible tool for addressing the need to maintain lands in wild and working landscapes. Conservation easements are either voluntarily sold or donated by private landowners and constitute legally binding agreements that limit certain types of uses or prevent development from taking place now and in the future (The Nature Conservancy 2006). Conservation easements are an incentive for private landowners to contribute to the conservation of natural resources while allowing them to retain certain property rights. A central attribute of easements is that their restrictions and terms can be designed to fit the needs of the underlying landowner and the easement holder so long as they retain a public purpose or intent (The Nature Conservancy 2006). Conservation easements have become an important incentive-based tool because they can address many of the challenges facing private landowners when it comes to contributing to conservation planning across landscapes while meeting economic objectives. The Nature Conservancy (TNC) alone reports conservation easement activity in 20 states across the United States (The Nature Conservancy 2006). Better understanding of the use and values of conservation easements is needed by private landowners, particularly by some in the ranching community who have seen these tools used by conservation organizations to leverage conservation concessions from landowners during times of economic difficulty, and may not recognize their value as voluntary tools.

Incentives to produce desired conditions on private lands within planning landscapes can take a variety of forms. Existing incentive programs need to be strengthened so that increased funding mechanisms are available to private landowners (Haufler and Kernohan 2001), particularly those willing to provide for biodiversity. A number of incentive programs exist to provide assistance in producing on-the-ground management, but are often inadequate to meet the demands for conservation incentives. Examples of incentive programs include various Farm Bill programs administered by Natural Resource Conservation Service (NRCS 2006) such as the Wildlife Habitat Incentive Program, the Environmental Quality Incentive Program, the Stewardship Incentive Program, the Grassland Reserve Program, and the Wetland Reserve Program; programs administered by the U.S. Fish and Wildlife Service (USFWS 2006) such as the Private Stewardship Program, the Partners Program, and the Landowner Incentive Program; programs administered by state agencies (e.g., Habitat Improvement Grants), programs administered by the National Fish and Wildlife Foundation, and other grants and programs administered by private organizations. Typically, the availability of incentive funds is predicated on a cost-share strategy (e.g., Stewardship Incentive Program [Natural Resource Conservation Service 2006]), which could be a limitation to a private landowner who may lack even the small amount of money needed to complete the program. Some programs, such as the Stewardship Incentive Program, require approved Forest Stewardship Plans (Natural Resource Conservation Service 2006) in order to receive funding.

A current limitation of the on-the-ground incentive programs is that few of these effectively link to conservation planning at landscape scales. For example, most Farm Bill practices (Natural Resource Conservation Service 2006) are targeted at individual farms or ranches, with little regard for conservation needs across a broader landscape. While the Conservation Security Program (Natural Resource Conservation Service 2006) is designed to look at conservation within watersheds, the bulk of the Farm Bill programs, while not precluding use of broader landscape planning, do not encourage such planning. Most other on-the-ground programs operate in a similar manner. New incentives are needed that place a premium for practices conducted that directly support conservation planning conducted at landscape scales. Increasing the effectiveness of incentive programs will require adjustments to funding mechanisms and increased availability of technical support for private landowners, both by funding private landowner-led planning and through increased agency technical support to landowners interested in conducting conservation planning.

Other Solutions to Private Lands Challenges

Several solutions to private land conservation challenges can be implemented by federal agencies. One of the challenges identified previously is the need to fund private landowner planning efforts. A few sources of funds are currently available, especially for those engaged in watershed planning, where various Environmental Protection Agency programs fund water-related projects. However, more funding directed at privately led conservation planning is needed. Along with this is the need to change the prevalent view in agencies that they are the ones that do the planning and that landowners should simply accept what they are told. While technical assistance from agencies will continue to play a critical role in conservation planning, the rights, ability, and importance of landowner-led conservation planning needs to be respected and supported.

The ability to enter into parallel agreements for private and public lands in mixed-ownership landscapes needs strengthening. While the need for regulations and guidelines for planning implementation on federal lands is recognized, processes must be developed to allow federal planning and private land agreements to proceed in a more coordinated manner. Current efforts at streamlining federal planning and project implementation have primarily focused on excluding various planning or projects from review processes, using tools such as categorical exclusions. Such exclusions do little to help coordinate planning in mixed-ownership landscapes. New ways of allowing closer links and faster processing for joint private and public land conservation planning in mixed-ownership landscapes are needed that do not preclude appropriate public input and review.

MODELING OF PRIVATE LANDS

In this chapter we have focused on the challenges of conducting conservation planning involving private lands in the United States. Models used in conservation planning such as ecosystem dynamics models or species assessment models (Beck and Suring, this volume; Larson et al., this volume) should operate equally well for any ownership. The challenge is to obtain the data from private lands required to drive such models, data that may be more difficult to obtain for the reasons discussed previously.

Assumptions used for modeling outputs concerning types of land uses or management may differ between public and private lands. For example, Spies et al. (2007) assessed some of the ecological and socioeconomic effects of recently enacted forest management policies in the 2.3 million ha Coast Range Physiographic Province of Oregon and made various assumptions about the different management that would occur on public and private lands. While such differences will occur, the focus of this chapter has been on approaches that allow private lands to be more effectively integrated into conservation plans.

Examples of Private Land Conservation Planning at Landscape Scales

Private Forestlands.—Addressing the challenges described in the preceding sections and creating workable solutions is not an easy task for private forestland owners. A diverse suite of entities including small woodlot owners, family-owned timber companies, timberland investment management organizations (TIMOs), public and private real estate investment trusts (REITs), and integrated forest products companies own private forestland in the United States. Regardless of who owns the forests, landscape considerations for conservation planning on private lands remain a challenge. Most examples of conservation planning at landscape scales in forested landscapes come from large, integrated forest products companies or large, forest investment/management companies (e.g., TIMOs and REITs) (Haufler and Kernohan 2001, National Council for Air and Stream Improvement 2001, Loehle et al. 2002). Forest products companies often undertake landscape planning to demonstrate the compatibility of forest management with ecological functions and to explore viable alternatives to restrictive regulations (Loehle et al. 2002). Forest products companies have commonly used habitat conservation plans at the federal or state level to accomplish these goals. Loehle et al. (2002) reported that three companies that developed management plans at the landscape level in the Pacific Northwest found their planning efforts expensive and difficult to develop and implement, with unclear benefits. Planning efforts, whether related to federal or state laws, should have clear achievable goals, be based on scientific principles and credible data, and be driven by realistic monitoring and adaptive management.

As an alternative to habitat conservation plans and other regulatory-based tools, Haufler and Kernohan (2001), National Council for Air and Stream Improvement (2001), and Haufler et al. (2002) described an approach to conservation planning at landscape scales developed by what was then Boise Cascade Corporation, a publicly traded integrated forest products company. Described as private landowner-led collaborative programs, these projects followed the steps of an ecosystem management process described by Haufler et al. (1996, 1999, 2002) and were established in central Idaho, south-central Washington, and northern Minnesota. All three projects used the ecosystem diversity approach described previously and incorporated collaboration considerations appropriate for projects in mixed-ownership landscapes (The Keystone Center 1996, Kernohan and Haufler 1999, Haufler and Kernohan 2001).

Creation of viable planning processes in mixed-ownership landscapes was achieved largely because of the leadership of a private company. By removing the tendency of public agencies to manage using a fine filter approach through regulatory action, multiple stakeholders were able to come together and conduct ecological assessments using an ecosystem diversity approach. The ecosystem approach allowed data to be summarized and presented in a manner that was not threatening to private landowners; private property rights were respected and proprietary information was controlled. In the case of the Minnesota project, the ecosystem approach described previously paved the way for the Minnesota Forest Resources Council (2006) to effectively implement a landscape-based planning and coordination program throughout the state of Minnesota that addressed many of the challenges facing conservation planning on private lands.

Thunder Basin Grasslands Prairie Ecosystem Association.—The Thunder Basin Grasslands Prairie Ecosystem Association (Association) is a nonprofit organization composed of landowners (ranchers and energy production companies) who manage over 300,000 acres within a delineated 945,000-acre landscape in eastern Wyoming (Fig. 6-1). The Association formed to develop a collaborative, responsible, commonsense, science-based approach to long-term management of their lands. The landscape is mixed-ownership, containing private ranchlands, energy production company lands, Thunder Basin National Grasslands, Wyoming state lands, and Bureau of Land Management (BLM) lands. The area, recognized as one of the most ecologically significant grasslands in the United States, supports some of the largest remaining populations of black-tailed prairie dogs (*Cynomys ludovicianus*), mountain plovers, ferruginous hawks (*Buteo regalis*), burrowing owls (*Athene cunicularia*), upland sandpipers, loggerhead shrikes (*Lanius ludovicianus*), long-billed curlews, and many other grassland-associated species. The area is also a potential site for reintroduction of black-footed ferrets (*Mustela nigripes*). The area has a solid population of sage grouse and has large herds of pronghorn antelope (*Antilocapra americana*), mule deer (*Odocoileus hemionus*), white-tailed deer (*Odocoileus virginianus*), and elk (*Cervus elaphus*) that support considerable recreational hunting. Management

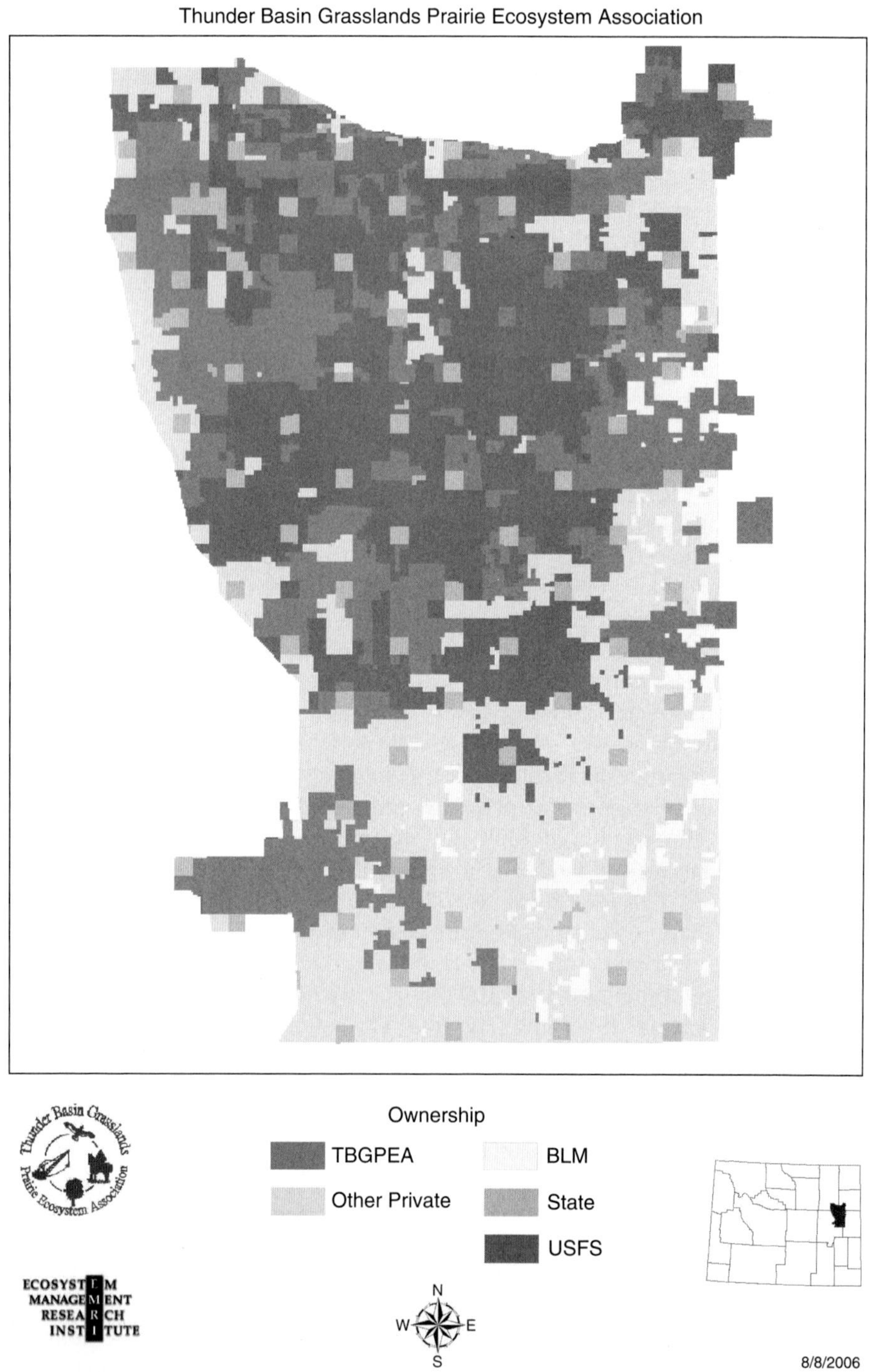

FIG. 6-1 (See color insert.)

The Thunder Basin Grasslands Prairie Ecosystem Association (TBGPEA) manages over 300,000 acres within a delineated 945,000-acre landscape in Converse, Campbell, Weston, and Niobrara Counties in eastern Wyoming, USA.

controversies including appropriate grazing practices, intensive and expanding energy developments, prairie dog management, and potential land conversions place the future productivity and ecological integrity of these lands at risk. The landowners of the Association were interested in conducting their own conservation planning for the landscape. They rejected offers from state and federal agencies as well as from conservation organizations to do the planning for them. However, they recognized that the mixed-ownership pattern necessitated cooperative management among private landowners and state and federal agencies if both effective conservation and economic vitality were to be achieved. The Association has cooperated with state and federal agencies in its efforts to produce long-term management plans. The Association has cooperative arrangements with the USFWS, U.S. Forest Service (USFS), BLM, NRCS, Wyoming Game and Fish, and Wyoming Department of State Lands. The Association has been developing an Ecosystem Diversity Plan and a Prairie Dog Plan that, when combined, could provide the basis for a Candidate Conservation Agreement with Assurances with the USFWS. The conservation measures that the landowners could provide could be the basis for receiving assurances that would allow them to continue with appropriate ranching and energy production activities, even if species occurring in the area are listed under the ESA. The USFS has been working on parallel management plans that are being implemented through Allotment Management Plans and a Grasslands Plan Amendment for the National Grasslands. The long-term objective would be to provide for the habitat needs of all species of concern in the planning landscape, while still maintaining ranching, energy developments, and recreational activities.

The Association recognized that addressing conservation concerns for one species at a time would not provide a comprehensive and consistent long-term plan for the landscape. The members of the Association also recognized that each landowner, working independently, would not be as effective as a collaborative effort that considered the cumulative contributions of all lands within the landscape for ecological, economic, and social objectives. Consequently, the Association focused its efforts on developing an ecosystem management plan that addressed the habitat needs of all species of concern while balancing these needs with sustainable economic and social activities. The ecosystem management plan, if implemented, would provide the science-based information and integration needed to meet these objectives and would provide the basis for either individual landowners or for a group of landowners to enter into appropriate conservation agreements.

The Association engaged the Ecosystem Management Research Institute (EMRI) to conduct an ecological assessment of the landscape that has characterized both a historical reference and existing ecosystem diversity. Ecosystem diversity was characterized as the native ecosystems that occurred in the area based on the influence of NRCS ecological sites (Natural Resource Conservation Service 2006) and the role of historical disturbances, specifically the influences of fire, grazing by native herbivores, and weather. The spatially explicit vegetation dynamics model SIMPPLLE (Chew et al. 2004) was used to model historical vegetation dynamics. Initial vegetation maps were generated by overlaying ecological site information (Natural Resource Conservation Service 2006) with information derived from General Land Office survey information recorded in the original surveyor logs. Existing ecosystem conditions were delineated through new 10 m resolution satellite mapping of grass and shrub dominated areas, coupled with on-the-ground sampling of vegetation conditions. Ecosystem diversity objectives, defined as representative levels of historical ecosystem diversity for the area, were identified. Many historically occurring ecosystems were found to be adequately represented in the planning area. Ecosystem diversity that was not well represented in the planning area included plant communities resulting from light grazing, as well as more heavily grazed areas containing an acceptably low level of exotic plant species, especially cheatgrass (*Bromus tectorum*). The Association has started on-the-ground practices to correct some of these deficiencies for both grassland and sagebrush ecosystems.

The proposed levels of ecosystem diversity that might be provided by the private landowners are being evaluated using a habitat-based species viability assessment (Roloff and Haufler 1997, 2002) for selected focal species. This assessment models home ranges of varying quality for selected species within a planning landscape. For example, output of the habitat-based species viability analysis shows over 6000 high-quality home ranges for the lark bunting under existing habitat conditions (Fig. 6-2). The viability assessment compares this number to the number of high-quality home ranges that would be present following full implementation of the potential management plan, and demonstrates that the plan would provide for sufficient high-quality habitat for this species to expect the population to remain viable in the landscape.

The assumption of the plan is that the habitat needs of all species will be provided through representation of all historically occurring ecosystems. The focal species are used to assess that this is a reasonable assumption. Monitoring of vegetation and corresponding species responses would be an ongoing component of the plan. The assumptions of the plan can be evaluated through this monitoring, and an active adaptive management plan is part of the implementation process (Franklin et al. 2007).

There are several advantages to using the ecosystem diversity approach in the Thunder Basin. This approach addresses the habitat needs of all species and provides the landowners with assurance that as new species of concern are identified, they will already be addressed in their area through the provision of ecosystem diversity. The landowners understand the reasons for the specific ecosystem diversity goals and can see the benefits of treating areas to obtain these conditions, both from the standpoint of species of concern and for improvements to rangeland conditions. The ecosystem diversity goals, stated as desired grassland or shrubland conditions, can be funded using a large number of possible sources, including several Farm Bill programs and associated rangeland practices. Finally, the ecosystem diversity approach can be used to

FIG. 6-2

Map of over 6000 high-quality home ranges for lark bunting determined using a habitat-based species viability approach (Roloff and Haufler 1997, 2002) within the Thunder Basin Grasslands Prairie Ecosystem Association (TBGPEA) landscape under existing habitat conditions.

address the habitat needs of species of concern, but presents goals recognized and supported by range conservationists working in the area. This approach can bring agency biologists and range conservationists to the same table where both can see and agree on the merits of the proposed actions.

The Association of private landowners also developed a conservation plan for prairie dogs with input from the USFS, BLM, Wyoming Game and Fish Department, and a number of conservation organizations. The prairie dog conservation plan was needed because prairie dogs are not limited by available habitat that is addressed by the ecosystem diversity plan. Rather they are limited by where prairie dogs are allowed to occur. The prairie dog conservation plan was designed to provide for not only prairie dogs, but also other associated species including sufficient conditions for the potential reintroduction of black-footed ferrets. The prairie dog plan was developed as a potential component of the assurances agreement between private landowners and the USFWS and it provided the basis for a Grasslands Plan Amendment by the USFS.

The Association needed to overcome a number of the challenges discussed previously to move this conservation planning initiative forward. The first obstacle that they encountered was the lack of available funding for conducting privately led conservation planning. No normal funding channels were available to provide the support needed to conduct an ecological assessment and develop the conservation plans. The Association was able to gain the support of the Wyoming Congressional delegation as well as the State Conservationist with NRCS to generate funds for this work, augmented with funding provided by a number of foundations, energy production companies, and the Association itself. However, the work required to obtain these funds slowed the overall conservation planning process by several years and highlighted the need for new funding programs to provide private landowner-led planning.

The ranching members of the Association were leery of providing information on the distribution and status of species of concern occurring on their lands, including black-tailed prairie dogs, mountain plovers, sage-grouse, and others. However, they recognized the need of having baseline information on these species. They agreed to allow access to their lands by EMRI as an independent institute. EMRI, funded by nongovernmental sources for the survey work so no federal or public claims could be made to the data, entered into individual agreements with each landowner, protecting the specific information collected on each landowner's property. EMRI was allowed to provide the Association and the general public summary information on the numbers and status of species of concern within the landscape. In this way, information needed for conservation planning was gathered while respecting the property rights and proprietary information concerns of the landowners.

The Association led all aspects of the conservation planning process. The Association was provided technical assistance by EMRI and state and federal agencies. However, they are determining what management they want to apply to their lands. One of the more contentious issues was management of prairie dogs. The ranching community was concerned about the potential spread of this species across the planning area, with federal lands having few options for managing this species. The conservation plan developed for the area identified areas on both public and private lands where prairie dogs are encouraged to remain and expand, but also identified where prairie dogs are not desired, and should be managed to reduce their occurrence.

Incentives are a key component of the long-term implementation of the plan. For example, incentive payments by the Sand County Foundation and the Landowner Incentive Program have been provided to landowners who voluntarily agree to have prairie dog colonies on their lands. These payments are designed to offset the reduction in grazing productivity caused by the prairie dogs. Incentives for ecosystem restoration may also be a key component. To date, incentive payments for on-the-ground treatments including prescribed burning, control of exotic species, and restoration of desired plant communities have been obtained from NRCS's Environmental Quality Incentives Program (EQIP), Peabody Energy Company, Wyoming Wildlife and Natural Resources Trust Fund, and the Landowner Incentive Program administered by the Wyoming Department of Game and Fish. In addition, state lands are being included in the restoration work, and the Wyoming Department of State Lands and Investments have allocated funds.

The work of the Thunder Basin Grassland Prairie Ecosystem Association demonstrates that many challenges to conservation planning at landscape scales can be met while incorporating private landowner concerns. This collaborative process brought together different landowners with a shared mission of landscape planning. Private property rights and proprietary information were respected, while information needed for planning was produced. Funding, while an obstacle to the effort, was obtained to allow the private landowners to lead the planning effort. Agencies were engaged as cooperators, but were not allowed to drive the planning process. Sound science provided the underlying basis of any adopted plans. An ecosystem diversity approach avoided the need to address a long list of individual species, but still demonstrated how the needs of focal species and species of concern could be incorporated and assessed. The ecosystem diversity approach addressed objectives of both biologists and range conservationists. Through this effort, landowners may gain assurances that they can continue to practice appropriate ranching and energy production activities even if various species of concern are listed under the ESA. Landowners are voluntarily involved, with their involvement encouraged through incentive programs that minimize economic impacts. Cooperative planning across public and private lands will result in greater conservation, continued economic vitality of the landscape, and reduced conflicts.

SUMMARY

Private lands offer some of the best opportunities and greatest challenges for conservation planning, especially at landscape scales. Nearly all large landscapes contain significant amounts of private lands that are often the most productive lands and offer some of the best potential for biodiversity conservation and other conservation objectives. Use of models for conservation planning of private lands at landscape scales generally parallel similar efforts on public lands. However, conservation planning on private lands presents various challenges not inherent in planning for public lands. Foremost are concerns for private property rights and proprietary information that complicate planning and model development. Distrust of planning conducted by agencies and lack of funding for planning conducted by private landowners unless paid for independently are formidable challenges. Mixed-ownership landscapes present additional challenges with respect to linking effective planning processes for both public and private lands. Various solutions exist, including the use of an ecosystem diversity focus in planning and linking planning processes and outputs to existing programs used by private landowners including various Farm Bill programs. Finally, generating interest in conservation planning by private landowners and identifying suitable incentives that can help balance conservation objectives with economic and other objectives of private lands holds promise. We presented two examples of private land conservation planning that incorporate these recommendations.

ACKNOWLEDGMENTS

This chapter was improved through input received from Josh Millspaugh, Frank Thompson, and two anonymous reviewers. We appreciated their assistance and suggestions.

LITERATURE CITED

Ben Wu, X., and F. E. Smeins. 2000. Multiple-scale habitat modeling approach for rare plant conservation. *Landscape and Urban Planning* 51:11–28.

Carignan, V., and M.-A. Villard. 2002. Selecting indicator species to monitor ecological integrity: A review. *Environmental Monitoring and Assessment* 78:45–61.

Carroll, C., R. F. Noss, and P. C. Paquet. 2001. Carnivores as focal species for conservation planning in the Rocky Mountain region. *Ecological Applications* 11:961–980.

Chew, J. D., C. Stalling, and K. Moeller. 2004. Integrating knowledge for simulating vegetation change at landscape scales. *Western Journal of Applied Forestry* 19:102–108.

Flather, C. H., K. R. Wilson, D. J. Dean, and W. C. McComb. 1997. Identifying gaps in conservation networks: Of indicators and uncertainty in geographic-based analyses. *Ecological Applications* 7:531–542.

Fleishman, E., C. J. Betrus, R. B. Blair, R. MacNally, and D. D. Murphy. 2002. Nestedness analysis and conservation planning: The importance of place, environment, and life history across taxonomic groups. *Oecologia* 133:78–89.

Fleishman, E., R. B. Blair, and D. D. Murphy. 2001. Empirical validation of a method for umbrella species selection. *Ecological Applications* 11:1489–1501.

Franklin, T. M., R. Helenski, and A. Manale. 2007. Using adaptive management to meet conservation goals. Pages 103–113 *in* J. B. Haufler, editor. *Fish and wildlife responses to Farm Bill conservation practices.* The Wildlife Society, Technical Review 07-1, Bethesda, Maryland, USA.

Freemark, K., D. Bert, and M.-A. Villard. 2002. Patch-, landscape-, and regional-scale effects on biota. Pages 58–83 *in* K. J. Gutzwiller, editor. *Applying landscape ecology in biological conservation.* Springer-Verlag, New York, New York, USA.

Groves, C. R. 2003. *Drafting a conservation blueprint: A practitioner's guide to planning for biodiversity.* Island Press, Washington, D.C., USA.

Groves, C. R., L. S. Kutner, D. M. Stoms, M. P. Murray, J. M. Scott, M. Schafale, and R. L. Pressey. 2000. Owning up to our responsibilities: Who owns lands important for biodiversity. Pages 275–300 *in* B. A. Stein, L. S. Kutner, and J. S. Adams. *Precious heritage: The status of biodiversity in the United States.* Oxford University Press, New York, New York, USA.

Haufler, J. B., R. K. Baydack, H. Campa, III, B. J. Kernohan, L. J. O'Neil, L. Waits, and C. Miller. 2002. *Performance measures for ecosystem management and ecological sustainability.* The Wildlife Society, Technical Review 02-01, Bethesda, Maryland, USA.

Haufler, J. B., and B. J. Kernohan. 2001. Ecological principles for land management across mixed ownerships: Private land considerations. Pages 73–94 *in* V. H. Dale and R. A. Haeuber, editors. *Applying ecological principles to land management.* Springer-Verlag, New York, New York, USA.

Haufler, J. B., C. A. Mehl, and G. J. Roloff. 1996. Using a coarse-filter approach with a species assessment for ecosystem management. *Wildlife Society Bulletin* 24:200–208.

Haufler, J. B., C. A. Mehl, and G. J. Roloff. 1999. Conserving biological diversity using a coarse-filter approach with a species assessment. Pages 107–125 *in* R. K. Baydack, H. Campa, III, and J. B. Haufler, editors. *Practical approaches to the conservation of biological diversity.* Island Press, Washington, D.C., USA.

Hughes, J. B., G. C. Daily, and P. H. Ehrlich. 2000. Conservation of insect diversity. *Conservation Biology* 14:1788–1797.

Kernohan, B. J., and J. B. Haufler. 1999. Implementation of an effective process for the conservation of biological diversity. Pages 233–249 *in* R. K. Baydack, H. Campa, III, and J. B. Haufler, editors. *Practical approaches to the conservation of biological diversity.* Island Press, Washington, D.C., USA.

Kintsch, J. A., and D. L. Urban. 2002. Focal species, community representation, and physical proxies as conservation strategies: A case study in the Amphibolite Mountains, North Carolina, USA. *Conservation Biology* 16:936–947.

Lambeck, R. J. 1997. Focal species: A multi-species umbrella for nature conservation. *Conservation Biology* 11:849–856.

Lawler, J. J., D. White, J. C. Sifneos, and L. L. Master. 2003. Rare species and the use of indicator groups for conservation planning. *Conservation Biology* 17:875–882.

Lindenmayer, D. B., R. B. Cunningham, C. F. Donnelly, and R. Lesslie. 2002. On the use of landscape surrogates as ecological indicators in fragmented forests. *Forest Ecology and Management* 159:203–216.

Louisiana Department of Wildlife and Fisheries. 2006, December 23. LDWF Home Page. <http://www.wlf.louisiana.gov/experience/lawildlife/nongame/rcw_safe_harbor.cfm>. Accessed 23 December 2006.

Loehle, C., J. G. MacCracken, D. Runde, and L. Hicks. 2002. Forest management at landscape scales: Solving the problems. *Journal of Forestry* 100:25–33.

Minnesota Forest Resources Council. 2006, May 22. MFRC Home Page. <http://www.frc.state.mn.us/Landscp/Landscape.html>. Accessed 2 January 2007.

National Council for Air and Stream Improvement. 2001. *Managing forested landscapes: Industry approaches, experiences, and perspectives in the Pacific Northwest.* NCASI, Technical Bulletin No. 0840, Research Triangle Park, North Carolina, USA.

Natural Heritage Data Network. 1993. *Perspectives on species imperilment.* The Nature Conservancy, Washington, D.C., USA.

Natural Resource Conservation Service. 2006, October 6. NRCS Home Page. <http://www.nrcs.usda.gov/PROGRAMS/>. Accessed 23 December 2006.

Nichols, W. F., K. T. Killingbeck, and P. V. August. 1998. The influence of geomorphological heterogeneity on biodiversity: II. A landscape perspective. *Conservation Biology* 12:371–379.

Niemi, G. J., J. M. Hanowski, A. R. Lima, T. Nicholls, and N. Weiland. 1997. A critical analysis on the use of indicator species in management. *Journal of Wildlife Management* 61:1240–1252.

Noon, B. R., and V. H. Dale. 2002. Broad-scale ecological science and its application. Pages 34–52 *in* K. J. Gutzwiller, editor. *Applying landscape ecology in biological conservation.* Springer-Verlag, New York, New York, USA.

Noon, B. R., D. D. Murphy, S. R. Beissinger, M. L. Shaffer, and D. DellaSala. 2003. Conservation planning for U.S. National Forests: Conducting comprehensive biodiversity assessments. *BioScience* 53:1217–1220.

Oliver, I., A. Holmes, J. M. Dangerfield, M. Gillings, A. J. Pik, D. R. Britton, M. Holley, M. E. Montgomery, M. Raison, V. Logan, R. L. Pressey, and A. J. Beattie. 2004. Land systems as surrogates for biodiversity in conservation planning. *Ecological Applications* 14:485–503.

Pearson, D. L., and S. S. Carroll. 1998. Global patterns of species richness: Spatial models for conservation planning using bioindicator and precipitation data. *Conservation Biology* 12:809–821.

Roloff, G. J., and J. B. Haufler. 1997. Establishing population viability planning objectives based on habitat potentials. *Wildlife Society Bulletin* 25:895–904.

Roloff, G. J., and J. B. Haufler. 2002. Modeling habitat-based viability from organism to population. Pages 673–686 *in* J. M. Scott, P. J. Heglund, M. L. Morrison, J. B. Haufler, M. G. Raphael, W. A. Wall, and F. B. Samson, editors. *Predicting species occurrences: Issues of accuracy and scale.* Island Press, Washington, D.C., USA.

Spies, T. A., K. N. Johnson, K. M. Burnett, J. L. Ohmann, B. C. McComb, G. H. Reeves, P. Bettinger, J. D. Kline, and B. Garber-Yonts. 2007. Cumulative ecological and socioeconomic effects of forest policies in coastal Oregon. *Ecological Applications* 17:5–17.

Stribley, J. M., and J. B. Haufler. 1999. Landscape effects on cowbird occurrences in Michigan: Implications to research needs in forests of the inland west. *Studies in Avian Biology* 18:68–72.

Su, J. C., D. M. Debinski, M. E. Jakubauskas, and K. Kindscher. 2004. Beyond species richness: Community similarity as a measure of cross-taxon congruence for coarse-filter conservation. *Conservation Biology* 18:167–173.

The Keystone Center. 1996. *The keystone national policy dialogue on ecosystem management: Final report.* Keystone, Colorado, USA.

The Nature Conservancy. 1982. *Natural heritage program operations manual.* The Nature Conservancy, Arlington, Virginia, USA.

The Nature Conservancy. 2006, December 23. TNC Home Page. <http://www.nature.org/aboutus/howwework/conservationmethods/privatelands/conservationeasements/>. Accessed 23 December 2006.

Thompson, J. R., M. D. Anderson, and K. N. Johnson. 2004. Ecosystem management across ownerships: The potential for collision with antitrust laws. *Conservation Biology* 18:1475–1481.

U.S. Department of Agriculture. 1997. *Agricultural resources and environmental indicators, 1996–1997.* USDA, Agricultural handbook number 712, Economic Research Service, Washington, D.C., USA.

U.S. Fish and Wildlife Service. 2006, December 19. USFWS LIP Home Page. <http://federalaid.fws.gov/lip/lip.html>. Accessed 23 December 2006.

van Jaarsveld, A. S., S. Freitag, S. L. Chown, C. Muller, S. Koch, H. Hull, C. Bellamy, M. Kruger, S. Endrödy-Younga, M. W. Mansell, and C. H. Scholtz. 1998. Biodiversity assessment and conservation strategies. *Science* 279:2106–2108.

Vos, C. C., H. Baveco, and C. J. Grashof-Bokdam. 2002. Corridors and species dispersal. Pages 84–104 *in* K. J. Gutzwiller, editor. *Applying landscape ecology in biological conservation.* Springer-Verlag, New York, New York, USA.

Wessels, K. J., S. Freitag, and A. S. van Jaarsveld. 1999. The use of land facets as biodiversity surrogates during reserve selection at a local scale. *Biological Conservation* 89:21–38.

Wiens, J. A. 2002. Central concepts and issues of landscape ecology. Pages 3–21 *in* K. J. Gutzwiller, editor. *Applying landscape ecology in biological conservation.* Springer-Verlag, New York, New York, USA.

CHAPTER

7 A Multiscale, Stepwise Approximation Approach for Wildlife Conservation Assessment and Planning

John R. Probst and Eric J. Gustafson

Comprehensive wildlife conservation will be more effective if it incorporates a large area perspective. Different species may have affinities for various ecosystem types, seral stages, and habitat elements at different scales, so conservation planning must consider diverse ecosystems and habitat conditions from a broad-scale perspective (Askins et al. 2007). A planning area must be sufficiently large to accommodate populations of the most wide-ranging species, maintain habitat for area-sensitive species, and provide interspersed habitat types for species that require them. Diverse ownership types and multiple natural resource objectives add to the complexity of wildlife conservation. Therefore, a major challenge for ecologists, resource managers, and policy makers is to develop comprehensive natural resource assessments that integrate current scientific knowledge of the requirements of multiple species. Here, we apply principles of landscape ecology to provide a context for landscape-level wildlife conservation planning and implementation. We begin by discussing limitations of models too specific for general application elsewhere, and review some literature on landscape ecology concepts that could be applied as a general framework for a variety of landscape-level modeling approaches. A broad-scale context is used to help set wildlife goals and objectives for landscape-level planning before choosing modeling objectives, methods, and evaluation criteria. We offer three examples of applying the concepts of multiple scales, stepwise levels of detail, or spatiotemporal heterogeneity in species' occurrences to help set or achieve conservation objectives at successively narrower scales.

There is growing acceptance of the importance of large spatial scales and multiscale interrelationships (i.e., macroecology) in ecology (e.g., May 1994, Brown 1995, Maurer 1999). Ecological systems exhibit domains of scale with hierarchical relationships that vary among levels of the hierarchy of scale (O'Neill et al. 1986), making it difficult to apply models across scales. Ecological heterogeneity within landscapes also makes it difficult to apply locally derived models at higher levels (Pickett and Cadenasso 1995). Many ecological phenomena are mediated by higher level processes (Urban et al. 1987, Wiens 1989), and one cannot predict broad-scale population dynamics by merely scaling up relationships known at finer scales (Bissonette 1997). Therefore, models developed at local scales may not be relevant at broader scales.

Another obstacle to comprehensive integrated assessment is that available information is typically narrow in scope and not uniform among species. Thus, there is an inadequate scientific basis to extrapolate narrowly focused survey and research results to other species, or spatial and temporal scales, geographic areas, or species (Johnson and Herring 1999, Swanson and Greene 1999). For example, even simple species occurrence models frequently fail to accurately predict presence and absence (e.g., Raphael and Marcot 1986), and these shortcomings are usually attributed to inadequate model development, refinement, interpretation, or application (Block et al. 1994, Morrison 2001). However, the problem is often that the models were applied at inappropriate scales or because species habitat associations (Collins 1983, Grzybowski et al. 1996) and ecosystems (Johnson and Herring 1999) are variable in space and time.

Researchers frequently sacrifice generality for precision when choosing to do high-resolution, local-scale research on a narrow, focused problem of limited geographical scope (e.g., May 1994, Brown 1995, Swanson and Greene 1999). Many high-resolution models have limited ability to make predictions for other times and places (e.g., Wiens et al. 1987, Van Horne and Wiens 1991, Levin 1992, Oreskes et al. 1994). The lack of appreciation of temporal variation in populations has been perpetuated because so few studies have been conducted for long time periods (Wiens 1981, Donner 2007). Combining highly detailed, single-species models to predict the cumulative effects of ecosystem change on multiple species becomes impossible because the models are usually parameterized with different variables measured at different spatial and temporal scales. Also, models are developed with varying degrees of rigor, so comparable models are rarely available for all species under consideration for landscape-level planning. Attempts to apply multispecies models are made difficult by several factors: (1) a model designed to optimize predictions from a given data set may lack useful generality; (2) integrating different model forms and data requirements of different species is difficult; (3) many models implicitly assume habitats support populations at carrying capacity, but populations of uncommon species are rarely at carrying capacity; and (4) most species distributions, associations, and biological processes are classified into categories, yet many ecological relationships are best represented by gradients (McGarigal and Cushman 2005).

Many wildlife models are correlative rather than mechanistic (Bissonette 1997), but some have argued that effort should be invested in developing mechanistic models instead (Morrison 2001, Wiens 2002). However, the mechanisms relating multiscale landscape patterns to wildlife ecology processes are not well known (but see Hansen et al. 1993, Thompson et al. 1993, White et al. 1997, Smallwood et al. 1998, Cogan 2002, Cushman et al. 2007), so it would be desirable to frame correlative models in a context of regional distribution or population variability to first separate regional from landscape-habitat effects. The clear majority of "multiscale" studies are actually done at only two scales (Wiens 2002), typically landscape and local or regional and local. Thus, what is needed to provide context for variable patterns in time and space is a series of general to specific models (Van Horne 2002) or a hierarchy of less detailed, or even qualitative models to organize thinking about processes (Rubin 1991). A related solution to these problems is to develop less detailed but more robust models for larger scales (e.g., presence-absence at habitat or landscape scales), and then develop subsequent nested models for smaller areas with more detail (i.e., abundance, productivity, survivorship). Such models can be framed within a hierarchy of general models that lead to an understanding of ecological processes (James and McCulloch 2002, Stauffer 2002, Van Horne 2002). Although coarser-resolution work is typically done at larger geographic scales, it is important to realize that levels of detail are not the same as a hierarchy of spatial scales.

Applying habitat models developed in one landscape to another landscape may be difficult when there is variability in ecosystem composition among landscapes, or when responses of various species to landscape composition and structure vary geographically. Generalizing habitat association models to other species, places, or times than those for which they were developed can be inappropriate because (1) species are distributed unevenly because there is considerable geographic variability in landscapes, ecosystems, biological community composition, and biophysical processes; (2) every species has unique niche and life history characteristics and species may perceive the "grain" of landscape patterns differently; (3) populations vary in time and space, so the assumption that there is always a useful and robust mathematical relationship between habitat and landscape conditions is often not realized; and (4) hierarchical relationships exist among species, ecosystems, and ecological processes. Clearly, ignoring these constraints when applying models violates some basic principles of ecology. If resources are unevenly distributed across gradients in time and space, and there is variation in species distributions and biophysical processes, then it is desirable to establish a multiscale strategy to describe the geographic variability of species' ecological attributes. We demonstrate how explicitly embracing the concept of hierarchies of scale (e.g., Root and Schneider 1995) and levels of resolution (e.g., Menges and Gordon 1996) allows practitioners to assess conservation issues by a process of stepwise approximation.

STEPWISE APPROXIMATION STRATEGY

We propose a multiple-level, stepwise strategy that links coarse and fine resolution information to balance the need for generality and specificity for integrated wildlife conservation assessments. By this strategy, local-scale, high-resolution research is interpreted within the context of larger scale, multispecies distribution patterns. The framework can be implemented using existing information and databases, producing time and cost savings, while guiding future research hypotheses and data collection to the most appropriate species and locations.

The key to considering all species while still directing more attention to the most critical issues is to take a general-to-specific perspective on population processes and wildlife habitat relationships. This is best done in a stepwise manner, proceeding from coarse to finer levels of detail (Freemark et al. 1993). We are not recommending a specific protocol for assessing species population dynamics, nor is there a rigid number of steps to an assessment, as each situation will have its own set of relevant scales (Wiens 1989). Nevertheless, we provide a hierarchical strategy for assessing population dynamics by describing seven discretionary levels of analytical resolution for inferring process from pattern based on literature synthesis and empirical data. The analytical levels for conducting a comprehensive population assessment at increasingly finer levels of detail are as follows: (1) delineate the ranges of all species of interest and their range overlaps to identify dissimilar distributions; (2) define or hypothesize species-habitat associations using gradients whenever possible; (3) test these habitat-gradient relationships with species occurrence surveys; (4) evaluate habitat quality by using abundance estimates and its temporal variation in abundance; (5) refine the habitat quality assessment by using indirect productivity inferences or direct productivity measurements, and/or estimate survivorship; (6) synthesize (levels 1–5) using maps and summary models to assess species viability, allowable harvest, conservation plans, or implications of changing conditions; and (7) test the synthesis and its extrapolation to other contexts in space and time through a well-designed strategic monitoring program. Not all levels of this strategy must be used in every conservation assessment, but each level adds understanding to the assessment process because ecological dynamics at each level are better understood when they are related to others (e.g., Urban et al. 1987, Pickett and Cadenasso 1995). Several of the levels can be developed simultaneously as long as generality is not prematurely sacrificed for local, fine-resolution precision. Additionally, adding sequential levels of detail is pragmatic when problem solving because it becomes an application of the scientific Principle of Parsimony (i.e., using the simplest explanations for observed patterns). The steps chosen for this chapter are best suited for wildlife conservation planning and implementation at the landscape level, but we argue that any broad context in space and time can be valuable for most wildlife and natural resource models and planning efforts.

To demonstrate how elements of any stepwise approximation process can be modified, adapted, and reduced for individual situations, we provide three examples of ongoing assessments: (1) a multispecies assessment conducted at multiple scales for prioritizing species for conservation action (Level 1); (2) a multispecies assessment that uses several levels of detail to estimate species' responses to openland management (Levels 1–5); and (3) a single-species range-wide assessment summary model (Level 6) that infers process from pattern using five levels of detail. These examples draw on different avian population studies within and among levels to demonstrate how unconnected research conducted at various scales can be linked within this stepwise framework.

Level of Resolution 1: Broad-Scale Multispecies Ranges

Understanding variability in population distribution among geographic areas and habitats is important when attempting to generalize data from one locality to another. The first prerequisite for aggregating distribution patterns of multiple species is knowledge of species occurrence at several scales to prevent local-scale research on population dynamics from being generalized to places where a species does not occur in its typical habitat. At continental scales, range-wide maps of species distributions are available in field guides or other reviews. Because the geographical extent and sampling intensity of surveys varies by taxa, using general geographic information provides some preliminary comparability among taxa by excluding some distribution details that are important only at finer scales. Large-scale, comprehensive distribution data are available for taxa such as birds (e.g., Price et al. 1995), fish, and trees. Atlases and other comprehensive distribution data are being assembled for other vertebrates, plants, and invertebrates (Johnson and Sargeant 2002).

Because only the most common species are found in all places within their range and suitable habitats (e.g., data in Brewer et al. 1991, Corace 2007), a tabulation of known species presence by regional landforms, ecosystems, cover types (e.g., forest, shrubland, grassland), or disturbance history is an important step in any broad assessment of species conservation status (e.g., Probst and Thompson 1996). At Level 1, species distributions are developed to a degree that might allow them to be matched to land covers from remote sensing and species distribution maps (e.g., Jennings 2000) without supplementary field observation or knowledge of habitat gradients within land cover types. At scales intermediate between regional and local, patterns of presence and absence within a species' geographical range is apparent, and some species may be absent where other species with the same or similar habitat preferences might be present. For example, upland sandpiper (*Bartramia longicauda*) is infrequent in southeast Michigan openlands, whereas grasshopper sparrow (*Ammodramus savannarum*) is not found in a majority of the counties in Michigan's Upper Peninsula (Fig. 7-1). These differences may be due to range limits, landscape structure, or more specific habitat preferences than just "openland" or "grassland" (see Level 2).

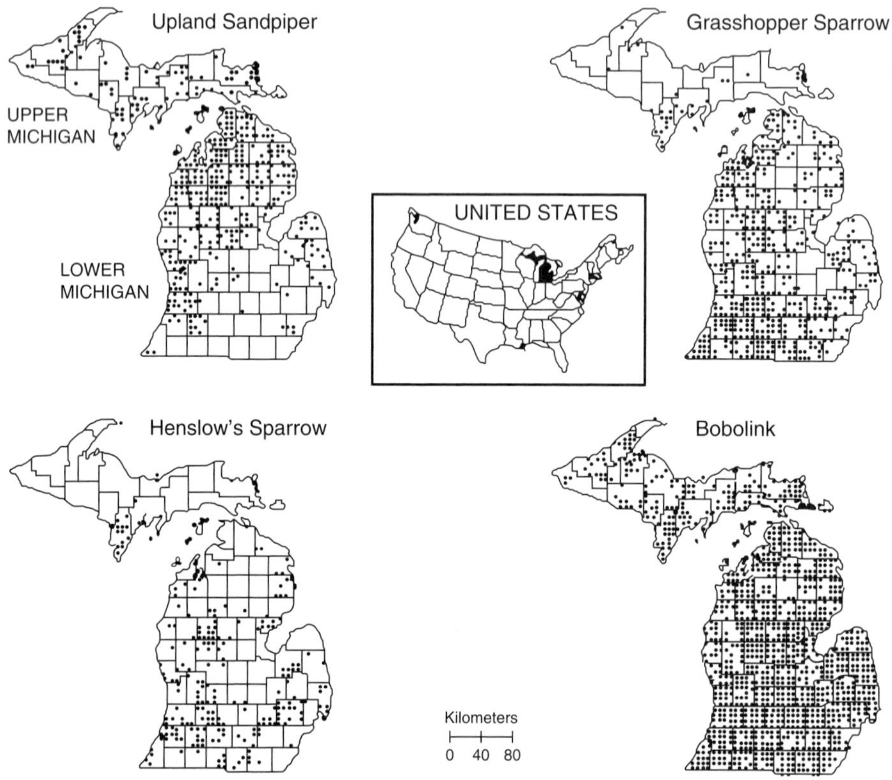

FIG. 7-1

Breeding Bird Atlas distribution of four species of openland birds in Michigan, USA (adapted from Brewer et al. 1991).

Multiscale Bird Distribution Assessment

As an example, we present a conceptual overview of a multiscale assessment of openland cover type and bird species distribution. The assessment is being developed to summarize species distributions at both regional and subregional (1,000–5,000 km^2) political and ecological scales to prioritize species for conservation efforts, and to allocate resources for land acquisition, preservation, restoration, management, mitigation, and easements (Corace 2007). At the Midwestern United States scale (Fig. 7-2), we see a disproportionately large area of grassland cover type in the four most western Midwestern states, and a disproportionately large area of hayfields in Kansas, Missouri, and the three Great Lakes States (Minnesota, Wisconsin, and Michigan). The areas in each cover

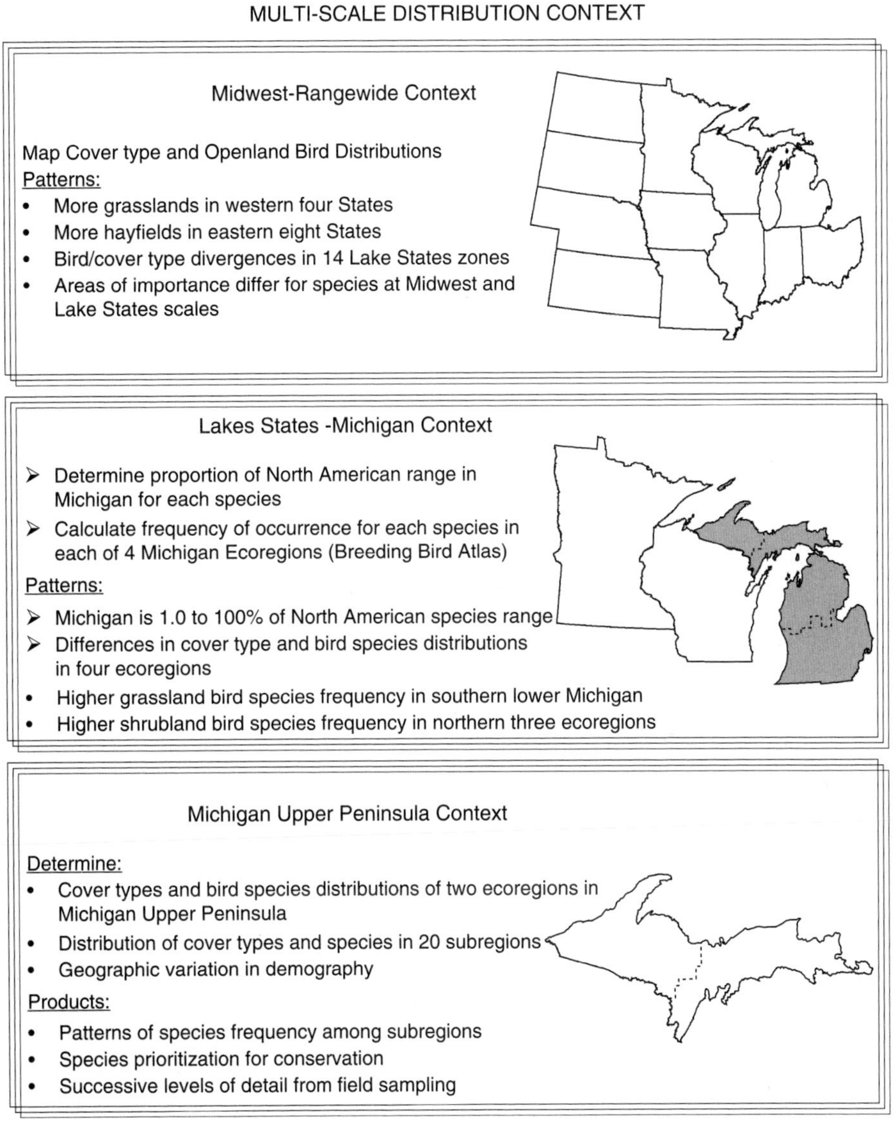

FIG. 7-2

A multiscale assessment of cover types and bird species distributions for species prioritization of conservation actions in the Midwestern USA.

type by state and ranked bird cover-type affinities were combined to produce a score for each species in each state. Bird species distribution centroids differ on east-west and north-south axes, and further divergences in cover type and bird species distribution are apparent within the Great Lakes States. These patterns guide initial conservation actions at the state and ecoregion scales within states by developing an initial land cover assessment for each species.

At the Michigan scale, the proportion of the North American range for each species in Michigan was calculated, and the frequency of occurrence of each species in each of four ecoregions in Michigan was tabulated from the Michigan Breeding Bird Atlas (Brewer et al. 1991). In general, grassland bird species were more frequent in southern Michigan and shrubland birds were more frequent in the three northern ecoregions of the state (Corace 2007). However, there were some exceptions that should affect conservation planning in each ecoregion that leads to considerations of grassland maintenance in the Upper Peninsula and management of shrublands in southern Michigan.

At the Michigan Upper Peninsula scale, conservation priority species were identified as (a) those whose proportion of North American range occurring in the Upper Peninsula is above the median (3.6%) for all Michigan species, or (b) species with higher Upper Peninsula breeding-bird atlas block frequency than in Michigan's Southern Peninsula. At this scale, not all species were found in all 20 subregions within the two Michigan Upper Peninsula ecoregions. Species were further prioritized by their breeding bird atlas block frequency by subregion, further refining the geographic specificity of prioritization. Finally, extrapolation errors were illustrated from presence-absence data by calculating the percent area of suitable cover type for a species where it was not found by the breeding-bird atlas, relative to the Upper Peninsula total area for that cover type. Although it is important to acknowledge that "cover types" may not accurately define suitable habitat for a given species, quantification of presence-absence patterns helps estimate errors when relating cover type areas to more detailed demographic attributes at subsequent levels of resolution (see following description). This simple example illustrates how conservation priorities can be identified without detailed population estimates, and the potential difficulties trying to estimate cumulative population sizes resulting from conservation actions.

At all scales, we find that important locations (as indicated by distribution and abundance) are different for each species, and all areas may be important for at least some species. The mix of cover types and bird species varies geographically, so the notion of a "suite of species" for conservation objectives has limited utility. In summary, the broad-scale patterns provide a context for species prioritization and generalization of field measurements, and the finer scales provide detail necessary to interpret the coarser scales. The broad-scale patterns can be used for modifying Partners in Flight prioritization and species rankings, and are being used for state-level landscape planning for Michigan Important Bird Areas. The context of spatial scales and distribution described here can also inform a stepwise approximation process that moves from less-detailed to more-detailed field measurements of bird species demographic parameters to identify specific conservation needs.

Level of Resolution 2: Refine Species Distributions in Terms of Habitat-Gradient Relationships

The next level in the strategy is to refine species distributions at finer scales of species-gradient relationships. At regional or multistate scales, the habitat distribution of vertebrates is available based on field guides, expert opinions, and limited inventories. However, there are questions and biases associated with application of expert opinion to distribution data (e.g., Johnson and Gillingham 2004), as discussed later in Level 3. Typically, species are grouped into habitat categories, vegetation associations or communities, physiographic units, or ecological units for research synthesis, assessment, and planning within the context of coarse filter biodiversity programs (e.g., Pregitzer et al. 2001). Additionally, we can combine information on species' geographic range and habitat affinities to describe geographic variability in continental or regional habitat distributions (e.g., Probst and Thompson 1996). However, individual species respond differently to environmental gradients or changes, so a community association of one species may be a poor predictor of other species distributions (e.g., Graham 1994; Noon et al., this volume). Further, some taxa are too poorly surveyed to reliably place them in either communities or gradients. Although the continuum concept of species (Gleason 1926, Curtis 1959) is basic to general ecology, it is often ignored in conservation plans that use community surrogacy, multispecies taxa, umbrella species (Simberloff 1997), or biophysical units as substitutes for a comprehensive enumeration of the species for which information is currently available.

Classifying species in terms of their response to gradients is a useful classification framework because gradients can accommodate changing species distributions and helps to integrate a variety of new or existing classification systems. Conversely, more arbitrary categorical classification systems can be an impediment to data integration, extrapolation of trends, or cross-agency assessment. For example, species distributions may be arranged on two or more ecological gradients rather than using arbitrary species associations or biological communities, such as habitat types or seral stages (e.g., Fig. 7-3). Gradients are important for understanding limits or boundary conditions (nested within Level 1) and provide context for more-detailed categorical comparisons or models (Van Horne 2002, Probst and DonnerWright 2003), which may be limited to only a part of the total range of potential suitable conditions.

Classification frameworks are an important part of assembling and interpreting distribution data for assessment. Classifying species based on ecological gradients allows scientists to predict changes in species distributions and ecosystem associations as a consequence of changes in the gradient factor(s). Furthermore, the gradient concept allows the formulation of varying management actions across the range of ecosystem conditions. It also allows one to make predictions about habitat suitability within ecosystems that have not been adequately surveyed, and surveys can be designed to test the predictions (Level 7).

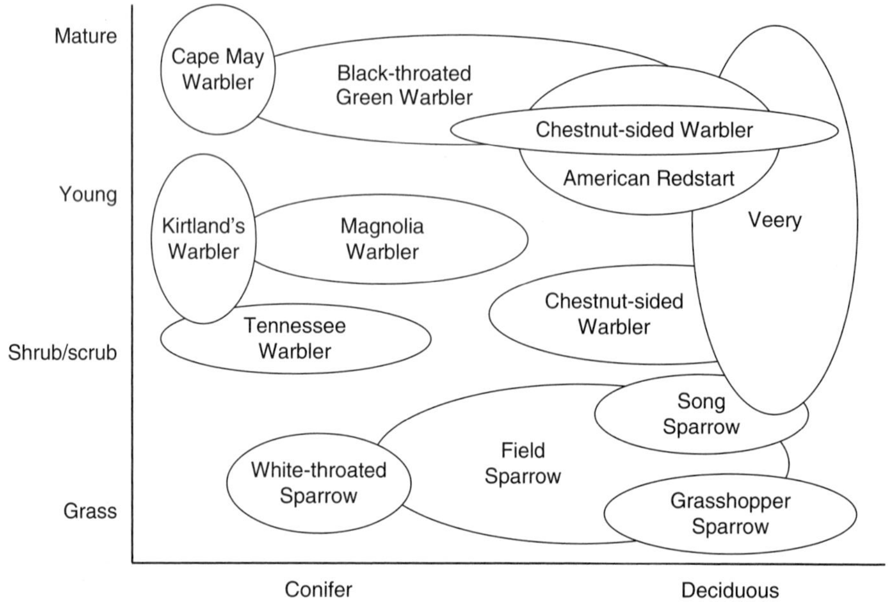

FIG. 7-3

Hypothetical ordination of avian habitat distribution on gradients of seral stage and tree life-form.

The simplest models are usually the best and most robust, and are the most useful to managers (Starfield 1997; Stauffer 2002; Shifley et al., this volume). A direct gradient approach (e.g., Whittaker 1956, Herkert 1991) to conservation biology has far more ability to interpret the effect of disturbances, biological cycles, or directional change than do relationships with indirect multivariate axes. The cumulative effects of global change or management on ecological processes are also more readily assessed by gradients than by arbitrary classification boundaries.

Level of Resolution 3: Test Species Habitat or Gradient Relationships Across Landscapes

Collecting frequency data on species occurrence is less labor intensive than taking measurements of density, biomass, and most demographic variables (Levels 4/5), so survey and monitoring can be planned and conducted in a stepwise manner to infer process from pattern and test hypotheses (Freemark et al. 1993). For example, surveys can be developed to test existing databases as part of stepwise approximation for confirmation of habitat gradient relationships of various species. Species habitat relationships and their spatio-temporal variation can be tested at low resolution (and low expense) by surveys for presence and absence. These surveys may show surprising departures from published field guides, expert opinion, or habitat models (e.g., Avery and van Riper 1990, Short and Hestback 1995) because presence-absence studies further refine our knowledge of species-habitat relationships. Species can be found where not predicted (omission error), absent from where predicted (commission error), or where variable in the same place through time (Raphael and Marcot 1986, Timothy and Stauffer 1991, Block et al. 1994). These surveys have been interpreted without reference to carrying capacity dynamics through time (e.g., Boone and Krohn 2000, Karl et al. 2000), and consequently, unreliable species-habitat model performance might actually be a reflection of populations too small to colonize all apparently suitable habitat (e.g., Donner 2007). Simply extrapolating habitat preference of a species (Levels 2/3) to the area of that habitat on maps can lead to serious population overestimates because most species are not found everywhere within preferred land covers or habitats (e.g., Raphael and Marcot 1986, Smith and Jenks 2002). For example, only 2 of 10 bird species most common in mature oak forests were found in more than 90% of the census plots in a study conducted in oak forests in central Pennsylvania (Probst 1979).

We can infer the dynamics of population processes and their underlying mechanisms when we conduct comparative surveys for presence-absence data in different geographic areas. Initial survey results can help direct subsequent surveys to illuminate the factors that may determine the population distribution. For example, species may have scattered distributions at range borders (Brown 1984) or in rare habitats that can be better understood by studying occurrence patterns in central versus peripheral areas of species' ranges. Species may also have disjunct populations because of area, isolation, or edge effects; so targeted surveys can be designed to document geographic variability in degrees of habitat occupancy and their change through time. Putative favorable landscape conditions can be tested for having higher frequency of occurrence than less favorable ones. Thus, frequency of habitat occupancy may be a useful coarse-resolution variable for evaluating the direction of population trends or geographic variability in populations and habitat preferences.

If planners and scientists have reasonably adequate data through Level 3, they may have enough information to proceed with an iterative planning process without incurring the time and expense of collecting additional field data. The spatial patterns identified by integrating available information may suggest some additional data collection needs, but the need may be for fewer species and locations than would be done in a comprehensive synthesis.

Level of Resolution 4: Evaluate Habitat Quality Through Abundance Estimates and Temporal Change

The next level of resolution considers population size and population variability in different ecosystems through time to make inferences about population dynamics. The focus at this level may be a subset of species and issues of concern identified at Levels 2 and 3. Abundance estimates may be added to presence-absence surveys if resources permit, and abundance data can be ranked to mitigate biases in sampling methods or species detection differences (Smith and Jenks 2002).

Temporal changes in frequency estimates can be used to make preliminary inferences about population dynamics across a species range. We can interpret the direction of population trends based on whether populations have expanded or retracted geographically. For example, variability in geographic distribution and local presence-absence data has been used to infer processes affecting bird population trends, such as sources and sinks (Howe et al. 1991, Probst and Weinrich 1993). Regional and landscape population changes first appear in sink populations, which tend to exhibit more variable dynamics (e.g., Boulinier et al. 1998). This means that degrees of habitat occupancy (i.e., frequency in a land cover or habitat) can be used to understand basic population dynamics and be applied to estimate a realistic carrying capacity for conservation planning purposes. Comparing population variability within breeding seasons (Zimmerman 1982, Lanyon and Thompson 1986, Probst 1988, Howe et al. 1991) or among years (e.g., Probst and Weinrich 1993, Probst et al. 2003) can also provide insight into metapopulation dynamics in relation to habitat quality and its regional population context. Low population variability in a habitat or landscape may imply that an area is a source, whereas annual fluctuations may be seen in overflow sinks. Consequently, monitoring programs (Level 7) could emphasize population variability such as source-sink comparisons, as one example, rather than tracking only relatively stable sources.

Completing data integration and field work through Level 4 for selected species and locations is a potential stopping point. The productivity and survivorship studies of Level 5 are very labor intensive and thus expensive and time consuming, and may not always be necessary.

Level of Resolution 5: Refine Habitat-Quality Evaluations with Productivity and Survivorship Estimates

Inferences about habitat quality and population dynamics developed from distribution patterns usually must be tested using data on other demographic factors such as productivity. Patterns detected from occurrence and abundance surveys can be used to suggest populations that may need research on reproduction or survivorship. Studies can then be initiated to refine habitat quality evaluations for the subset of species and locations identified as problematic by the surveys. Many demographic variables can be approximated for larger areas and smaller sample sizes before doing more standard, fine-resolution studies, such as reproductive studies. These productivity studies can be designed at intermediate resolution to provide indirect productivity measures across a range of species and locations. For example, relative natality can be estimated across space and time from reproductive condition, percentage of adults, the proportion of adults mated or paired (e.g., Verner 1964, Probst and Hayes 1987, Gibbs and Faaborg 1990, Villard et al. 1993), and the proportion of immature individuals to adults during the post-breeding period (e.g., Howe et al. 1991, Bollinger and Linder 1994, Bart et al. 1999). By conducting less intensive relative productivity studies of greater scope and geographical breadth, planners will have a better context for more detailed studies because they will have knowledge of geographic patterns in species' presence, density, and indirect productivity. Conversely, a nested hierarchy of resolution can be used to more reliably extrapolate detailed population dynamics data to a broader multiscale geographic context. Detailed productivity studies will be more effective and efficient if they are designed to test hypotheses developed during coarser-resolution surveys and indirect productivity studies.

Because not all individuals of many species attempt to breed, studies on productivity (Level 5) cannot be directly extrapolated to population density (Level 4), but can be initially approximated by the number of breeding pairs. In cases where productivity is measured directly, abundance is not necessarily well correlated with productivity (Pidgeon et al. 2003). In all situations, failure to understand the context of the varying geographical distribution and abundance developed at Levels 2–4 could lead to gross overestimates of populations and productivity. All field studies described here may be done sequentially in a single field season, so the levels of data need not be confounded by temporal variation across years.

Studies on survivorship are difficult and may take many years to complete, especially for motile organisms. It is often impossible to separate dispersal from mortality (e.g., Brewer and Harrison 1975, Greenwood and Harvey 1982), especially for migratory species (e.g., Probst 1986) or species with long dispersal distances. Often, an estimated survivorship can be calculated from productivity and total population change data (e.g., Probst and Hayes 1987) and tested directly at a more advantageous time.

Openland Bird Assessment Example (Levels 1–5)

This approach was applied to a hierarchical population assessment of openland birds in the Upper Midwest for multispecies, interregional conservation planning (Fig. 7-4). Researchers summarized where species or habitats were common, uncommon, or both, at three scales. At each level, a context was developed to facilitate interpretation at the next finer level of detail, and the number of species under consideration was narrowed based on the patterns observed (e.g., species rarity or distribution of habitats used).

Bird species occurrence and openland area (shrub-grasslands, hayfields, etc.) were compared at three spatial scales: 12 Midwestern states, 3 Upper Great Lakes states (Michigan, Wisconsin, and Minnesota) and 14 subregions (40,000–50,000 ha) (Level 1). Results confirmed that species' distributions and large-scale abundance patterns differed geographically, suggesting that many species-habitat models could not be transferred reliably to ecoregions with different habitat

STEPWISE CONSERVATION ASSESSMENT: OPENLAND BIRDS[A]

Upper Midwest (Level 1): Document species' ranges (4 X 10^6 km^2)

- Michigan (Level 1): Record frequency distribution using 10.4 km^2 breeding bird atlas blocks (37 species)
  - Northern Lower Michigan Barrens[B] (Level 3): Record frequency distribution at landscape scale using road point count transect (20,000 km^2) (20 species)
    - Landscape Scale (Level 4): Determine abundance in 1-4 year clear-cuts by walking transects through landscapes (1000 km^2) (12 species)
      - Landscape-local (Level 5): Determine number of pairs present (9 species)
      - Landscape-local (Level 5): Determine pair productivity by finding fledged young (6 species)

Result: Broad distribution context of most openland birds established for regional population assessment. Documentation of abundance and status of selected species, and further refined by productivity estimates at the landscape-local level.

[A] Grass-shrubland birds

[B] Xeric shrublands and clearcuts

FIG. 7-4

A framework for stepwise population assessment of openland birds in the Upper Great Lakes States, USA. At finer scales level, a successively smaller pool of species is examined at more detailed levels of resolution based on information available at the previous levels.

relationships or population levels. The 4 Northern Great Plains states contained most of the openlands and grasslands in particular. The 893,000 km^2 of open lands in the Great Lakes states represented only 16% of the openlands in the 12 states context, but represented 25% of the hayfield habitat, and included the majority of the total range for some bird species of population concern

(Crow et al. 2006, Corace 2007). Published landscape-level abundance information (Price et al. 1995) at the mid-scale 3-states level was used to identify zones of species presence and absence. At the scale of one state (approximately 250,000 km^2), existing frequency data were tabulated from a Breeding Bird Atlas (Brewer et al. 1991) for all openland types (Corace 2007) and within barrens and shrublands only (Fig. 7-4).

The assessment documented which species occur within subdivisions of their geographic range with Breeding Bird Atlases (e.g., Fig. 7-1), but focused on just two habitat-types (barrens and xeric shrublands) within an age-gradient in a single ecosystem (Level 2) to document multiple bird species' use of young, jack pine (*Pinus banksiana*) regeneration. Field surveys documented landscape-level (1500–6000 ha) spatial variability in species-habitat relationships (Level 3). Not all species found in Levels 1 or 2 were found in the sampled habitats of landscapes, so the pool of species under consideration was reduced at each level. At the landscape level, it was possible to choose study sites where a species of interest, especially rare species, actually occurred, before conducting finer resolution investigations. In this example, we selected nine species of conservation interest, five of which were area-sensitive openland species. Thus, the landscape-local studies on abundance (Level 4) and reproduction (Level 5) could be generalized to larger-scale areas of habitat more reliably, because it was documented beforehand where species actually occur in respect to geography, hierarchical ecosystems, and habitat. Ultimately, we should achieve an understanding of multispecies response to openland management at the levels of frequency, abundance, and indirect productivity measurements. Such information has been incorporated into state-level planning such as Important Bird Areas and to document multispecies use of an ongoing endangered species management program for Kirtland's warblers (*Dendroica kirtlandii*).

APPLICATIONS TO PLANNING

Level of Resolution 6: Synthesize Levels 1–5 with Maps and Summary Models

One major application of syntheses of population dynamics is in the development of conservation plans, including species viability assessments (e.g., Beissinger and Westphal 1998). Once ecological processes and their multiscale interactions are better understood, range-wide or other broad-scale assessments are more useful. Many times viability is inferred based on minimum population size needed to withstand demographic uncertainty, or to conserve genetic diversity (e.g., Shaffer and Samson 1985), but rarely is range-wide or even regional information considered. For example, an exhaustive habitat assessment for the northern goshawk (*Accipiter gentilis*; Reynolds et al. 1992) initially did not address the entire species' range, but a more comprehensive geographic scope has been added and is ongoing. Conversely, a range-wide study on black-capped vireo (*Vireo atricapillus*) showed regional differences in habitats and their relative use by different age classes of birds (Grzybowski et al. 1996), which is vital to understand how geographic differences may affect species viability throughout its range. Thus, habitat models using information from part of the species' range could not be transferred uncritically to other places in the range. In the initial northern spotted owl (*Strix occidentalis*) conservation plans, the single species assessment was followed by, rather than preceded by, the multispecies considerations in the Forest Ecosystem Management Assessment Team (Thomas 1994). Because the context for application was lacking in the initial single-species focus, the management plan for Pacific Northwest forests had to be modified.

Similarly, comprehensive assessments of neotropical migrant bird populations have proposed programs focused on subareas of species ranges, with planning conducted for subregional physiographic units (Williams and Pashley 2000). We advocate improving conservation planning by beginning with less-detailed, broader-scale steps at the level of many physiographic units (e.g., Probst and Thompson 1996; Fig. 7-2), before examining subregional or landscape issues. For example, assessment across the entire known range of Kirtland's Warbler has given practitioners a realistic framework of possibilities and limitations for the total habitat and population potential and its distribution in time and space. This understanding was generated by a general-to-specific examination of patterns (Fig. 7-5), where each level was constrained by the level before. In particular, Kirtland's Warbler cannot be expected to occupy habitat classes predictably when the context of the regional population and habitat area are so dynamic. The regional population may simply be insufficient to fill one or more habitat types. Thus, a temporal component is often necessary to evaluate the significance of current species population densities (Probst et al. 2003). Although the Kirtland's Warbler has a smaller range than that of many bird species, a stepwise approach can still be used for larger-scaled multispecies assessments (Fig. 7-2). A grassland plan for prairie grouse (Vodehnal and Haufler 2007) incorporates most of the ranges of four species and can interface with state-level planning for these species and others in or adjacent to the interregional planning area.

FUTURE DIRECTION

Level of Resolution 7: Testing Syntheses: Integrating Monitoring into Science

Traditionally, monitoring is conducted principally to document trends, while research is done to explain mechanisms determining the trends. In contrast, we propose that monitoring be used to test hypotheses so that the scientific method becomes an integral part of monitoring conservation strategies. From this perspective, monitoring is logically delayed within a stepwise approach

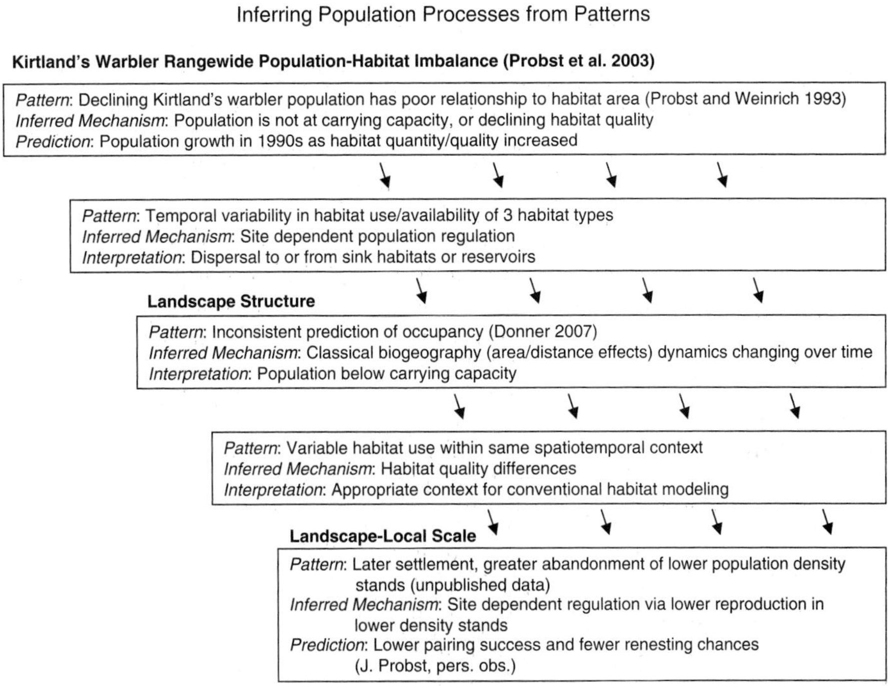

FIG. 7-5

Stepwise inferences about mechanisms controlling patterns of Kirtland's Warbler habitat occupancy. Each inference about process leads to examination of another pattern, so detailed habitat modeling is deferred until processes are elucidated.

until early inferences are made from survey data. Initial surveys are essentially the observation phase of the scientific method, and targeted surveys and long-term monitoring data structured in time and space are used to test hypotheses. Sequential surveys (Levels 3, 4) can target processes such as source-sink interactions before long-term monitoring is established.

Past approaches to resource management monitoring tended to consider only a few species or resources as indicators. However, the indicator approach is unlikely to cover the range of biodiversity concerns, and surrogate representation of other species by ecological indicators is questionable (Mannan et al. 1984; Verner 1984; Landres et al. 1988; Noon et al., this volume). Monitoring groups of species, or guilds, may moderate some of the problems of an indicator approach (Verner 1984, Tilghman and Verner 1989), but even monitoring several representative guilds may not provide a comprehensive evaluation of cumulative management effects. Another improvement is to consider as many species as possible at lower resolution before initiating intensive field studies. Once population sizes are estimated initially and population dynamics and interrelationships are postulated using a stepwise approximation strategy, developing monitoring plans to test hypotheses becomes possible.

Survey and monitoring applications in past approaches are frequently separated from the scientific process. Some established monitoring work should certainly be continued to maintain long-term baselines, but new surveys could be planned within a conceptual framework designed to document resource conditions by stepwise approximation. Such a framework is likely to direct surveys to specific, contrasting locations rather than a random, systematic, or stratified sampling scheme. Any monitoring that is distributed randomly or evenly may be suboptimal if not based on less-intensive sampling. Surveys might be short-term initially, but can be replaced by specific higher resolution work as needed. Indeed, sequential surveys, carefully planned in space and time, can be powerful tools for understanding the mechanisms of environmental change and drivers of population dynamics. In fact, there is a need to develop surveys, targeted toward areas or species groups that are dynamic in time, and that are not permanent monitoring plots and projects. Such mid-resolution surveys could be extremely incisive, using lower sampling intensity, and might best be assembled by interagency partnerships.

CONCLUSIONS

Researchers and managers often approach integrated wildlife management by using bottom-up applications of an indicator species approach, a pilot project, an ecosystem management demonstration area, or other narrowly focused but detailed approaches. Most detailed analyses must be repeated when placed in a broader context of space, time, and other resource issues. In contrast, solving problems using stepwise approximation strategies with nested, general models (Rubin 1991, Van Horne 2002) is a useful first step to achieve large-scale, integrated objectives, especially across academic disciplines.

In summary, the key concepts of a stepwise approximation approach for linking coarse scale and finer scale species population patterns are

1. Employ more than two spatial scales or several levels of resolution over a large geographic spatial context.
2. Consider most well-known species or several taxa at some common level of resolution before conducting detailed research or modeling.
3. Set ranges of environmental conditions (i.e., gradients) to establish context before doing categorical comparisons in research or modeling.
4. Initially use less precise models to facilitate geographic comparisons and integration across spatial scales.
5. Infer processes from presence-absence patterns in space and time to generate and test hypotheses about mechanisms driving observed patterns.

6. Assume that species are not at carrying capacity so that spatiotemporal heterogeneity can be captured and evaluated for its explanatory value.
7. Employ a hierarchy of simple or even qualitative models rather than overly precise predictive or correlative models until there is an adequate understanding of controlling processes or finer-scale mechanisms.

The stepwise approximation strategy can help produce better, more integrated conservation plans for multiple species at reduced cost, by deferring more detailed analyses until it has been determined that it is necessary.

SUMMARY

A major challenge for ecologists, resource managers, planners, and policy makers is the development of comprehensive wildlife conservation assessments that synthesize current scientific knowledge. We presented a stepwise approximation approach that incorporated a broad consideration of population patterns by emphasizing a larger geographic context before integrating more detailed studies at successively finer scales. Although there are various approaches to establishing context for assessments, we described one that integrates multiple levels of resolution in a stepwise manner to (1) delineate multispecies ranges to identify range overlaps and dissimilar distributions; (2) refine species distributions in terms of habitat-gradient relationships; (3) test species habitat-gradient relationships with species occurrence surveys; (4) evaluate habitat quality through abundance estimates and temporal change; (5) further refine habitat-quality evaluations with productivity and survivorship estimates; (6) synthesize levels 1–5 using maps and summary models to assess viability, conservation plans, or changing conditions; and (7) design monitoring strategies to test the validity of the synthesis. Stepwise approximation can connect coarse and fine scales using general to detailed species population patterns and can help produce more integrated conservation plans for multiple species at reduced cost, by deferring more detailed analyses until it has been determined that it is necessary.

ACKNOWLEDGMENTS

We thank Greg Corace, Tom Crow, Russell Graham, Peter Landres, Pam Nankervis, Anna Pidgeon, Brian Sturtevant, and several anonymous referees for critical reviews of the manuscript drafts. Deahn Donner edited and improved numerous manuscript and figure drafts and contributed valuable ideas. Sue Lietz provided technical assistance. The approaches and strategies presented here benefited from interactions with practitioners and land managers from government, industrial, and conservation organizations.

LITERATURE CITED

Askins, R. A., F. Chávez-Ramirez, B. C. Dale, C. A. Haas, J. R. Herkert, F. L. Knopf, and P. D. Vickery. 2007. Conservation of grassland birds in North America: Understanding ecological processes in different regions. *Ornithological Monographs* 64:1-46.

Avery, M. L., and C. van Riper, III. 1990. Evaluation of wildlife-habitat relationships database for predicting bird community composition in central California chaparral and blue oak woodlands. *California Fish and Game* 76:103-117.

Bart, J., C. Kepler, P. Sykes, and C. Bocetti. 1999. Evaluation of mist-net sampling as an index to productivity in Kirtland's Warblers. *Auk* 116:1147-1151.

Beissinger, S. R., and M. I. Westphal. 1998. On the use of demographic models of population viability in endangered species management. *Journal of Wildlife Management* 63:821-841.

Bissonette, J. A. 1997. Scale-sensitive ecological properties: Historical context, current meaning. Pages 3-31 *in* J. A. Bissonette, editor. *Wildlife and landscape ecology: Effects of pattern and scale.* Springer-Verlag, New York, New York, USA.

Block, W. M., M. L. Morrison, J. Verner, and P. N. Manley. 1994. Assessing wildlife habitat relationships models: A case study with California oak woodlands. *Wildlife Society Bulletin* 22:549-561.

Bollinger, E. K., and E. T. Linder. 1994. Reproductive success of neotropical migrants in a fragmented Illinois forest. *Wilson Bulletin* 106:46-54.

Boone, R. B., and W. B. Krohn. 2000. Predicting broad-scale occurrences of vertebrates in patchy landscapes. *Landscape Ecology* 15:63-74.

Boulinier, T., J. D. Nichols, J. E. Hines, J. R. Sauer, C. H. Flather, and K. H. Pollock. 1998. Higher temporal variability of forest breeding bird communities in fragmented landscapes. *Proceedings of the National Academy of Sciences* 95:7497-7501.

Brewer, R., and K. G. Harrison. 1975. The time of habitat selection by birds. *Ibis* 117:521-522.

Brewer, R., G. A. McPeek, and R. J. Adams, Jr., editors. (1991). *The atlas of breeding birds of Michigan.* Michigan State University Press, East Lansing, Michigan, USA.

Brown, J. H. 1984. On the relationship between abundance and distribution of species. *American Naturalist* 124:255-279.

Brown, J. H. 1995. *Macroecology.* University of Chicago Press, Chicago, Illinois, USA.

Cogan, C. B. 2002. Biodiversity conflict analysis at multiple spatial scales. Pages 229-240 *in* J. M. Scott, P. J. Heglund, M. L. Morrison, J. B. Haufler, M. G. Raphael, W. A. Wall, and F. B. Samson, editors. *Predicting species occurrences: Issues of accuracy and scale.* Island Press, Washington, D.C., USA.

Collins, S. L. 1983. Geographic variation in habitat structure of the black-throated green warbler (*Dendroica virens*). *Auk* 100:382-389.

Corace, R. G., III. 2007. *Using multiple spatial scales to prioritize openland bird conservation in the Midwest.* Dissertation, Michigan Technological University, Houghton, Michigan, USA.

Crow, T. R., D. T. Cleland, D. M. DonnerWright, E. J. Gustafson, D. E. Lytle, L. R. Parker, J. Probst, L. A. Schulte, B. R. Sturtevant, and P. A. Zollner. 2006. Managing Midwestern landscapes using ecological principles. Pages 251-281 *in* J. Chen, S. C. Saunders, K. D. Brosofske, and T. R. Crow, editors. *Linking ecology to landscape hierarchies.* Nova Science Publishers, New York, New York, USA.

Curtis, J. T. 1959. *The vegetation of Wisconsin: An ordination of plant communities.* University of Wisconsin Press, Madison, Wisconsin, USA.

Cushman, S. A., D. McKenzie, D. L. Peterson, J. Little, and K. S. McKelvey. 2007. *Research agenda for integrated landscape modeling.* U.S. Forest Service, Rocky Mountain Forest Research Station, General Technical Report RMRS-GTR-194, Fort Collins, Colorado, USA.

Donner, D. M. 2007. *Spatiotemporal response of the male Kirtland's Warbler population to changing landscape structure over 26 years.* Dissertation, University of Wisconsin, Madison, Wisconsin, USA.

Freemark, K. E., J. R. Probst, J. B. Dunning, and S. J. Hejl. 1993. Adding a landscape ecology perspective to conservation and management planning. Pages 346-352 *in* D. M. Finch and P. W. Stangel, editors. *Status and management of neotropical migrant birds.* U.S. Forest Service, Rocky Mountain Forest and Range Experiment Station, General Technical Report RM-229, Fort Collins, Colorado, USA.

Gibbs, J. P., and J. Faaborg. 1990. Estimating the viability of ovenbird and Kentucky warbler populations in forest fragments. *Conservation Biology* 4:193–196.

Gleason, H. A. 1926. The individualistic concept of the plant association. *American Midland Naturalist* 21:92–110.

Graham, R. W. 1994. Late Pleistocene faunal changes as a guide to understanding effects of greenhouse warming on the mammalian fauna of North America. Pages 76–87 *in* R. L. Peters and T. E. Lovejoy, editors. *Global warming and biological diversity.* Yale University Press, New Haven, Connecticut, USA.

Greenwood, P. J., and P. H. Harvey. 1982. The natal and breeding dispersal of birds. *Annual Review of Ecology and Systematics* 13:1–21.

Grzybowski, J. A., D. J. Tazik, and G. D. Schnell. 1996. Regional analysis of black-capped vireo breeding habitats. *Condor* 96:512–544.

Hansen, A. J., S. L. Garman, B. Marks, and D. L. Urban. 1993. An approach for managing vertebrate diversity across multiple-use landscapes. *Ecological Applications* 3:481–496.

Herkert, J. R. 1991. Prairie birds of Illinois: Population response to two centuries of habitat change. *Illinois Natural History Survey Bulletin* 34:393–399.

Howe, R. W., G. J. Davis, and V. Mosca. 1991. The demographic significance of sink populations. *Biological Conservation* 57:239–255.

James, F. C., and C. F. McCulloch. 2002. Predicting species presence and abundance. Pages 461–466 *in* J. M. Scott, P. J. Heglund, M. L. Morrison, J. B. Haufler, M. G. Raphael, W. A. Wall, and F. B. Samson, editors. *Predicting species occurrences: Issues of accuracy and scale.* Island Press, Washington, D.C., USA.

Jennings, M. D. 2000. Gap analysis: Concepts, methods and recent results. *Landscape Ecology* 15:5–20.

Johnson, C. J., and M. P. Gillingham. 2004. Mapping uncertainty: Sensitivity of wildlife habitat ratings to expert opinion. *Journal of Applied Ecology* 41:1032–1041.

Johnson, D. H., and G. A. Sargeant. 2002. Toward better atlases: Improving presence-absence information. Pages 391–398 *in* J. M. Scott, P. J. Heglund, M. L. Morrison, J. B. Haufler, M. G. Raphael, W. A. Wall, and F. B. Samson, editors. *Predicting species occurrences: Issues of accuracy and scale.* Island Press, Washington, D.C., USA.

Johnson, K. N., and M. Herring. 1999. Understanding bioregional assessments. Pages 341–376 *in* K. N. Johnson, F. Swanson, M. Herring, and S. Greene, editors. *Bioregional assessments: Science at the crossroads of management and policy.* Island Press, Washington, D.C., USA.

Karl, J. W., P. J. Heglund, E. O. Garton, J. M. Scott, N. M. Wright, and R. L. Hutto. 2000. Sensitivity of species habitat-relationship model performance to factors of scale. *Ecological Applications* 10:1690–1705.

Landres, P. B., J. Verner, and J. W. Thomas. 1988. Ecological uses of vertebrate indicator species: A critique. *Conservation Biology* 2:316–328.

Lanyon, S. M., and C. F. Thompson. 1986. Site fidelity and habitat quality as determinants of settlement patterns in male painted buntings. *Condor* 88:206–210.

Levin, S. A. 1992. The problem of pattern and scale in ecology. *Ecology* 73:1943–1967.

Mannan, R. W., M. L. Morrison, and E. C. Meslow. 1984. Comment: The use of guilds in forest bird management. *Wildlife Society Bulletin* 12:426–430.

Maurer, B. 1999. *Untangling ecological complexity: The macroscopic perspective*. University of Chicago Press, Chicago, Illinois, USA.

May, R. M. 1994. The effects of spatial scale on ecological questions and answers. Pages 1–17 *in* P. J. Edwards, R. M. May, and N. R. Webib, editors. *Large scale ecology and conservation biology.* 35th Symposium of the British Ecological Society with the Society for Conservation Biology. Blackwell Scientific Publications, London, United Kingdom.

McGarigal, K., and S. A. Cushman. 2005. The gradient concept of landscape structure. Pages 112–119 *in* J. A. Wiens and M. R. Moss, editors. *Issues and perspectives in landscape ecology.* Cambridge University Press, Cambridge, United Kingdom.

Menges, E. S., and D. R. Gordon. 1996. Three levels of monitoring intensity for rare plant species. *Natural Areas Journal* 16:227–237.

Morrison, M. L. 2001. A proposed research emphasis to overcome the limits of wildlife-habitat relationship studies. *Journal of Wildlife Management* 65:613–623.

O'Neill, R. V., D. L. DeAngelis, J. B. Waide, and T. F. H. Allen. 1986. *A hierarchical concept of ecosystems.* Princeton University Press, Princeton, New Jersey, USA.

Oreskes, N., K. Shrader-Frechette, and K. Belitz. 1994. Verification, validation and confirmation of numerical models in the earth sciences. *Science* 263:641–646.

Pickett, S. T. A., and M. L. Cadenasso. 1995. Landscape ecology: Spatial heterogeneity in ecological systems. *Science* 269:331–334.

Pidgeon, A. M., V. C. Radeloff, and N. E. Mathews. 2003. Landscape-scale patterns of black-throated sparrow (*Amphispiza bilineata*) abundance and nest success. *Ecological Applications* 13:530–542.

Pregitzer, K. S., P. C. Goebel, and T. B. Wigley. 2001. Evaluating forestland classification schemes as tools for maintaining biodiversity. *Journal of Forestry* 99:33–40.

Price, J., S. Droege, and A. Price. 1995. *The summer atlas of North American birds.* Academic Press, San Diego, California, USA.

Probst, J. R. 1979. Oak forest bird communities. Pages 80–88 *in* R. M. Degraaf and K. E. Evans, technical coordinators. *Proceedings of the workshop, management of North Central and Northeastern forests for nongame birds.* U.S. Forest Service, North Central Forest Experiment Station, General Technical Report NC-51, St. Paul, Minnesota, USA.

Probst, J. R. 1986. A review of factors limiting the Kirtland's Warbler on its breeding grounds. *American Midland Naturalist* 116:87–100.

Probst, J. R. 1988. Kirtland's Warbler breeding biology and habitat management. Pages 28–35 *in* T. W. Hoekstra and J. Capp, compilers. *Integrating forest management for wildlife and fish: 1987 Society of American Foresters National Convention.* U.S. Forest Service, North Central Forest Experiment Station, General Technical Report NC-122, St. Paul, Minnesota, USA.

Probst, J. R., D. M. Donner, C. I. Bocetti, and S. Sjogren. 2003. Population increase in Kirtland's Warbler and summer range expansion to Wisconsin and Michigan's Upper Peninsula. *Oryx* 37:365–375.

Probst, J. R., and D. M. DonnerWright. 2003. Fire and shade effects on ground cover structure in Kirtland's Warbler habitat. *American Midland Naturalist* 149:320–334.

Probst, J. R., and J. P. Hayes. 1987. Pairing success of Kirtland's Warbler in marginal versus suitable habitat. *Auk* 104:234–241.

Probst, J. R., and F. R. Thompson, III. 1996. A multi-scale assessment of the geographic and ecological distribution of Midwestern neotropical migratory birds. Pages 22–40 *in* F. R. Thompson, III, editor. *Management of Midwestern landscapes for the conservation of neotropical migratory birds.* U.S. Forest Service, North Central Forest Experiment Station, General Technical Report NC-187, St. Paul, Minnesota, USA.

Probst, J. R., and J. Weinrich. 1993. Relating Kirtland's Warbler population to changing landscape composition and structure. *Landscape Ecology* 8:257–271.

Raphael, M. G., and B. G. Marcot. 1986. Validation of a wildlife habitat relationships model: Vertebrates in a Douglas-fir sere. Pages 129–138 *in* J. Verner, M. L. Morrison, and C. J. Ralph, editors. *Wildlife 2000: Modeling habitat relationships of terrestrial vertebrates.* University of Wisconsin Press, Madison, Wisconsin, USA.

Reynolds, R. T., R. T. Graham, and M. H. Reiser. 1992. *Management recommendations for the northern goshawk in the southwestern United States.* U.S. Forest Service, Rocky Mountain Forest Research Station, General Technical Report RM-217, Fort Collins, Colorado, USA.

Root, T. L., and S. H. Schneider. 1995. Ecology and climate research: Strategies and implications. *Science* 269:334–341.

Rubin, E. S. 1991. Benefit-cost implications of acid rain controls: An evaluation of NAPAP integrated assessment. *Journal of Air and Waste Management Association* 41:914–921.

Shaffer, M. L., and F. B. Samson. 1985. Population size and extinction: A note on determining critical population sizes. *American Naturalist* 125:144–152.

Short, H. L., and J. B. Hestbeck. 1995. National biotic resource inventories and gap analysis. *BioScience* 45:535–539.

Simberloff, D. 1997. Flagships, umbrellas, keystones: Is single-species management passé in the landscape era?. *Biological Conservation* 83:247–257.

Smallwood, K. S., B. Wilcox, R. Leidy, and K. Yarris. 1998. Indicators assessment for habitat conservation plan of Yolo County, California, USA. *Environmental Management* 22:947–958.

Smith, V. J., and J. A. Jenks. 2002. Ranked modeling of small mammals based on capture data. Pages 441–446 *in* J. M. Scott, P. J. Heglund, M. L. Morrison, J. B. Haufler, M. G. Raphael, W. A. Wall, and F. B. Samson, editors. *Predicting species occurrences: Issues of accuracy and scale.* Island Press, Washington, D.C., USA.

Starfield, A. M. 1997. A pragmatic approach to modeling for wildlife management. *Journal of Wildlife Management* 61:261–270.

Stauffer, D. F. 2002. Linking populations in habitats: Where have we been? Where are we going? Pages 53–62 *in* J. M. Scott, P. J. Heglund, M. L. Morrison, J. B. Haufler, M. G. Raphael, W. A. Wall, and F. B. Samson, editors. *Predicting species occurrences: Issues of accuracy and scale.* Island Press, Washington, D.C., USA.

Swanson, F. J., and S. Greene. 1999. Perspectives on scientists and science. Pages 56–68 *in* K. N. Johnson, F. Swanson, M. Herring, and S. Greene, editors. *Bioregional assessments: Science at the crossroads of management and policy.* Island Press, Washington, D.C., USA.

Thomas, J. W. 1994. Forest ecosystem management assessment team: Objectives, process and options. *Journal of Forestry* 92:12–19.

Thompson, F. R., III, J. R. Probst, and M. G. Raphael. 1993. Silvicultural options for neotropical migratory birds. Pages 353–362 *in* D. M. Finch and P. W. Stangel, editors. *Status and management of neotropical migratory birds.* U.S. Forest Service, Rocky Mountain Forest and Range Experiment Station, General Technical Report RM-229, Fort Collins, Colorado, USA.

Tilghman, N. G., and J. Verner. 1989. Featured species, guilds and species richness: Three approaches to wildlife habitat management in northeastern forests. Pages 66–83 *in* J. C. Finley and M. C. Brittingham, editors. *Proceedings of the 1989 Penn State Forest Resources Issues Conference*. State College, Pennsylvania, USA.

Timothy, K. G., and D. F. Stauffer. 1991. Reliability of selected avifauna information in a computerized information system. *Wildlife Society Bulletin* 19:80–88.

Urban, D. L., R. V. O'Neill, and H. H. Shugart, Jr. 1987. Landscape ecology. *Bioscience* 37:119–127.

Van Horne, B. 2002. Approaches to habitat modeling: The tensions between pattern and process and between specificity and generality. Pages 63–72 *in* J. M. Scott, P. J. Heglund, M. L. Morrison, J. B. Haufler, M. G. Raphael, W. A. Wall, and F. B. Samson, editors. *Predicting species occurrences: Issues of accuracy and scale.* Island Press, Washington, D.C., USA.

Van Horne, B., and J. A. Wiens. 1991. *Forest bird habitat suitability models and the development of general habitat models*. U.S. Fish and Wildlife Service, Fish and Wildlife Research 8, Washington, D.C., USA.

Verner, J. 1964. Evolution of polygamy in the long-billed marsh wren. *Evolution* 18:252–261.

Verner, J. 1984. The guild concept applied to management of bird populations. *Environmental Management* 8:1–14.

Villard, M.-A., P. R. Martin, and C. G. Drummond. 1993. Habitat fragmentation and pairing success in the ovenbird (*Seiurus aurocapillus*). *Auk* 110:759–768.

Vodehnal, W. L., and J. B. Haufler, compilers. 2007. *A grassland conservation plan for prairie grouse*. North American Grouse Partnership. Fruita, Colorado, USA.

Whittaker, R. H. 1956. Vegetation of the Great Smoky Mountains. *Ecological Monographs* 26:1–80.

White, D., P. G. Minotti, J. J. Barczak, J. C. Sifneos, K. E. Freemark, M. V. Santelmann, C. F. Steinitz, A. R. Kiester, and E. M. Preston. 1997. Assessing risks to biodiversity from future landscape change. *Conservation Biology* 11:349–360.

Wiens, J. A. 1981. Single-sample surveys of communities: Are the revealed patterns real? *American Naturalist* 117:90–98.

Wiens, J. A. 1989. Spatial scaling in ecology. *Functional Ecology* 3:385–397.

Wiens, J. A. 2002. Predicting species occurrences: Progress, problems and prospects. Pages 739–750 *in* J. M. Scott, P. J. Heglund, M. L. Morrison, J. B. Haufler, M. G. Raphael, W. A. Wall, and F. B. Samson, editors. *Predicting species occurrences: Issues of accuracy and scale.* Island Press, Washington, D.C., USA.

Wiens, J. A., J. T. Rotenberry, and B. Van Horne. 1987. Habitat occupancy patterns of North American shrubsteppe birds: The effects of spatial scale. *Oikos* 48:132–147.

Williams, E. J., and D. N. Pashley. 2000. Designation of conservation planning units. Pages 44–47 *in* R. Bonney, D. N. Pashley, R. J. Cooper, and L. Niles, editors. Strategies for bird conservation: The Partners in Flight planning process. *Proceedings of the Third Partners in Flight Workshop.* U.S. Forest Service, Rocky Mountain Forest Research Station, RMRS-P-16, Ogden, Utah, USA.

Zimmerman, J. L. 1982. Nesting success of dickcissels (*Spiza americana*) in preferred and less preferred habitats. *Auk* 99:292–298.

CHAPTER

An Emerging Agency-Based Approach to Conserving Populations Through Strategic Habitat Conservation

8

Rex R. Johnson, Charles K. Baxter, and Michael E. Estey

We describe a framework for strategic habitat conservation (SHC) that enables the efficient maintenance of wildlife populations at objective levels through protection of existing habitat, habitat restoration, and habitat manipulation. The conventional model of habitat conservation by fish and wildlife agencies has for many species, at best, slowed the rate of long-term population decline. This is largely due to insufficient conservation resources compared to ever increasing human pressures on natural systems and to insufficient regulatory authorities. We believe that both are primarily due to three factors: lack of explicit and socially accepted conservation objectives; lack of clear compelling conservation strategies that describe why populations have declined and what may be done to aid their recovery; and a limited ability to demonstrate the population effects of our conservation actions. Collectively, these factors contribute to a lack of awareness by the public, elected officials, and representatives of other government agencies, which reduces the credibility and influence of wildlife conservation agencies.

The traditional approach to conservation in many areas can be characterized as an agency operating with limited awareness of the goals and the potentially beneficial or adverse activities of other agencies working in the same landscapes. Planning is often viewed as onerous and the plans themselves as static documents with limited value. Research and monitoring may be perceived to be expensive luxuries with little relevance to making conservation decisions.

Conversely, the SHC approach is planning intense; requires the integration of planning, conservation delivery, monitoring and research; and benefits from inter-agency collaboration and coordination. The approach is essentially a business model, and the concept of a conservation business model is gaining acceptance (Keen and Qureshi 2006). Successful businesses must articulate their purpose, develop products, identify target markets and marketing strategies, and create feedback loops that ensure product quality and continued viability in a competitive environment (Prahalad and Hamel 1990, Drucker 1994, Keen and Qureshi 2006). These elements are developed into a business strategy that includes communication and marketing tools designed to inspire investor confidence. A conservation strategy serves the same purposes.

The idea of inspiring investor confidence may initially sound like an odd concept for government agencies; however, the competition for public funding may be as intense as competition in the marketplace. Inspiring investor confidence requires that agencies demonstrate their ability to efficiently achieve results. This may be the greatest failing of the traditional conservation paradigm. Even small budget increases are often accompanied by an implicit expectation that perceptible benefits will result. Failure to produce these perceptible benefits reduces public and policy maker (i.e., investor) confidence. Although the general magnitude of the challenge of conserving populations at objective levels may be intuitive to conservation professionals, most of the public and elected officials are lay people who routinely lack this understanding.

For wildlife conservation agencies to be more successful and increase support by the public, it is imperative that these agencies be more explicit about objectives, strategies, and estimated costs of attaining objectives. Furthermore, we believe that wildlife conservation agencies must more fully adopt the role of stewards and purveyors of the biological foundation for conservation, seeking to influence the actions of other government agencies and inform policy makers and public perceptions. We believe that developing and communicating explicit, science-based habitat conservation strategies are critical to building this support and that the concepts we present in this chapter can help remedy current deficiencies.

The framework we describe places models in a useful context of the larger conservation enterprise. It is based on our personal experiences in attempting to meet the information needs of managers in government wildlife conservation agencies, and it is not a synthesis of the extensive literature on theories of conservation biology. Although many of the concepts we describe are well known within the scientific community, their application is still novel within most conservation agencies. As agencies implement this framework, they will be more efficient, transparent, and accountable and ultimately more credible and effective in informing the actions of policy makers and other agencies. We believe that if scientists understand the state of agency-based strategic conservation, they may recognize their role in facilitating it.

AN OVERVIEW OF SHC

We define SHC as an iterative process of, first, setting explicit objectives for populations and systematically figuring out how to achieve them *most efficiently* using agency resources and by working with partners. Or, more specifically, it is a process of developing and refining a conservation strategy, making

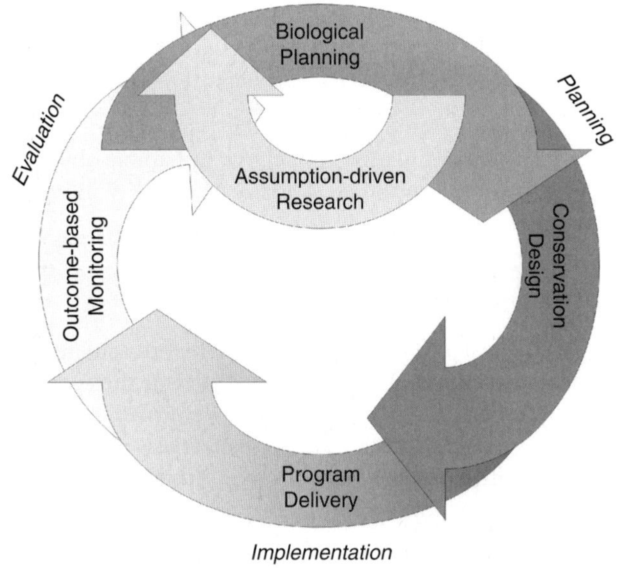

FIG. 8-1

The strategic habitat conservation cycle.

efficient conservation decisions based on that strategy, and using research and monitoring to assess accomplishments and inform future iterations of the conservation strategy (Fig. 8-1). SHC is a form of adaptive resource management (Walters 1986, Walters and Holling 1990, Williams 2003) wherein habitat conservation at multiple spatial scales is the primary form of intervention.

The goal of SHC is to make natural resource conservation agencies more efficient and transparent and, in part, thereby making them more credible and wide-reaching in effect (Johnson et al. 2006). *Conservation efficiency may be thought of as the ratio of population impacts to conservation costs*. Science-based habitat conservation strategies are developed to increase efficiency over random or haphazard conservation delivery. This approach presumes that sites vary in their potential to affect populations in a predictable fashion, and habitat managers are able and willing to prioritize their actions (i.e., operate strategically).

A comprehensive habitat conservation strategy should address the following questions:

1. Why have long-term average populations declined?
2. What do we want to achieve and how can we achieve it?
    a. What are our objectives for populations?
    b. What factors are acutely limiting populations below objective levels?
    c. What conservation treatments are available to overcome these limiting factors?

3. Where should we apply these conservation treatments to effect the greatest change in populations at the lowest possible total monetary and nonmonetary costs to conservation agencies and society?
4. How much of a particular type of conservation will be necessary to reach our population objectives (a habitat objective—a minimum estimate, but useful nonetheless for reasons we will describe)?
5. What are the key uncertainties in the answers to questions 1–4 and what assumptions were made in developing the strategy that will guide our research and monitoring activities?

In the case of federal and state fish and wildlife conservation agencies, it is usually most appropriate to ask and answer these questions in terms of populations; however, these basic questions are equally applicable to other ecosystem functions. Other agencies and organizations with different mandates may focus on these other functions by applying the same basic concepts.

Efficient conservation requires that agencies strategically apportion their resources at broad scales. This commonly means that agencies must undertake SHC in multiple regions, since the relationship of a species to its habitats is likely to vary among major ecoregions. Strategic habitat conservation will be more efficient when it is applied to ecoregions for which species of concern, population-habitat relationships, including limiting factors, and possible future threats to habitats are relatively homogeneous. This enables the use of strategies tailored to a particular part of a species' range and to a particular season of the year, if necessary, and it also enables more reliable inferences from research and monitoring. Conducting SHC within ecologically based regions such as Bird Conservation Regions (U.S. Fish and Wildlife Service 1999; Sauer et al. 2003; Fitzgerald et al., this volume) is a logical way to apply SHC across a state, country, or continent.

SHC TECHNICAL ELEMENTS

We focus on the technical elements of SHC—*biological planning*, *conservation design*, *assumption-driven research*, and *mission-based monitoring* (Fig. 8-1). These elements are not a rigid, linear sequence of events (Fig. 8-2). Biological planning, conservation design, and research and monitoring blend together in an iterative process. However, the process achieves its full value only when all five elements, including conservation delivery (Fig. 8-1), are in place.

Biological Planning

Biological planning is the systematic application of scientific knowledge about species and habitat conservation. It includes articulating measurable population objectives for selected species, considering what may be limiting populations to

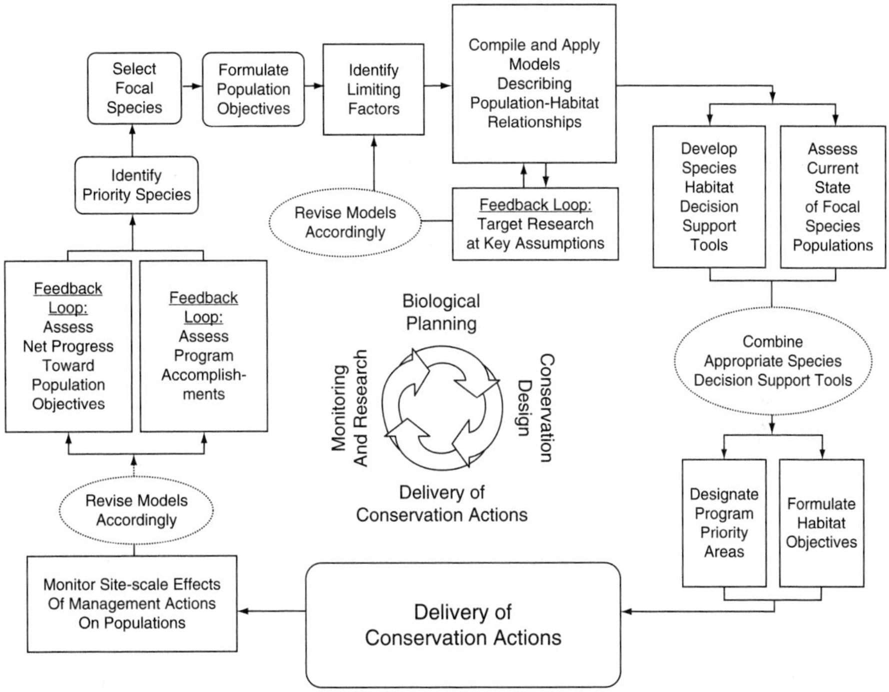

FIG. 8-2

Important elements in the iterative, strategic habitat conservation approach to conservation.

less than objective levels, and compiling models that describe how populations are expected to respond to specific habitat conservation actions.

Select Focal Species.— Strategic habitat conservation is focused on priority species whose populations are at less than desired levels. Ideally, we would model species-habitat relationships and spatial patterns in conservation potential for every priority species. The use of focal species, however, is usually necessary because trying to integrate information about too many species representing key ecological processes can become overwhelming.

Focal species are used as surrogates for the needs of larger guilds of species that use habitats and respond to conservation similarly (Noon et al., this volume); however, focal species may be more sensitive to patch characteristics, landscape context, or habitat conservation (Lambeck 1997, 2002). Other focal species may have unique habitat needs (e.g., some threatened and endangered species) or may be keystone species and therefore important determinants of ecosystem function (Mills 2007). Hagan and Whitman (2006) provided a valuable overview of the use of indicator species. They recommended selecting 5–15 species that are sensible indicators of the ecological communities and processes stakeholders value most. Of course, the assumption that other species and ecological processes will respond as predicted to habitat conservation for focal species must be evaluated (Lambeck 2002).

Lambeck (1997, 2002) and Lindenmayer et al. (2002) suggested that the use of multiple focal species will typically be more satisfactory than the use of a single umbrella species. There is no single prescription for selecting focal species or the number of focal species (Hagan and Whitman 2006, Mills 2007). Focal species may be selected for biological, socioeconomic, programmatic, or political reasons. One useful method for selecting focal species may be to start by assigning species to guilds based on their basic habitat needs and response to conservation. One or more focal species may be selected from each guild (Appendix A).

Because one outcome of SHC is an objective for each general habitat type, it will often be important to also select focal species with large enough population objectives to ensure adequate habitat to meet public demand for these species. Often these will be high-profile game species that are actually less limited in their habitat use than some other species. The U.S. Fish and Wildlife Service (USFWS) may be better served by selecting focal species that help us make better decisions about managing our trust resource responsibilities. Likewise, partners may select the focal species that best meet their needs. This does not preclude a continuous dialogue with partners, but agencies with different trust responsibilities plan separately for focal species and *then* integrate the outcomes of the biological planning processes.

Set Population Objectives.—Unlike some past approaches to conservation, which have tended to view activities like wetland restoration or reforestation as objectives, SHC requires explicit objectives for populations because most agencies are charged with the conservation of populations—not habitat. Efficient conservation strategies can be developed only after unambiguous mission-based objectives are established.

If an agency *was* simply to conserve habitats, an objective like "restore 5,000 ha of wetlands in the Great Lakes ecoregion of the United States each year" might be adequate. However, an activity-based objective like this does not promote accountability because no explicit relationship has been established between the habitat objective and the mandate to conserve populations. It is an objective without a clear ending point and without benchmarks for success (i.e., the objective is to do more wetland restoration each year). Of significant concern, a habitat objective without a clearly articulated set of predicted population outcomes provides no justification for increased resources for conservation because there are no explicit predictions for the public or policy makers about the consequences of succeeding or failing to attain the objective.

Strategic habitat conservation is founded on objectives expressed as desired population states, such as "maintain an average annual capacity to produce 1.7 million duck recruits per year in the Great Lakes ecoregion of the United States." These are "mission-based objectives." Efficient attainment of a mission-based objective requires that we know the current state of the system relative to the objective, make informed assumptions about environmental factors that are limiting populations below objective levels, determine where and how conservation can most effectively remediate these limiting factors, and monitor population state relative to the objective. Furthermore, we acknowledge that site and landscape-scale factors interact to affect the population impacts of conservation. Thus, *where* we deliver conservation is an important determinant of *how much* habitat is required to sustain populations at objective levels.

Population objectives may be more useful if they are composed of desired abundance and a performance indicator. For convenience, we will refer to these as P_1 and P_2 subobjectives, respectively. Examples of hypothetical population objectives might be

1. Maintain a population of 1250 moose (*Alces alces*) (P_1) in northwestern Minnesota with a mean annual calf:cow ratio of 0.84 (P_2);
2. Increase king rail (*Rallus elegans*) density 300% (P_1) at marsh bird survey sites and maintain a mean annual nesting success of 60% (P_2) in the southeastern coastal plain ecoregion; or
3. Maintain 25 distinct stream segments (P_1) with stable or increasing (P_2) breeding populations of lake sturgeon (*Acipenser fulvescens*) in Michigan.

In each case, the P_1 subobjective enables us to estimate how much habitat we need to conserve based on model-based abundance estimates, or where information on a species is more limited, predictions of relative habitat suitability, territory size, or average density in suitable habitat. Above minimum viable population sizes, P_1 subobjectives are value-based expressions of how many individuals of a species we want, or, more accurately, that we believe the public wants and will support. Ecoregional-scale P_1 objectives should be stepped down from range-wide objectives when these broad-scale goals exist; doing so links local conservation actions to state, national, or continental strategies and vice versa.

P_2 subobjectives, which are commonly vital rates, describe how we want to affect the population. If we believe that some habitats yield higher productivity or density than others, the P_2 subobjective should help us decide how to configure or treat the habitats we conserve. In practice, it will often be necessary to express P_2 subobjectives as assumptions about the effects of conservation.

Although vital rates are notoriously difficult to estimate, monitoring both P_1 and P_2 subobjectives paints a much clearer picture of how we are influencing focal species populations and ecological function than we would get from monitoring abundance alone because estimating short-term trends from annual abundance data often requires unrealistically intensive monitoring. For some species, P_1 and P_2 subobjectives may be combined, as in the previous Great Lakes duck example, in terms of number of recruits produced, rather than a P_1 subobjective for number of a breeding pairs and a P_2 subobjective for recruitment rate.

Identify Limiting Factors and Appropriate Conservation Treatments.— The purpose of habitat conservation is to relieve the constraints limiting factors impose on population size. "The presence and success of an organism or group of organisms depends upon a complex of conditions. Any condition which approaches or exceeds the limits of tolerance is said to be a limiting condition or a limiting factor...first and primary attention should be given to factors that are operationally significant to the organism at some time during its life cycle" (Odum 1971, pp. 110–111). One purpose of SHC is to identify areas where these limiting factors can be most efficiently alleviated, i.e., areas where potential population impacts are relatively high, conservation costs are relatively low, and tactics are socially acceptable.

Limiting factors are often related to the appropriate area, type, quality, or configuration of habitat necessary to sustain a population at objective levels. For example, consider a hypothetical example in which low reproductive success in small forests limits populations of a species of interior forest breeding bird. There are not enough large patches to sustain the population at objective levels of abundance. Individuals that settle in small patches fail to recruit young into the population, so the population must be maintained by birds that are able to settle in large patches. Once we understand the limiting factor, several potential conservation treatments designed to increase recruitment or survival may be considered:

1. Use reforestation to create more large patches;
2. Focus on increasing nonbreeding survival;
3. Use nest predator and nest parasite control in small patches; and
4. Raise birds in a hatchery and release them into the wild.

Generally, one or two conservation treatments will be most practical and compatible with our goals for the ecosystem and the other species that inhabit it. In this case, managers would likely choose reforestation as the preferred conservation treatment—coalescing small patches where recruitment is low into larger patches where recruitment is higher. If survival remains the same and reproductive success increases in response to increasing patch size, the population will grow toward objective levels. Hence, a primary purpose of the conservation strategy for the guild of interior forest breeding birds in this ecoregion would be strategic targeting of reforestation to most efficiently increase the number or area of large patches.

Develop and Apply Models.— Developing an efficient conservation strategy requires that we understand the relationship between populations and limiting factors. A defining feature of SHC is the application of models to spatial data to target specific conservation treatments. Models are simply a means of organizing our science to aid in understanding how a system functions by expressing real relationships in simplified terms (Starfield and Bleloch 1991; Millspaugh et al., this volume).

Whether aware of it or not, almost all managers use models to predict the probable outcomes of applying a particular conservation practice at a particular site. The difference between this intuitive approach to modeling and the more deliberate use of models in SHC is that, in the latter, models are stated in explicit, measurable terms. The advantages of explicitly stating and systematically applying models are that

1. Models and the products of applying them are useful for communicating the scientific foundation for actions, decisions, and recommendations, thereby yielding greater transparency and credibility;
2. The process of explicitly stating a model enables critical evaluation of uncertainties and assumptions and thus
    a. Determine how confident we should be about our predictions; and
    b. Target critical uncertainties with research to make future predictions more reliable;
3. Models may be used to report accomplishments expressed as estimated population effects.

Model predictions must be expressed in the same terms as population objectives to (1) estimate the amount of habitat conservation necessary to attain population objectives; and (2) to facilitate estimates of project, program, or agency accomplishments and net progress toward population objectives. The implications are that the information available to create models will affect the form of model predictions, which in turn affect the expression of population objectives. Thus, data collection, model development, and population objectives are iterative within the overall cycle of SHC. Although it is tempting to focus on using models to make maps of conservation priority areas, these other benefits of using models are often just as important.

For our hypothetical focal species of forest breeding bird example, we believe that it is really the ratio of patch edge:area that limits recruitment rate; as patches become larger and blockier, recruitment rate goes up. Thus, answering the questions of *where* and *how much* requires that we use models that describe the relationship between the ratio of perimeter:area and recruitment rate (Fig. 8-3). In this example, we see that after the perimeter:area ratio exceeds 0.1 (about 50 ha for a square patch, larger for irregularly shaped patches), further increases in recruitment rate begin to slow down. We have reached the point of diminishing returns. A strategic approach to attaining our objectives, as informed by this model, would indicate that once we have reached a ratio of 0.1, we should move on to a new patch rather than continue to make the same patch bigger and bigger for less and less additional benefit.

The value of a model is measured by the extent to which it adds useful information to the conservation of focal species. Generally speaking, as model

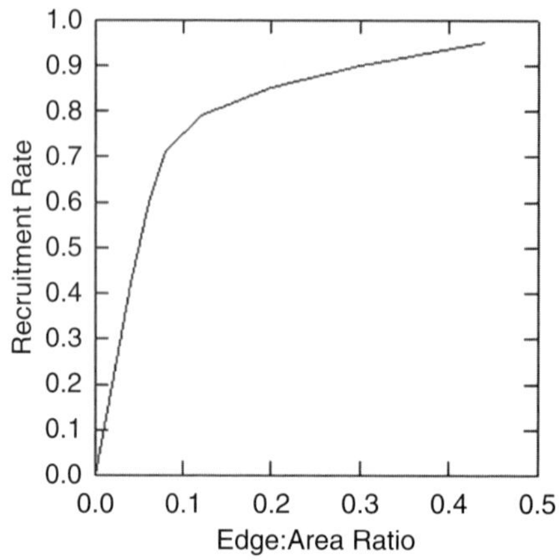

FIG. 8-3

A hypothetical relationship between interior forest breeding bird recruitment and the ratio of forest patch edge:area.

complexity goes up, so does the added value for decision making because model predictions move beyond our capacity for intuition. Advances over the past two decades in spatial data management enable depiction of complex multidimensional biological models in two-dimensional (map) form that contribute to a better understanding of how conservation potential varies among landscapes. However, models and the maps derived by applying them to spatial data have inherent uncertainties that should be explicitly acknowledged, but this is rarely the case. Burgman et al. (2005) recommended using multiple competing models in pursuit of robust conservation strategies that are likely to result in tolerable outcomes, despite uncertainties.

Numerous types of models are described in this book. We describe the most basic dichotomy among types of models as data-based (empirical) and experience-based (conceptual) models. Niemuth et al. (this volume) present empirical models for breeding duck access to grasslands, sora (*Porzana carolina*) use of wetlands, and empirical and conceptual models for marbled godwits (*Limosa fedoa*).

Both empirical and conceptual models may be used to predict factors (in increasing sophistication) such as probability of occurrence or apparent habitat suitability, abundance or density, and demographic rates such as productivity or survival (Fig. 8-4). Each may be estimated in relative or absolute terms. Generally, models tend to be more data-driven and less experience-based as the sophistication of their predictions increases. For example, although conceptual modeling like that for marbled godwits in Niemuth et al. (this volume) is useful for predicting relative apparent habitat suitability, the outcome of estimating abundance using a purely conceptual approach would be less certain. However, if apparent habitat suitability is all we are able to reliably predict, we may still

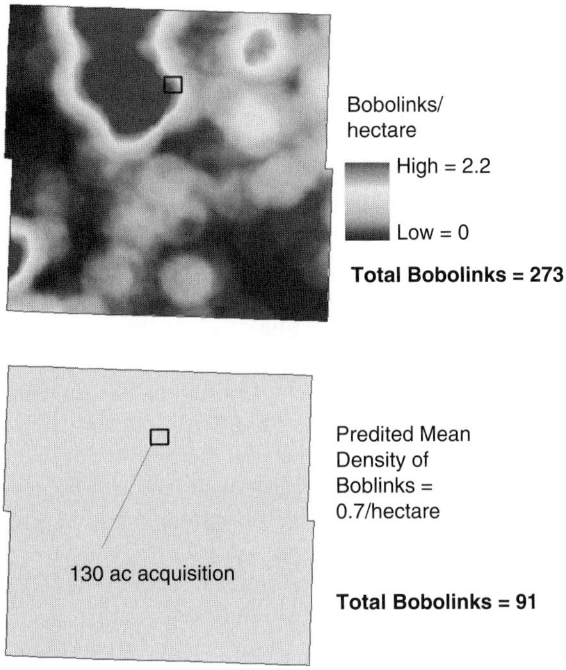

FIG. 8-4 (See color insert.)

The predicted relative abundance of bobolinks (*Dolichonyx oryzivorus*) in Grant County, Minnesota, assessed for 1 ha pixels (top) and at a county scale (bottom) and a hypothetical 52 ha grassland acquisition. County or large watershed-scale assessments make the unreasonable assumption that the entire area has homogeneous conservation potential.

predict abundance using empirically derived average density estimates from "suitable" versus "unsuitable" or "less suitable" sites.

Estimating the effects of habitat conservation on population vital rates is an ideal that is presently impossible for most species because appropriate data for model development do not exist. However, models could be constructed to estimate probability of occurrence, relative abundance, or habitat suitability for most species although many would have to be experience-based and may contain untested assumptions.

Conservation Design

Conservation Design is predicated on the belief that the potential to affect populations varies in space in response to site characteristics and landscape context. If not, it matters little where we manage habitat. The development of maps predicting patterns in the ecosystem is the outstanding feature of Conservation

Design. Maps that are not based on the systematic application of science can be misleading and may impede conservation success. Maps used in SHC are the product of applying empirical or experience-based models to spatial data. Hence, we propose the use of the phrase "spatially explicit models" (SEMs) in lieu of maps to emphasize that developing and applying models relating a species to limiting habitat factors is the essence of SHC.

Assess the Current State of the Ecosystem.— A conservation strategy is a route between the current state of the system and the objective state. Models used to create SEMs also may be used to estimate the current state of the system. The current state of the system must be expressed in the same units as population objectives. The objective state minus the current state represents a conservation deficit to be made up as efficiently as possible. Note that the deficit is expressed in terms of populations, not area of habitat.

Develop Species-Specific SEMs.— Spatially explicit models will generally be specific to a focal species and a conservation treatment that affects that species (e.g., targeting a particular conservation treatment like reforestation or wetland protection to address factors limiting populations below objective levels). For example, the ornate box turtle (*Terrapene ornata*) is listed as threatened or endangered in a number of midwestern states. Population declines are primarily attributable to the loss of sandy-soil grasslands and road-related mortality within remnant populations. Conservation treatments, therefore, include strategic grassland restoration on sandy sites (away from roads) and road signage placed around known or suspected populations. In this situation, an SEM for ornate box turtles may be based on a simple empirical or experience-based model with only two variables: land cover and soil type. Spatially explicit models derived from these models may be combined with data on the distribution of roads to identify areas with existing or potential turtle populations for population surveys, potential population restoration sites, and sites to erect road signage.

The resolution of SEMs should match or be smaller than the scale at which conservation occurs. Maps of large geographic units like counties or major watersheds may be deceptive because they implicitly include the unreasonable assumption that conservation anywhere within the county or watershed will yield the same outcomes (i.e., they are simply too coarse to reflect site and landscape effects on potential conservation outcomes). The geographic units assessed using models and portrayed in SEMs should be small parcels that more or less match, or are finer than the typical scale of conservation (i.e., as fine as possible but generally <256 ha) (Fig. 8-4).

Spatially explicit models typically include an assessment of the potential of every part of the ecoregion to impact a population or set of populations. This means that geographic units with high, moderate, and even low potential to affect populations are included. This is important because managers typically only deal with willing landowners, and it is not always possible to limit conservation to the highest priority sites; and a conservation action with a lower predicted biological impact may still be efficient if costs are low enough.

Formulate Habitat Objectives.— Habitat objectives are developed for habitat types, not species. The size of an objective for a particular habitat type depends on the diversity of species that depend on it, their population objectives, and on their range of responses to conservation. For example, grassland habitat objectives for an ecoregion will be smaller if every priority species prefers idled grasslands than if some prefer idled and some disturbed habitat because the potential for aggregate population impacts is greater for each acre.

Habitat objectives may be expressed for the total area of habitat in public and private, protected and unsecured status, or they may be defined more specifically, such as the number of hectares to be restored and placed in the conservation estate. In theory, if we know the capacity of every hectare to contribute toward our population objective for each focal species, we can simply tally up the smallest area (cost per ha being equal) that overcomes the aggregate conservation deficit. Of course, this is an absolute minimum estimate of the amount of habitat that will actually be required to achieve population objectives, since it is almost never possible to work exclusively in the areas with the greatest potential to affect populations.

Since potential to affect populations varies in response to site characteristics and landscape context, the relative efficiency with which we make up the conservation deficit and attain our population objectives depends on our ability to act strategically by operating at sites with the greatest potential to affect each focal species' populations and reconcile potential management conflicts. Because managers cannot typically work exclusively within the highest priority sites, estimates of the amount of habitat needed to attain our population objectives will likely be underestimates. Nonetheless, explicit habitat objectives based on population-habitat relationships enable us to convey to policy makers and stakeholders the extent of actions required to conserve populations. While some deviation from our strategy is inevitable, close adherence to it by limiting our conservation actions to high-priority sites will help ensure that our habitat objectives, while minimal, come close to providing the anticipated population response. Timely adjustments to habitat objectives can be made based on recent conservation accomplishments, new scientific information, and other influences on habitat due to policy changes and socioeconomic factors.

Designate Priority Areas.— Priority areas can only be delineated in the context of explicit objectives. "Show me the best areas for conservation" is not a satisfactory question on which to base conservation assessment. Because no site is likely to actually have high value for every species, some interpretation of relative priority is necessary. "Show me the set of sites with the greatest conservation potential to affect species X" and "Show me the set of sites with the greatest aggregate potential to affect species X, Y, and Z" are more appropriate questions.

Multiple species-specific SEMs may be integrated to assess the relative potential of each unit of the landscape to yield aggregate population benefits consistent with unique program, agency, and partner priorities. Caution must be used in combining SEMs because prediction errors propagate in the overlay process and because not all species that could occur at a site have compatible habitat needs or responses to site scale management (e.g., burning). Before combining SEMs, we need to (1) know what species or environmental benefits a program emphasizes the most; (2) know what treatments can be employed under a program; and (3) thoughtfully integrate SEMs based on management compatibility.

Different partners will often be most interested in benefiting different combinations of species. Thus, while it may be possible to designate a single set of priority areas for a specific program, it is seldom practical for conservation partnerships. This is why developing a portfolio of focal species by treatment SEMs is important. Once created, SEMs can be rapidly combined to match the unique priorities of programs, agencies, and partners (i.e., a portfolio of SEMs provides a rapid response capability to inform conservation). Optimization models have been used to select areas for conserving species richness or the occurrence of selected species and to factor in costs associated with acquisition or management (see Flather et al., this volume; Haight and Gobster, this volume), but potentially could be used to select areas to meet population objectives for a group of focal species.

Biodiversity and Species Richness Maps.— Although no single standard definition exists for biodiversity, it is commonly interpreted as the totality of genes, species, and ecosystems of a *region*. Thus, concepts of biodiversity conservation have little utility at the pixel, patch, or local scales at which conservation actions actually occur. Instead, conserving biodiversity requires balancing the area and configuration of habitats needed by the full array of species within an ecoregion. Biodiversity indices are often implicitly emphasized over species-based approaches to strategic conservation (Simberloff 1998). However, rather than a one-size-fits-all approach to program delivery, the appropriate approach is to conserve and manage tracts such that ecoregional biodiversity is conserved, with each agency contributing to biodiversity conservation consistent with its specific conservation mandate and priorities.

Contributing to the conservation of biodiversity is undeniably a high priority; however, SHC is founded on being explicit, measurable, and communicable. Unless a measurable and universally acceptable definition of biodiversity can be developed, it cannot be described in a mission-based objective. Because explicit definitions of biodiversity are elusive (Wilson 1997), other measures like species richness are often equated to biodiversity conservation potential. Maps of species richness are easy to produce using modern geographic information system techniques. Species richness maps are commonly based on data such as range maps or species occurrence. Abstract goals such as maximizing species richness at patch scales are inappropriate, as implementing plans that emphasize high local diversity can reduce overall (gamma) diversity (Noss 1987) and are of little use for programs that typically have a more narrowly defined purpose when they are established. Maps of species richness are likely to identify ecotones, mountains, and river corridors as priority areas because they have greater habitat diversity although they are often poor habitat for many priority species. We have the following concerns about maps of species richness as they are commonly portrayed:

1. Occurrence data are notoriously subject to errors, particularly errors of omission (e.g., where no one has looked for a species). This is particularly true for uncommon, candidate, or listed species;
2. The approach is not founded on explicit objectives or predictions of population response; there are normally no benchmarks against which to assess accomplishments;
3. The approach makes limited use of the biological foundation available for many species including factors that are limiting populations and thus
   a. Provides little information about how and where conservation can be effectively used for species recovery, especially using habitat restoration;
   b. Provides no means of estimating conservation effects on populations, which is critical for targeting conservation and for estimating accomplishments; and
   c. Provides no foundation for assumption-driven research;
4. Habitat heterogeneity is often the most important factor in determining species richness. Number of species and habitat heterogeneity are often poor predictors of the importance of a site for conservation;
5. Conservation compatibility is often not explicitly considered. For example, both American woodcock (*Scolopax minor*) and cerulean warbler (*Dendroica cerulea*) may be assigned to mixed deciduous forest tracts although the two species respond very differently to stand age and common forest management practices;
6. Estimates of species richness are scale-dependent and common scales of assessment (e.g., large hexagons, hydrologic units, or counties) are much larger than the scale at which conservation decisions are routinely implemented. The implicit message is that habitat conservation anywhere within a large geographic unit will provide equal benefits to the full array of species. This assumption is usually unwarranted. Inferences resulting from assessment at fine scales (e.g., 30 m pixels, 16 ha parcels) can be generalized to larger geographic units, but coarse scale assessments cannot be broken down to make fine-scale inferences.

For these reasons, maps of species richness within hydrologic units or counties are not useful tools for SHC. Nevertheless, maps of species richness are often compelling, as is the misperception that they are surrogate predictions of biodiversity. As such, they can inadvertently impede more sophisticated approaches to biological planning and conservation design, based on a critical assessment of trust responsibilities, program authorities and priorities, population objectives, limiting factors, management compatibility, and spatial scales. Although single-species planning and conservation seem to be falling out of favor in the scientific literature, developing a portfolio of species-specific assessment products enables a rapid response to requests to designate priority areas tailored to a program's unique authorities and priorities, including species priorities.

Evaluation

Although our knowledge of ecological systems will always be incomplete, agencies must still make conservation decisions using the best information to guide their actions. Models force us to make assumptions about limiting factors and their effects on populations, and this highlights uncertainties in the biological foundation for conservation. The advantages of an iterative process of SHC are two-fold with respect to reconnecting management and science. On one hand, the overall process is a systematic means of applying the existing, albeit incomplete, biological foundation about how species relate to habitats and management at local and landscape-scales. However, science is primarily a means of learning. The scientific method is founded on articulating hypotheses (assumptions in the planning process) and then setting out to try to disprove them (evaluation through research). Without monitoring and research, SHC is not an iterative process by which managers learn and increase their effectiveness.

Assumption-Driven Research.— Not all assumptions made in biological planning are equally important. Two important criteria for evaluating assumptions are (1) how uncertain is the assumption, and (2) to what extent would better information affect conservation decisions. Assumptions that are both tenuous and high impact are priorities for research. For example:

Scenario 1—Research shows that soybean fields are used extensively by greater prairie chicken (*Tympanuchus cupido*) broods, even in the presence of adjacent native grasslands. Soybeans are superabundant at this time in the vicinity of grasslands used by prairie chickens, but soybean distribution varies annually.

Assumption 1: Soybeans are a preferred habitat for greater prairie chicken broods.

Conclusion: Limited uncertainty with little decision-making value of better information because of high but annually variable soybean abundance driven by market forces.

Scenario 2—Ornate box turtles are known to burrow extensively in sandy soils, but surveys are limited. There are presently no plans for box turtle releases or reintroductions.

Assumption 2: Ornate box turtles have a relative density at sites with sandy loam soils that is 200% greater than their density at sites with clay soils.

Conclusions: Considerable uncertainty but little value of additional information unless long-range conservation plans suggest releases or reintroductions will be necessary to sustain populations.

Scenario 3—Dabibling duck daily nest survival rates have been shown to vary with percent grass (+) and cropland (–) in the landscape (Greenwood et al. 1995, Reynolds et al. 2001). Unfortunately, the relationship is highly variable, and its exact nature has been difficult to ascertain.

Assumption 3: Waterfowl nesting success increases linearly with the percent grass within a 2 mi radius of a nest site (Reynolds et al. 2001).

Conclusions: Considerable uncertainty and considerable value of better information because millions of dollars are spent annually to protect grassland for upland nesting ducks, and millions more are spent to restore grasslands through programs like the Conservation Reserve Program.

Among the three hypothetical assumptions, Assumption 3 is the highest priority for research because of its degree of uncertainty and the potential benefits to conservation of obtaining better information. Assumption 3 may be restated as at least four competing hypotheses (Fig. 8-5):

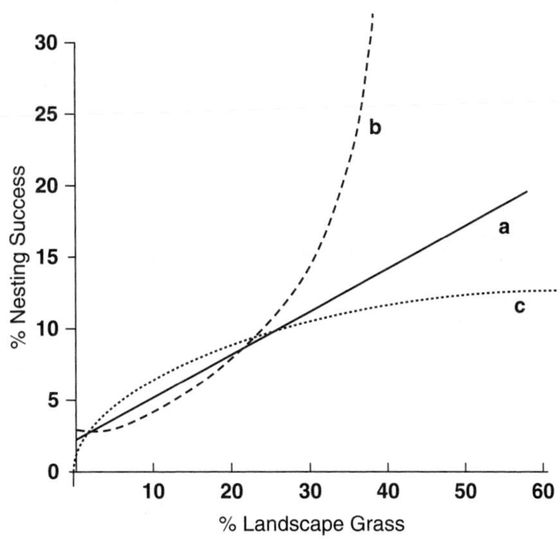

FIG. 8-5

Alternative relationships that describe a general trend of increasing waterfowl nesting success with increasing amounts of grass in a 3.2 km radius landscape. Each implies the need for a different grassland restoration strategy.

H_O: Nesting success and percent grassland are independent;

H_{A1}: Nesting success and percent grassland are positively and linearly related (the current assumption);

H_{A2}: Nesting success and percent grassland are positively related but the relationship is exponential; and

H_{A3}: Nesting success and percent grassland are positively related but the relationship is nonlinear and reaches an asymptote at about 40% grassland in the landscape.

If you are working exclusively within landscapes with <25% grasslands, the value of better information is minimal because all three hypotheses predict similar nesting success. However, the implications of obtaining better information about this relationship when working in landscapes with 25% or more grassland are huge. If the relationship is linear (curve a), restoration of grass in any location will yield the same incremental increase in nesting success. If curve b more accurately describes the relationship, an agency should invest all its grassland protection and restoration resources in a few sites until the entire landscape is grassland or nest success approaches 100%, whichever comes first. If curve c is the best fit, an agency should add grass to locations within landscapes with 25–40% grassland. Above 40% we should move on to other areas because additional grassland restoration will have less and less effect on increasing nesting success. If the null hypothesis (H_O) cannot be disproved, grassland protection and restoration would not seem to be a very effective treatment for increasing nesting success.

When research priorities are established as an outcome of biological planning, we are targeting mission-critical research, not simply indulging our intellectual curiosity. Thus, model-based biological planning helps an agency articulate its research priorities. Moreover, model-based biological planning is the means by which research results find their way into conservation decisions in the iterative SHC framework.

Outcome-Based Monitoring.— Conservation agencies should evaluate their actions based on (1) the effects of specific conservation actions on habitats and individuals; (2) program and agency accomplishments expressed in terms of population impacts; and (3) net progress toward population objectives.

Assessing the Effects of Conservation Actions.— To evaluate whether conservation actions have the predicted consequences, we need to monitor actual outcomes. For example, did the conservation action yield the expected habitat response, and did the change in habitat evoke the expected species response? Answers to the first question enable managers to adjust their tactics to more consistently achieve desired habitat conditions. The second question is the means whereby we compare observed and predicted population response at the site scale and refine our models of species-habitat relationships. This means that monitoring programs should be structured around the same ecoregions as biological planning to ensure efficient model updating. It may not be necessary to monitor the outcome of every conservation action, but monitoring outcomes with repeated counts at a valid sample of sites is essential.

Assessing Program and Agency Accomplishments.—Populations vary in space and time in response to a variety of short-term, uncontrollable environmental and anthropogenic factors. Population status and trend estimates tend to have high variances because of limited sample sizes and short-term environmental variations. Consequently, except for intensive sampling to assess long-term trends, actual counts of individuals often have little utility for assessing annual accomplishments. Rather than using highly variable counts of individuals, we can use models used to target conservation actions to estimate population impacts of conservation that actually occurred (Fig. 8-6). The sum of the estimated impacts of each individual conservation action is an agency's

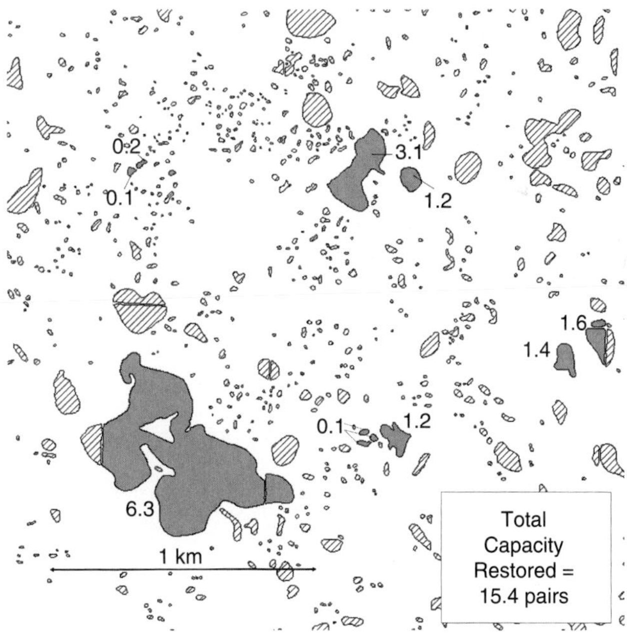

FIG. 8-6

Models, such as regression models, which can predict numbers of breeding ducks (Neimuth et al., this volume) can be used to target habitat conservation practices such as wetland restoration. They can also be used to estimate accomplishments in terms of population impacts. Crosshatched features are drained wetlands, solid features are restored wetlands, and numbers are the predicted increase in capacity to support breeding pairs in a 10 km^2 western Minnesota landscape.

accomplishments in accordance with the science built into the models. In other words, population monitoring at a sample of sites used to assess the effects of conservation actions is used to indirectly assess accomplishments, with model refinement and estimation as the intermediate step. This approach to accomplishment reporting has two important implications:

1. Managers should report their annual accomplishments in terms of predicted aggregate population effects. Overall program and agency accomplishments are the sum of the output of individual managers; and

2. Program and agency accomplishments will be expressed in the same terms as mission-based population objectives.

Assessing Net Progress Toward Population Objectives.—Net progress toward population objectives is a function of habitat gains versus losses, both of which may be driven more by socioeconomic or long-term environmental factors than by agency accomplishments. Just like assessing agency accomplishments, assessing net progress toward population objectives is a model-driven process. Essential field data collection consists of site-scale data on species response to habitat codified in models, as noted previously; and ecoregional, national, or continental data on habitat abundance, distribution, and quality (e.g., from regularly updated land cover data). Most broad-scale (national or continental) population monitoring has not compiled data on habitats, with little effort to systematically monitor population responses to habitat at site-scales. Continued broad-scale surveillance monitoring of populations is still warranted, because if model-based predictions do not match observations of species status or trends, it is likely that an important limiting factor has been overlooked.

CONCLUSIONS

We believe that SHC can make conservation planning

1. More efficient at habitat conservation because of the ability to estimate biological benefits relative to conservation costs;

2. More transparent and defensible because actions are based on a systematic application of the best available science;

3. More strategic in allocating limited research and monitoring funds;

4. More compelling at communicating the magnitude and nature of the conservation challenges and the strategies proposed to address them;

5. More accountable;

6. More wide-reaching in informing agencies and policy makers, contributing to greater leadership in the conservation.

Although from time to time the focus may shift from one element to another, SHC is a continuous iterative process of overlapping elements that occur both sequentially and simultaneously: biological planning, conservation design, conservation delivery, assumption-driven research, and outcome-based monitoring. Conservation strategies are dynamic suites of objectives, tactics, and tools that change as new factors or information influence the system. The very act of doing assumption-driven research and monitoring implies a commitment to continuous replanning using better information about how a species responds to its habitat and conservation actions. Furthermore, external forces operate on habitats and populations, and our strategies must acknowledge their effects on the attainment of our objectives. The SHC framework is designed to promote learning about populations and how they respond to habitat. By following the adaptive cycle of planning, doing and evaluating, and replanning described as SHC, we continuously move toward more and more reliable conservation decisions. The elements of conservation strategies—objectives, tactics, spatially explicit models of priority areas, monitoring programs, etc.—are all subject to change as new information becomes available or new forces operate on the system.

SUMMARY

We described a framework for strategic habitat conservation (SHC)—biological planning, conservation design, conservation delivery and monitoring, and research—that enables efficient conservation of wildlife populations. Strategic habitat conservation is gaining greater acceptance among conservation agencies that historically tended to manage habitat opportunistically, often without regard to site and landscape heterogeneity and the magnitude of potential population responses. We described the basic elements of SHC as they are being communicated within and among agencies, including most importantly establishing explicit, outcome-based objectives and the use of models relating populations to limiting habitat factors. Although many of the concepts we described are well known within the scientific community, their application is still novel within most conservation agencies. We believe that as agencies implement this framework, they will be more efficient, transparent, and accountable and ultimately more credible and effective in informing the actions of policy makers and other agencies. We present this chapter with the expectation that if scientists understand the state of agency-based strategic conservation, they may recognize their role in facilitating it.

ACKNOWLEDGMENTS

A number of individuals have contributed to refining these concepts and helped promote their understanding within conservation agencies. We especially wish to acknowledge the contributions of Neal Niemuth, Bill Uihlein, Ron Reynolds, Randy Wilson, and Seth Mott of the U.S. Fish and Wildlife Service, and Patricia Heglund and Wayne Thogmartin, U.S. Geological Survey (USGS). Members of the USFWS-USGS National Ecological Assessment Team and the SHC Technical Advisory Team are gratefully acknowledged for their contributions to this chapter.

LITERATURE CITED

Burgman, M. A., D. B. Lindenmayer, and J. Elith. 2005. Managing landscapes for conservation under uncertainty. *Ecology* 86:2007–2017.

Drucker, P. F. 1994. The theory of the business. *Harvard Business Review* 70:96–104.

Greenwood, R. J., A. B. Sargeant, D. H. Johnson, L. M. Cowardin, and T. L. Shaffer. 1995. Factors associated with duck nest success in the prairie pothole region of Canada. *Wildlife Monographs* 128:1–57.

Hagan, J. M., and A. A. Whitman. 2006. Biodiversity indicators for sustainable forestry: Simplifying complexity. *Journal of Forestry* 104:203–210.

Johnson, R. R., S. E. Mott, and D. A. Smith. 2006. Biological planning for bird conservation at the landscape scale: Developing shared conservation strategies in the United States. Pages 711–716 *in* G. C. Boere, C. A. Galbraith, and D. A. Stroud, editors. *Waterbirds around the world.* The Stationery Office, Edinburgh, United Kingdom.

Keen, P., and S. Qureshi. 2006. *Organizational transformation through business models: A framework for business model design*. Proceedings of the 39th Hawaii International Conference on System Sciences. CD-ROM ISBN 0-7695-2507.

Lambeck, R. J. 1997. Focal species: A multi-species umbrella for nature conservation. *Conservation Biology* 11:849–856.

Lambeck, R. J. 2002. Focal species and restoration ecology: RESPONSE to Lindenmayer et al. *Conservation Biology* 16:549–551.

Lindenmayer, D. B., A. D. Manning, P. L. Smith, H. P. Possingham, J. Fischer, I. Oliver, and M. A. McCarthy. 2002. The focal species approach and landscape restoration: A critique. *Conservation Biology* 16:338–345.

Mills, L. S. 2007. *Conservation of wildlife populations: Demography, genetics and management.* Blackwell Publishing, Malden, Massachusetts, USA.

Noss, R. F. 1987. Do we really want diversity? *Whole Earth Review* 1987:126–128.

Odum, E. P. 1971. *Fundamentals of Ecology.* WB Saunders Co., Philadelphia, Pennsylvania, USA.

Prahalad, C. K., and G. Hamel. 1990. The core competences of the organization. *Harvard Business Review* 68:79–93.

Reynolds, R. E., T. L. Shaffer, R. W. Renner, W. E. Newton, and B. D. Batt. 2001. Impact of the Conservation Reserve Program on duck recruitment in the U.S. Prairie Pothole Region. *Journal of Wildlife Management* 65:765–780.

Sauer, J. R., J. E. Fallon, and R. R. Johnson. 2003. Use of North American Breeding Bird Survey data to estimate population change for Bird Conservation Regions. *Journal of Wildlife Management* 67:372–389.

Simberloff, D. 1998. Flagships, umbrellas, and keystones: Is single-species management passé in the landscape era? *Biological Conservation* 83:247–257.

Starfield, A. M., and A. L. Bleloch. 1991. *Building models for conservation and wildlife management.* Second edition. Interaction Book Company, Edina, Minnesota, USA.

U. S. Fish and Wildlife Service. 1999. *A proposed framework for delineating ecologically-based planning, implementation and evaluation units for cooperative bird conservation in the U.S.* U.S. Fish and Wildlife Service, Washington, D. C., USA.

Walters, C. J. 1986. *Adaptive management of renewable resources*. MacMillan Publishing, New York, New York, USA.

Walters, C. J., and C. S. Holling. 1990. Large-scale management experiments and learning by doing. *Ecology* 71:2060–2068.

Williams, B. K. 2003. Policy, research, and adaptive management in avian conservation. *Auk* 120:212–217.

Wilson, E. O. 1997. Introduction. Pages 1–3 *in* M. L. Reaka-Kudla, D. E. Wilson, and E. O. Wilson, editors. *Biodiversity II: Understanding and protecting our biological resources.* Joseph Henry Press, Washington, D.C., USA.

APPENDIX A

Forested ecosystems may be characterized by stand composition and age structure. In this simple example, we describe stand composition as deciduous, coniferous, or mixed, and stand age as young or old. Species occur in one or more of these forest community types. We may start by constructing a matrix of forest types by age and assigning species (represented by letters) to guilds.

	Forest Type		
Stand Age	**Deciduous**	**Coniferous**	**Mixed**
Young	A,B,E,H,K,M,N	A,C,F,G	A,B,D,L,K
Old	A,B,I,J	A,C,F,G	A,B,D,Q,R

Note that species A is a habitat generalist that uses all our forest habitats, making it unsuitable as a focal species unless there are other compelling reasons to use it in the planning process. Note also that the species composition is the same in young and old age stands of conifers. Consequently, we will combine the two age classes in the planning process. Species C, F, and G require coniferous forest, but F is the most sensitive to patch size and landscape context.

Species E and H occur only in young deciduous stands; however, H is an interior forest breeding species requiring large block habitats, while E is area independent. We will use H as a focal species because its habitat needs are more restrictive. Similarly, species I and J require mature deciduous forests, but I is believed to be highly sensitive to disturbance along roads and trails, which J is not.

Lastly, species L occurs only in young mixed forests, and Q only in old-age mixed stands. Furthermore, species L is a popular hunted species with a high population objective. This factor alone recommends it as a focal species because it requires large amounts of habitat to attain population goals.

Thus, through the selection of focal species, planning for the conservation of 16 priority species has been consolidated into the development and application of models for five species: F, H, I, L, and Q. Of course, continued monitoring is necessary to ensure that populations of other species in the same guilds are responding as predicted. If not, they must be brought more directly into the conservation planning process.

CHAPTER

A Review of Methods for Quantifying Wildlife Habitat in Large Landscapes

9

Michael A. Larson, Joshua J. Millspaugh, and Frank R. Thompson, III

Spatially explicit information about land use and vegetation composition and structure is now available for landscapes (i.e., areas of large spatial extent) around the world. Advancements in remote sensing and related technologies have increased the resolution and quantity of landscape data, and efforts have been made to increase the availability of landscape data (e.g., national Gap Analysis Program; Scott et al. 1993, Jennings 2000). Advancements also have been made in desktop computers and software to model vegetation dynamics [e.g., LANDIS (He et al. 1999, 2005; Mladenoff and He 1999), FVS (Dixon 2002)] and quantify spatial patterns [e.g., FRAGSTATS (McGarigal and Marks 1995, McGarigal et al. 2002)]. Geographic information system (GIS) software allows managers and researchers to study the spatially explicit effects of management decisions and other disturbances on vegetation at large spatial scales and allows researchers to include complex spatial processes in models of wildlife-habitat relationships (Roloff et al., this volume).

For practitioners to take advantage of relatively recent advances in information about landscapes and the means of processing, they must select from among the many different methods of quantifying habitat quality in large landscapes. Fortunately, part of the reason so many options exist is that methods have been developed for a variety of uses related to the management of land and wildlife (Beck and Suring, this volume). Some methods are designed to provide a general index of habitat quality, whereas others attempt to use measures of habitat quality to predict animal presence, population density, or population viability. Also, methods have been developed to incorporate increasingly complex interactions between wildlife and landscapes and to reduce various technical limitations, such as statistical assumptions.

Papers addressing the topic of quantifying wildlife habitat in large landscapes occur in a wide range of primary literature, from wildlife and other ecology journals to those in remote sensing and urban planning, so reviews of this material are needed. Roloff et al. (2001*b*) provided an overview of models used in wildlife management and included guides to help managers select an appropriate category of habitat model for a specific use. Guisan and Zimmermann (2000) reviewed the steps of predictive habitat distribution modeling, including theoretical and technical aspects of model formulation and calibration, and they reviewed some specific statistical methods. Guisan and Thuiller (2005) discussed the history of, recent advances in, and methodological issues (e.g., how to select an appropriate spatial scale) related to species distribution modeling. None of these reviews, however, focused on the variety of GIS-based methods for quantifying the quality and quantity of wildlife habitat in large landscapes.

Our primary goal is to provide a synopsis of available techniques to help managers and other practitioners become aware of the many alternatives that exist. We provide a context for understanding the various techniques for quantifying the quality and quantity of habitat in landscapes by organizing them in a progression of increasing complexity in terms of their objectives, minimum data requirements, and key assumptions. These three criteria are useful when considering which modeling approach might be best for a specific management question or problem. We also provide citations for numerous examples of different modeling approaches.

COMPLEXITY GRADIENT

We classified methods of modeling wildlife habitat into five general approaches that correspond to common objectives in habitat studies (Table 9-1). These approaches generally follow a complexity gradient that represents increasing levels of potential realism in the models. More complex habitat models also tend to be more mechanistic rather than descriptive, and they often rely on more data to estimate parameter values for the purported wildlife–habitat relationships (Fig. 9-1). More complex models, however, are not necessarily better than less complex models (Starfield 1997; Millspaugh et al., this volume); the utility of a model depends only on how well suited it is for its intended use.

The order in which we discuss the five approaches follows the admittedly imperfect complexity gradient with one exception. Models for predicting the presence or absence of wildlife can be quite simple (e.g., a dichotomous habitat quality index) or much more complex (e.g., logistic regression), but we chose to discuss predicting presence in the same section as the probability of occurrence. Likewise, the correlation between model complexity and the modeler's reliance on data is somewhat contrived because the degree to which wildlife–habitat relationships in a model are supported by data rather than expert opinion can vary. The conceptual framework represented in Fig. 9-1, however, provides a practical starting point for considering the options currently available for habitat modeling in a GIS.

Table 9-1 Specific Techniques and Examples for Five General Approaches to Evaluating the Quality of Wildlife Habitat Over Large Landscapes Using a Geographic Information System (GIS). Some Main Strengths of Each Approach Include the Potential Applications, Whereas Some Main Weaknesses Include the Minimum Data Requirements and Assumptions.

General Approach	Objective	Specific Techniques	Potential Applications[a]	Minimum Data Requirements[b]	Key Assumptions[c]	Citations for Examples
Presence or absence	Predict presence Estimate species richness	Designate only 2 habitat quality categories Logistic regression	Mapping species distributions Identifying biodiversity hotspots	Expert opinion supporting putative wildlife–land cover associations	Habitat influences animal distribution	White et al. 1997 Smith et al. 1998
Habitat quality indexing	Quantify habitat quality	Habitat evaluation procedures revised for GIS Weighted linear combination procedures Object-oriented programming	Development mitigation or compensation Evaluating site quality for restoration	Expert opinion supporting putative wildlife–habitat relationships	All significant habitat variables are included	Rickers et al. 1995 Hepinstall et al. 1996 Rickel et al. 1998 Clevenger et al. 2002
Probability of occurrence	Predict presence Estimate species richness Quantify probability of occurrence	Logistic regression Occupancy estimation Discrete choice Mahalanobis distance Resource selection functions (RSF) Artificial neural networks	Mapping important resources Evaluating interspecific interactions	Habitat measures at used and often random locations	Habitat influences animal distribution Animal use estimates are accurate All significant habitat variables are included	Clark et al. 1993 Cooper and Millspaugh 1999 Mace et al.1999 Özesmi and Özesmi 1999 Cooper and Millspaugh 2001 Fleishman et al. 2001 MacKenzie et al. 2002

continues

Table 9-1 Specific Techniques and Examples for Five General Approaches to Evaluating the Quality of Wildlife Habitat Over Large Landscapes Using a Geographic Information System (GIS). Some Main Strengths of Each Approach Include the Potential Applications, Whereas Some Main Weaknesses Include the Minimum Data Requirements and Assumptions. *cont...*

General Approach	Objective	Specific Techniques	Potential Applications[a]	Minimum Data Requirements[b]	Key Assumptions[c]	Citations for Examples
Density	Predict abundance Predict density	Home range or territory mapping Application of RSF Pattern recognition Linear regression	Estimating carrying capacity Evaluating land management alternatives	Habitat measures at locations with varying animal density Density estimates at the locations	Animal density estimates are accurate Population and habitat quality are in equilibrium All significant habitat variables are included	Roloff and Haufler 1997 Bellamy et al. 1998 Boyce and McDonald 1999 McClain and Porter 2000 Penhollow and Stauffer 2000
Population viability	Determine if area is large enough for a viable population Evaluate the level of viability	Quantify availability of high-quality habitat Bayesian belief networks Demographic simulation Estimate population growth rates directly	Identifying limiting factors Endangered species risk assessment or recovery planning	Habitat measures at locations with individuals of varying fitness Fitness estimates of individuals at the locations	Animal fitness estimates are accurate Population and habitat quality are in equilibrium All significant habitat variables are included	Liu et al. 1995 Akçakaya and Raphael 1998 Edelmann et al. 1998 Moilanen and Hanski 1998 Marcot et al. 2001

[a] *Potential applications are examples that are not necessarily exclusive to a single general approach.*
[b] *All listed techniques require a digital map of each habitat characteristic. The presence or absence approach may require only a land cover map.*
[c] *All listed techniques require an assumption that the habitat maps are accurate.*

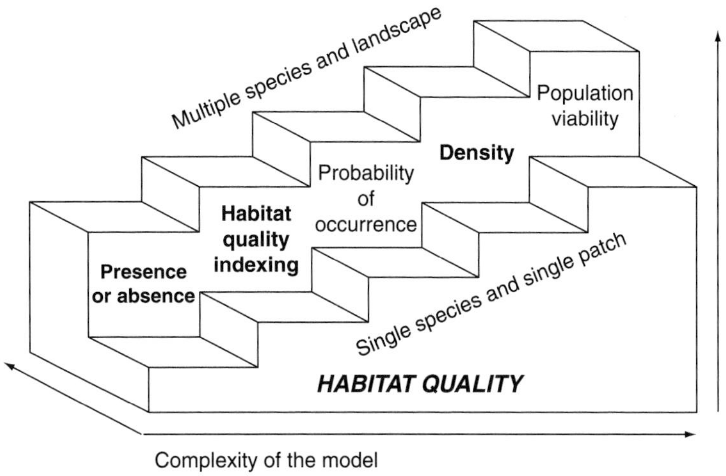

FIG. 9-1

A conceptual framework for GIS-based techniques used to evaluate the quality of wildlife habitat over large landscapes. Habitat quality forms the foundation of all analyses, and as the complexity of a model increases, so does the reliance on data to specify appropriate wildlife–habitat relationships. Additional methodological advancement is necessary for multispecies analyses, which are more complex than their single-species analogs.

Indexing Habitat Quality

Habitat Suitability Index (HSI) models are the basis of the Habitat Evaluation Procedure (HEP) developed by the U.S. Fish and Wildlife Service (1980, 1981) and constitute a basic mathematical technique for quantifying the quality of wildlife habitat. Expert opinion (Crance 1987) and published literature are used to develop functional relationships between the value of habitat attributes (e.g., % canopy cover) and index values between 0 (= not habitat) and 1 (= habitat of highest quality). In the HEP, which was the most common practice for modeling habitat until the 1990s, it was expected that a biologist applying the model collected field-based data to calculate HSI scores. This labor-intensive approach might limit the size of study sites and historically resulted in a focus on microhabitat characteristics (e.g., percent visual obstruction, density of woody stems). Now, however, it is common to use aerial photographs or satellite images of potentially large areas (e.g., 10s to 1000s of km^2) to quantify habitat attributes for use in HSI models (Breininger et al. 1991, Rempel et al. 1997, White et al. 1997, McClain and Porter 2000). Such remotely sensed data have also been used to directly define areas that provide suitable habitat and areas that do not (White et al. 1997). Models of wildlife–habitat relationships based on remotely sensed data often emphasize patch- and landscape-level characteristics rather than smaller-scale characteristics, but the distinction between microhabitat and landscape approaches is becoming less clear. Remotely sensed attribute data can be of high resolution (e.g., <1 m), field measurements of microhabitat characteristics from a sample can be extrapolated across a landscape based on cover types defined in a GIS, and habitat models can contain variables from different scales (Mitchell et al. 2001).

Geographic information system technology has transformed HSI modeling by allowing biologists to easily incorporate spatial structure (e.g., minimum area requirements, effects of habitat edges, juxtaposition of or distance to multiple life requisite resources, proportion of different land cover types), which is often appropriate and desirable (Donovan et al. 1987, Rickers et al. 1995, McClain and Porter 2000). For example, Gustafson et al. (1994) developed a proximity index ([$size_i$/nearest-neighbor $distance_i$] for all patches i=1 to n within a 300 m buffer) to incorporate the spatial configuration of forest patches in a habitat model for wild turkeys (*Meleagris gallopavo*). Another common technique to incorporate spatial structure is to evaluate habitat quality in a portion of the landscape (i.e., an analysis window) and then move the window across the landscape systematically (Riitters et al. 1997). Assuming a raster GIS is used, the HSI score for each window can be assigned to the central cell (= map pixel) or to all cells in the window, and the window can be moved 1 or many cells at a time. Window size (Roloff and Haufler 1997, Ortega-Huerta and Medley 1999, McClain and Porter 2000), percent window overlap, method of calculating the mean or interpolating HSI scores between window centers, and treatment of landscape boundaries are important considerations that may influence the results of a moving window habitat analysis (Hepinstall et al. 1996), but have not been studied comprehensively. We think window size should be ecologically meaningful, usually the scale at which the animal is believed to perceive, evaluate, and be influenced by habitat quality (e.g., home range size; Hildén 1965). Multiple window sizes relating to different aspects of habitat selection (e.g., nest site, foraging) could be used within a model for a single species. Given the high processing rate of modern desktop computers, we recommend moving windows 1 cell at a time because moving them any greater distance (e.g., with only 50% window overlap) requires interpolation and introduces unnecessary error in results of the moving window analysis.

Weighted linear combination (WLC) procedures for multicriteria evaluations (Voogd 1983) provide an alternative framework to HEP for habitat quality indexing (HQI). When WLC procedures are implemented in raster GIS (Eastman et al. 1995), values in maps of habitat attributes are standardized to the same unitless scale (e.g., 0–1). Then a matrix of pairwise comparisons between habitat variables is created from knowledge of or hypotheses about the relative importance of the individual variables to overall habitat quality (Saaty 1977). The principal eigenvector of the matrix contains weights that sum to 1. A habitat quality map is created by calculating the weighted mean of the standardized habitat variable scores in each raster cell. The advantages of the WLC approach are that pairwise-comparison matrices can simplify model development, most GIS software provides tools for implementing such models, and WLC is easily incorporated in a formal decision analysis (Eastman et al. 1995). Clevenger et al. (2002) used WLC procedures to incorporate expert opinion from individuals and published literature into assessments of black bear (*Ursus americanus*) habitat in Banff National Park, Alberta, Canada. They found that the literature-based model performed acceptably and similarly to a discriminant function when compared to independent data about where bears crossed a road, but the expert opinion model performed poorly.

An empirical alternative for indexing habitat quality involves calculating the Mahalanobis distance (D^2), or lack of similarity, which can be recoded to a 0–1 scale, between a vector of attributes of a raster cell (x) and the vector of mean attribute values from locations where animals have been observed (u), often obtained by radio-tracking (Clark et al. 1993). A main benefit of D^2 relative to other multivariate methods is that it incorporates the covariance matrix to standardize across variables and account for correlations among them. Knick and Rotenberry (1998), however, discovered that it was inappropriate to apply u and its covariance matrix in landscapes that were not sampled because differences in habitat characteristics (e.g., patch size) always result in greater D^2, even if a difference is in the direction of better habitat. They proposed as a solution using only the partitioned components of D^2 that vary the least and therefore represent a minimum rather than purportedly optimum set of habitat requirements (Rotenberry et al. 2002). Thatcher et al. (2006) and Browning (2005) provided recent examples of both approaches. As with other statistical models, it is important to include only habitat attributes that have hypothesized relationships to habitat quality to avoid spurious correlations.

The use of fuzzy systems and object-oriented programming (OOP) is a more sophisticated approach that allows for high-level abstractions of wildlife–habitat relationships and may represent the real world better than procedural programming (Booch et al. 1999). Fuzzy logic also can be used to incorporate the accuracy of spatial classification into habitat assessments. Although OOP is used to model animals and habitat patches as interacting objects, the approach is similar to other HQI modeling techniques because degree of membership scores, ranging from 0 to 1, are assigned to habitat patches based on the presumed quality of their habitat characteristics (Rickel et al. 1998). Furthermore, the main output, mean degrees of membership for a habitat patch, is analogous to an HSI score, which is usually a weighted mean of multiple suitability index scores on a 0–1 scale. A fundamental difference exists, however, between the theories underlying habitat models that are based on discrete categories of index values and those based on ambiguous, or fuzzy, categories (Hill and Binford 2002).

Predicting Presence and Probability of Occurrence

In some cases, models for predicting the presence or absence of a species in an area are less complex and require less data than an HQI model (Fig. 9-1) because one can simply define only two categories on the landscape: areas that provide habitat (= present) and areas that do not (= absent). Category definitions often are based on land use or land cover classifications (Aspinall and Veitch 1993, White et al. 1997, Smith et al. 1998), but they could be based on a minimum habitat quality threshold below which it is presumed the species will not occur (Table 9-1). For example, Browning et al. (2005) maximized the predictive gain by using as a threshold the index value at which the vertical distance between cumulative frequency distributions for percentage of study area and percentage of used sites was greatest. Liu et al. (2005) used common model assessment indices (e.g., sensitivity, overall prediction success) to compare 12 different approaches for determining a threshold value, assuming errors due to false positive species occurrences and false negative occurrences were equally important. They determined that the observed prevalence of species occurrence, mean predicted suitability (or probability of occurrence), and several joint functions of sensitivity and specificity (e.g., maximizing their sum) performed best. Maximizing the Kappa statistic performed worse, and subjective thresholds were worst. Regardless of category definitions and thresholds, using mutually exclusive categories to simplify habitat assessments can be problematic because true wildlife responses are unlikely to match the categories.

The theory and practice of modeling the presence of wildlife and estimating the probability of a site being occupied has advanced recently (Scott et al. 2002, MacKenzie et al. 2006). Logistic regression models (Coker and Capen 1995, Bellamy et al. 1998, Penhollow and Stauffer 2000, Fleishman et al. 2001), resource selection function (RSF; Manly et al. 2002) models (Mace et al. 1999), multiple-visit detection-nondetection surveys (MacKenzie et al. 2002), and artificial neural network models (Anderson 1995, Özesmi and Özesmi 1999) can be used to predict probability of occurrence or use from vegetation, landscape, and other habitat characteristics. Johnson and Gillingham (2005) directly compared Mahalanobis distance, RSF, and ecological niche models using the same training, habitat, and validation data. They also evaluated a qualitative HSI model that was based on expert opinion. Correlations between model predictions and validation data were high for the quantitative models ($r > 0.88$, $p \leq 0.02$) and lower for the HSI model ($r = 0.51, p = 0.30$). The quantitative models differed in how well they predicted areas of high versus low probabilities of occurrence, which resulted in low similarity among models in the spatial distribution of mapped habitats (Kappa $\leq$ 0.19). Dettmers et al. (2002) also compared multiple modeling methods (i.e., logistic regression, Mahalanobis distance, classification and regression tree [CART; Breiman et al. 1984], and discriminant function models), with a few qualifications mostly related to how the different models were applied to produce comparable results. Which model(s) performed best differed among the six bird species considered, but in general the discriminant function models performed well on the original study site but not when tested with independent data from another site; logistic regression and Mahalanobis distance models performed well at predicting probabilities of occurrence but performed worse at predicting presence, likely due to the need to apply a threshold value to define presence; and CART models were best at predicting bird presence.

Whereas the simplest models of presence and HQI models may be based solely on expert opinion, predicting probability of occurrence or use generally requires statistical models that are fit to data (Table 9-1). It is difficult to determine the absence of a species in a particular area, and presence is not perfectly related to habitat quality or quantity because the existence of habitat is a necessary but not sufficient condition for animal presence. Information about the presence or absence of a species, regardless of the modeling approach used for prediction, may be desired when species richness, rather than density or viability of any single species, is the response variable of interest (if so, see "Multispecies Approaches" later).

Estimating Population Density

When data exist or assumptions are justified to predict population density from habitat variables or index values, all the methods discussed in the preceding sections, especially RSF modeling, can be used to estimate density (Boyce and McDonald 1999). Even if seemingly appropriate data exist, several key assumptions are necessary to link habitat attributes to population density (or even presence). One must assume that knowledge of all limiting factors is included in the model, empirical data accurately reflect wildlife–habitat relationships, animals always select habitat perfectly, temporal fluctuations in habitat quality are adequately incorporated in the model, and the population has equilibrated with the habitat (Boyce and McDonald 1999). It also might be important to account for spatial autocorrelation in statistical models because different spatial scales of variation in wildlife abundance and habitat variables can affect inferences about wildlife–habitat relationships (Keitt et al. 2002). Relationships that allow the use of HQI models or models of species presence to make inferences about animal density (Gaston et al. 2000, Royle and Nichols 2003), however, should not be assumed to exist. In fact, the processes and variables that determine presence and density may differ, and even if positive relationships between presence and density exist, these relationships may change over time (Nielsen et al. 2005).

Various methods have been used to model densities of wildlife. Bellamy et al. (1998) used a 6-point scale of availability of large hardwoods to define suitable and unsuitable habitat for nuthatches (*Sitta europaea*) in woodlots of eastern Britain. Then they applied a nonlinear regression model developed for nuthatches in the Netherlands to predict expected number of breeding pairs from the area of suitable habitat in the woodlot and the surrounding landscape. Pereira and Itami (1991) multiplied the land area in each of several probability-of-occurrence categories by the mean density of squirrel activity areas in those categories to estimate habitat equivalents (U.S. Fish and Wildlife Service 1980), which represented the potential density of squirrels in portions of the landscape. McClain and Porter (2000) predicted population-density potential for white-tailed deer (*Odocoileus virginianus*) directly using pattern recognition (PATREC) methods (Williams et al. 1978), which seek patterns in habitat variables that are associated with areas of high and low animal abundance. Deer harvest data corrected for land accessibility, snow depth, and variation in hunting intensity at the scale of townships were used to develop the PATREC model and as a source of independent data for model validation. The PATREC model explained approximately twice as much of the variability in the independent deer harvest data as did an HSI model containing the same variables but parameterized based on a published study conducted in the same region. An advantage of the HSI model, however, was its greater spatial resolution.

Relative abundance of wildlife can be modeled using count data. Penhollow and Stauffer (2000) used multiple linear regression models to relate the mean number of detections of 23 bird species during point counts to GIS-based habitat characteristics, including forest age, overstory cover type, and eight metrics from FRAGSTATS. The species-specific models had adjusted R^2 values of 0.17–0.77, demonstrating how variable the success of applying a particular type of model can be among species. Royle et al. (2002) developed a statistically rigorous method for extrapolating information from counts of animals at points to a map of relative abundance over an entire region. The method accounts for spatial correlation in counts, incorporates habitat and other covariates, and allows for a spatially explicit assessment of uncertainty, which can be used to improve the allocation of subsequent survey effort. Thogmartin et al. (2004) applied this method to explore habitat relationships and map the relative abundance of cerulean warblers (*Dendroica cerulea*).

Roloff and Haufler (1997) developed a method of mapping home ranges that results in an estimate of population density. In their approach, home ranges are delineated using a GIS to accumulate a minimum number of habitat units (i.e., the area of a patch multiplied by its HSI score, summed over all patches of interest). The process begins with cells of the highest habitat quality and essentially follows an ideal free distribution (Fretwell and Lucas 1970). Furthermore, the mean HSI score of the focal cells for a home range must exceed a predefined habitat quality threshold of viability (Roloff and Haufler 1997). The population density is the product of the number of viable home ranges and the mean number of individuals in a home range divided by the total area of interest. Roloff and Haufler (2002) applied the same approach to the mapping of bird territories.

Evaluating Population Viability

Approaches to evaluating the viability of a population using habitat data fall into four categories: assessing availability of high-quality habitat, Bayesian belief networks (BBNs; Oliver and Smith 1990), population simulation, and estimating population growth rates directly. Availability of high-quality habitat can be estimated with home range mapping procedures (Roloff and Haufler 1997). First, a minimum viable population size must be specified. An area is then deemed sufficient to support a viable population if it is large enough to contain the minimum number of adequate home ranges. This approach may be useful for assessing the suitability of a reserve of limited size for animals that have large home ranges or territories. It can also be used to evaluate more subtle changes in viability due to changes in habitat quality (Roloff and Haufler 2002).

A BBN "depicts the logical or causal relations among ecological factors that influence the likelihood of outcome states of some parameter(s) of interest" (Marcot et al. 2001:30). Habitat-based BBN models can be used to predict virtually any habitat or population performance measure (i.e., from an index of habitat quality to population viability). To evaluate wildlife population viability, Marcot et al. (2001) specified input (e.g., habitat) variables and assigned them prior probabilities of being in a given state. The probability that intermediate (e.g., life requisite) variables were in a given state was conditional on (i.e., linked to) input and other intermediate variables that affected them. The final, posterior probability was the likelihood of a population being in one of five qualitative viability categories. The BBN models developed by Marcot et al. (2001) were integrated with GIS and were used to evaluate habitat at multiple scales.

Simulating population growth is a common method for population viability analysis (PVA; Beissinger and Westphal 1998; Beissinger et al., this volume; Akçakaya and Brook, this volume). Models for PVA range from deterministic, single population matrix projections (Caswell 2001) to stochastic, spatially explicit metapopulation simulations (Hanski and Simberloff 1997), with increasing complexity and data requirements. Although each type of model can be used to evaluate the effects of habitat quality and quantity on vital rates and the effects of vital rates on population viability, it is often easiest to account for spatial variability in habitat quality and vital rates in metapopulation models. Incidence function models (Hanski 1994) predict whether or not metapopulation units (i.e., patches) are occupied, usually from individual extinction and colonization probabilities (E_i and C_i) for each patch. Incidence function models often base E_i and C_i predominantly on patch size and interpatch distances, respectively. Moilanen and Hanski (1998) used patch-specific habitat quality information to influence the effective patch size and interpatch distances in their incidence function model. More specifically, they multiplied E_i and C_i by parameters that were, in turn, defined by third order polynomial functions of habitat quality variables. Donovan and Thompson (2001) used a modified matrix model to estimate metapopulation growth rates of a generalized migratory songbird in landscapes composed of various proportions of high- and low-quality habitat that differed in the reproductive success rates of birds nesting in them. Akçakaya and Raphael (1998) used a home range mapping procedure to define habitat patches (and their carrying capacity for northern spotted owls [*Strix occidentalis*]), which were then used as population units in a spatially explicit, habitat-based metapopulation simulation model (RAMAS GIS; Akçakaya 2000*a*, 2006; Akçakaya et al. 2004*a*). Nickelson and Lawson (1998) used regression models to estimate potential production and survival rates of coho salmon (*Oncorhynchus kisutch*) smolts for specific stream reaches from habitat data. Then they combined the vital rates for smolts with estimates of other population parameters in a reach-specific metapopulation simulation model. Liu et al. (1995) used habitat-specific fecundity in a third type of spatially explicit population simulation model, an individual-based model, to evaluate the population viability of Bachman's sparrows (*Aimophila aestivalis*). Liu (1993*a*) also discussed how the initial distribution of individual sparrows among patches that vary in habitat quality and the length of time each patch provides habitat can influence population viability.

As an alternative to data- and computer-intensive population simulations, population growth rates can be estimated directly using habitat quality data. Growth rates are indicators of viability, but they may not account for factors such as density dependence when they are estimated from habitat data. Fortunately, models relating habitat quality to population growth rates are flexible enough to overcome many potential drawbacks. Edelmann et al. (1998) developed regression models to estimate sage grouse (*Centrocercus urophasianus*) vital rates for each cell in a raster GIS from habitat data. A second type of regression model was used to estimate population growth rates (λ) from the cell-specific vital rates. Evaluation of population viability was based on the frequency distribution of λ in all cells. Similarly, Moore et al. (2000) used a quadratic function of stand age and basal area to model λ directly for a hypothetical forest bird metapopulation. They simulated stand-specific population sizes, recalculated λ at each time step because habitat quality changed over time, and incorporated source-sink dynamics (Pulliam 1988).

ADDITIONAL CONSIDERATIONS

Selecting and applying methods for quantifying the quality and quantity of wildlife habitat in landscapes involves consideration of more than just the gradient of model complexity. Additional considerations should include whether to use a multispecies model rather than multiple single-species models, how to make comparisons of habitat quality among different landscapes, and how to evaluate the habitat model.

Multispecies Approaches

Several investigators have incorporated >1 species in their habitat evaluations, but most deal with each species separately rather than quantifying habitat quality for groups of species (Kliskey et al. 1999; Marzluff et al. 2002; Noon et al., this volume). For example, a computer program developed by Li et al. (2000) contains habitat models for hundreds of species, but effects of differences among landscapes are not integrated across species. Van Horne and Wiens (1991) discussed difficulties they encountered in their evaluation of the feasibility of combining HSI models for 16 forest bird species into a single, more general model. An approach proposed by Hansen et al. (1999:1461) "integrates aspects of species prioritization, dynamic habitat modeling, and PVA" in an effort to balance coarse filter and fine filter methods under time and budget constraints imposed on managers. Using this approach, species are ranked by risk of extinction at continental to watershed scales based on population status and threats to habitat. A few species with the highest ranks (i.e., lowest viability) are targeted for local demographic research and quantitative PVA. Habitat quality models are developed for several species with the next highest rankings. Then alternative management strategies based on species rankings are evaluated (Hansen et al. 1999). Fine filter models for ecological indicator species (Morrison et al. 1998:337) can also be used to evaluate coarse filter habitat quality for multiple members of a community.

One way to evaluate habitat availability for multiple species directly is to link habitat characteristics to the species richness or diversity of wildlife. White et al. (1997) evaluated the diversity of terrestrial vertebrates based on habitat relationships, and they summarized their results in terms of species richness and habitat abundance (i.e., area of suitable habitat summed across species). A habitat model tested by Flather et al. (1992) predicted avian species richness in landscapes of the eastern United States. Similarly, Penhollow and Stauffer (2000) developed linear regression models to relate landscape characteristics to avian species richness and assemblage index values (e.g., Shannon-Wiener). Fleishman et al. (2002) modeled species richness as well as occurrence and persistence in their analysis of butterfly habitat in the Great Basin of the western United States. However, a concern with modeling species richness for conservation planning is that all species are treated equally, and common species tend to contribute the most to richness (Flather et al., this volume).

Occupancy models (MacKenzie et al. 2004) and artificial neural networks (ANN; Anderson 1995) have potential for multispecies habitat modeling. Both types of models can predict probabilities of occurrence for >1 species simultaneously, thereby incorporating interactions among them. Unlike occupancy models, in which probabilities of occurrence are calculated jointly for all species, ANN models calculate the probabilities separately for each species, so the output is summarized as species richness. Özesmi and Özesmi (1999) developed a two-species ANN model for bird nest site habitat and implemented it in a GIS, but it was based on microhabitat variables that typically are not available for large areas (e.g., height and density of herbaceous stems).

Akçakaya (2000*b*:S80) proposed a unique way to use habitat-based, single-species models of population viability to produce a "multispecies conservation value" map. Each cell in the map contains the weighted mean of HSI scores for all species at that location. The weighting factor accounts for the probability of extinction or decline of each species (presumably, any index of species importance could be substituted) and the contribution of the location to the viability of each species, which is the difference in extinction probability between a model that includes the location and one that does not (Akçakaya 2000*b*). The weighted linear combination procedure described by Eastman et al. (1995) could be used to weight species importance for subsequent inclusion in Akçakaya's (2000*b*) multispecies conservation map or for combining habitat quality maps for multiple species in the absence of quantitative viability information.

In addition to maximizing species richness or a weighted mean HSI score, Hof and Raphael (1993) recognized two other multispecies objectives. The main objective of a management plan could be to maximize the minimum probability of viability among all wildlife species of interest or to maximize the joint probability of viability across all species of interest. Hof and Raphael (1993) assumed in their model that viability was linearly or logistically related to relative abundance. They demonstrated, however, that results differed depending on which objective was satisfied.

Comparing Landscapes

Spatial Comparisons.—Habitat analysis can be used to evaluate a single landscape, but often it is used to compare $\geq$2 landscapes across spatial or temporal (Rickers et al. 1995) intervals. Habitat suitability index values have been summarized across a landscape by calculating the mean (Donovan et al. 1987, Rickers et al. 1995), multiplying patch size by HSI score to obtain habitat units (Rickers et al. 1995, Marzluff et al. 2002), or providing the quantity of area in distinct HSI value categories (Herr and Queen 1993; Hepinstall et al. 1996; Riitters et al. 1997; Kliskey et al. 1999; Ortega-Huerta and Medley 1999; Dijak and Rittenhouse, this volume). Further summarization may also be useful, such as calculating the ratio of "optimal" to "marginal" patch area, which may influence some populations (Lidicker 1988).

It would be more informative to provide the entire cumulative frequency distribution of HSI values in the landscape (Gustafson 1998). The generic GIS model developed by Ortigosa et al. (2000) to calculate values for HSI models creates a frequency table of mutual occurrence (based on a cell-by-cell comparison) of HSI categories, which is analogous to comparing cumulative distribution functions of HSI values. These simple summaries fail to make full use of available GIS technology because they do not fully characterize spatial differences among landscapes. Riitters et al. (1997) used landscape metrics (e.g., connectivity, contagion, fractal dimension) and mean patch size of suitable habitat to elaborate their summary of habitat quality, and Trani (2002) demonstrated the effect of spatial resolution on landscape metrics. Mitchell (1997:129) used Moran's I coefficient (see Cliff and Ord 1973, Cressie 1993) as an index of spatial continuity (i.e., autocorrelation) in HSI values for his landscape comparisons. More recently, Johnson et al. (2004) used a local quadrat variance method to identify spatial patterns in land cover, and similar methods could be used to quantify patterns among patches of habitat. Perhaps most comprehensively, elaborations of the Kappa statistic can be used to compare both the quantity and location of differences in model predictions at multiple spatial resolutions in raster maps of categorical data (Pontius 2000, 2002; Hagen 2003), but see the critique and an alternative for presence-absence comparisons by Allouche et al. (2006).

Predicted animal densities or metrics of population viability can be used to compare landscapes when habitat-based models of density or viability are used. Data relating HSI values to animal density also can be used to summarize landscape quality in these terms (Breininger et al. 1991). Once the quality or quantity of habitat or the density or viability of a wildlife population is summarized for a landscape, spatial comparisons among landscapes can be made using the corresponding summary statistics of greatest interest.

Temporal Comparisons.—A further level of sophistication in landscape comparisons involves simulating the habitat or population-level effects of vegetation dynamics and alternative management decisions over time. Several studies provide examples of simulating forest growth and harvest followed by or linked with an evaluation of wildlife habitat quality. Kliskey et al. (1999) used a GIS-based HSI model to compare habitat suitability for marten (*Martes americana*) and woodland caribou (*Rangifer tarandus*) among four timber harvest scenarios for a large watershed in British Columbia. Marzluff et al. (2002) used a similar approach to evaluate habitat suitability for three nongame species under five management scenarios for a 566 ha forest in western Washington. Bettinger et al. (1996) modeled the effect of several timber management options on a habitat effectiveness index for elk (*Cervus elaphus*; Ager and Hitchcock 1994) in a watershed in northeastern Oregon. LEEMATH evaluates alternative forest management strategies in terms of habitat potential for many bird, reptile, and amphibian species (Li et al. 2000). Some of these species models also incorporated other landscape planning considerations such as stream fish habitat and economics.

Temporal comparisons of habitat values also can be made at higher levels of the complexity gradient. Olson and Orr (1999) developed a model that tracks the presence and absence of wildlife species as a function of the size, density, and species composition of trees within the forest stands of a timber growth and yield model (FREIGHTS; Krumland 1990). Boyce and McDonald (1999:271) advocated the use of RSFs to compare wildlife population sizes at different points in time, but they acknowledged that "RSF coefficients might change as resource availability changes."

Few models of which we are aware simultaneously simulate habitat dynamics (e.g., forest succession and harvest) and habitat-dependent population viability. The model developed by Moore et al. (2000) incorporated an optimization framework that defined timber harvest strategies (i.e., a series of decisions about where, when, and how much to harvest) to maximize bird abundance at the end of a planning horizon. Birds within a stand were treated as a population whose growth rate was a quadratic function of stand age and basal area. ECOLECON (Liu 1993*b*) used habitat-specific fecundity to directly link an individual-based population simulation with a forest growth and yield subroutine in a spatially explicit landscape (Liu et al. 1995). Similarly, RAMAS Landscape (Applied

Biomathematics, Setauket, New York), which is available commercially, integrates a landscape vegetation model, LANDIS (He et al. 2005), with a wildlife metapopulation model, RAMAS GIS (Akçakaya 2006; Akçakaya and Brook, this volume). RAMAS Landscape has been used to evaluate the effect of timber harvest regimes on the viability of sharp-tailed grouse (*Tympanuchus phasianellus*) populations in Wisconsin (Akçakaya et al. 2004*b*) and the effect of different fire-return intervals on the viability of Bell's sage sparrow (*Amphispiza belli*) in California (Akçakaya et al. 2005). Larson et al. (2004) modeled the viability of ovenbird (*Seiurus aurocapillus*) populations in Missouri using LANDIS and RAMAS GIS separately.

Model Evaluation

Habitat models are developed to satisfy a variety of objectives. Each model, therefore, should be evaluated by the degree to which it accomplishes the specified objectives, not by its complexity, or the degree to which it represents reality (Millspaugh et al., this volume). A basic objective of most habitat models is to predict some aspect of a wildlife population (e.g., presence, density, survival), so assessing predictive ability is a critical component of model validation (Vaughan and Ormerod 2005). This requires wildlife-use data that are independent of those from which the model was developed. Some evaluation procedures are related to specific methods for habitat modeling (e.g., Boyce et al. 2002, Fielding 2002) or the type of data available for model building and testing (e.g., Ottaviani et al. 2004). It is informative not only to evaluate model predictions with new observations from the original study site but also to evaluate predictions in new geographic areas (Mladenoff et al. 1999). Vanreusel et al. (2007) argue that models based on functional resources (e.g., specific foods) rather than environmental surrogates (e.g., topography, climate) are more likely to be transferable to new areas. True validation also addresses other components of the modeling process (Roloff and Kernohan 1999), such as a logical analysis of model formulation (Guisan and Zimmermann 2000).

Beutel et al. (1999) discussed six assumptions inherent in models of presence or abundance—habitat influences animal distribution, the predicted distribution is adequately modeled, habitat is adequately measured, distribution is adequately measured, measured distribution equates with habitat quality, and habitat quality is adequately measured—and ways to test their validity. Roloff and Kernohan (1999) provided a checklist for HSI model validation studies and emphasized that surrogates of fitness (e.g., reproductive rates, survival), rather than presence or abundance, should be used. One of their main concerns was that authors often ignored the variability of input data and its impact on model outputs. Although the presence of sampling error in habitat attribute data gathered in the field is well known, the measurement error associated with remotely sensed data and other GIS databases may not be as widely appreciated (Stoms et al. 1992). Monte Carlo simulation can be used to calculate confidence intervals for HSI scores from uncertainty in input variables (Bender et al. 1996), and fuzzy math (Ferson et al. 1998) can be used to calculate reliability bounds on HSI scores from both statistical and structural uncertainty in the model (Burgman et al. 2001).

When appropriate data exist, validation of habitat models is straightforward in a GIS framework because correlations can be computed between values in a map of population parameters of interest and values in a habitat quality map (Duncan et al. 1995). Roloff et al. (2001*a*) used the volume of intersection index (Millspaugh et al. 2004) to compare the utilization distribution predicted by a habitat effectiveness model for elk and the utilization distributions of elk herds, which were based on radio-tracking data. Mladenoff et al. (1999) used compositional analysis (Aebischer et al. 1993), comparing habitat use and availability, to validate their logistic regression model of gray wolf (*Canis lupus*) habitat in Wisconsin. Resampling methods (e.g., cross-validation, bootstrapping) provide an alternative to collecting new, independent data for evaluating correct classification rates for models used to predict presence and absence (Verbyla and Litvaitis 1989).

Reducing the resolution of a model or its output (see "Comparing Landscapes" earlier for examples of how to summarize habitat model outputs) for validation purposes may be helpful if independent data about wildlife populations (e.g., harvest indexes, regional surveys) are more readily available at a landscape scale (e.g., township, county, watershed) rather than a patch scale (i.e., an area of relatively homogeneous vegetation). Furthermore, the scale at which land management objectives are most relevant, often the landscape, is also the most relevant scale at which to evaluate model performance. Model validity, however, is currently limited by a lack of information about the spatial components of wildlife habitat (e.g., minimum patch size) and relationships between habitat quality and landscape indices (Li et al. 2000).

Furthermore, model validation alone is not sufficient for determining whether a given model is better or worse than possible alternatives (Conroy and Moore 2002). Adaptive management provides a framework for using repeated collections of new data to reduce the uncertainty about which model is best for predicting the consequences of management decisions (Williams et al. 2002).

CONCLUSIONS

Recent technological advancements in the collection and analysis of spatially explicit data for large geographic areas have facilitated the development of new methods for evaluating wildlife habitat. Currently, our ability to incorporate wildlife population objectives in land management plans is limited more by a lack of knowledge about the relationships between the quality and quantity of habitat and animal use and demographics. We recommend, therefore, that a focus remain on improving our understanding of wildlife–habitat relationships (Morrison 2001), which will require studies of resource use by animals across a wide range of spatial scales. Such information would be most useful if it were summarized and made widely available in published databases (e.g., Wisdom et al. 2000).

Habitat modeling should be encouraged, even when knowledge is imperfect or data are scant, for several reasons. Developing models compels us to formalize our hypotheses about wildlife-habitat relationships. Habitat models organize existing knowledge in a format useful to managers. They also help us identify gaps in existing knowledge and prioritize future research objectives for filling the gaps. Managers and researchers also should be aware of the potential drawbacks of misusing habitat models. All models are imperfect representations of reality, and inferences based on them should be treated as hypotheses, or best estimates, rather than facts. Furthermore, habitat models are more useful for making relative comparisons of habitat quality among landscapes than they are for making accurate predictions about habitat quality at a given place and time.

Given the broad range of objectives for which wildlife habitat models are developed, we expect that the methods and approaches for developing such models will continue to expand rather than narrow to just a few that might be deemed best, or universally applicable. Some specific methodological advancements seem particularly important. Biologists should strive for a comprehensive approach to summarizing the quality, quantity, and spatial structure of habitat over large areas (Gustafson 1998). Summarizing and analyzing changes in those statistics over time may require a different approach. Furthermore, quantifying the uncertainty in the output of models should be more commonly applied. Such practices are necessary to make complete and valid comparisons among management alternatives. We also need to develop additional options for considering >1 species simultaneously at all levels of habitat modeling complexity (see Noon et al., this volume).

A general approach we think continues to hold promise was mentioned by Wiens (2002) as he speculated about the future of habitat modeling: focusing on process. Several processes at the individual and population levels link wildlife populations to the way they use space, and understanding those processes will improve our conceptualization of wildlife habitat. When we consider the processes that produce the patterns we observe, habitat models are more meaningful and the causes of discrepancies between predicted and realized patterns of use can be logically investigated.

Progress in habitat modeling techniques is likely to continue to be driven, or at least accompanied, by increasing quantitative and scientific rigor. The demand for skilled modelers with a sound understanding of the best available methods also may increase. We hope the future also holds the true integration of science and decision making. Not only are the expertise and experience of both researchers and managers necessary for developing most habitat models, resources spent on model development are largely wasted if the model is not used to inform management decisions.

SUMMARY

Recent technological advancements in the collection and analysis of spatially explicit data for large geographic areas have facilitated the development of new methods for evaluating wildlife habitat. We reviewed various methods that have been developed or revised for geographic information systems to evaluate the quality of wildlife habitat or to predict how wildlife populations respond to spatially explicit changes in habitat caused by land management decisions. Our goal was to provide a synopsis of available techniques and organize the methods along a complexity gradient with increasing levels of potential realism in the models. We also discussed multispecies approaches, methods of comparing landscapes, validation of habitat models, and several applications of each method. Further research is needed to advance knowledge of wildlife-habitat relationships; to develop multispecies methods; and to provide a comprehensive approach to summarizing the quality, quantity, and spatial structure of wildlife habitat over large spatial and temporal scales.

ACKNOWLEDGMENTS

We thank J. A. Hepinstall, B. J. Kernohan, G. J. Roloff, and several anonymous referees for providing critical reviews that helped us improve earlier drafts of this manuscript. We also appreciate the support of our employers, the Minnesota Department of Natural Resources, the University of Missouri, and the U.S. Forest Service.

LITERATURE CITED

Aebischer, N. J., P. A. Robertson, and R. E. Kenward. 1993. Compositional analysis of habitat use from animal radio-tracking data. *Ecology* 74:1313–1325.

Ager, A., and M. Hitchcock. 1994. HEICALC: Elk habitat-effectiveness index software. *Wildlife Society Bulletin* 22:126–128.

Akçakaya, H. R. 2000*a*. Viability analyses with habitat-based metapopulation models. *Population Ecology* 42:45–53.

Akçakaya, H. R. 2000*b*. Conservation and management for multiple species: Integrating field research and modeling into management decisions. *Environmental Management* 26:S75–S83.

Akçakaya, H. R. 2006. *RAMAS GIS: Linking spatial data with population viability analysis. Version 5.0*. Applied Biomathematics, Setauket, New York, USA.

Akçakaya, H. R., M. A. Burgman, O. Kindvall, C. C. Wood, P. Sjögren-Gulve, J. S. Hatfield, and M. A. McCarthy, editors. 2004*a*. *Species conservation and management: Case studies*. Oxford University Press, New York, New York, USA.

Akçakaya, H. R., J. Franklin, A. D. Syphard, and J. R. Stephenson. 2005. Viability of Bell's sage sparrow (*Amphispiza belli* spp. *belli*): Altered fire regimes. *Ecological Applications* 15:521–531.

Akçakaya, H. R., V. C. Radeloff, D. J. Mladenoff, and H. S. He. 2004*b*. Integrating landscape and metapopulation modeling approaches: Viability of the sharp-tailed grouse in a dynamic landscape. *Conservation Biology* 18:526–537.

Akçakaya, H. R., and M. G. Raphael. 1998. Assessing human impact despite uncertainty: Viability of the northern spotted owl metapopulation in the northwestern USA. *Biodiversity and Conservation* 7:875–894.

Allouche, O., A. Tsoar, and R. Kadmon. 2006. Assessing the accuracy of species distribution models: Prevalence, kappa and the true skill statistic (TSS). *Journal of Applied Ecology* 43:1223–1232.

Anderson, J. A. 1995. *An introduction to neural networks*. MIT Press, Cambridge, Massachusetts, USA.

Aspinall, R., and N. Veitch. 1993. Habitat mapping from satellite imagery and wildlife survey data using a Bayesian modeling procedure in a GIS. *Photogrammetric Engineering and Remote Sensing* 59:537–543.

Beissinger, S. R., and M. I. Westphal. 1998. On the use of demographic models of population viability in endangered species management. *Journal of Wildlife Management* 62:821–841.

Bellamy, P. E., N. J. Brown, B. Enoksson, L. G. Firbank, R. J. Fuller, S. A. Hinsley, and A. G. M. Schotman. 1998. The influences of habitat, landscape structure and climate on local distribution patterns of the nuthatch (*Sitta europaea L.*). *Oecologia* 115:127–136.

Bender, L. C., G. J. Roloff, and J. B. Haufler. 1996. Evaluating confidence intervals for habitat suitability models. *Wildlife Society Bulletin* 24:347–352.

Bettinger, P., K. N. Johnson, and J. Sessions. 1996. Forest planning in an Oregon case study: Defining the problem and attempting to meet goals with a spatial-analysis technique. *Environmental Management* 20:565–577.

Beutel, T. S., R. J. S. Beeton, and G. S. Baxter. 1999. Building better wildlife-habitat models. *Ecography* 22:219–223.

Booch, G., R. Martin, and J. Newkirk. 1999. *Object-oriented analysis and design with applications*. Third edition. Addison Wesley, Boston, Massachusetts, USA.

Boyce, M. S., and L. L. McDonald. 1999. Relating populations to habitats using resource selection functions. *Trends Ecology and Evolution* 14:268–272.

Boyce, M. S., P. R. Vernier, S. E. Nielsen, and F. K. A. Schmiegelow. 2002. Evaluating resource selection functions. *Ecological Modelling* 157:281–300.

Breiman, L., J. H. Friedman, R. A. Olshen, and C. J. Stone. 1984. *Classification and regression trees*. Wadsworth International Group, Belmont, California, USA.

Breininger, D. R., M. J. Provancha, and R. B. Smith. 1991. Mapping Florida scrub jay habitat for purposes of land-use management. *Photogrammetric Engineering and Remote Sensing* 57:1467–1474.

Browning, D. M., S. J. Beaupré, and L. Duncan. 2005. Using partitioned Mahalanobis $D^2(k)$ to formulate a GIS-based model of timber rattlesnake hibernacula. *Journal of Wildlife Management* 69:33–44.

Burgman, M. A., D. R. Breininger, B. W. Duncan, and S. Ferson. 2001. Setting reliability bounds on habitat suitability indices. *Ecological Applications* 11:70–78.

Caswell, H. 2001. *Matrix population models: Construction, analysis, and interpretation*. Second edition. Sinauer Associates, Sunderland, Massachusetts, USA.

Clark, J. D., J. E. Dunn, and K. G. Smith. 1993. A multivariate model of female black bear habitat use for a geographic information system. *Journal of Wildlife Management* 57:519–526.

Clevenger, A. P., J. Wierzchowski, B. Chruszcz, and K. Gunson. 2002. GIS-generated, expert-based models for identifying wildlife habitat linkages and planning mitigation passages. *Conservation Biology* 16:503–514.

Cliff, A. D., and J. K. Ord. 1973. *Spatial autocorrelation*. Pion, London, United Kingdom.

Coker, D. R., and D. E. Capen. 1995. Landscape-level habitat use by brown-headed cowbirds in Vermont. *Journal of Wildlife Management* 59:631–637

Conroy, M. J., and C. T. Moore. 2002. Wildlife habitat modeling in an adaptive framework: The role of alternative models. Pages 205–218 *in* J. M. Scott, P. J. Heglund, M. L. Morrison, J. B. Haufler, M. G. Raphael, W. A. Wall, and F. B. Samson, editors. *Predicting species occurrence: Issues of accuracy and scale.* Island Press, Washington, D.C., USA.

Cooper, A. B., and J. J. Millspaugh. 1999. The application of discrete choice models to wildlife resource selection studies. *Ecology* 80:566–575.

Cooper, A. B., and J. J. Millspaugh. 2001. Accounting for variation in resource availability and animal behavior in resource selection studies. Pages 243–273 *in* J. J. Millspaugh and J. M. Marzluff, editors. *Radio tracking and animal populations.* Academic Press, San Diego, California, USA.

Crance, J. H. 1987. Guidelines for using the Delphi technique to develop habitat suitability index curves. Biological Report 82(10.134). U.S. Fish and Wildlife Service, Washington, D.C., USA.

Cressie, N. A. C. 1993. *Statistics for spatial data.* Revised edition. John Wiley and Sons, New York, New York, USA.

Dettmers, R., D. A. Buehler, and J. B. Bartlett. 2002. A test and comparison of wildlife-habitat modeling techniques for predicting bird occurrence at a regional scale. Pages 607–615 *in* J. M. Scott, P. J. Heglund, M. L. Morrison, J. B. Haufler, M. G. Raphael, W. A. Wall, and F. B. Samson, editors. *Predicting species occurrence: Issues of accuracy and scale.* Island Press, Washington, D.C., USA.

Dixon, G. E. 2002. *Essential FVS: A user's guide to the Forest Vegetation Simulator.* U.S. Forest Service, Forest Management Service Center, Fort Collins, Colorado, USA. Last revised: April 2007.

Donovan, M. L., D. L. Rabe, and C. E. Olson, Jr. 1987. Use of geographic information systems to develop habitat suitability models. *Wildlife Society Bulletin* 15:574–579.

Donovan, T. M., and F. R. Thompson, III. 2001. Modeling the ecological trap hypothesis: A habitat and demographic analysis for migrant songbirds. *Ecological Applications* 11:871–882.

Duncan, B. W., D. R. Breininger, P. A. Schmalzer, and V. L. Larson. 1995. Validating a Florida scrub jay habitat suitability model, using demography data on Kennedy Space Center. *Photogrammetric Engineering and Remote Sensing* 61:1361–1370.

Eastman, J. R., W. Jin, P. A. K. Kyem, and J. Toledano. 1995. Raster procedures for multi-criteria/multi-objective decisions. *Photogrammetric Engineering and Remote Sensing* 61:539–547.

Edelmann, F. B., M. J. Ulliman, M. J. Wisdom, K. P. Reese, and J. W. Connelly. 1998. *Assessing habitat quality using population fitness parameters: A remote sensing/GIS-based habitat-explicit population model for sage grouse* (Centrocercus urophasianus). Idaho Forest, Wildlife and Range Experiment Station, University of Idaho, Technical Report 25, Moscow, USA.

Ferson, S., W. Root, and R. Kuhn. 1998. *RAMAS/RiskCalc: Risk assessment with uncertain numbers.* Applied Biomathematics, Setauket, New York, USA.

Fielding, A. H. 2002. What are the appropriate characteristics of an accuracy measure? Pages 271–280 *in* J. M. Scott, P. J. Heglund, M. L. Morrison, J. B. Haufler, M. G. Raphael, W. A. Wall, and F. B. Samson, editors. *Predicting species occurrence: Issues of accuracy and scale.* Island Press, Washington, D.C., USA.

Flather, C. H., S. J. Brady, and D. B. Inkley. 1992. Regional habitat appraisals of wildlife communities: A landscape-level evaluation of a resource planning model using avian distribution data. *Landscape Ecology* 7:137–147.

Fleishman, E., R. MacNally, J. P. Fay, and D. D. Murphy. 2001. Modeling and predicting species occurrence using broad-scale environmental variables: An example with butterflies of the Great Basin. *Conservation Biology* 15:1674–1685.

Fleishman, E., D. D. Murphy, and P. Sjögren-Gulve. 2002. Modeling species richness and habitat suitability for taxa of conservation interest. Pages 507–517 *in* J. M. Scott, P. J. Heglund, M. L. Morrison, J. B. Haufler, M. G. Raphael, W. A. Wall, and F. B. Samson, editors. *Predicting species occurrence: Issues of accuracy and scale.* Island Press, Washington, D.C., USA.

Fretwell, S. D., and H. L. Lucas, Jr. 1970. On territorial behavior and other factors influencing habitat distribution in birds. I. Theoretical development. *Acta Biotheoretica* 19:16–36.

Gaston, K. J., T. M. Blackburn, J. J. D. Greenwood, R. D. Gregory, R. M. Quinn, and J. H. Lawton. 2000. Abundance-occupancy relationships. *Journal of Applied Ecology* 37:39–59.

Guisan, A., and W. Thuiller. 2005. Predicting species distribution: Offering more than simple habitat models. *Ecology Letters* 8:993–1009.

Guisan, A., and N. E. Zimmermann. 2000. Predictive habitat distribution models in ecology. *Ecological Modelling* 135:147–186.

Gustafson, E. J. 1998. Quantifying landscape spatial pattern: What is the state of the art? *Ecosystems* 1:143–156.

Gustafson, E. J., G. R. Parker, and S. E. Backs. 1994. Evaluating spatial pattern of wildlife habitat: A case study of the wild turkey (*Meleagris gallopavo*). *American Midland Naturalist* 131:24–33.

Hagen, A. 2003. Fuzzy set approach to assessing similarity of categorical maps. *International Journal of Geographical Information Science* 17:235–249.

Hansen, A. J., J. J. Rotella, M. P. V. Kraska, and D. Brown. 1999. Dynamic habitat and population analysis: An approach to resolve the biodiversity manager's dilemma. *Ecological Applications* 9:1459–1476.

Hanski, I. 1994. A practical model of metapopulation dynamics. *Journal of Animal Ecology* 63:151–162.

Hanski, I., and D. Simberloff. 1997. The metapopulation approach, its history, conceptual domain, and application to conservation. Pages 5–26 *in* I. Hanski and M. Gilpin, editors. *Metapopulation biology: Ecology, genetics, and evolution*. Academic Press, San Diego, California, USA.

He, H. S., W. Li, B. R. Sturtevant, J. Yang, B. Z. Shang, E. J. Gustafson, and D. J. Mladenoff. 2005. *LANDIS, a spatially explicit model of forest landscape disturbance, management, and succession—LANDIS 4.0 User's Guide*. U.S. Forest Service, North Central Research Station, General Technical Report NC-263, St. Paul, Minnesota, USA.

He, H. S., D. J. Mladenoff, and J. Boeder. 1999. An object-oriented forest landscape model and its representation of tree species. *Ecological Modelling* 119:1–19.

Hepinstall, J. A., L. P. Queen, and P. A. Jordan. 1996. Application of a modified habitat suitability index model for moose. *Photogrammetric Engineering and Remote Sensing* 62:1281–1286.

Herr, A. M., and L. P. Queen. 1993. Crane habitat evaluation using GIS and remote sensing. *Photogrammetric Engineering and Remote Sensing* 59:1531–1538.

Hildén, O. 1965. Habitat selection in birds. *Annales Zoologici Fennici* 2:53–75.

Hill, K. E., and M. W. Binford. 2002. The role of category definition in habitat models: Practical and logical limitations of using Boolean, indexed, probabilistic, and fuzzy categories. Pages 97–106 *in* J. M. Scott, P. J. Heglund, M. L. Morrison, J. B. Haufler, M. G. Raphael, W. A. Wall, and F. B. Samson, editors. *Predicting species occurrence: Issues of accuracy and scale*. Island Press, Washington, D.C., USA.

Hof, J. G., and M. G. Raphael. 1993. Some mathematical programming approaches for optimizing timber age-class distributions to meet multispecies wildlife population objectives. *Canadian Journal of Forestry Research* 23:828–834.

Jennings, M. D. 2000. Gap analysis: Concepts, methods, and recent results. *Landscape Ecology* 15:5–20.

Johnson, C. J., M. S. Boyce, R. Mulders, A. Gunn, R. J. Gau, H. D. Cluff, and R. L. Case. 2004. Quantifying patch distribution at multiple spatial scales: Applications to wildlife-habitat models. *Landscape Ecology* 19:869–882.

Johnson, C. J., and M. P. Gillingham. 2005. An evaluation of mapped species distribution models used for conservation planning. *Environmental Conservation* 32:117–128.

Keitt, T. H., O. N. Bjørnstad, P. M. Dixon, and S. Citron-Pousty. 2002. Accounting for spatial pattern when modeling organism-environment interactions. *Ecography* 25:616–625.

Kliskey, A. D., E. C. Lofroth, W. A. Thompson, S. Brown, and H. Schreier. 1999. Simulating and evaluating alternative resource-use strategies using GIS-based habitat suitability indices. *Landscape and Urban Planning* 45:163–175.

Knick, S. T., and J. T. Rotenberry. 1998. Limitations to mapping habitat use areas in changing landscapes using the Mahalanobis distance statistic. *Journal of Agricultural, Biological, and Environmental Statistics* 3:311–322.

Krumland, B. 1990. The California timber supply model: Design and implementation. Report submitted to the California Department of Forestry and Fire Protection, Agreement Number 8CA74741, Sacramento, California, USA.

Larson, M. A., F. R. Thompson, III, J. J. Millspaugh, W. D. Dijak, and S. R. Shifley. 2004. Linking population viability, habitat suitability, and landscape simulation models for conservation planning. *Ecological Modelling* 180:103–118.

Li, H. B., D. I. Gartner, P. Mou, and C. C. Trettin. 2000. A landscape model (LEEMATH) to evaluate effects of management impacts on timber and wildlife habitat. *Computers and Electronics in Agriculture* 27:263–292.

Lidicker, W. Z., Jr. 1988. Solving the enigma of microtine "cycles." *Journal of Mammalogy* 69:225–235.

Liu, C., P. M. Berry, T. P. Dawson, and R. G. Pearson. 2005. Selecting thresholds of occurrence in the prediction of species distributions. *Ecography* 28:385–393.

Liu, J. 1993*a*. Discounting initial population sizes for prediction of extinction probabilities in patchy environments. *Ecological Modelling* 70:51–61.

Liu, J. 1993*b*. ECOLECON: An ECOLogical-ECONomic model for species conservation in complex forest landscapes. *Ecological Modelling* 70:63–87.

Liu, J., J. B. Dunning, Jr., and H. R. Pulliam. 1995. Potential effects of a forest management plan on Bachman's sparrows (*Aimophila aestivalis*): Linking a spatially explicit model with GIS. *Conservation Biology* 9:62–75.

Mace, R. D., J. S. Waller, T. L. Manley, K. Ake, and W. T. Wittinger. 1999. Landscape evaluation of grizzly bear habitat in western Montana. *Conservation Biology* 13:367–377.

MacKenzie, D. I., L. L. Bailey, and J. D. Nichols. 2004. Investigating species co-occurrence patterns when species are detected imperfectly. *Journal of Animal Ecology* 73:546–555.

MacKenzie, D. I., J. D. Nichols, G. B. Lachman, S. Droege, J. A. Royle, and C. A. Langtimm. 2002. Estimating site occupancy rates when detection probabilities are less than one. *Ecology* 83:2248–2255.

MacKenzie, D. I., J. D. Nichols, J. A. Royle, K. H. Pollock, L. L. Bailey, and J. E. Hines. 2006. *Occupancy estimation and modeling: Inferring patterns and dynamics of species occurrence*. Academic Press, Burlington, Massachusetts, USA.

Manly, B. F. J., L. L. McDonald, D. L. Thomas, T. L. McDonald, and W. P. Erickson. 2002. *Resource selection by animals: Statistical design and analysis for field studies*. Second edition. Kluwer Academic Publishers, Dordrecht, The Netherlands.

Marcot, B. G., R. S. Holthausen, M. G. Raphael, M. M. Rowland, and M. J. Wisdom. 2001. Using Bayesian belief networks to evaluate fish and wildlife population viability under land management alternatives from an environmental impact statement. *Forest Ecology and Management* 153:29–42.

Marzluff, J. M., J. J. Millspaugh, K. R. Ceder, C. D. Oliver, J. Withey, J. B. McCarter, C. L. Mason, and J. Comnick. 2002. Modeling changes in wildlife habitat and timber revenues in response to forest management. *Forest Science* 48:191–202.

McClain, B. J., and W. F. Porter. 2000. Using satellite imagery to assess large-scale habitat characteristics of Adirondack Park, New York, USA. *Environmental Management* 26:553–561.

McGarigal, K., S. A. Cushman, M. C. Neel, and E. Ene. 2002. *FRAGSTATS: Spatial pattern analysis program for categorical maps. Version 3*. University of Massachusetts, Amherst, USA.

McGarigal, K., and B. J. Marks. 1995. *FRAGSTATS: Spatial pattern analysis program for quantifying landscape structure*. U.S. Forest Service, Pacific Northwest Research Station, General Technical Report 351, Portland, Oregon, USA.

Millspaugh, J. J., R. A. Gitzen, B. J. Kernohan, M. A. Larson, and C. L. Clay. 2004. Comparability of three analytical techniques to assess joint space use. *Wildlife Society Bulletin* 32:148–157.

Mitchell, M. S. 1997. *Optimum home ranges: Models and application to black bears*. Dissertation, North Carolina State University, Raleigh, USA.

Mitchell, M. S., R. A. Lancia, and J. A. Gerwin. 2001. Using landscape-level data to predict the distribution of birds on a managed forest: Effects of scale. *Ecological Applications* 11:1692–1708.

Mladenoff, D. J., and H. S. He. 1999. Design and behavior of LANDIS, an object-oriented model of forest landscape disturbance and succession. Pages 125–162 *in* D. J. Mladenoff and W. L. Baker, editors. *Advances in spatial modeling of forest landscape change: Approaches and applications.* Cambridge University Press, Cambridge, United Kingdom.

Mladenoff, D. J., T. A. Sickley, and A. P. Wydeven. 1999. Predicting gray wolf landscape recolonization: Logistic regression models vs. new field data. *Ecological Applications* 9:37–44.

Moilanen, A., and I. Hanski. 1998. Metapopulation dynamics: Effects of habitat quality and landscape structure. *Ecology* 79:2503–2515.

Moore, C. T., M. J. Conroy, and K. Boston. 2000. Forest management decisions for wildlife objectives: System resolution and optimality. *Computers and Electronics in Agriculture* 27:25–39.

Morrison, M. L. 2001. A proposed research emphasis to overcome the limits of wildlife-habitat relationship studies. *Journal of Wildlife Management* 65:613–623.

Morrison, M. L., B. G. Marcot, and R. W. Mannan. 1998. *Wildlife-habitat relationships: Concepts and applications*. Second edition. University of Wisconsin Press, Madison, USA.

Nickelson, T. E., and P. W. Lawson. 1998. Population viability of coho salmon, *Oncorhynchus kisutch*, in Oregon coastal basins: Application of a habitat-based life cycle model. *Canadian Journal of Fisheries and Aquatic Sciences* 55:2383–2392.

Nielsen, S. E., C. J. Johnson, D. C. Heard, and M. S. Boyce. 2005. Can models of presence-absence be used to scale abundance? Two case studies considering extremes in life history. *Ecography* 28:197–208.

Oliver, R. M., and J. Q. Smith, editors. 1990. *Influence diagrams, belief nets and decision analysis.* Wiley, Chichester, United Kingdom.

Olson, C. M., and B. Orr. 1999. Combining tree growth, fish and wildlife habitat, mass wasting, sedimentation, and hydrologic models in decision analysis and long-term forest land planning. *Forest Ecology and Management* 114:339–348.

Ortega-Huerta, M. A., and K. E. Medley. 1999. Landscape analysis of jaguar (*Panthera onca*) habitat using sighting records in the Sierra de Tamaulipas, Mexico. *Environmental Conservation* 26:257–269.

Ortigosa, G. R., G. A. DeLeo, and M. Gatto. 2000. VVF: Integrating modelling and GIS in a software tool for habitat suitability assessment. *Environmental Modelling and Software* 15:1–12.

Ottaviani, D., G. J. Lasinio, and L. Boitani. 2004. Two statistical methods to validate habitat suitability models using presence-only data. *Ecological Modelling* 179:417–443.

Özesmi, S. L., and U. Özesmi. 1999. An artificial neural network approach to spatial habitat modelling with interspecific interaction. *Ecological Modelling* 116:15–31.

Penhollow, M. E., and D. F. Stauffer. 2000. Large-scale habitat relationships of neotropical migratory birds in Virginia. *Journal of Wildlife Management* 64:362–373.

Pereira, J. M. C., and R. M. Itami. 1991. GIS-based habitat modeling using logistic multiple regression: A study of the Mt. Graham red squirrel. *Photogrammetric Engineering and Remote Sensing* 57:1475–1486.

Pontius, R. G., Jr. 2000. Quantification error versus location error in comparison of categorical maps. *Photogrammetric Engineering and Remote Sensing* 66:1011–1016.

Pontius, R. G., Jr. 2002. Statistical methods to partition effects of quantity and location during comparison of categorical maps at multiple resolutions. *Photogrammetric Engineering and Remote Sensing* 68:1041–1049.

Pulliam, H. R. 1988. Sources, sinks, and population regulation. *American Naturalist* 132:652–661.

Rempel, R. S., P. C. Elkie, A. R. Rodgers, and M. J. Gluck. 1997. Timber-management and natural-disturbance effects on moose habitat: Landscape evaluation. *Journal of Wildlife Management* 61:517–524.

Rickel, B. W., B. Anderson, and R. Pope. 1998. Using fuzzy systems, object-oriented programming, and GIS to evaluate wildlife habitat. *AI Applications* 12:31–40.

Rickers, J. R., L. P. Queen, and G. J. Arthaud. 1995. A proximity-based approach to assessing habitat. *Landscape Ecology* 10:309–321.

Riitters, K. H., R. V. O'Neill, and K. B. Jones. 1997. Assessing habitat suitability at multiple scales: A landscape-level approach. *Biological Conservation* 81:191–202.

Roloff, G. J., and J. B. Haufler. 1997. Establishing population viability planning objectives based on habitat potentials. *Wildlife Society Bulletin* 25:895–904.

Roloff, G. J., and J. B. Haufler. 2002. Modeling habitat-based viability from organism to population. Pages 673–686 *in* J. M. Scott, P. J. Heglund, M. L. Morrison, J. B. Haufler, M. G. Raphael, W. A. Wall, and F. B. Samson, editors. *Predicting species occurrence: Issues of accuracy and scale.* Island Press, Washington, D.C., USA.

Roloff, G. J., and B. J. Kernohan. 1999. Evaluating reliability of habitat suitability index models. *Wildlife Society Bulletin* 27:973–985.

Roloff, G. J., J. J. Millspaugh, R. A. Gitzen, and G. C. Brundige. 2001*a*. Validation tests of a spatially explicit habitat effectiveness model for Rocky Mountain elk. *Journal of Wildlife Management* 65:899–914.

Roloff, G. J., G. F. Wilhere, T. Quinn, and S. Kohlmann. 2001*b*. An overview of models and their role in wildlife management. Pages 512–536 *in* D. H. Johnson and T. A. O'Neil, editors. *Wildlife-habitat relationships in Oregon and Washington.* Oregon State University Press, Corvallis, USA.

Rotenberry, J. T., S. T. Knick, and J. E. Dunn. 2002. A minimalist approach to mapping species' habitat: Pearson's planes of closest fit. Pages 281–289 *in* J. M. Scott, P. J. Heglund, M. L. Morrison, J. B. Haufler, M. G. Raphael, W. A. Wall, and F. B. Samson, editors. *Predicting species occurrence: Issues of accuracy and scale.* Island Press, Washington, D.C., USA.

Royle, J. A., W. A. Link, and J. R. Sauer. 2002. Statistical mapping of count survey data. Pages 625–638 *in* J. M. Scott, P. J. Heglund, J. B. Haufler, M. L. Morrison, M. G. Raphael, W. B. Wall, and F. Samson, editors. *Predicting species occurrences: Issues of accuracy and scale.* Island Press, Washington, D.C., USA.

Royle, J. A., and J. D. Nichols. 2003. Estimating abundance from repeated presence-absence data or point counts. *Ecology* 84:777–790.

Saaty, T. L. 1977. A scaling method for priorities in hierarchical structures. *Journal of Mathematical Psychology* 15:234–281.

Scott, J. M., F. Davis, B. Csuti, R. Noss, B. Butterfield, C. Groves, H. Anderson, S. Caicco, F. D'Erchia, T. C. Edwards, Jr., J. Ulliman, and R. G. Wright. 1993. GAP analysis: A geographic approach to protection of biological diversity. *Wildlife Monographs* 123:225–250.

Scott, J. M., P. J. Heglund, J. B. Haufler, M. L. Morrison, M. G. Raphael, W. B. Wall, and F. Samson, editors. 2002. *Predicting species occurrences: Issues of accuracy and scale*. Island Press, Washington, D.C., USA.

Smith, J. L. D., S. C. Ahearn, and C. McDougal. 1998. Landscape analysis of tiger distribution and habitat quality in Nepal. *Conservation Biology* 12:1338–1346.

Starfield, A. M. 1997. A pragmatic approach to modeling for wildlife management. *Journal of Wildlife Management* 61:261–270.

Stoms, D. M., F. W. Davis, and C. B. Cogan. 1992. Sensitivity of wildlife habitat models to uncertainties in GIS data. *Photogrammetric Engineering and Remote Sensing* 58:843–850.

Thatcher, C. A., F. T. Van Manen, and J. D. Clark. 2006. Identifying suitable sites for Florida panther reintroduction. *Journal of Wildlife Management* 70:752–763.

Thogmartin, W. E., J. R. Sauer, and M. G. Knutson. 2004. A hierarchical spatial model of avian abundance with application to cerulean warblers. *Ecological Applications* 14:1766–1779.

Trani, M. K. 2002. The influence of spatial scale on landscape pattern description and wildlife habitat assessment. Pages 141–155 *in* J. M. Scott, P. J. Heglund, M. L. Morrison, J. B. Haufler, M. G. Raphael, W. A. Wall, and F. B. Samson, editors. *Predicting species occurrence: Issues of accuracy and scale.* Island Press, Washington, D.C., USA.

U.S. Fish and Wildlife Service. 1980. *Habitat evaluation procedures (HEP)*. USFWS, Division of Ecological Services, Manual 102, Washington, D.C., USA.

U.S. Fish and Wildlife Service. 1981. *Standards for the development of habitat suitability index models for use in the habitat evaluation procedure*. USFWS, Division of Ecological Services, Manual 103, Washington, D.C., USA.

Van Horne, B., and J. A. Wiens. 1991. *Forest bird habitat suitability models and the development of general habitat models*. Fish and Wildlife Research 8. U.S. Fish and Wildlife Service, Washington, D.C., USA.

Vanreusel, W., D. Meas, and H. Van Dyck. 2007. Transferability of species distribution models: A functional habitat approach for two regionally threatened butterflies. *Conservation Biology* 21:201–212.

Vaughan, I. P., and S. J. Ormerod. 2005. The continuing challenges of testing species distribution models. *Journal of Applied Ecology* 42:720–730.

Verbyla, D. L., and J. A. Litvaitis. 1989. Resampling methods for evaluating classification accuracy of wildlife habitat models. *Environmental Management* 13:783–787.

Voogd, H. 1983. *Multicriteria evaluation for urban and regional planning*. Pion, London, United Kingdom.

White, D., P. G. Minotti, M. J. Barczak, J. C. Sifneos, K. E. Freemark, M. V. Santelmann, C. F. Steinitz, A. R. Kiester, and E. M. Preston. 1997. Assessing risks to biodiversity from future landscape change. *Conservation Biology* 11:349–360.

Wiens, J. A. 2002. Predicting species occurrences: Progress, problems, and prospects. Pages 739–749 *in* J. M. Scott, P. J. Heglund, M. L. Morrison, J. B. Haufler, M. G. Raphael, W. A. Wall, and F. B. Samson, editors. *Predicting species occurrence: Issues of accuracy and scale.* Island Press, Washington, D.C., USA.

Williams, B. K., J. D. Nichols, and M. J. Conroy. 2002. *Analysis and management of animal populations: Modeling, estimation, and decision making*. Academic Press, San Diego, California, USA.

Williams, G. L., K. R. Russell, and W. K. Seitz. 1978. Pattern recognition as a tool in the ecological analysis of habitat. Pages 521–531 *in* A. Marmelstein, editor. *Classification, inventory, and analysis of fish and wildlife habitat.* U.S. Fish and Wildlife Service, Office of Biological Services, FWS/OBS-78/76, Washington, D.C., USA.

Wisdom, M. J., R. S. Holthausen, B. C. Wales, C. D. Hargis, V. A. Saab, D. C. Lee, W. J. Hann, T. D. Rich, M. M. Rowland, W. J. Murphy, and M. R. Eames. 2000. *Source habitats for terrestrial vertebrates of focus in the interior Columbia basin: Broad-scale trends and management implications*. U.S. Forest Service, Pacific Northwest Research Station, General Technical Report 485, Portland, Oregon, USA.

CHAPTER

Wildlife Habitat-Relationships Models: Description and Evaluation of Existing Frameworks

10

Jeffrey L. Beck
and Lowell H. Suring

Wildlife habitats are areas of land that provide resources such as food, cover, and water and environmental conditions such as precipitation and soil types that affect occupancy of individuals or populations of species, allowing those species to survive and reproduce (Morrison et al. 2006). Changing requirements in the 1970s to evaluate and report the effects of land management activities on wildlife habitats and associated populations led to a need for new analysis techniques. Wildlife habitat-relationships models were first developed in the mid-1970s (Salwasser et al. 1980) to provide practitioners with tools to evaluate habitat quality for selected species. The underlying goal of many habitat-relationships modeling frameworks is to evaluate habitat quality for wildlife populations. Habitat quality was described by Hall et al. (1997:178) as "the ability of the environment to provide conditions appropriate for individual and population persistence."

Habitat capability models provide an estimate of the area within which resources for a modeled species can be found, or ranking an area based on the capability of that area to support a species based on a few important environmental variables (Morrison et al. 2006:337). Habitat effectiveness models rank resources in an area to the degree that maximum use or carrying capacity can be met (Morrison et al. 2006:337), with effectiveness often tempered to reflect the constraints of human activities on the area actually usable by animals (Lyon and Christensen 1992, Merrill et al. 1999). Throughout our chapter, we generally refer to habitat-relationships modeling frameworks, while recognizing that frameworks have been developed under a variety of structures including species-habitat matrices, habitat suitability, habitat capability, and habitat effectiveness (Morrison et al. 2006). We define frameworks as conceptual modeling structures including modeling shells (e.g., expert systems) and general modeling approaches (e.g., artificial neural networks, Bayesian belief networks, spatial optimization) within which models are constructed that are similar in purpose and function.

Two general approaches have been developed to assess habitat quality for wildlife populations. Under species-habitat matrix frameworks, the starting point is a classification of vegetation within which each classification unit is assigned a value describing its value as habitat for one or more wildlife species (Morrison et al. 2006). Frameworks that use guilds often are structured as a species-habitat matrix, because guilds represent aggregates of species needs typically including generalizations of habitat needs. Work by Thomas (1979) in the Blue Mountains of northeastern Oregon and southeastern Washington, Hoover and Willis (1984) in Colorado forests, and DeGraaf et al. (1992) in New England forests are examples of species-habitat matrix modeling frameworks. The second approach to modeling wildlife habitat quality includes frameworks that begin with the habitat requirements of a species and then quantifies these requirements through specific vegetation and other variables to evaluate how an area provides the various required requirements. The Habitat Evaluation Procedures (HEPs) developed by the U.S. Fish and Wildlife Service (1981) established the underpinnings for this approach from which many other modeling frameworks have been developed.

Habitat-relationships modeling frameworks have increased in number and complexity since the mid-1970s. Consequently, selecting a modeling framework to match the objectives of a wildlife conservation program that appropriately consider data availability and the analytical abilities of practitioners can be difficult. The purpose of our review was to describe the structure, uses, output, and operation of wildlife habitat-relationships modeling frameworks to provide practitioners with a basis for selecting frameworks. Our specific objectives were to (1) identify wildlife habitat-relationships modeling frameworks that are currently available for use; and (2) provide a descriptive analysis of frameworks to assist practitioners in selecting approaches to modeling wildlife-habitat relationships that best fit their objectives.

METHODS

Identifying and Rating Habitat-Relationships Modeling Frameworks

To focus our search for modeling frameworks, we bounded our definition of wildlife habitat-relationships modeling frameworks with four criteria that were based on the modeling objectives of each framework. We (1) considered frameworks that were designed to evaluate habitat for terrestrial wildlife species; (2) considered frameworks that have the potential for multispecies applications, thus avoiding approaches designed solely for one species (e.g., Gutiérrez et al. 1992); (3) avoided statistical modeling techniques (e.g., logistic regression, discriminant function analysis, resource selection functions) designed to quantify selection of habitat by a species, although we considered modeling frameworks that incorporate statistical or other analytical concepts to describe habitat relationships (e.g., artificial neural networks, Bayesian belief networks, expert systems, fuzzy logic, spatial optimization); and (4) considered only frameworks that were operational, avoiding those that are currently being conceptualized or were otherwise incomplete.

In many cases, the recently developed wildlife-habitat relationships frameworks we identified were improvements of earlier, more general frameworks. For instance, several newer frameworks including ArcHSI (Juntti and Rumble 2006), HABIT@ (McGarigal and Compton 2003), HCI (McComb et al. 2002), HQI (Rickel 1997), Landscape HSI (Larson et al. 2003, 2004; Dijak et al. 2007; Rittenhouse et al. 2007), and LMS (Marzluff et al. 2002; Oliver et al., this volume) retain elements of the original 1981 HSI framework, but provide more sophistication through incorporation of advancements such as GIS and spatially explicit analyses. Consequently, we retained newer frameworks that were built on the platforms of older frameworks as independent observations because their advancements allow them to function in different ways than the previously described frameworks. In other cases, frameworks were stand-alone, not based on previously described frameworks. To be consistent, however, in each case we adhered to the four criteria to identify frameworks according to their modeling objectives.

After identifying the major habitat-relationships modeling frameworks that fit the above four criteria, we rated each according to 10 nominal- and 5 ordinal-scale criteria to quantify our evaluation (Table 10-1). Nominal criteria included (1) whether the breadth of application of the framework could consider a wide range of species in a wide range of environments or was limited to certain taxa or a single environment; (2) whether the frameworks linked habitat conditions with population demographics or surrogates; (3) whether the frameworks were included in comprehensive landscape modeling systems; (4) availability of input data; (5) whether at least one individual species model based on a particular framework had been validated with field data; (6) capability of frameworks to examine habitat relationships at single or multiple scales; (7) whether multi-scaled frameworks required linkage information among scales to function; (8) whether the frameworks had attained scientific credibility through publication or application of results suggesting acceptance by an array of professionals; (9) the spatial application of the framework (i.e., does the framework use geographic data [spatial framework]?; does the framework examine spatial relationships in habitat data at specific locations or coordinates [spatially explicit]?; or, does the framework not rely on geographic or spatial data [aspatial])?; and (10) whether vegetation and its attributes were applied in the framework as the basis for a species-habitat matrix or as variables to assess habitat relationships for wildlife species (Table 10-1). Ordinal criteria included (1) whether documentation was adequate to clearly understand and apply the modeling frameworks; (2) ease of application; (3) whether output was well defined and measurable; (4) whether frameworks were well suited for the scales they were developed to examine; and (5) transparency of the frameworks' structure (Table 10-1). We conducted two independent reviews of each framework and then reached consensus on criteria ratings that differed.

Table 10-1 Nominal- and Ordinal-Scale Criteria Used to Rate Wildlife Habitat-Relationships Modeling Frameworks

Criteria	Definition	Rating Scale
Nominal criteria		
Breadth of application	Can the framework be used to define habitat relationships for a wide range of species in a wide range of environments?	0 = only suited for a single species or environment 1 = suited for a wide range of species in a wide range of environments
Habitat–population linkage	Does the modeling framework incorporate vital rates (e.g., production, survival), other demographic parameters (e.g., density, population size); surrogates (e.g., quality of home ranges, habitat conditions in critical reproductive habitats, presence/absence) of population demographic parameters; or does the modeling framework model habitat conditions without specific consideration of wildlife population parameters?	0 = does not rely on population demographics or surrogates of modeled species 1 = relies on surrogates for population demographic parameters or framework; can utilize population demographics if desired, but is not dependent on them 2 = specifically relies on population demographics of modeled species
Independence	Is the framework part of a larger landscape modeling system?	0 = a component of a larger landscape modeling system 1 = stands alone and is not part of a larger landscape modeling system
Input requirements	Is the required input data (e.g., GIS coverages, stand and wildlife inventory data) readily available in agency inventories?	0 = not readily available 1 = readily available
Model validation	Has output from at least 1 model developed within a framework been validated with field data?	0 = no validation known or validation impossible 1 = model validated
Scale application	Is the framework limited to 1 scale or can it explicitly examine differences in habitat conditions at a range of spatial scales?	1 = limited to 1 scale 2 = capable of examining habitat conditions at more than 1 scale (e.g., forest and region)
Scale linkage	If the framework is multiscaled, are the scales linked?	0 = scales are not linked 1 = scales are linked
Scientific credibility	Has the framework gained credibility through publication of results, application of results, or other mechanisms to suggest acceptance by an array of professionals?	0 = limited credibility 1 = at least 1 publication of results using this framework, or other application of the modeling framework

continues

Table 10-1 Nominal- and Ordinal-Scale Criteria Used to Rate Wildlife Habitat-Relationships Modeling Frameworks *cont...*

Criteria	Definition	Rating Scale
Spatial application	Does the framework: not rely on geographic data (aspatial); examine geographic data (spatial framework); or examine spatial relationships in habitat data at specific locations or coordinates as part of its structure (spatially explicit)?	1 = aspatial 2 = spatial 3 = spatially explicit
Vegetation application	How does the framework apply vegetation and its attributes in modeling?	0 = applied as the basis for a wildlife species-habitat matrix 1 = applied as habitat variables to assess wildlife–habitat relationships
Ordinal criteria		
Documentation	Is there sufficient documentation (e.g., a user's manual or website) to clearly understand the modeling framework?	0 = limited 1 = marginal 2 = sufficient
Ease of application	Is the model difficult to parameterize, run, and understand the output?	1 = difficult 2 = moderate 3 = easy
Output definition	Is the output well defined and will it translate to something that can be measured?	1 = difficult 2 = moderate 3 = easy
Scale definition	Is the framework well suited for the scales it is defined to examine?	0 = not well suited 1 = moderately well suited 2 = very well suited
Transparency	Is the structure of the framework clear (i.e., is the flow of the framework apparent)?	1 = difficult 2 = moderate 3 = easy

Description of Habitat-Relationships Modeling Frameworks

To depict trends in development of wildlife habitat-relationships modeling frameworks, we plotted nominal criteria as proportions across the three decades encompassing our review (1980s, 1990s, and 2000s), with the final decade covering 2000–2006. Because California wildlife habitat relationships (Salwasser et al. 1980), pattern recognition (Williams et al. 1977), and wildlife habitat quality (Roller 1978) modeling frameworks were developed in the mid- to late-

1970s, we included these frameworks with those described in the 1980s. We developed narratives for each framework summarizing the origins of the framework, capabilities of the framework including data inputs and outputs, and related information (e.g., availability of software).

We conducted cluster analyses to better understand relationships among frameworks and to identify frameworks with similar characteristics. We used agglomerative hierarchical cluster methods to identify groupings of habitat-relationships modeling frameworks based on dissimilarity distance between each framework (PROC CLUSTER; SAS Institute 2003). Our input data for cluster analyses were the criteria ratings for each framework. Because our ratings consisted of nominal and ordinal data, we computed Gower's similarity coefficients (Gower 1971) between each pair of frameworks. We then computed Gower's dissimilarity coefficient (1 - Gower's similarity coefficient) in PROC DISTANCE (SAS Institute 2003) to base clustering on heterogeneity within the data ratings between frameworks. We used the average linkage cluster method, which is an unweighted pair-group method that uses arithmetic averages of dissimilarity coefficients to compute distance between clusters (PROC CLUSTER; SAS Institute 2003). We used an R^2-type measure of total within-cluster heterogeneity to evaluate the proportion of variance accounted for by joining each cluster. When each framework is in a cluster by itself, $R^2 = 1$ because there is no within-cluster variability; as frameworks are grouped into clusters, within-cluster variability increases from 0 and R^2 decreases from 1. We plotted R^2 values for each cluster in a hierarchical tree diagram (PROC TREE; SAS Institute 2003) and used a cutoff value of $R^2 = 0.60$ to define cluster groupings. We computed Gower's dissimilarity coefficients within each identified cluster group to evaluate within-cluster variability and report the mean and range in these coefficients for each cluster (PROC MEANS; SAS Institute 2003). Because Gower's dissimilarity coefficients range from 0 to 1, higher values indicate greater within-cluster heterogeneity. Lastly, we described attributes of each cluster group to better understand common patterns.

RESULTS

Identifying and Rating Habitat-Relationships Modeling Frameworks

We identified 40 modeling frameworks (Table 10-2); 13 frameworks developed through the 1980s, 12 frameworks developed in the 1990s, and 15 developed since 2000. Ten (0.25) frameworks exist within a larger landscape assessment system (ALCES, BOREAL, CompPATS, EMDS, HCI, LEAM, LEEMATH, LMS, SESI, and SIMFOR). Although HCI was developed as a component of the Coastal Landscape Analysis and Modeling System (CLAMS; Spies et al. 2002), it can model wildlife-habitat relationships outside this system (B. C. McComb, University of Massachusetts, personal communication, 2006). Eight (0.20) frameworks (Arc-Habcap, BEST, BIRDHAB, CompPATS, CWHR, HABSCAPES, PATCH, and SHM) apply vegetation and its

Table 10-2 Summary of 40 Habitat-Relationships Modeling Frameworks

Framework	Description	Primary References
A Landscape Cumulative Effects Simulator (ALCES)	ALCES quantifies economic contributions of land use practices, identifies associated environmental and industrial issues, and assists in development of mitigation strategies. The availability and quality of habitat for specific wildlife species is determined by tracking the area and area-weighted value of different vegetation and landscape types.	Schneider et al. 2003, ALCES 2005
Animal, Landscape and Man Simulation System (ALMaSS)	ALMaSS predicts the effect of changing landscape structure or management on key wildlife species. It incorporates detailed species-specific life history information and is agent-based, allowing each individual to interact with other individuals and the environment.	Topping et al. 2003
Artificial Neural Network (ANN)	Neural network models are inspired by natural physiology and mimic the neurons and synaptic connections of the brain. Once trained for a given task, a network can be applied by providing suitable data on the network inputs. Published applications used habitat variables to model nesting habitat for red-winged blackbirds, marsh wrens, and northern bobwhite quail.	Özesmi and Özesmi 1999, Lusk et al. 2002, Özesmi et al. 2006
Arc-Habcap	Arc-Habcap is a deterministic GIS-based wildlife habitat model that originated from a spreadsheet-based habitat capability (Habcap) model. The model in Benkobi et al. (2004) predicts effectiveness of forage, cover, and cover-forage proximity, as well as effects of roads, on elk distributions. The Arc-Habcap framework can be used to model habitat for any terrestrial vertebrate based on association with vegetation structural stages.	Benkobi et al. 2004
Arc Habitat Suitability Index (ArcHSI)	ArcHSI is a GIS-based model that estimates the ability of an area to meet the food and cover requirements of an animal species. The components and parameters of the model occur in tables and can be easily edited or otherwise modified. ArcHSI runs on personal computers with the full installation of ArcGIS. Also see ArcView HABCAP (U.S. Forest Service 2005).	Juntti and Rumble 2006

continues

Table 10-2 Summary of 40 Habitat-Relationships Modeling Frameworks *cont...*

Framework	Description	Primary References
Bayesian Belief Networks (BBN)	BBNs depict probabilistic relations among variables and use Bayesian statistics to calculate probabilities of outcomes, such as population presence, given conditions of input variables (e.g., condition of habitat).	Marcot et al. 2001, Raphael et al. 2001, Marcot 2006
Biodiversity Expert System Tool (BEST)	BEST uses data from the U.S. Geological Survey's Gap Analysis Program (GAP) and other data in a GIS environment. This tool provides predictions of conflict between proposed land uses and biotic elements and is intended for use at the start of a development review process.	Crist et al. 2000
BIRDHAB	BIRDHAB is a wildlife habitat relationships model developed for national forests in the Southern Region to assist in assessment of proposed management actions. It is written as an ArcInfo GIS program that accesses stand inventory data and a species-habitat matrix to describe the relative quality of habitat for 271 species of birds.	U.S. Forest Service 1994, Kilgo et al. 2002
BOREAL	BOREAL is a tactical planning decision support system that predicts the effects of alternative forest management strategies on forest product yields, revenues, and habitat area and distribution. This framework uses readily available inventory data and provides tabular, graphical, and map output.	Puttock et al. 1998
Computerized Project Analysis and Tracking System (CompPATS)	CompPATS evaluates the effects of forest management on wildlife habitat, sedimentation, visual quality, timber yield, and net revenue. Wildlife values describe habitat capacity, not an estimate of animal abundance.	Ouachita National Forest 1988, Keller et al. 1994
California Wildlife Habitat Relationships (CWHR)	CWHR is maintained by the California Department of Fish and Game. Habitat suitability indices may be calculated for land use planning assessments using GIS and fuzzy logic.	Salwasser et al. 1980, Raphael and Marcot 1986, Block et al. 1994, California Department of Fish and Game 2005
Effective Area Model (EAM)	EAM is an empirically based spatial model that incorporates patch size and shape, composition of matrix habitats, and species-specific edge responses to predict the organization of animal assemblages occupying heterogeneous landscapes. Specifically, it predicts the effects of matrix habitats on species abundances in habitat patches.	Sisk et al. 1997, Brand et al. 2006

continues

Table 10-2 Summary of 40 Habitat-Relationships Modeling Frameworks *cont...*

Framework	Description	Primary References
Ecosystem Management Decision Support (EMDS)	EMDS v. 2.0 is an application framework for knowledge-based decision support of ecological assessments that is designed for use at any geographic scale. The system integrates GIS and knowledge-based reasoning technologies in the Microsoft Windows® environment.	Reynolds 1999a, b; Reynolds 2001, Stoms et al. 2002
Expert Systems	Expert systems are a formalized method of organizing and applying information and opinion which utilize quantitative information when available, but usually rely primarily on expert opinion. Results may be expressed in terms of conditional states or probabilities.	Marcot 1986
FORHAB	FORHAB is a deciduous forest stand simulation model that may be used to predict changes in available breeding habitat for birds.	Smith et al. 1981
HABIT@	HABIT@ evaluates habitat at multiple, interconnected scales through indices that represent the quality of selected variables with numerous options for summarizing, combining, and/or comparing model variables (e.g., arithmetic mean, product, geometric mean, minimum).	McGarigal and Compton 2003
HABSCAPES	HABSCAPES uses spatial databases to map the predicted occurrence of all terrestrial vertebrate and aquatic amphibian species relative to landscape pattern over large geographic areas. Spatial databases describing the landscape are linked to databases containing wildlife habitat relationships and life history characteristics using custom FORTRAN programs and PARADOX scripts.	Huff et al. 2001; Mellen et al. 1995, 2001
HABSIM	HABSIM tracks vegetation seral stages, quantifies the change in vegetation structure and composition for each seral stage over time, and relates this information to potential carrying capacity for the species of interest.	Raedeke and Lehmkuhl 1986
Habitat-Based Species Viability (HBSV) Model	With HBSV, areas of high quality habitat for a species are assumed to support individuals in smaller home ranges, with higher rates of survival, and with higher reproductive success. The number of individual home ranges of different quality habitat for an individual species are mapped and quantified to assess the potential viability of the species.	Roloff and Haufler 1997, 2002

continues

Table 10-2 Summary of 40 Habitat-Relationships Modeling Frameworks *cont...*

Framework	Description	Primary References
Habitat Capability Index (HCI)	HCI estimates the capability of a landscape patch and its surrounding neighborhood to provide conditions important to a species survival and reproduction. These values are based on vegetation and physical conditions over a range of scales on the landscape.	McComb et al. 2002, Spies et al. 2002
Habitat Effectiveness Index (HEI)	HEI originated through the development of models to evaluate cumulative effects and is computed as the difference between analogues of death and birth rates, which yields a measure of habitat suitability. An index of human activity may be used as an analogue of death rates. An index of habitat quality, potentially described by vegetation, food availability, and abiotic factors is often used as an analogue of birth rate.	Thomas et al. 1988, Merrill et al. 1999
Habitat Quality (HQ)	The HQ framework measures habitat interspersion (Is) and juxtaposition (Jx) through GIS processes and incorporates it with limiting factors (RDF) that are essential for the species of interest. The form of the relationship is HQ = (0.2*Is/8) + (0.6*Jx/12) + (0.2*RDF) resulting in values from 0.0 to 1.0.	Roy et al. 1995
Habitat Quality (HQI) and Habitat Quality Plus (HQI+)	This is a GIS (ArcView) PC application that was developed to provide information for development of forest plans (HQI for single species analyses; HQI+ for multiple species analyses). An index value from 0.0 to 1.0 is assigned to habitat patches based on cover type, canopy, tree size, and season.	Rickel 1997
Habitat Suitability Index (HSI)	HSI indices are a composite (often a geometric mean) of individual suitability index (SI) scores reflective of habitat variables that represent cover types, life requisites, and life stages for habitats of individual species, each scaled 0 (unsuitable habitat) to 1 (optimum habitat). SI scores range from 0 to 1 and are computed as a ratio of a value of interest (i.e., estimate or measure of habitat conditions) divided by a standard of comparison (i.e., optimum habitat condition). HSI models assume a linear relationship between the index value and carrying capacity for the species of interest.	U.S. Fish and Wildlife Service 1981

continues

Table 10-2 Summary of 40 Habitat-Relationships Modeling Frameworks *cont...*

Framework	Description	Primary References
Landscape HSI	Landscape HSI applies a 0–1 habitat suitability index to large landscapes through the use of GIS-based modeling of raster data (e.g., tree species and age) across entire landscapes. Landscape HSI has also incorporated other programming to facilitate evaluation of spatially explicit landscape attributes (e.g., LANDIS) and wildlife population fitness parameters (e.g., RAMAS).	Larson et al. 2003, 2004, Shifley et al. 2006, Dijak et al. 2007, Rittenhouse et al. 2007
Land Use Evolution and Impact Assessment Model (LEAM)	The LEAM model determines the location of habitat patches likely to sustain populations of species of interest, estimates population size, and assesses the degree of connectivity and potential gene flow between patches. When applied to a changing landscape, the results of the model indicate changes in species-specific patch connectivity and determine the impact of land-use change on population isolation and habitat fragmentation.	Aurambout et al. 2005
Landscape Evaluation Effects of Management Effects on Timber and Habitat (LEEMATH)	LEEMATH is a spatially and temporally explicit tool that integrates habitat attributes, habitat suitability, stand growth, spatial habitat attributes, and landscape characteristics. Model input is a management regime defined by a timber harvest schedule, a silvicultural treatment plan, the spatial distribution of stands, and the target wildlife species. Outputs include timber growth and harvest (e.g., total basal area), habitat attributes (e.g., mean habitat patch size) and habitat suitability (e.g., total habitat area).	Li et al. 2000
Landscape Management System (LMS)	LMS is a computerized system that integrates landscape-level spatial information, stand-level inventory data, and distance-independent individual tree growth models to project changes through time in tree growth and snag decay across forested landscapes. Management scenarios are evaluated in terms of wildlife habitat and timber revenue.	Marzluff et al. 2002, Oliva et al. (this volume)
Program to Assist in Tracking Critical Habitat (PATCH)	PATCH is a spatially explicit, individual-based, life history simulator designed to project populations of territorial terrestrial vertebrate species through time. Inputs include habitat maps, specifications for habitat use (territory size and habitat affinity), vital rates (survival and reproduction), and descriptions of species' movement behavior. Outputs include spatial estimates of habitat occupancy rate and source-sink characteristics.	Schumaker 1998, Schumaker et al. 2004

continues

Table 10-2 Summary of 40 Habitat-Relationships Modeling Frameworks *cont...*

Framework	Description	Primary References
Pattern Recognition (PATREC)	PATREC is a modeling framework that relies on Bayesian statistical inference, which requires that habitat conditions be expressed as conditional probabilities (i.e., 1 or more of the habitat conditions under consideration is much more probable [occurs more frequently] than the others). Expected densities of animals can be computed based on knowledge of densities and habitat conditions.	Williams et al. 1977, Grubb 1988
Point Specific Estimator (PSE)	PSE estimates quality of habitat from single variable databases (e.g., vegetation maps) in terms of interspersion, juxtaposition, and spatial diversity. Input requirements include cover type and values of cover types to wildlife species. Outputs for raster-based maps are possible through application of the spatial diversity index values to each grid cell.	Mead et al. 1981, Lyon et al. 1987
RAMAS Landscape	RAMAS Landscape integrates the LANDIS landscape model with the RAMAS GIS habitat-based metapopulation model to provide predictions about the viability, recovery, and growth of species based on predicted changes in landscapes.	Akçakaya et al. 2004, 2005
Spatially Explicit Species Index (SESI)	SESI models are similar to HSI models in that population response is predicted by a set of habitat relationships and in that habitat quality is quantified by an index value. However, SESI models can focus either on one part of a life cycle, such as breeding or foraging, or whole life cycles. They incorporate temporal changes in the environment, can be used to model the responses of any species in the system, and provide a landscape index map rather than just a single index or set of indices.	Curnutt 2000
SIMFOR	SIMFOR evaluates the response of forest vegetation to management or natural disturbances, and calculates potential landscape and wildlife habitat conditions. By matching wildlife species requirements with projected habitat attributes, SIMFOR estimates species-specific habitat suitability. Simple landscape metrics based on seral stage, patch size, and edge characteristics are also calculated.	Wells et al. 1999, Wells and May 2002, Seely et al. 2004

continues

Table 10-2 Summary of 40 Habitat-Relationships Modeling Frameworks *cont...*

Framework	Description	Primary References
Spatially Neutral Bayesian Model (SNBM)	The simplest potential distribution of a wildlife species is a random distribution where all sites have equal probabilities. A more ecologically appropriate potential spatial distribution accounts for environmental variation. This expected distribution is called a spatially neutral model, because it is generated without hypothesizing spatial factors that regulate the distribution of resources or organisms.	Milne et al. 1989
Spatial Optimization	Spatial optimization is not a habitat modeling framework, per se, but provides a framework within which the results of habitat modeling may be applied to obtain habitat configurations to best meet specific management objectives. Optimization of landscapes aims to identify landscape and land-use patterns, which support certain ecosystem functions in an optimal way. The chosen performance criteria are based on the ecosystem functions considered for optimization.	Hof and Bevers 1998
Species-area Relationship (SPPAREA)	Species-area curves are computed as $S = cAz$, where S = number of species, c = a constant that varies with taxon and geographic region, A = area, and z = a constant measuring the slope of the line relating S and A. Species-habitat area relationships were first explored on islands, but have been extended to a wide variety of habitats.	Schroeder 1996
Species-Habitat Matrices (SHM)	Species-habitat matrices are databases used to predict the presence or relative abundance of species within geographic areas or within sera1 stages of vegetation types. More detailed predictions include ratings for life requisites of species such as reproduction, feeding, and cover. Most species-habitat matrices rely on previously published information and expert opinion as the basis for their entries.	Thomas 1979, Hoover and Willis 1984, DeGraaf et al. 1992, Scott et al. 1993, Karl et al. 2000
Species Sorting Algorithm (SSA)	SSA derives data from a spatial landscape analysis and from published species life-histories to evaluate the full suite of species that could occur on a landscape. The SSA identifies and concentrates attention on species that have, due to ecological factors such as habitat specificity or negative response to management activities, the potential to be affected by proposed land management.	Reed et al. 2001, Higdon et al. 2005, 2006

continues

Table 10-2 Summary of 40 Habitat-Relationships Modeling Frameworks *cont...*

Framework	Description	Primary References
Wildlife Habitat Quality (WHQ)	WHQ generates numerical ratings of habitat quality based on an analysis of digital habitat maps and associated information. Information on vegetation and terrain (as they affect availability of food and cover), habitat interspersion, and habitat juxtaposition are integrated to provide a score from 0 to 100 to quantify habitat quality.	Roller 1978

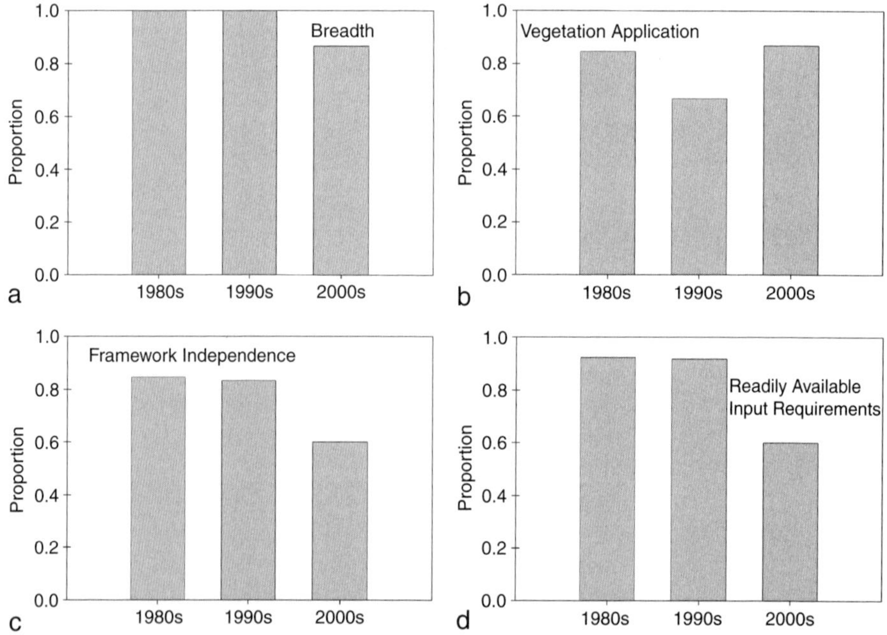

FIG. 10-1

Proportion of wildlife habitat-relationships modeling frameworks developed by decade (A) suited for a wide range of species in a wide range of environments; (B) where vegetation was applied as habitat variables to assess wildlife-habitat relationships; (C) that are standalone frameworks, not a component of a landscape modeling system; and (D) with input requirements that are readily available in agency inventories.

attributes as the basis for evaluating wildlife-habitat relationships within species-habitat matrices.

Since development of wildlife-habitat relationship models began, most frameworks have defined habitat relationships for a wide range of species in a wide range of environments (Fig. 10-1A). During the 1990s, more (0.33) frameworks

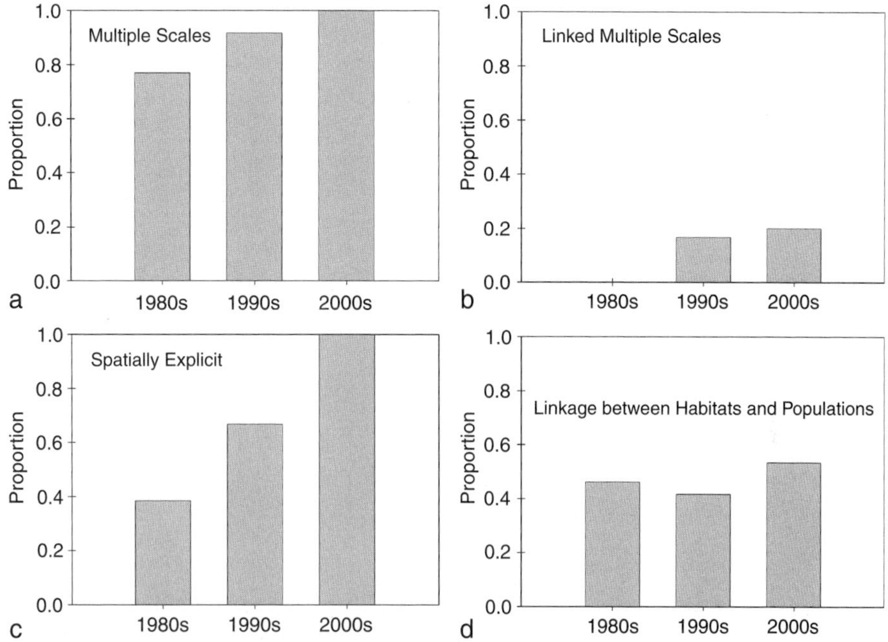

FIG. 10-2

Proportion of wildlife habitat-relationships modeling frameworks developed by decade that (A) examine habitat relationships at multiple scales; (B) provide linkage between scales if multiscaled; (C) are spatially explicit; and (D) use population demographics or surrogates of population demographics to model habitat relationships.

applied vegetation attributes within the context of species-habitat matrices than other decades (Fig. 10-1B). The proportion of frameworks that are not components of larger landscape modeling systems (Fig. 10-1C) and that use input data that are typically readily available in natural resource agency inventories declined from 1980 through 2006 (Fig. 10-1D). The proportion of frameworks that examine habitat relationships at multiple scales (Fig. 10-2A), link scales when multiscaled (Fig. 10-2B), and that are spatially explicit (Fig. 10-2C) increased from the 1980s through 2006. The proportion of frameworks that use population demographics or surrogates generally increased from the 1980s through 2006 (Fig. 10-2D). Over time, the proportion of frameworks where at least one species model based on that framework has been validated through comparing predictions to observed data, reserving data to use in validation, or other techniques never exceeded 0.58 (Fig. 10-3A), but the proportion of frameworks that have received scientific credibility through peer-reviewed publication or application of results or other mechanisms has consistently remained >0.83 (Fig. 10-3B).

Only two (0.05) frameworks (ALMaSS and LEEMATH) were limited to a single environment (Table 10-3). Of the total, three (0.08) frameworks were aspatial (Expert Systems, HABSIM, CompPATS; Table 10-3). Four (0.10; ANN, CompPATS,

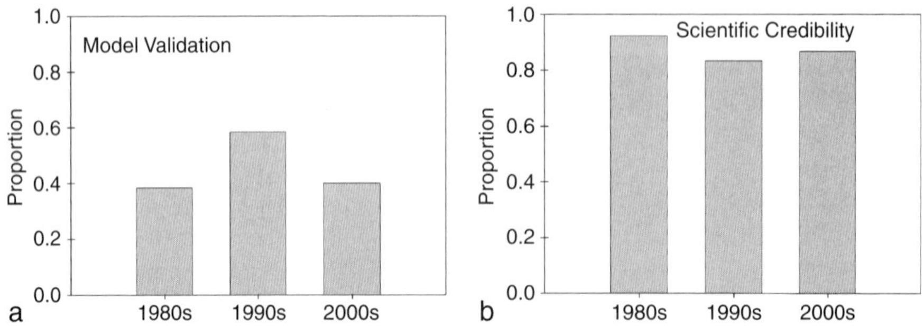

FIG. 10-3

Proportion of wildlife habitat-relationships modeling frameworks developed by decade (A) where at least one model developed within that framework has been validated with field data; and (B) that have attained scientific credibility through publication of results, application of results, or other mechanisms to suggest acceptance by an array of professionals.

SPPAREA, and WHQ) of the reviewed frameworks considered habitat relationships at a single spatial scale (Table 10-3). Five of the 36 (0.14) multiscale frameworks (BBN, HABIT@, HCI, EMDS, and PATCH) provided linkage between scales (Table 10-3). Nineteen (0.48) frameworks incorporated population demographics or surrogates into modeling. Twenty-seven (0.68) frameworks have the ability to incorporate spatially explicit characteristics (Table 10-3).

Description of Habitat-Relationships Modeling Frameworks

Total heterogeneity between CompPATS, HABSCAPES, and other frameworks was $R^2 \geq 0.60$, indicating these two frameworks were different from other frameworks based on our criteria so they were not included in any clusters (Fig. 10-4). Heterogeneity was lowest between frameworks for the cluster formed by HEI and HBSV ($R^2 = 1.000$) and highest ($R^2 = 0.000$) between CompPATS, HABSCAPES, and all clusters (Fig. 10-4). Thirty-eight frameworks were apportioned within 7 clusters, each cluster containing an average of 5.4 (range = 2–10) frameworks. Mean dissimilarity between all modeling frameworks was 0.352 (range: 0.034–0.753), indicating average heterogeneity was low-to-moderate, yet the range in heterogeneity between frameworks was broad.

Cluster 1.—Cluster 1 consisted of HSI and nine other frameworks ($R^2 = 0.739$) that rely on emerging analysis techniques (ANN, CWHR, HEI, HBSV, PATCH, and PATREC) and/or evaluate wildlife-habitat relationships within the context of species-habitat matrices (Arc-Habcap, BIRDHAB, CWHR, PATCH, and SHM; Fig. 10-4; Table 10-3). Mean dissimilarity between all frameworks was 0.241 (range: 0.071–0.429), indicating that frameworks within the cluster were rather similar in their characteristic abilities (i.e., how they fit our evaluation criteria). Input for all frameworks in Cluster 1 was readily available in natural resource

Table 10-3 Ratings for Criteria Used to Assess Wildlife Habitat-Relationships Modeling Frameworks

	Nominal Criteria[a]										Ordinal Criteria[b]				
Framework	**Breadth**	**Habitat Pop-Link**	**Indep**	**Input Req**	**Model Valid**	**Scale**	**Scale Link**	**Credible**	**Spatial Appl**	**Veg Appl**	**Document**	**Ease**	**Output**	**Scale Def**	**Trans**
Cluster 1															
ANN	W	S	I	RA	V	S	NL	C	S	HV	S	M	E	VWS	D
Arc-Habcap	W	S	I	RA	V	M	NL	C	SE	SHM	L	M	E	VWS	E
BIRDHAB	W	No	I	RA	V	M	NL	C	S	SHM	S	E	E	VWS	E
CWHR	W	S	I	RA	V	M	NL	C	S	SHM	S	M	E	VWS	M
HBSV	W	S	I	RA	V	M	NL	C	SE	HV	S	M	E	VWS	E
HEI	W	S	I	RA	V	M	NL	C	SE	HV	S	E	E	VWS	E
HSI	W	No	I	RA	V	M	NL	C	S	HV	S	E	E	MWS	E
PATCH	W	S	I	RA	V	M	L	C	SE	SHM	S	M	E	VWS	E
PATREC	W	S	I	RA	V	M	NL	C	S	HV	S	E	E	VWS	E
SHM	W	No	I	RA	V	M	NL	C	S	SHM	S	M	E	MWS	E
Cluster 2															
ALCES	W	No	NI	NRA	NV	M	NL	C	SE	HV	S	M	M	VWS	D
BOREAL	W	No	NI	RA	NV	M	NL	C	S	HV	L	D	M	VWS	D
EMDS	W	No	NI	RA	NV	M	L	C	SE	HV	S	D	E	VWS	D
HCI	W	No	NI	RA	V	M	L	C	SE	HV	S	M	E	VWS	M
LEAM	W	S	NI	NRA	NV	M	NL	C	SE	HV	M	M	M	VWS	M
LEEMATH	S	No	NI	NRA	V	M	NL	C	SE	HV	M	D	E	VWS	D
LMS	W	No	NI	RA	V	M	NL	C	SE	HV	S	D	E	VWS	M

continues

Table 10-3 Ratings for Criteria Used to Assess Wildlife Habitat-Relationships Modeling Frameworks *cont...*

	Nominal Criteria[a]										Ordinal Criteria[b]				
Framework	**Breadth**	**Habitat Pop-Link**	**Indep**	**Input Req**	**Model Valid**	**Scale**	**Scale Link**	**Credible**	**Spatial Appl**	**Veg Appl**	**Document**	**Ease**	**Output**	**Scale Def**	**Trans**
SESI	W	No	NI	RA	V	M	NL	C	SE	HV	M	D	M	VWS	M
SIMFOR	W	No	NI	RA	V	M	NL	C	SE	HV	S	M	E	VWS	D
Cluster 3															
EAM	W	P	I	RA	V	M	NL	C	SE	HV	M	M	E	VWS	M
Expert Systems	W	No	I	RA	NV	M	NL	C	A	HV	S	M	E	VWS	M
HABSIM	W	P	I	RA	NV	M	NL	C	A	HV	M	M	E	VWS	M
HQ	W	No	I	RA	NV	M	NL	C	SE	HV	M	E	E	VWS	M
Landscape HSI	W	No	I	RA	NV	M	NL	C	SE	HV	S	E	E	VWS	E
RAMAS Landscape	W	P	I	RA	NV	M	NL	C	SE	HV	S	M	E	VWS	D
SNBM	W	S	I	RA	V	M	NL	C	SE	HV	M	D	E	VWS	D
Spatial Optimization	W	S	I	RA	NV	M	NL	C	SE	HV	S	D	E	VWS	D
SPPAREA	W	No	I	RA	NV	S	NL	C	SE	HV	S	E	E	VWS	E
WHC	W	No	I	RA	NV	S	NL	C	SE	HV	M	M	M	VWS	M
Cluster 4															
ArcHSI	W	No	I	RA	NV	M	NL	NC	SE	HV	S	E	E	MWS	E
HQI	W	No	I	RA	NV	M	NL	NC	S	HV	S	E	M	MWS	M
PSE	W	No	I	RA	NV	M	NL	C	SE	HV	S	M	D	MWS	D

Cluster 5															
BBN	W	S	I	NRA	V	M	L	C	SE	HV	S	E	E	VWS	M
HABIT@	W	S	I	RA	NV	M	L	NC	SE	HV	S	E	M	VWS	M
Cluster 6															
BEST	W	S	I	RA	NV	M	NL	C	S	SHM	L	M	M	MWS	M
FORHAB	W	S	I	NRA	NV	M	NL	C	S	HV	M	D	E	MWS	D
Cluster 7															
ALMASS	S	P	I	NRA	NV	M	NL	C	SE	HV	S	D	E	VWS	M
SSA	W	P	I	NRA	NV	M	NL	C	SE	HV	M	M	M	MWS	E
Nonclustered frameworks															
HABSCAPES	W	No	I	NRA	NV	M	NL	NC	SE	SHM	S	D	E	VWS	D
COMPATS	W	No	NI	RA	NV	S	NL	NC	A	SHM	S	E	M	MWS	D

[a]*Definitions for nominal criteria ratings:*
Breadth of application (Breadth) = suited for a single species or one environment (S) or for a wide range of species in a wide range of environments (W).
Habitat–population linkage (Habitat pop-link) = does the framework rely on population demographic parameters (P), surrogates of population demographic parameters (S), or does not rely on population demographics or surrogates (No) of modeled species.
Independence (Indep) = framework is independent of (I) or a part of a larger landscape modeling system (NI).
Input requirements (Input req) = not readily available (NRA) or readily available (RA) in agency inventories.
Model validation (model valid) = at least 1 model based on each framework not validated (NV) or validated (V) with field data.
Scale = is the framework limited to 1 scale (S) or is it capable of examining habitat relationships at more than 1 scale (M).
Scale linkage (Scale link) = scales in multiscaled frameworks are not linked (NL) or linked (L).
Scientific credibility (Credible) = framework has gained credibility (C) or not (NC) through publication or application of results.
Spatial application (Spatial appl) = Does the framework: solely examine aspatial (A) data, evaluate geographic data (spatial [S]), or examine spatial relationships in habitat data at specific locations or coordinates as part of its structure (spatially explicit [SE]).
Vegetation application (Veg appl) = within the framework, vegetation is applied as the basis for a wildlife species-habitat matrix (SHM) or vegetation is applied as habitat variables that are used to assess habitat relationships for wildlife species (HV).
[b]*Definitions for ordinal criteria ratings:*
Documentation (Document) = is documentation limited (L), marginal (M), or sufficient (S) to understand the modeling framework.
Ease = framework is difficult (D), moderate (M), or easy (E) to parameterize, run, and understand the output.
Output = difficult [D], moderate [M]), or easy [E] to define and measure.
Scale definition (Scale def) = is the framework not well suited (NWS), moderately well suited (MWS), or very well suited (VWS) to examine the scales it is defined to examine.
Transparency (Trans) = is the structure of the framework difficult (D), moderate (M), or easy (E) to understand.

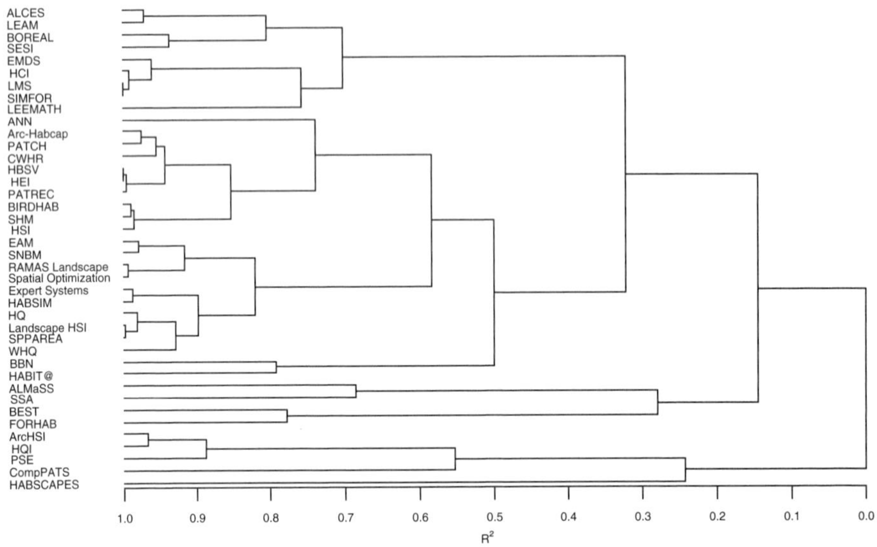

FIG. 10-4

Hierarchical tree diagram depicting heterogeneity between clusters of 40 wildlife habitat-relationships modeling frameworks evaluated in 2007.

agency inventories. Output was easy to define and measure for all frameworks in Cluster 1 (Table 10-3). Species-specific models for each framework in Cluster 1 have been validated; each framework was suited for a wide range of species in a wide range of environments and has attained scientific credibility (Table 10-3). Among the three largest clusters, Cluster 1 was highest (0.70) for frameworks that relied on population demographics or surrogates. All frameworks in Cluster 1 were moderate or easy to parameterize, run, and understand the output and 0.90 were moderate or easily transparent. With the exception of Arc-Habcap, all frameworks in Cluster 1 had sufficient documentation to clearly understand the framework (Table 10-3).

Cluster 2.—Cluster 2 included all frameworks (R^2 = 0.703), with the exception of CompPATS, that were components of larger landscape modeling systems (ALCES, BOREAL, EMDS, HCI, LEAM, LEEMATH, LMS, SESI, and SIMFOR; Table 10-3; Fig. 10-4). Mean dissimilarity between all nine frameworks was 0.302 (range: 0.119–0.500), indicating that most frameworks within the cluster were similar in their characteristic abilities. All the frameworks in Cluster 2 have received scientific credibility through publication, and all but BOREAL were spatially explicit (Table 10-3). However, data inputs were not readily available in agency inventories for three of nine of the frameworks; species-specific models for 4 of 9 frameworks have not been validated; each framework is moderate or difficult to parameterize, run, and understand the output; and transparency in model structure was moderate or difficult for every framework (Table 10-3).

Documentation for four frameworks was limited or marginal. None of the frameworks in Cluster 2 used population demographics, although LEAM used surrogates of population demographics (Table 10-3).

Cluster 3.—Cluster 3 consisted of 10 frameworks (EAM, expert systems, HABSIM, HQ, Landscape HSI, RAMAS Landscape, SNBM, spatial optimization, SPPAREA, and WHQ; Fig. 10-4; $R^2 = 0.887$). Mean dissimilarity between all frameworks within the cluster was 0.239 (range: 0.071–0.429), indicating that most frameworks within the cluster were similar in their characteristic abilities. Cluster 3 was characterized by frameworks that were generally well documented; have attained scientific credibility; used readily accessible input data; had output that is well defined and measurable; but tended to be difficult to run, parameterize, and understand the output (Table 10-3). Half of these frameworks emphasized population demographics or surrogates; the structure of only two frameworks in Cluster 3 was easily transparent; 8 of 10 frameworks do not have species-specific models that have been validated; two frameworks (SPPAREA and WHQ) considered habitat relationships at a single spatial scale; and all frameworks, except expert systems and HABSIM, were spatially explicit. In addition, all frameworks were very well suited to examine the scales they were designed for (Table 10-3).

Cluster 4.—Cluster 4 included three frameworks (ArcHSI, HQI, and PSE; Fig. 10-4) that had the lowest within-cluster variability ($R^2 = 0.887$) of all clusters. Mean dissimilarity between all frameworks within Cluster 4 was 0.256 (range: 0.154–0.308), further indicating that frameworks within this cluster were similar in their characteristic abilities. All the frameworks in Cluster 4 used readily available input data, had sufficient documentation to understand the framework, and were moderately well suited to examine the multiple scales they were designed to evaluate (Table 10-3). None of the frameworks in Cluster 4 used population demographics or surrogates or have been validated through species-specific models. These frameworks are mixed (difficult, moderate, and easy; Table 10-3) relative to our assessment of practitioners being able to measure model output and understand framework transparency.

Cluster 5.—Cluster 5 consisted of two spatially explicit frameworks (BBN, HABIT@), which were both linked to the multiple scales they were very well suited to examine (Fig. 10-4). Within-cluster heterogeneity was $R^2 = 0.791$ and within-cluster dissimilarity was 0.364. Both frameworks had sufficient documentation; were easy to parameterize, run, and provided understandable output; used surrogates of population demographics; and were ranked moderate in transparency (Table 10-3). BBN, but not HABIT@, attained model validation and scientific credibility (Table 10-3).

Cluster 6.—Cluster 6 included two scientifically credible, spatial frameworks (BEST and FORHAB; Fig. 10-4) that were moderately well suited for the multiple scales they were designed to examine (Table 10-3). Within-cluster heterogeneity was $R^2 = 0.778$. Dissimilarity between frameworks was 0.429, indicating that the frameworks forming this cluster were relatively more dissimilar than frameworks in the other clusters. Both frameworks incorporated surrogates of population demographics; were capable of modeling a wide range of species in a wide range of environments; but did not have examples of validated models developed within the frameworks. However, other characteristic abilities based on rating criteria differed. BEST used readily available data from natural resource agency inventories and incorporated vegetation and its attributes within a species-habitat matrix.

Cluster 7.—Cluster 7 included two spatially explicit, credible frameworks (ALMaSS and SSA; Fig. 10-4), which specifically relied on population demographics to evaluate wildlife-habitat relationships (Table 10-3). Within-cluster heterogeneity was highest in this cluster when compared among all seven clusters (R^2 = 0.686), and within-cluster dissimilarity (0.400) was second highest among clusters. Input data for both frameworks were not readily available in natural resource agency inventories, and neither framework has attained validation through a species-specific model. ALMaSS was suited for a single environment (i.e., temperate Europe); was moderately transparent in understanding model structure; was very well suited to examine the scales for which it was designed; was difficult to run, parameterize, and understand its output; but has detailed documentation (Table 10-3). Although marginally well documented, the structure of SSA was easily transparent; however, it was rated moderate for all other ordinal-scale criteria (Table 10-3).

DISCUSSION

Development of model components through the past three decades has coincided with technological advancements including landscape modeling applications, statistical techniques, and computing capabilities (Capen 1981, Scott et al. 2002, Stauffer 2002). Developments in ecological theory have also influenced habitat-relationships modeling. For instance, newer frameworks often consider wildlife habitat relationships from a landscape viewpoint by including fragmentation or patch size effects on wildlife populations (e.g., LEAM [Aurambout et al. 2005]), grouping terrestrial species into guilds based on expected responses to different amounts and distributions of habitat across landscapes (HABSCAPES [Mellen et al. 2001]), integrating landscape and metapopulation models to predict demographic responses based on predicted landscape changes (RAMAS Landscape [Akçakaya et al. 2004, 2005]); and predicting the effects of matrix habitats, including edge responses of species, on species abundances in habitat patches (EAM [Sisk et al. 1997, Brand et al. 2006]).

Habitat suitability under HEP was defined as a 0–1 index of habitat quality ranging from unsuitable to optimal (U.S. Fish and Wildlife Service 1981). Many newer modeling frameworks (e.g., ArcHSI [Juntti and Rumble 2006], HABIT@ [McGarigal and Compton 2003], HCI [McComb et al. 2002], HQ [Roy et al. 1995], HQI [Rickel 1997], and Landscape HSI [Larson et al. 2003, 2004, Dijak et al. 2007; Rittenhouse et al. 2007]) follow this convention by defining habitat capability or suitability in 0–1 index form. This approach provides an easily interpretable basis to compare current habitat conditions or suitability of sites to optimal habitat conditions at sites for a given species.

Habitat Evaluation Procedures suggested that population variables should not usually be included in a habitat model because they are costly to obtain, difficult to predict, and often not indicative of habitat suitability (U.S. Fish and Wildlife Service 1981). Even though including population variables in habitat-relationships modeling may have been avoided in the past, we considered this criterion in our evaluations of modeling frameworks because the value of habitats to wildlife populations is better understood when population parameters can be linked with habitat conditions (Van Horne 1983). The results of habitat-relationships modeling are increasingly reported within a population context, including available breeding bird habitat (Smith et al. 1981), habitat effectiveness (Merrill et al. 1999), potential population density (Mattson and Merrill 2004), presence or relative abundance (Scott et al. 1993), and viable home ranges (Roloff and Haufler 1997).

Since their inception, wildlife habitat-relationships modeling frameworks have incorporated additional characteristic abilities such as application at multiple scales, linking scales when multi scaled, and incorporation of population demographics or surrogates. Our evaluation provides practitioners with information that will be useful in selecting frameworks to meet specific needs. In the following sections, we examine scenarios in which frameworks in each cluster have potential application. We also provide a key to assist practitioners in selecting the most appropriate framework for potential applications (Table 10-4).

Potential Applications

Cluster 1.—Frameworks forming Cluster 1 provide many characteristics that practitioners may find desirable including data inputs that are readily available, field validation, scientific credibility, transparency, and the added benefit of using population demographics or surrogates to model habitat relationships. Although Cluster 1 included frameworks that evaluate wildlife habitat quality within the simplistic context of species-habitat matrices, as compared to frameworks that rely on more complex emerging analysis techniques, the characteristic abilities of frameworks using these approaches were similar. A practical application of species-habitat matrix frameworks is their use when conducting environmental impact assessments, where the quality of habitat for various species within impacted or nonimpacted habitats or habitat structural stages is of more importance than predicting occurrence or abundance (Kilgo et al. 2002). Although they provide interpretable output, frameworks that use emerging analysis techniques may require technical support to parameterize and interpret model output. For instance, to model habitat relationships, ANN uses artificial neural networks (Özesmi and Özesmi 1999, Lusk et al. 2002,

Table 10-4 Key to Assist Practitioners in Selecting the Most Appropriate Framework for Potential Applications from Among 40 Identified Wildlife Habitat-Relationships Modeling Frameworks

1.		Large landscape modeling system is not desired	2
1.		Large landscape modeling system is desired	
	A.	Framework with scientific credibility is desired	Cluster 2
	B.	Framework with scientific credibility is not important	COMPATS
2.		Input data must be readily available from agency databases	3
2.		Not critical that input data be readily available from agency databases	5
3.	A.	Framework where output from 1 model has been validated is desired	Cluster 1
	B.	Framework where output from 1 model has not been validated is acceptable	4
4.		Frameworks are very well suited for the scales they are designed for	Cluster 3
4.		Frameworks are moderately well suited for the scales they are designed for	Cluster 4
5.		The use of population demographics or surrogates is not an objective	HABSCAPES
5.		Framework which uses population demographics or surrogates is desired	6
6.	A.	The spatial application of the framework simply uses geographical data	Cluster 6
	B	Spatially explicit applications by the framework are desired	7
7.	A.	Framework that uses surrogates of population demographics is desired	Cluster 5
	B.	Framework that uses population demographics is desired	Cluster 7

Özesmi et al. 2006); PATREC uses Bayesian probabilities (Williams et al. 1977, Grubb 1988); CWHR provides an option to apply fuzzy logic to calculate habitat suitability indices (California Department of Fish and Game 2005); and HBSV is a habitat-based approach to population viability modeling (Roloff and Haufler 1997, 2002). The original HSI framework provides advantages in ease of interpretability and has many completed models that have been validated. In addition, techniques are available to evaluate the reliability in HSI model inputs, providing a means to infer differences between HSI scores (Bender et al. 1996, Burgman et al. 2001). Those wishing to select a framework that uses surrogates or population demographics to link with habitat conditions should also consider Cluster 1. In comparison, frameworks in Cluster 4 do not incorporate a habitat-population linkage, and fewer frameworks in Clusters 2 and 3 provide these options as compared to Cluster 1.

Cluster 2.—All the modeling frameworks comprising Cluster 2 are scientifically credible components of larger landscape modeling systems. Thus, practitioners may want to consider selecting these frameworks only if they are going to be involved in a comprehensive assessment of a large landscape and therefore are willing to devote the effort necessary to parameterize and run the more comprehensive landscape model. It may be advisable for practitioners to establish a dialogue with the developers of these systems prior to initiating modeling; without establishing such dialogue, it would be difficult for practitioners to independently implement these frameworks. LEEMATH was developed to evaluate alternative management strategies for multiple species in industrial forest landscapes in the southeastern United States (Li et al. 2000); however, all other frameworks in Cluster 2 are suitable for a wide range of species in a wide range of landscapes. Major weaknesses of Cluster 2 are that only LEAM uses surrogates of population demographics, and without consultation with framework developers, transparency of the structure of frameworks is moderate at best. An advantage of several frameworks in Cluster 2 is that websites have been provided that detail their application (i.e., ALCES, EMDS, HCI [via CLAMS; Spies et al. 2002], LEAM, LMS, SESI, SIMFOR). Limitations associated with availability of input data, documentation, model parameterization, and transparency for frameworks in this cluster are largely related to the fact that these frameworks are components of larger landscape modeling systems. However, the value of understanding the influences of landscape processes and management activities such as logging on wildlife habitat quality makes consideration of these frameworks advantageous over those in other clusters.

Cluster 3.—Each framework in Cluster 3 was scientifically credible and used readily available input data, but only EAM and SNBM had models that have been field verified. Frameworks forming Cluster 3 approach habitat modeling under the context of a modeling shell (expert systems and spatial optimization), a GIS-based modeling system (Landscape HSI, RAMAS Landscape), or a modeling framework that uses a diversity of techniques to model habitat relationships. For instance, EAM utilizes a variety of spatially explicit analyses to predict the effects of matrix habitats on species abundances in habitat patches (Sisk et al. 1997, Brand et al. 2006), and SNBM generates expected distributions for wildlife species without hypothesizing spatial factors that regulate the distribution of resources or organisms (Milne et al. 1989). Spatial optimization allows one to apply the results of habitat modeling to optimize habitat configurations. However, implementation of habitat modeling with spatial optimization requires strong quantitative skills. RAMAS Landscape (Akçakaya et al. 2004, 2005; Bekessy et al., this volume) provides practitioners with a useful website and integrates a landscape model (LANDIS; He et al. 1999; He, this volume) with a metapopulation model (RAMAS GIS; Akçakaya 1998). Expert systems offer modelers the ability to structure models with expert opinion and quantitative data, often within the structure of a modeling shell (e.g., Sodja et al. 2002). A major advantage of frameworks in Cluster 3 compared to other clusters is the flexibility in modeling through modeling shells, GIS-based modeling systems, and other innovative techniques. A disadvantage of several frameworks in the cluster (i.e., EAM, HABSIM, HQ, SNBM, and WHQ) is marginal documentation.

Cluster 4.—Major strengths of frameworks in Cluster 4 are input data that are readily available in agency databases, abilities to evaluate spatial or spatially explicit data, and sufficient documentation to clearly understand each modeling framework. A major advantage of frameworks in Cluster 4 is their simple approach to evaluate habitat quality. ArcHSI and HQI are more sophisticated versions of the original HSI framework, are easy to parameterize and understand model output, and were developed for use within a GIS. PSE uses simple landscape metrics to evaluate habitat quality with single variable databases (Mead et al. 1981, Lyon et al. 1987). Although frameworks in Cluster 4 use simple approaches to model habitat quality, they are limited by their inability to link habitats with populations, and only PSE has achieved scientific credibility.

Cluster 5.—Cluster 5 is the only cluster where all frameworks link multiple scales. In addition, unlike the linked multiscale frameworks in Cluster 3, HABIT@ and BBN use surrogates of population demographics in assessing wildlife habitat quality. BBN provides practitioners with endless opportunities to evaluate habitat quality through depicting probabilistic relations among variables (Marcot et al. 2001, Raphael et al. 2001, Marcot 2006). HABIT@ represents one of the most innovative frameworks because it evaluates linked, spatially explicit habitat attributes at local, home range, and population scales (McGarigal and Compton 2003).

Cluster 6.—Cluster 6 is characterized by spatial frameworks that predict changes in habitats. FORHAB predicts changes in bird breeding habitats (Smith et al. 1981), while BEST is based on a species-habitat matrix that provides predictions of where land uses may conflict with the conservation of biotic elements of the landscape (Crist et al. 2000). In addition to predictive abilities, other strengths of frameworks in Cluster 6 include scientific credibility and linkage between habitats and populations. Limitations of frameworks in Cluster 6 include limited or marginal documentation, no model validation, and models where functional transparency is marginal or difficult to understand.

Cluster 7.—Frameworks in Cluster 7 provide predictive tools that are useful in assessing impacts of land management activities on species and habitats. These predictive frameworks are stronger than those in Cluster 6 because they are spatially explicit and directly use population demographics to evaluate habitat quality. ALMaSS addresses policy questions regarding effects of changing landscape or management scenario on selected wildlife species; however, it was specifically developed to model wildlife habitats in temperate Europe (Topping et al. 2003) and may have limited application elsewhere. SSA focuses on species that have the potential to be adversely affected by proposed land management due to specific habitat requirements or characteristic responses to management activities (Reed et al. 2001; Higdon et al. 2005, 2006). Weaknesses of frameworks in Cluster 7 include input data are not readily available in agency databases, models have not been validated, and frameworks are difficult or marginal to parameterize and understand the output.

FUTURE DIRECTIONS

Many recently developed modeling frameworks incorporate linkages between habitats and populations at multiple scales and link those scales, while incorporating spatially explicit data. We suggest that developers of new frameworks consider incorporating these components because the ecological concepts addressed often provide a better understanding of wildlife-habitat relationships and management implications. An emerging trend in wildlife habitat-relationships modeling is for frameworks to be components of larger landscape modeling systems. Although we view this trend as potentially problematic for practitioners not involved in comprehensive landscape assessments, many contemporary frameworks still allow independent applications.

Habitat suitability index models were originally developed to assist in quantifying and evaluating the effects of management actions on wildlife populations and their habitats (U.S. Fish and Wildlife Service 1981). Since the development of HEP, many other habitat-relationships modeling frameworks have also focused on evaluating land management actions on wildlife habitats. For instance, some frameworks have been developed to evaluate prescriptions for harvesting timber on wildlife habitats (e.g., BOREAL [Puttock et al. 1998], LEEMATH [Li et al. 2000]), whereas others consider influences of a variety of perturbations and ecological and industrial issues in conjunction with wildlife habitats (e.g., ALCES [ALCES 2005], CompPATS [Ouachita National Forest 1988], LMS [Marzluff et al. 2002], SESI [Curnutt et al. 2000], SIMFOR [Seely et al. 2004]). Future frameworks that focus on evaluations of management practices or perturbations on wildlife habitats will be more widely applied if they address a variety of management questions (e.g., energy development, transportation corridors).

A current trend in framework development is to incorporate spatially explicit procedures when evaluating wildlife-habitat relationships. We suggest all future frameworks for wildlife conservation in large landscapes be able to evaluate habitat conditions under explicit spatial contexts. Spatially explicit habitat modeling frameworks provide practitioners with the ability to evaluate habitat in relation to conditions in adjoining parcels, according to configurations of resources, and in relation to habitat features such as roads that may influence animal movements or other behaviors (McGarigal and Compton 2003).

Emerging frameworks that show promise for describing wildlife-habitat relationships and that may be considered by developers include Petri nets, which are mathematical tools that are useful for modeling concurrent, distributed, asynchronous behavior in a system (e.g., Gronewold and Sonnenschein 1998). Also, qualitative modeling (e.g., loop analysis [Justus 2006]) may be more practical as a framework than quantitative modeling because qualitative models require fewer resources and less modeling experience.

Developers of frameworks have consistently attained scientific credibility through published manuscripts describing the development or applications of models developed within their frameworks, but a major weakness for many frameworks continues to be a lack of validation (Raphael and Marcot 1986, Block et al. 1994, Roloff and Kernohan 1999). Model validation is critical so that models developed within any framework can be used with confidence (Shifley et al., this volume). Therefore, we recommend that models be validated through independent field study or by reserving some data used in model development. Of particular interest is the need to validate frameworks. Although some frameworks have been validated (e.g., BIRDHAB [Kilgo et al. 2002], CWHR [Block et al. 1994], EAM [Sisk et al. 1997], SHM (Karl et al. 2000]), validation has typically been applied to individual species models developed within the structure of frameworks. Both frameworks and models need validation; a framework may work well conceptually, while a specific habitat-relationships model developed within the framework may not. Although we focused on evaluating whether at least one species-specific model within a framework had been validated, we suggest that the need to validate frameworks is of even greater importance.

We suggest developers of future frameworks carefully consider the capability of practitioners to develop and apply models. Specifically, developers of new frameworks should consider using input data that are readily available in agency inventories, and develop frameworks with transparent structure and adequate documentation so that practitioners may clearly understand and apply the framework. We remind practitioners that if available data are poor quality or fail to adequately describe variables critical to the habitat requirements of a species, then only poor quality outputs will result. Thus, obtaining quality input data is paramount in modeling activities. A particularly important consideration for new frameworks is ensuring the availability of documentation, either online or printed user's manuals that clearly describe application of models developed within the framework, present examples of model applications, offer other resources such as descriptions of input and output data, document assumptions and functional forms (i.e., equations), and provide schematic descriptions of framework structures to enhance understanding of the model applications by practitioners.

As model frameworks become more sophisticated, users will increasingly face the issue of parameterizing complex models for species whose ecological relationships may not be well understood. For instance, the current understanding of spatial relationships and even basic habitat associations is poor for many vertebrates (e.g., U.S. Forest Service 2006). Therefore, it will be important to retain the ability within potentially complicated frameworks to develop simple models that reflect the level of ecological understanding for particular species.

SUMMARY

Wildlife habitat-relationships models were first developed in the mid-1970s to provide practitioners with tools to evaluate habitat quality. We identified and described the structure, uses, output, and operation of major habitat-relationships modeling frameworks. We defined frameworks as conceptual modeling structures such as modeling shells and general modeling approaches within which models are constructed that are similar in purpose and function. Frameworks provide the foundation for building models for a wide array of animals in almost any environmental setting. We also provided a descriptive analysis of frameworks to assist practitioners in selecting approaches that fit specific operational objectives. We identified 40 frameworks (13 through the 1980s, 12 in the 1990s, and, 15 since 2000) and grouped them according to 10 nominal- and 5 ordinal-scale criteria. The proportion of frameworks that are not components of larger landscape modeling systems using input data readily available in natural resource agency inventories declined from 1980 through 2006. The proportion of frameworks that examine habitat relationships at multiple scales, link scales when multiscaled, and that are spatially explicit increased from the 1980s through 2006. The proportion of frameworks that have received scientific credibility through publication or application of results or other mechanisms has remained above 0.83, but the proportion of frameworks where output from at least one model developed within a framework has been validated with field data never exceeded 0.58. We used agglomerative hierarchical cluster methods to identify groupings of habitat-relationships modeling frameworks based on dissimilarity distance between each framework according to criteria ratings. CompPATS and HABSCAPES did not meet our cluster grouping criteria, but the remaining 38 frameworks were apportioned among seven clusters, each containing an average of 5.4 (range = 2–10) frameworks. Each cluster was characterized by specific strengths and limitations that practitioners should assess prior to selecting a framework that best meets their modeling objectives. Cluster 1 included HSI and nine other frameworks that were based on species-habitat matrices or newly emerging analysis techniques. Cluster 2 was characterized by frameworks that were components of larger landscape modeling systems. Cluster 3 approached habitat modeling through modeling shells, GIS-based modeling systems, or a diversity of other techniques to model habitat relationships. Frameworks in Cluster 4 use simple approaches to evaluate habitat quality, often developed for use within a GIS. Both frameworks in Cluster 5 link multiple-scales to evaluate habitat quality. Frameworks in Cluster 6 predict changes in habitats. Frameworks in Cluster 7 provide predictive tools that are useful in assessing impacts of land management activities on species and habitats. Our evaluation provides conceptual information for practitioners evaluating how well wildlife habitat-relationships frameworks may achieve modeling objectives. To assist developers of future wildlife habitat-relationships modeling frameworks, we provided insights to the development of rigorous yet practical frameworks that follow current trends in wildlife-habitat relationships modeling and suggestions to overcome limitations in existing frameworks.

ACKNOWLEDGMENTS

J. L. Beck initiated work leading to this chapter while conducting postdoctoral research in the Department of Zoology and Physiology at the University of Wyoming. J. B. Haufler and G. J. Roloff provided expert reviews of our initial compilations of modeling frameworks. R. M. King, U.S. Forest Service, Rocky Mountain Research Station, provided statistical assistance. J. B. Haufler, G. D. Hayward, B. G. Marcot, J. J. Millspaugh, and F. R. Thompson, III, reviewed earlier drafts of our chapter. We thank them for their insightful reviews. Funding for our work was provided by the U.S. Forest Service; Washington Office; Watershed, Fish, Wildlife, Air, and Rare Plant Staff (University of Wyoming Project Number USDAFS45321).

LITERATURE CITED

Akçakaya, H. R. 1998. *RAMAS GIS: Linking landscape data with population viability analysis, version 3.0.* Applied Biomathematics, Setauket, New York, USA.

Akçakaya, H. R., V. C. Radeloff, D. J. Mladenoff, and H. S. He. 2004. Integrating landscape and metapopulation modeling approaches: Viability of the sharp-tailed grouse in a dynamic landscape. *Conservation Biology* 18:526–537.

Akçakaya, H. R., J. Franklin, A. D. Syphard, and J. R. Stephenson. 2005. Viability of Bell's sage sparrow (*Amphispiza belli* ssp *belli*): Altered fire regimes. *Ecological Applications* 15:521–531.

ALCES. 2005. ALCES: A landscape cumulative effects simulator. An integrated landscape management tool. Forem Technologies, Bragg Creek, Alberta, Canada. <http://www.foremtech.com>. Accessed 16 November 2006.

Aurambout, J. P., A. G. Endress, and B. M. Deal. 2005. A spatial model to estimate habitat fragmentation and its consequences on long-term persistence of animal populations. *Environmental Monitoring and Assessment* 109:199–225.

Bender, L. C., G. J. Roloff, and J. B. Haufler. 1996. Evaluating confidence intervals for habitat suitability models. *Wildlife Society Bulletin* 24:347–352.

Benkobi, L., M. A. Rumble, G. C. Brundige, and J. J. Millspaugh. 2004. *Refinement of the Arc-Habcap model to predict habitat effectiveness for elk.* U.S. Forest Service, Rocky Mountain Research Station, Research paper RMRS-RP-51, Fort Collins, Colorado, USA.

Block, W. M., M. L. Morrison, J. Verner, and P. N. Manly. 1994. Assessing wildlife-habitat-relationships models: A case study with California oak woodlands. *Wildlife Society Bulletin* 22:549–561.

Brand, L. A., B. R. Noon, and T. D. Sisk. 2006. Predicting abundance of desert riparian birds: Validation and calibration of the Effective Area Model. *Ecological Applications* 16:1090–1102.

Burgman, M. A., D. R. Breininger, B. W. Duncan, and S. Ferson. 2001. Setting reliability bounds on habitat quality indices. *Ecological Applications* 11:70–78.

California Department of Fish and Game. 2005. *California Interagency Wildlife Task Group. User's manual for version 8.1 of the California Wildlife Habitat Relationships System and Bioview.* California Fish and Game Department, Sacramento, California, USA.

Capen, D. E., editor. 1981. *The use of multivariate statistics in studies of wildlife habitat.* U.S. Forest Service, Rocky Mountain Forest and Range Research Station, General Technical Report RM- GTR-87, Fort Collins, Colorado, USA.

Crist, P. J., T. W. Kohley, and J. Oakleaf. 2000. Assessing land-use impacts on biodiversity using an expert systems tool. *Landscape Ecology* 15:47–62.

Curnutt, J. L., J. Comiskey, M. P. Nott, and L. J. Gross. 2000. Landscape based spatially explicit species index models for Everglades restoration. *Ecological Applications* 10:1849–1860.

DeGraaf, R. M., M. Yamasaki, W. B. Leak, and J. W. Lanier. 1992. *New England wildlife: Management of forested habitats*. U.S. Forest Service, Northeastern Forest Experiment Station, General Technical Report NE-144, Radnor, Pennsylvania, USA.

Dijak, W. D., C. D. Rittenhouse, M. A. Larson, F. R. Thompson, III, and J. J. Millspaugh. 2007. Landscape habitat suitability index software. *Journal of Wildlife Management* 71:668–670.

Gower, J. C. 1971. A general coefficient of similarity and some of its properties. *Biometrics* 27:857–871.

Gronewold, A., and M. Sonnenschein. 1998. Event-based modelling of ecological systems with asynchronous cellular automata. *Ecological Modelling* 108:37–52.

Grubb, T. G. 1988. *Pattern recognition—A simple model for evaluating wildlife habitat*. U.S. Forest Service, Rocky Mountain Research Station, Research Note RM-487, Fort Collins, Colorado, USA.

Gutiérrez, R. J., J. Verner, K. S. McKelvey, B. R. Noon, G. N. Steger, D. R. Call, W. S. LaHaye, B. B. Bingham, and J. S. Senser. 1992. Habitat relations of the California spotted owl. Pages 79–98 *in* J. Verner, K. S. McKelvey, B. R. Noon, R. J. Gutiérrez, G. I. Gould, Jr., and T. W. Beck, technical coordinators. *The California Spotted Owl: A technical assessment of its current status*. U.S. Forest Service, Pacific Southwest Research Station, General Technical Report PSW-GTR-133, Albany, California, USA.

Hall, L. S., P. R. Krausman, and M. L. Morrison. 1997. The habitat concept and a plea for standard terminology. *Wildlife Society Bulletin* 25:173–182.

He, H. S., D. J. Mladenoff, and J. Boeder. 1999. An object-oriented forest landscape model and its representation of tree species. *Ecological Modeling* 119:1–19.

Higdon, J. W., D. A. MacLean, J. M. Hagan, and J. M. Reed. 2005. Evaluating vertebrate species risk on an industrial forest landscape. *Forest Ecology and Management* 204:279–296.

Higdon, J. W., D. A. MacLean, J. M. Hagan, and J. M. Reed. 2006. Risk of extirpation for vertebrate species on an industrial forest in New Brunswick, Canada: 1945, 2002, and 2027. *Canadian Journal of Forest Research* 36:467–481.

Hof, J., and M. Bevers. 1998. *Spatial optimization for managed ecosystems*. Columbia University Press, New York, New York, USA.

Hoover, R. L., and D. L. Willis, editors. 1984. *Managing forested lands for wildlife*. Colorado Division of Wildlife in cooperation with USDA Forest Service, Rocky Mountain Region, Denver, Colorado, USA.

Huff, M., T. K. Mellen, and R. Hagetedt. 2001. Case study 2: A model to assess potential vertebrate habitat at landscape scales: HABSCAPES. Pages 544–549 *in* D. H. Johnson and T. A. O'Neil, managing directors. *Wildlife-habitat relationships in Oregon and Washington*. Oregon State University Press, Corvallis, USA.

Juntti, T. M., and M. A. Rumble. 2006. *Arc Habitat Suitability Index computer software*. U.S. Forest Service, Rocky Mountain Research Station, General Technical Report RMRS-GTR-180WWW, Fort Collins, Colorado, USA.

Justus, J. 2006. Loop analysis and qualitative modeling: Limitations and merits. *Biology and Philosophy* 21:647–666.

Karl, J. W., P. J. Heglund, E. O. Garton, J. M. Scott, N. M. Wright, and R. L. Hutto. 2000. Sensitivity of species habitat-relationship model performance to factors of scale. *Ecological Applications* 10:1690–1705.

Keller, D., D. Belcher, and M. Robertson. 1994. *CompPATS Users Guide (version 3.2)—A systems approach to project level implementation of forest plans*. U.S. Forest Service, Southern Region, Atlanta, Georgia, USA.

Kilgo, J. C., D. L. Gartner, B. R. Chapman, J. B. Dunning, Jr., K. E. Franzels, S. A. Gauthreaux, C. H. Greenberg, D. J. Levey, K. V. Miller, and S. F. Pearson. 2002. A test of an expert-based bird-habitat relationship model in South Carolina. *Wildlife Society Bulletin* 30:783–793.

Larson, M. A., W. D. Dijak, F. R. Thompson, III, and J. J. Millspaugh. 2003. *Landscape-level habitat suitability models for twelve wildlife species in southern Missouri*. U.S. Forest Service, North Central Research Station, General Technical Report NC-233, St. Paul, Minnesota, USA.

Larson, M. A., F. R. Thompson, III, J. J. Millspaugh, W. D. Dijak, and S. R. Shifley. 2004. Linking population viability, habitat suitability, and landscape simulation models for conservation planning. *Ecological Modelling* 180:103–118.

Li, H., D. I. Gartner, P. Mou, and C. C. Trettin. 2000. A landscape model (LEEMATH) to evaluate effects of management impacts on timber and wildlife habitat. *Computers and Electronics in Agriculture* 27:263–292.

Lusk, J. J., F. S. Guthery, and S. J. DeMaso. 2002. A neural network model for predicting northern bobwhite abundance in the rolling Red Plains of Oklahoma. Pages 345–355 *in* J. M. Scott, P. J. Heglund, M. L. Morrison, J. B. Haufler, M. G. Raphael, W. A. Wall, and F. B. Samson, editors. *Predicting species occurrences: Issues of accuracy and scale.* Island Press, Washington, D.C., USA.

Lyon, J. G., and A. G. Christensen. 1992. *A partial glossary of elk management terms.* U.S. Forest Service, Intermountain Research Station, General Technical Report INT-288, Ogden, Utah, USA.

Lyon, J. G., J. T. Heinen, R. A. Mead, and N. E. G. Roller. 1987. Spatial data for modeling wildlife habitat. *Journal of Surveying Engineering* 113:88–100.

Marcot, B. G. 1986. Use of expert systems in wildlife-habitat modeling. Pages 145–150 *in* J. Verner, M. L. Morrison, and C. J. Ralph, editors. *Wildlife 2000: Modeling habitat relationships of terrestrial vertebrates.* University of Wisconsin Press, Madison, Wisconsin, USA.

Marcot, B. G. 2006. Habitat modeling for biodiversity conservation. *Northwestern Naturalist* 87:56–65.

Marcot, B. G., R. S. Holthausen, M. G. Raphael, M. M. Rowland, and M. J. Wisdom. 2001. Using Bayesian belief networks to evaluate fish and wildlife population viability under land management alternatives from an environmental impact statement. *Forest Ecology and Management* 153:29–42.

Marzluff, J. M., J. J. Millspaugh, K. R. Ceder, C. D. Oliver, J. Whithey, J. B. McCarter, C. L. Mason, and J. Comnick. 2002. Modeling changes in wildlife habitat and timber revenues in response to forest management. *Forest Science* 48:191–202.

Mattson, D. J., and T. Merrill. 2004. A model-based appraisal of habitat conditions for grizzly bears in the Cabinet-Yaak region of Montana and Idaho. *Ursus* 15:76–89.

McComb, W. C., M. T. McGrath, T. A. Spies, and D. Vesely. 2002. Models for mapping potential habitat at landscape scales: An example using northern spotted owls. *Forest Science* 48:203–216.

McGarigal, K., and B. W. Compton. 2003. *An introduction to modeling wildlife habitat with HABIT@—working draft* (12 June 2003). Landscape Ecology Program, Department of Natural Resources Conservation, University of Massachusetts, Amherst, USA.

Mead, R. A., T. L. Sharik, S. P. Prisley, and J. T. Heinen. 1981. A computerized spatial analysis system for assessing wildlife habitat from vegetation maps. *Canadian Journal of Remote Sensing* 7:34–40.

Mellen, K., M. Huff, and R. Hagestedt. 1995. "HABSCAPES" Interpreting landscape patterns: A vertebrate habitat relationships approach. Pages 135–145 *in* J. Thompson, compiler. *Analyis in support of ecosystem management.* U.S. Forest Service, Ecosystem Management Analysis Center, Washington, D.C., USA.

Mellen, K., M. Huff, and R. Hagestedt. 2001. *"HABSCAPES" reference manual and users guide.* U.S. Forest Service, Pacific Northwest Region, Mt. Hood National Forest, Portland, Oregon, USA.

Merrill, T., D. J. Mattson, R. G. Wright, and H. B. Quigley. 1999. Defining landscapes suitable for restoration of grizzly bears *Ursus arctos* in Idaho. *Biological Conservation* 87:231–248.

Milne, B. T., K. M. Johnston, and R. T. T. Forman. 1989. Scale-dependent proximity of wildlife habitat in a spatially-neutral Bayesian model. *Landscape Ecology* 2:101–110.

Morrison, M. L., B. G. Marcot, and R. W. Mannan. 2006. *Wildlife-habitat relationships: Concepts and applications.* Third edition. Island Press, Washington, D.C., USA.

Ouachita National Forest. 1988. *User's manual: Computerized project analysis and tracking system.* U.S. Department of Agriculture, Ouachita National Forest, Land Management Planning, Hot Springs, Arkansas, USA.

Özesmi, S. L., and U. Özesmi. 1999. An artificial neural network approach to spatial habitat modelling with interspecific interaction. *Ecological Modelling* 116:15–31.

Özesmi, U., C. O. Tan, S. L. Özesmi, and R. J. Robertson. 2006. Generalizability of artificial neural network models in ecological applications: Predicting nest occurrence and breeding success of the red-winged blackbird *Agelaius phoeniceus*. *Ecological Modelling* 195:94–104.

Puttock, G. D., I. Timossi, and L. S. Davis. 1998. BOREAL: A tactical planning system for forest ecosystem management. *The Forestry Chronicle* 74:413–420.

Raedeke, K. J., and J. F. Lehmkuhl. 1986. A simulation procedure for modeling the relationships between wildlife and forest management. Pages 377–381 *in* J. Verner, M. L. Morrison, and C. J. Ralph, editors. *Wildlife 2000: Modeling habitat relationships of terrestrial vertebrates.* University of Wisconsin Press, Madison, Wisconsin, USA.

Raphael, M. G., and B. G. Marcot. 1986. Validation of a wildlife-habitat-relationships model: Vertebrates in a Douglas-fir sere. Pages 129–138 *in* J. Verner, M. L. Morrison, and C. J. Ralph, editors. *Wildlife 2000: Modeling habitat relationships of terrestrial vertebrates.* University of Wisconsin Press, Madison, Wisconsin, USA.

Raphael, M. G., M. J. Wisdom, M. M. Rowland, R. S. Holthausen, B. C. Wales, B. G. Marcot, and T. D. Rich. 2001. Status and trends of habitats of terrestrial vertebrates in relation to land management in the interior Columbia river basin. *Forest Ecology and Management* 153:63–88.

Reed, J. M., J. M. Hagan, and A. A. Whitman. 2001. *A process for identifying species at risk in forested landscapes.* Manomet Center for Conservation Sciences Mosaic, Science Notes #2001-1, Brunswick, Maine, USA.

Reynolds, K. M. 1999*a*. *EMDS users guide (version 2.0): Knowledge-based decision support for ecological assessment.* U.S. Forest Service, Pacific Northwest Research Station, General Technical Report PNW-GTR-470, Portland, Oregon, USA.

Reynolds, K. M. 1999*b*. *NetWeaver for EMDS version 2.0 user guide: A knowledge base development system.* U.S. Forest Service, Pacific Northwest Research Station, General Technical Report PNW-GTR-471, Portland, Oregon, USA.

Reynolds, K. M. 2001. *Fuzzy logic knowledge bases in integrated landscape assessment: Examples and possibilities.* U.S. Forest Service, Pacific Northwest Research Station, General Technical Report PNW-GTR-521, Portland, Oregon, USA.

Rickel, B. W. 1997. *Habitat quality index model.* U.S. Forest Service, Southwestern Region, Albuquerque, New Mexico, USA.

Rittenhouse, C. D., W. D. Dijak, F. R. Thompson, III, and J. J. Millspaugh. 2007. *Development of landscape-level habitat suitability models for ten wildlife species in the central hardwoods region.* U.S. Forest Service, Northern Research Station, General Technical Report NRS-4, Newtown Square, Pennsylvania, USA.

Roller, N. E. G. 1978. Quantitative evaluation of deer habitat. Pages 137–146 *in Proceedings of Pecora IV symposium on applications of remote sensing data to wildlife management.* M. E. Berger, D. W. Carneggie, M. Fletcher, A. Marmelstein, G. A. Thorley, and G. Watson, coordinators. National Wildlife Federation, Sioux Falls, South Dakota, USA.

Roloff, G. J., and J. B. Haufler. 1997. Establishing population viability planning objectives based on habitat potentials. *Wildlife Society Bulletin* 25:895–904.

Roloff, G. J., and J. B. Haufler. 2002. Modeling habitat-based viability from organism to population. Pages 673–686 *in* M. J. Scott, P. J. Heglund, M. L. Morrison, J. B. Haufler, M. G. Raphael, W. A. Wall, and F. B. Samson, editors. *Predicting species occurrence: Issues of accuracy and scale.* Island Press, Washington, D.C., USA.

Roloff, G. J., and B. J. Kernohan. 1999. Evaluating reliability of habitat suitability index models. *Wildlife Society Bulletin* 27:973–985.

Roy, P. S., S. A. Ravan, N. Rajadnya, K. K. Das, A. Jain, and S. Singh. 1995. Habitat suitability analysis of *Nemorhaedus goral*—A remote sensing and geographic information system approach. *Current Science* 69:685–691.

Salwasser, H., J. C. Capp, H. Black, and J. F. Hurley. 1980. The California Wildlife Habitat Relationships Program: An overview. Pages 369–378 *in* R. M. DeGraaf, editor. *Workshop proceedings:*

Management of western forests and grasslands for nongame birds. U.S. Forest Service, General Technical Report INT-86, Ogden, Utah, USA.

SAS Institute. 2003. *SAS/STAT user's guide, release 9.1.* SAS Institute, Cary, North Carolina, USA.

Schneider, R. R., J. B. Stelfox, S. Boutin, and S. Wasel. 2003. Managing the cumulative impacts of land uses in the Western Canadian Sedimentary Basin: A modeling approach. *Conservation Ecology* 7(1), <http://www.consecol.org/vol7/iss1/art8/>. Accessed 16 November 2006.

Schroeder, R. L. 1996. *Wildlife community habitat evaluation using a modified species-area relationship.* U.S. Army Engineer Waterways Experiment Station, Technical Report WRP-DE-12, Vicksburg, Mississippi, USA.

Schumaker, N. H. 1998. *A users guide to the PATCH model.* U.S. Environmental Protection Agency, Environmental Research Laboratory, EPA/600/R-98/135, Corvallis, Oregon, USA.

Schumaker, N. H., T. Ernst, D. White, J. Baker, and P. Haggerty. 2004. Projecting wildlife responses to alternative future landscapes in Oregon's Willamette Basin. *Ecological Applications* 14:381-400.

Scott, J. M., F. Davis, B. Csuti, R. Noss, B. Butterfield, C. Groves, H. Anderson, S. Caicco, F. D'erchia, T. C. Edwards, Jr., J. Ulliman, and R. G. Wright. 1993. Gap analysis: A geographic approach to protection of biological diversity. *Wildlife Monographs* 123:1-41.

Scott, J. M., P. J. Heglund, M. L. Morrison, J. B. Haufler, M. G. Raphael, W. A. Wall, and F. B. Samson, editors. 2002. *Predicting species occurrences: Issues of accuracy and scale.* Island Press, Washington D.C., USA.

Seely, B., J. Nelson, R. Wells, B. Peter, M. Meitner, A. Anderson, H. Harshaw, S. Sheppard, F. L. Bunnell, K. Kimmins, and D. Harrison. 2004. The application of a hierarchical, decision-support system to evaluate multi-objective forest management strategies: A case study in northeastern British Columbia. *Forest Ecology and Management* 199:283-305.

Shifley, S. R., F. R. Thompson III, W. D. Dijak, M. A. Larson, and J. J. Millspaugh. 2006. Simulated effects of forest management alternatives on landscape structure and habitat suitability in the Midwestern United States. *Forest Ecology and Management* 229:361-377.

Sisk, T. D., N. M. Haddad, and P. R. Ehrlich. 1997. Bird assemblages in patchy woodlands: Modeling the effects of edge and matrix habitats. *Ecological Applications* 7:1170-1180.

Smith, T. M., H. H. Shugart, and D. C. West. 1981. FORHAB: A forest simulation model to predict habitat structure for nongame bird species. Pages 114-123 *in* D. E. Capen, editor. *The use of multivariate statistics in studies of wildlife habitat.* U.S. Forest Service, Rocky Mountain Forest and Range Research Station, General Technical Report RM- GTR-87, Fort Collins, Colorado, USA.

Sodja, R. S., J. E. Cornely, and A. E. Howe. 2002. Development of an expert system for assessing trumpeter swan breeding habitat in the Northern Rocky Mountains. *Waterbirds* 25(Special Publication 1):313-319.

Spies, T. A., G. H. Reeves, K. M. Burnett, W. C. McComb, K. N. Johnson, G. Grant, J. L. Ohmann, S. L. Garman, and P. Bettinger. 2002. Assessing the ecological consequences of forest policies in a multi-ownership province in Oregon. Pages 179-207 *in* J. Liu and W. W. Taylor, editors. *Integrating landscape ecology into natural resource management.* Cambridge University Press, Cambridge, United Kingdom.

Stauffer, D. F. 2002. Linking populations and habitats: Where have we been? Where are we going? Pages 53-61 *in* M. J. Scott, P. J. Heglund, M. L. Morrison, J. B. Haufler, M. G. Raphael, W. A. Wall, and F. B. Samson, editors. *Predicting species occurrence: Issues of accuracy and scale.* Island Press, Washington, D.C., USA.

Stoms, D. M., J. M. McDonald, and F. W. Davis. 2002. Environmental assessment: Fuzzy assessment of land suitability for scientific research reserves. *Environmental Management* 29:545-558.

Thomas, J. W. 1979. *Wildlife habitats in managed forests: The Blue Mountains of Oregon and Washington.* U.S. Forest Service, Agricultural Handbook number 553, Washington, D.C., USA.

Thomas, J. W., D. A. Leckenby, M. Henjum, R. J. Pedersen, and L. D. Bryant. 1988. *Habitat-effectiveness index for elk on Blue Mountain winter ranges.* U.S. Forest Service, Pacific Northwest Experiment Station, General Technical Report PNW-GTR-218, Portland, Oregon, USA.

Topping, C. J., T. S. Hansen, T. S. Jensen, J. U. Jepsen, F. Nikolajsen, and P. Odderskær. 2003. ALMaSS, an agent-based model for animals in temperate European landscapes. *Ecological Modelling* 167:65-82.

U.S. Fish and Wildlife Service. 1981. *Standards for the development of habitat suitability index models for use in the Habitat Evaluation Procedures*. USFWS, Division of Ecological Services, ESM 103, Washington, D.C., USA.

U.S. Forest Service. 1994. *BIRDHAB: GIS bird habitat evaluation for resource managers, users guide. Beta version 0.999*. USFS, Southern Region, Atlanta, Georgia, USA.

U.S. Forest Service. 2005. ArcView HABCAP extension: documentation and users's guide. USFS, Rocky Mountain Region, Denver, Colorado, USA.

U.S. Forest Service. 2006. Species conservation program: Species conservation assessments. USFS, Rocky Mountain Region. Denver, Colorado, USA. <http://www.fs.fed.us/r2/projects/scp/assessments/index.shtml>. Accessed 25 April 2006.

Van Horne, B. 1983. Density as a misleading indicator of habitat quality. *Journal of Wildlife Management* 47:893-901.

Wells, R. W., and A. May. 2002. *SIMFOR version 3.01 User manual*. Center for Applied Conservation Research. University of British Columbia, Vancouver, Canada.

Wells, R. W., E. Valdal, C. Steeger, and P. Vernier. 1999. *SIMFOR habitat analysis of two FSSIM harvest scenarios in the Rocky Mountain trench*. Invermere Forest District, Enhanced Forest Management Report 19. Invermere, British Columbia, Canada.

Williams, G. L., K. R. Russell, and W. K. Seitz. 1977. Pattern recognition as a tool in the ecological analysis of habitat. Pages 521-531 *in Proceedings of the symposium on classification, inventory, and analysis of fish and wildlife habitat*. A. Marmelstein, general chairman. U.S. Fish and Wildlife Service, Office of Biological Services, Washington, D.C., USA.

CHAPTER

Lessons Learned from Using GIS to Model Landscape-Level Wildlife Habitat

11

Gary J. Roloff, Michael L. Donovan, Daniel W. Linden, and Marshall L. Strong

Wildlife-habitat relationships models include spatially explicit models that "keep track of the exact locations of plants and animals" (Kareiva and Wennergren 1995:299) or "have a structure that specifies the location of each object of interest" (Dunning et al. 1995:4). Application of spatially explicit habitat models has been facilitated by readily available aerial and satellite imagery, global positioning systems (GPS), increasingly comprehensive field inventories, and geographic information systems (GIS) (O'Neil et al. 2005; Dijak and Rittenhouse, this volume; Fitzgerald et al., this volume). These tools, particularly GIS, are common in landscape-level habitat assessments (O'Neil et al. 2005), though their use is not without concern (Stoms et al. 1992, Corsi et al. 2000). Numerous GIS-based landscape-level habitat models have been published in the ecological literature, with most authors emphasizing habitat model construction and performance while ignoring how inherent limitations in GIS data, habitat classification schemes, or data processing assumptions may have influenced their findings. It is our goal to increase awareness of these issues when conducting large-scale wildlife-habitat assessments using GIS.

Models benefit wildlife-habitat relationships studies by offering a framework for integrating uncertainty and error and by identifying complex and sometimes obscure relationships (Anderson and Gutzwiller 2005). Model applications are common at all levels of resource decision making ranging from operational, site-specific evaluations to strategic, large-scale evaluations. Modeling wildlife habitat over this range of scales requires many assumptions about the relationships between wildlife population metrics and habitat occurrence, quality, and spatial distribution (Beutel et al. 1999). Standard modeling protocol is to (1) explicitly state all assumptions early in the process; (2) substantiate those assumptions with field data, published information, or expert opinion; (3) hypothesize the relationships among wildlife and their habitat; and (4) use the modeling framework to evaluate sensitivities and produce output.

One critical assumption underlying this protocol is that habitat is accurately characterized at ecologically relevant scales to the organism(s) of interest.

Projects involving the testing of spatially explicit wildlife-habitat models often emphasize wildlife model mechanics and outputs, with less consideration given to how GIS maps and associated attribute information interface with model performance. This is an erroneous approach to understanding model performance. The first question to ask and integrate into spatially explicit wildlife model evaluations is whether supporting GIS data provide an accurate representation of habitat. It is important to remember that GIS data are often derived from models having their own assumptions that influence how data should be interpreted and used. In many cases, wildlife-habitat modelers have little or no involvement in creating GIS data that support their projects and thus may not be aware of data-specific limitations.

The combined experience of the chapter authors as GIS data developers and users in support of wildlife-habitat modeling provides the basis for this chapter. In recounting some of our experiences, we hope to facilitate scientifically rigorous application of GIS to wildlife habitat assessments (also see Corsi et al. 2000 and Bissonette and Storch 2003). We divided our experiences and commentary into six general categories: (1) analysis structure; (2) data abundance; (3) vegetation change analyses; (4) classification systems; (5) spatial and temporal scale issues; and (6) technological considerations. These categories are not independent; rather they are often inseparable in GIS modeling projects. We also stress that our presentation should not be viewed as a complete treatise on GIS's role in characterizing wildlife habitat across landscapes. Textbooks have been written on individual components in our chapter, and the literature cited should help in identifying these sources. Instead, we offer our experiences to stimulate critical thinking during the process of conducting landscape-level habitat assessments using GIS. We suggest that wildlife modelers with GIS skills tend to have a good understanding of ecological issues, but few have a complete understanding of GIS technology and data limits and potential relationships to analysis and interpretation.

ANALYSIS STRUCTURE

At the core of all good wildlife-habitat analyses is a well-thought-out issue or problem statement and an analytical process to address it (Starfield 1997, Morrison et al. 1998:141, Corsi et al. 2000). If the issue statement is erroneous or incomplete, even the best analysis will be inadequate or potentially wrong (Hammond et al. 1999). We have witnessed firsthand how this basic tenet of good decision making and analysis can be compromised by data availability or use restrictions. In most situations, the issue or problem statement retains its integrity and the path to the "best" answer is altered. In extreme situations, even the issue or problem statement can change.

Consider the use of Forest Inventory and Analysis (FIA) plots in wildlife-habitat modeling. The FIA program is a collection of related surveys designed to focus on different aspects of America's forested ecosystems. A portion of this program includes a temporally continuous forest inventory based on a systematic grid of plots across the country. The exact locations of FIA plots are legally protected from the public (U.S. Forest Service 2005). This is not an uncommon practice, as numerous worldwide examples exist of governments restricting access to high-resolution "science-quality" data (Estes and Mooneyhan 1994). Individual landowners can acquire their FIA plot locations through the Forest Service because landscape-level habitat assessments often span multiple ownerships. However, use of geo-referenced FIA data is restricted to those with proprietary permission. Thus, the utility of FIA data is limited for wildlife modelers external to the Forest Service wishing to map habitats across large landscapes using algorithms that require geo-referenced point data (e.g., Frescino et al. 2001). One solution to this limitation has been to work with the Forest Service on using FIA data to attribute habitat classification schemes that hide or blur specific plot locations.

Linden (2006) recently used this approach to map habitat for Canada lynx (*Lynx canadensis*) across Michigan's Upper Peninsula. Linden initially proposed using a clustering algorithm (k-means; MacQueen 1967) that required geo-referenced inventory data for mapping forest structure. The k-means approach was selected as the best procedure for fulfilling project objectives because it generated a robust forest structure map based on attributes deemed important to lynx along with classification error estimates. However, to comply with the proprietary FIA data issues, Linden (2006) abandoned the k-means approach and instead, working with the Forest Service, associated FIA plots to strata in a preconceived classification. The preconceived classification was ecologically based but developed for uses other than lynx habitat assessments. Thus, although Linden identified a preferable method for mapping and attributing lynx habitat across Michigan's Upper Peninsula, this methodology was modified because of data use constraints. The effect of using this alternative methodology on lynx habitat assessment accuracy is unknown and, at the study area extent may be negligible. However, Linden's habitat assessment serves as an example of how certain scale(s) (in this case a less resolute classification system) can be forced onto projects by data use restrictions.

Recommendations

Wildlife-habitat modelers should cautiously adjust study designs or analyses in response to available data constraints. One must ensure that the original problem or issue statement can be addressed. A large part of framing appropriate analyses is developing a sound understanding of input data and the intended target scales (e.g., geographic scale, temporal scale, attribute detail). We have experienced situations in which tool or data availability determined how a modeling application proceeds, sometimes compromising the original question. Failure to understand the driving purpose for conducting an analysis can result in inapplicable results, wasted resources, wrong directions of applications, or a combination of these.

Part of developing a sound issue statement and project objectives is consideration of how scale and accuracy affect model outputs. These considerations tend to be ignored during project development stages, and modelers find their results are less useful or corrupted by these factors. We encourage modelers to identify their scalar needs (e.g., data grain, extent, positional accuracy, classification error) while formulating their project objectives. We are not necessarily advocating for explicit inclusion of scalar needs in the issue statement or project objectives. Rather, these needs should be used to evaluate the likelihood of fulfilling project expectations. An early understanding of scalar needs provides one metric for evaluating data set utility.

DATA ABUNDANCE

In the age of Internet data libraries, user-friendly software, desktop computers, and high-speed Internet access, data are abundant and readily available. Thorough data documentation is critical in all analysis situations, but many poorly documented data sets are publicly available, often used, and transferred among colleagues. The tendency for data users to manipulate GIS layers (e.g., change resolution, conduct reclassifications) and not document these manipulations exacerbates potential misuses. In general, data developers tend to provide complete documentation. However, that completeness often erodes as users modify data to fulfill their needs. Improvements in data documentation software (e.g., Earth Systems Research Institute's [ESRI's] metadata management tools) have helped alleviate some documentation issues, but we contend that the majority of GIS users fail to consistently document their data sets. Part of the scientific process associated with using GIS data includes understanding its correct use. Poor or incomplete data documentation hinders this process.

In some situations habitat modelers may simply be overwhelmed by data and be forced to rank the utility of those resources (O'Neil et al. 2005). When available, GIS metadata offer a source of information that can help identify pertinent GIS layers. Metadata are referred to as "data about data" and can be thought of as detailed data descriptions. Users should initially evaluate metadata for publication date, abstract, purpose, use constraints, and positional and attribute accuracy (e.g., significant figures of measurement) statements to ensure that the data are aligned with project objectives. In data selection situations in which metadata are lacking or a quantitative evaluation of available data is needed, we favor exploratory data analyses. Exploratory data analyses are designed to identify important GIS layers, where importance is defined by the layer's association with patterns of wildlife species occurrence, abundance, or habitat use. Numerous tools exist for conducting exploratory data analysis ranging from simple evaluation of statistical distributions and correlation matrices to more complex multivariate analyses (e.g., cluster analysis, factor analysis; Hirzel et al. 2002, Zaniewski et al. 2002, Engler et al. 2004, among others). Tools also are available for determining significance of GIS layers once they are selected including generalized linear models (e.g., Pereira and Itami 1991, Bian and West 1997, Mladenoff et al. 1999, Osborne et al. 2001, Luoto and Seppälä 2002, Brotons et al. 2004), GIS-based simulation (Wu, F., 2004; Wu, J., 2004), and others (see review in Guisan and Zimmerman 2000, Segurado and Araújo 2004).

Although there is often an abundance of data available to support wildlife-habitat modeling, it seems we seldom can find exactly what we need. This leads to another potential problem related to data abundance and the ease with which even novice GIS users can manipulate data. Spatial data are often provided in different data structures, formats, and resolutions (Garbrecht et al. 2001). Data processing and conversion into a consistent spatial reference system, format, and resolution are often needed for practical applications (Garbrecht et al. 2001). Also, because we rarely have exactly what we want, we often manipulate data to meet our needs. Simple GIS processes (such as "intersect") that generate new combinations of polygons (a map data structure based on areas closed by lines) or rasters (a map data structure based on cells) from multiple input layers are commonly used. These combinations can result in data integrity degradation (e.g., formation of "sliver" polygons, unrealistic attribute combinations) and a divergence from the original metadata. We recognize the value of manipulating data to meet specific project needs; however in doing so users should check the resulting data layers for common errors. These include boundary mismatches (e.g., shorelines between two data sets may not match depending on water levels at the time of original mapping), polygons or patches below the minimum mapping unit, and unrealistic attribute combinations (e.g., a forest cover type that overlays a lake).

Recommendations

When considering data resources developed external to a project or by others, first and most importantly, one must seek out and read any metadata, supporting reports, or other detailed background information which helps in using the data correctly. Sometimes this assessment requires map displays of source information, and during this phase visualization tools can facilitate understanding (Ramsey and Strong 2000). One needs to pay close attention to issues of scale and accuracy that will influence the ability to make inferences on wildlife-habitat relationships. Failing to conduct this initial step is often the first mistake in a modeling project and will ultimately impact all other steps. When manipulating data, one needs to closely document analysis steps and incorporate processes and decisions into the resulting layer's metadata (see the Federal Geographic Data Committee metadata standard as reviewed in O'Neil et al.

[2005] or the international standards reviewed by Moellering et al. [2005] for guidance on building or editing metadata). One must ensure that subsequent users of data and analyses can replicate the work. Data abundance offers the opportunity to explore multiple relationships and develop better modeling designs; however, we caution against the tendency to make models overly complex (Burnham and Anderson 1998).

VEGETATION CHANGE ANALYSES

Data limitations in modeling vegetation (or habitat) change over multiple time periods are frequently ignored. Vegetation change analyses often involve comparing GIS data derived from different sources and processes. Our general concern is that classification processes, resolution, and accuracy limitations of time series data are often ignored or downplayed in change analyses. This concern applies to any spatial time series but is perhaps most exemplified by analyses using presettlement vegetation.

It is not uncommon for modelers to rely on presettlement vegetation maps as a baseline for assessing habitat or cover type changes (e.g., Rodgers and Anderson 1979, Van Deelen et al. 1996, Cowell 1998, Radeloff et al. 1999, Farley et al. 2002, Pidgeon et al. 2005, Schulte et al. 2005). These analyses are extremely useful for understanding broad-scale changes in vegetation types and associated ecological processes (Manies and Mladenoff 2000, Delcourt and Delcourt 2006); however, presettlement data have several limitations that influence their utility and proper application (reviewed by Wang 2005). In our experience, modelers often embark on using presettlement data without fully understanding three important limitations related to scale, boundary inaccuracies, and underrepresentation of certain vegetation types.

Scale-Related Limitations

Although presettlement data are most appropriately used for broad-scale, regional assessments (Schulte and Mladenoff 2001, Manies and Mladenoff 2000, Delcourt and Delcourt 2006), the data are frequently used for finer scale change analyses. For example, simulation data from Michigan suggest that inference to scales <65 ha should be avoided when using presettlement vegetation maps derived from the General Land Office (GLO) survey (Delcourt and Delcourt 2006). This scale roughly corresponds to a quarter-section, one of the smallest scales at which GLO data were consistently collected. It is common practice to ignore this scale-related limitation when conducting change analyses between presettlement and current data. For example, a simple change analysis using GLO and more current (circa 2000) satellite-derived data from Michigan suggests a noticeable change in spatial distribution and patch sizes of aspen (*Populus* spp.) (Fig. 11-1). Often, the validity of an analysis like that presented in Fig. 11-1 is not questioned.

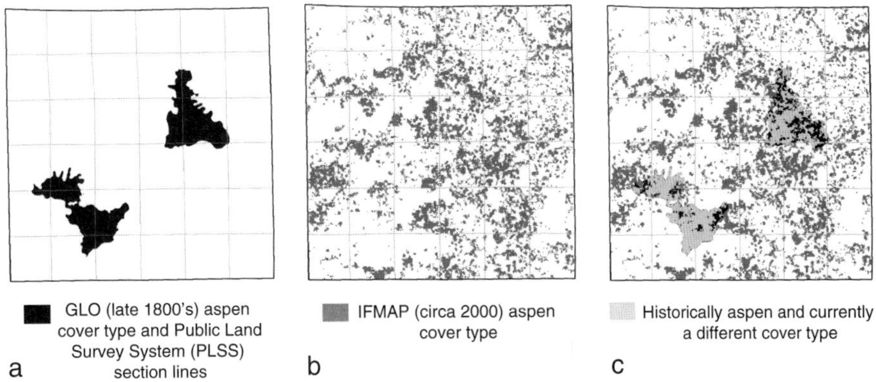

FIG. 11-1

Scale limitation of General Land Office (GLO) settlement data for conducting change analyses on aspen cover type in a northern lower peninsula of Michigan township, Michigan, USA. A) GLO aspen cover type; B) Michigan Department of Natural Resources (MDNR) Integrated Forest Monitoring Assessment and Prescription (IFMAP) aspen cover type for the same township; and C) a change analysis showing historic, current, and areas of overlap.

However, it is important to remember that GLO data were not collected in a manner that would have detected the numerous small aspen patches not intersecting section or quarter-section corners or section lines (Fig. 11-1). Some modelers have explored techniques for extracting finer scale information from the GLO, but these procedures have generally been viewed as qualitative and not easily reproducible (reviewed in Schulte and Mladenoff 2001). The coarse resolution of GLO data makes extrapolation to small-scale vegetation associations problematic except in cases of extremely homogenous vegetation (Cowell 1995).

There are also two issues related to GLO data temporal scale that are relevant to change analyses: survey timing and observed phenomenon age (Wang 2005). Although GLO data collection often spanned several decades (Haines and Sando 1969, Wang 2005), the data are generally accepted as a single ecological period of vegetation development (Brown 1998*b*, Schulte and Mladenoff 2001). GLO surveyors did not specifically record the age of observed events, but they did collect enough information to permit inference on disturbance frequencies and successional processes (reviewed by Wang 2005). For example, in Michigan a commonly used presettlement map is based on GLO survey records dating 1816 to 1856 (Comer et al. 1995). In vegetation change analyses, these 40 years are often viewed as a snapshot in time (e.g., Fig. 11-1). By today's standards of mapping vegetation dynamics, this would not be an acceptable practice, particularly for operational or tactical decision making.

Recommendations.—When conducting vegetation change analyses, GIS users should restrict inference to the coarsest minimal map unit for their input data layers. In Michigan's case, presettlement data patches <65 ha in size are erroneously included in comparisons. A better representation of vegetation change could be derived by simulating GLO data collection processes and patch mapping algorithms on the current vegetation data layer. This would help standardize spatial scales.

We caution modelers conducting vegetation change analyses over multiple time periods to consider how mismatches in temporal scales among data sources may influence their results. For example, understand how vegetation community dynamics, particularly for short-lived communities, relate to the temporal scale of measurement. Consider the temporal scales of disturbance regimes and how disturbances may affect the ability to detect certain vegetation types. Vegetation change analyses are most appropriately conducted on data derived using the same sources (e.g., satellite sensor), techniques, and spatial and temporal scales.

Boundary Inaccuracy Limitation

Boundary mapping errors should be considered when evaluating changes in spatial extent or boundary locations among time periods. For example, Michigan's presettlement vegetation maps were derived from an ecologist's interpretation of GLO surveyor's notes along survey section lines (Public Land Survey System, PLSS). Boundaries between different vegetation types that occurred within section boundaries were interpolated using USGS 7.5 minute quadrangles and should be considered an approximation (Comer et al. 1995). Brown (1998*b*) suggested that presettlement boundaries are best portrayed as fuzzy, indicative of the vagueness associated with the mapping process. Additionally, Wang (2005) noted that many Public Land Survey Records (of which the GLO is a subset in the United States) contain three inherent positional accuracy issues: mislocation of corners, mislocation of landscape features, and incorrect positions of bearing trees. These inaccuracies reinforce the importance of using presettlement data for coarse-level assessments and not inferring an artificial sense of accuracy to the representations.

Boundary limitations are not unique to comparisons using GLO. We conducted a simple, visual inspection of wetland maps delineated by the Michigan Department of Natural Resources (MDNR; 2001) (based on 1998, 1:12,000 aerial photography) and National Wetlands Inventory (NWI) maps (based on 1971, 1:58,000 scale aerial photography) (Fig. 11-2). The MDNR wetland polygons were digitized by a photo interpreter. In contrast, the NWI maps for Michigan were transformed by hand onto 1:24,000 U.S. Geological Survey (USGS) quadrangles and then scanned to produce a digital form. Even though two significantly different processes were used to generate the wetlands maps, an uninformed user may simply compare the 1971 NWI and 1998 MDNR data to calculate the aerial extent of wetland change over that time period. For wetlands A and C in Fig. 11-2, that analysis would suggest 0.8 and 0.5 ha increases,

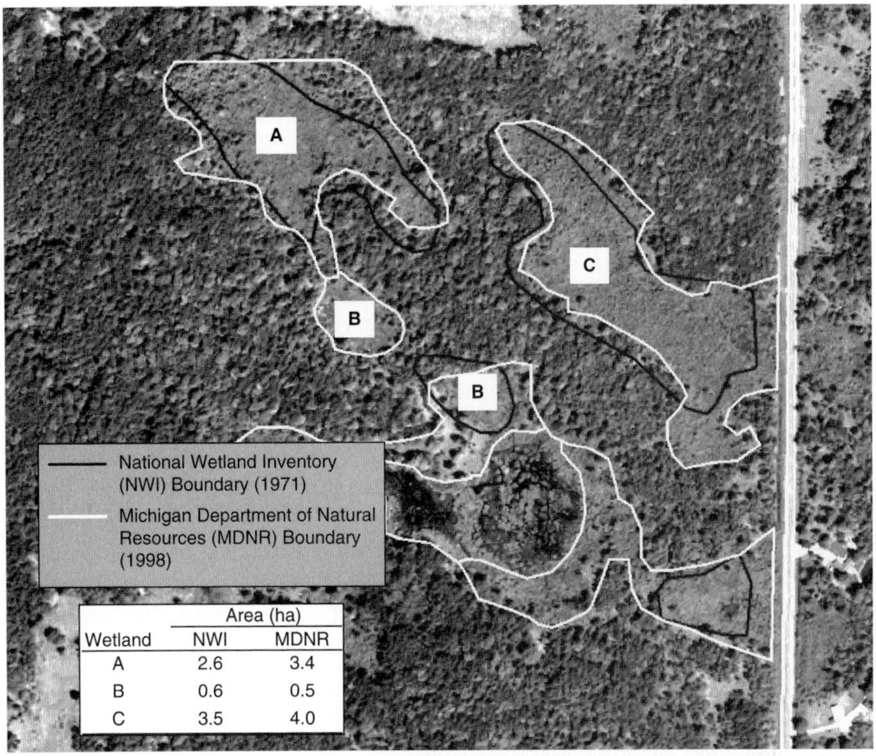

FIG. 11-2

Boundary delineations for National Wetlands Inventory (NWI) (1971) and Michigan Department of Natural Resources (MDNR) wetlands (1998) overlaid on National Aerial Imagery Program (NAIP); USDA (2000).

respectively. Although the areas of wetland B were similar (0.1 ha difference), note the apparent positional error between the data sources (Fig. 11-2). Boundary discrepancies occurred between the NWI and MDNR data, and from a simple comparison one might conclude that wetlands A and C are increasing in size and all wetland boundaries are dynamic. Is this a real ecological phenomenon or an artifact of boundary delineation processes? Without evaluating the metadata to understand differences in how boundaries were delineated between the two data sets, a modeler may incorrectly attribute the increase in wetland size as an ecologically significant event.

Recommendations.—Evaluate metadata to understand methods for boundary delineation. Quantify potential boundary delineation error among map sources and establish confidence in the boundary locations. Then, if the shifts in boundary locations fall outside confidence estimates, there is more reason to hypothesize that the differences are ecologically based.

Underrepresentation of Vegetation Types Limitation

The third limitation potentially influencing correct use of historical data for vegetation change analysis involves underrepresentation of vegetation communities that historically occurred as small patches (Barbour et al. 1999), such as aspen-birch (*Betula* spp.) and some wetlands in Michigan. Additionally, those vegetation types located on complex topography may not have been as thoroughly surveyed as other areas (Liegel 1982, Manies and Mladenoff 2000, Black and Abrams 2001). Biases in bearing tree selection in the GLO also influence its utility for portraying some vegetation types, but this bias can be somewhat alleviated by also including the section line notes (Wang 2005). Underrepresentation of certain vegetation types and GLO survey biases warrant caution when comparing the frequency and abundance of vegetation types between presettlement data and more recent cover type data acquired through air photo or satellite interpretation.

Recommendations.—It is important to question whether the cover type(s) of interest had similar detection probabilities between the different sampling approaches prior to drawing strong inference from a change analysis. If detection probabilities are unequal due to the scale of source data, then a data bias exists. Recognizing these limitations among data collected using different technologies and from different time periods will increase scientific rigor in conducting vegetation change analyses.

CLASSIFICATION SYSTEMS

Classification Language

Wildlife habitat models rely on a description of habitat. There are two general mapping models used to portray habitat in a GIS environment: (1) a patch-corridor-matrix model and (2) a continuum model. The patch-corridor-matrix model (Forman and Godran 1981, 1986) relies on our ability to denote "patches" of relatively homogenous communities, identify connecting "corridors" between those patches, and map the "matrix" that forms the background for patches and corridors (Forman and Godran 1981). The patch-corridor-matrix model is most commonly used in wildlife applications that can be portrayed in vector or raster data structures in the GIS. This type of discrete vegetation model has been recognized as inadequate where representations of spatial gradients or spatial patterns of boundary uncertainty are desired (Brown 1998*a*), but in terms of practical application, the patch-corridor-matrix model is preferred because it aligns with the ecological community concept and is easier to visualize on the ground. In contrast, the continuum model relies on the fact that gradients often exist among natural communities, and thus, definitive boundaries are difficult to accurately portray (e.g., Brown 1998*a*). The continuum model portrays communities on some continuous gradient (e.g., soil moisture, elevation) and most typically results in raster data structure (e.g., Zhu et al. 2001, Lehmann 2004). The use of continuum models is gaining popularity in wildlife-habitat relationships modeling as statistical techniques evolve, though their interpretation and application are often cumbersome to practitioners.

Patch-corridor-matrix GIS models are often based on some type of ecological classification system (see review in Grossman et al. 1999). A classification system relies on language that describes vegetation or habitat entities. Language resolution becomes an important determinant of predictive capability for any models that use the classification system. By necessity, most classification systems provide a limited vocabulary (or class names) to describe a landscape, and the amount of variability within a class can potentially be high (e.g., Roloff et al. 1994). In some situations we have observed premature rejection of habitat models based on "poor performance" without consideration of how the GIS classification system may have influenced the results. A better approach would be to denote that the model performed poorly in the data environment for which it was applied. A comparison of two systems currently used to support MDNR management decisions will help expose classification limitations of each system for modeling wildlife habitat.

Michigan Department of Natural Resources' (2001) Operations Inventory (OI) system for classifying cover types has a limited vocabulary of 26 class names. In contrast, MDNR's Integrated Forest Monitoring Assessment and Prescription (IFMAP) cover-type classification system is hierarchical and at the most resolute level has 137 classes. Both OI and IFMAP rely on remotely sensed imagery interpretation and ground data collection to produce cover type maps, but the decisions that lead to class assignments and polygon boundary locations differ between the two systems. Operations Inventory is based on a forester's interpretation of vegetation composition, structure, and site variables. It is also partially based on future management objectives, which can result in major differences in boundary delineation and class membership. Because OI is based on a qualitative process, it is sometimes difficult to establish the primary determinants of stand boundary location and class membership. Stand polygons in the IFMAP system were delineated based on patterns in aerial imagery. Ground verification was used to verify or edit those boundaries based on specific measurements of vegetation species composition. Class assignments were made using a computer algorithm that interprets these canopy measurements.

What are the ramifications of these classification processes to wildlife modelers using OI and IFMAP? Consider a modeler assessing habitat for a wildlife species that relies on aspen. Aspen communities in Michigan occur in two general forms: (1) as monotypic stands and (2) as associates in a variety of forest types. Aspen is represented in the OI system by a single class and in IFMAP by nine classes. If the modeler wants to identify relatively monotypic aspen stands as important patches for this species, he or she will need to determine which classes from OI and IFMAP are likely to represent this condition. Under the OI scenario, the modeler would likely be forced to accept the lone aspen class and ignore potential bias of management intent (i.e., the forester's perception) in class determination. Using IFMAP, the modeler could evaluate the decision rules associated with multiple aspen classes and determine which were appropriate. Selection of the "correct" data set to use relates back to how the analysis was framed. If the modeler requires quantifiable, repeatable spatial characteristics of aspen stands, then knowledge of the decision rules used for boundary determinations would be important and IFMAP would be preferred. If a modeler were to use OI to address this analysis, he or she may add additional uncertainty and nonrepeatability to habitat model results. Conversely, if the analysis calls for a general assessment of monotypic aspen acreage, the OI may suffice.

The importance of classification system resolution for conducting wildlife habitat assessments cannot be overstated. Results of wildlife-habitat relationships models can vary substantially depending solely on the classification system used (e.g., Lawler et al. 2004). Consider the habitat model results in Fig. 11-3. These results represent brown creeper (*Certhia americana*) habitat suitability, on a scale of 0 (nonhabitat) to 100 (optimal habitat), for a 390 km^2 landscape in central Idaho. Here, a brown creeper habitat suitability index model was applied using three commonly used ecological classification schemes: ecological land units (Haufler et al. 1996), structural stages (Johnson et al. 1994) within a Daubenmire plant series (sensu; Daubenmire 1952), and "alliances" in the National Vegetation Classification (Grossman et al. 1998).

Ecological land units provided the most resolute classification scheme, often used to describe and map combinations of vegetation successional stage and site potential in the context of major disturbance regimes (Haufler et al. 1996).

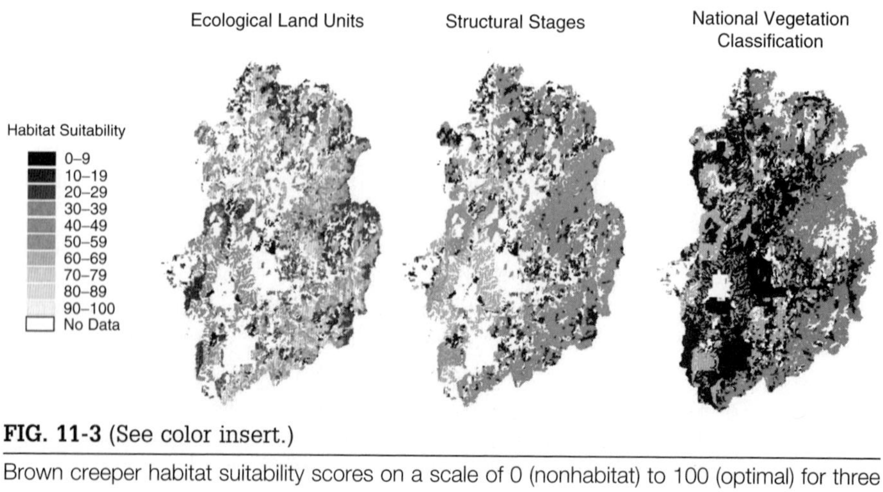

FIG. 11-3 (See color insert.)

Brown creeper habitat suitability scores on a scale of 0 (nonhabitat) to 100 (optimal) for three ecological classification systems: ecological land units (Haufler 1996), structural stages (Johnson et al. 1994) with Daubenmire plant series (sensu Daubenmire 1952), and the National Vegetation Classification (Grossman et al. 1998).

Structural stages provided an intermediate resolution that was based on percent canopy cover in five tree size classes (Johnson et al. 1994). The structural stage classification roughly corresponds to the mid-scale land classification used by the Interior Columbia River Basin Ecosystem Management Process (U.S. Forest Service 1996). The National Vegetation Classification uses both physiognomy and floristics to classify existing vegetation into "alliances" (Grossman et al. 1998). Alliances are roughly equivalent to the Society of American Forester's cover types (Eyre 1980) and were defined as physiognomically uniform groups of plant associations sharing one or more dominant or diagnostic species which are generally found in the upper-most stratum of vegetation (Grossman et al. 1998).

A detailed, geo-referenced vegetation inventory was available for the study area and these data were stratified according to each classification scheme. Average vegetation structure values required by the brown creeper model were calculated by stratum. The results of this analysis demonstrated that habitat model output can vary substantially depending on the classification scheme (Fig. 11-3). Depending on the classification system used, one could draw completely different inferences from the habitat model results. With ecological land units, the model suggests that a relatively broad range of habitat conditions are distributed across the landscape and higher quality patches of habitat are well dispersed (Fig. 11-3). With the National Vegetation Classification, the habitat model suggests that habitat quality is poor to marginal across the landscape, with virtually no high-quality patches (Fig. 11-3).

Recommendations.—Classification language and how map attributes are portrayed can significantly affect wildlife-habitat model output. It is not uncommon for practitioners to dismiss a perfectly good wildlife habitat model on the pretext that "it doesn't do a good job of mapping potential habitat" when in fact the issue resides with classification system resolution and variability. This often raises a dilemma for wildlife modelers. Should we use the model output and describe the sources of error and caveats for use (most typical strategy), should the habitat model be adjusted to accommodate the resolution and errors of available data (less typical strategy), or should new data be collected that directly meet the needs of the habitat assessment (least typical strategy)? Unfortunately, we often tend to ignore error descriptions and caveats associated with habitat maps because of project deadlines and budgetary constraints. Plus, something about a GIS "map" instills a sense of reality and faith in habitat model output. We caution modelers and users of GIS products to understand and convey data limitations to decision makers.

To help account for classification system effects on wildlife habitat model output, we advocate aligning scales of species habitat selection (which are multiscalar; Johnson 1980, Morris 1987) to classification resolution (also see Barry and Elith 2006). Rather than rely on a preconceived classification system and assume that specific vegetation structures are present, we support the recent trend of mapping individual structures as continuous inputs to habitat models

(Edwards et al. 2003, Holmström and Fransson 2003, Tuominen et al. 2003, Lu et al. 2004, Tuominen and Pekkarinen 2005). There are advantages to using continuous versus discrete GIS data in habitat modeling (Conner et al. 2003, Théau et al. 2005, but see Dussault et al. 2005), if the continuous data layers are derived from an adequate set of spatially explicit data. For example, it is a common practice to use geo-referenced sample points (e.g., FIA data) to interpolate continuous data surfaces. We caution that the utility of a continuous data layer depends on the number and distribution of sample points. Users of interpolated continuous data layers should understand the important scales of their project objectives and ensure that the density of supporting data support inference at the correct scale(s).

In regards to using continuous data surfaces in wildlife-habitat modeling, practitioners may argue that they still need to make management decisions based on stands or patches. We contend that stand boundaries that were most likely derived for some purpose (like resource management or planning) other than species habitat modeling can be overlaid on structure-based continuous habitat model output. The issue of classification language resolution is removed in a continuous map, but interpolation errors can be high if few data points are available. Our criticisms of class-based GIS maps should not be used to negate the utility of wildlife-habitat models that rely on this data type as they provide useful conservation information, especially at larger scales (Raphael and Marcot 1986, Edwards et al. 1996, Bolger et al. 1997, Karl et al. 2000, Pearlstine et al. 2002). We encourage modelers to evaluate both continuous and class-based GIS data and use the format that best fulfills their project objectives.

Attribute Variability

Attribute data portray what is known about a spatial location in the GIS. For example, a GIS data layer of vegetation patches may be attributed with vegetation cover type, percent canopy cover, plant species lists, plant structure, area of the patch, and a host of other attributes that explicitly describe an individual patch type. In most situations, attribute data represent field-collected observations, GPS point collections, or GIS-calculated information. In the case of field-collected information, the data contain sample errors. For GIS-calculated information, the data contain potential positional and classification errors. Variation in attribute data should not be ignored when conducting wildlife-habitat assessments using GIS. The concept of fuzzy logic (i.e., a system of logic dealing with the concept of partial truth with values ranging from completely true to completely false) has been incorporated into some habitat modeling efforts (e.g., Silvert 1997, Robinson 2003, Bojórquez-Tapia et al. 2004, Zhang et al. 2004, Cheung et al. 2005) to help account for GIS data uncertainties. Here we focus on GIS attribute data variability as it relates to potential effects on wildlife modeling.

As demonstrated in the analysis conducted to produce Fig. 11-3, habitat elements are often portrayed using map stratum averages. Within-stratum variability is frequently ignored in wildlife-habitat modeling (Roloff and Kernohan 1999), even though it can have significant effects on model interpretation (e.g., Verbyla and Litvaitis 1989, Bender et al. 1996, Hess and Bay 2004). Linden (2006) examined the effects of simulating within-stratum variability in habitat attributes for an ecological land classification as an alternative to using stratum averages. Frequency distributions of habitat characteristics were estimated from field surveys for each stratum and linked to the occurrence of cell values in a GIS raster. The spatial distribution of cell values within each stratum was randomized through multiple iterations, and habitat quality for Canada lynx and snowshoe hare (*Lepus americanus*) was examined. This evaluation indicated a significant effect on modeled habitat quality for those strata with high attribute variability. In these highly variable strata, habitat quality would have been inaccurately portrayed by the use of averages. Linden (2006) noted that as the scale of analysis increased, differences in habitat quality decreased. This observation is consistent with known relationships on how variance responds to changes in map extent, with larger areas tending to dampen the variance extremes (Stoms 1994, Wolock and Price 1994, Wu, F., 2004; Wu, J., 2004). Linden (2006) concluded that the decision to explicitly include attribute variability in a habitat modeling project depends on map data resolution and how it relates to the scale at which focal organisms perceive their habitats and the spatial extent of the landscape. In Linden's example, raster cell variability between iterated maps was less important for Canada lynx than for snowshoe hare because the scales of modeled habitat selection behavior (i.e., the probability that habitat would support a home range) substantially differed between the species. In other words, habitat suitability for snowshoe hare home ranges was modeled as more sensitive to the habitat structure of an individual patch compared to lynx, which was modeled as less sensitive to individual patches. Linden's analysis demonstrates the importance of understanding the effects of scale and attribute variability on habitat model output.

Recommendations.—Attribute error should be incorporated into habitat models. Recent advances in resampling statistics (Manly 1997, Simon 1997) and computing capabilities have offered mechanisms for bounding habitat model outputs with confidence estimates, but computer processing time across large landscapes still prohibits incorporation of these tools into many projects. It is important not to confuse this source of habitat model error with that portrayed by some quantitative modeling procedures. In quantitative models, the error estimates represent a composite of system errors (e.g., error in animal location and perhaps the variation in land classification system, particularly if the continuum GIS data model is used). A quantitative model derived from GIS layers based on stratum averages (e.g., a cover type map with average tree densities per class) does not account for within-stratum variability in model error estimates. As Bender et al. (1996) demonstrated, ignoring this source of error in wildlife-habitat relationships modeling can lead to erroneous conclusions.

SPATIAL AND TEMPORAL SCALE ISSUES

Correct determination of the appropriate scale is the cornerstone of habitat analysis and model development (Morrison et al. 1998:141, Morrison 2002). Here, we offer advice on some common spatial scale issues that we have encountered in our habitat modeling efforts. Readers should refer to Turner et al. (1989), Wiens (1989), Levin (1992), and Hunsaker et al. (2001) for more complete treatises of this topic.

Spatial Scale Considerations

There are at least four meanings of spatial scale in remote sensing and GIS (partially taken from Cao and Lam [1997], also see Morrison et al. [1998:241]):

1. *Cartographic (or map) scale:* the relationship (proportion) of map distance to ground distance (e.g., 1 cm on the map = 1 km on the ground);
2. *Geographic (or observational) scale:* the size or spatial extent of a study;
3. *Operational scale:* the geographic extent at which certain processes operate within the environment. This could be the scale associated with nutrient-cycling in a wetland;
4. *Measurement scale:* the smallest distinguishable part of an object. Examples include the cell size in a remotely sensed image or the sampling interval in an ecological study.

In wildlife-habitat modeling, most explicit references to scale use the second (geographic or observational) or third (operational) meanings of the term. The phenomena of interest in most ecological studies are observed at operational scale (s). One of the basic challenges in using GIS data to model wildlife habitats across landscapes is to understand how observations at the operational scale(s) are influenced by observational and measurement scales.

Observational Scale.—It is not uncommon for significant ecological information to be lost or compromised because habitat modelers are forced to use data from some arbitrarily determined study area boundary (Johnson 1980, Porter and Church 1987, Morrison 2002). Other authors have documented the importance of identifying an appropriately scaled study area (Brennan et al. 2002, Morrison 2002), especially in the context of resource selection studies (e.g., Aebischer et al. 1993, Erickson et al. 2001). The recommended approach is to ensure that phenomena measured at operational scales can be fully encompassed by the observational scale (i.e., the study area boundary; Brennan et al. 2002). Morrison (2002) concluded that study area size and its relationship to wildlife population data and how the location of a study area fits the geographic range of species are seldom studied or incorporated into analyses. Unfortunately, the scale(s) at which ecological processes operate may not become evident until data are collected. This was the case in Roloff et al. (2001) in which radio-collared elk (*Cervus elaphus*) were used to test the reliability of a habitat potential model. Detailed habitat maps were generated for the study area, which was defined by the boundary of Custer State Park, South Dakota. All radio-tagged elk were located within the study area boundary, but one sub-herd tended to seasonally use habitats outside this area. As a result, Roloff et al. (2001) needed to censor that sub-herd from the habitat model evaluation. Fortunately, some of the sub-herds that were radio-tagged used habitats entirely within the study boundary, so the analysis could still occur; however, sample size was reduced because of a study boundary constraint. This example illustrates the significance that observational scale can have on wildlife habitat modeling results.

Measurement Scale.—Different measurement scales are inherent in either vector or raster GIS data structures. The usefulness of each data format for characterizing habitat and landscape patterns has been presented elsewhere (e.g., Johnson 1990, Haines-Young et al. 1994, Corsi et al. 2000, Wade et al. 2003). The best scenario for deciding which type of data to use in a GIS habitat modeling project is to properly frame the analysis and select the data format that best fulfills project objectives. As noted previously, that process often breaks down in that modelers are often forced into using whatever data are available to generate a product within time and budgetary constraints. Though conversions between GIS data structures are available, the processing artifacts can have significant impacts on habitat model output.

The most basic example of how measurement scale in vector and raster GIS data can influence habitat model output is in the simple calculation of patch area and edge. Consider the example in Fig. 11-4. Here, the same land cover data

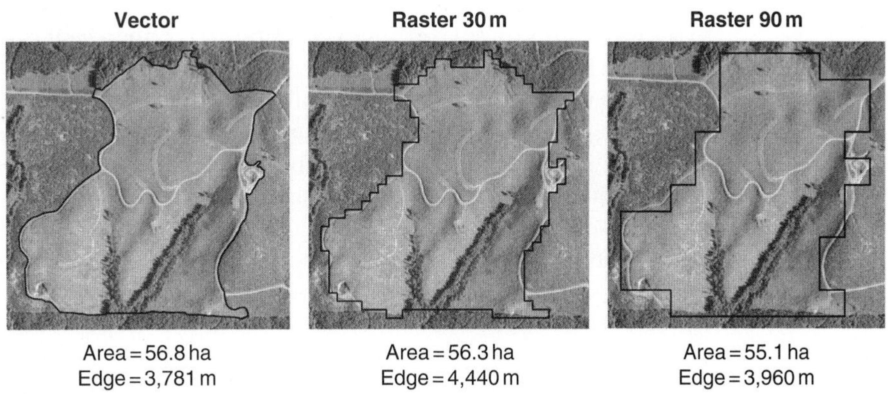

FIG. 11-4

Vector and raster representations with area and edge GIS measurements of a forested stand in Washington, USA. Background photography from the National Agriculture Imagery Program (NAIP), USDA.

for a forested stand in Washington are portrayed in vector and raster formats. These data represent typical stand boundaries used in resource management and planning. The vector data were developed using on-screen digitizing with National Agriculture Imagery Program (NAIP) as the background. The raster data represent the same vector polygon at 30 m and 90 m cell resolution. Area estimates differ depending on data structure and resolution. Edge measurements differed substantially (by almost 700 m between the extremes) depending on the data type. Which representation of the stand in Fig. 11-4 is the best? We argue that the vector-based representation is the most accurate in this case because the layer was generated directly from the air photo and because this stand had relatively hard edges (i.e., a continuous line was a good representation of the real boundary). In wildlife-habitat assessments that evaluate edge effects based on linear edge distances or boundaries between patch types, raster-based data appear to overestimate the absolute amount of edge. This data artifact is of concern only if the absolute value is important in a habitat model, i.e., relative evaluation of edge distances help alleviate this bias among data types.

An additional cautionary note on using raster data structure in wildlife habitat modeling projects warrants mention here. It is generally well accepted that square cells in rasters may not meaningfully represent space for animals (Tischendorf 1997). Tischendorf's (1997) argument is that animal movement analyses are constrained by the resolution of square cells and the fact that a diagonal movement is longer than a perpendicular movement. Thus, vector-based animal movement data can only be portrayed as a "jump" from cell to cell, and the distance of those jumps can vary. Solutions to this problem have included using a raster with cells of different sizes (Tischendorf 1997), resampling GIS data to a small resolution relative to animal movement distances, and portraying raster GIS data as hexagon-shaped cells (the idea being that a hexagon most closely resembles the least biased geometric shape, a circle, that can be portrayed in a nonoverlapping arrangement) (e.g., Schumaker 1996, Lawler and Schumaker 2004, Lawler et al. 2004, Jackson et al. 2005, Fernández et al. 2006). All these solutions have potential problems, particularly with data management, simulation times, and loss of descriptive capabilities as resolution is compromised (Gough and Rushton 2000, Hunsaker et al. 1994).

The advantages of using hexagonal shapes to describe landscapes and allocate sample plots were noted by McCollum and Cochran (2003) in their description of the previously described FIA program as (1) spatial compactness; (2) uniform spatial coverage; (3) flexibility for altering grid density; (4) reduced likelihood of correspondence to anthropogenic landscape components; and (5) reduced variance estimates in comparison to a random sample. Despite these advantages, the most common technique for characterizing landscapes in current wildlife and ecological literature is to use rasters with square cells, most likely because that geometry corresponds to how most remotely sensed GIS data are collected and made available, and the structure is amenable to common GIS software. We caution habitat modelers to at least question whether square cell biases may significantly alter their study designs or inference space prior to project implementation.

Regardless of whether raster or vector data structures are used in a habitat modeling project, one of the most important influences on habitat model interpretation is minimum mapping unit or map grain (which often corresponds to measurement scale as described previously). Minimum map unit is defined as the smallest size aerial entity to be mapped as a discrete entity (Saura 2002). It is well known that in heterogeneous landscapes between-patch variance decreases as resolution decreases (i.e., as minimum mapping unit or grain size increases, between-patch differences tend to decrease; Turner et al. 1989, Wu, F., 2004; Wu, J., 2004). In other words, coarser resolutions tend to homogenize landscapes into the dominant patch types (Saura 2002). It follows that map accuracy tends to increase as minimum map unit increases (Knight and Lunetta 2003). Sensitivity of most landscape-level data to minimum map unit further exemplifies the importance of understanding minimum map unit or measurement scale in the context of processes being modeled (i.e., operational scales) for the study area (i.e., observational scale). This same concept applies to selecting a sample plot or quadrat size. It is not uncommon for researchers to arbitrarily select a plot size without considering variation in the underlying vegetation. In some situations the natural variation in vegetation patches may provide more reliable sampling units than arbitrarily chosen plot sizes (Bowyer et al. 1996, Stohlgren et al. 1997).

Temporal Scale Considerations

Issues surrounding the effects of temporal scale on wildlife-habitat relationships modeling are equally as important as spatial scale issues but have received less attention (Morrison 2002, Chapter 14). Temporal scales can categorically be viewed similar to spatial scales, with observational temporal scale referring to the duration of the study, operational temporal scale referring to the time that certain processes operate in the environment, and measurement temporal scale referring to the smallest unit of time for which we have a measurement. With the advent of technologies for making historical map and tabular data spatially explicit, more researchers and modelers are conducting time series comparisons of habitat change.

Wildlife researchers recognize the importance of temporal variation in wildlife populations, and most try to account for this phenomenon by collecting data over multiple years or seasons. In wildlife-habitat relationships studies, less attention is given to short-term temporal variation of habitat elements, unless the wildlife response variable is explicitly linked to a habitat element known to exhibit frequent changes (e.g., Boyd 1996, Whitehead 1996, Fortin et al. 2002, McCarty et al. 2002). One standard approach used in wildlife studies is to collect population data for 2–4 years, develop a habitat GIS layer that reasonably corresponds to the time period of population data collection, and then assume that habitat remains static. Most researchers track the occurrence of major habitat altering events during their studies, but few incorporate the subtle effects of short-term fluctuations in habitat elements (e.g., vegetation reproductive status, growth, mortality, and successional processes).

We caution that habitat model components be evaluated for their sensitivity to short-term variations. For example, some of the authors (G. Roloff and D. Linden) recently completed a habitat modeling project for Canada lynx in which outputs from the model were used to influence a 5-year forest management plan. We used a standard approach for depicting habitat (see Linden 2006, Appendix D, for a description of the process used to estimate attributes for the lynx model). Because the lynx model relied on vegetation cover provided by tree branches and boles, we were concerned that 5 years of vegetation succession may have a significant influence on model results that would go undetected if we simply assumed no major vegetation changes. We tested this potential effect by simulating vegetation changes for 10 common vegetation types in the landscape using a well-established tree growth and yield model. According to the growth and yield model and Linden's (2006) method of calculating understory cover, horizontal cover differences within a stand varied from -7% to 7% over the 5-year period (Table 11-1). At first glance this change seems negligible relative to how lynx may use habitats; however, the magnitude and direction of change must be viewed in the context of how the habitat model uses the information. Our lynx project used a habitat suitability index model that was based on linear relationships with definitive thresholds. If the ±7% applied to a stand with vegetation conditions close to a critical threshold, habitat model output could be significantly affected. This example illustrates the importance of considering short-term vegetation changes in wildlife-habitat relationships modeling.

Recommendations.—Spatial scale considerations are one of the most important in characterizing and analyzing wildlife habitat for landscapes. A basic challenge faced by all wildlife-habitat modelers that use GIS is to understand how their characterization of operational processes is influenced by observational and measurement scales. We stress the importance of understanding your project objectives in the context of observational and measurement scales. Consider how GIS data formats can influence your modeling work. Make sure that data format (i.e., vector or raster) does not introduce hidden biases into your descriptions of operational processes. Minimum map unit is a critical consideration for habitat modeling projects. Modelers should ensure that operational processes are occurring at scales ≥ minimum map unit. If this is not the case, make sure that assumptions used to infer smaller scale habitat characteristics to map units are sound and accurately portray habitat dynamics in the landscape. Subtle temporal variations in habitat elements are often ignored in modeling projects, and we caution that these changes can influence habitat model output. We reiterate the recommendation of previous authors (Corsi et al. 2000, Maurer 2002, Morrison 2002) on the importance of evaluating research designs or habitat modeling projects in the context of influential spatial and temporal scales.

Table 11-1 Horizontal Cover Values in Three Height Strata Over a 5-Year Period as Modeled for a Canada Lynx Habitat Suitability Index Model to Demonstrate the Potential Effects of Short-Scale Temporal Changes Caused by Vegetation Succession on Habitat Model Output

	Horizontal Cover (%) Year 0			Horizontal Cover (%) Year 5			Horizontal Cover (%) Difference Between Years 0 and 5		
	Height Strata			Height Strata			Height Strata		
Stand	**0–0.99 m**	**1–1.99 m**	**2–2.99 m**	**0–0.99 m**	**1–1.99 m**	**2–2.99 m**	**0–0.99 m**	**1–1.99 m**	**2–2.99 m**
1	27	21	22	23	28	24	−4	7	2
2	26	37	31	25	38	38	−1	1	7
3	32	42	46	33	39	49	1	−3	3
4	19	23	31	19	25	29	0	2	−2
5	28	32	43	30	31	40	2	−1	−3
6	16	27	24	15	30	30	−1	3	6
7	41	71	66	39	64	69	−2	−7	3
8	19	25	28	19	25	30	0	0	2
9	45	60	69	45	56	66	0	−4	−3
10	18	23	31	20	25	29	2	2	−2

The influence of these scales on wildlife populations is inseparable, and we recommend evaluating the organization of your wildlife modeling project with this in mind (see Corsi et al. 2000; Fig. 11-3 for a useful conceptual model). Consistent with other authors, we recommend conducting habitat assessments at multiple scales (at least one scale above and below the process of interest) to ensure that the operational processes are being characterized appropriately (Orians and Wittenberger 1991, Wu and Loucks 1995, Bowyer et al. 1996, Jelinski and Wu 1996, Qi and Wu 1996, Bennetts and Kitchens 1997, Maurer 2002, Morrison 2002).

TECHNOLOGICAL CONSIDERATIONS

Another important consideration when using GIS to model wildlife habitat relates to the technology used for deriving land classification systems. Consider the common practice of automatically classifying satellite imagery to portray landscapes (e.g., Osborne et al. 2001, Betts et al. 2003, Kerr and Ostrovsky 2003, Jeganathan et al. 2004). We will focus our discussion on two principal sources of error in the process of creating a classified landscape from satellite imagery (also see O'Neil et al. 2005:433–443). The first source of error, which we will call "mixed-pixel error," is associated with image acquisition and spectral response (the range of light reflected by an object and detected by a sensor) on the Instantaneous Field of View (IFOV). The IFOV is the scan spot size or instantaneous geographic coverage of a satellite sensor. The area of the IFOV varies depending on the satellite sensor and most commonly ranges from 1 m^2 to 1 km^2. For example, the IFOV for Landsat Thematic Mapper sensor (one type of satellite sensor that produces data commonly used in characterizing landscapes) is collected at 30 m $\times$ 30 m (0.09 ha) resolution. Within that area, spectral response is integrated for all objects into a single cell value ranging from 0–255. If an IFOV is composed of a single, homogenous object on the ground (e.g., a slab of concrete), then the remote sensing analyst can be confident that the recorded cell value is associated with concrete. However, large IFOVs are rarely composed of a single, homogenous object, even in simplified ecosystems. Larger IFOVs are composed of multiple objects such as tree canopies of different species, ground cover between trees, water, or manmade structures. The ability to consistently associate a cell value with objects of interest on the ground is confounded by integration of spectral response from multiple objects over the entire area of the IFOV. In some situations mixed-pixel error may have negligible effects on habitat interpretations (e.g., Wickham and Riitters 1995). Nonetheless, it is important for the wildlife modeler to consider how habitat elements of interest relate to the size of an IFOV's integrated measure of spectral response. Habitat elements that are significantly smaller than an IFOV, and therefore "mixed" with many objects not of interest, will be much harder to accurately map from satellite imagery than those that are significantly larger than the IFOV.

We experienced the effect of IFOV in working with the data for Michigan's Gap Analysis Program (GAP; Donovan et al. 2004). Michigan's GAP project relied on a cover type map derived from 30 m Landsat Thematic Mapper imagery. For those wildlife species that required specialized habitat elements (e.g., snags, rock outcroppings, vernal ponds), "mixed-pixel error" was accounted for by assuming that these features are consistently associated with specific cover types (Donovan et al. 2004). This adjustment, along with other error accumulated in the GAP process (Dean et al. 1997, Laba et al. 2002), results in commission (i.e., erroneously designating an entity something it is not) and omission errors (i.e., erroneously not designating something it truly is), with a tendency for higher commission errors. These sources of error tend to diminish at larger scales consistent with the intended use of GAP products for regional-level conservation assessments (Edwards et al. 1996). The importance of accounting for these errors in conservation decision making cannot be lost because those wildlife species with specialized habitat requirements frequently are those in greatest conservation need (Short and Hestbeck 1995).

Another source of error in classifying a landscape from satellite data that affects wildlife habitat modeling in GIS relates to "mixed-pixel error" and the level of detail (or number of classes) desired in the final classification system. We will refer to this error as "spectral-distance error." A remote sensing analyst generally embarks on a mapping project with a predefined classification system or some idea as to what constitutes a class in the final map. In satellite image interpretation, the only information available to the analyst is the measure of spectral response in the IFOV over multiple wavelength bands provided by the sensor. For example, Landsat Thematic Mapper has seven wavelength bands available for transformation into map classes. The process for classifying imagery usually involves plotting the spectral response of different vegetation classes in multidimensional spectral space with the hope that identifiable and clearly separable clusters are present. This is rarely the case; usually there are multiple classes of interest that intersect or cluster very near each other (Fig. 11-5). The closer classes are in spectral space, the more they will be confused with each other in the classification process (Fig. 11-5). This level of confusion constitutes "spectral-distance error."

Both "mixed-pixel error" and "spectral-distance error" can have substantial effects on the inference drawn from landscape-level wildlife habitat models. For those classification processes that include accuracy assessments, these sources of error are typically portrayed in classification error matrices, though these matrices also contain other error sources (Congalton and Green 1993, Stehman 1997, Congalton and Green 1999, McGwire and Fisher 2001). Nonetheless, classification error matrices should be consulted prior to using GIS layers for habitat modeling to ensure that data classes required by the habitat model can be portrayed with acceptable accuracy levels. Also, habitat modelers should ensure that the classification error matrix contains data from all ecologically significant portions of their assessment landscape. Without well-dispersed classification

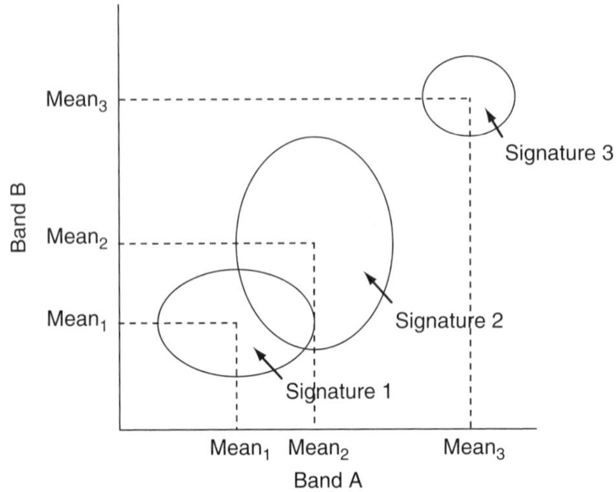

FIG. 11-5

Spectral-distance error and its effect on classification accuracy (based on ERDAS 1982). Means correspond to multidimensional means in spectral space and correspond to average classification values (i.e., the values used to map classes).

test data, important nuances in the classification process or the classification system may be overlooked. For example, one of the co-authors (M. Donovan) recently completed a project that used a wildlife-habitat relationships model to estimate statewide sharp-tailed grouse (*Tympanuchus phasianellus*) habitat potential in Michigan (Fig. 11-6). Landsat derived cover types were used in the model, and a good understanding of classification error was based on dispersed training plots throughout the state (Space Imaging Solutions 2001). The model identified several potential areas for sharp-tailed grouse habitat management and restoration. In the Upper Peninsula the predicted habitats corresponded well to known grouse locations. However, the model also predicted substantial habitat potential in the thumb of the Lower Peninsula. This area was visited by field biologists, and they concluded that although the vegetation was typed correctly, habitat potential for sharp-tailed grouse was low because of current land management practices. In this example, classification accuracy for cover types important to sharp-tailed grouse was high, but the classification scheme lacked an important component of habitat: current land management status. Without field visits by biologists to Michigan's thumb area, potential sharp-tailed grouse habitat would have been significantly overpredicted. This example also points to the utility of using predictive models even if classification data or schemes are less than perfect. Here, the model helped direct field biologists to a specific area in a large landscape.

There is a common misconception among wildlife-habitat modelers that high resolution is always better. This is not always the case. Often, modelers focus on

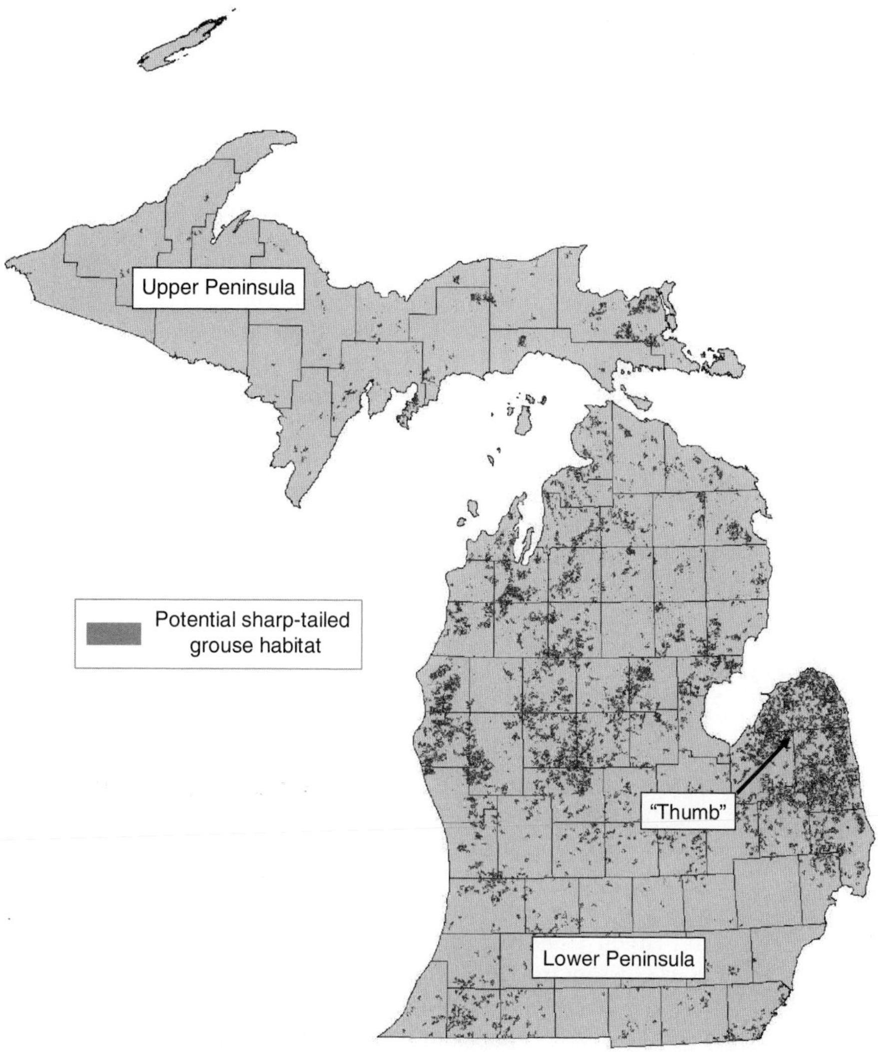

FIG. 11-6

Potential sharp-tailed grouse habitat in Michigan, USA, as estimated by a wildlife-habitat relationships model.

spatial resolution as the primary determinant of data utility, but spectral contrast of habitats in relation to their surroundings is also an important consideration. Trade-offs between spatial and spectral resolutions of data should be evaluated in the context of importance to the modeling effort. This decision depends on habitat element size, with higher spatial resolution not necessary for coarser habitat elements (e.g., mature tree canopies). For coarse habitat elements, it may prove more advantageous to favor higher spectral resolution. Most modelers have a limited choice in available imagery and thus may not spend much time in evaluating spatial and spectral resolution trade-offs.

Recommendations

We cannot overstress the importance of evaluating the classification error matrix as a means to understand how "mixed-pixel" and "spectral-distance" errors can potentially influence habitat model results (see Foody 2002). Conducting error simulations (e.g., Fleming et al. 2004, Hines et al. 2005) holds great promise for understanding not only class errors but also spatial location errors when using satellite imagery. Remember that classification accuracy errors compound as data layers are overlaid in a GIS, and care should be exercised when conducting operations that modify polygons or combine raster cells. Always incorporate known errors into metadata descriptions and in caveats on how habitat model results should be used. Avoid the misconception that higher resolution data are always better. There is a trade-off between spatial and spectral resolution, and the best approach is to align data resolution with the project objectives.

SUMMARY

Habitat maps generated by a GIS are often viewed by decision makers and the general public as absolute truth regardless of accuracy assessments, summaries of data use limitations, or spatial and temporal scale considerations. We caution wildlife-habitat relationships modelers to consider how the inherent properties of data sets and uncertainty from all these sources should be included and portrayed in their analyses. Texts have been devoted specifically to some of these topics such as accuracy assessments (Goodchild and Gopal 1989, Congalton and Green 1999, Lowell and Jaton 1999), spatial and temporal scale considerations (Bissonette 1997, Waring and Running 1998), and integrating uncertainty into GIS analyses (Chiles and Delfiner 1999, Mowrer and Congalton 2000, Hunsaker et al. 2001), and our intention was not to repeat that information. Rather, we offered some of our experiences in using GIS to model habitats at landscape levels as a means of drawing attention to commonly overlooked issues with data analysis and interpretation.

Successful development and implementation of wildlife models over large landscapes is a complex endeavor. The process often involves using landscape-level data derived from a variety of technologies and by analysts not necessarily associated with the wildlife modeling project. GIS software has empowered analysts and modelers to rapidly manipulate data, sometimes at the expense of evaluating data and processing limitations. During our experience in diverse GIS applications and projects, we have repeatedly witnessed poor planning or processing methodologies resulting in extra work to correct problems found in the results. In our opinion, a small amount of time spent understanding the data, the planned analysis, the limitations of both, and project needs will facilitate completion of project goals. We echo the sentiments of O'Neil et al. (2005) who noted that spatial technologies are only as accurate and reliable as the underlying data. Spatial tools alone cannot improve accuracy, precision, and bias of information. It is our hope that by sharing our experiences using GIS for landscape-level wildlife habitat modeling, we encouraged critical thinking about establishing project objectives, data limitations and structures, and issues of temporal and spatial scale.

ACKNOWLEDGMENTS

We thank the Michigan Department of Natural Resources (MDNR), Wildlife Division for their support in writing this chapter. Dr. Patrick Lederle with the MDNR, Wildlife Division saw the value in this chapter and encouraged M. Donovan and M. Strong to engage in this effort. The patience and editorial advice offered by Joshua Millspaugh and Frank Thompson are greatly appreciated. Comments from William Dijak and one anonymous reviewer greatly improved the manuscript.

LITERATURE CITED

Aebischer, N. J., P. A. Robertson, and R. E. Kenward. 1993. Compositional analysis of habitat use from animal radio-tracking data. *Ecology* 74:1313–1325.

Anderson, S. H., and K. J. Gutzwiller. 2005. Wildlife habitat evaluation. Pages 489–502 *in* C. E. Braun, editor. *Techniques for wildlife investigations and management.* The Wildlife Society, Bethesda, Maryland, USA.

Barbour, M. G., J. H. Burk, W. D. Pitts, F. S. Gilliam, and M. W. Schwartz. 1999. *Terrestrial plant ecology.* Third edition. Addison Wesley Longman, Menlo Park, California, USA.

Barry, S., and J. Elith. 2006. Error and uncertainty in habitat models. *Journal of Applied Ecology* 43:413–423.

Bender, L. C., G. J. Roloff, and J. B. Haufler. 1996. Evaluating confidence intervals for habitat suitability models. *Wildlife Society Bulletin* 24:347–352.

Bennetts, R. E., and W. M. Kitchens. 1997. Population dynamics and conservation of snail kites in Florida: The importance of spatial and temporal scale. *Colonial Waterbirds* 20:324–329.

Betts, M. G., S. E. Franklin, and R. G. Taylor. 2003. Interpretation of landscape pattern and habitat change for local indicator species using satellite imagery and geographic information system data in New Brunswick, Canada. *Canadian Journal of Forest Research* 33:1821–1831.

Beutel, T. S., R. J. S. Beeton, and G. S. Baxter. 1999. Building better wildlife-habitat models. *Ecography* 22:219–223.

Bian, L., and E. West. 1997. GIS modeling of elk calving habitat in a prairie environment with statistics. *Photogrammetric Engineering and Remote Sensing* 63:161–167.

Bissonette, J. A. 1997. *Wildlife and landscape ecology: Effects of pattern and scale*. Springer-Verlag, New York, New York, USA.

Bissonette, J. A., and I. Storch. 2003. *Landscape ecology and resource management: Linking theory with practice*. Island Press, Washington, D.C., USA.

Black, B. A., and M. D. Abrams. 2001. Analysis of temporal variation and species-site relationships of witness tree data in southeastern Pennsylvania. *Canadian Journal of Forest Research* 31:419–429.

Bojórquez-Tapia, L. A., L. Juárez, and G. Cruz-Bello. 2002. Integrating fuzzy logic, optimization, and GIS for ecological impact assessments. *Environmental Management* 30:418–433.

Bolger, D. T., T. A. Scott, and J. T. Rotenberry. 1997. Breeding bird abundance in an urbanizing landscape in coastal southern California. *Conservation Biology* 11:406–421.

Bowyer, R. T., J. G. Kie, and V. Van Ballenberghe. 1996. Sexual segregation in black-tailed deer: Effects of scale. *Journal of Wildlife Management* 60:10–17.

Boyd, I. L. 1996. Temporal scales of foraging in a marine predator. *Ecology* 77:426–434.

Brennan, J. M., D. J. Bender, T. A. Conteras, and L. Fahrig. 2002. Focal patch landscape studies for wildlife management: Optimizing sampling effort across scales. Pages 68–91 *in* J. Liu and W. W. Taylor, editors. *Integrating landscape ecology into natural resource management.* Cambridge University Press, Cambridge, United Kingdom.

Brotons, L., W. Thuiller, M. B. Araújo, and A. H. Hirzel. 2004. Presence-absence versus presence-only modelling methods for predicting bird habitat suitability. *Ecography* 27:437–448.

Brown, D.G 1998*a*. Classification and boundary vagueness in mapping presettlement forest types. *International Journal of Geographical Information Science* 12:105–129.

Brown, D.G 1998*b*. Mapping historical forest types in Baraga County Michigan, USA as fuzzy sets. *Plant Ecology* 134:97–111.

Burnham, K. P., and D. R. Anderson. 1998. *Model selection and inference: A practical information-theoretic approach*. Springer-Verlag, New York, New York, USA.

Cao, C., and N. S.-N. Lam. 1997. Understanding the scale and resolution effects in remote sensing and GIS. Pages 57–72 *in* D. A. Quattrochi and M. F. Goodchild, editors. *Scale in remote sensing and GIS.* CRC Press, Boca Raton, Florida, USA.

Cheung, W. W. L., T. J. Pitcher, and D. Pauly. 2005. A fuzzy logic expert system to estimate intrinsic vulnerabilities of marine fishes to fishing. *Biological Conservation* 124:97–111.

Chiles, J.-P., and P. Delfiner. 1999. *Geostatistics: Modeling spatial uncertainty.* Wiley, New York, New York, USA.

Comer, P. J., D. A. Albert, H. A. Wells, B. L. Hart, J. B. Raab, D. L. Price, D. M. Kashian, R. A. Corner, and D. W. Schuen. 1995. *Michigan's native landscape, as interpreted from the General Land Office Surveys 1816–1856.* Report to the U. S. Environmental Protection Agency Water Division and Michigan Department of Natural Resources Wildlife Division, Michigan Natural Features Inventory, Lansing, Michigan, USA.

Congalton, R. G., and K. Green. 1993. A practical look at the sources of confusion in error matrix generation. *Photogrammetric Engineering and Remote Sensing* 59:641–644.

Congalton, R. G., and K. Green. 1999. *Assessing the accuracy of remotely sensed data: Principles and practices.* Lewis Publishers, Boca Raton, Florida, USA.

Conner, L. M., M. D. Smith, and L. W. Burger. 2003. A comparison of distance-based and classification-based analyses of habitat use. *Ecology* 84:526–531.

Corsi, F., J. de Leeuw, and A. Skidmore. 2000. Modeling species distribution with GIS. Pages 389–434 *in* L. Boitani and T. K. Fuller, editors. *Research techniques in animal ecology: Controversies and consequences.* Columbia University Press, New York, New York, USA.

Cowell, C. M. 1995. Presettlement piedmont forests: Patterns of composition and disturbance in central Georgia. *Annals of the Association of American Geographers* 85:65–83.

Cowell, C. M. 1998. Historical change in vegetation and disturbance on the Georgia Piedmont. *American Midland Naturalist* 140:78–89.

Daubenmire, R. 1952. Forest vegetation of Idaho and adjacent Washington, and its bearing on concepts of vegetation classification. *Ecological Monographs* 22:301–330.

Dean, D. J., K. R. Wilson, and C. H. Flather. 1997. Spatial error analysis of species richness for a gap analysis map. *Photogrammetric Engineering and Remote Sensing* 63:1211–1217.

Delcourt, H. R., and P. A. Delcourt. 2006. Presettlement landscape heterogeneity: Evaluating grain of resolution using General Land Office survey data. *Landscape Ecology* 11:363–381.

Donovan, M. L., G. M. Nesslage, J. J. Skillen, and B. A. Maurer. 2004. *The Michigan gap analysis project final report*. Michigan Department of Natural Resources, Wildlife Division, Lansing, Michigan, USA.

Dunning, J. B., Jr., D. J. Stewart, B. J. Danielson, B. R. Noon, T. L. Root, R. H. Lamberson, and E. E. Stevens. 1995. Spatially explicit population models: Current forms and future uses. *Ecological Applications* 5:3–11.

Dussault, C., J.-P. Ouellet, and R. Courtois. 2005. A comparison of distance-based and classification-based analyses of habitat use: Comment. *Ecology* 86:3119–3125.

Edwards, T. C., Jr., E. T. Deshler, D. Foster, and G.G Moisen. 1996. Adequacy of wildlife habitat relation models for estimating spatial distributions of terrestrial vertebrates. *Conservation Biology* 10:263–270.

Edwards, T. C., Jr., G.G Moisen, T. S. Frescino, and J. J. Lawler. 2003. Modeling multiple ecological scales to link landscape theory to wildlife conservation. Pages 153–176 *in* J. A. Bissonette and I. Storch, editors. *Landscape ecology and resource management: Linking theory with practice.* Island Press, Washington, D.C., USA.

Engler, R., A. Guisan, and L. Rechsteiner. 2004. An improved approach for predicting the distribution of rare and endangered species from occurrence and pseudo-absence data. *Journal of Applied Ecology* 41:263–274.

ERDAS. 1982. *ERDAS field guide*. Fourth edition. ERDAS, Atlanta, Georgia, USA.

Erickson, W. P., T. L. McDonald, K. G. Gerow, S. Howlin, and J. W. Kern. 2001. Statistical issues in resource selection studies with radio-marked animals. Pages 209–242 *in* J. J. Millspaugh and J. M. Marzluff, editors. *Radio tracking and animal populations.* Academic Press, San Diego, California, USA.

Estes, J. E., and D. W. Mooneyhan. 1994. Of maps and myths. *Photographic Engineering and Remote Sensing* 60:517–524.

Eyre, F. H. 1980. *Forest cover types of the United States and Canada*. Society of American Foresters, Washington, D.C., USA.

Farley, S. C., R. E. Masters, and D. M. Engle. 2002. Riparian landscape change in central Oklahoma 1872–1991. *Proceedings of the Oklahoma Academy of Sciences* 82:57–71.

Fernández, N., M. Delibes, and F. Palomares. 2006. Landscape evaluation in conservation: Molecular sampling and habitat modeling for the Iberian lynx. *Ecological Applications* 16:1037–1049.

Fleming, K. K., K. A. Didier, B. R. Miranda, and W. F. Porter. 2004. Sensitivity of a white-tailed deer habitat-suitability index model to error in satellite land-cover data: Implications for wildlife habitat-suitability studies. *Wildlife Society Bulletin* 32:158–168.

Foody, G. M. 2002. Status of land cover classification accuracy assessment. *Remote Sensing of Environment* 80:185–201.

Forman, R. T. T., and M. Godron. 1981. Patches and structural components for a landscape ecology. *BioScience* 31:733–740.

Forman, R. T. T., and M. Godron. 1986. *Landscape ecology.* John Wiley and Sons, New York, New York, USA.

Fortin, D., J. M. Fryxell, and R. Pilote. 2002. The temporal scale of foraging decisions in bison. *Ecology* 83:970–982.

Frescino, T. S., T. C. Edwards, Jr., and G. G. Moisen. 2001. Modeling spatially explicit forest structural attributes using generalized additive models. *Journal of Vegetation Science* 12:15–26.

Garbrecht, J., F. L. Ogden, P. A. DeBarry, and D. R. Maidment. 2001. GIS and distributed watershed models. I: Data, coverages, and sources. *Journal of Hydrologic Engineering* 6:506–514.

Goodchild, M. F., and S. Gopal, editors. 1989. *The accuracy of spatial databases.* Taylor & Francis, Bristol, Pennsylvania, USA.

Gough, M. C., and S. P. Rushton. 2000. The application of GIS-modelling to mustelid landscape ecology. *Mammal Review* 30:197–216.

Grossman, D. H., P. Bourgeron, W.-D. Busch, D. Cleland, W. Platts, G. C. Ray, C. R. Robins, and G. Roloff. 1999. Principles for ecological classification. Pages 353–393 *in* R. C. Szaro, N. C. Johnson,

W. T. Sexton, and A. J. Malk, editors. *Ecological stewardship: A common reference for ecosystem management*. Volume 2, Elsevier Science, Oxford, United Kingdom.

Grossman, D. H., D. Faber-Langendoen, A. S. Weakley, M. Anderson, P. Bourgeron, R. Crawford, K. Goodin, S. Landaal, K. Metzler, K. Patterson, M. Pyne, M. Reid, and L. Sneddon. 1998. *International classification of ecological classification: Terrestrial vegetation of the United States*. Volume I. The National Vegetation Classification System: Development, status, and applications, The Nature Conservancy, Arlington, Virginia, USA.

Guisan, A., and N. E. Zimmermann. 2000. Predictive habitat distribution models in ecology. *Ecological Modelling* 135:147–186.

Haines, D. A., and R. W. Sando. 1969. *Climatic conditions preceding historically great fires in the north central region*. U.S. Forest Service, North Central Forest Experiment Station, Research Paper NC-34, St. Paul, Minnesota, USA.

Haines-Young, R., D. R. Green, and S. H. Cousins, editors. 1994. *Landscape ecology and GIS*. Taylor & Francis, Philadelphia, Pennsylvania, USA.

Hammond, J. S., R. L. Keeney, and H. Raiffa. 1999. *Smart choices: A practical guide to making better decisions*. Harvard Business School Press, Boston, Massachusetts, USA.

Haufler, J. B., C. A. Mehl, and G. J. Roloff. 1996. Using a coarse-filter approach with species assessment for ecosystem management. *Wildlife Society Bulletin* 24:200–208.

Hess, G. R., and J. M. Bay. 2004. Generating confidence intervals for composition-based landscape indexes. *Landscape Ecology* 12:309–320.

Hines, E. M., J. Franklin, and J. R. Stephenson. 2005. Estimating the effects of map error on habitat delineation for the California spotted owl in southern California. *Transactions in GIS* 9:541–559.

Hirzel, A. H., J. Hausser, D. Chessel, and N. Perrin. 2002. Ecological-niche factor analysis: How to compute habitat-suitability maps without absence data. *Ecology* 83:2027–2036.

Holmström, H., and J. E. S. Fransson. 2003. Combining remotely sensed optical and radar data in kNN-estimation of forest variables. *Forest Science* 49:409–418.

Hunsaker, C. T., M. F. Goodchild, M. A. Friedl, and T. J. Case, editors. 2001. *Spatial uncertainty in ecology: Implications for remote sensing and GIS applications*. Springer-Verlag, New York, New York, USA.

Hunsaker, C. T., R. V. O'Neill, B. L. Jackson, S. P. Timmins, D. A. Levine, and D. J. Norton. 1994. Sampling to characterize landscape pattern. *Landscape Ecology* 9:207–226.

Jackson, V. L., L. L. Laack, and E. G. Zimmerman. 2005. Landscape metrics associated with habitat use by ocelots in south Texas. *Journal of Wildlife Management* 69:733–738.

Jeganathan, P., R. E. Green, K. Norris, I. N. Vogiatzakis, A. Bartsch, S. R. Wotton, C. G. R. Bowden, G. H. Griffiths, D. Pain, and A. R. Rahmani. 2004. Modelling habitat selection and distribution of the critically endangered Jerdon's courser *Rhinoptilus bitorquatus* in scrub jungle: An application of a new tracking method. *Journal of Applied Ecology* 41:224–237.

Jelinski, D. E., and J. Wu. 1996. The modifiable areal unit problem and implications for landscape ecology. *Landscape Ecology* 11:129–140.

Johnson, C. G., Jr., R. R. Clausnitzer, P. J. Mehringer, and C. D. Oliver. 1994. *Biotic and abiotic processes of eastside ecosystems: The effects of management on plant and community ecology, and on stand and landscape vegetation dynamics*. U.S. Forest Service, Pacific Northwest Research Station, General Technical Report PNW-GTR-322, Portland, Oregon, USA.

Johnson, D. H. 1980. The comparison of usage and availability measurements for evaluating resource preference. *Ecology* 61:65–71.

Johnson, L. B. 1990. Analyzing spatial and temporal phenomena using geographical information systems: A review of ecological applications. *Landscape Ecology* 4:31–43.

Kareiva, P., and U. Wennergren. 1995. Connecting landscape patterns to ecosystem and population processes. *Nature* 373:299–302.

Karl, J. W., P. J. Heglund, E. O. Garton, J. M. Scott, N. M. Wright, and R. L. Hutto. 2000. Sensitivity of species habitat-relationship model performance to factors of scale. *Ecological Applications* 10:1690–1705.

Kerr, J. T., and M. Ostrovsky. 2003. From space to species: Ecological applications for remote sensing. *Trends in Ecology and Evolution* 18:299–305.

Knight, J. F., and R. S. Lunetta. 2003. An experimental assessment of minimum mapping unit size. *IEEE Transactions on GeoScience and Remote Sensing* 41:2132–2134.

Laba, M., S. K. Gregory, J. Braden, D. Ogurcak, E. Hill, E. Fegraus, J. Fiore, and S. D. DeGloria. 2002. Conventional and fuzzy accuracy assessment of the New York gap analysis project land cover map. *Remote Sensing of Environment* 81:443–455.

Lawler, J. J., R. J. O'Connor, C. T. Hunsaker, K. B. Jones, T. R. Loveland, and D. White. 2004. The effects of habitat resolution on models of avian diversity and distributions: A comparison of two land-cover classifications. *Landscape Ecology* 19:515–530.

Lawler, J. J., and N. H. Schumaker. 2004. Evaluating habitat as a surrogate for population viability using a spatially explicit population model. *Environmental Monitoring and Assessment* 94:85–100.

Lehmann, A. 2004. GIS modeling of submerged macrophyte distribution using generalized additive models. *Plant Ecology* 139:113–124.

Levin, S. A. 1992. The problem of pattern and scale in ecology. *Ecology* 73:1943–1967.

Liegel, C. 1982. The pre-European settlement vegetation of the Aldo Leopold Memorial Reserve. *Transactions of the Wisconsin Academy of Science, Arts and Letters* 70:13–26.

Linden, D. W. 2006. *Modeling current and historic habitat for Canada lynx (*Lynx canadensis*) in the upper peninsula of Michigan*. Thesis, Michigan State University, East Lansing, USA.

Lowell, K., and A. Jaton, editors. 1999. *Spatial accuracy assessment: Land information uncertainty in natural resources.* Ann Arbor Press, Chelsea, Michigan, USA.

Lu, D. S., P. Mausel, E. Brondízio, and E. Moran. 2004. Relationships between forest stand parameters and Landsat TM spectral responses in the Brazilian Amazon Basin. *Forest Ecology and Management* 198:149–167.

Luoto, M., and M. Seppälä. 2002. Modelling the distribution of palsas in Finnish Lapland with logistic regression and GIS. *Permafrost and Periglacial Processes* 13:17–28.

MacQueen, J. 1967. Some methods for classification and analysis of multivariate observations. Pages 281–297 *in* L. M. L. Cam and J. Neyman, editors. *Proceedings of the Fifth Berkeley symposium on mathematical statistics and probability.* Volume 1, University of California Press, Berkeley, USA.

Manies, K. L., and D. J. Mladenoff. 2000. Testing methods to produce landscape-scale presettlement vegetation maps from the U.S. General Land Office survey records. *Landscape Ecology* 15:741–754.

Manly, B. F. J. 1997. *Randomization, bootstrap and Monte Carlo methodology in biology.* Second edition. CRC Press, Boca Raton, Florida, USA.

Maurer, B. A. 2002. Predicting distribution and abundance: Thinking within and between scales. Pages 125–139 *in* J. M. Scott, P. J. Heglund, M. L. Morrison, J. B. Haufler, M. G. Raphael, W. W. Wall, and F. B. Samson, editors. *Predicting species occurrences: Issues of accuracy and scale.* Island Press, Washington, D.C., USA.

McCarty, J. P., D. J. Levey, C. H. Greenberg, and S. Sargent. 2002. Spatial and temporal variation in fruit use by wildlife in a forested landscape. *Forest Ecology and Management* 164:277–291.

McCollum, J. M., and J. K. Cochran. 2003. Forest inventory and analysis and forest health monitoring: Piecing the quilt. Pages 175–182 *in* R. E. McRoberts, G. A. Reams, P. C. Van Deusen, and W. H. McWilliams, editors. *Proceedings of the Fifth Annual Forest Inventory and Analysis Symposium.* U.S. Forest Service, General Technical Report WO-69, Washington, D.C., USA.

McGwire, K. C., and P. Fisher. 2001. Spatially variable thematic accuracy: Beyond the confusion matrix. Pages 308–329 *in* C. T. Hunsaker, M. F. Goodchild, M. A. Friedl, and T. J. Case, editors. *Spatial uncertainty in ecology: Implication for remote sensing and GIS applications.* Springer-Verlag, New York, New York, USA.

Michigan Department of Natural Resources. 2001. *Current use inventory classification system definitions*. MDNR, Michigan Resource Information System, Land Resource Programs Division, Lansing, Michigan, USA.

Mladenoff, D. J., T. A. Sickley, and A. P. Wydeven. 1999. Predicting gray wolf landscape recolonization: Logistic regression models vs. new field data. *Ecological Applications* 9:37–44.

Moellering, H., H. J. G. L. Aalders, and A. Crane, editors. 2005. *World spatial metadata standards.* Elsevier Press, San Diego, California, USA.

Morris, D. W. 1987. Ecological scale and habitat use. *Ecology* 68:362–369.

Morrison, M. L. 2002. Role of temporal and spatial scale. Pages 123–124 *in* J. M. Scott, P. J. Heglund, M. L. Morrison, J. B. Haufler, M. G. Raphael, W. W. Wall, and F. B. Samson, editors. *Predicting species occurrences: Issues of accuracy and scale.* Island Press, Washington, D.C., USA.

Morrison, M. L., B. G. Marcot, and R. W. Mannan. 1998. *Wildlife-habitat relationships: Concepts and applications.* Second edition. University of Wisconsin Press, Madison, Wisconsin, USA.

Mowrer, H. T., and R. G. Congalton. 2000. *Quantifying spatial uncertainty in natural resources: Theory and applications for GIS and remote sensing.* Ann Arbor Press, Chelsea, Michigan, USA.

O'Neil, T. A., P. Bettinger, B. G. Marcot, B. W. Luscombe, G. T. Koeln, H. J. Bruner, C. Barrett, J. A. Pollock, and S. Bernatas. 2005. Application of spatial technologies in wildlife biology. Pages 418–447 *in* C. E. Braun, editor. *Techniques for wildlife investigations and management.* The Wildlife Society, Bethesda, Maryland, USA.

Orians, G. H., and J. F. Wittenberger. 1991. Spatial and temporal scales in habitat selection. *American Naturalist* 137:S29–S49.

Osborne, P. E., J. C. Alonso, and R. G. Bryant. 2001. Modelling landscape-scale habitat use using GIS and remote sensing: A case study with great bustards. *Journal of Applied Ecology* 38:458–471.

Pearlstine, L. G., S. E. Smith, L. A. Brandt, C. R. Allen, W. M. Kitchens, and J. Stenberg. 2002. Assessing state-wide biodiversity in the Florida Gap analysis project. *Journal of Environmental Management* 66:127–144.

Pereira, J. M. C., and R. M. Itami. 1991. GIS-based habitat modeling using logistic multiple regression: A study of the Mt Graham red squirrel. *Photogrammetric Engineering and Remote Sensing* 57:1475–1486.

Pidgeon, A. M., L. A. Schulte, and D. J. Mladenoff. 2005. *Pre-Euroamerican distribution of pine warblers* (Dendroica pinus) *in Wisconsin: A tool for setting regional conservation priorities.* U.S. Forest Service, Pacific Southwest Research Station, General Technical Report PSW-GTR-191, Berkeley, California, USA.

Porter, W. F., and K. E. Church. 1987. Effects of environmental pattern on habitat preference analysis. *Journal of Wildlife Management* 51:681–685.

Qi, Y., and J. Wu. 1996. Effects of changing spatial resolution on the results of landscape pattern analysis using spatial autocorrelation indices. *Landscape Ecology* 11:39–49.

Radeloff, V. C., D. J. Mladenoff, H. S. He, and M. S. Boyce. 1999. Forest landscape change in the northwestern Wisconsin pine barrens from pre-European settlement to the present. *Canadian Journal of Forest Research* 29:1649–1659.

Ramsey, E. W., III, and M. L. Strong. 2000. Conveying multiple, complex themes and classes for natural resource assessments. *Geocarto International* 15:29–35.

Raphael, M. G., and B. G. Marcot. 1986. Validation of a wildlife-habitat-relationships model: Vertebrates in a Douglas-fir sere. Pages 129–138 *in* J. Verner, M. Morrison, and C. J. Ralph, editors. *Wildlife 2000: Modeling habitat relationships of terrestrial vertebrates.* University of Wisconsin Press, Madison, USA.

Robinson, V. B. 2003. A perspective on the fundamentals of fuzzy sets and their use in geographic information systems. *Transactions in GIS* 7:3–30.

Rodgers, C. S., and R. C. Anderson. 1979. Presettlement vegetation of two prairie peninsula counties. *Botanical Gazette* 140:232–240.

Roloff, G. J., and B. J. Kernohan. 1999. Evaluating reliability of habitat suitability index models. *Wildlife Society Bulletin* 27:973–985.

Roloff, G. J., J. J. Millspaugh, R. A. Gitzen, and G. C. Brundige. 2001. Validation tests of a spatially explicit habitat effectiveness model for Rocky Mountain elk. *Journal of Wildlife Management* 65:899–914.

Roloff, G. J., M. C. Reynolds, and J. B. Haufler. 1994. Variability in the structure and composition of regenerating aspen with implications for ruffed grouse management. *Michigan Academician* 26:75–82.

Saura, S. 2002. Effects of minimum mapping unit on land cover data spatial configuration and composition. *International Journal of Remote Sensing* 23:4853–4880.

Schulte, L. A., and D. J. Mladenoff. 2001. The original U.S. public land survey records: Their use and limitations in reconstructing pre-European settlement vegetation. *Journal of Forestry* 99:5–10.

Schulte, L. A., A. M. Pidgeon, and D. J. Mladenoff. 2005. One hundred fifty years of change in forest bird breeding habitat: Estimates of species distributions. *Conservation Biology* 19:1944–1956.

Schumaker, N. H. 1996. Using landscape indices to predict habitat connectivity. *Ecology* 77:1210–1225.

Segurado, P., and M. B. Araújo. 2004. An evaluation of methods for modelling species distributions. *Journal of Biogeography* 31:1555–1568.

Short, H. L., and J. B. Hestbeck. 1995. National biotic resource inventories and GAP analysis: Problems of scale and unproven assumptions limit a national program. *BioScience* 45:535–539.

Silvert, W. 1997. Ecological impact classification with fuzzy sets. *Ecological Modelling* 96:1–10.

Simon, J. L. 1997. *Resampling: The "new statistics."* Resampling Statistics, Arlington, Virginia, USA.

Space Imaging Solutions. 2001. *Integrated forest monitoring assessment and prescription, IFMAP. Review of remote sensing technologies for the IFMAP project*. Report prepared for the Michigan Department of Natural Resources, Space Imaging Solutions, Ann Arbor, Michigan, USA.

Starfield, A. M. 1997. A pragmatic approach to modeling for wildlife management. *Journal of Wildlife Management* 61:261–270.

Stehman, S. V. 1997. Selecting and interpreting measures of thematic classification accuracy. *Remote Sensing of Environment* 62:77–89.

Stohlgren, T. J., G. W. Chong, M. A. Kalkhan, and L. D. Schell. 1997. Multiscale sampling of plant diversity: Effects of minimum mapping unit size. *Ecological Applications* 7:1064–1074.

Stoms, D. M. 1994. Scale dependence of species richness maps. *The Professional Geographer* 46:346–358.

Stoms, D. M., F. W. Davis, and C. B. Cogan. 1992. Sensitivity of wildlife habitat models to uncertainties in GIS data. *Photogrammetric Engineering and Remote Sensing* 58:843–850.

Théau, J., D. R. Peddle, and C. R. Duguay. 2005. Mapping lichen in a caribou habitat of northern Quebec, Canada, using an enhancement-classification method and spectral mixture analysis. *Remote Sensing of Environment* 94:232–243.

Tischendorf, L. 1997. Modelling individual movements in heterogeneous landscapes: Potentials of a new approach. *Ecological Modelling* 103:33–42.

Tuominen, S., S. Fish, and S. Poso. 2003. Combining remote sensing, data from earlier inventories, and geostatistical interpolation in multisource forest inventory. *Canadian Journal of Forest Research* 33:624–634.

Tuominen, S., and A. Pekkarinen. 2005. Performance of different spectral and textural aerial photograph features in multi-source forest inventory. *Remote Sensing of Environment* 94:256–268.

Turner, M. G., R. V. O'Neill, R. H. Gardner, and B. T. Milne. 1989. Effects of changing spatial scale on the analysis of landscape pattern. *Landscape Ecology* 3:153–162.

U.S. Forest Service. 1996. *Status of the interior Columbia basin: Summary of scientific findings*. U.S. Forest Service, Pacific Northwest Research Station, General Technical Report PNW-GTR-385, Portland, Oregon, USA.

U.S. Forest Service. Forest inventory and analysis program. <http://fia.fs.fed.us/>. Accessed 5 January 2007.

Van Deelen, T. R., K. S. Pregitzer, and J. B. Haufler. 1996. A comparison of presettlement and present-day forests in two northern Michigan deer yards. *American Midland Naturalist* 135:181–194.

Verbyla, D. L., and J. A. Litvaitis. 1989. Resampling methods for evaluating classification accuracy of wildlife habitat models. *Environmental Management* 13:783–787.

Wade, T. G., J. D. Wickham, M. S. Nash, A. C. Neale, K. H. Riitters, and K. B. Jones. 2003. A comparison of vector and raster GIS methods for calculating landscape metrics used in environmental assessments. *Photogrammetric Engineering and Remote Sensing* 69:1399-1405.

Wang, Y.-C. 2005. Presettlement land survey records of vegetation: Geographic characteristics, quality and modes of analysis. *Progress in Physical Geography* 29:568-598.

Waring, R. H., and S. W. Running. 1998. *Forest ecosystems: Analysis at multiple scales*. Second edition. Academic Press, San Diego, California, USA.

Whitehead, H. 1996. Variation in the feeding success of sperm whales: Temporal scale, spatial scale and relationship to migrations. *Journal of Animal Ecology* 65:429-438.

Wickham, J. D., and K. H. Riitters. 1995. Sensitivity of landscape metrics to pixel size. *International Journal of Remote Sensing* 16:3585-3594.

Wiens, J. A. 1989. Spatial scaling in ecology. *Functional Ecology* 3:385-397.

Wolock, D. M., and C. V. Price. 1994. Effects of digital elevation model map scale and data resolution on a topography-based watershed model. *Water Resources Research* 30:3041-3052.

Wu, F. 2004. GIS-based simulation as an exploratory analysis for space-time processes. *Journal of Geographical Systems* 1:199-218.

Wu, J. 2004. Effects of changing scale on landscape pattern analysis: Scaling relations. *Landscape Ecology* 19:125-138.

Wu, J., and O. L. Loucks. 1995. From balance-of-nature to hierarchical patch dynamics: A paradigm shift in ecology. *Quarterly Review of Biology* 70:439-466.

Zaniewski, A. E., A. Lehmann, and J. M. Overton. 2002. Predicting species spatial distributions using presence-only data: A case study of native New Zealand ferns. *Ecological Modelling* 157:261-280.

Zhang, L., C. Liu, C. J. Davis, D. S. Solomon, T. B. Brann, and L. W. Caldwell. 2004. Fuzzy classification of ecological habitats from FIA data. *Forest Science* 50:117-127.

Zhu, A. X., B. Hudson, J. Burt, K. Lubich, and D. Simonson. 2001. Soil mapping using GIS, expert knowledge, and fuzzy logic. *Soil Science Society of America Journal* 65:1463-1472.

CHAPTER

A Review of LANDIS and Other Forest Landscape Models for Integration with Wildlife Models

12

Hong S. He

The essential goal of wildlife management is to conserve wildlife populations. This goal is often evaluated with wildlife habitat models or population models. These wildlife models often assume that wildlife habitats remain static (Akçakaya 2001). Such an assumption can be problematic when models span long time periods and large spatial extents. Habitat abundance, quality, and distribution will change due to many natural and anthropogenic processes, including spatial and landscape processes. Forest landscape processes are a set of spatially contagious processes such as fire, wind, seed dispersal, insect and disease propagation, and forest harvest. They operate at large spatial (10^3–10^7 ha) and temporal scales (10^1–10^3 years), overlapping with the scales required to study habitat and population change. Forest landscape processes are key factors affecting forest landscape dynamics and consequently habitat, abundance, quality, and distribution for wildlife. Such processes can be addressed with forest landscape models. Integrating forest landscape modeling and wildlife modeling provides a viable tool to fully address forest wildlife habitats and population under various natural and anthropogenic regimes (Akçakaya et al. 2004, Wintle et al. 2005, Pichancourt et al. 2006, Shifley et al. 2006).

Over the past 15 years, there has been a rapid development in the field of forest landscape modeling, fueled by both technological and theoretical advances (Noon et al., this volume). From a technological perspective, forest landscape models have benefited greatly from increasing computing capacity and the development of GIS, remote sensing, and software engineering (Roloff et al., this volume). Incorporating ecological processes into forest landscape models and digitally representing these processes and their interactions can be facilitated through the use of well-designed computer software (He et al. 1999, He et al. 2002). The discipline of landscape ecology studies the interaction of spatial pattern and ecological processes under various spatiotemporal scales and theories of disturbance, equilibrium, and nonequilibrium approaches of vegetation and ecosystems. Landscape ecology provides a strong theoretical and conceptual basis for forest landscape modeling. The general background of forest landscape model development is reviewed by Sklar and Costanza (1990), Gardner et al. (1999), Mladenoff and Baker (1999), Mladenoff (2004), Scheller and Mladenoff (2007), and He (2008).

Forest landscape models share two common features: (1) They simulate forest vegetation response at large spatial and temporal scales; and (2) they simulate outcomes of repeated, stochastic landscape processes (e.g., seed dispersal, fire, wind, insects, diseases, harvests, and fuel treatments). Depending on the model purpose and design limitations, forest landscape models may differ in the key ecological processes incorporated, the extent to which mechanistic details are simulated for each process, and the type and scope of applications. These differences render certain forest landscape models more suitable for wildlife modeling than others.

I review approaches used to simulate site-level vegetation dynamics in forest landscape modeling because different approaches have different potentials and limitations for use with wildlife modeling. I use the LANDIS model to illustrate a general framework of integrating a forest landscape model with habitat suitability models or population models. I show that (1) habitat suitability can be evaluated under a combination of forest management and natural disturbance scenarios; (2) habitat suitability is dynamic, driven by succession, disturbance, and management; (3) habitat maps can be derived using relevant species and age class combinations targeted specifically to individual wildlife species; and (4) aggregations and disaggregations of vegetation classes using raster data simulated in the LANDIS model allow habitats to be mapped at multiple scales. Finally, I discuss limitations of the LANDIS model in this coupled modeling framework.

FOREST LANDSCAPE MODELS AND THEIR POTENTIALS IN WILDLIFE MODELING

A forest landscape model simulates spatiotemporal characteristics (distribution, shape, abundance, etc.) of at least one recurrent landscape process in a spatial context. Under this definition, a forest landscape model (1) is a simulation model where the model objects of time t are derived from the model objects of time t-1; (2) simulates recurrence of one or more landscape processes; and (3) operates at a large spatial and temporal extent that is adequate to simulate the landscape process (He 2008).

Forest landscape models have been developed using diverse approaches largely driven by research or application. Two features that are keys to integrating forest landscape models with wildlife models are the number and types of landscape processes simulated and the type of vegetation data tracked by each model entity (i.e., basic modeling unit such as pixel or polygon).

Most forest landscape models employ simplified approaches to simulate site-level vegetation dynamics, under the premise that fine-scale, site-level dynamics can be aggregated, whereas modeled landscape-scale objects are relatively less affected (Rastetter et al. 1992). Three approaches are generally used to simulate site-level vegetation dynamics, each approach having its own potentials and limitations for wildlife modeling.

Landscape Process Models

Landscape process models do not explicitly simulate site-level vegetation dynamics; rather, variables representing simulated landscape processes are used as surrogates for site-level succession dynamics (He 2008). For example, the variable time since last fire is used to represent stand age in DISPATCH (Baker et al. 1991) and ONFIRE (Li et al. 1997), time since last treatment is used to represent the amount of fuel accumulated in FIRESCAPE (Cary 1998), and time since last harvest is used to represent stand age in HARVEST (Gustafson and Crow 1996), while the actual site-level vegetation or succession is not simulated. These models are either highly theoretical or tend to work in the systems where dominant landscape processes may override site-level succession dynamics. For example, in boreal forests in Canada (Li et al. 1997), western coniferous forests in the United States (Romme and Despain 1989), chaparral shrub lands in southern California (Franklin et al. 2005), and Eucalypt forests in Australia (Gott 2005), fires tend to be stand replacing, and once they occur, they can reset succession to the initial stage.

For this group of models, inferences about vegetation are usually derived using empirical relationships based on the variables representing the landscape process such as time since last disturbance or harvest recorded at each site (pixel or polygon). The inferred vegetation information is usually general, such as stand development stage (e.g., young or old forests) or seral stages such as early, mid, or late successional vegetation. The inferred vegetation characteristics and these simulated spatial patterns can then be used to model wildlife habitats or populations. For example, HARVEST is a timber harvest allocation model developed by Gustafson and Crow (1994). It does not track actual timber or vegetation for each site and therefore does not simulate site-level dynamics. Rather, it allows the input of specific rules of clearcutting to generate landscape patterns that reflect the "look and feel" of managed landscape. The output of HARVEST can be used to determine the effect of variation in harvest size, rotation, and total harvest area on the spatial pattern of a forested landscape. The simulated spatial patterns were further assessed for a generalized neotropical migrant forest bird using a GIS model (Gustafson and Crow 1994).

These models usually simulate one landscape process, either fire or harvest, because simulating multiple landscape processes generally requires simulating site-level succession. Thus, application of these models is limited primarily to situations in which there is interest in the effects of one dominant landscape process, such as fire or timber harvest, but not both.

Succession Pathway Models

Succession pathway models track vegetation types for each model unit (pixel or polygon) and use state-and-transition models to represent succession by linking vegetation types or development stages to the transition time. Succession proceeds along pathways until it reaches a climax or stable vegetation type. Succession pathways are deterministic, but stochastic characteristics such as transition time and transition probability can be built in using Markov modeling (e.g., Gardner et al. 1999, Hargrove et al. 2000). Landscape processes interact with the pathway by forwarding or rewinding succession stages. It is possible to incorporate a landscape process of different forms (e.g., fire with different intensities) or multiple landscape processes (e.g., fire, insect, disease, and harvesting) into succession pathways.

Succession pathway models are highly empirical; the transition time and direction are often quantified from empirical knowledge through field observation. Succession pathways can be as simple as vegetation development stages (e.g. seedling, sapling, young forest, and old forest) or as specific as major vegetation types of different seral stages such as those developed by Keane et al. (2004) for mountain pine beetle.

Single pathway models are usually associated with one landscape process, such as fire in the model by EMBYR (Gardner et al. 1999, Hargrove et al. 2000). In EMBYR, vegetation is interpreted as fuel types and updated per iteration via a predefined transition probability. Multiple pathway models are associated with multiple landscape processes or one model object with multiple forms; examples are LANDSUM (Keane et al. 2002), SIMPPLLE (Chew et al. 2004), and LADS (Wimberly et al. 2000). In LANDSUM and LADS, fire can have multiple forms in terms of intensity, such as stand replacement fire (high severity) and nonlethal surface fire (low severity). Vegetation can have multiple predefined pathways under these fire severities.

The potential to integrate succession pathway models with wildlife models varies depending on the succession pathways defined. The advantage of succession pathway models is that when empirical knowledge is available, succession pathway approaches have the flexibility to incorporate multiple landscape processes. Multiple succession pathways may provide alternatives of disturbance and management, under which more realistic examination of wildlife response can be achieved. A limitation of integrating succession pathway models with wildlife models is the predefined pathways. Since succession pathways are predefined in these models, wildlife habitat types are also predefined. Thus, succession pathway models may be of limited value for studying wildlife habitats that are not included in the predefined succession pathways, which are driven by disturbance and succession stages.

Vital Attribute Models

Vital attribute models use vital attributes defined as a set of autecological characteristics necessary to predict plant species' behavior in environments of recurrent disturbance (Nobel and Slatyer 1980). The vital attributes generally reflect plant species succession (longevity, sexual maturity, sprouting), competition (shade tolerance), dispersal, and their tolerance to disturbance. They can be defined either for individual species or functional groups of species (Roberts and Betz 1999). Vegetation dynamics on an individual site are simulated as competitive processes driven by species' longevity, maturity, seeding and resprouting capability, and environmental suitability, defined by species' vital attributes. Without disturbance, more shade-tolerant species will outcompete less shade-tolerant species to reach climax or a stable state. Species' vital attributes can also interact with disturbances through species' tolerance to disturbance. Postdisturbance response is driven by a combination of species' longevity, maturity, seeding capability, sprouting capability, and environmental adaptability. Roberts (1996) first implemented the vital attribute approach in LANDSIM, a polygon-based model, and this approach is used by the LANDIS models (Mladenoff 2004).

Vital attribute approaches are more mechanistic, less deterministic, and more flexible in deriving vegetation or habitat types than succession pathway approaches because vegetation types are not predefined. Vital attribute approaches generally have a greater potential for integration with wildlife models than do landscape process or successional pathway models. The reason is that vegetation type is an emergent property determined by the interactions of plant species, environment, disturbance, and/or management. Thus, wildlife habitat requirements can be defined specifically based on a focal plant species or a group of plant species, and habitats can be generated from model outputs such as plant species and age class.

THE LANDIS MODEL

Overall Model Design and Structure

LANDIS is a vital-attribute, raster-based forest landscape model (Mladenoff et al. 1996, Mladenoff and He 1999) (Fig. 12-1). Each raster unit or cell tracks (1) the presence or absence of 1- to 10-year age cohorts of individual species; (2) fuel accumulation level; (3) the time since last disturbance (e.g., fire, wind, insects, and disease) and the time since last management (e.g., harvest and fuel treatment); and (4) species establishment ability in particular land types. For each cell, species birth, growth, death, regeneration, random mortality, and vegetative reproduction are simulated at 1- to 10-year time steps. At landscape scales, seed dispersal, disturbances, and management are simulated each iteration. To simulate heterogeneous landscapes, one uses land types derived from other

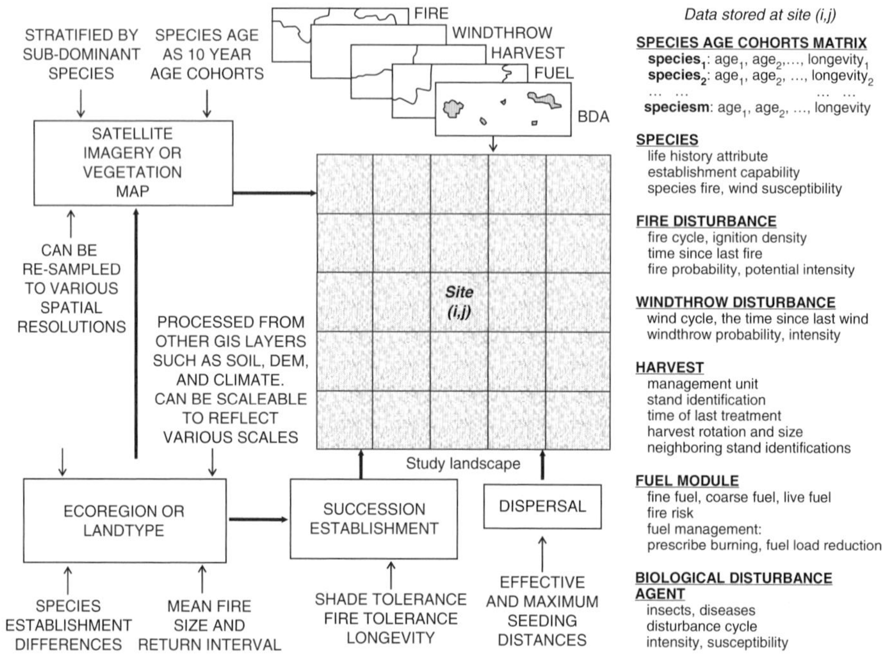

FIG. 12-1

LANDIS model structure (duplicated from Figure 1 in He et al. 2005). In LANDIS, a landscape is divided into equal-sized individual cells or sites. Each *site* (*i, j*) resides on a certain land type and includes a unique list of species present and their associated age cohorts. The species/age cohort information varies via establishment, succession, and seed dispersal, and interacts with disturbances. For disturbance heterogeneity (except for wind and biological disturbances), LANDIS stratifies the heterogeneous disturbance regimes using disturbance regime maps. Within-regime heterogeneity is further simulated by the stochastic process of each disturbance regime, and pixel-level heterogeneity is simulated through the interaction of disturbance and vegetation in the particular pixel.

climate and soil GIS data layers to stratify the landscape into smaller homogenous land units. At a given focal resolution such as within each ecoregion, physical conditions are assumed homogeneous, as are some characteristics such as fuel accumulation and decomposition rates, and species establishment (He et al. 1999, Mladenoff and He 1999). LANDIS simulates succession as a nonspatial process and seed dispersal, fire, wind, insect and disease, timber harvesting, and fuel treatment as spatial processes (He et al. 2005).

Succession

Succession is a result of birth establishment, growth, death, and vegetative reproduction of individual species and competition among species. Species' competitive ability is determined based on simple logical rules by the combination of life history attributes and land type suitability (Mladenoff and He 1999). Shade-intolerant species (species with lower shade-tolerance class) cannot establish on a site where more shade-tolerant and mature species are present. On the other hand, the most shade-tolerant species are delayed in establishment on an open site until specified years of shade creation are met. A shade-checking algorithm defines shade by the most shade-tolerant species cohort present that is also sexually mature. Species cohorts younger than the minimum seed-producing age are ignored in this shade-checking algorithm. This approach was implemented as a surrogate for crown closure. Without disturbance, shade-tolerant species will tend to dominate the landscape if other attributes are not highly limiting and land types (reflected as species establishment coefficients) are generally suitable.

In LANDIS, vegetation heterogeneity is modeled at multiple hierarchical levels from the landscape to the pixel. Land type captures the highest level (coarse grain) of spatial vegetation-heterogeneity caused by various environmental controls. A somewhat uniform suite of ecological conditions that results in similar species establishment patterns is assumed for each land type. Within a land type, stochastic processes such as seed dispersal can result in intermediate-level heterogeneity of a species distribution. At an individual site or pixel level, succession, competition, and probabilistic establishment may result in heterogeneity of species presence and age cohorts even among pixels that were initially identical.

Natural Disturbance

To simulate different types of disturbances and their effects on tree species composition and landscape pattern is central to the design of LANDIS. LANDIS can simulate three different types of natural disturbances (fire, wind, and biological), which can be applied in any combination. While each disturbance module follows its own set of rules, there are some basic designs common to all disturbance modules that are described here.

A disturbance event can be simplified into three steps: (1) selecting sites to be disturbed; (2) determining disturbance intensity; and (3) removing susceptible and intolerable species-age cohorts (e.g., disturbance-caused mortality or effects). Disturbance site selection involves both the starting site (e.g., ignition) and spread algorithms that vary by disturbance types. For example, a wind or fire event spreads across a subset of sites forming disturbance patches. In each case, stochastic (e.g., ignition and some components in spread) and deterministic processes (mortality) determine the final shape and form of the disturbance event.

Disturbance intensity classes in LANDIS approximate the relative strength of the simulated disturbance event, and their specific calculation varies by disturbance type. Vulnerability of species and/or age cohorts to a given disturbance type can vary; in LANDIS, tolerance class defines the relative vulnerability of a species to a given disturbance type and intensity, and susceptibility class defines the relative vulnerability of a species age group to a given disturbance type and intensity (He and Mladenoff 1999). For example, fire is simulated, in general, as a bottom-up disturbance, in which the youngest age cohorts are most susceptible to mortality. However, a low intensity fire may not kill species of high fire tolerance class even if its age cohorts are young. Removal of intolerant species and susceptible age classes is resultant from the interaction of disturbance intensity and species tolerance and susceptibility at each site and calculated for each species age cohort present on that site to determine which species age cohorts are removed by the disturbance.

Timber Harvest and Fuel Treatment

The LANDIS harvest and fuel treatment modules use a hierarchical approach that reflects the typical decisions made and execution of timber harvest and fuel treatment activities in managed forests. In both modules, a landscape is divided into forest management areas that represent spatial zones to which specific management goals are assigned. Within each management area, the landscape is divided into stands of various forest types. Stands are basic implementation units where harvest prescriptions are implemented. Each stand contains a group of grid cells, and each cell is populated with a specific combination of species and age cohorts. Management practices are the combination of temporal, spatial, and species components (Gustafson et al. 2000) and specify (1) when or how often to harvest or treat fuel for a stand; (2) how to allocate harvest and fuel treatment based on stand ranking, which in turn is based on user-specified criteria such as economic value or fuel loading; and (3) how to harvest a species age cohort (shelterwood, selection, and clear-cutting) or conduct a particular fuel treatment (prescribed burn or coarse woody debris reduction). The combination of these three components covers most forest management practices currently being used across a wide spectrum of ownership (Gustafson et al. 2000; He et al. 2004; Shang et al. 2004, 2007).

Operational Design of LANDIS

LANDIS has been continually updated and improved (Mladenoff and He 1999, He et al. 2005, Scheller and Mladenoff 2007). LANDIS uses a component-based approach to conduct simulation, which breaks the monolithic program into multiple small, standalone, and functionally more specific components (He et al. 2002, Scheller and Mladenoff 2007). The model is complex in terms of algorithms, interactions, and parameters. However, due to the modular design (He et al. 1999, 2002), LANDIS is relatively easy to use because users need to parameterize only the components of interests while turning off all modules not of interest.

Operationally, the model is a free-standing program. Implemented processes such as wind, fire, insect and disease, harvest, and fuel management have their

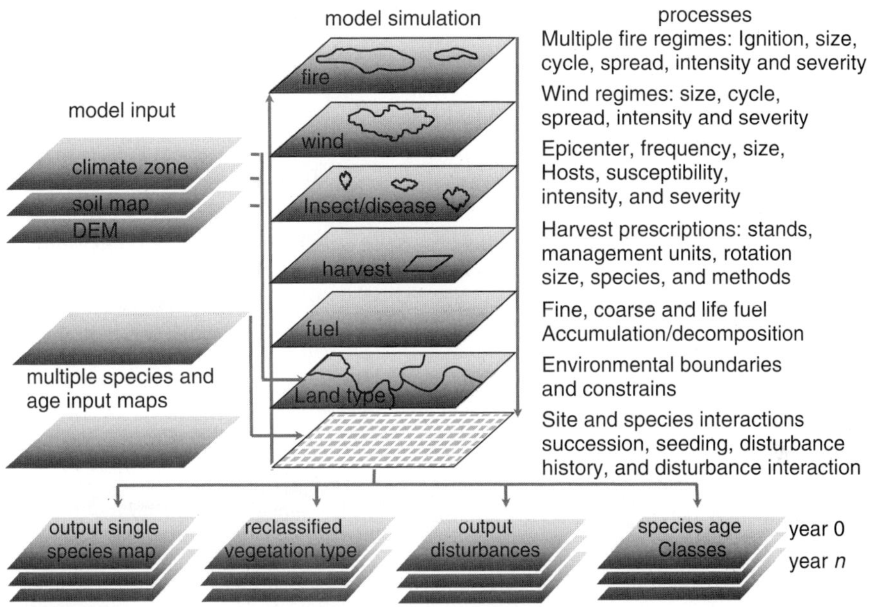

FIG. 12-2

The LANDIS model is a free-standing program that operates on a raster GIS format. Implemented processes such as harvest, fuel management, and disturbance can have their own respective input maps. The model can output maps of individual species, individual species age cohorts, forest type, and various disturbance and management history maps.

own respective input maps. The model can output maps of individual species, individual species age cohorts, forest type, and various disturbance and management history maps (Fig. 12-2).

GENERAL FRAMEWORK FOR INTEGRATING WILDLIFE MODELS AND LANDIS

One approach to integrating a wildlife model with LANDIS is to conduct scenario analysis. The lack of management experience at landscape scales and the limited feasibility of conducting landscape-scale experiments have resulted in the increasing use of scenario modeling to analyze the effects of different management actions on focal forests or wildlife species. Model scenarios are created by altering input parameters to reflect changes in climate, disturbance, and/or fuel or harvest alternatives, whereas the built-in model relationships remain unchanged. Comparing results from different model scenarios provides relative measurements regarding the direction and magnitude of changes within the simulated landscape.

Multiple simulation scenarios are derived by using the same set of spatial input and varied nonspatial input representing different simulation scenarios. Spatial input includes species composition and age classes representing the initial or current vegetation conditions, land type or ecoregion map that reflects the environmental heterogeneity, disturbance regimes maps, stand map, and management unit map for harvesting and fuel treatment. Nonspatial input includes parameters of each disturbance regime and management alternative as well as species vital attributes driving vegetation succession dynamics. Each scenario is independently simulated, and the output for each scenario contains time series maps of individual species, age classes, vegetation types, and disturbance and management effects (Fig. 12-3).

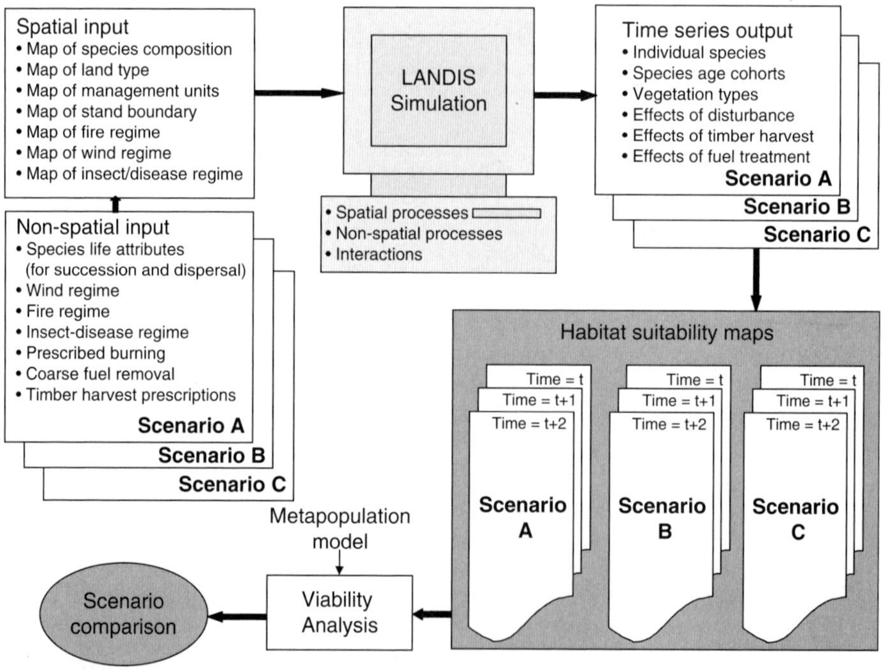

FIG. 12-3

Wildlife models can be integrated with multiple scenarios representing different management and disturbance alternatives that were simulated in LANDIS. The output for each scenario from LANDIS contains time series maps of individual species, age classes, vegetation types, and disturbance and management effects. Habitat suitability index (HSI) models can be used to create a time series of habitat (patch) maps from the LANDIS outputs, and population models can be used to assess viability. The habitat abundance, quality, and spatial structure over time via an HSI model and species viability via a metapopulation model can be compared among scenarios to evaluate the effects of disturbance and management alternatives on wildlife focal species or a suite of species.

For each scenario, LANDIS generates a time series of output maps of individual tree species and their age classes (Fig. 12-2). These output maps, along with other GIS layers, can be used as inputs for wildlife models, such as habitat suitability (HSI) models. This is an independent, post-LANDIS simulation process that is often performed in a GIS environment (Larsen et al. 2003) or with standalone software (Dijak et al. 2007; Dijak and Rittenhouse, this volume). This process often involves moving or sliding window GIS techniques, and within each window, HSI score, a measure of the quality of the habitat, is calculated based on plant species composition and age classes. Window sizes usually reflect the biological requirement of the focal species (e.g., home range or minimum territory requirement). The habitat abundance, quality, and spatial structure over time can be compared among scenarios to evaluate the effects of disturbance and management alternatives on wildlife habitat for a focal or a suite of species (Shifley et al. 2006).

Wildlife habitat models provide an assessment of habitat quality, but they do not provide specific information on populations. Population models can be applied to the time series of outputs from LANDIS, or habitat maps produced from these, to simulate population trajectories and viability. Thus, the effects of modeled scenarios of disturbance and management on wildlife population can be more specifically compared and evaluated (Akçakaya et al. 2004; Wintle et al. 2005; Akçakaya and Brook, this volume; Bekessy et al., this volume). Linking population models with LANDIS is first based on habitats delineated from habitat modeling (Fig. 12-3). The habitat quality (e.g., HSI score) of each pixel can be further used to derive habitat patches and their maximum carrying capacity (Akçakaya et al. 2005). This often provides the initialization of wildlife population for each pixel at the beginning of the LANDIS simulation year. Other demographic factors such as sex ratio, reproduction rate, fecundity, and survival rate are either measured directly in the field or parameterized from published sources to support the subsequent demographic modeling, and can also be linked to habitat quality. Typical results of demographic modeling coupled with LANDIS include species abundance and viability under various simulated disturbance and management regimes (Akçakaya et al. 2004, 2005). The coupling of LANDIS and a wildlife model can be loose or seamless. In the former case, LANDIS is run independently, and the simulated results are separately analyzed with wildlife habitat or population models (Larson et al. 2004, Shifley et al. 2006). Seamless simulations using LANDIS and the population model RAMAS can be run with the RAMAS-LANDIS model (Akçakaya et al. 2004; Bekessy et al., this volume).

Akçakaya et al. (2004) demonstrated the use of RAMAS-LANDIS in assessing the effects of forest management scenarios on sharp-tailed grouse (*Tympanuchus phasianellus*) in the northern Wisconsin Pine Barrens. The region has been severely altered since human settlement, resulting in relatively old red pine (*Pinus resinosa*) and lack of jack pine (*Pinus banksiana*) forests that affect sharp-tailed grouse, which persisted in fire-generated openings of presettlement times (Radeloff et al. 2006). In this work, Akçakaya et al. (2004) simulated eight management scenarios using LANDIS. These scenarios contained silvicultural parameters ranging from small to large clearcut sizes of jack pine, as well as several clearcut sizes and minimum cutting age combinations under red pine management (Radeloff et al. 2006). They showed that different timber harvest scenarios result in different amounts of available habitat, measured by the total carrying capacity of all habitat patches. Scenarios with the largest amount of habitat, however, were the worst scenarios in terms of population viability. Such results suggest that ranking management options only in terms of the habitat they provide for threatened or declining species, while ignoring the demography of species, may be misleading. Also, approaches that ignore changes in landscape may overestimate viability and give results that are too optimistic compared with the more realistic simulations that incorporate landscape dynamics.

Wintle et al. (2005) also used the approach developed by Akçakaya et al. (2004) and examined the effects of eight management and disturbance scenarios on brown creeper (*Certhia americana*) in a managed, boreal landscape in north-central Ontario, Canada (see Bekessy et al., this volume). Disturbance scenarios include the current fire regime under fire suppression (long fire return interval, small mean fire size, and higher fire intensity) and natural fire regime (short fire return interval, large mean fire size, and lower fire intensity). The management scenarios ranged from no timber harvesting to natural disturbance-emulation harvesting to intensive harvesting with fire suppression and salvage logging in burned forests under altered fire regime. Compared with using the metapopulation model alone, results from the integrated model showed that trajectories for the brown creeper under alternative management scenarios differed from the base-model, with declines predicted as the intensity of disturbance increased, and under most scenarios the predicted minimum population size was not in direct proportion to the change of carry capacity over the simulation. Their results suggested that population processes, beyond simple habitat availability, influenced model results.

Larson et al. (2004) combined all three components of a habitat-based population viability analysis for land management planning, including landscape simulation using LANDIS, quantifying wildlife habitat quality using HSI models, and population viability analysis using RAMAS GIS. They demonstrated this application for ovenbirds (*Seiurus aurocapillus*) in two simulation scenarios: (1) no harvest, in which forest growth is only disturbed by fire and windthrow; and (2) even-aged management on a 100-year rotation, in which forest growth and succession are disturbed by fire, windthrow, and a clearcut of 10% of the area each decade. They found that ovenbird habitat quality in the study area differed between the no harvest and even-aged harvest scenarios during the first 100-year period, but was similar during the second 100-year period, since natural tree mortality and wind and fire disturbance in the later stage of the simulation increased. Their results further showed that the viability of ovenbird populations was noticeably lower under the even-aged management scenario. These results cannot be derived using habitat suitability models or population models alone. However, the trade-off of involving all three models is increased uncertainties, which are difficult to evaluate due to the complexity of models.

ADVANTAGES AND LIMITATIONS

The main advantages of linking wildlife models to landscape vital-attribute models such as LANDIS are (1) the rich data of individual tree species and age classes recorded for each pixel, and (2) the ability to simulate multiple landscape processes that impact landscapes and wildlife. Wildlife species have a wide range of habitat requirements that can be assessed using basic vegetation information such as vegetation type and age. For example, ovenbirds prefer late successional forest habitat, prairie warblers (*Dendroica discolor*) prefer early successional forest habitat, and hooded warblers (*Wilsonia citrina*) prefer forest gaps. The respective habitat for each species can be derived from the LANDIS tree age map by interpreting 0–20-year-old forest as gaps or patches of early successional forest and forest >40 years old as mature and late successional forest (Shifley et al. 2006). Pine warblers (*Dendroica pinus*), for example, prefer mature coniferous forest, so LANDIS tree age and tree species maps can be used to assess habitat suitability (Larson et al. 2003). Because LANDIS records vegetation information at individual species and age class levels, vegetation information from the same set of simulations can be used to evaluate habitat for multiple wildlife species. Such flexibility is difficult to obtain with succession pathway models that use predefined vegetation or habitat types.

LANDIS is one of the few comprehensive forest landscape models that explicitly simulates silvicultural level timber harvest and common fuel treatments such as prescribed burning and coarse woody debris removal, along with simulation of fire, windthrow, and insect and disease disturbance. The capacity of including multiple landscape processes makes it possible to evaluate the effects of multiple disturbances on wildlife habitat and population dynamics (Shifley et al. 2008). The most common processes used to integrate LANDIS and wildlife models are timber harvesting and fire disturbance (e.g., Akçakaya et al. 2004, Wintle et al. 2005), whereas Larson et al. (2004) and Shifley et al. (2006) simulated all above disturbances except insect and disease in their studies. Common scenario comparisons include fire scenarios such as natural, current, historic, or scenarios with suppression (e.g., Wimberly et al. 2000, Keane et al. 2002); timber management scenarios representing combinations of harvest rotation, size, and even and uneven age harvesting (e.g., Shifley et al. 2000, Zollner et al. 2005); fuel treatments representing combinations of prescribed burning and coarse woody debris removal at various intensities and rotations (e.g., Shang et al. 2007); and current climate versus warming climate scenarios (e.g., Scheller and Mladenoff 2005).

Advantages of integrating LANDIS with HSI models or demographic models also include the flexible spatial scales that LANDIS accommodates. Because LANDIS tracks only the presence-absence of species age cohorts, not individual trees, the essential information of presence-absence is relatively independent of cell size, and therefore LANDIS is capable of simulating forest succession at cell sizes ranging from 10 to 500 m (He et al. 1999). Tracking presence and absence of species age cohorts greatly reduces the computational loads and allows the model to simulate much larger landscapes than models that track biomass for each age cohort. Raster-based modeling approaches employed in LANDIS allow habitat patches to be aggregated and disaggregated depending on the requirement of the wildlife species being studied. The aggregation and disaggregation can reflect the appropriate scale of interests. Furthermore, raster models support more complicated computation tasks such as proximity to water and home range sizes, which are often required to derive the habitat map.

There are several limitations to applying wildlife models with LANDIS, many of which are due to the LANDIS data structure. LANDIS records dead biomass as fine and coarse woody debris classes and cannot differentiate standing dead trees from other dead biomass. The model does not record the vigor of live biomass such as cavity trees. Both standing dead and cavity trees are unique habitat requirements, which have to be derived separately. For example, Fan et al. (2003, 2004) developed a model of cavity tree abundance for LANDIS output. Another limitation related to LANDIS data structure is that LANDIS does not directly record forest density and thus does not directly simulate canopy closure, a common requirement for many wildlife species. Independent functions in habitat models, however, can be used to derive tree density based on tree species, age, and land type, which are mapped in LANDIS (Larson et al. 2003).

Early versions of LANDIS used 10-year time intervals to simulate succession, which generally suits studies of long-term effects (e.g., >150 years). However, forest management plans of National Forests are required to be revised at 10- to 15-year intervals, and most forest management prescriptions (fuel treatment, timber harvest, etc.) are processed annually. Thus, a 10-year time step poses a limitation of examining short-term effects (e.g. <20 years). A version of LANDIS with 1-year time steps has been developed and tested (Syphard et al. 2007) for fire and succession modules. Additional developments are underway for the 1-year time step model to work with harvest, fuel treatment, and insect and disease modules. The development of LANDIS II with variable time steps sheds light on addressing the coarse temporal interval issue (Scheller et al. 2007).

Error propagation and uncertainty are obvious in such integrated models because of the numerous parameters and procedures involved in such modeling frameworks. Moreover, there is a lack of formal procedures to analyze error propagation and uncertainties in forest landscape models. In general, uncertainties embedded in model parameters are subjective uncertainties that are often related to measurement, observation, and synthesis. Uncertainties associated with random algorithms built in the model are stochastic uncertainties.

Stochastic uncertainties larger than subjective uncertainties suggest that input model parameters play little role in model outcome (Xu et al. 2004, 2005).

SUMMARY

The quality of wildlife habitat is often evaluated using habitat suitability index models or more sophisticated population viability models. An emerging approach for assessing long-term management effects or landscape changes on wildlife is to link landscape models with wildlife models. In this framework, habitat dynamics under various natural and anthropogenic disturbances are simulated with a landscape model while impacts on wildlife, such as habitat abundance, quality, structure, and population size or trend, are evaluated with habitat models or population models. Over the past 15 years there has been rapid development in forest landscape modeling, fueled by both technological and theoretical advances. Here, I reviewed and classified forest landscape models in three groups, discussed their suitability for integration with wildlife models, and then discussed in detail the model LANDIS. Vital attribute models are generally the most flexible in deriving habitat types from the output of species and age classes. They are more suitable for integration with wildlife models than successional pathway models and landscape process models that do not simulate vegetation dynamics. LANDIS is a raster-based, vital attribute model that has been used with wildlife models. Applying LANDIS to habitat mapping and wildlife population modeling has the following advantages: (1) Habitat suitability can be evaluated under a combination of forest management and natural disturbance scenarios that are study site specific; (2) habitat suitability is dynamic, driven by succession, disturbance, and management; (3) habitat maps can be derived using relevant species and age class combinations targeting specific wildlife species; and (4) aggregations and disaggregations of vegetation classes using raster data simulated in the LANDIS model allow habitat mapping to be accomplished at multiple scales. In this chapter I also discussed LANDIS model structure and examples of habitat suitability mapping using LANDIS as well as limitations of integrating LANDIS with wildlife models.

ACKNOWLEDGMENTS

This work was funded by the Missouri GIS Mission Enhancement Program and the U.S. Forest Service, Northern Research Station. LANDIS can be freely downloaded at www.missouri.edu/~landis. I thank Joshua Millspaugh, Frank Thompson, and an anonymous reviewer for their helpful comments that improved this manuscript.

LITERATURE CITED

Akçakaya, H. R. 2001. Linking population-level risk assessment with landscape and habitat models. *Science of the Total Environment* 274:283–291.

Akçakaya, H. R, J. Franklin, A. D. Syphard, and J. R. Stephenson. 2005. Viability of Bell's sage Sparrow (*Amphispiza Belli* spp. Belli): Altered fire regimes. *Ecological Applications* 15:521–531.

Akçakaya, H. R., V. C. Radeloff, D. J. Mladenoff, and H. S. He. 2004. Integrating landscape and metapopulation modeling approaches: Viability of the sharp-tailed grouse in a dynamic landscape. *Conservation Biology* 18:526–537.

Baker, W. L., S. L. Egbert, and G. F. Frazier. 1991. A spatial model for studying the effects of climatic change on the structure of landscapes subject to large disturbances. *Ecological Modelling* 56:109–125.

Cary, G. J. 1998. *Predicting fire regimes and their ecological effects in spatially complex landscapes*. Dissertation. The Australian National University, Canberra, Australia.

Chew, J. D., C. Stalling, and K. Moeller. 2004. Integrating knowledge for simulating vegetation change at landscape scales. *Western Journal of Applied Forestry* 19:102–108.

Dijak, W. D., C. D. Rittenhouse, M. A. Larson, F. R. Thompson, III, and J. J. Millspaugh. 2007. Landscape habitat suitability index software. *Journal of Wildlife Management* 71:668–670.

Fan, Z., D. R. Larsen, S. R. Shifley, and F. R. Thompson. 2003. Estimating cavity tree abundance by stand age and basal area, Missouri, USA. *Forest Ecology and Management* 179:231–242.

Fan, Z., S. R. Shifley, F. R. Thompson, III, and D. R. Larsen. 2004. Simulated cavity tree dynamics under alternative timber harvest regimes. *Forest Ecology and Management* 193:399–412.

Franklin, J., A. D. Syphard, H. S. He, and D. J. Mladenoff. 2005. Altered fire regimes affect landscape patterns of plant succession in the foothills and mountains of southern California. *Ecosystems* 8:885–898.

Gardner, R. H., W. H. Romme, and M. G. Turner. 1999. Predicting forest fire effects at landscape scales. Pages 163–185 *in* D. J. Mladenoff and W. L. Baker, editors. *Spatial modeling of forest landscape change: Approaches and applications.* Cambridge University Press, Cambridge, United Kingdom.

Gott, B. 2005. Aboriginal fire management in south-eastern Australia: Aims and frequency. *Journal of Biogeography* 32:1203–1208.

Gustafson, E. J., and T. R. Crow. 1994. Modeling the effects of forest harvesting on landscape structure and the spatial distribution of cowbird brood parasitism. *Landscape Ecology* 9:237–248.

Gustafson, E. J., and T. R. Crow. 1996. Simulating the effects of alternative forest management strategies on landscape structure. *Journal of Environmental Management* 46:77–94.

Gustafson, E. J., S. R. Shifley, D. J. Mladenoff, K. K. Nimerfro, and H. S. He. 2000. Spatial simulation of forest succession and timber harvesting using LANDIS. *Canadian Journal of Forest Research* 30:32–43.

Hargrove, W. W., R. H. Gardner, M. G. Turner, W. H. Romme, and D. G. Despain. 2000. Simulating fire patterns in heterogeneous landscapes. *Ecological Modelling* 135:243–263.

He, H. S. 2008. Forest landscape models, definition, characterization, and classification. *Forest Ecology and Management.* 254:484–498.

He, H. S., D. R. Larsen, and D. J. Mladenoff. 2002. Exploring component-based approaches in forest landscape modeling. *Environmental Modelling and Software* 17:519–529.

He, H. S., W. Li, B. R. Sturtevant, J. Yang, Z. B. Shang, E. J. Gustafson, and D. J. Mladenoff. 2005. *LANDIS 4.0 users guide. LANDIS: A spatially explicit model of forest landscape disturbance, management, and succession*. U.S. Forest Service, North Central Research Station, General Technical Report NC-263, St. Paul, Minnesota, USA.

He, H. S., and D. J. Mladenoff. 1999. Spatially explicit and stochastic simulation of forest-landscape fire disturbance and succession. *Ecology* 80:81–99.

He, H. S., D. J. Mladenoff, and J. Boeder. 1999. An object-oriented forest landscape model and its representation of tree species. *Ecological Modelling* 119:1–19.

He, H. S., Z. B. Shang, T. R. Crow, E. J. Gustafson, and S. R. Shifley. 2004. Simulating forest fuel and fire risk dynamics across landscapes—LANDIS fuel module design. *Ecological Modelling* 180:135–151.

Keane, R. E., G. J. Cary, I. D. Davies, M. D. Flannigan, R. H. Gardner, S. Lavorel, J. M. Lenihan, C. Li, and T. S. Rupp. 2004. A classification of landscape fire succession models: Spatial simulations of fire and vegetation dynamics. *Ecological Modelling* 179:3–27.

Keane, R. E., R. A. Parsons, and P. F. Hessburg. 2002. Estimating historical range and variation of landscape patch dynamics: Limitations of simulation approach. *Ecological Modelling* 151:29–49.

Larson, M. A., W. D. Dijak, F. R. Thompson, III, and J. J. Millspaugh. 2003. *Landscape-level habitat suitability models for twelve wildlife species in Southern Missouri*. U.S. Forest Service, North Central Research Station, General Technical Report NC-233, St. Paul, Minnesota, USA.

Larson, M. A., F. R. Thompson, III, J. J. Millspaugh, W. D. Dijak, and S. R. Shifley. 2004. Linking population viability, habitat suitability, and landscape simulation models for conservation planning. *Ecological Modelling* 180:103–118.

Li, C., M. Ter-Mikaelian, and A. Perera. 1997. Temporal fire disturbance patterns on a forest landscape. *Ecological Modelling* 99:137–150.

Mladenoff, D. J. 2004. LANDIS and forest landscape models. *Ecological Modelling* 180:7–19.

Mladenoff, D. J., and W. L. Baker, editors. 1999. *Spatial modeling of forest landscape change: Approaches and applications.* Cambridge University Press, Cambridge, United Kingdom.

Mladenoff, D. J., and H. S. He. 1999. Design, behavior and application of LANDIS, an object-oriented model of forest landscape disturbance and succession. Pages 125–162 *in* D. J. Mladenoff and W. L. Baker, editors. *Spatial modeling of forest landscape change: Approaches and applications.* Cambridge University Press, Cambridge, United Kingdom.

Mladenoff, D. J., G. E. Host, J. Boeder, and T. R. Crow. 1996. LANDIS: A spatial model of forest landscape disturbance, succession, and management. Pages 175–180 *in* M. F. Goodchild, L. T. Steyaert, and B. O. Parks, editors. *GIS and environmental modeling: Progress and research issues.* GIS World Books, Fort Collins, Colorado, USA.

Nobel, I. R., and R. O. Slatyer. 1980. The use of vital attributes to predict successional changes in plant communities subject to recurrent disturbances. *Vegetation* 43:5–21.

Pichancourt, J.-B., F. Burel, and P. Auger. 2006. Assessing the effect of habitat fragmentation on population dynamics: An implicit modelling approach. *Ecological Modelling* 192:543–556.

Radeloff, V. C., D. J. Mladenoff, E. J. Gustafson, R. M. Scheller, P. A. Zollner, H. S. He, and R. H. Akçakaya. 2006. Modeling forest harvesting effects on landscape pattern in the Northwest Wisconsin Pine Barrens. *Forest Ecology and Management* 236:113–126.

Rastetter, E. B., A. W. King, B. J. Cosby, G. M. Hornberger, R. V. O'Neill, and J. E. Hobibie. 1992. Aggregating fine-scale ecological knowledge to model coarser-scale attributes of ecosystems. *Ecological Applications* 2:55–70.

Roberts, D. W. 1996. Landscape vegetation modelling with vital attributes and fuzzy systems theory. *Ecological Modelling* 90:175–184.

Roberts, D. W., and D. W. Betz. 1999. Simulating landscape vegetation dynamics of Bryce Canyon National Park with the vital attributes/fuzzy systems model VAFS/LANDSIM. Pages 99–124 *in* D. J. Mladenoff and W. L. Baker, editors. *Spatial modeling of forest landscape change: Approaches and applications.* Cambridge University Press, Cambridge, United Kingdom.

Romme, W. H., and D. G. Despain. 1989. The Yellowstone fires. *Scientific American* 261:37–46.

Scheller, R. M., J. B. Domingo, B. R. Sturtevant, J. S. Williams, A. Rudy, E. J. Gustafson, and D. J. Mladenoff. 2007. Design, development, and application of LANDIS-II, a spatial landscape simulation model with flexible temporal and spatial resolution. *Ecological Modelling* 201:409–419.

Scheller, R. M., and D. J. Mladenoff. 2005. A spatially interactive simulation of climate change, harvesting, wind, and tree species migration and projected changes to forest composition and biomass in northern Wisconsin, USA. *Global Change Biology* 11:307–321.

Scheller, R. M., and D. J. Mladenoff. 2007. An ecological classification of forest landscape simulation models: Tools and strategies for understanding broad-scale forested ecosystems. *Landscape Ecology* 22:491-505.

Shang, Z. B., H. S. He, T. R. Crow, and S. R. Shifley. 2004. Fuel load reductions and fire risk in central hardwood forests of the United States: A spatial simulation study. *Ecological Modelling* 180:89-102.

Shang, Z. B., H. S. He, D. E. Lytle, S. R. Shifley, and T. C. Crow. 2007. Modeling the long-term effects of fire suppression on central hardwood forests in Missouri Ozarks, using LANDIS. *Forest Ecology and Management* 242:776-790.

Shifley, S. R., F. R. Thompson, III, W. D. Dijak, and Z. Fan. Forecasting landscape-scale, cumulative effects of forest management on vegetation and wildlife habitat: A case study of issues, limitations, and opportunities. *Forest Ecology and Management: In press.*

Shifley, S. R., F. R. Thompson, III, W. D. Dijak, M. A. Larson, and J. J. Millspaugh. 2006. Simulated effects of forest management alternatives on landscape structure and habitat suitability in the Midwestern United States. *Forest Ecology and Management* 229:361-377.

Shifley, S. R., F. R. Thompson, III, D. R. Larsen, and W. D. Dijak. 2000. Modeling forest landscape change in the Missouri Ozarks under alternative management practices. *Computers and Electronics in Agriculture* 27:7-27.

Sklar, F. H., and R. Costanza. 1990. The development of dynamic spatial models for landscape ecology: A review and synthesis. Pages 239-288 *in* M. G. Turner and R. H. Gardner, editors. *Quantitative methods in landscape ecology.* Springer-Verlag, New York, New York, USA.

Syphard, A. D., J. Yang, J. Franklin, H. S. He, and J. E. Keeley. 2007. Calibrating a forest landscape model to simulate frequent fire in Mediterranean-type shrublands. *Environmental Modelling and Software* 22:1641-1653.

Wimberly, M. C., T. A. Spies, C. J. Long, and C. Whitlock. 2000. Simulating historical variability in the amount of old forests in the Oregon coast range. *Conservation Biology* 14:167-180.

Wintle, B. A., S. A. Bekessy, L. A. Venier, J. L. Pearce, and R. A. Chisholm. 2005. Utility of dynamic-landscape metapopulation models for sustainable forest management. *Conservation Biology* 19:1930-1943.

Xu, C., H. S. He, Y. Hu, Y. Chang, D. R. Larsen, X. Li, and R. Bu. 2004. Assessing the effect of cell level uncertainty on a forest landscape model simulation in northeastern China. *Ecological Modelling* 180:57-72.

Xu, C., H. S. He, Y. Hu, Y. Chang, X. Li, and R. Bu. 2005. Latin hypercube sampling and geostatistical modeling of spatial uncertainty in a spatially explicit forest landscape model simulation. *Ecological Modelling* 185:255-269.

Zollner, P. A., E. J. Gustafson, H. S. He, V. C. Radeloff, and D. J. Mladenoff. 2005. Modeling the influence of dynamic zoning of forest harvesting on ecological succession in a northern hardwoods forest. *Environmental Management* 35:410-425.

CHAPTER

Simulating Landscape Change Using the Landscape Management System

13

Chadwick D. Oliver, James B. McCarter, Jeffrey M. Comnick, Kevin Ceder, and Christopher S. Nelson

Management of forested landscapes is becoming increasingly complex as people are demanding more and diverse values from these landscapes. Different wildlife species can require different forest conditions that often are mutually exclusive. For example, some owl and woodpecker species may require closed, old forests, whereas butterflies and some songbirds in the same forest types may require openings with sparse tree cover. In addition, there are increasing demands for values other than timber such as scenic beauty, fire protection, and carbon sequestration.

Forests are constantly changing through growth and disturbances (Fig. 13-1). Consequently, managers need to anticipate how different forest stands will change and how they can manage the change to ensure that a diversity of values is provided at all times. Because of the mutual exclusivity of different habitats and other values, managers must also make trade-offs among these values. Although necessary, such trade-offs can be minimized with informed management.

Effective landscape management requires rapid processing of large amounts of information on site-specific, local levels. Forest growth models, mapping systems/geographic information systems (GIS), and similar tools are helpful; however, these tools need to be integrated. Proper integration requires methods that allow data to be converted to appropriate formats, analyzed thoroughly, synthesized, and communicated effectively. Such methods and tools need to be based on management and natural sciences. Furthermore, they must also be robust enough to be adjustable for local geographic and administrative variations (Oliver et al. 1992, 1993).

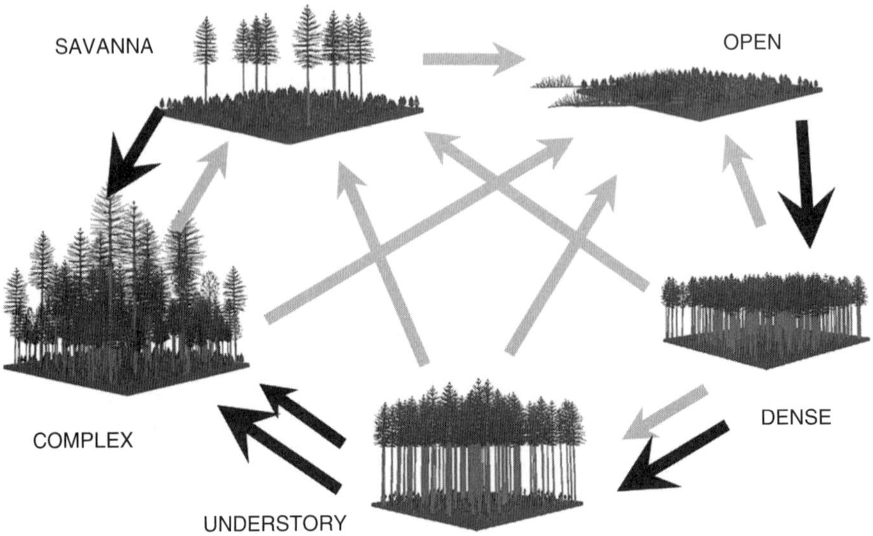

FIG. 13-1 (See color insert.)

Forests exist in many structures, such as the five-structure classification system shown here (Oliver and Larson 1996). The stands change constantly among structures through growth and disturbances. Each structure provides habitat for some species. *(Copyright, C. Oliver, Yale University, 2006).*

SCIENTIFIC BASIS

In this chapter, we first describe the scientific basis for managing wildlife and other values across forested landscapes. Then we describe the organization of the Landscape Management System (LMS) and how its modularity allows it to be improved and integrated with other systems. We address details of the LMS by describing the methods for applying sequential components followed by results of using those components. These details are described in two parts: the basic tool and how it can be used to make and implement effective decisions. Throughout the chapter, we try to emphasize wildlife habitat as part of a suite of values to be managed across a landscape.

Stand and Landscape Dynamics

The systems we describe herein are based on the natural science understanding that forests are dynamic and constantly changing (Botkin and Sobel 1975, Oliver 1981, Kimmins 1987). Some stand conditions, and their associated values to people, diminish, whereas others increase as stands change (Fig. 13-1). Management is the task of guiding stand changes in coordination with other stands to provide a targeted suite of benefits, the management objectives, across the landscape as a whole. A diversity of stand structures is needed to provide a mixture of noncommodity and commodity objectives (Oliver 1992, Boyce 1995). Using existing forest growth models, GIS, and forest inventories, managers try to predict these stand and landscape changes, and the resulting changes in values, both spatially and temporally; they then try to coordinate these changes using various natural and human activities to achieve the targeted objectives.

The dynamic nature of forests means a stand with a given species, or species mix, can be managed by different silvicultural pathways (Oliver and O'Hara 2004) and produce different consequences. A silvicultural pathway is herein defined as a unique change in the stand over time caused by a combination of growth and specific silvicultural operations at specific times. For example, the same stand would follow different silvicultural pathways if it was managed by the clearcut, shelterwood, or selection method. Different pathways would also result from timing of and selection of treatments (e.g., thinning, prescribed burning, planting to different species and/or densities). Subsequent operations, or lack thereof, will further define the silviculture pathway.

Wildlife, Structures and Functions, Other Values, and Trade-Offs

Managing landscapes can enhance their value to humans by providing appropriate habitats for desired wildlife species. This habitat management is an important determinant of wildlife presence and abundance in addition to hunter harvest of the target species, its prey, or its predators. Habitats often consist of specific stand structures, combinations of stand structures, or such structures within certain proximities to each other or to rock outcrops or to water or beyond certain distances from roads. Forest habitats are managed at two levels: "coarse level" and "fine level" (see Noon et al., this volume). At the coarse level, all habitats (all stand structures in the case of forests) are maintained across the landscape to ensure that most species can survive. For species of conservation concern, more targeted, fine-level management is done to ensure the individual species has the habitat features required (coarse and fine filters; Noon et al., this volume).

Complexity and Systems Theory

Systems approaches are being developed to address complexity in forest management and research by grouping similar entities and then addressing the behavior within a group and among interacting groups (Checkland 1999, Oliver and Twery 1999). Systems approaches integrate the holism and reductionism philosophical approaches and make use of quantitative methods developed in management sciences (Wilson 1887, Taylor 1911, Krick 1962, Cleland and King 1968, von Bertalanffy 1968, Dieter 1991). Grouping naturally leads to variation, which is recognized as inherent in systems. Variation is managed by using summary or (emergent) values of groups and by encouraging the user to develop a more complex "mental model" (i.e., conceptual model, accumulated knowledge) for the level of concern (Senge 1990)—a tree, stand, landscape, or region. We utilize systems theory and these simplistic summary values when developing computer models such as the LMS; however, we expect the user to form more complex mental models using these computer models as frameworks.

Hierarchies and Top-Down Planning

Grouping of stands can form a hierarchy. For example, individual plants and animals can form the base of a hierarchy that is grouped upward to stands, landscapes, watersheds, regions, and even higher (Litterer 1965, Bare 1996, Clegg et al. 1996, Oliver et al. 1999). Inherent variability can be compounded upward through each level of the hierarchy. The compounding error is overcome by managing in a way that avoids excessive centralized planning, especially at lower levels of the hierarchy. Emergent values are targeted at higher levels, but the details of management to accomplish them are left to the lowest possible level. Similarly, variability is addressed in science by avoiding theories (or models) that try to explain too many hierarchical levels in great detail at once (see Probst and Gustafson, this volume).

Decision Analysis

Multiple objectives often cannot be completely achieved when managing a forest landscape; therefore, trade-offs among different amounts of the different objectives that will be provided must be made. Two robust decision approaches have been developed to address trade-offs: the noniterative and iterative approaches (Morgan and Henrion 1990, Rauscher 1995, Oliver and Twery 1999). Noniterative approaches predetermine the relative trade-offs and commonly entail use of optimization programs. Iterative approaches present a variety of management alternatives and display them to enable the decision maker to view the trade-offs among objectives. The iterative approach, often in a matrix format, seems most appropriate for the complexities of modern forest ecosystem management (Krick 1962, Roberts 1979, Oliver and Lippke 1995, Oliver et al. 1997, Oliver and Twery 1999). We use the iterative approach in the "toggle" tool within the LMS Framework.

Overview of the Landscape Management System and Associated Tools

The overall management system consists of two sets of tools: the Landscape Management System containing the LMS tool and the Decision Analysis System

(DAS) Tools containing the "Scope & Group" and "Toggle" tools. All tools and tutorials can be downloaded from: <http//lms.cfr.washington.edu/>.

The LMS and DAS tools can be conceptually organized into seven components (Fig. 13-2). NED (Twery et al. 2005), SILVAH (Marquis and Ernst 1992), and other landscape models have similar organizations. Various efforts are underway to link and integrate these systems, as appropriate. LMS can be used independently of the DAS Tools to display and implement a developed forest management plan or to develop a simple plan. The DAS Tools can supplement LMS for developing complex plans of large areas and many objectives. Computer specifications of the tools are described in the following paragraphs.

Land Management System.—The Landscape Management System (McCarter et al. 1998) coordinates the activities of approximately 50 computer programs to facilitate the evaluation of alternative management approaches. The LMS software package uses a point-and-click graphical user interface (GUI) with dropdown menus to interact with information for a specific landscape. LMS 2.x works on computers running Microsoft Windows 98 (and newer). LMS 3.1

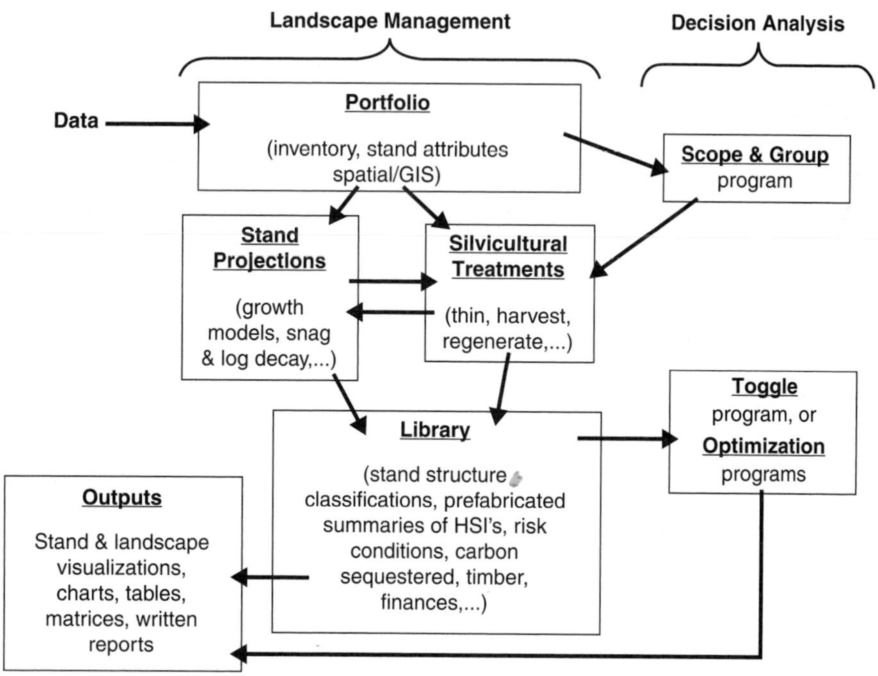

FIG. 13-2

Modular nature of the Landscape Management System, and similar models such as NED (Twery et al. 2005) and SILVAH (Marquis and Ernst 1992). *(Copyright, C. Oliver, Yale University, 2006).*

(released in November 2006) runs on Windows 2000 (and newer). Landscape Management System consists of the following components:

1. Portfolio: the stand, inventory, and spatial information needed to run LMS;
2. Stand Projection: the tree growth models and snag/down log decomposition models;
3. Silvicultural Treatments: the manipulations done to individual stands to simulate either silvicultural operations or natural perturbations and natural regeneration; and
4. Outputs and Library: the display of information through stand and landscape visualizations, graphs, tables, and spreadsheet templates. The visualizations, tables, and spreadsheet templates represent a collection of algorithms for determining structural stages, risk conditions, habitat, carbon, timber, finances, and other values.

Scope & Group.— Scope & Group is a Microsoft Excel® spreadsheet file (~6 MB) used to classify a landscape into groups of similar stands. The version available on the Web is calibrated to Pack Forest and is named "PackScope.xls."

Toggle.—Toggle is currently an Excel spreadsheet file (~10 MB), but is being converted to Visual Basic format to increase user-friendliness. This tool is used to allocate the landscape iteratively among different silvicultural pathways for each group of similar stands. It provides immediate feedback to the user on the consequences of management objectives of having certain stands follow certain pathways. The version available on the Web is calibrated to Pack Forest and is named "PackToggle.xls."

GEOGRAPHIC AREA OF STUDY AND APPLICATION

To date, we have converted the inventory information (and in some cases spatial information) of over 50 separate landscapes (~ 700,000 hectares) to LMS portfolios that they can be used with LMS. These landscapes are in the United States and Canada: Alaska, Arizona, British Columbia, California, Colorado, Connecticut, Florida, Michigan, Mississippi, Missouri, New Brunswick, New Hampshire, North Carolina, Ontario, Oregon, Utah, Washington, and West Virginia. In addition, LMS is being applied to landscapes in China and Ukraine.

We have applied and tested versions of the LMS tool for more than 10 years at the University of Washington, the Yale University School of Forestry and Environmental Studies, Pennsylvania State University, and elsewhere. Additionally, we have used LMS, Scope & Group, and Toggle tools and techniques in 10 mid-career professional development courses and three graduate-level courses using simulated management situations based on real landscapes and management plans. Finally, we have implemented pilot projects using these tools on landscapes in various locations throughout the United States and in other online and offline tests described throughout this chapter.

The most complete application to date was developed for Charles L. Pack Experimental Forest of the University of Washington, College of Forest Resources (1,740 hectares in the western Cascade Range, near Eatonville, Washington, USA). The Pack Forest Landscape Plan (<http://www.packforest.org/plan/>) was developed using LMS and customized versions of Scope & Group and Toggle spreadsheets.

LANDSCAPE MANAGEMENT SYSTEM: APPLICATION METHODS AND RESULTS

A portfolio in LMS consists of a group of stands that are combined into a larger planning unit or landscape. Any user can create a portfolio by using minimal inventory and stand attribute information about individual stands or polygons. Menus within LMS facilitate the performance of functions such as growth, stand treatments (thinning, planting, or mimicking natural disturbances such as mortality from a fire), and visualization of stands and landscapes. Finally, Microsoft Access® and/or Microsoft Excel® tables can be generated for further analyses outside LMS.

Creating an LMS Portfolio

Methods.— Three sources of data are used to create an LMS portfolio for a given landscape. If the data are in digital (computerized) format, application-specific programs can convert them for use within LMS. The sources of data are as follows:

1. *Stand attribute information:* Attributes for each stand such as stand area, site index, habitat type, latitude, slope, aspect, and elevation are inputs required by some growth models and are stored in the portfolio data set. These values are used to localize estimates from the growth models.
2. *Inventory information:* An inventory tree list is required, with species, diameter, heights (if available), and number of trees per record, summarized to the average acre. Missing tree heights are calculated by the embedded growth models and subsequently added to the inventory data set by LMS. If only stand average data are available, the user can expand these data to a tree list. When inventory information is missing from some stands, the user can extrapolate data from similar stands. For all cases, the user is responsible for limitations of the data.
3. *Spatial/GIS information:* Digital elevation models (DEMs) and stand boundary, road, and stream layers from GIS can be used by LMS if management criteria include spatially explicit objectives or if landscape visualization is desired. Currently, LMS supports and interfaces with ESRI software (ArcView and ArcGIS) and files.

Results to Date.— A common issue is the landowner's organization of data, a management concern not limited to forestry. The issue occurs when information about stand inventories, management practices, or treatments are diligently collected but not effectively organized and catalogued. Often a large amount of information is collected, but no easy method for using the information exists, allowing ineffective organization to go unnoticed. The Landscape Management System provides a centralized location for data. Common problems from underutilized data include multiple copies of the master file with conflicting information, mistyped data, and poor definitions of codes and fields. When constructing portfolios, LMS examines the stand and inventory information to find mistyped data and various data conflicts. The user must rectify poorly defined fields and codes prior to using LMS. Additional data organizational issues can be corrected while applying Continuous Quality Improvement (CQI) techniques at the end of the first iteration of the LMS management cycle. LMS does not solve all information management issues but can provide a beginning focus to improve the process.

Treating and Projecting Stands

Methods.— Each LMS portfolio is configured to use one growth model per portfolio. LMS supports the U.S. Forest Service's Forest Vegetation Simulator (FVS; Stage 1973) and Oregon State University's Organon (Hann et al. 1995) growth models. The Forest Vegetation Simulator (U.S. Forest Service 2007) offers variants for most forest types and regions in the United States and portions of Canada. Organon (Organon Growth and Yield Project 2007) provides three variants for use in the Pacific Northwest west of the Cascades. Both models can be used for projecting inventories of individual stands or entire landscapes to a user-specified year in multiples of the time step chosen (i.e., 5- or 10-year increments). Future projected inventories for each time step are stored within the LMS portfolio.

Any stand can be treated at any time step by manipulating the stand inventory. These treatments will be directly reflected in the inventory files; future projections will be adjusted accordingly. The treatment menu allows any specification of trees to be harvested and retained (thinned from above, below, proportional; by species or diameter; and to a specified number of trees, basal area, or stand density index). The user designates the expected regeneration, which will appear in the following time step. The regeneration editor allows the user to create, name, and save regeneration files within LMS for current and future use. Common treatments can also be named and stored in LMS for future use. Treatments to be done at different years to many stands in a landscape can be created as a "scenario" and stored for later use. Disturbances can be modeled using the treatment procedures to simulate the effects of disturbances on stands.

All inventory information, including harvested tree information, is stored within the LMS portfolio for each stand and time step. These changes over time form a silvicultural pathway for the stand (Fig. 13-3). Trees that the growth

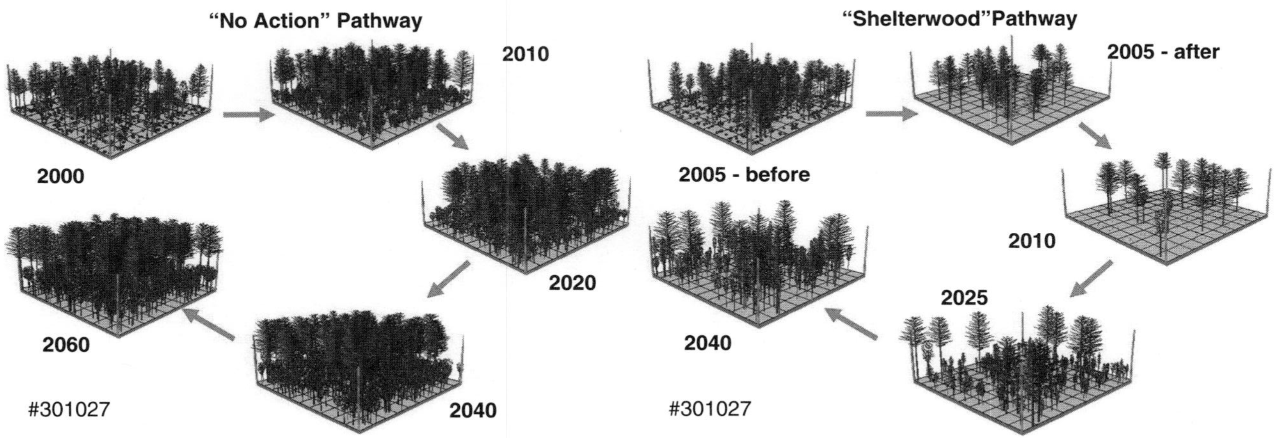

FIG. 13-3

Silvicultural pathways, such as the two shown here for Pacific Northwest mixed conifer forests in the USA, consist of intended changes in a stand with time as a result of growth and silvicultural operations. They can be planned, projected, attributes assessed, and reported (Fig. 13-4), and visualized, as shown here. *(Copyright, C. Oliver, Yale University, 2006).*

model predicts to die, as well as trees intentionally treated to create snags and other coarse woody material, are tracked separately.

Results to Date.— Projections of standing volume for long intervals (e.g., 50 years) using different growth projections inside LMS reveal differences that are inherent in the growth models. For example, systematic differences in projected volumes are observed when the same portfolio is projected using FVS (PN variant) and Organon because of differences in the growth models resulting from differences in the modeling databases used to develop each growth model. Although important, local managers can generally interpret the variations in growth model outputs while the models are being improved, once they are aware of the different models' tendencies.

Outputs and Library: Analyzing and Displaying Objectives

Methods.—The Land Management System provides three types of output:

1. *Inventory Information:* Inventory data can be exported to Excel® spreadsheet, database, or ASCII formats from LMS. In these familiar formats, the data can be manipulated to calculate measurable criteria for management objectives. The results of these queries can then be displayed as tables, graphs, and charts.

2. *Summarized Information:* LMS provides a variety of prefabricated tables that present summarized information to address specific management issues. Some of these tables calculate common forest information summaries, such as canopy layers and species and diameter distributions (Baker and Wilson 2000). Other tables calculate less traditional summaries such as stand structural stages (Oliver and Larson 1996) and habitat suitability for a specific species (Marzluff et al. 2002). These tables' algorithms are typically written in the Python programming language; additional tables can be inserted by a user knowledgeable in Python. The tables can be exported to spreadsheet and database formats. These data can then be pasted into user-defined templates that show outputs over time in graph and/or chart form.

3. *Visualizations:* Visualizations can be made for any year, any stand, and from any viewpoint for both stands and entire landscapes using the included U.S. Forest Service Stand Visualization System (SVS; McGaughey 1997) and EnVision (McGaughey 2000) programs. Stand-level visualizations utilize the LMS tree lists and are displayed on a unit area basis. Landscape visualization uses the LMS tree list and such GIS information as stand boundaries and digital elevation models (DEMs). Realism of the visualization can be adjusted from stick figure trees to photorealistic images (Wilson and McGaughey 2000).

We have developed local interpretations and measurable values for most "Criteria for Sustainable Forestry" of the Montreal Process (Forestry Commission 1998). These are displayed within many LMS tables. These criteria are as follows:

1. *Criterion # 1: Biological Diversity:* Several considerations relative to biological diversity entail analysis of both "coarse filter" and "fine filter" biodiversity. Measures of coarse filter biodiversity (Hunter 1990) on the landscape can be calculated by using any of the seven stand structure classification systems within LMS (O'Hara et al. 1996, Oliver and Larson 1996, Carey et al. 1999, Johnson and O'Neil 2001). Most of these classification systems use mutually exclusive categories that include "sharp boundaries." Each uses a hierarchical classification system to assign each stand to a specific stand structure at each time step. The hierarchy means that all stands are analyzed for their "fit" into the criteria of the different structures sequentially; a "default" structure accepts stands that do not fit any of the other classes. Usually, the "open" structure is the default. Fine filter biodiversity measures can be calculated as the amount of a specific structure in the landscape or species-specific Habitat Suitability Indexes (HSIs). Some habitat suitability indices have been incorporated into LMS and others can be added (Ceder and Comnick 2001, Raedeke 2001, Marzluff et al. 2002). Another technique being developed for calculating carrying capacity for species in mixed forest/grassland areas is to integrate rangeland analyses of animal unit months (AUMs) from forage conditions with assessments of hiding and thermal cover (Han 2006). By transferring data between GIS and LMS, the user can assign spatial attributes to the different stands and lengths of contiguous boundaries between stands. These additional data enable calculations of animal habitats that require different stand structures to be within a certain proximity or that require an edge between two specific structures. Finally, changes in the total area and the conditions (structure, density, species composition, and others) of stands held in reserve can be monitored over time.

2. *Criterion # 2: Productive Capacity:* Tree volume growth, harvest, and standing volume can be calculated and displayed over time to evaluate the sustainability of timber harvest. Volumes by tree sizes, species, and log grades can be calculated for any stands or the landscape and for each time period of the simulation by using locally adaptable log sort tables within LMS.

3. *Criterion # 3: Forest Health:* The user can calculate wind, fire, and insect susceptibility using LMS and/or interfacing LMS with GIS information (Wilson and Baker 1998, Wilson et al. 1998, Wilson and Oliver 2000).

4. *Criterion # 4: Soil and Water Protection:* Soil protection measures have been developed as a combination of the silvicultural operations (e.g., for soil compaction), the type of soil and slope, and the time between

operations. Water protection measures have been developed as analyses of riparian zones (e.g., direct sunlight, coarse woody debris, and particulate organic matter reaching streams).

5. *Criterion # 5: Global Carbon Sequestration:* Carbon sequestration in the standing forest, in nondecomposed wood products, and in unburned fossil fuel when forest products replace more fossil fuel-consuming substitutes (e.g., steel, concrete) have been developed after Lippke et al. (2004) and are available as summarized tables and charts in LMS.

6. *Criterion # 6: Socio-Economic Benefits:* Economic benefits can be analyzed from a Financial Analysis module within LMS. This module allows the user to define costs and prices, and it displays the costs, returns, and cash flow for stands or the landscape for any or all time steps. Discounted cash flow and internal rate of return can be analyzed by exporting the tables to spreadsheets. Social benefit measures relate employment to various silvicultural activities and to the amount and quality of outputs (Oliver and Lippke 1995).

7. *Criterion # 7: Legal, Institutional, Economic Framework:* The Landscape Management System and Decision Analysis System methods and tools can be used to analyze legal, institutional, research, and economic policy alternatives.

Results to Date.— We describe examples of specific interpretations and measurements for different landscapes in <http://lms.cfr.washington.edu/> under "Case Studies." Specific concerns regarding wildlife habitats are as follows:

1. The algorithms currently used to calculate stand structures (e.g., for use in "coarse filter" biodiversity analyses) create an artifact of sharp boundaries between structures within a classification because of the binary nature of the current classification systems. Consequently, if a stand does not meet all minimum criteria for a certain structure, it is completely excluded from being considered to have that structure. Alternative classification systems are being investigated that allow more "fuzzy" boundaries.

2. We have compared habitat suitability indices within LMS with other methods of analysis by Marzluff et al. (2002) and found them to be consistent.

   Specific analyses of other outputs from LMS to date suggest the following results as well:

3. Currently, LMS can merchandize logs by sizes into grades using a simple bucking algorithm. The bucking algorithm is being enhanced to allow better handling of additional species, especially hardwoods, and more flexibility in log lengths, merchandizing options, and output volumes.

4. Projections of stand structures appear relatively insensitive to growth models.

5. Projections of height/diameter (Ht/DBH) ratios of the tallest 247 trees per hectare using different growth models indicate that measures of wind susceptibility are highly sensitive to the growth model used (Wilson and Oliver 2000).
6. Carbon sequestration has been calibrated for the Pacific northwestern and southeastern United States coniferous forests (Lippke et al. 2004); however, they are less reliable in other forest types, especially hardwood forests.

We suggest three ways to address the issues described here. First, local managers can adjust these output variances within their "mental model" of the expected behavior of the forest. For example, they would know that a low height/diameter ratio projected using the FVS (PN variant) model does not necessarily indicate low wind risk, nor does a high height/diameter ratio projected in the Organon (SWO variant) model necessarily indicate high wind risk. Second, over time the user can calibrate these regional growth models to simulate local conditions better. Last, it is understood that over time the specific growth models will improve. The integrative LMS tool now allows growth models to be used in ways for which they were not originally developed. Already, developers of growth models are improving the predictive capabilities of current and new growth models.

Applying LMS to Management

Methods.— The Land Management System can be used to evaluate an existing management plan, or it can be used to develop a new management plan. If applied to an existing plan, LMS can display the planned treatments and expected results in a variety of tabular, graphical, and visual ways, and on both temporal and spatial scales. The Decision Analysis System in conjunction with LMS can also help develop complex plans for large landscapes, as described later in this chapter.

The management plan can be divided into two time frames: the "planning horizon" and the "management cycle." The planning horizon is the total time for which management treatments and expected results are projected, commonly 50, 80, or 100 years. While accuracy is likely to diminish with longer growth periods, the projection is useful for several reasons: (1) to ensure the temporal sustainability of the different values; (2) to compare long-term differences and trade-offs in alternative management approaches; (3) to estimate future trends in operations, equipment, and labor needs for strategic planning and policy development; and (4) to estimate future trends in availability of commodity and noncommodity products (e.g., habitats for certain species, products of certain dimensions).

The management cycle is the first projection period of the growth model used within LMS (user-defined, but usually 5 or 10 years). The chosen management plan results from specific treatments having been assigned to each stand in

LMS. The user projects all treatments in LMS as if they were implemented at the beginning of each growth cycle.

During implementation, the treatments are done opportunistically within the management cycle, allowing the local manager to respond to variations in markets, weather, and other operational opportunities/constraints. Allowing this variation is consistent with the systems approach of addressing variation within a system by not externally specifying inputs/outputs too precisely (e.g., through central planning).

The user is expected to correct deviations caused by variations in treatment times by updating the plan at the end of each 5- or 10-year management cycle. At this update, the manager can revisit and refine any objectives. The objectives are revisited and the analytical techniques can be refined using the Continuous Quality Improvement process or similar processes (Deming 1982, Ishikawa 1982, Feigenbaum 1983, Juran and Gryna 1988, Dieter 1991).

For the management cycle, the user can generate a treatment-outcome list from LMS tables showing each stand name, treatment (if any), and expected outcome for a variety of objectives (Fig. 13-4). The manager can use this list as a business portfolio (Oliver 1994, Wilson and Baker 2001). Knowing the intended inputs and outputs for the current management cycle, the manager can "bundle" high- and low-value products to ensure they are sold, sell marginal products (e.g., floral greens) by being specific about where they are to be found, sell wildlife access by demonstrating the number and location of target species, sell products at optimum market or labor cycles, and otherwise balance the "cash flow" for the different treatments.

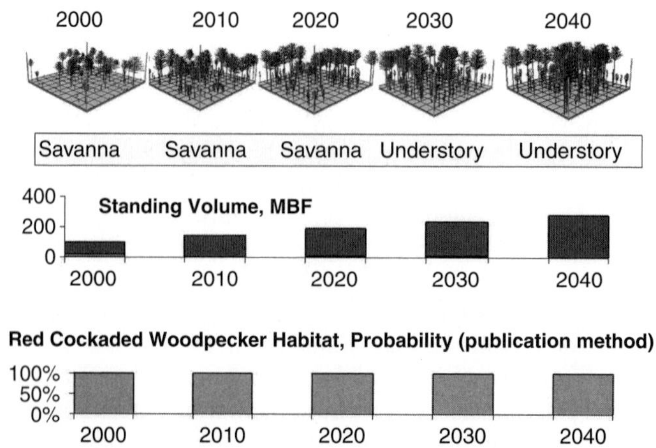

FIG. 13-4

Many attributes for a given stand at each time period can be assessed and displayed in different ways, as shown here for longleaf pine stands in Florida, USA. *(Copyright, C. Oliver, Yale University, 2006).*

People want assurances that the forests will actually provide the values as stated in the management plans. The Land Management System can help provide that assurance. Using stand and landscape visualizations and GIS, it can identify and display how any stand or landscape will be treated, and where expected conditions (e.g., stand structures or habitats for specific animals) will be found at each future time (Figs. 13-4 and 13-5).

Results to Date.— One test plan at Pack Forest has been carried through the step of developing treatment-outcome lists for the management cycle and each subsequent cycle in the planning horizon. The treatment list for the current management cycle is proving helpful in the test by giving managers the information as a "portfolio," with the certainty and flexibility helpful in a business environment. The treatment-outcome lists for subsequent cycles enable the managers to begin activities (e.g., order seedlings, upgrade roads) needed for operations to be implemented in the more distant future. The LMS treatment lists and visualizations are proving helpful for communicating with the public, other managers, loggers, and silviculture contractors. Wilson and McGaughey

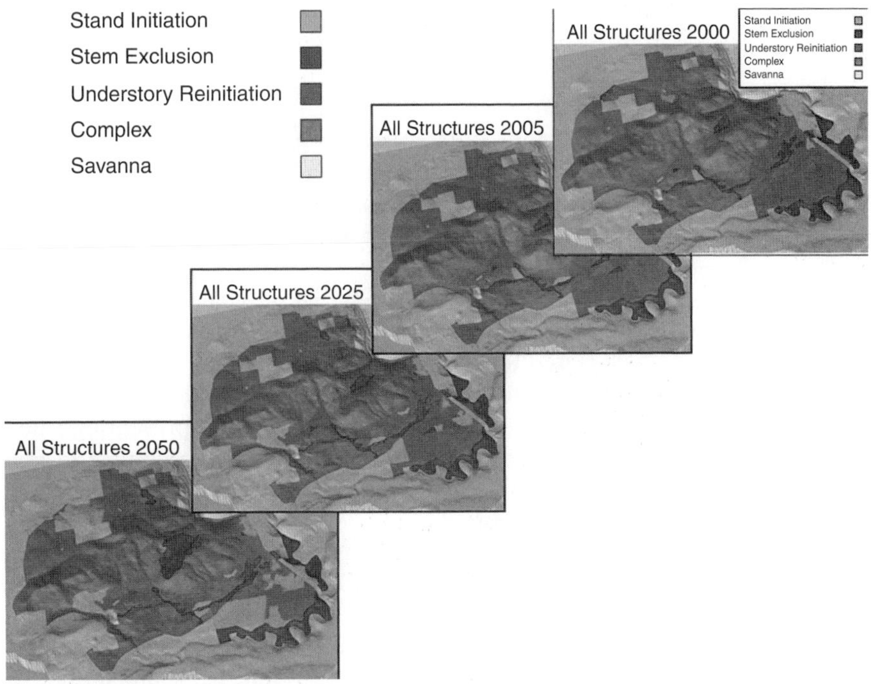

FIG. 13-5 (See color insert.)

The anticipated results of different attributes can be visually displayed across the landscape over time, as shown here for stand structures (Fig. 13-1) in the Pacific Northwest Pack Forest Plan, USA. *(Copyright, C. Oliver, Yale University, 2006).*

(2000) evaluated the use of photorealistic visualizations and the potential for misleading the intended audience. No management plan using LMS has yet completed its first management cycle; therefore, the monitoring and CQI system using LMS has not been fully tested online. Offline testing of simulated plans on real landscapes in mid-career short courses suggests the techniques described previously will be effective.

DECISION ANALYSIS SYSTEM: APPLICATION METHODS AND RESULTS

Forest planning over large spatial scales (10,000 to 100,000 acres), long time periods (50–100 years), and for multiple objectives is a complex problem. As the number of stands in a landscape increases, the ability of the manager to analyze and understand the trade-offs among many alternatives becomes limited. A number of analytical approaches have been developed to address this problem. Each approach requires simplification of real-world complexity but emphasizes different forest dynamics (e.g., spatial patterns, stand growth, and response to treatments).

Comparing different approaches is facilitated by identifying the most common simplifications used in forest modeling. Many models retain the spatial complexity of the landscape but simplify the data used to describe each stand and the response of stands to treatment and growth. Simplification is accomplished by reducing the number of unique forest types used to represent the forest, the sophistication of the within-stand growth projections, or the detail of the stand data. Alternatively, models can retain the stand data at the expense of spatial complexity.

The LMS is designed to use the most detailed forest inventory data available, to link to growth models, and to function with a built-in library of outputs. It is very well suited to analyzing both forest growth and responses to various treatments. A Decision Analysis System (DAS) was developed to aid in the most common analysis tasks needed for developing and presenting landscape alternatives and their consequences. It does not choose or recommend any alternative. After describing the DAS, we will contrast it with two other models: LANDIS (He, this volume) and Woodstock.

Systems approaches reduce complexity by grouping, or stratifying, many entities into a manageable number of groups. The utility of the system is based on how well the groups are formed, with a desire for much lower variation within groups than between groups. The systems approach can be applied to planning large landscapes by grouping similar stands managing within and among groups. We have developed the "Scope & Group" program to facilitate effective grouping and the "Toggle" program to examine the effects of alternative management actions within and among groups.

Grouping Stands and Developing Alternative Silvicultural Pathways

Methods.— The Scope & Group program facilitates the process of grouping stands and selecting a representative stand for each group. When select tables generated in LMS are imported into the Scope & Group program, multiple graphs and charts are immediately generated that show the landscape's area distribution by combinations of dominant species, age classes, site indexes, slopes, and aspects. The user then stratifies the landscape into ecologically similar groups of stands based on this information, and the stands within each group are listed. The user can then choose a stand to represent each group based on the key variables (e.g., site index, slope, aspect, average DBH, trees per acre, and density measures) and local knowledge of the area.

The user then develops several silvicultural pathways for each group and applies them to the representative stand using LMS. Fifteen independent pathways have commonly been used for each group; however, more or fewer can be applied at the user's discretion. One pathway is usually a no-action pathway, reflecting the stand's development and values provided in the absence of all natural and anthropogenic disturbances.

Using LMS, the user can create a separate portfolio for the representative stand for each group by using the *subset* command under the *tools* drop-down menu. This stand is copied and renamed to create numerous "stands" of identical initial characteristics. A different pathway is applied to each initially identical stand within a group's portfolio and projected for the planning horizon. The results are displayed in output tables showing how the different objectives would change over time under each silvicultural pathway.

Results to Date.— We have applied the techniques and program described in the preceding paragraph to six landscapes in eastern and western Washington and Oregon. For all landscapes, we readily transferred the information from LMS to the Scope & Group program and displayed the information graphically. We then divided each landscape into six or more groups based on combinations of the following criteria chosen by the users: age, species composition, site index, slope, aspect, and/or elevation.

We also developed silvicultural pathways for each group in each of these six landscapes. Experience to date suggests that the results are effective, but that considerable silvicultural expertise is necessary to develop realistic silvicultural pathways.

Developing Alternative Landscape Plans

Methods.— Two methods can be used to develop a landscape plan that optimizes trade-offs among values. One method is to use an optimization program, such as LINDO (LINDO Systems Inc., Chicago, Illinois, USA). Another method is to use an iterative approach guided by the user. Both methods entail similar processing of data by grouping and developing silvicultural alternatives for each group.

A program named Toggle was developed with LMS to facilitate the iterative approach. The user transfers output tables of the alternative pathways for each group into the Toggle program. This program allows the user to view the effects on all objectives when specified proportions of each group's area are allocated to different pathways (Fig. 13-6). Inside the Toggle program, the user can change the proportion of a group allocated to each pathway rapidly. The outputs for a pathway are weighted by the proportion of the group area allocated and the total area in that group. The area-weighted outputs for all pathways across all groups are automatically summed. The Toggle program displays the summary outputs for the entire landscape in a series of charts showing the changing values over time. The user can develop different landscape alternatives by changing the proportions of any pathways in any group and can immediately understand the effects on the multiple objectives by viewing the changing graphs. The different landscape alternatives can be saved for later use.

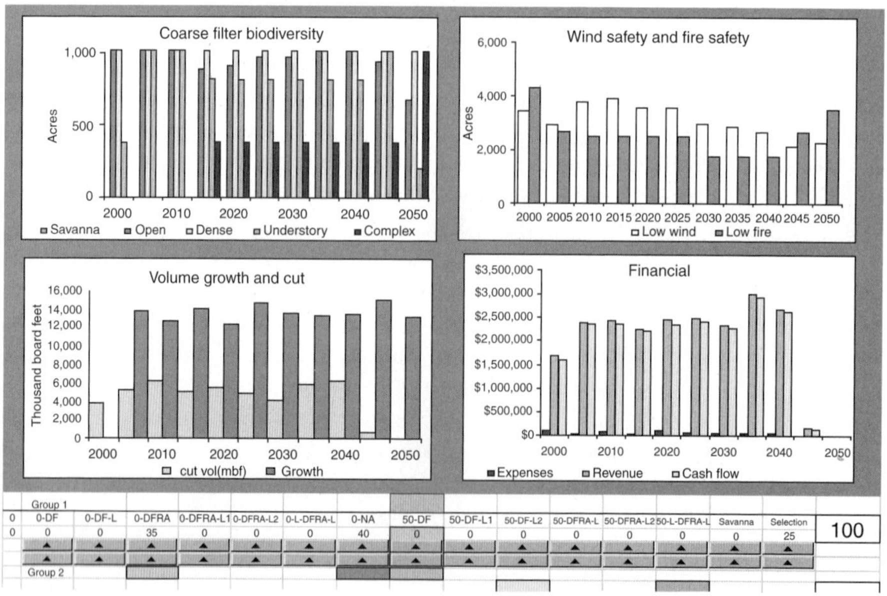

FIG. 13-6

The flow over time of many attributes can be viewed in the Toggle program. By pressing the "spinner" buttons at the bottom, the user changes the proportion of each group of stands that follows each silvicultural pathway. The values of all graphs change instantaneously, so the user can develop a management alternative consisting of a series of silvicultural pathways that provide a given trade-off of many values over time. *(Copyright, C. Oliver, Yale University, 2006).*

The Toggle program also develops a summary value for each management objective, a single number for each objective generalizing how well that objective is achieved over time. The summary value number for each objective also changes as the proportions treated by the different pathways change, showing the summary effects of changing management alternatives. Decision makers first gain an overview of the effects of different management alternatives through summary values. Then the decision maker examines promising alternatives in more depth by viewing the charts underlying the summary values. The user can store potentially interesting summary values, as well as the pathway proportions that generated them, in a matrix of objectives and alternatives (Fig. 13-7) inside the Toggle program.

The Toggle program facilitates manual (but rapid) completion of common steps required to develop multiple (nonspatial) landscape alternatives. While these steps can be accomplished with optimization tools, the iterative process is an important element of the analysis, helping the user develop a mental model of the full effects of the alternatives.

Results to Date.— In its present form, we have applied the Toggle program to six landscapes of 1,000 to 40,000 acres in western and eastern Washington and Oregon. Up to 13 graphs projected one or more objectives over time, and 18 summary values have been displayed dynamically in the Toggle program.

In two cases, Pack Forest (McKinley 2002), and Satsop Forest (Ceder 2001), the Toggle program was calibrated with groups, silviculture pathways, and objectives for the targeted forest landscape, and the calibrated program was given to representative stakeholders. These stakeholders aided in developing management alternatives. A general consensus was that the Toggle was a useful method for displaying trade-offs among objectives and for generating alternative management scenarios.

Generating Stand-Specific Prescriptions from the Groups

Methods.— After the designated decision maker chooses an alternative from the matrix described previously, local managers use their understanding of variations among stands within groups to assign specific pathways to specific stands. When managers are allocating stands to pathways, the objective is to have the same proportion of the group's area in each pathway as was assigned by the chosen alternative inside the Toggle program. Local managers can use local knowledge, matrixes generated in spreadsheets, maps, road layout and harvest scheduling programs, and GIS tools to assist in these assignments. Those stands to be treated will then receive considerable measurement attention. At the same time, the managers may be less concerned with the measurement precision of stands that will not be targeted for treatment for several decades. Consequently, an intensive inventory may not be needed on those stands to be treated much later.

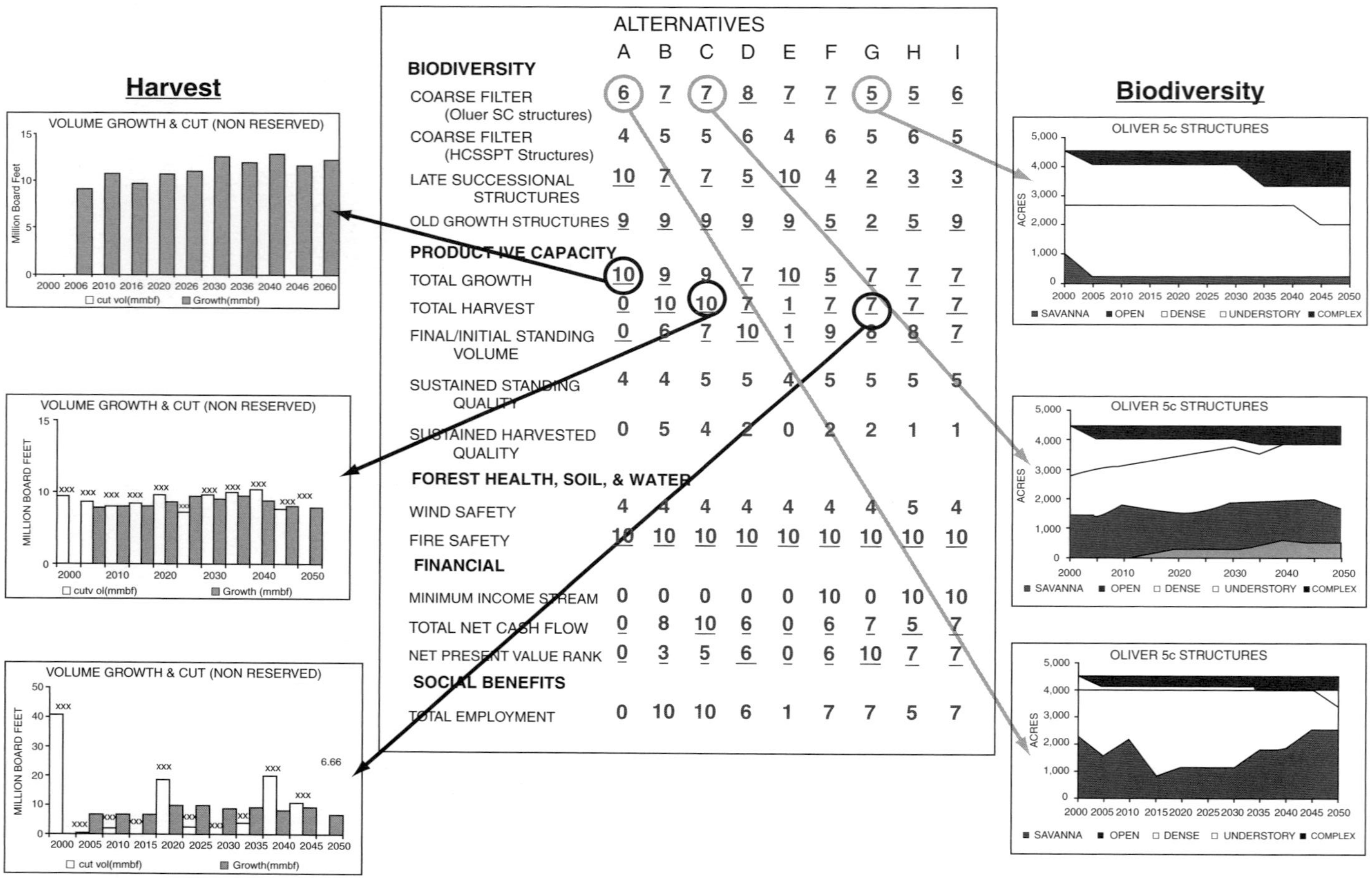

FIG. 13-7

The flow of each attribute following each management alternative can be converted to a number in the Toggle program and put into a matrix, allowing the decision maker to view a summary of the effects of a variety of management alternatives on a suite of management objectives. *(Copyright, C. Oliver, Yale University, 2006).*

Results to Date.— McKinley (2002) chose a management alternative for Pack Forest using the Toggle program and assigned stands to specific pathways using the local managers' knowledge aided by a matrix of all stands in each group and the targeted percentages of area to be treated by each pathway for the group. He ran a scenario containing the targeted prescription for each stand in LMS for the planning horizon and then compared the resulting flows of objectives over time with the chosen alternative from the Toggle program. He further modified stand assignments and again projected the alternative with LMS. After two iterations of such modifications and reprojections, the flow of objectives over time was very similar to those of the chosen alternative.

COMPARISON OF LMS WITH OTHER FOREST PLANNING MODELS

Comparing LMS with other forest models allows relative strengths and weaknesses to be identified; though one model may be more appropriate for a particular task, ultimately they are complementary, and will be discussed in that context.

LMS and LANDIS

LANDIS is a landscape disturbance and succession model that emphasizes spatial processes and patterns at broad spatial scales (Scheller et al. 2007; He, this volume). Landscapes are represented as a grid of interacting cells, with vegetation described by attributes of age-defined cohorts (longevity, shade tolerance, disturbance tolerance). LANDIS is designed as an extensible simulation environment that allows scientists to develop new forest models (i.e., new ecoregions or new disturbance modules) within the established model structure.

LANDIS simulates over larger areas and longer time periods than LMS, with a focus on spatial processes. Compared to LMS, LANDIS may help the manager understand the interaction between landscape management alternatives and disturbance agents such as fire or wind. Translating desired landscape patterns from LANDIS to silvicultural prescriptions for specific stands is better suited to LMS. The relationship between the models could also work in reverse, with the robustness of landscape alternatives developed in LMS tested against disturbance agents in LANDIS.

An important distinction should be made between the LMS and LANDIS modeling environments and potential analysts. LANDIS models are developed using a programming language, requiring higher technical skills by the analyst. LMS utilizes a graphical user interface to facilitate analysis by a wider range of individuals, including field personnel. While all analysts must understand the assumptions and limitations of a model (enforced by LANDIS), local knowledge and practical management expertise can be important elements of the analysis.

LMS and Spatial Woodstock

Spatial Woodstock (Remsoft Inc., Fredericton, New Brunswick, Canada) is a spatial harvest scheduling system with a spatial (~100,000 acres) and temporal (~100 years) planning scale similar to LMS. The model uses a hierarchical approach to forest planning, first producing long-term nonspatial landscape alternatives and then suggesting short-term spatial solutions (Walters et al. 1999). Spatial Woodstock is commonly used in places in the United States and Canada where a primary objective is timber production.

Like LANDIS, Woodstock requires a relatively sophisticated analyst to develop a forest model within a programming environment. Woodstock can directly access growth models such as those used in LMS but also commonly uses preprocessed stand growth and treatment simulations for a limited number of forest types. The LMS could be useful for stratifying the forest and simulating growth as an input to Woodstock. Though Woodstock can be run as a simulator, forest management alternatives are often developed through optimization. Optimization is a more efficient and powerful technique than the iterative approach suggested with the LMS Decision Analysis System but may be less informative for analysts and transparent for other interested parties. Hall (2001) compared the results of an optimization program (LINDO) and Toggle and noted that the optimization program appears slightly more accurate but less easy to use and less informative to the user.

Differences between LMS and Woodstock may correspond to the flow of information within an organization. Forest management plans would generally be developed in Woodstock at a central office (such as a corporate office). These plans could be resimulated in LMS for viewing by personnel in field offices. Local managers can then account for inherent system variation with their "mental models." For example, local foresters and wildlife biologists often need to refine generalized treatments from broader plans. For wildlife management, snag creation or minor vegetation conditions may influence silvicultural prescriptions, which could not be previously accounted for (especially if snag, log, and minor vegetation data are limited). Refined analysis with LMS that is augmented with local knowledge can then inform future modeling efforts in Woodstock.

DISCUSSION

The LMS and companion Decision Analysis Tools (Scope & Group and Toggle) allow wildlife habitats and other management objectives to be managed across forested landscapes of different sizes. Recognition of the strengths and weaknesses of LMS is important to use the technology appropriately and effectively.

The strengths of LMS include that it (1) incorporates modern concepts of decision analysis, systems approaches, forest dynamics, and wildlife habitat identification; (2) enables the user to perform detailed analyses with a user-friendly, point-and-click interface; (3) provides the user with multiple options for user-friendly display; (4) allows the user to adjust the input data in many ways; (5) can be applied to landscapes in many places by using relatively standard data; and (6) is available on IBM-compatible PCs.

On the other hand, there remain a number of challenges and weaknesses, which include (1) the classification algorithms classify stands into stand structures ("coarse filter" biodiversity) that have "sharp boundaries" and give no credit for being close to a desired structural classification; (2) LMS does not account well for within-stand heterogeneity or for edge effects on stand growth and therefore does not represent some of the existing variation in our landscapes; and (3) LMS visualizations and charts may give the novice user a greater sense of accuracy with the projections than is warranted.

The LMS is more useful in projecting trends than in providing exact values of wildlife habitat, timber volume, carbon sequestration, fire and insect risks, and other values. The inaccuracies involved with data collection and growth model projection reduce its effectiveness in predicting exact outcomes. On the other hand, the systematically repeated nature of the precise measurements and projections means that the errors will be systematic in nature and therefore makes the system suitable for investigating and identifying trends.

This system does not simulate stochastic stand changes because it does not simulate disturbances such as windstorms or insect outbreaks. It can, however, provide estimates of the relative susceptibility of each stand to disturbances during each time step. LMS can be used to estimate risk. Then the users can determine the level of susceptibility of disturbance they are comfortable with when managing a specific landscape. The users can simulate any disturbance with the treatment tools and project the consequences of specific events.

We anticipate both incremental and large changes in the LMS. Incrementally, we expect continuous improvement of the system as problems are identified and corrected. We also anticipate improvements to the Toggle system. We plan to develop stand structure classification algorithms that address the problem with sharp boundaries between classes. We also expect to include more prefabricated summaries of objectives (e.g., inclusion of habitat suitability indices for more species).

The LMS was designed with a modular architecture so that its components can be improved, replaced, and/or integrated with similar systems (e.g., NED and SILVAH). Improved integration with existing GIS systems is also planned. We are receiving strong interest to apply LMS to different parts of the world and to include more growth models. For this, we need to develop a stronger methodology for improving projections, perhaps through the Continuous Quality Improvement process.

Our method of inventory may change as spectral imagery and Light Detection and Ranging (LiDAR) systems become more common and further development of feature extraction occurs. In the future, it may be possible to take a "census" of all upper stratum trees in a forest, rather than the current field-based sampling methodology. This census will lead to changes in growth models, visualizations systems, and many other functions within the LMS.

CONCLUSIONS

The LMS, and similar tools, can revolutionize management from the policy to the application levels. At the policy level, an easily understood demonstration that the forest changes without human actions can avoid a presumption of stability in natural systems and a realization that wildlife habitats and risks of fire and insect damage change with time. The tools can also make managers' actions transparent, since they can demonstrate what they expect of the forest before any manipulations for management. If the forest does not appear as projected, the managers can be held accountable.

Environmental services such as carbon sequestration, wildlife habitat, fire protection, and forest cover can be more easily marketed because the buyer can know the amounts, times, and places where the services are provided. The tools allow the data to be understood and analyzed in different degrees of analytical depth, from simple visualizations to in-depth analysis of the variations within and between stands. The tools enable Continuous Quality Improvement techniques to be applied, since the projections can be compared with future monitoring and the differences corrected iteratively in projections.

A major constraint of the tools is their potential inaccuracy. This inaccuracy is caused by inaccuracies in the collection of data and can be compounded by inaccuracies in growth projections and prefabricated classifications such as habitat suitability indices. On the other hand, the precision and consistency of treatment of the data allow trends to be projected quite well. Such trends include directions of change of habitats for various wildlife species, of fire or insect risk, of standing timber volume, of carbon sequestration, and others.

SUMMARY

The Landscape Management System includes a suite of tools (current forest inventory, growth models, treatment and disturbance simulation, habitat suitability indices, hazard risk measures, carbon sequestration and economic analyses, and visualization tools) for assessing changes in tree inventory information in stands across landscapes over time. It automates many of the routine steps necessary for simulating many stands over several decades. It does not offer prescriptions for predefined goals, nor does it automatically include stochastic events such as fire or windstorms; instead, it provides estimates of susceptibility to these events. The projected tree inventory information is converted to habitat, hazard, carbon, financial, timber, and other measures. It can then be analyzed and visualized to determine current and future stand and landscape conditions. Because the users define their own objectives, they can evaluate the landscape for timber, habitat, conservation, scenic, and other objectives.

ACKNOWLEDGMENTS

Development of LMS system has been partially supported by the U.S. Forest Service.

LITERATURE CITED

Baker, P. J., and J. S. Wilson. 2000. A quantitative technique for the identification of canopy stratification in tropical and temperate forests. *Forest Ecology and Management* 127:77–86.

Bare, B. B. 1996. Hierarchical forest planning: Some general observations. Proceedings of a workshop on hierarchical approaches to forest management in public and private organizations. Pages 164–165 *in* Information Report PI-X-124, Petawawa National Forestry Institute, Canadian Forestry Service. University of Toronto, May 25–29, 1992.

Botkin, D. B., and M. T. Sobel. 1975. Stability in time-varying ecosystems. *The American Naturalist* 109:625–646.

Boyce, S. G. 1995. *Landscape forestry.* John Wiley and Sons, New York, New York, USA.

Carey, A. B., B. R. Lippke, and J. Sessions. 1999. Intentional systems management: Managing forests for biodiversity. *Journal of Sustainable Forestry* 9:83–125.

Ceder, K. R. 2001. *Using silviculture to sustain wildlife habitat: Assessing changes and trade-offs in forest habitats using a habitat evaluation procedure within the landscape management system.* Thesis, University of Washington, Seattle, USA.

Ceder, K. R., and J. M. Comnick. 2001. Potential Roosevelt elk populations on the Sol Duc Adaptive Management Area under various management prescriptions: Simulation and analysis. *In* M. E. Ferry, T. S. Peterson, and J. C. Calhoun, technical editors. *Status of elk populations on the Olympic Peninsula. Olympic Natural Resources Center Conference Proceedings.* University of Washington, Olympic Natural Resources Center, Forks, Washington, USA.

Checkland, P. 1999. *Systems thinking, systems practice.* John Wiley and Sons, New York, New York, USA.

Clegg, S. R., C. Hardy, and W. R. Nord, editors. 1996. *Handbook of organization studies.* Sage Publications, London, United Kingdom.

Cleland, D. I., and W. R. King. 1968. *Systems analysis and project management.* McGraw-Hill, New York, New York, USA.

Deming, W. E. 1982. *Out of the crisis.* Massachusetts Institute of Technology, Center for Advanced Engineering Study, Cambridge, Massachusetts, USA.

Dieter, G. E. 1991. *Engineering design.* Second edition. McGraw-Hill, New York, New York, USA.

Feigenbaum, A. V. 1983. *Total quality control.* Third edition. McGraw-Hill, New York, New York, USA.

Forestry Commission. 1998. *The UK Forestry Standard: The government's approach to sustainable forestry.* Forestry Commission, Edinburgh, United Kingdom.

Hall, J. S. 2001. *Forest management planning: A comparison between linear programming and new tools in the landscape management system.* Thesis, University of Washington, Seattle, USA.

Han, X. 2006. *Assessing the Amur tigers' current habitat and evaluating the habitat fluctuations within the dynamic forests in Northeast China and the Russian Far East.* Ph.D. Prospectus. Yale University School of Forestry and Environmental Studies, New Haven, Connecticut, USA.

Hann, D. W., A. S. Hester, and C. L. Olsen. 1995. *Organon user's manual: Edition 5.0*. Department of Forest Resources, Oregon State University, Corvallis, USA.

Hunter, M. L., Jr. 1990. *Wildlife, forests, and forestry*. Prentice Hall, Englewood Cliffs, New Jersey, USA.

Ishikawa, K. 1982. *Guide to quality control*. Asian Productivity Organization. Quality Resources. White Plains, New York, USA.

Johnson, D. H., and T. A. O'Neil. 2001. *Wildlife habitat relations in Oregon and Washington*. Oregon State University Press, Corvallis, USA.

Juran, J. M., and F. M. Gryna, editors. 1988. *Juran's quality control handbook*. Fourth edition. McGraw-Hill, New York, New York, USA.

Kimmins, J. P. 1987. *Forest ecology*. Macmillan Publishing, New York, New York, USA.

Krick, E. V. 1962. *Methods engineering: Design and measurement of work methods*. John Wiley and Sons, New York, New York, USA.

Lippke, B., J. Wilson, J. Perez-Garcia, J. Bowyer, and J. Meil. 2004. CORRIM: Life-cycle environmental performance of renewable building materials. *Forest Products Journal* 54:8–19.

Litterer, J. A. 1965. *The analysis of organizations*. John Wiley and Sons, New York, New York, USA.

Marquis, D. A., and R. L. Ernst. 1992. *User's guide to SILVAH: Stand analysis, prescription, and management simulator program for hardwood stands of the Alleghenies*. U.S. Forest Service, Northeastern Forest Experiment Station, General Technical Report NE-162, Newtown Square, Pennsylvania, USA.

Marzluff, J. M., J. J. Millspaugh, K. R. Ceder, C. D. Oliver, J. Withey, J. B. McCarter, C. L. Mason, and J. Comnick. 2002. Modeling changes in wildlife habitat and timber revenues in response to timber management. *Forest Science* 48:191–202.

McCarter, J. B., J. S. Wilson, P. J. Baker, J. L. Moffett, and C. D. Oliver. 1998. Landscape management through integration of existing tools and emerging technologies. *Journal of Forestry* 96:17–23.

McGaughey, R. J. 1997. Visualizing forest stand dynamics using the stand visualization system. Pages 248–257 *in Proceedings of the 1997 ACSM/ASPRS Annual Convention and Exposition and American Society for Photogrammetic and Remote Sensing* 4:248–257.

McGaughey, R. J. 2000. Seeing the forest and the trees: Visualizing stand and landscape conditions. Pages 105–109 *in* H. Gucinski, C. Miner, and B. Bittner, editors. *Proceedings of Views from the Ridge: Considerations for planning at the landscape level.* U.S. Forest Service, Pacific Northwest Research Station, General Technical Report PNW-GTR-596, Portland, Oregon, USA.

McKinley, M. 2002. *Final development of Pack Forest landscape management plan*. Thesis, University of Washington, Seattle, USA.

Morgan, M. G., and M. Henrion. 1990. *Uncertainty: A guide to dealing with uncertainty in quantitative risk and policy analysis*. Cambridge University Press, Cambridge, United Kingdom.

O'Hara, K. L., P. A. Latham, P. Hessburg, and B. G. Smith. 1996. A structural classification for inland northwest forest vegetation. *Western Journal of Applied Forestry* 11:97–102.

Oliver, C., D. Adams, T. Bonnicksen, J. Bowyer, F. Cubibage, N. Sampson, S. Schlarbaum, R. Whaley, H. Wiant, and J. Sebelius. 1997. *Report on forest health of the United States by the Forest Health Science Panel.* A panel chartered by Charles Taylor, Member, United States Congress, 11th District, North Carolina. (Submitted April 7, 1997), Also available as reprint through University of Washington College of Forest Resources CINTRAFOR RE43 (main document) and RE43A (summary).

Oliver, C. D. 1981. Forest development in North America following major disturbances. *Forest Ecology and Management* 3:153–168.

Oliver, C. D. 1992. A landscape approach: Achieving and maintaining biodiversity and economic productivity. *Journal of Forestry* 90:20–25.

Oliver, C. D. 1994. A portfolio approach to landscape management: An economically, ecologically, and socially sustainable approach to forestry. Pages 66–76 *in* C. R. Bamsey, editor. *Proceedings of Innovative Silvicultural Systems in Boreal Forests.* Clear Lake Limited, Edmonton, Alberta, Canada.

Oliver, C. D., D. R. Berg, D. R. Larsen, and K. L. O'Hara. 1992. Integrating management tools, ecological knowledge, and silviculture. Pages 361–382 *in* R. Naiman, editor. *New perspectives for watershed management.* Springer-Verlag, New York, New York, USA.

Oliver, C. D., M. Boydak, G. Segura, and B. B. Bare. 1999. Forest organization, management, and policy. Pages 556–596 *in* M. L. Hunter, Jr., editor. *Maintaining biodiversity in forest ecosystems.* Oxford University, United Kingdom.

Oliver, C. D., W. H. Knapp, and R. Everett. 1993. A system for implementing ecosystem management. Pages 377–388 *in* M. E. Jensen and P. S. Bourgeron, editors. *Eastside Forest Ecosystem Health Assessment: Volume 2: Ecosystem management: Principles and applications.* U.S. Forest Service, Pacific Northwest Research Station, General Technical Report PNW-GTR-318, Portland, Oregon, USA.

Oliver, C. D., and B. C. Larson. 1996. *Forest stand dynamics.* Update edition. John Wiley and Sons, New York, New York, USA.

Oliver, C. D., and B. R. Lippke. 1995. Wood supply and other values and ecosystem management in western Interior Forests. *In* R. L. Everett and D. M. Baumgartner, editor. *Ecosystem management in western interior forests.* Washington State University, Pullman, Washington, USA.

Oliver, C. D., and K. L. O'Hara. 2004. Effects of restoration at the stand level. Pages 31–59 *in* J. A. Stanturf and P. Marsden, editors. *Restoration of boreal and temperate forests.* CRC Press, New York, New York, USA.

Oliver, C. D., and M. Twery. 1999. Decision support systems: Models and analyses. Pages 661–685 *in* N. C. Johnson, A. J. Malk, W. T. Sexton, and R. C. Szaro, editors. *Ecological stewardship: A common reference for ecosystem management.* Elsevier Science Ltd., Oxford, United Kingdom.

Organon Growth and Yield Project. 2007 November 20. Organon Growth and Yield Project Home Page. <http://www.cof.orst.edu/cof/fr/research/organon/index.htm>. Accessed 20 November 2007.

Raedeke, K. 2001. Model approach to habitat conditions influencing elk populations: An approach to habitat modeling. *In* M. E. Ferry, T. S. Peterson, and J. C. Calhoun, technical editors. *Status of elk populations on the Olympic Peninsula.* Olympic Natural Resources Center Conference Proceedings, University of Washington, Olympic Natural Resources Center, Forks, Washington, USA.

Rauscher, H. M. 1995. Natural resource decision support: Theory and practice. *AI Applications* 9:1–2.

Roberts, J. C. 1979. Principles of land use planning. Pages 47–63 *in Planning the uses and management of land.* ASA-CSSA-SSSA, Madison, Wisconsin, USA.

Scheller, R. M., J. B. Domingo, B. R. Sturtevant, J. S. Williams, A. Rudy, E. J. Gustafson, and D. J. Mladenoff. 2007. Design, development, and application of LANDIS-II, a spatial landscape simulation model with flexible temporal and spatial resolution. *Ecological Modeling* 201:409–419.

Senge, P. S. 1990. *The fifth discipline: The art and practice of the learning organization.* Currency Doubleday, New York, New York, USA.

Stage, A. R. 1973. *Prognosis model for stand development.* U.S. Forest Service, Intermountain Forest and Range Experiment Station, Research Paper INT-137, Ogden, Utah, USA.

Taylor, F. W. 1911. *The principles of scientific management.* Harper and Brothers, New York, New York, USA.

Twery, M. J., P. D. Knopp, S. A. Thomasma, H. M. Rauscher, D. E. Nute, W. D. Potter, F. Maier, J. Wang, M. Dass, H. Uchiyama, A. Glende, and R. E. Hoffman. 2005. NED-2: A decision support system for integrated forest ecosystem management. *Computers and Electronics in Agriculture* 49:24–43.

U.S. Forest Service. 2007. USFS Forest Vegetation Simulator Home Page. <http://www.fs.fed.us/fmsc/fvs/>. Accessed 20 November 2007.

von Bertalanffy, L. 1968. *General systems theory: Foundations, development, applications.* George Braziller, New York, New York, USA.

Walters, K. R., U. Feunekes, A. Cogswell, and E. Cox. 1999. *A forest planning system for solving spatial harvest scheduling problems.* CORS National Conference. Toronto. 7–9 June 1999.

Wilson, J. S., and P. J. Baker. 1998. Mitigating fire risk to late-successional forest reserves on the east slope of the Washington Cascade Range, USA. *Forest Ecology and Management* 110:59–75.

Wilson, J. S., and P. J. Baker. 2001. Flexibility in forest management: Managing uncertainty in Douglas-fir forests of the Pacific Northwest. *Forest Ecology and Management* 145:219–227.

Wilson, J. S., E. S. Isaac, and R. I. Gara. 1998. Impacts of mountain pine beetle (*Dendroctonus ponderosae*) infestation on future landscape susceptibility to the western spruce budworm (*Choristoneura occidentalis*) in north central Washington. *Journal of Applied Entomology* 122:239–245.

Wilson, J. S., and R. McGaughey. 2000. Presenting landscape-scale forest information: What is sufficient and what is appropriate? *Journal of Forestry* 98:21–27.

Wilson, J. S., and C. D. Oliver. 2000. Stability and density management in Douglas-fir plantations. *Canadian Journal of Forest Research* 30:910–920.

Wilson, W. 1887. The study of administration. *Political Science Quarterly* 2:197–222.

CHAPTER

Development and Application of Habitat Suitability Models to Large Landscapes

14

William D. Dijak and Chadwick D. Rittenhouse

Technological advancements in data collection, data processing, analytical procedures, and geographic information systems (GIS) facilitate the use of spatially explicit data for modeling landscape-level wildlife-habitat relationships (Larson et al., this volume; Roloff et al., this volume). Correspondingly, there is a variety of software programs that may be used to model wildlife-habitat relationships. These programs include species presence or probability of occurrence models such as BIOCLIM (Busby 1991), HABITAT (Walker and Cocks 1991), DOMAIN (Carpenter et al. 1993), and BIOMAPPER (Hirzel et al. 2002, 2006). Carpenter et al. (1993), Guisan and Zimmermann (2000), and Hirzel and Arlettaz (2003) reviewed or discussed differences among these programs. An important distinction is made between classification-based approaches (e.g., BIOCLIM, HABITAT) that describe the extent of a species distribution (i.e., niche) based on species' presence-absence information versus multivariate, distance-based approaches that describe both the extent and probability of species occurrence within that extent (e.g., DOMAIN, BIOMAPPER). Habitat Suitability Index (HSI) models provide an alternative approach that quantifies habitat quality, as opposed to species presence or probability of occurrence directly. The suitability relationships may be defined by empirical data, literature review, or expert opinion, or a combination of these. Software programs for HSI models include Landscape Scripting Language (Kushneriuk and Rempel 2007), Landscape HSImodels (Dijak et al. 2007), and VVF (Ortigosa et al. 2000). Also, HSI models have been developed directly within GIS software (Nichols et al. 2000, Juntti and Rumble 2006, Tirpak et al. 2007). Additionally, RAMAS GIS (Akçakaya 1998) provides the means to link wildlife demographic response (i.e., population viability and metapopulation dynamics) to habitat suitability (e.g., Larson et al. 2004; Bekessy et al., this volume). Regardless of which program is used, the goals are similar: to quantify wildlife response in terms of the quality, quantity, or spatial structure of habitat.

Roloff et al. (this volume) discussed some of the issues related to the use of GIS to model landscape-level wildlife habitat. In this chapter, we continue the discussion in the context of habitat suitability model development and application to large landscapes. We refer readers to Beissinger et al. (this volume) and Akçakaya et al. (this volume) for extension of HSI models to viability modeling. We begin with an overview of HSI models and HSI model development, discuss emerging issues regarding data availability and populating landscapes from different data sources, and conclude with a case study that illustrates the use of Landscape HSImodels software to the Hoosier National Forest, Indiana.

HSI MODELS

Habitat Suitability Index modeling is an outgrowth of the Habitat Evaluation Procedures (HEPs) developed during the early 1980s (U.S. Fish and Wildlife Service 1980, 1981). The purpose of HSI models is to numerically quantify wildlife habitat quality. In their original form, HSI models were based on measurements of habitat components at a local scale, which were numerically scored (e.g., suitability indices) and combined into an overall habitat suitability value for selected wildlife species. Collection of detailed local scale information becomes impractical when evaluating habitat at the landscape scale. Surrogates of local scale habitat components may be utilized to provide information about habitat components at the landscape scale. In addition, spatial relationships of habitat components such as area sensitivity, edge sensitivity, the interspersion and quantity of life requisite habitats, and distance to resource become important when modeling wildlife-habitat relationships at the landscape scale (Morrison et al. 1998).

As with any modeling endeavor, the development of HSI models is best accomplished when following an established protocol that outlines the philosophy, assumptions, data sources, analytical approaches, validation procedures, and appropriate applications for the models. The philosophy underlying HSI models is that each species requires a distinctive set of physical environmental factors used for survival and reproduction (e.g., habitat; Block and Brennan 1993). In its most general sense, these factors include food, cover, and, in the case of birds, nest locations for reproduction (Hildén 1965). Often, these environmental factors are associated with specific vegetative communities (e.g., habitat types) and with increasing level of detail, to vegetation structure, species composition, and vegetation age or succession stage. Habitat Suitability Index models hypothesize a functional relationship between the quantity of a resource and its suitability value (or quality). The value of each of these suitability indices (SIs) range from 0 (low or nonsuitable habitat) to 1 (highly suitable) for a specific resource attribute. A composite HSI value is formed by combining multiple SIs in an HSI equation.

Increasingly, HSI models are being developed and applied within a spatial framework (Roseberry and Woolf 1998; Juntti and Rumble 2006; Dijak et al. 2007; Tirpak et al. 2007; Fitzgerald et al., this volume). The application of HSI models to large geographic areas, use of landscape-level data sources, and inclusion of spatial attributes of wildlife-habitat relationships facilitates the transition away from field-based, measurement-intensive HSI models. Large-scale HSI models have a variety of uses, from facilitating evaluation of alternative management strategies in the development of a natural resource management plan (see following description), to identification of priority areas for management activities, and estimation of population viability (Larson et al. 2004). We acknowledge that large-scale HSI models are not immune to some challenges. For example, population density is sometimes used as a surrogate for habitat quality, and some HSI model validations use density as a measure of habitat quality (Duncan et al. 1995, Breininger et al. 1998, Kroll and Haufler 2006). Because HSI models do not account for intra- or inter-specific interactions such as competition and predation, behavioral responses to changes in resource conditions (i.e., changes in space use, movements, or resource selection), nor the error associated with the HSI value (Van Horne 1983, Roloff and Kernohan 1999, Morrison et al. 2006), interpretation and validation of HSI models can be difficult for some species (Shifley et al., this volume). Despite these concerns, the relatively simple conceptual framework of HSI models, availability of GIS data layers, and use of output maps as visual aids elevate the utility of large-scale HSI models and may enhance communication between managers, planners, biologists, and stakeholders.

MODEL DEVELOPMENT

Suitability, Abundance, or Viability?

The first consideration when developing large-scale HSI models is to state the model assumptions. HSI models predict habitat suitability, which is generally assumed to be related to probability of occurrence, population density, or population viability. In other words, habitat with a high suitability value will have high population density or maintain viable wildlife populations. The degree to which this assumption is met can depend on intra- and inter-specific interactions such as competition and predation, seasonal differences in habitat use, and temporal unpredictability in resource distribution or abundance; these factors affect whether abundance is an indicator of viability (Van Horne 1983). Habitat suitability models might also not predict abundance well if regional populations exist well below carrying capacity—that is, if habitat is not limiting. For some species it may be possible to incorporate SIs that explicitly address factors influencing population density or viability. For many species, however, we lack the empirical data or knowledge to support such relationships. When such information is available, an additional consideration is whether or not to mix factors by including suitability relationships for different types of demographic responses. For example, models designed to predict habitat suitability for breeding birds may contain suitability relationships for factors influencing territory density (e.g., patch area) as well as factors affecting nest success (e.g., distance to edge) (Rittenhouse et al. 2007). Habitat Suitability Index models for habitat specialists (e.g., yellow-breasted chat [*Icteris virens*]) may perform better than HSI models for habitat generalists (e.g., wood thrush [*Hylocichla mustelina*]), particularly when the same factors influence density and nest success (Rittenhouse 2008). If HSI models contain suitability relationships for different types of demographic responses, the most appropriate use of the models may be as indicators of probability of occurrence as opposed to specific demographic response(s). We recommend adherence to the most basic assumptions of HSI models: (1) Habitat influences animal distributions; (2) HSI models predict habitat suitability (not occurrence or abundance); and (3) all significant habitat variables are included in the model.

Geographic Extent

The second consideration when developing large-scale HSI models is to explicitly define the purpose of the model and the geographic extent of application. Probably the most common purpose of HSI models for avian species is to evaluate breeding habitat suitability, since it is the most studied portion of the avian life cycle. However, many migratory avian species have spatially distinct breeding, migration, and over-wintering habitat that span multiple ecoregional domains (Bailey 1983). For these species, we recommend using an ecoregional classification system such as Bailey (1983, 1996) to establish the geographic area for model application. For example, we developed our large-scale HSI models to predict breeding habitat suitability for the Central Hardwoods Region (Rittenhouse et al. 2007), which we defined as the Hot Continental Division (220) located within the Humid Temperate Domain, excluding the mountainous portions (M220), and including the eastern portion of the Prairie Division (250; Bailey 1996). The forested areas within the Central Hardwoods Region contain primarily oak (*Quercus* spp.) and hickory (*Carya* spp.), with some maple (*Acer* spp.) and beech (*Fagus* spp.), and lesser amounts of pine (*Pinus* spp.) and cedar (*Juniperus virginiana*). This definition restricts the application of our models to the area defined; application to other regions should not occur without modification to site-specific conditions.

Spatial Grain and Extent

The third consideration is to define the spatial scale of model application. Spatial scale has two attributes: grain and extent. Grain defines the lower limit of resolution for the landscape map and is often synonymous with patch or cell size (Wiens 1989). Typically, grain is established by the size of the cells in the available GIS layers, such as the digital elevation model (DEM) or land cover type. The concept of grain may also be used in a biological context. For example, biological grain may be defined as the resolution at which an animal perceives and responds to habitat cues. In large-scale HSI models, biological grain is often expressed at the size of the average home range; however, biological grain may range from micro-habitat to a forest stand to a landscape depending on the habitat cue. Spatial extent refers to the size and location of the study area or landscape (Wiens 1989).

We define large-scale HSI models as those applicable to landscapes >1000 ha in size. Often, the goal is to apply large-scale HSI models to landscapes with high resolution (e.g., small cell size) across large spatial extents. To do this, one needs to define life requisites at multiple spatial scales within a GIS. The ability to do this for a given species is often limited by the data available.

Data Sources

Habitat Suitability Index models are relatively unique among modeling approaches in that they use both empirical data, existing knowledge (based on literature review), and expert opinion. Expert opinion may be invaluable for species with limited empirical data or to describe complex relationships. While expert opinion has great utility, it may be difficult to quantify. For example, many experts and some empirical data support the importance of canopy gaps for cerulean warblers (*Dendroica cerulea*; Burhans et al. 2001). When translating the importance of canopy gaps into a suitability relationship, one needs not only to quantify this relationship in terms of the size, distribution (i.e., random, clumped), density, or position of gaps on the landscape (e.g., bottomland gaps versus upland gaps), but also to associate some metric of cerulean warbler response to canopy gaps (e.g., nest success, population density, or survival during the breeding season). The key is quantifying the resource in terms of its attributes—size, area, quantity, density, age, type, and distribution—and have some metric of animal response to the resource (i.e., demographic, resource use, movements/space use). The transition from a purely qualitative relationship to a quantitative one not only improves the suitability relationship and overall HSI, but also identifies data needs and directions for future studies.

Ideally, empirical data would be available from multiple studies across the geographic extent of interest at multiple spatial scales that affect habitat quality. Literature searches are valuable for identifying data sources, key habitat relationships (factors), and the form of the suitability response (e.g., linear or nonlinear). The context of a study is important: The study design, methods, and analysis should be appropriate for the intended application. One should not assume that the conclusions made from studies conducted at a particular spatial scale are applicable to relationships expressed at a different spatial scale (McCarty et al. 1956). Landscape-level data are often limiting because most empirical studies have been conducted at high resolution for small geographic extents (e.g., micro-habitat or patch-level studies). The strength of the suitability relationship may be improved if it is based on studies conducted at multiple spatial scales or replicated at a single scale across multiple habitat types, study sites, or ecoregions. Another consideration when evaluating empirical data is whether the study was experimental or correlative. Experimental studies are optimal because they can identify the specific mechanisms underlying wildlife-habitat relationships; however, correlative studies are valuable when conducted across habitat gradients.

Suitability Functions

Habitat type and structure.—Landscape-scale HSI models will generally have an SI that is based on a species preference for a habitat type. Habitat types are often inferred from land cover or land use data, classified aerial photography, or stand inventory data where available. For forest species this often includes knowledge of the suitability of tree species, tree species groups (e.g., red oaks, white oaks, pine/cedar, and maple) or forest land cover type (e.g., deciduous, coniferous, mixed). We usually begin HSI models for forest species with an SI that identifies tree species, species groups, forest type, or land cover type associations (Larson et al. 2003, Rittenhouse et al. 2007). For example, we evaluate the dominant tree species (group) for each cell on the landscape and assign SI = 1.0 if the cell contains the resource or SI = 0.0 if it does not. We also typically incorporate successional stage, tree size, or age class, as an indicator of structure, in a second SI or in combination with tree species in the first SI. These functions establish the maximum extent and quantity of potentially suitable habitat.

Area sensitivity.—Additional SIs may be incorporated to address spatial relationships such as area or edge sensitivity, or the composition of habitat within a specified area (e.g., average home range size). Many avian species are considered area sensitive, meaning that a minimum area of contiguous habitat is required before occupancy or breeding occurs. We estimate an SI for area sensitivity using a patch-definition algorithm (Larson et al. 2003, Dijak et al. 2007, Rittenhouse et al. 2007). Prior to applying the algorithm, we assigned suitability based on tree age, tree species, ecological land type, or land cover type as described previously. We used the patch-definition algorithm to join adjacent (i.e., horizontal, vertical, or diagonal) cells of suitable habitat. We then used an SI to assign values to cells based on the size of the habitat patch formed by aggregation. We determined the suitability value by plotting probability of occupancy, density, or nest success on the y-axis and patch size on the x-axis. We assigned the maximum suitability value (SI = 1.0) to the patches with the highest occupancy, density, or nest success and rescaled the y-axis to range from 0 to 1. We assigned the minimum suitability value (SI = 0.0) to patches equal to the cell size (e.g., 0.09 ha for 30 m $\times$ 30 m cells) or the minimum patch size at which occupancy, density, or nest success is nonzero. The form of the function depends on the species response and may be linear or nonlinear. We fit a logistic function to the suitability by patch size data and assigned suitability to all patches using this function.

Distance.—The distance to resources can have a positive or negative effect on habitat quality. For example, bats need water within their home range in order to survive, and roost sites are often clustered around water holes (Adams and

Thibault 2006). As the distance to water increases, the energy expended to utilize the resource increases and the quality of the habitat declines. For black bears (*Ursus americanus*), habitat quality increases as distance from roads increases (Tietje and Ruff 1983). This relationship could be expressed as habitat within 200 m of a road has a value of 0. Between 200 m and 1000 m habitat would gradually increase as expressed by the formula 0.00125 * DISTANCE – 0.25 and habitat greater than 1000 m from a road is assigned a value of 1.0 (Larson et al. 2003).

Edge effects.—Another common spatial relationship is edge sensitivity. Edge sensitivity varies by the type of edge and species' response to edges. We define habitat edges as a change in land cover type (e.g., forest to grassland) or tree age and its associated structure (e.g., early successional forest to mature forest). Species response to edges may be positive if different habitat types are used to meet life requisites. For example, in the Central Hardwoods Region, northern bobwhites (*Colinus virginianus*) nest in grasslands, forage in croplands, and use woody edges for escape cover (Stoddard 1931, Roseberry and Klimstra 1984, Roseberry and Sudkamp 1998, Williams et al. 2001). Suitable habitat contains all three habitat types within a biologically relevant area, such as the average bobwhite home range size. Species response to edges may be negative if the edge decreases the probability of occupancy, survival, or nest success.

A moving window approach can be used to model edge effects. The size of the neighborhood of cells represents the distance to which an edge effect penetrates the interior of a habitat. For example, if we have a 5 cell × 5 cell circular moving window and the raster cells are 30 m × 30 m resolution, the edge effect would extend a distance of 60 m, the maximum distance any cell in the neighborhood is away from the center cell. If any of the cells within the moving window create an edge that increases or decreases the value of the habitat represented by the center cell, the center cell value of the SI would be assigned the increase or decrease in habitat quality.

Landscape composition.—We quantify the landscape context through a more computationally and data-intensive approach. We compute the percent of a particular cover type (i.e., forest) within a moving window (Larson et al. 2003, Dijak et al. 2007, Rittenhouse et al. 2007). A moving window approach requires knowledge of habitat quality as a function of percent cover type and the effective landscape size in which to evaluate the percent cover type. The size of the moving window may be based on the biology of the species (e.g., maximum dispersal or movement distance) or a large value based on landscape size or attributes needed to support a population (e.g., 1, 5, or 10 km).

Landscape composition is the relative amount of individual habitat components found within a biologically relevant area, such as an animal's home range. The habitat components must be available in the correct proportions within the specified area to achieve optimal habitat. As the proportions deviate from the ideal, habitat quality declines. We use a circular moving window to process portions of the landscape equal in size to a typical home range for a species as the area within which habitat composition would be evaluated (Fig. 14-1). The moving window for a raster cell operates by evaluating the neighboring cells

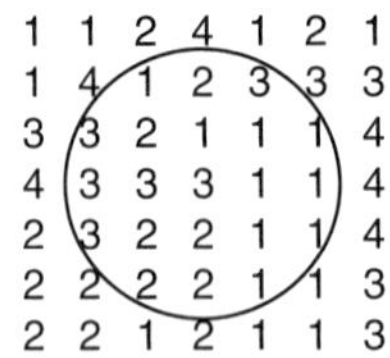

FIG. 14-1

Habitat composition determined through the use of a circular moving window. Numbers represent one of four habitats located at each raster cell. Composition of this window is 0.43 habitat 1, 0.28 habitat 2, 0.28 habitat 3, and 0.00 habitat 4.

within a given area, in this case the area of a home range, and calculates the proportion of the neighboring cells for each habitat. These proportions are then compared to proportions thought to provide the best habitat, and an SI value determined from the comparison is then assigned to the center cell and the process continues by moving to the next adjoining cell and repeating the process. Optimum landscape composition receive a value of 1.0, and values decline as composition deviates from the optimum. In the example shown in Table 14-1,

Table 14-1 Habitat Composition Table for Two Habitat Components, A and B. Axes Represent the Observed Proportions of Habitats A and B. Table Values are the Suitability Index Values Resulting from Differences Between the Observed and Ideal Compositions of Habitats A and B. Optimum Suitability would be Achieved with Observed Proportion of A at 0.2 and Proportion of B at 0.8

Proportion of Habitat Component B	Proportion of Habitat Component A										
	0.0	0.1	0.2	0.3	0.4	0.5	0.6	0.7	0.8	0.9	1.0
0.0	0.00	0.00	0.00	0.00	0.00	0.00	0.00	0.00	0.00	0.00	0.00
0.1	0.00	0.27	0.30	0.27	0.24	0.21	0.18	0.15	0.12	0.09	
0.2	0.00	0.36	0.40	0.36	0.32	0.28	0.24	0.20	0.16		
0.3	0.00	0.45	0.50	0.45	0.40	0.35	0.30	0.25			
0.4	0.00	0.54	0.60	0.54	0.48	0.42	0.36				
0.5	0.00	0.63	0.70	0.63	0.56	0.49					
0.6	0.00	0.72	0.80	0.72	0.64						
0.7	0.00	0.81	0.90	0.81							
0.8	0.00	0.90	1.00								
0.9	0.00	0.81									
1.0	0.00										

the optimum composition would be 20% component A and 80% component B. The values derived for the table values not equal to 0 use the equation (1-|optimum proportion A - observed proportion A|) * (1-|optimum proportion B - observed proportion B|). Both habitat components must be present to be considered suitable habitat, so if the proportion of either component equals 0, the composition is equal to 0. If the decline in habitat quality is thought to be more severe as the proportions deviate from the ideal, one or both terms can be squared or cubed. Other formulas are possible including the geometric or arithmetic mean of the two terms.

INPUT DATA LAYERS

Various data layers are used to provide information on landscape characteristics such as landform, land cover, and DEMs, which are important in defining habitat suitability. A basic landform map (Fig. 14-2) can be derived from a DEM using a topographic position index (Jenness 2006, Tirpak et al. 2007). Topographic position index is calculated as the elevation of a particular cell minus the mean elevation of cells in a moving window neighborhood divided by the standard deviation of the mean cell elevation within the window. Slope and aspect layers are created from the DEM. The slope layer and two moving windows of different sizes representing a large and small scale are used to evaluate a cell's elevation compared to the large-scale variation and small-scale variation in elevation to define landform classes. Decision rules (Table 14-2) provide an example of how different landform classes are determined. Landform in some instances can be used to identify ecological land types (ELTs; Van Kley 1994).

Input layers that we commonly use for forest species include a general land cover map (i.e., forest, croplands, water, etc.), a landform map (i.e., ridge, bottomlands, etc.), a dominant tree species map (i.e., white oak, maple, etc.), and a dominant tree age class map. Age class maps can be replaced with maps defining areas of similar forest structure if age classes are unavailable. Land cover maps are available from a variety of sources. The national land cover data (NLCD) map provides land cover based on classified satellite imagery with a resolution of 30 m that spans the United States. Many states have developed their own land cover classification as part of the national Geographic Approach to Planning for Biological Diversity (GAP) project. Some states have classified satellite imagery from the National Agriculture Statistics Service (NASS). Most states have National Aerial Imagery Program (NAIP) data that can be used as the basis to digitize land cover layers for smaller regions of interest. These land cover data layers offer general land cover classes that can be the foundation of land cover data used in HSI modeling.

The land cover data can be augmented with data from Forest Inventory and Analysis (FIA) data (Miles et al. 2001) and landform data to create spatially representative forest type maps, forest species composition maps, and forest age class

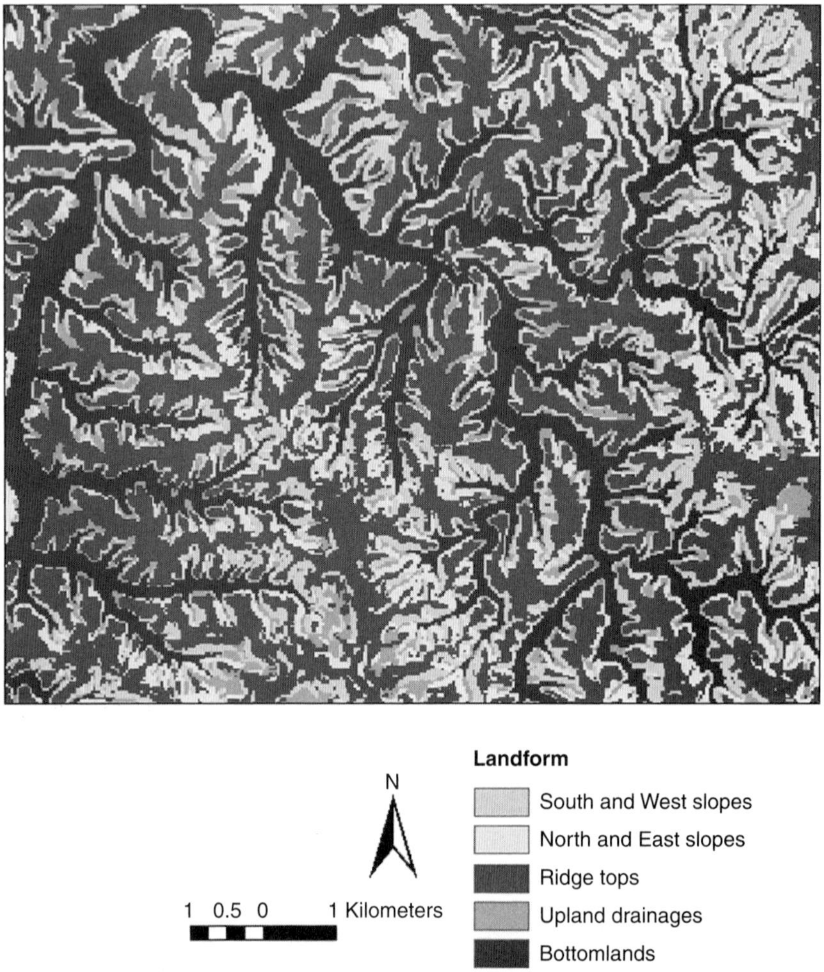

FIG. 14-2

Map of landforms created using topographic position index calculated from a digital elevation map.

maps. These maps will not be spatially accurate (i.e., placing a tree of specific age and species on an exact point in the landscape) but will be spatially representative of forests within a region. Tree species and age maps are created using land cover maps to separate forested lands from nonforested lands. If information in the land cover maps also separates deciduous, coniferous, and mixed forests, location of forest types within those land covers becomes more spatially representative of the true condition. The land cover data are combined with landform data to create patches representing the different combinations of forest land covers and landforms. These patches are surrogates for forest stands. Forest Inventory Analysis data from each state is broken up into geographic regions

Table 14-2 Criteria used to Assign Landform Classes for Landscape-Scale Habitat Suitability Models Based on Topographic Position Index

	Topographic Position Index (TPI)			
Landform	**240 m Radius Window**	**750 m Radius Window**	**% slope**	**Aspect**
Bottomland	TPI < 1 SD	TPI ≤ –1 SD		
Upland drainage	TPI ≤ –1 SD	–1 SD < TPI		
S&W slopes	–1 SD < TPI < 1 SD	–1 SD < TPI < 1 SD	slope > 5%	135 > aspect < 315
N&E slopes	–1 SD < TPI < 1 SD	–1 SD < TPI < 1 SD	slope > 5%	135 ≤ aspect ≥ 315
Ridges	–1 SD < TPI	–1 SD < TPI	slope ≤ 5%	
Ridges	TPI ≥ 1 SD			

called units; FIA data should be used from the unit that corresponds to the geographic extent of the landscape being created. Forest Inventory Analysis data for the unit is summarized to represent the proportions of forest types and age classes by landform and forest land cover (deciduous, coniferous, etc.). Forest Inventory Analysis age classes are converted to size classes, and a forest type and forest size class are randomly assigned to each patch based on the proportions of forest types and size classes found in the FIA data. The next step is to create the tree species and tree age maps. All subplots are pooled for each combination of landform, forest type, and size class plots. Subplots are assigned to a raster cell based on the raster cells' landform, forest type, and size class. The subplot data contain the list of tree species and diameters found on the subplot. The tree diameters are converted to tree ages, and the dominant tree species and age are assigned to cells in the dominant tree species and age maps. Though this process is tedious, it retains the patchy nature of forest stands by first assigning forest types and size class but includes the heterogeneity of species and age classes found within forest stands.

Another source of base map information is forest stand inventory data collected by national and state forests. These inventory data layers provide information about the forest type, forest structure, size class, and/or age of forest stands within sampling areas within a state. Using FIA subplot data and a land form data layer as in our previous example, one can create tree composition data layers that reflect the tree species and ages of trees typical to the forest stands. Similar methods of assigning forest structure parameters from FIA to forest patches are discussed by Tirpak et al. (2007).

The methods discussed in the preceding paragraphs describe ways to develop spatially representative data layers of forest tree species and structure.

Advancements are being made to directly measure these landscape attributes to create spatially exact rather than spatially representative landscapes.

Light Detection and Ranging (LiDAR) technology is leading the way in providing direct structural measurement of forests from remotely sensed technology. Forest structure has been shown to be important to a variety of species of birds (MacArthur and MacArthur 1961, James 1971, Rotenberry and Wiens 1980). LiDAR uses a pulsed laser beam emitted from an airplane or helicopter flying a specified route. The time it takes the light beam to reflect back to the aircraft can be used to determine the elevation of an object on the ground. Light beams that pierce the canopy and reflect from the ground are used to determine surface elevation. LiDAR data are often collected at submeter resolution and can have a vertical accuracy of 15 cm. Forest structure such as mean tree canopy height, dominant tree height, mean diameter, stem number, basal area, timber volume (Naesset 2002), canopy density (Lefsky et al. 1999, Maier et al. 2006), and quadratic mean canopy height (Lefsky et al. 1999) can be calculated from LiDAR in certain forest types. As the costs of acquiring LiDAR declines and the potential of the data to solve questions increases, LiDAR is becoming an essential data layer in many projects. For example, the U.S. Army Corps of Engineers and the U.S. Natural Resources Conservation Service are in the process of acquiring 2900 square miles of LiDAR data along the Missouri River for preliminary design of agricultural practices such as terracing, grade stabilization, and vegetative condition. It is expected that the data will enable them to perform detailed land cover mapping and vegetative species identification along the flood plain. Similar acquisitions of data are occurring across the United States. In an effort to expand the availability and utilization of LiDAR, the first U.S. National LiDAR Initiative meeting was held in February 2007 in Reston, Virginia.

Advancements are also being made in image classification of remotely sensed data for nondiscrete habitat classes. For example, texture analysis allows for the classification of habitats where there is high structural diversity but little distinct change from one habitat type to the next, such as what might occur in semi-arid regions and grasslands. The process evaluates more than the values of an individual raster cell. It bases the classification on repeated patterns occurring in a neighborhood of raster cells. Texture is defined by Hawkins (1969, p. 347) as (1) "some local 'order' is repeated over a region, which is large in comparison to the order's size"; (2) "the order consists in a nonrandom arrangement of elementary parts"; and (3) "the parts are roughly uniform entities having approximately the same dimensions everywhere within the textured region." For other definitions of texture and methods to determine texture, see Haralick et al. (1973) and Tuceryan and Jain (1998). Applications of image texture analysis include predicting avian species richness (Hepinstall and Sader 1997, Knick and Rotenberry 2000, St-Louis et al. 2006) and mapping nesting habitat (Pasher et al. 2007).

Software has been developed to perform object-oriented classification of imagery and LiDAR data including textured areas through a process known as segmentation. One such software, Definiens eCognition software (Definiens 2003), has been used to classify satellite and LiDAR data simultaneously to create land cover polygons of agricultural lands (Manakos et al. 2000). Their object-oriented classification outperformed the traditional ISODATA pixel classification approach. Levick and Rogers (2006) used object-oriented classification of color aerial photography and LiDAR data to monitor the spatio-temporal changes of savanna woody vegetation in Kruger Park, South Africa.

HSI EQUATIONS

An HSI value is a combination of individual SIs. The functional response by a species to a resource attribute can take many forms, but the most commonly used form for SIs is a linear relationship. More complex forms may be appropriate when supported by empirical data or expert opinion. These include sigmoid, exponential, and piecewise-regression functions. We recommend using a sigmoid function when there is uncertainty about the endpoints of the hypothesized relationship. Piecewise regression may be used to estimate the breakpoints (i.e., thresholds) of nonlinear suitability relationships, such as a species' response to edge effects (Toms and Lesperance 2003). However, these equations are data hungry and computationally intensive.

The form of the HSI equation varies depending on whether an SI represents a critical or limiting resource, or modifies a resource based on a spatial attribute such as size, proximity to edge, or composition. We used geometric, arithmetic, and logical relationships to calculate HSI scores depending on the number and type of species' life requisites (Larson et al. 2003, Dijak et al. 2007, Rittenhouse et al. 2007). We used a geometric mean when all habitat characteristics were necessary for habitat suitability:

$$HSI = \sqrt[3]{SI_1 \times SI_2 \times SI_3}.$$

With a geometric mean, the HSI value is zero if any suitability index is zero. We used an arithmetic mean when habitat characteristics were substitutable. In other words, the HSI value is greater than zero when at least one SI is nonzero.

Suitability indexes may be included as modifiers to decrease habitat quality. For example, we included an SI for fire in our worm-eating warbler (*Helmitheros vermivorum*) HSI model (Rittenhouse et al. 2007). The final habitat suitability value was the geometric mean of deciduous habitat (SI_1), tree age by ELT (SI_2), and deciduous patch size (SI_3), multiplied by SI_4 to account for reduced suitability due to fire:

$$HSI = \left(\sqrt[3]{SI_1 \times SI_2 \times SI_3}\right) \times SI_4.$$

Logical relationships are useful when a species' life requisites cannot occur in a single cell. Recall the northern bobwhite example earlier, where bobwhites use woody edges for escape cover, grasslands for nesting, and croplands for forage.

In this situation, the suitability value of a given cell represents only one of the life requisites. We used a maximum function to identify the greatest contribution to habitat suitability among the three requisites (Rittenhouse et al. 2007). The final habitat suitability value was the sum of (1) the maximum value of grassland (SI_1), cropland (SI_2), and woody cover (SI_3); and (2) the product of habitat composition (SI_4) and a modifier to reduce the suitability of roads and urban areas within the moving window for habitat composition (SI_5):

$$HSI = Maximum(Maximum(SI_1, SI_2), SI_3) + (SI_4 \times SI_5).$$

We used an additive HSI equation instead of a geometric mean because we recognized that grassland, cropland, or woody cover provided bobwhite habitat; however, the highest suitability value occurred when at least two of the three habitat types were present within a bobwhite's home range. Alternatively, a minimum function can be used when a suitability index represents a limiting factor.

LANDSCAPE HSImodels SOFTWARE

We developed Landscape HSImodels software (Dijak et al., 2007) to provide a user-friendly interface to evaluate the spatial relationships of wildlife habitat at the landscape scale. Version 2.1.1 contains models for 21 species of wildlife, including American woodcock (*Scolopax minor*), black bear, bobcat (*Lynx rufus*), cerulean warbler, eastern wild turkey (*Meleagris gallopovo silvestris*), gray squirrel (*Sciurus carolinensis*), Henslow's sparrow (*Ammodramus henslowii*), hooded warbler (*Wilsonia citrina*), Indiana bat (*Myotis sodalis*), northern bobwhite, northern long-eared bat (*Myotis septentrionalis*), ovenbird (*Seiurus aurocapilla*), pine warbler (*Dendroica pinus*), prairie warbler (*Dendroica discolor*), red bat (*Lasiurus borealis*), ruffed grouse (*Bonasa umbellus*), southern redback salamander (*Plethodon serratus*), Timber rattlesnake (*Crotalus horridus*), wood thrush, worm-eating warbler, and yellow-breasted chat. A generic model is also included so that suitability relationships from different species models can be recombined into models for a species not represented in the software. We created models using literature review, expert opinion, and from previous local-scale models (see Larson et al. [2003] and Rittenhouse et al. [2007]). Each species model contains an interface that guides the user through the calculation of each SI (Fig. 14-3). The individual SIs are combined into an overall HSI by an equation specified by the user (Fig. 14-3). All models come with default parameters and equations developed for the Central Hardwoods Region of the United States (Larson et al. 2003, Rittenhouse et al. 2007) but can be modified to fit habitat relationships that occur in other parts of a species range.

Input and output data formats are ASCII rasters, which may be created in ArcView 3.x by exporting a data source, in ArcGIS using ArcToolBox, and in ArcInfo by issuing the gridascii command. ASCII rasters created in other GIS software packages need to follow the Environmental Systems Research Institute (ESRI, Redlands, California, USA) format for header lines (ESRI, ArcGIS, ArcView,

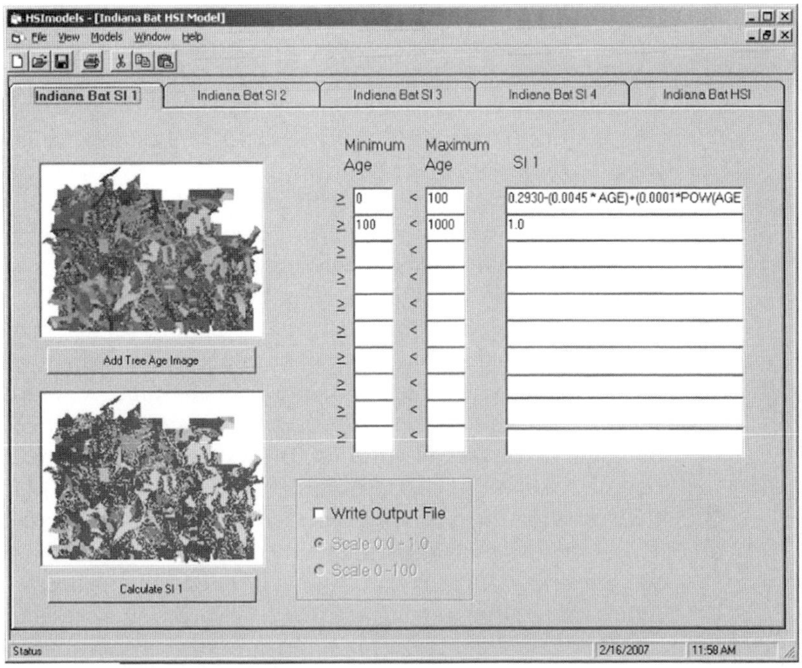

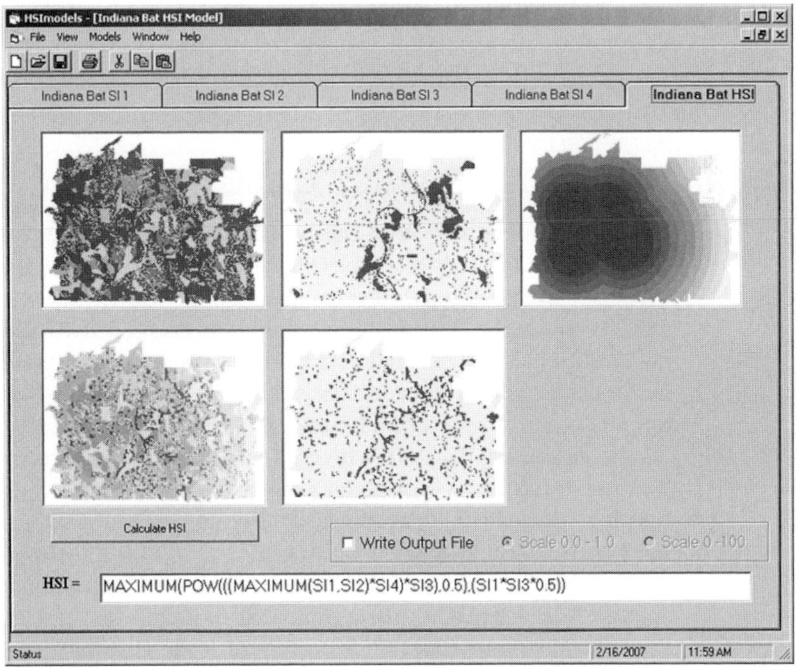

FIG. 14-3 (See color insert.)

Examples of the Indiana bat model in Landscape HSImodels software. The top window is a map of tree age (light green for young forest and dark green for older forest) and the resulting suitability index values ranging from light yellow (SI 1 = 0.0) to dark red (SI 1 = 1.0). The lower window is a map of the overall Habitat Suitability Index (lower, left map) and four suitability index maps.

ArcToolBox, ArcInfo are trademarks, registered trademarks, or service marks of ESRI in the United States, the European Community, or certain other jurisdictions, Environmental Services Research Institute, Redlands, California, USA). Input layers as well as SI and HSI layers are displayed as the user works through the model. All SI and HSI layers can be exported from the software and imported back into GIS software for further analysis.

The minimum computer system recommended is a PC with a 1.7 gigahertz (GHz) processor and 500 megabytes of random access memory (RAM). We also recommend using a 17-inch or larger monitor. Computers with faster processors and more RAM will reduce model processing time. A computer with the above configuration was successfully used to process a 1200 row by 1200 column landscape. A landscape with 2000 rows and 3000 columns was modeled on a computer with a 3.0 GHz processor and 2 gigabytes of RAM. The maximum size of a landscape that can be processed will vary from model to model based on the number of individual suitability indices incorporated into the model and the complexity of the calculations that need to be processed within the model. Models using large moving windows on large landscapes take several hours to complete. The limitation in landscape size is controlled by the amount of RAM the operating system is capable of utilizing. At the time of this printing, none of the HSI models have been validated, and the authors recognize the importance of validation. The software has been applied to districts of the Hoosier and Mark Twain National Forests (Shifley et al. 2006).

HOOSIER NATIONAL FOREST CASE STUDY

Working cooperatively with the personnel of the Hoosier National Forest (HNF) (Fig. 14-4), and in support of the HNF management plan, we applied Landscape HSI-models to five proposed forest management plan alternatives. Alternative 1 was the current plan and was mostly focused on uneven-aged management using single tree and group selection harvesting of timber with only a small percentage of the forest being harvested per decade. Alternative 2 had no harvesting, no maintenance of openings, and no prescribed burning. Alternative 3 had greater levels of uneven-aged management than alternative 1 and included a moderate amount of prescribed burning. Alternative 4 had even-aged management and a high level of prescribed burning. Alternative 5 was similar to alternative 1 but provided for a focal area that used even-aged management to provide for wildlife species needing early successional forest. Alternatives 3 and 4 also included this focal area for early successional species. The alternative plans were first modeled through LANDIS, a forest landscape simulation model (Mladenoff et al. 1996, He et al. 1999, Mladenoff and He 1999) that applies forest management practices and natural disturbance to current conditions to produce maps of future forest age class patterns and forest species composition. Methods similar to those described above were used to build input layers for LANDIS with the exception that multiple species and age cohorts were

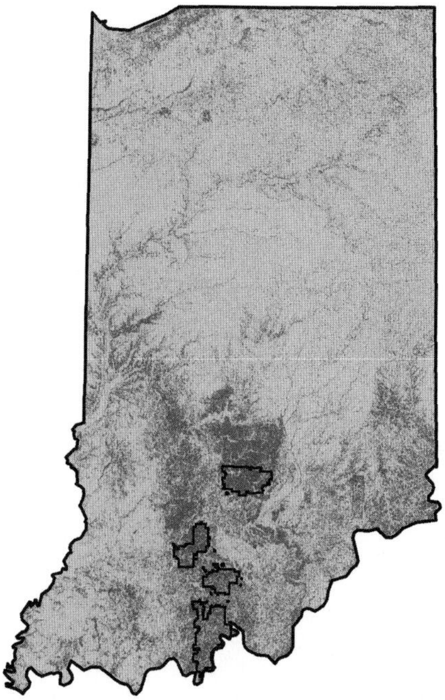

FIG. 14-4

Hoosier National Forest (outlined in black) located in south central Indiana, USA. Dark gray depicts forested; and light gray, nonforested areas of the state.

assigned to each cell of the current condition map, since LANDIS uses this information in forecasting future forest landscapes. Landscapes were modeled at 10 m resolution representing the size of a mature tree crown so that single tree selection harvesting could be modeled. The forest landscapes were modeled through 15 decades of each management alternative producing sets of forest landscape maps at each decade. LANDIS output maps were converted to ASCII rasters, and nine wildlife models were then applied to the current conditions as well as maps forecasting forest conditions at 10, 50, and 150 years of age.

We modeled the effects of alternatives on American woodcock, cerulean warbler, Henslow's sparrow, Indiana bat, northern bobwhite, ruffed grouse, wood thrush, worm-eating warbler, and yellow-breasted chat; these represented species that were disturbance dependent, area sensitive, edge sensitive, fire sensitive, mast dependent, game species, species dependent on specific forest ages or structures, and species of special concern. By selecting a suite of species that respond in different ways to varying management methods, we were able to evaluate the trade-offs in habitats for each species. Changes from current condition HSI values occurred over time and between competing alternatives. Habitat Suitability Index maps (Fig. 14-5) were produced as well as tabular summaries and charts (Fig. 14-6). Alternatives 1 and 2 did not

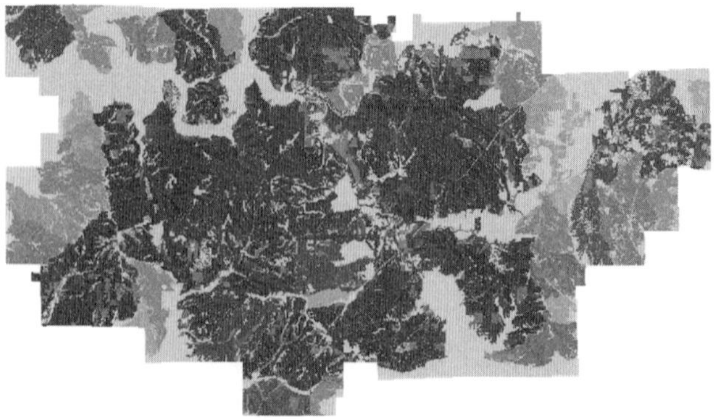

FIG. 14-5

Worm-eating warbler Habitat Suitability Index map for current condition on the Pleasant Run district of the Hoosier National Forest, Indiana, USA. Values range from 0.0 (light gray) to 1.0 (dark gray).

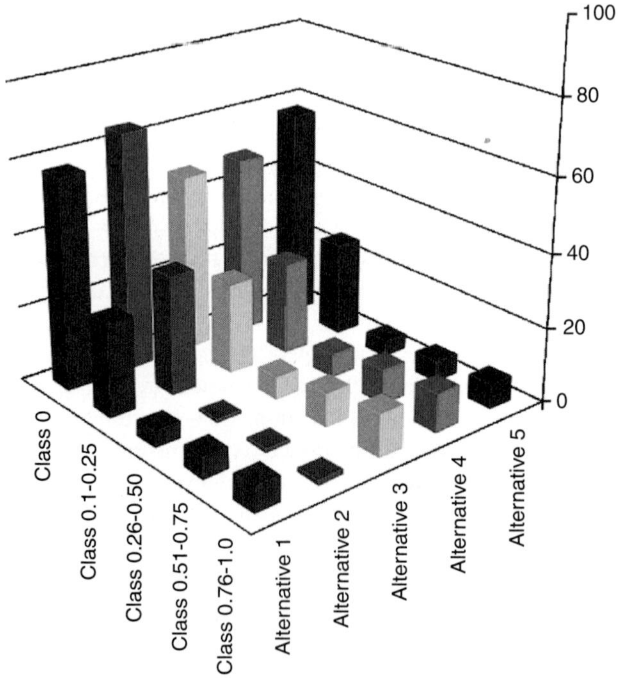

FIG. 14-6

The proportion of the Hoosier National Forest, Indiana, USA, that falls within five different habitat suitability classes for ruffed grouse after 50 years of forest management under five different management alternatives.

provide adequate habitat for American woodcock, ruffed grouse, and yellow-breasted chat. Alternative 2 did not provide adequate habitat for Henslow's sparrow. Alternatives 3, 4, and 5 all provided adequate habitat in varying degrees to all species. This information was included with the proposed plan alternatives to provide managers and stakeholders with information on the cumulative effects over time of all the proposed management alternatives.

SUMMARY

Extending HSI modeling to the landscape scale allows for the evaluation of habitat quality for larger geographic areas based on our knowledge of spatial wildlife-habitat relationships. When used with landscape forest simulation models, they provide a method of evaluating temporal changes, including proposed management activities. Landscape-level planning and management of populations requires knowledge of habitat quality at the landscape scale.

Suitability indices can be developed to represent habitat relationships based on habitat type and structure and landscape patterns such as patch size, distance to features, edge effects, and landscape composition. Input layers in the form of GIS layers can be developed from a variety of remote sensing products or large-scale field inventories to calculate suitability index values based on landform, land cover, forest type or tree species, forest age class, etc. By varying the values of SIs and varying the methods used to combine SIs into an HSI, we can examine the effects of individual habitat components on overall habitat suitability to help us to determine which habitat components are most lacking for a species.

Methods of deriving landscape information and monitoring landscape changes are improving quickly, and the availability of software such as Landscape HSImodels (Dijak et al. 2007) further facilitates the use of large-scale HSI models. Better and more concise models can be developed as our knowledge of habitat components increases at the landscape scale, but management cannot and should not wait for the perfect model. We contend that applying the best current knowledge is better than waiting until all wildlife-habitat relationships are thoroughly defined.

ACKNOWLEDGMENTS

We wish to thank Michael Larson, Joshua Millspaugh, and Frank Thompson for their insightful comments on this chapter.

LITERATURE CITED

Akçakaya, H. R. 1998. *RAMAS GIS: Linking landscape data with population viability analysis. Version 3.0.* Applied Biomathematics, Setauket, New York, USA.

Adams, R. A., and K. M. Thibault. 2006. Temporal partitioning by bats at water holes. *Journal of Zoology* 270:466–472.

Bailey, R. G. 1983. Delineation of ecosystem regions. *Environmental Management* 7:365–373.

Bailey, R. G. 1996. *Ecosystem geography.* Springer-Verlag, New York, New York, USA.

Block, W. M., and L. A. Brennan. 1993. The habitat concept in ornithology: Theory and applications. Pages 35–91 *in* D. M. Powers, editor. *Current ornithology.* Volume 11. Plenum Press, New York, New York, USA.

Breininger, D. R., V. L. Larson, B. W. Duncan, and R. B. Smith. 1998. Linking habitat suitability to demographic success in Florida scrub-jays. *Wildlife Society Bulletin* 26:118–128.

Burhans, D. E., W. D. Dijak, and F. R. Thompson, III. 2001. *Conservation assessment: Cerulean warbler* Dendroica cerulea. USDA Forest Service, North Central Research Station, Columbia, Missouri, USA.

Busby, J. R. 1991. BIOCLIM—a bioclimate analysis and prediction system. Pages 64–68 *in* C. R. Margules and M. P. Austin, editors. *Nature conservation: Cost effective biological surveys and data analysis.* CSIRO, Melbourne, Australia.

Carpenter, G., A. N. Gillison, and J. Winter. 1993. DOMAIN—A flexible modeling procedure for mapping potential distributions of plants and animals. *Biodiversity and Conservation* 2:667–680.

Definiens. 2003. *Definiens, eCognition, object oriented image analysis, Version 3.* Definiens Imaging, Munich, Germany.

Dijak, W. D., C. D. Rittenhouse, M. A. Larson, F. R. Thompson, III, and J. J. Millspaugh. 2007. Landscape habitat suitability index software. *Journal of Wildlife Management* 71:668–670.

Duncan, B. W., D. R. Breininger, P. A. Schmalzer, and V. L. Larson. 1995. Validating a Florida scrub-jay habitat suitability model, using demography data on Kennedy Space Center. *Photogrammetric Engineering & Remote Sensing* 61:1361–1370.

Guisan, A., and N. E. Zimmermann. 2000. Predictive habitat distribution models in ecology. *Ecological Modelling* 135:147–186.

Haralick, R. M., K. Shanmugan, and I. Dinstein. 1973. Textural features for image classification. *IEEE Transactions on Systems, Man, and Cybernetics SMC* 3:610–621.

Hawkins, J. K. 1970. Textural properties for pattern recognition. *In* B. S. Lipkin and A. Rosenfeld, editors. *Picture processing and psychopictorics* pp. 347–370. Academic Press, New York, New York, USA.

He, H. S., D. J. Mladenoff, and J. Boeder. 1999. An object-oriented forest landscape model and its representation of tree species. *Ecological Modelling* 119:1–19.

Hepinstall, J. A., and S. A. Sader. 1997. Using Bayesian statistics, thematic mapper satellite imagery, and breeding bird survey data to model bird species probability of occurrence in Maine. *Photogrammetric Engineering & Remote Sensing* 63:1231–1237.

Hildén, O. 1965. Habitat selection in birds, a review. *Annales Zoologici Fennici* 2:53–75.

Hirzel, A. H., and R. Arlettaz. 2003. Modelling habitat suitability for complex species distributions by the environmental-distance geometric mean. *Environmental Management* 32:614–623.

Hirzel, A. H., J. Hausser, and N. Perrin. 2002. *Biomapper 2.0.* Division of Conservation Biology, University of Switzerland, Bern. <http://www.unil.ch/biomapper>.

Hirzel, A. H., G. Le Lay, V. Helfer, C. Randin, and A. Guisan. 2006. Evaluating the ability of habitat suitability models to predict species presences. *Ecological Modelling* 199:142–152.

James, F. C. 1971. Ordination of habitat relationships among breeding birds. *Wilson Bulletin* 83:215–236.

Jenness, J. 2006. Topographic Position Index (tpi_jen.avx) extension for ArcView 3.x version 1.2. Jenness Enterprises website. <http://www.jennessent.com/arcview/tpi.htm>. Accessed 15 January 2008.

Juntti, T. M., and M. A. Rumble. 2006. *Arc habitat suitability index computer software*. U.S. Forest Service, Rocky Mountain Research Station. General Technical Report RMRS-GTR-180WWW, Fort Collins, Colorado, USA.

Knick, S. T., and J. T. Rotenberry. 2000. Ghosts of habitats past: Contribution of landscape change to current habitats used by shrubland birds. *Ecology* 81:220–227.

Kroll, A. J., and J. B. Haufler. 2006. Development and evaluation of habitat models at multiple spatial scales: A case study with the dusky flycatcher. *Forest Ecology and Management* 229:161–169.

Kushneriuk, R. S., and R. S. Rempel. 2007. *LSL: Landscape Scripting Language*. Centre for Northern Forest Ecosystem Research, Ontario Ministry of Natural Resources, Thunder Bay, Ontario, Canada.

Larson, M. A., W. D. Dijak, F. R. Thompson, III, and J. J. Millspaugh. 2003. *Landscape-level habitat suitability models for 12 wildlife species in southern Missouri*. U.S. Forest Service, North Central Research Station, General Technical Report NC-233, St. Paul, Minnesota, USA.

Larson, M. A., F. R. Thompson, III, J. J. Millspaugh, W. D. Dijak, and S. R. Shifley. 2004. Linking population viability, habitat suitability and landscape simulation models for conservation planning. *Ecological Modelling* 180:103–118.

Lefsky, M. A., D. Harding, W. B. Cohen, G. Parker, and H. H. Shugart. 1999. Surface LiDAR remote sensing of basal area and biomass in deciduous forests of eastern Maryland, USA. *Remote Sensing of the Environment* 67:83–98.

Levick, S. R., and K. H. Rogers. 2006. LiDAR and object-based image analysis as tools for monitoring the structural diversity of savanna vegetation. *In* S. Lang, T. Blaschke, and E. Schöpfer, editors. *Proceedings of the 1st International Conference on object-based image analysis.* International Archives of Photogrammetry, Remote Sensing and Spatial Information Sciences, Volume 34, Part XXX.

MacArthur, R. H., and J. W. MacArthur. 1961. On bird species diversity. *Ecology* 42:594–598.

Maier, B., D. Tiede, and L. Dorren. 2006. Assessing mountain forest structure using laser scanning and landscape metrics. *In* S. Lang, T. Blascbke, and E. Schöpfer, editors. *Proceedings of the 1st International Conference on object-based image analysis*. International Archives of Photogrammetry, Remote Sensing and Spatial Information Sciences, Volume XXX VI4/C42, Saizburg, Austria.

Manakos, I., T. Schneider, and U. Ammer. 2000. A comparison between the ISODATA and the eCognition classification methods on basis of field data. *International Archives of Photogrammetry, Remote Sensing and Spatial Information Sciences,* Volume XXXIII (Supplement B7): 133–139.

McCarty, H. H., J. C. Hook, and D. S. Knos. 1956. *The measurement of association in industrial geography.* Department of Geograpy Report 1, State University of Iowa, Iowa City, USA.

Miles, P. D., G. J. Brand, C. L. Alerich, L. F. Bednar, S. W. Woudenberg, J. F. Glover, and E. N. Ezell. 2001. *The forest inventory and analysis database: Database description and users manual version 1.0*. U.S. Forest Service, North Central Research Station, General Technical Report NC-218, St. Paul, Minnesota, USA.

Mladenoff, D. J., and H. S. He. 1999. Design, behavior and application of LANDIS, an object-oriented model of forest landscape disturbance and succession. Pages 125–162 *in* D. J. Mladenoff and W. L. Baker, editors. *Spatial modeling of forest landscape change: Approaches and applications.* Cambridge University Press, Cambridge, United Kingdom.

Mladenoff, D. J., G. E. Host, J. Boeder, and T. R. Crow. 1996. LANDIS: A spatial model of forest landscape disturbance, succession, and management. Pages 175–180 *in* M. F. Goodchild, L. T. Steyaert, B.O Parks, C. Johnston, D. Maidment, M. Crane, and S. Glendinning, editors. *GIS and environmental modeling: Progress and research issues.* GIS World Books, Fort Collins, Colorado, USA.

Morrison, M. L., B. G. Marcot, and R. W. Mannan. 1998. *Wildlife-habitat relationships: Concepts and applications*. Second edition. University of Wisconsin Press, Madison, Wisconsin, USA.

Morrison, M. L., B. G. Marcot, and R. W. Mannan. 2006. *Wildlife-habitat relationships: Concepts and applications*. Third edition. Island Press, Washington, D.C., USA.

Naesset, E. 2002. Predicting forest stand characteristics with airborne scanning laser using a practical two-stage procedure and field data. *Remote Sensing of Environment* 80:88-99.

Nichols, J. M., S. A. Husain, and C. Papadas. 2000. *Integrating GIS technology with forest management and habitat assessment efforts on our national forests.* U.S. Forest Service, North Central Research Station, General Technical Report NC-205, St. Paul, Minnesota, USA.

Ortigosa, G. R., G. A. De Leo, and M. Gatto. 2000. VVF: Integrating modelling and GIS in a software tool for habitat suitability assessment. *Environmental Modelling & Software* 15:1-12.

Pasher, J., D. King, and K. Lindsay. 2007. Modelling and mapping potential hooded warbler (*Wilsonia citrina*) habitat using remotely sensed imagery. *Remote Sensing of Environment* 107:471-483.

Rittenhouse, C. D., W. D. Dijak, F. R. Thompson, III, and J. J. Millspaugh. 2007. *Development of landscape-level habitat suitability models for ten wildlife species in the Central Hardwoods Region.* U.S. Forest Service, Northern Research Station, General Technical Report NRS-4, Newtown Square, Pennsylvania, USA.

Roloff, G. J., and B. J. Kernohan. 1999. Evaluating reliability of habitat suitability index models. *Wildlife Society Bulletin* 27:973-985.

Roseberry, J. L., and W. D. Klimstra. 1984. *Population ecology of the bobwhite.* Southern Illinois University Press, Carbondale, Illinois, USA.

Roseberry, J. L., and S. D. Sudkamp. 1998. Assessing the suitability of landscapes for northern bobwhite. *Journal of Wildlife Management* 62:895-902.

Roseberry, J. L., and A. Woolf. 1998. Habitat-population density relationships for white-tailed deer in Illinois. *Wildlife Society Bulletin* 26:252-258.

Rotenberry, J. T., and J. A. Wiens. 1980. Habitat structure, patchiness, and avian communities in North American steppe vegetation: A multivariate approach. *Ecology* 61:1228-1250.

Shifley, S. R., F. R. Thompson, III, W. D. Dijak, M. A. Larson, and J. J. Millspaugh. 2006. Simulated effects of forest management alternatives on landscape structure and habitat suitability in the Midwestern United States. *Forest Ecology and Management* 229:361-377.

St-Louis, V., A. M. Pidgeon, V. C. Radeloff, T. J. Hawbaker, and M. K. Clayton. 2006. High-resolution image texture as a predictor of bird species richness. *Remote Sensing of Environment* 105:299-312.

Stoddard, H. L. 1931. *The bobwhite quail: Its life history and management.* Charles Scribner's Sons, New York, New York, USA.

Tietje, W. D., and R. L. Ruff. 1983. Responses of black bears to oil development in Alberta. *Wildlife Society Bulletin* 11:99-112.

Tirpak, J. M., D. T. Jones-Farrand, F. R. Thompson, III, D. J. Twedt, M. D. Nelson, and W. B. Uihlein, III. 2008. Predicting bird habitat quality from a geospatial analysis of FIA data. Pages xx-xx *in* R. E. McRoberts, G. A. Reams, P. C. Van Duesen, and W. H. McWilliams, eds. *Proceedings of the 8th annual forest inventory and analysis symposium.* U.S. Forest Service, General Technical Report WO-xx, Newtown Square, Pennsylvania, USA.

Toms, J. D., and M. L. Lesperance. 2003. Piecewise regression: A tool for identifying ecological thresholds. *Ecology* 84:2034-2041.

Tuceryan, M., and A. K. Jain. 1998. Texture analysis. Pages 207-248 *in* C. H. Chen, L. F. Pau, and P. S. P. Wang, editors. *The handbook of pattern recognition and computer vision.* Second edition,. World Scientific Publishing, River Edge, New Jersey, USA.

U.S. Fish and Wildlife Service. 1980. *Habitat evaluation procedures (HEP).* U.S. Fish and Wildlife Service, Division Ecological Services, Manual 102, Washington, D.C., USA.

U.S. Fish and Wildlife Service. 1981. *Standards for development of habitat suitability index models for use in the habitat evaluation procedure.* U.S. Fish and Wildlife Service, Division Ecological Services, Manual 103, Washington, D.C., USA.

Van Horne, B. 1983. Density as a misleading indicator of habitat quality. *Journal of Wildlife Management* 47:893-901.

Van Kley, J. E., G. R. Parker, D. P. Franzmeier, and J. C. Randolph. 1994. *Field guide: Ecological classification of the Hoosier National Forest and surrounding areas of Indiana*. U.S. Department of Agriculture, Forest Service, Hoosier National Forest, Bedford, Indiana, USA.

Walker, P. A., and K. D. Cocks. 1991. HABITAT—A procedure for modeling a disjoint environmental envelope for a plant or animal species. *Global Ecology and Biogeography Letters* 1:108–118.

Wiens, J. A. 1989. Spatial scaling in ecology. *Functional Ecology* 3:385–397.

Williams, B. K., J. D. Nichols, and M. J. Conroy. 2001. *Analysis and management of animal populations*. Academic Press, San Diego, California, USA.

CHAPTER

Modeling Understory Vegetation and Its Response to Fire

15

Donald McKenzie, Crystal L. Raymond, and Samuel A. Cushman

The understory is an oft-neglected element in landscape modeling. Most landscape models focus on the dominant vegetation and how it responds over successional time to climate, competitive interactions, and disturbance (Keane et al. 2004, Cary et al. 2006). Even forest stand-level models rarely consider understory components other than seedlings, saplings, and downed wood (Pacala et al. 1993, He and Mladenoff 1999, Gratzer et al. 2004), except in special cases such as the need for estimating surface fuels for fire modeling (Rebain 2006).

Understory vegetation has important ecological functions on many landscapes. For example, in many coniferous forests, which are relatively depauperate in tree species, the understory, including shrubs, forbs, grasses, and nonvascular plants, accounts for most plant diversity (Halpern and Spies 1995, Gilliam and Roberts 2003, Halpern et al. 2005). Habitat quality for both arboreal and ground-dwelling wildlife is often controlled by understory characteristics such as regeneration and subcanopy tree layers, shrub cover and density, dead and downed wood, and edible herbaceous species (Hansen et al. 1995, Block and Morrison 1998, Pearman 2001, MacFaden and Capen 2002, Manning and Edge 2004).

In contrast to canopy vegetation composition and structure, which can be quantified reasonably well from remotely sensed images in both forested or nonforested landscapes, understory vegetation is largely invisible to remote sensing. Our ability to characterize it accurately at broad spatial scales depends on inferences about its relationships with observable characteristics like overstory structure or species composition. In this chapter, we review understory models—qualitative or quantitative frameworks for estimating understory composition, diversity, structure, and spatial pattern. Understory models are valuable as a baseline for predicting and managing other ecosystem components that depend on vegetation, including wildlife and wildlife habitat, insects, specifically population dynamics of defoliators and beetles, fungi, and fire hazard and fire effects (Holt et al. 1995, Wisdom et al. 2002). We identify three types of models:

1. *Empirical models,* which predict understory characteristics from a set of independent variables, using statistical or expert-system approaches, or

both. A leading edge of research in this area studies hierarchical models that incorporate multiple contingencies;

2. *Process-based models,* which simulate understory development, usually within a mechanistic ecosystem-modeling paradigm that focuses on photosynthesis, element cycling, mortality, and decomposition; and
3. *Qualitative or knowledge-based models,* which infer understory characteristics indirectly using a combination of logic and expert opinion.

We concentrate on empirical modeling because of the relative wealth of literature on this paradigm compared to the others. We then discuss the need and available methods for extrapolating understory models across the large landscapes that are the focus of this book. Disturbance plays a key role in forest understory dynamics, and its effects are rarely included in understory models. Fire is the principal disturbance in western North America, which is also a focal area for wildlife conservation. In fire-adapted ecosystems such as dry forests, shrublands, and grasslands, surface fuels (i.e., understory) determine fire behavior and severity (Raymond and Peterson 2005, Vaillant et al. 2006, Wright and Prichard 2006). Some landscape models assign fire behavior fuel models (Deeming et al. 1978, Anderson 1982) to dominant vegetation types to predict fire behavior and fire effects, but these are derived heuristically from the dominant vegetation and do not provide details on understory vegetation. We therefore examine the fire-effects literature, as it can inform understory modeling by identifying key variables that control fire effects on understory vegetation.

Lastly, we look to the future of understory models and recommend areas in need of further research. We emphasize the inherent uncertainty associated with estimating understory characteristics, especially in the context of a rapidly changing climate (Bonan et al. 2003, Neilson et al. 2005), changes in disturbance regimes (McKenzie et al. 2004, Gedalof et al. 2005), and the ubiquity of invasive species that respond to those changes (Keeley 2003, 2006; Brooks et al. 2004; Emery and Gross 2005).

UNDERSTORY MODELING PARADIGMS

An ideal understory model should:

1. Capture the range of variability across space and time associated with a particular ecosystem type (Landres et al. 1999), and also, where possible, specify the error structure of quantitative models. For example, given a range in the percent canopy cover, what is the range of shrub cover expected? The ranges of variability of most statistics in natural resources are at least as useful, and generally less biased, than predicted means. Several types of intervals around the mean are important: confidence intervals for predicted mean values, prediction intervals for the estimates of individual

observations, and tolerance intervals for proportions of new observations that fall within a specific range (Vardeman 1992).

2. Be dynamic, as opposed to giving only a snapshot in time. Because ecosystems can change rapidly, ecological data from stand inventories to coarse-scale GIS layers can quickly become outdated.
3. Incorporate disturbance types associated with the ecosystem; for example, fire frequency, grazing, and invasive species in rangelands, or fire severity and frequency, insect outbreaks, and logging in forests.
4. Provide a means for robust validation of results and for extrapolation outside the range of conditions within which the model was built. It should also identify limits to extrapolation. For example, a statistical model to predict total shrub cover might be invalid outside the range of overstory densities used to build it, whereas a model to predict herb-layer composition might be more sensitive to overstory cover type. Crucial to this validation is accounting for various sources of uncertainty, such as errors in measurement, model specification, or parameter estimates, or an incomplete specification of the spatial and temporal domain of the model.
5. For specific applications, focus on key variables for management or for predicting other ecosystem components. For example, management for a particular wildlife species that needs a specific shrub for food or cover would need an accurate understory model for that species more than a model for total herbaceous cover, seedling density, or coarse woody debris.

In practice, each type of understory model (described next) falls short of these ideals in different ways.

EMPIRICAL MODELS

Most models that explicitly estimate understory characteristics are of the empirical model type. Predictive vegetation mapping (Franklin 1995) is an active area of research that has produced special issues of journals (Guisan et al. 2002, Moisen et al. 2006), comparisons of multiple methods (Bolliger et al. 2000, Elith et al. 2002, Moisen and Frescino 2002, Leathwick et al. 2006), and broad theoretical investigation (Oksanen and Minchin 2002, Austin et al. 2006). Forest understory models are the most problematic, particularly at broad spatial scales, because canopy cover interferes with direct remote sensing of the understory.

Modeling techniques abound, but they fall broadly into two types: (1) machine-learning or expert-system approaches and (2) gradient-based methods, whether parametric or nonparametric (Cushman et al. 2007). The former category includes artificial neural networks (Ripley 1996), genetic algorithms (Stockwell 1999, 2006), Bayesian classification (Termansen et al. 2006), and recursive partitioning, the best known of which is classification and regression trees (Breiman et al. 1984) and its offshoots such as adaptive regression splines (Friedman 1991) and random forests (Breiman 2001). The latter category includes both univariate and multivariate methods. The most widely used univariate methods are ordinary multiple regression, generalized linear models (GLMs; McCullagh and Nelder 1989), generalized additive models (GAMs; Hastie and Tibshirani 1990), and embellishments of these to include spatial dependence in the response variable. Of these latter, two important developments are generalized linear mixed models (GLMMs; Hooten et al. 2003, Gelfand et al. 2005), which are particularly applicable in a landscape context, and multivariate gradient models, which are either eigenvector-based (e.g., canonical correspondence analysis [CCA]; ter Braak 1986) or distance-based (e.g., multidimensional scaling [NMDS]; Faith et al. 1987), and can also incorporate spatial dependencies (Gelfand et al. 2005).

Cushman et al. (2007) identify strengths and weaknesses of machine learning versus gradient modeling for vegetation modeling. Machine-learning methods excel at pattern matching—separating signal from noise in a data set and thereby optimizing prediction accuracy. These methods often yield better classification accuracy than gradient modeling (Moisen and Frescino 2002). In contrast, gradient modeling invokes driver-response mechanisms more directly, making it more robust to ecological interpretation and to extrapolation beyond conditions associated with the model database. Typical predictors for both methods are either surrogate variables for ecological mechanisms, such as elevation, geographic coordinates, and satellite spectra, or more direct drivers such as climate or climate-derived variables (e.g., snowpack depth or soil moisture), and biotic variables such as stand structure or composition (Fig. 15-1). For forest understories, overstory variables (Fig. 15-1) are often the strongest predictors because they modify the direct effects of environmental factors (Alaback 1982, Riegel et al. 1995, Klinka et al. 1996, McKenzie et al. 2000).

Given the vast literature on predictive vegetation models that use empirical approaches, we provide a few illustrative examples of understory shrub models from western North America. A comprehensive review is found in Guisan and Zimmerman (2000).

Binary Data

Data on the presence-absence of a species, life form, plant association, etc., should be analyzed with models that restrict predicted values to (0,1), either probabilistically or by allowing only presence or absence to be predicted. Generalized linear models or generalized additive models of the binomial family are typical gradient-based methods, whereas classification trees and Genetic Algorithms for Rule-set Prediction (GARP; Stockwell 2006) are standard machine-learning methods. Franklin (1998) used GLMs, GAMs, and classification trees to predict presence-absence of 20 chaparral and coastal sage shrub species in southwestern California, USA, from climate and terrain variables.

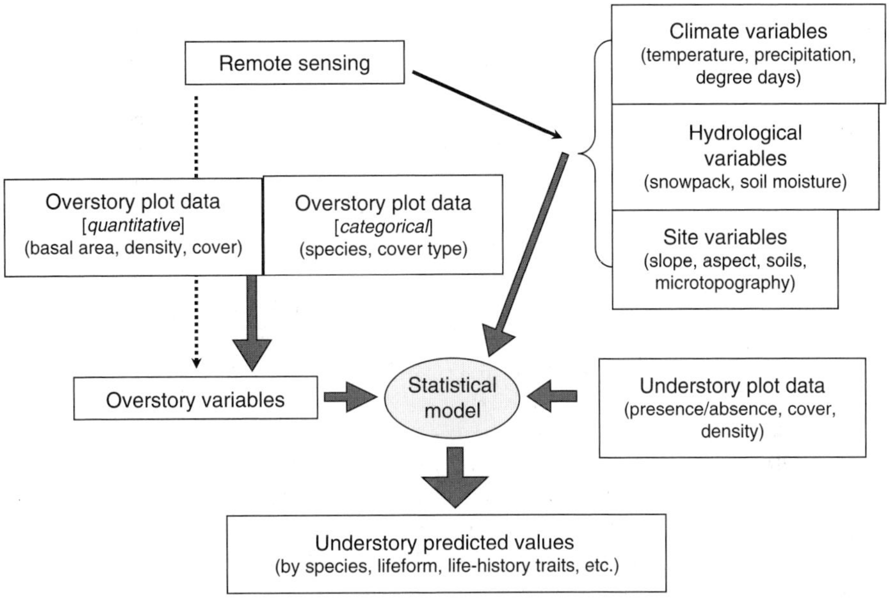

FIG. 15-1

Key elements of empirical models for predicting understory vegetation. Climate, hydrological, and site variables may or may not be predicted from remote sensing. Overstory plot data are preferred for generating overstory predictors, but remote sensing is also used to generate these directly when plot data are lacking.

Error misclassification rates, more appropriate for a binary response than deviance explained, were between 5–30% for GLMs, 2–27% for GAMs, and 3–25% for classification trees. Franklin (1998) notes that the gradient models (GLMs, GAMs) were easier to interpret ecologically than the tree-based models, an observation supported by the analysis of Cushman et al. (2007).

Abundance Data

Data on plant cover or density require models that predict a nonnegative response, either continuous or discrete. Generalized linear models or GAMs of the gamma and Poisson families, respectively, are gradient-based methods (note that standard linear regression is ill advised because it can predict negative values). Regression trees are a standard machine-learning method for abundance data, whereas multinomial models (count data) can be fit via feed-forward neural networks (Venables and Ripley 2002). Kerns and Ohmann (2004) used regression-tree models to predict shrub cover at a regional scale in managed coastal forests of Oregon, USA, after initially trying multiple linear regression. Forest structural variables were the best predictors, and shrub cover was lowest

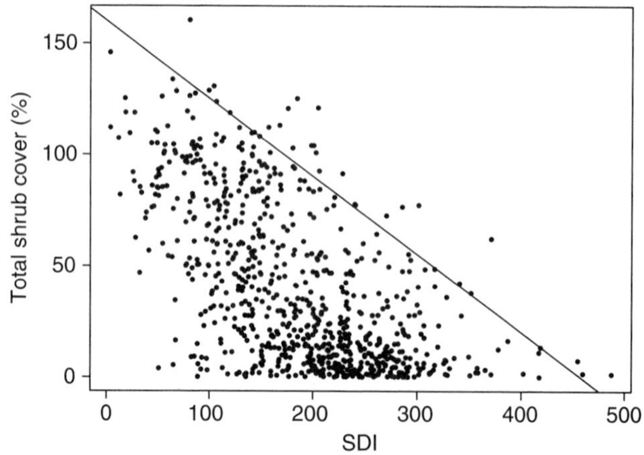

FIG. 15-2

Response of total shrub cover to stand density index (sqrt(basal_area * trees_per_ha)). Models of the mean response are noisy; models of maximum response produce a much better fit. Adapted from McKenzie et al. (2000).

during periods associated with stem exclusion, but explanatory power was compromised by the influence of human-caused disturbance.

McKenzie et al. (2000) examined the response of shrub cover to stand variables graphically (Fig. 15-2) and with regression-tree analysis and then elected to look at methods for estimating maximum abundance as a function of limiting factors (Guo et al. 1998, Scharf et al. 1998). This method is distinct from quantile regression (Cade et al. 1999, 2005; Cade and Guo 2000; Koenker and Hallock 2001) in that it uses a semiqualitative procedure to identify a constraint line as a surrogate for a limiting factor, rather than developing a full statistical model with associated error structure with confidence and prediction intervals. Proportion of deviance explained from nonlinear maximum-abundance models was 0.84–0.92 for shrub and herb response variables, as opposed to 0.49–0.80 from regression-tree models for estimating mean abundance (McKenzie et al. 2000). Maximum-abundance models clearly facilitate ecological interpretation but are not applicable to the prediction of mean values, nor is the error structure well defined in current implementations (Scharf et al. 1998, McKenzie et al. 2000).

Quantile regression is a feasible alternative to GLMs or regression trees for estimating means (i.e., 50% quantiles) or other quantiles (e.g., 5% and 95% quantiles, which can be surrogates for minimum and maximum response). It also provides a more rigorous means of quantifying confidence intervals, error structure, and goodness of fit (Koenker and Hallock 2001) than other maximum-response techniques. Extensions of quantile regression can accommodate asymmetrical distributions, nonlinear parametric models, and nonparametric models such as GAMs.

Multivariate Data

The presence-absence or abundance, or rarely for vegetation modeling, compositional data (proportions that sum to 1) for multiple species can be fit by gradient modeling. We are unaware of any examples of multivariate vegetation data modeled via neural networks or other machine-learning methods. Ohmann and Spies (1998) used gradient modeling (CCA) at the regional scale to predict the abundance and spatial pattern of woody plant species, including understory shrubs, across forests of Oregon, USA. Explanatory power was relatively low, not surprisingly, given that predictor variables (climate being the best) likely affected plant abundance at different scales and interacted in complex ways.

Data Requirements and Limitation

Measurement error is the initial uncertainty associated with any modeling effort. Obviously, every effort should be made to assure data quality (as with any research). Some additional desirable attributes of data for empirical models are (1) capture as much of each associated ecological gradient as possible in data collection, for the response and predictor variables; (2) avoid surrogate variables where possible (for example, elevation and latitude-longitude often turn out to be good predictors for ecological responses but represent ecological mechanisms poorly, if at all); and (3) in only apparent contrast to #2, try to collect at least a surrogate variable for each expected limiting factor in an ecological process being modeled. For example, energy and water are universal requirements for vegetation, understory or other. A model of understory response will likely miss much explanatory power if it does not represent both these elements, with variables such as soil temperature, degree days, or solar radiation (energy), or soil moisture, precipitation, or snowpack (water).

ECOSYSTEM DYNAMICS MODELS

In contrast to empirical models, which begin by assembling data, models in this paradigm begin by specifying key ecological mechanisms to be simulated (Fig. 15-3). Ecosystem dynamics models use mechanistic approaches to simulate plant growth, regeneration, and mortality, decomposition, and nutrient cycling (Neilson and Running 1996, Landsberg and Gower 1997). Waring and Running (1998) distinguish between biogeochemical (BGC) models and "gap-phase" (gap) models. BGC-type models emphasize physiology and biogeochemistry, whereas gap models emphasize life-cycle dynamics.

Gap models simulate growth, death, regeneration, and stand structure based on initial vegetation and biophysical conditions, and can in theory simulate explicit understory attributes such as shrub and herbaceous cover, density, and composition. In practice, however, they focus almost exclusively on forests

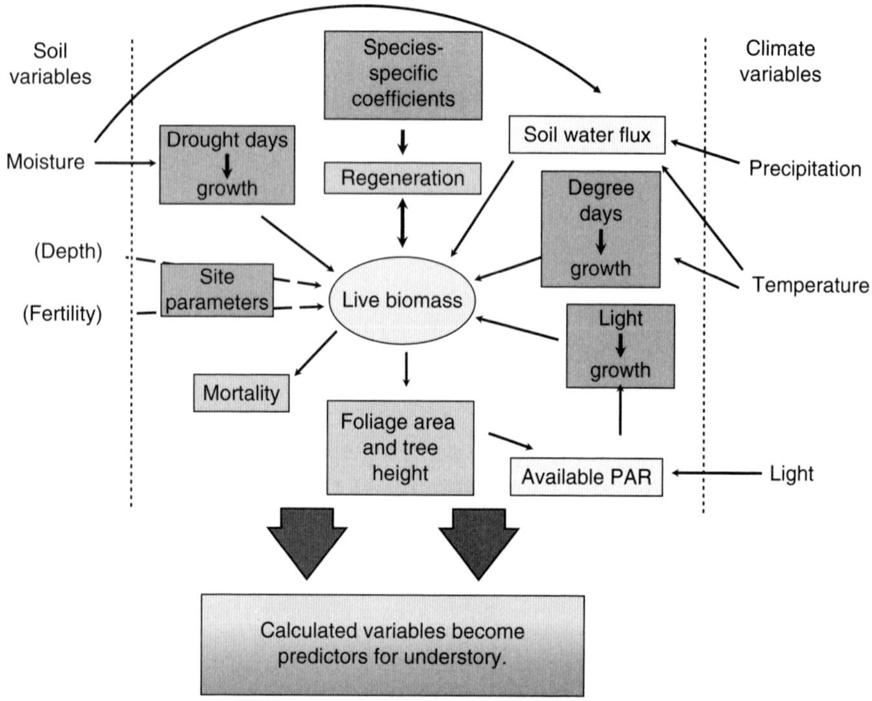

FIG. 15-3

Key elements of ecosystem models for predicting understory vegetation. At each time step, tree regeneration is simulated explicitly and other understory vegetation is predicted from intermediate calculations. Adapted from Cushman et al. (2007).

and the tree layer, with tree regeneration being the only understory component simulated (Urban and Shugart 1992, Pacala et al. 1993, Bugmann 1996). We know of no simulation model to date within the gap-phase mechanistic paradigm that includes nontree understory components (Bugmann 2001). However, shrubs could be modeled as trees in models such as LANDIS (He et al., this volume) which considers attributes such as species shade tolerance and longevity.

The Forest Vegetation Simulator (FVS; Dixon 2002) provides an example of how nontree understory attributes could be extracted from a gap-phase model. It is based on empirical growth equations rather than mechanistic modeling, updating a "tree list" at each time step with respect to growth, mortality, and regeneration. The FVS has two add-on modules: COVER (Laursen 1984, Moeur 1985), which calculates shrub cover based on empirical equations, and

FVS-FFE (Fire and Fuels Extension; Reinhardt and Crookston 2003, Rebain 2006), which calculates understory fuels. The allometry of these calculations is based on empirical models like those we discussed previously. Much as ecosystem dynamics models include stochastic routines for regeneration and mortality, they could include empirical models for understory components until such time as an explicit mechanistic framework is developed.

Data Requirements and Limitations

As with empirical-statistical models, the weak link in process-based models is often the lack of databases for initializing and calibrating model algorithms. An oft-neglected limitation of these models is the reliability of the sources of the "physical" processes modeled. For example, as noted previously, the FVS model is based on empirical growth equations throughout. More subtly, however, many process-based simulators have empirical equations at their core. It is the specific use of these equations to represent processes as mechanistically as possible, rather than some embedded association with ecological mechanisms, that justifies their being part of process-based modeling.

QUALITATIVE MODELING

When data to inform empirical or process-based models are of poor quality, at the wrong scale, or simply lacking, qualitative reasoning, based on logic, expert opinion, or both provide an alternative paradigm. Knowledge-based systems are often used successfully in natural resource applications (Puccia and Levins 1985, Robertson et al. 1991, Schmoldt and Rauscher 1996 and references therein). For example, successional pathways provide an experiential logic for estimating changes in overstory and understory structure and composition over time (Cattellino et al. 1979, Beukema et al. 2003, Kipfmueller and Kupfer 2005). They can stand alone as predictive models or be embedded in simulation systems or larger qualitative frameworks (Keane and Long 1998, Hemstrom et al. 2001). However, much care is warranted in applying qualitative models, as they are by definition a formalization of assumptions and are neither derived from nor usually tested with empirical data. In some cases, including expert opinion as explanatory factors in empirical models has produced worse performance and lower predictive success (Seoane et al. 2005).

As with process-based simulations, few successional-pathway models provide an explicit description, let alone quantification, of the understory apart from tree regeneration. Understory composition and structure generally are inferred indirectly. For example, Raymond et al. (2006) developed a classification of fuels, in both overstory and understory, by combining overstory structure and composition with successional changes, a comprehensive inventory database, and expert opinion of local fire managers.

Data Requirements and Limitations

Data requirements can be just as substantial as for empirical-statistical models or process-based models. Qualitative inferences do not necessarily imply a lack of available data, although it is often the case. When the statistical properties of the data are not evident or weakly specified, that is where qualitative inference often comes in. The main limitation of this approach is the lack of a rigorous quantitative framework to guide inferences and extrapolations to different landscapes or future conditions. In general, extrapolations will depend on a new set of qualitative inferences, rather than evolving out of the model structures themselves.

CAN WE BUILD THE IDEAL UNDERSTORY MODEL?

At best, empirical models can satisfy four of our five requirements for an ideal understory model. Because each model is a snapshot of data collected at specific places and times, it does not capture the transient dynamics associated with ecological mechanisms. Time lags, nonequilibrium dynamics, and mismatch of temporal scale between responses and drivers reduce the effectiveness of equilibrium models. In contrast, process-based modeling with explicit time steps can more directly relate organism responses to the action of specific mechanisms and address temporal disequilibria and transient dynamics (Neilson 1995, Keane et al. 1996, Waring and Running 1998). The ideal understory model will use robust empirical models to inform parameter choices in process-based simulations (green-shaded components in Fig. 15-3), while incorporating temporal dynamics, either via successional pathways or more quantitatively—for example, with state-transition components to a gap model (Acevedo et al. 1996). It should also account for stochastic perturbations (disturbances, human-caused or natural).

Which model paradigm is best? We suggest that this comparison is valid only within the context of particular studies, and that none is best globally. In general, however, the more clearly defined the statistical properties of the model database, the more empirical-statistical models should be favored over qualitative models. Similarly, the better understood the temporal dynamics of the system, the better the argument for using process-based simulations. For dynamic landscape modeling of understory vegetation, we suggest linking empirical and process-based models so that each can do what it does best.

PREDICTING UNDERSTORY CHARACTERISTICS OF LARGE LANDSCAPES

We suggested previously that process-based simulations, with due attention to choices of parameters, can project empirical relationships through time, particularly those from gradient models. For management of large landscapes, we would also like to project these relationships across space. With a representative sample of inventory plots and robust statistical relationships across scales between explanatory and response variables (Fig. 15-1), predicted values for response variables can be assigned to new observational units, often cells or polygons rather than plots. This process, termed "imputation," is distinct from specifying distributions for missing data at the same scale as existing observations, and has received much recent attention because of increasing emphasis on large-scale management of vegetation and disturbance. Nearest-neighbor algorithms (University of Minnesota 2006), which impute values at a new location from those at one or more nearby locations (either in geographic or parameter space), preserve much of the explicit covariance structure of the empirical data in imputed responses, and are therefore superior to strict interpolation methods such as Kriging (Isaaks and Srivastava 1989) or inverse-distance weighting (Hessl et al. 2007). Unlike interpolation, however, imputation methods require explanatory variables to be available, usually as GIS layers, at the spatial scale and resolution for which predictions are to be made. For an understory model, overstory plot data may have to be imputed from coarser-scale predictors before the understory response is specified.

There are three standard approaches to nearest-neighbor imputation in vegetation analysis. Most-similar-neighbor (MSN; Moeur and Stage 1995) is the simplest conceptually. A distance measure of choice, usually multivariate, is applied to the GIS layer(s) of explanatory variables. The response variables at each unsampled location are then assigned the values at the sampled location for which the multivariate distance to the unsampled location is smallest. In its simplest form, MSN is simple to use and draws on the full range of values from sampled locations but cannot assign any new values to unsampled locations.

K-nearest-neighbor imputation (KNN; Dale 2002) draws on multiple ("k") nearest neighbors and applies a weighting scheme of choice to impute new values of the response variable at unsampled locations. As k increases, however, the averaging process decreases the variance among locations compared to that from the original samples (Pierce and Ohmann 2006). KNN algorithms are based in machine-learning and can be computationally intensive, so approximate techniques are useful (Finley et al. 2006).

Gradient-nearest-neighbor (GNN; Ohmann and Gregory 2002; Fig. 15-4) takes advantage of nonlinearities in species-environment relationships and is particularly useful across large landscapes in which gradients are long, species turnover is high, and there are many zeros in response variables (e.g., understory species presence-absence). The relationship between GNN and KNN is analogous to that between gradient-based and machine-learning methods in empirical modeling. Gradients in the ordination space of GNN models (Fig. 15-4) are interpretable in the same way as gradients in the empirical models that are the basis for imputation, and can reinforce our understanding of driver-response relationships and limiting factors that affect understory composition, structure, and

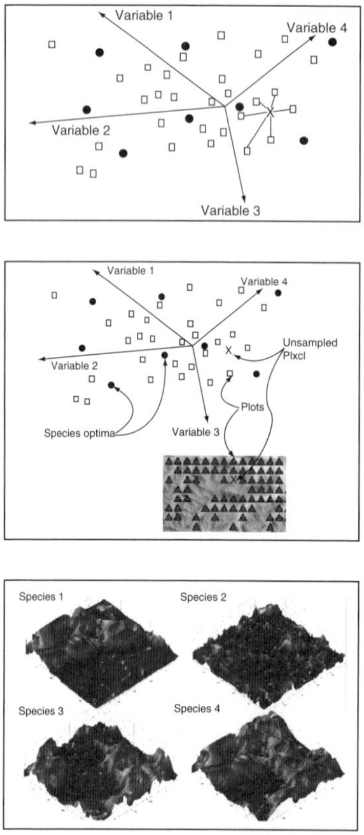

(1) Independent variables at a range of spatial scales are used for a nonlinear gradient model of niche structure for each species. This example uses CCA. Axes are orthogonal dimensions of ecological space. Species optima locate the center of their environmental niche with respect to the measured predictor variables.

(2) The niche model uses ecological space, but we can make inferences about suitability in geographic space using *GNN.* Each unsampled location (cell) is projected iteratively into the environmental space defined by the niche model. By imputing the value of the species at the nearest sampled plot (or a weighted average of nearby plots) we estimate the suitability of the new location for that species.

(3) Gradient imputation allows the explicit translation from quantitative models in environmental space to suitability maps in geographic space. Normalized values for multiple species at each location provide equilibrium estimates of relative competitive ability, useful for defining parameters in ecosystem dynamics (gap) models.

FIG. 15-4 (See color insert.)

Gradient imputation example using canonical correspondence analysis (ter Braak 1986) and gradient-nearest-neighbor (Ohmann and Gregory 2002). Adapted from Cushman et al. (2007).

abundance (Cushman et al. 2007). In contrast, machine-learning methods (e.g., combining regression-tree models with KNN imputation) are more difficult to interpret in terms of ecological mechanisms.

On landscapes affected by warming temperatures, changing disturbance regimes, invasive species, and accelerated land-use change, the most robust models will be those that can be interpreted in terms of ecological fundamentals even under new environmental conditions. We suggest that gradient-based imputation techniques such as GNN, perhaps with hierarchical elements to account for processes at multiple scales (Cushman and McGarigal 2003, Gehringer 2006), are in general the best choice for extrapolating understory models to large landscapes.

FIRE AND UNDERSTORY MODELS

Fire is a key disturbance worldwide and universally affects understory vegetation, whether fire regimes are of low or high severity. To inform our understanding of understory composition, structure, and dynamics, we briefly review fire effects because information on understory response to fire can refine variables used in understory models and enable better representation of disturbances and their effects on succession.

Fire, and disturbance in general, can reset or drastically alter successional pathways, especially in the presence of invasive species. Understory vegetation drives forest succession in the boreal forest by influencing tree seedling regeneration and below-ground nutrient cycling processes, and wildfire is the main determinant of understory vegetation in this ecosystem (Nilsson and Wardle 2005). Frequent burning favors invasive species, and time-since-fire is an important predictor of invasive species presence in mixed-conifer forests, blue oak savannahs, and chaparral ecosystems (Keeley 2003). Sagebrush ecosystems are particularly sensitive to invasive species like cheatgrass (*Bromus tectorum*) because grasses increase fuel cover and continuity, thereby increasing the risk of large frequent fires that destroy habitat for the endangered sage grouse (*Centrocercus urophasianus;* Miller and Rose 1999).

Fire is often a key to wildlife habitat quality. Some species are fire-dependent, whereas others are fire-sensitive. Sometimes a change in fire regime, particularly increased fire severity, can change habitat quality (e.g., nesting, forage, protection from predators). In many ecosystems of the United States (and worldwide), fire management is central to ecosystem resilience, restoration, and maintenance. Land management in western North America is increasingly focused on wildfire mitigation and fuel reduction, including mechanical removal of biomass and reintroduction of fire. These treatments alter understory structure and composition, which can affect wildlife habitat and forage. Understory models can be useful tools for predicting the effects of management, and conversely, empirical data from management activities can be used to refine simulation and empirical models and contribute knowledge to qualitative models.

About half (17 of 35) of the fire effects studies we reviewed were conducted in pine forests, pine/oak woodlands, or pine grasslands; four were in boreal forests and the remainder included mixed conifer, woodlands, and sagebrush ecosystems. Studies of fire effects typically evaluate initial and short-term responses (less than 5 years); only six are long-term studies in which fire effects were studied for >5 years postdisturbance. Understory response to fire varies with species, fire, and site characteristics, but in the majority of studies reviewed, fire significantly increased understory diversity (Griffis et al. 2001, Wang and Kemball 2005), richness (Griffis et al. 2001, Huisinga et al. 2005), abundance (Sparks et al. 1998, Lloret et al. 2003), and cover (Keeley 2003, Huisinga et al. 2005). In some cases, fire had no or minimal effect on understory (Rego et al. 1991, Fulé et al.

2005) or initially decreased abundance (Schwartz and Heim 1996), diversity (Wang and Kemball 2005), and richness (Metlen et al. 2004).

The following factors influence fire effects on species structure and composition and represent a set of variables that may be relevant for understory models that include a disturbance component. Fire effects can vary with hillslope position (ridge, mid, valley). In a burned pine/oak forest of eastern North America, species diversity in the understory layer increased on ridges, decreased on the mid-slope, and did not change on the low slope (Elliot et al. 1999). Season of burned in mixed-oak forests, dormant or growing, can also alter the effects of burning and may be a key management choice for achieving desired effects on understory composition (Hutchinson et al. 2005). Variability in fire intensity, frequency, and severity affects overstory basal area, creating an indirect pathway by which fire affects understory vegetation through an inverse relationship between overstory basal area and understory production and composition (Bataineh et al. 2006) and richness (Beckage and Stout 2000). Fire severity can also directly affect understory vegetation and alter the relevant abundance of vegetative forms. For example, severe fire in the boreal forest initially decreases species diversity and richness and favors herbaceous and nonvascular plants over woody plants (Wang and Kemball 2005). Fire frequency can directly affect understory vegetation because the relative abundance of invading and residual species changes with time since disturbance (Halpern 1989).

Fire effects on understory can, in part, be predicted by known conditions prior to disturbance. Understory species respond to fire in a way consistent with life history traits (Halpern 1989, Lloret et al. 2003). For example, in mixed oak/pine forests, more frequent fires increase the abundance of sprouting grasses, shrubs, and hardwoods, whereas seeding species reach greatest abundance with intermediate fire frequencies (Lloret et al. 2003). Incorporating life history traits into process-based models should therefore improve predictions of understory vegetation response to disturbance. Site conditions such as prefire overstory structure and fuel loads (Fulé et al. 2005) and water table depths (Blank et al. 2003) can mitigate the effects of fire and explain the variability in species richness and diversity that is observed postfire. Climate can alter fire effects because vegetation response following fire is driven in part by inter-annual climatic variability in ecosystems susceptible to drought stress (Fulé et al. 2005).

Fire extent, frequency, and severity are likely to increase in a warming climate (Flannigan et al. 1998, McKenzie et al. 2004, Gedalof et al. 2005, Westerling et al. 2006), increasing the influence of fire on understory composition, structure, and succession. Our ideal understory model (described previously) clearly should incorporate fire effects dynamically. A less desirable alternative, though certainly easier to implement, would be to represent fire as a "snapshot" variable, e.g., time-since-fire. State transition (successional pathway) models use fire events to reset succession and vegetation development (Keane and Long 1998, Hemstrom et al. 2001), and empirical models use time-since-fire as a predictor for understory richness and diversity (Chipman and Johnson 2002, Laughlin and Grace 2006).

Ecosystem dynamics (gap phase) models can incorporate fire dynamically. For example, Miller and Urban (1999, 2000) and Miller (2003) integrated fire into a gap model to examine landscape patterns and the effects of climatic change, but the understory was only represented in terms of fuel loadings and fuel-bed connectivity. A dynamic fire-succession model with an understory component is still in the future, but clearly within the range of current modeling paradigms.

FUTURE DIRECTIONS

We see four research tasks ahead to improve our ability to model and manage the understory component of large landscapes.

Incorporate the Range of Variability of Understory Vegetation into Models

Even the best statistical models of understory vegetation are noisy (Franklin 1998, Ohmann and Spies 1998, McKenzie and Halpern 1999, McKenzie et al. 2000), reflecting the considerable range of variability in understories even under strong abiotic and biotic controls. Imputation to large landscapes should take advantage of the ability of statistical models to quantify uncertainty, rather than treat variation as noise to be overcome by more precise models driven by more advanced algorithms. For example, suppose we are imputing the results from 1,000 sampled locations (plots) to 20,000 unsampled locations (cells in a GIS layer). The algorithms MSN, GNN, or KNN populating those cells will assign each either a fitted value from a model (MSN, GNN) or a weighted average (KNN), but iteration of this process disguises the uncertainties associated with the original models. Instead, imputation procedures could draw on the full distributions associated with predicted values, depending on the model type (e.g., binomial for presence-absence data, Poisson for count data, etc.).

For example, suppose a binomial GLM were built to estimate the probability of presence of an understory species given environmental conditions. Estimates would have standard errors associated with them, so imputed values could be drawn from the associated binomial distribution, not just assigned the means (fitted values). Alternatively, a hierarchical Bayesian model might be used to quantify the variance structure at different scales (Hooten et al. 2003), providing a direct approach to probabilistic imputation. As spatial scales of inference broaden, procedures like these are more appropriate than they would be for inferences at single points, whether for mean values or single observations, and more representative of landscape variability.

Integrate Space with Time in the Modeling Domain

Empirical gradient models, imputation to landscapes, and (process-based) ecosystem dynamics models need to be integrated to combine spatial patterns and temporal dynamics. Cushman et al. (2007) propose a framework for doing this, which we outline here:

1. Environmental data and stand inventories are used to build empirical gradient models;
2. Gradient imputation is used to populate landscapes with predicted values from gradient models;
3. Gradient models can supply parameters to ecosystem models; for example, unimodal response functions of organisms, both mature trees and understory vegetation, to climate variables (Miller and Urban 1999, 2000; McKenzie et al. 2003); and
4. Ecosystem models simulate succession over time at each cell.

Incorporate Disturbance Quantitatively into Understory Models

We have seen in our brief review of fire effects that in many ecosystems, fire is a pervasive influence on understory composition, structure, and dynamics. Fire has been incorporated into mechanistic vegetation models (Keane et al. 1999, Miller and Urban 1999) and broader-scale stochastic vegetation models (LANDIS; He, this volume), but to our knowledge the successional-pathway approach is the only instance of including fire in temporal dynamics of the understory (other than tree regeneration). However, given the ubiquity of inventory databases, satellite-based models, and fire observations, at least on public lands in the United States (Hicke et al. 2002, Jenkins et al. 2003, National Interagency Fire Center 2006), there is now an opportunity to better quantify the response of understory nontree vegetation to fire in both statistical and simulation models.

Integrate over Multiple Scales

Aggregating information to broader spatial scales produces multiple uncertainties and can propagate and magnify errors in unknown ways (Rastetter et al. 1992, McKenzie et al. 1996). We have proposed gradient imputation as a means of preserving much of the range of variability across landscapes as represented in inventory data. Controls on understory vegetation operate at multiple scales, however, from biome-scale (e.g., climate) down to biotic interactions between individual organisms. Ideally, understory models could reflect multiple-scale controls. For example, Cushman and McGarigal (2003, 2004) built hierarchical models of avian species-environment relationships in the Oregon Coast Range, USA, and these methods could be transferred to understory vegetation. Wagner (2004) developed methods for multiscale ordination with CCA, and in theory, this technique could be used in GNN (Ohmann and Gregory 2002) or other forms of gradient imputation. Gehringer (2006) showed how nearest-neighbor imputation could be applied hierarchically. In general, hierarchical methods partition variance into scale-specific components, improving our ability to preserve the range of variability of individual data points when extrapolating to large landscapes.

As a possible alternative to these (non-Bayesian) hierarchical methods, Wikle (2003) developed a more mathematically unified, though less detailed, approach to combining spatial and temporal processes in ecological models. The temporal process, in this case movement of birds across the eastern United States, was specified as a diffusion process and nested in a hierarchy of conditional probabilities to be estimated by Markov Chain Monte Carlo procedures. If made accessible to working ecologists, and shown to account for cross-scale interactions (Wagner 2004, Gehringer 2006), this procedure could complement the Cushman et al. (2007) approach, particularly in cases where more rigorous statistical inferences were desired.

CONCLUSIONS

Understory modeling presents unique problems. Particularly in forests, understories are opaque to remote sensing, so inferences about understories must be made either from fine-scale plot data or indirectly through models. Indirect methods are also required to extrapolate fine-scale data to large landscapes; these methods propagate the considerable uncertainties associated with most fine-scale understory models and add more of their own. By incorporating these uncertainties directly into broad-scale predictions, focusing on the range of variability in understory response and on aggregate measures appropriate to broad scales, we can minimize biases that lead to poor management decisions.

We also propose that natural disturbance, particularly fire, be incorporated dynamically into understory models. Because understory succession is often rapid, on a scale of years rather than decades, understanding disturbance regimes *per se* (e.g., frequency, severity, and extent), will lead to better models of understory dynamics. Lastly, we look to active research in scaling and hierarchical models, whether incorporating scale explicitly (Cushman and McGarigal 2003, 2004; Wagner 2004; Li and Wu 2006) or of the Bayesian variety (Wikle 2003, Clark 2007), and integrating space with time in vegetation models (Cushman et al. 2007), to provide new insights and methods for modeling understories across large landscapes.

SUMMARY

Canopy vegetation composition and structure, whether in forested or nonforested landscapes, can be quantified reasonably well from remotely sensed images. In contrast, understory vegetation is largely invisible to remote sensing. Our ability to characterize it accurately at broad spatial scales depends on inferences about relationships to observable characteristics like overstory structure. We reviewed three modeling approaches to predicting understory vegetation from observable quantities: environmental variables, canopy vegetation, or disturbance history. Empirical models predict understory characteristics such as species composition or abundance from statistical relationships with predictors. Process-based simulation models use ecophysiological or biogeochemical algorithms to predict ecosystem properties such as rates of biomass accumulation or decomposition. Knowledge-based or expert-system approaches use qualitative reasoning, often when there is a dearth of empirical data for modeling. The optimal approach to understory modeling depends on both availability of data and the state of knowledge. Where a good understanding of ecological mechanisms exists, process-based models may be superior, whereas rich inventory data sets suggest the empirical approach. In all cases, understory models need to be scalable to be applied to large landscapes. We reviewed methods of extrapolation, focusing on gradient imputation, which preserves the covariance structure of models at their original scale. Fire is a ubiquitous disturbance across North America and an important control on understory structure. Conversely, live understory vegetation and dead woody fuels are the principal determinants of the severity and effects of understory fire. We reviewed the literature on fire effects on understories and suggested how fire might be incorporated into understory modeling. We concluded by offering recommendations for choosing the optimal method(s) for understory modeling of particular landscapes and pointed to directions for future research.

LITERATURE CITED

Acevedo, M. F., D. L. Urban, and H. H. Shugart. 1996. Models of forest dynamics based on roles of tree species. *Ecological Modelling* 87:267–284.

Alaback, P. A. 1982. Dynamics of understory biomass in Sitka spruce-western hemlock forests of southeast Alaska. *Ecology* 63:1932–1948.

Anderson, H. A. 1982. *Aids to determining fuel models for estimating fire behaviour.* U.S. Forest Service, Intermountain Research Station, General Technical Report INT-122. Ogden, Utah, USA.

Austin, M. P., L. Belbin, J. A. Meyers, M. D. Doherty, and M. Luoto. 2006. Evaluation of statistical models used for predicting plant species distributions: Role of artificial data and theory. *Ecological Modelling* 199:197–216.

Bataineh, A. L., B. P. Oswald, M. M. Bataineh, H. M. Williams, and D. W. Coble. 2006. Changes in understory vegetation of a ponderosa pine forest in northern Arizona 30 years after a wildfire. *Forest Ecology and Management* 235:283–294.

Beckage, B., and I. J. Stout. 2000. Effects of repeated burning on species richness in a Florida pine savanna: A test of the intermediate disturbance hypothesis. *Journal of Vegetation Science* 11:113–122.

Beukema, S. J., W. A. Kurz, C. B. Pinkham, K. Milosheva, and L. Frid. 2003. *Vegetation Dynamics Development Tool, User's Guide, Version 4.4c.* Prepared by ESSA Technologies Ltd., Vancouver, British Columbia, Canada.

Blank, R. R., J. C. Chambers, and D. Zamudio. 2003. Restoring riparian corridors with fire: Effects on soil and vegetation. *Journal of Range Management* 56:388-396.

Block, W. M., and M. L. Morrison. 1998. Habitat relationships of amphibians and reptiles in California oak woodlands. *Journal of Herpetology* 32:51-60.

Bolliger, J., F. Kienast, and H. Bugmann. 2000. Comparing models for tree distributions: Concept, structures, and behaviour. *Ecological Modelling* 134:89-102.

Bonan, G. B., S. Levis, S. Sitch, M. Vertenstein, and K. W. Oleson. 2003. A dynamic global vegetation model for use with climate models: Concepts and description of simulated vegetation dynamics. *Global Change Biology* 9:1543-1566.

Breiman, L. 2001. Random forests. *Machine Learning* 45:5-32.

Breiman, L., J. H. Friedman, R. A. Olshen, and C. J. Stone. 1984. *Classification and regression trees*. Wadsworth International Group, Belmont, California, USA.

Brooks, M. L., C. M. D'Antonio, D. M. Richardson, J. B. Grace, J. E. Keeley, J. M. DiTomaso, R. J. Hobbs, J. E. Keeley, M. Pellant, and D. Pyke. 2004. Effects of invasive alien plants on fire regimes. *Bioscience* 54:677-688.

Bugmann, H. 2001. A review of forest gap models. *Climatic Change* 51:259-305.

Bugmann, H. K. M. 1996. A simplified forest model to study species composition along climate gradients. *Ecology* 77:2055-2074.

Cade, B. S., and Q. Guo. 2000. Estimating effects of constraints on plant performance with regression quantiles. *Oikos* 91:245-254.

Cade, B. S., B. R. Noon, and C. H. Flather. 2005. Quantile regression reveals hidden bias and uncertainty in habitat models. *Ecology* 86:786-800.

Cade, B. S., J. W. Terrell, and R. L. Schroeder. 1999. Estimating effects of limiting factors with regression quantiles. *Ecology* 80:311-323.

Cary, G. J., R. E. Keane, R. H. Gardner, S. Lavorel, M. D. Flannigan, I. D. Davies, C. Li, J. M. Lenihan, T. S. Rupp, and F. Mouillot. 2006. Comparison of the sensitivity of landscape-fire-succession models to variation in terrain, fuel pattern, climate and weather. *Landscape Ecology* 21:121-137.

Chipman, S. J., and E. A. Johnson. 2002. Understory vascular plant diversity in the mixedwood boreal forest of western Canada. *Ecological Applications* 12:588-601.

Clark, J. S. 2007. *Models for ecological data: An introduction*. Princeton University Press, New Jersey, USA.

Cushman, S. A., and K. McGarigal. 2003. Hierarchical, multi-scale decomposition of species-environment relationships. *Landscape Ecology* 17:637-646.

Cushman, S. A., and K. McGarigal. 2004. Hierarchical analysis of forest bird species-environment relationships in the Oregon Coast Range. *Ecological Applications* 14:1090-1105.

Cushman, S. A., D. McKenzie, D. L. Peterson, J. Littell, and K. S. McKelvey. 2007. *Research agenda for integrated landscape modeling*. U.S. Forest Service, Rocky Mountain Research Station, Research Paper RMRS-GTR-194, Fort Collins, Colorado, USA.

Dale, M. R. T. 2002. *Spatial pattern analysis in plant ecology*. Cambridge University Press, Cambridge, United Kingdom.

Deeming, J. E., R. E. Burgan, and J. D. Cohen. 1978. *The National Fire-Danger Rating System—1978*. U.S. Forest Service, Intermountain Forest and Range Experiment Station, General Technical Report INT-39, Ogden, Utah, USA.

Dixon, G. E. compiler. 2002. *Essential FVS: A user's guide to the Forest Vegetation Simulator*. U.S. Forest Service, Forest Management Service Center, Internal Report, Fort Collins, Colorado, USA.

Elith, J., M. A. Burgman, and H. M. Regan. 2002. Mapping epistemic uncertainties and vague concepts in predictions of species distribution. *Ecological Modelling* 157:313-329.

Elliott, K. J., R. L. Hendrick, A. E. Major, J. M. Vose, and W. T. Swank. 1999. Vegetation dynamics after a prescribed fire in the southern Appalachians. *Forest Ecology and Management* 114:199-213.

Emery, S. M., and K. L. Gross. 2005. Effects of timing of prescribed fire on the demography of an invasive plant, spotted knapweed *Centaurea maculosa*. *Journal of Applied Ecology* 42:60–69.

Faith, D. P., P. R. Minchin, and L. Belbin. 1987. Compositional dissimilarity as a robust measure of ecological distance. *Vegetation* 69:57–68.

Finley, A. O., N. L. Crookston, and A. R. Ek. 2006. *Approximate nearest neighbors search and* yaImpute; *new weapons in the battle against "Big N."* Paper presented at the 2006 Nearest Neighbors Workshop, University of Minnesota, Minneapolis, Minnesota, USA.

Flannigan, M. D., Y. Bergeron, O. Engelmark, and B. M. Wotton. 1998. Future wildfire in circumboreal forests in relation to global warming. *Journal of Vegetation Science* 9:469–476.

Franklin, J. 1995. Predictive vegetation mapping: Geographic modelling of biospatial patterns in relation to environmental gradients. *Progress in Physical Geography* 19:474–499.

Franklin, J. 1998. Predicting the distribution of shrub species in southern California from climate and terrain-derived variables. *Journal of Vegetation Science* 9:733–748.

Friedman, J. H. 1991. Multivariate adaptive regression splines (with discussion). *Annals of Statistics* 19:1–141.

Fulé, P. Z., D. C. Laughlin, and W. W. Covington. 2005. Pine-oak forest dynamics five years after ecological restoration treatments, Arizona, USA. *Forest Ecology and Management* 218:129–145.

Gedalof, Z., D. L. Peterson, and N. Mantua. 2005. Atmospheric, climatic, and ecological controls on extreme wildfire years in the northwestern United States. *Ecological Applications* 15:154–174.

Gehringer, K. R. 2006. *Constructing a virtual forest: An hierarchical nearest neighbors method for generating simulated tree lists.* Paper presented at the 2006 Nearest Neighbors Workshop, University of Minnesota, Minneapolis, Minnesota, USA.

Gelfand, A. E., A. M. Schmidt, S. Wu, J. A. Silander, Jr., A. Latimer, and A. G. Rebelo. 2005. Modelling species diversity through species level hierarchical modelling. *Journal of the Royal Statistical Society: Series C (Applied Statistics)* 54:1–20.

Gilliam, F. S., and M. R. Roberts, editors. 2003. *The herbaceous layer in forests of Eastern North America.* Oxford University Press, Oxford, United Kingdom.

Gratzer, G., C. Canham, U. Dieckmann, A. Fischer, Y. Iwasa, R. Law, M. J. Lexer, H. Sandmann, T. A. Spies, B. E. Splechtna, and J. Szwagrzyk. 2004. Spatio-temporal development of forests: Current trends in field methods and models. *Oikos* 107:3–15.

Griffis, K. L., J. A. Crawford, M. R. Wagner, and W. H. Moir. 2001. Understory response to management in northern Arizona ponderosa pine forests. *Forest Ecology and Management* 146:239–245.

Guisan, A., T. C. Edwards, Jr., and T. Hastie. 2002. Generalized linear and generalized additive models in studies of species distributions: Setting the scene. *Ecological Modelling* 157:89–100.

Guisan, A., and N. E. Zimmermann. 2000. Predictive habitat distribution models in ecology. *Ecological Modelling* 135:147–186.

Guo, Q., J. H. Brown, and B. J. Enquist. 1998. Using constraint lines to characterize plant performance. *Oikos* 83:237–245.

Halpern, C. B. 1989. Early successional patterns of forest species: Interactions of life history traits and disturbance. *Ecology* 70:704–720.

Halpern, C. B., D. McKenzie, S. A. Evans, and D. A. Maguire. 2005. Initial responses of forest understories to varying levels and patterns of green-tree retention. *Ecological Applications* 15:175–195.

Halpern, C. B., and T. A. Spies. 1995. Plant species diversity in natural and managed forests of the Pacific Northwest. *Ecological Applications* 5:913–934.

Hansen, A. J., W. C. McComb, R. Vega, M. G. Raphael, and M. Hunter. 1995. Bird habitat relationships in natural and managed forests in the West Cascades of Oregon. *Ecological Applications* 5:555–569.

Hastie, T. J., and R. J. Tibshirani. 1990. Generalized additive models. *Monographs on statistics and applied probability* 43.

He, H. S., and D. J. Mladenoff. 1999. The effects of seed dispersal on the simulation of long-term forest landscape change. *Ecosystems* 2:308–319.

Hemstrom, M. A., J. J. Korol, and W. J. Hann. 2001. Trends in terrestrial plant communities and landscape health indicate the effects of alternative management strategies in the interior Columbia River basin. *Forest Ecology and Management* 153:105–125.

Hessl, A., J. Miller, J. Kernan, and D. McKenzie. 2007. Mapping paleo-fire boundaries from binary point data: Comparing interpolation approaches. *Professional Geographer* 59:87–104.

Hicke, J. A., G. P. Asner, J. T. Randerson, C. Tucker, S. Los, R. Birdsey, J. C. Jenkins, and C. Field. 2002. Trends in North American net primary productivity derived from satellite observations, 1982–1998. *Global Biogeochemical Cycles* 16:10.1029/2001GB001550.

Holt, R. D., S. W. Pacala, T. W. Smith, and J. Liu. 1995. Linking contemporary vegetation models with spatially explicit animal population models. *Ecological Applications* 5:20–27.

Hooten, M. B., D. R. Larsen, and C. K Wikle. 2003. Predicting the spatial distribution of ground flora on large domains using a hierarchical Bayesian model. *Landscape Ecology* 18:487–502.

Huisinga, K. D., D. C. Laughlin, P. Z. Fulé, J. D. Springer, and C. M. McGlone. 2005. Effects of an intense prescribed fire on understory vegetation in a mixed conifer forest. *Journal of the Torrey Botanical Society* 132:590–601.

Hutchinson, T. F., R. E. J. Boerner, S. Sutherland, E. K. Sutherland, M. Ortt, and L. R. Iverson. 2005. Prescribed fire effects on the herbaceous layer of mixed-oak forests. *Canadian Journal of Forest Research* 35:877–890.

Isaaks, E. H., and R. M. Srivastava. 1989. *An introduction to applied geostatistics*. Oxford University Press, Oxford, United Kingdom.

Jenkins, J. C., D. C. Chojnacky, L. S. Heath, and R. A. Birdsey. 2003. National-scale biomass estimators for United States tree species. *Forest Science* 49:12–35.

Keane, R. E., G. J. Cary, I. D. Davies, M. D. Flannigan, R. H. Gardner, S. Lavorel, J. M. Lenihan, C. Li, and T. S. Rupp. 2004. A classification of landscape fire succession models: Spatial simulations of fire and vegetation dynamics. *Ecological Modelling* 179:3–27.

Keane, R. E., and D. G. Long. 1998. A comparison of coarse scale fire effects simulation strategies. *Northwest Science* 72:76–90.

Keane, R. E., P. Morgan, and S. W. Running. 1996. *FIRE-BGC—A mechanistic ecological process model for simulating fire succession on coniferous forest landscapes of the Northern Rocky Mountains*. U.S. Forest Service, Intermountain Research Station, Research Paper INT-484, Ogden, Utah, USA.

Keane, R. E., P. Morgan, and J. D. White. 1999. Temporal patterns of ecosystem processes on simulated landscapes in Glacier National Park, Montana, USA. *Landscape Ecology* 14: 311–329.

Keeley, J. E. 2003. Fire and invasive plants in California ecosystems. *Fire Management Today* 63:18–19.

Keeley, J. E. 2006. Fire management impacts on invasive plant species in the western United States. *Conservation Biology* 20:375–384.

Kerns, B. K., and J. L. Ohmann. 2004. Evaluation and prediction of shrub cover in coastal Oregon forests (USA). *Ecological Indicators* 4:83–98.

Kipfmueller, K. F., and J. A. Kupfer. 2005. Complexity of successional pathways in subalpine forests of the Selway-Bitterroot Wilderness Area, USA. *Annals of the Association of American Geographers* 95:495–510.

Klinka, K., H. Y. H. Chen, Q. Wang, and L. de Montigny. 1996. Forest canopies and their influence on understory vegetation in early-seral stands on west Vancouver Island. *Northwest Science* 70:193–200.

Koenker, R., and K. F. Hallock. 2001. Quantile regression. *Journal of Economic Perspectives* 15:143–156.

Landres, P. B., P. Morgan, and F. J. Swanson. 1999. Overview of the use of natural variability concepts in managing ecological systems. *Ecological Applications* 9:1179–1188.

Landsberg, J. J., and S. T. Gower. 1997. *Applications of physiological ecology to forest management*. Academic Press, San Diego, California, USA.

Laughlin, D. C., and J. B. Grace. 2006. A multivariate model of plant species richness in forested systems: Old-growth montane forests with a long history of fire. *Oikos* 114:60–70.

Laursen, S. B. 1984. *Predicting shrub community composition and structure following management disturbance in forest ecosystems of the Intermountain West*. Dissertation, University of Idaho, Moscow, USA.

Leathwick, J. R., J. Elith, and T. Hastie. 2006. Comparative performance of generalized additive models and multivariate adaptive regression splines for statistical modelling of species distributions. *Ecological Modelling* 199:188–196.

Li, H., and J. Wu. 2006. Uncertainty analysis in ecological studies: An overview. Pages 45–66 *in* J. Wu, K. B. Jones, H. Li, and O. L. Loucks, *Scaling and uncertainty analysis in ecology*. Springer, Dordrecht, The Netherlands.

Lloret, F., J. G. Pausas, and M. Vilà. 2003. Responses of Mediterranean plant species to different fire frequencies in Garraf National Park (Catalonia, Spain): Field observations and modelling predictions. *Plant Ecology* 167:223–235.

MacFaden, S. W., and D. E. Capen. 2002. Avian habitat relationships at multiple scales in a New England forest. *Forest Science* 48:243–253.

Manning, J. A., and W. D. Edge. 2004. Small mammal survival and downed wood at multiple scales in managed forests. *Journal of Mammalogy* 85:87–96.

McCullagh, P., and J. A. Nelder. 1989. Generalized linear models. *Monographs on statistics and applied probability* 37.

McKenzie, D., Z. Gedalof, D. L. Peterson, and P. Mote. 2004. Climatic change, wildfire, and conservation. *Conservation Biology* 18:890–902.

McKenzie, D., and C. B. Halpern. 1999. Modeling the distributions of shrub species in Pacific Northwest forests. *Forest Ecology and Management* 114:293–307.

McKenzie, D., C. B. Halpern, and C. R. Nelson. 2000. Overstory influences on herb and shrub communities in mature forests of western Washington, USA. *Canadian Journal of Forest Research* 30:1655–1666.

McKenzie, D., D. L. Peterson, and E. Alvarado. 1996. Extrapolation problems in modeling fire effects at large spatial scales: A review. *International Journal of Wildland Fire* 6:165–176.

McKenzie, D., D. W. Peterson, D. L. Peterson, and P. E. Thornton. 2003. Climatic and biophysical controls on conifer species distributions in mountain forests of Washington State, USA. *Journal of Biogeography* 30:1093–1108.

Metlen, K. L., C. E. Fielder, and A. Youngblood. 2004. Understory response to fuel reduction treatments in the Blue Mountains of northeastern Oregon. *Northwest Science* 78:175–185.

Miller, C. 2003. Simulation of effects of climatic change on fire regimes. Pages 69–94 *in* T. T. Veblen, W. L. Baker, G. Montenegro, and T. W. Swetnam, *Fire and climatic change in temperate ecosystems of the western Americas*. Springer, New York, New York, USA.

Miller, C., and D. Urban. 1999. A model of surface fire, climate and forest pattern in Sierra Nevada, California. *Ecological Modelling* 114:113–135.

Miller, C., and D. Urban. 2000. Modeling the effects of fire management alternatives on Sierra Nevada mixed-conifer forests. *Ecological Applications* 10:85–94.

Miller, R. F., and J. A. Rose. 1999. Fire history and western juniper encroachment in sagebrush steppe. *Journal of Range Management* 52:550–559.

Moeur, M. 1985. *COVER: A user's guide to the CANOPY and SHRUBS extension of the Stand Prognosis model*. U.S. Forest Service, Intermountain Research Station, General Technical Report INT-190. Ogden, Utah, USA.

Moeur, M., and A. R. Stage. 1995. Most similar neighbor: An improved sampling inference procedure for natural resource planning. *Forest Science* 41:337–359.

Moisen, G. G., T. C. Edwards, Jr., and P. E. Osborne. 2006. Further advances in predicting species distributions. *Ecological Modelling* 199:129–131.

Moisen, G. G., and T. S. Frescino. 2002. Comparing five modelling techniques for predicting forest characteristics. *Ecological Modelling* 157:209–225.

National Interagency Fire Center. 2006. Wildland Fire Statistics. NIFC. <http://www.nifc.gov/fire_info/fire_stats.htm/>.

Neilson, R. P. 1995. A model for predicting continental-scale vegetation distribution and water balance. *Ecological Applications* 5:362–385.

Neilson, R. P., L. F. Pitelka, A. M. Solomon, R. Nathan, G. F. Midgley, J. M. V. Fragoso, H. Lischke, and K. Thompson. 2005. Forecasting regional to global plant migration in response to climate change. *BioScience* 55:749–759.

Neilson, R. P., and S. W. Running. 1996. Global dynamic vegetation modelling: Coupling biogeochemisry and biogeography models. Pages 451–465 *in* B. Walker and W. Steffen, *Global change and terrestrial ecosystems.* Cambridge University Press, Cambridge, United Kingdom.

Nilsson, M.-C., and D. A. Wardle. 2005. Understory vegetation as a forest ecosystem driver: Evidence from the northern Swedish boreal forest. *Frontiers in Ecology and the Environment* 3:421–428.

Ohmann, J. L., and M. J. Gregory. 2002. Predictive mapping of forest composition and structure with direct gradient analysis and nearest-neighbor imputation in coastal Oregon, USA. *Canadian Journal of Forest Research* 32:725–741.

Ohmann, J. L., and T. A. Spies. 1998. Regional gradient analysis and spatial pattern of woody plant communities in Oregon forests. *Ecological Monographs* 68:151–182.

Oksanen, J., and P. R. Minchin. 2002. Continuum theory revisited: What shape are species responses along ecological gradients. *Ecological Modelling* 157:119–129.

Pacala, S. W., C. D. Canham, and J. A. Silander. 1993. Forest models defined by field measurements: I. The design of a northeastern forest simulator. *Canadian Journal of Forest Research* 23:1980–1988.

Pearman, P. B. 2001. The scale of community structure: Habitat variation and avian guilds in tropical forest understory. *Ecological Monographs* 72:19–39.

Pierce, K. B., and J. L. Ohmann. 2006. *Variance of similar neighbors compared to random imputation.* Paper presented at the 2006 Nearest Neighbors Workshop, University of Minnesota, Minneapolis, Minnesota, USA.

Puccia, C. J., and R. Levins. 1985. *Qualitative modeling of complex systems.* Harvard University Press, Cambridge, Massachusetts, USA.

Rastetter, E. B., A. W. King, B. J. Cosby, G. M. Hornberger, R. V. O'Neill, and J. E. Hobibie. 1992. Aggregating fine-scale ecological knowledge to model coarser-scale attributes of ecosystems. *Ecological Applications* 2:55–70.

Raymond, C. L., L.-K. B. Kellogg, and D. McKenzie. 2006. Mapping fuels on the Okanogan and Wenatchee National Forests. Pages 293–304 *in* P. L. Andrews and B. W. Butler, compilers. *Fuels management—How to measure success.* U.S. Forest Service, Rocky Mountain Research Station, Proceedings RMRS-P-41, Fort Collins, Colorado, USA.

Raymond, C. L., and D. L. Peterson. 2005. Fuel treatments alter the effects of wildfire in a mixed-evergreen forest, Oregon, USA. *Canadian Journal of Forest Research* 35:2981–2995.

Rebain, S. 2006. *The fire and fuels extension to the forest vegetation simulator.* U.S. Forest Service, Rocky Mountain Research Station, addendum to General Technical Report RMRS-GTR-116, Ogden, Utah, USA.

Reinhardt, E. D., and N. L. Crookston, technical editors. 2003. *The fire and fuels extension to the forest vegetation simulator.* U. S. Forest Service, Rocky Mountain Research Station, General Technical Report RMRS-GTR-116, Ogden, Utah, USA.

Rego, F. C., S. C. Bunting, and J. M. da Silva. 1991. Changes in understory vegetation following prescribed fire in maritime pine forests. *Forest Ecology and Management* 41:21–31.

Riegel, G. M., R. F. Miller, and W. C. Krueger. 1995. The effects of aboveground and belowground competition on understory species composition in a *Pinus ponderosa* forest. *Forest Science* 41:864–889.

Ripley, B. D. 1996. *Pattern recognition and neural networks.* Cambridge University Press, Cambridge, United Kingdom.

Robertson, D., A. Bundy, R. Muetzelfeldt, M. Haggith, and M. Uschold. 1991. *Eco-logic: Logic-based approaches to ecological modelling.* The MIT Press, Cambridge, Massachusetts, USA.

Scharf, F. S., F. Juanes, and M. Sutherland. 1998. Inferring ecological relationships from the edges of scatter diagrams: Comparison of regression techniques. *Ecology* 79:448–460.

Schmoldt, D. L., and H. M. Rauscher. 1996. *Building knowledge-based systems for natural resource management*. Chapman and Hall, New York, New York, USA.

Schwartz, M. W., and J. R. Heim. 1996. Effects of prescribed fire on degraded forest vegetation. *Natural Areas Journal* 16:184–191.

Seoane, J., J. Bustamante, and R. Díaz-Delgado. 2005. Effect of expert opinion on the predictive ability of environmental models of bird distribution. *Conservation Biology* 19:512–522.

Sparks, J. C., R. E. Masters, D. M. Engle, M. W. Palmer, and G. A. Bukenhofer. 1998. Effects of late growing-season and late dormant-season prescribed fire on herbaceous vegetation in restored pine-grassland communities. *Journal of Vegetation Science* 9:133–142.

Stockwell, D. R. B. 1999. Genetic algorithms II. Pages 123–144 *in* A. H. Fielding, editor. *Machine learning methods for ecological applications*. Kluwer Academic Publishers, Boston, Massachusetts, USA.

Stockwell, D. R. B. 2006. Improving ecological niche models by data mining large environmental datasets for surrogate models. *Ecological Modelling* 192:188–196.

Termansen, M., C. J. McClean, and C. D. Preston. 2006. The use of genetic algorithms and Bayesian classification to model species distributions. *Ecological Modelling* 192:410–424.

ter Braak, C. J. F. 1986. Canonical correspondence analysis: A new eigenvector technique for multivariate direct gradient analysis. *Ecology* 67:1167–1179.

University of Minnesota. 2006. Nearest Neighbors Workshop, August 28–30, 2006. Minneapolis, Minneapolis, Minnesota, USA.

Urban, D. L., and H. H. Shugart. 1992. Individual-based models of forest succession. Pages 249–292 *in* D. C. Glenn-Lewin, R. K. Peet, and T. T. Veblen, *Plant succession: Theory and prediction*. Chapman and Hall, London, United Kingdom.

Vaillant, N. M., J. Fites-Kaufman, and S. L. Stephens. 2006. Effectiveness of prescribed fire as a fuel treatment in California coniferous forests. Pages 465–476 *in* P. L. Andrews and B. W. Butler, *Fuels management—how to measure success*. U.S. Forest Service, Rocky Mountain Research Station, Fort Collins, Colorado, USA Proceedings RMRS-P-41.

Vardeman, S. B. 1992. What about the other intervals? *American Statistician* 46:193–197.

Venables, W. N., and B. D. Ripley. 2002. *Modern applied statistics with S*. Fourth edition. Springer-Verlag, Dordrecht, The Netherlands.

Wagner, H. H. 2004. Direct multi-scale ordination with canonical correspondence analysis. *Ecology* 85:342–351.

Wang, G. G., and K. J. Kemball. 2005. Effects of fire severity on early development of understory vegetation. *Canadian Journal of Forest Research* 35:254–262.

Waring, R. H., and S. W. Running. 1998. *Forest ecosystems: Analysis at multiple scales*. Second edition. Academic Press, San Diego, California, USA.

Westerling, A. L., H. G. Hidalgo, D. R. Cayan, and T. W. Swetnam. 2006. Warming and earlier spring increase western U.S. forest wildfire activity. *Science* 313:940–943.

Wisdom, M. J., M. M. Rowland, B. C. Wales, M. A. Hemstrom, W. J. Hann, M. G. Raphael, R. S. Holthausen, R. A. Gravenmier, and T. D. Rich. 2002. Modeled effects of sagebrush-steppe restoration on greater sage grouse in the Interior Columbia Basin, U.S.A. *Conservation Biology* 16:1223–1231.

Wikle, C. 2003. Hierarchical Bayesian models for predicting the spread of ecological processes. *Ecology* 84:1382–1394.

Wright, C. S., and S. J. Prichard. 2006. Biomass consumption during prescribed fires in big sagebrush ecosystems. Pages 489–500 *in* P. L. Andrews and B. W. Butler, compilers. *Fuels management—how to measure success*. U.S. Forest Service, Rocky Mountain Research Station, Proceedings RMRS-P-41 , Fort Collins, Colorado, USA.

CHAPTER

Validation of Landscape-Scale Decision Support Models That Predict Vegetation and Wildlife Dynamics

16

Stephen R. Shifley, Chadwick D. Rittenhouse, and Joshua J. Millspaugh

Models have long been used to assess habitat quality and to predict how wildlife populations would respond to changes or manipulations in habitat (Verner et al. 1986, Morrison et al. 1998). In general, most modern approaches assess the value of habitat by relating a species' needs of food, cover, and other necessities to characteristics of vegetative cover types and other landscape features (e.g., distance to edge, patch size, interspersion of habitat features). Habitat suitability models evaluate the resource attributes considered important to a species' presence, abundance, survival, or reproduction and often result in predictive maps that illustrate suitability (Fig. 16-1), which may then be used to make relative comparisons across management alternatives (Gustafson et al. 2001, Marzluff et al. 2002). Ideally, habitat quality would be summarized in terms of demographic processes (Van Horne 1983, Johnson 2007), but this is often not the case due to insufficient data to measure and model demographic processes.

Models can incorporate the effects of spatial patterns of habitat quality on wildlife population viability by modeling demographic parameters of one or more populations such as carrying capacity, population size, and fecundity as a function of characteristics of the patches or landscape they occupy (Akçakaya 2001, 2002, Larson et al. 2004). However, the locations of habitat patches, populations, or individual home ranges change over time due to processes such as succession, natural disturbance, or anthropogenic alteration (Akçakaya et al. 2004). Therefore, changes in wildlife habitat over time in response to management actions or inaction are important determinants of wildlife population viability. Such broad-scale assessments require the use of dynamic landscape models

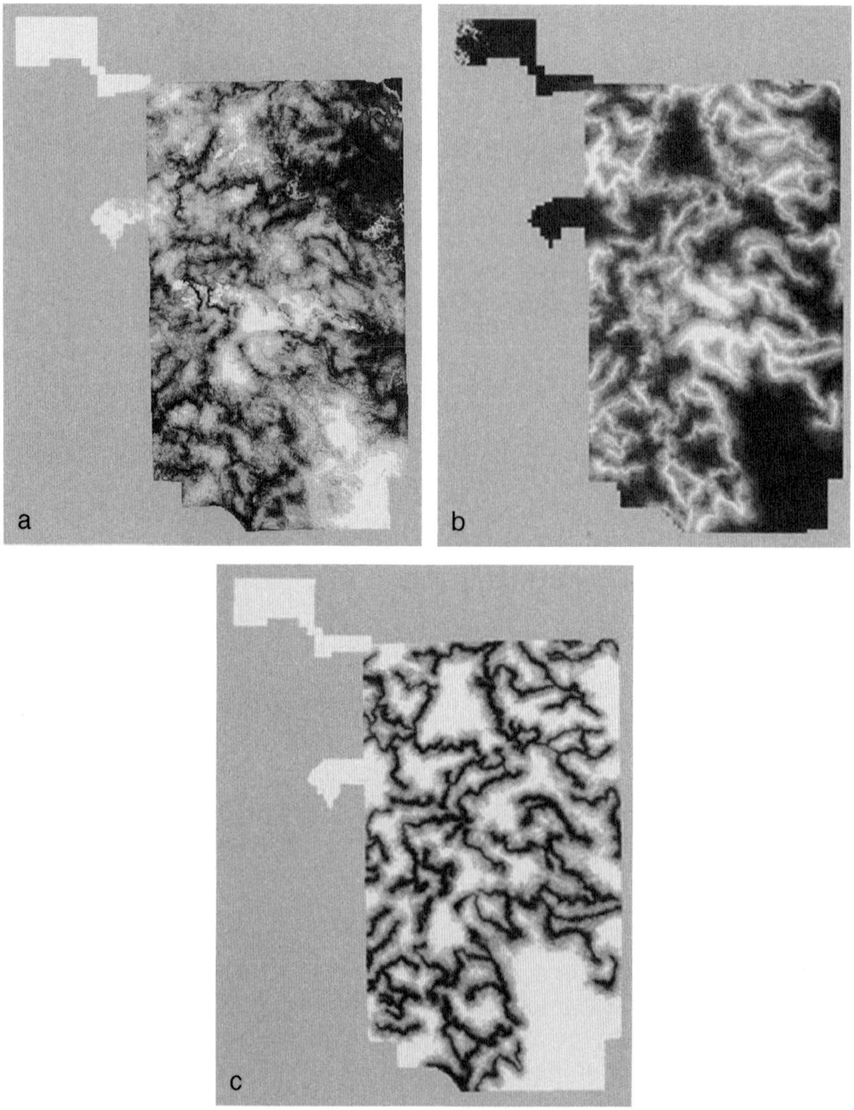

FIG. 16-1

Resource selection functions of (A) elk (*Cervus elaphus*), (B) mule deer (*Odocoileus hemionus*), and (C) white-tailed deer (*Odocoileus virginianus*) in Custer State Park, South Dakota, USA, during summer as determined by radiotelemetry techniques and logistic regression (Woeck 2003).

capable of projecting future habitat conditions (Akçakaya and Brook, this volume; Bekessy et al., this volume; He, this volume).

Landscape models, like all models, are abstractions of reality and should be validated to determine their utility. Under the best circumstances, models will capture the most important features and processes of the real system (Gentil and Blake 1981). But all models are imperfect; they are constrained by imperfect knowledge of the system and limited by resources (e.g., data, computational, human, and financial) for model building or execution (Miller et al. 1976). The challenge is finding or creating a model or group of models that strike a practical balance between usability and capability for one's intended purpose (Millspaugh et al., this volume). Box (1979) stated that "all models are wrong, some models are useful." The essence of landscape model validation and verification is not to determine if a model is wrong—we already know that. Rather we are interested in knowing whether the model is good enough for an intended purpose and whether it is superior to the alternative models that are available (Conroy and Moore 2002). In the case of wildlife habitat models, they represent our best knowledge of animal-habitat relationships. However, the form of the relationships between habitat quality and life history attributes may be unknown. Thus, it may be difficult to validate wildlife habitat suitability models or even to understand what constitutes validation (Van Horne 2002).

Much of our current understanding about simulation model development and validation in ecology is derived from methods originating in engineering and physics. Often model applications in those fields have outcomes that are straightforward and easier to test than those from ecological models (Cartwright 1983). For example, a beam that buckles, a bridge that collapses, or parts that fail with unacceptable frequency give unambiguous feedback on the utility of the models used in their design. The landscape-scale vegetation and wildlife models that are addressed in this book typically lack such clear indicators of outcome. It is inherently difficult to quantify habitat quality for multiple species across a large landscape at a single point in time, particularly when there is no obvious currency (or metric) and methodology for each species. It is even more difficult to quantify how well a model predicts changes in habitat in response to management for those same species. Consequently, with landscape-scale vegetation and wildlife models, we rarely have simple outcomes corresponding to a simple pass or fail. Rather we accumulate strengths and weaknesses or domains where the model appears to be useful and domains where it does not (Starfield and Bleloch 1991).

Evaluation of landscape-scale models is a complex process (Scott et al. 2002). Numerical or statistical prediction accuracy (Mayer and Butler 1993) is only one component of model evaluation (Hamming 1975, Hurley 1986). There are practical and theoretical considerations associated with model evaluation that may eliminate a model from consideration regardless of its statistical accuracy and precision. In this chapter we discuss the important factors that should be considered when evaluating a model for its intended purpose. Model evaluation proceeds with imperfect information (Starfield 1997) and is an iterative process; over time our understanding of model capabilities, data needs, and limitations improves. Models become more useful when applied in an adaptive management framework (Millspaugh et al., this volume). We describe a general process and present specific examples from the literature to help guide evaluation of landscape-scale vegetation and wildlife habitat suitability models. There is not a single recipe suitable for all applications, but the evaluation process will help model developers and model users arrive at a realistic assessment of a model's strengths, limitations, and utility for a stated purpose (Rykiel 1996). Related issues of model transparency, repeatability, and incorporating uncertainty in data are addressed elsewhere in the book (Millspaugh et al., this volume).

GENERAL CONCEPTS AND TERMINOLOGY

Three concepts should be clearly understood in the model validation process. First, evaluating a model in the absence of plainly stated objectives is meaningless. Landscape vegetation and wildlife suitability models can be used to gain insight into the theoretical underpinnings of system processes; to make general predictions of change over large areas and long time frames (Shifley et al. 2006); to make specific predictions for a small area (Marzluff et al. 2002); to address a single species of concern (Akçakaya et al. 2005); to evaluate trade-offs among multiple species (Marzluff et al. 2002, Root et al. 2003, Noon et al., this volume); to make a quick estimate based on readily available data; to establish the critical parameters for a long-term monitoring program; or for countless other purposes. Only when the objectives of model application are known can a model be evaluated and compared with alternatives (Johnson 2001).

Comparison among alternative models is a second core concept in model validation (Conroy and Moore 2002). There are always alternative models. However, alternative models are not always quantitative, and that can make comparisons difficult. For example, "conceptual models" (i.e., mental models) are always an initial point of comparison. Experienced managers may have highly refined conceptual models that work well in specific situations or for some species of wildlife. When conceptual models based on experience are all we have, they invariably are used to make management decisions because they are better than nothing. And "nothing" is the ultimate starting point for model comparisons. With GIS models applied in large landscapes, we sometimes even create a "nothing" or "null" landscape with random patterns and use that as a starting point for model comparison.

Finally, when one is quantitatively evaluating model performance, the currency (i.e., the primary metric) of the model output must be explicitly stated, because it must be measured objectively and compared to observations (Conroy and Moore 2002). When "habitat suitability" is defined in only vague terms, it becomes difficult to validate a model because a variety of competing independent data sources could be used, each offering a different view of model utility. There is an important difference in validating models that predict suitability and those that predict animal abundance. For example, failure of predicted suitability to relate to animal use or abundance does not necessarily mean the model is performing badly if the animal population is well below carrying capacity.

Consequently, additional knowledge about population status is often helpful. Validation with vital rates is sometimes better than with presence-absence data.

The literature relevant to model validation arises from numerous disciplines, and definitions of terms like model validation, verification, or evaluation vary among authors and applications (Morrison and Hall 2002). Also, criteria for successful validation, verification, or evaluation differ among authors (Marcot et al. 1983). Some authors contend that models can never be validated; they can only be falsified as in the context of a statistical hypothesis (e.g., Holling 1978). Others suggest that validation leads to a binomial outcome (true vs. false or good enough vs. not good enough). Others see validation as an ongoing iterative process of constantly refining a model and documenting its strengths and weaknesses (Starfield and Bleloch 1991). Rykiel (1996) provides the most comprehensive and coherent overview of ecological model validation that we have encountered; it is essential reading for anyone engaged in evaluating ecological models. We utilize his terminology whenever possible.

1. *Evaluation*—a nontechnical umbrella term for an assessment of the strengths, weaknesses, and utility of a model for a stated purpose. Evaluation is a broad concept that takes into account whether the model predictions are good enough (Mankin et al. 1979), and whether it is practical to utilize the model with the resources (people, data, money) available for the problem (Johnson 2001). Johnson (2001) argues that "evaluation" is a more appropriate term than "validation" because value is relative and valid is an absolute term. In the context of predicting vegetation and wildlife suitability, a model might not be valid (i.e., it is incorrect), yet still have value for its intended purpose.

2. *Validation*—involves "demonstration that a model within its domain of applicability possesses a satisfactory range of accuracy consistent with the intended application" (Rykiel 1996). Validation includes a specific context of applicability (e.g., species, geographic region, associated disturbance processes). In wildlife habitat suitability modeling, we sometimes extend this traditional definition to consider how well the model performs under conditions that are different from those used to develop the model (Conroy and Moore 2002).

3. *Verification*—demonstrating (1) that the ecological processes embodied in the model (e.g., vegetation response to simulated disturbance or predicted habitat quality change in response to vegetation change) are correctly and sufficiently embodied in the model's equations or algorithms; and (2) determining that the computer code used to implement the equations and algorithms is correct (*sensu* Rykiel 1996).

4. *Calibration*—improves the agreement between model output and observed data through estimation and adjustment of model parameters. Calibration data are most often distinguished from validation data by partitioning the data in time or space.

THREE ELEMENTS OF MODEL EVALUATION

Model evaluation can be partitioned into three areas: (1) the application environment, (2) conceptual design, and (3) quantitative performance (Buchman and Shifley 1982) (Fig. 16-2). Each of these areas addresses a potentially large number of questions, issues, and concerns associated with a given model. The numerical order of the three elements is relevant only in that it represents increasing levels of investment of resources (time, data, intellectual energy) to evaluate model performance. Therefore, it is usually most efficient to see if a model passes the tests related to the application environment and design before pursuing to the data-intensive or quantitative performance considerations.

Application Environment

Evaluation of the application environment simply determines whether the model is compatible with one's particular problem and available resources. Landscape-scale simulation models, such as those discussed in this book, are notoriously demanding of spatial data and computing power. A number of questions should be asked at this stage. Do you have (or can you acquire) the required data to apply the model to your region of interest? Do you have the necessary computing resources and the technical expertise to apply the model? Do you have the time it takes to acquire, learn, and apply the model? These and the related questions in Fig. 16-2 are practical considerations. Deficiencies in the application environment can be difficult to overcome and can quickly eliminate an otherwise suitable model from consideration. Alternatives to simulation models such as conceptual models or expert opinions may have serious limitations in the other two elements of model evaluation, but they usually offer a favorable application environment (i.e., they are widely available, they are fast, and they work with the available data).

Every model presents a unique application environment, but we have some practical observations relevant to landscape-scale simulation models.

1. These models require a great deal of data. Data for model initialization or for regional recalibration are inevitably limited or partially missing. Although there are ways to estimate missing data, they are usually time-consuming. Our general rule of thumb is that it will take longer to assemble the data and implement the model than anticipated—probably twice as long and maybe more. Despite the availability of some software specifically developed for large-scale modeling activities (Beck and Suring, this volume; Dijak and Rittenhouse, this volume; He, this volume; Larson et al., this volume), compiling the data necessary for landscape modeling is difficult.

2. Model initialization and calibration are the most time-consuming stages of landscape simulation modeling, especially for real landscapes. Our

Application Environment

User support
Is the system thoroughly and clearly documented?
Are user guides and sample projection runs available?
Is training available?
Who is available to answer questions or address problems?
Who maintains the system?
Is the system open-source with options for user modification?
Is there an established user group?
Does the system provide on-line help and user prompts?

Data
What data are required to use the system?
Are the available data compatible with those requirements?
Are there established methods to estimate missing data?
Does the system check for erroneous data values or ranges?
Is model output in a format suitable for the intended use?
Can output easily be reformatted or further processed?
Can user-supplied data be used to test model performance or recalibrate the model?

Computing
What computing capacity is required?
How quickly does the system produce results?
What specialized skills or knowledge are required of a user?
Can the system be reprogrammed if necessary?
Does the system require additional software for implementation or for analysis of results?
Have the program logic and coding been verified?

Design

Flexibility
Can the system address a wide array of issues or problems?
Can it be readily modified for new problems?
Is the system built of modules that can be revised or replaced as necessary?
Can models for other resources readily be added?
Can a wide array of management practices and other disturbance processes be modeled?
Can the system easily be recalibrated to new conditions or new locations?

Bio-logic
Are projection models formulated to incorporate basic biological and ecological theories of change?
Are appropriate feedback mechanisms and other system controls incorporated?
Do system component interact logically (e.g. vegetation change affects wildlife habitat quality and herbivory affects vegetation change)?
Over long projection periods, do predictions approach reasonable limits?
Are there real or hypothetical conditions that cause the model to predict results that are obviously unreasonable?

Performance

Entire system
How accurately and how precisely can the system forecast the quantities of interest?
- Wildlife habitat amount and suitability
- Wildlife population viability
- Vegetation species composition and size structure
- The spatial arrangement of vegetation and other landscape features
- The spatial and temporal distribution of disturbance events

Do forecasts show a systematic bias or loss of precision associated with length of projections, with certain species of wildlife or certain vegetation types, by ecological land type, or by landscape location?

System components
Is it possible to test individual components of the system?
- Wildlife habitat models
- Wildlife population viability models
- Vegetation change models
- Disturbance models

Do individual components show a systematic bias or loss of precision associated with length of projections, with certain species of wildlife or certain vegetation types, by ecological land type or landscape location, or other relevant variables?

FIG. 16-2

Considerations in model evaluation. The Application Environment issues often determine whether or not a model can be applied in a given situation. Design issues address factors that can make a landscape model adaptable to new places and new problems. Performance issues address the quantitative and qualitative evaluations that are typically associated with model validation and verification. This table is reproduced from Buchman and Shifley (1982) with only minor modifications and reordering criteria. Over the past 25 years, capabilities to access and manipulate data and to forecast landscape change and wildlife response have increased exponentially. However, the basic questions associated with model evaluation have changed little.

experience is that roughly 75% of the effort goes into gathering base data and setting up a large problem, 5% goes into running the actual simulation scenarios, and the remaining 20% goes toward summarizing, interpreting, and reporting results.

3. The large investment typically required to initialize a landscape simulation model and apply it on a realistic landscape is a significant barrier to model implementation. However, the models are usually amenable to addressing a wide range of issues. When the first application is completed, the models can be applied repeatedly to address other issues.
4. Purchasing more computing and data storage capacity is cheap relative to the cost of implementing the model. However, with landscape-scale simulation models, it is surprisingly easy to exceed all the computing and data handling capacity available.
5. We advise automating the computing processes to the greatest extent possible. Repeating a few hundred hours of simulation is easier to tolerate when it is fully automated than when frequent operator intervention is required.
6. Mistakes in modeling will occur. We inevitably rerun the simulation analysis to correct errors—usually many times.
7. Expertise in applying GIS, manipulating large amounts of data, and computer programming is a practical necessity for any large-scale application of vegetation and wildlife models (Roloff et al., this volume).

Conceptual Design

This set of criteria determines the extent to which the model appropriately incorporates the underlying theories of ecology, biology, vegetation dynamics, and wildlife population dynamics. Does the model account for the ecosystem processes necessary to address the issues at hand? How do wildlife respond to landscape features and predicted landscape change? This evaluation requires a careful look inside the model to ensure computations are correct (i.e., examination of equations or algorithms in the model); it also requires a detailed look at the model assumptions in the context of the specific application of the model. For some landscape models this step is difficult because the user interface and software documentation do not provide all the information needed. We encourage users to fully investigate the inner-workings of "black box" models to ensure they are comfortable with the assumptions being made. It is tempting to proceed with modeling without fully understanding the modeling components, only to find out later that critical assumptions are not appropriate for the intended application. Users should not be tempted to select software based solely on ease of use.

Theoretical derivation of processes and interactions affecting landscape change is a useful exercise because it provides a structured way to think about the problem of landscape and wildlife change in response to endogenous and exogenous factors and feedback loops. Proper theoretical formulation of the model is essential, but it is not entirely sufficient. Implementation of an elegant conceptual design can suffer from missing data, ill-defined causal pathways, and processes that are difficult to quantify. A model that is well designed from a theoretical standpoint may be impossible to implement and, thus, useless for practical purposes (Hurley 1986). Evaluation of model design must explicitly consider the qualitative trade-offs between design, data requirements, and utility for the intended purpose. All ecological models compromise theoretical detail and complexity for application utility; the important consideration is whether or not the model remains useful for its intended purpose. Well-designed models find a suitable balance among theoretical and practical considerations. Models often continue to be refined to incorporate more theoretical detail as practical issues like data availability or computing limitations are resolved over time.

Quantitative Performance

A difficult part of evaluating landscape-scale models is the quantitative assessment of the model predictions. In many cases it seems nearly intractable to complete a true validation, particularly when considering the availability of appropriate vegetation and wildlife data. When attempting to validate wildlife habitat suitability maps using field data, we often face issues with nonconstant detection probabilities of surveyed wildlife (McKenzie et al. 2006), seasonal movements and migration, variability in wildlife use among seasons and years, difficulties with sampling small populations (Thompson 2004), and technological limitations. Forest vegetation models, because they deal with stationary populations, avoid some of these issues, but they face other constraints such as slow rates of change over time (e.g., with regard to tree species composition). Despite these barriers to model validation, there are a few techniques, although imperfect, that can address the quantitative performance of landscape-scale decision support models.

Data are invariably a limiting factor in quantitative validation of landscape model predictions. For example, if we want to compare model predictions of change for real landscapes to observed change for those landscapes, where do we get the data? Clearly, we do not have a reference map for future time periods. Therefore, validation can be performed only for past or present conditions. If we are interested in a small landscape, roughly 1,000 ha in size, can we find a dozen landscapes where we have observational data on initial conditions, disturbance events, vegetation change over time, and wildlife change over time? Do we have such results for several decades of observation, or for a century, which is a typical planning horizon for forest landscapes? Usually, we do not. In a few cases, it is possible to test individual species over the long term with appropriate data sets. For example, one might use the breeding bird survey as a long-term data source (Newson et al. 2005, Somershoe et al. 2006, Freeman et al. 2007). One can also substitute space for time using widespread vegetation inventories such as Forest Inventory and Analysis (FIA) data (U.S. Forest Service 2008*a*) although limitations exist (Roloff et al., this volume). Testing with other detailed, species-specific options such as expert panels is also possible (e.g., Holthausen et al. 1994).

There are a number of procedures to quantitatively evaluate model performance. Sensitivity analysis is commonly used to evaluate the contribution of individual parameters to model performance (Rykiel 1996), but does not constitute validation (Johnson 2001). By systematically varying model parameters and comparing results across a wide range of landscape conditions, sensitivity analysis helps model users understand how robust the model is to small changes in the modeled relationships. That helps model users understand how much confidence to place in model results. Because of the interactive effects of multiple input parameters in complex models, it is advisable during sensitivity analysis to alter more than one parameter at a time. As Rykiel (1996) suggests, the most important parameters should be estimated well. Sensitivity analysis, although useful, should be complemented by other quantitative evaluation methods. Statistical procedures, including Receiver Operating Characteristic (ROC) curves (Zweig and Campbell 1993, Pearce and Ferrier 2000), confusion matrix-derived measures, such as the Kappa statistic (Cohen 1960, Fielding and Bell 1997), and cross-validation techniques (Boyce et al. 2002), are commonly used in validation tests. The appropriateness of these procedures is dependent on assumptions of the model, the data available, and intended purpose of the model.

We now turn our focus to a review and discussion of the methods used to evaluate models that make quantitative predictions of changes to vegetation over time and habitat suitability on large landscapes.

EVALUATION OF FOREST VEGETATION MODELS

Types of Forest Vegetation Models

Unlike wildlife, forest vegetation is rooted in place. Except as seeds, trees do not move beyond the reach of their roots or crowns in search of suitable habitat. Thus, observing changes over time for individual trees or communities of forest vegetation is relatively easy when compared to similar operations for wildlife. Consequently, models to predict vegetation change for trees and forest stands are well established (e.g., Ek et al. 1988, Vanclay 1994, Dixon 2003, Husch et al. 2003) and are widely used to predict forest change and develop management prescriptions (Miner et al. 1988, Mowrer 1997, Dixon 1998, Twery et al. 2005).

Methods for validating short-term predictions (e.g., up to 50 years) of tree or stand change are documented in numerous sources (e.g., Buchman and Shifley 1982, Reynolds 1984, Miner et al. 1988, Brand and Holdaway 1989, Vanclay 1994, Vanclay and Skovsgaard 1997). These typically compare model predictions with changes observed from remeasured forest inventory plots. The metrics of interest are usually (1) tree size or growth (e.g., diameter, height, and volume), (2) tree mortality, and (3) stand change per hectare (density, volume, diameter distribution). Validation generally consists of measuring the departure of the observed from predicted vegetation change and reporting bias and precision of estimates by species, per hectare, and over time. Occasionally, validation results for forest change also include estimated prediction intervals—expected errors if the model is applied at new locations (Reynolds 1984). In virtually all situations, model predictions and validation procedures for forests are limited to tree species and exclude other vegetation layers that lack abundant inventory data.

Software for implementing tree- and stand-scale tree growth models is comprehensive and widely available (Ek et al. 1988, Dixon 2003). For some landscapes it is now possible to apply tree- or stand-scale models to all stands across the landscape and, thus, forecast change over time for the entire landscape. This approach provides great detail in predicted forest vegetation structure and composition at all locations on the landscape, it is a highly intuitive methodology, and in some cases integrated software is available to visualize landscape vegetation change over time. Validation for the component vegetation change models are typically reported with the documentation of model calibration procedures (e.g., Brand and Holdaway 1989, U.S. Forest Service 2008*b*), so initial estimates of model precision and bias are available in most cases. Also, long-term predicted changes in stand size structure can be evaluated from a theoretical perspective using multidimensional, graphical analysis of modeled changes in stand density, height, mean tree size, volume, and site quality (Leary 1996). Although the tree- and stand-scale vegetation models usually provide a mechanism to model forest regeneration and species succession, forecasting long-term species succession is not their strength (for the same reasons outlined for other categories of landscape-scale models discussed later in this section).

The Forest Vegetation Simulator (FVS) is the most widely utilized system of tree and stand growth prediction models in the United States. It includes automated procedures for comparing predicted tree and stand changes with actual changes observed from remeasured forest inventory plots. It also provides automated procedures for recalibrating the prediction equations based on changes observed from local inventory data (Dixon 2003). In situations in which the landscape extent and the data on initial landscape vegetation conditions are compatible with model requirements, FVS or similar models can be an excellent choice for forecasting forest vegetation with known (or readily determined) levels of precision and bias. Moreover, it is possible to visualize projected forest landscape change over time as three-dimensional renderings using software such as the Landscape Management System (LMS; Millspaugh et al., this volume; Oliver et al., this volume; University of Washington 2008). A typical application would be for a landscape composed of a large, contiguous public or private ownership supported with a current stand-level inventory of forest conditions for each stand or each ecological land type.

Limitations to applying tree- and stand-level models in the specific context of large-scale landscape decision support modeling are the detailed data requirements for initial landscape conditions (e.g., forest inventory data observed or estimated for each site), the burden of carrying for each site highly detailed data that may be unnecessary for evaluating the final objectives (e.g., for estimating wildlife habitat suitability), and a limited capacity to model natural regeneration and species succession over long time periods. Thus, for landscapes larger than about 25,000 ha and for analyses approaching or exceeding the duration of a forest management rotation, the general tendency has been to model vegetation structure and composition using landscape decision support systems that carry far less detail about forest vegetation at each site. These fall into two broad categories: (1) raster-based models that track forest age class and/or tree species presence on sites ranging from 0.01 ha to 1 km^2 mosaiced across the entire landscape (e.g., LANDIS and HARVEST) (Gustafson and Crow 1999; He et al. 1999, 2005; Mladenoff and He 1999; Mladenoff 2004; Gustafson and Rasmussen 2005; Scheller et al. 2007; He et al. this volume), and (2) polygon-based models that track the progression of each landscape polygon (e.g., forest stand) through a finite number of habitat classes defined by the species composition and size structure of the dominant vegetation (e.g., LANDSUM, SIMPPLE, and VDDT/TELSA) (Chew 1995; Keane et al. 1997, 2002; Beukema and Kurz 1998; Barrett 2001; Chew et al. 2007). In both cases the primary emphasis is on predicting long-term (a century or more) changes in species composition and forest size structure in response to succession, harvest, fire, severe weather, insects, and disease.

Validation and Evaluation Considerations

For all landscape vegetation models, quantitative validation of predicted forest change is hampered by (1) a lack of long-term data documenting patterns of tree species succession and (2) a lack of data describing pre- and post-disturbance vegetation for sites affected by harvest, fire, severe weather, insects, or disease. Consequently, quantitative, data-driven validation estimates of predicted changes in species composition over time (with or without disturbance) are problematic. However, there are qualitative procedures for evaluating predicted species dynamics for landscape-scale models of vegetation change. Verification of the predicted rate and pattern of disturbance processes is also required.

Evaluation of landscape-scale forest vegetation models can be subdivided into three parts: (1) validation of forest structure change in the absence of disturbance; (2) validation of the rate of exogenous disturbance; and (3) evaluation of species dynamics as affected by disturbance type (e.g., harvest, fire, wind, none) and by ecological land type. The three elements do not exhaust the range of what could be (and ideally should be) done to evaluate a landscape-scale model of forest vegetation used to support decision making, but they provide an essential first iteration.

Validation of Forest Structure Change.—Validation of forest structure change in the absence of disturbance has strong ties to validation processes developed for traditional forest growth and yield models, and it is the validation component that typically has the best supporting data. In the absence of disturbance, established forest stands (or sites) go through predictable stages of development that can be described in terms of tree size and density, in terms of stand age, or in terms of structural stages (e.g., stand establishment stage, stem exclusion stage, understory reinitiation stage, old-growth stage; Oliver and Larson 1996).

For tree- and stand-based models of forest vegetation change (e.g., FVS) validation estimates typically encompass two to five decades of observed change for disturbed and undisturbed forests. For robust evaluation of model performance, the observed forest changes in the independent validation data set should cover a wide range of forest age, species composition, site quality, and ecological land types; these readily measured attributes are known to influence forest change. Examples and recommendations for validation of forest size and structure change based on inventory data can be found in Reynolds (1984), Brand and Holdaway (1989), Vanclay (1994), and Vanclay and Skovsgaard (1997). However, other types of landscape-scale vegetation change models track less detail about each site and produce outputs that are not directly comparable with remeasured forest inventory data. These models include LANDIS (He et al. 1999, 2005; Mladenoff and He 1999; Mladenoff 2004; Scheller et al. 2007; He et al., this volume), HARVEST (Gustafson and Crow 1999, Gustafson and Rasmussen 2005), LANDSUM (Keane et al. 1997, 2002), SIMPPLE (Chew 1995, Chew et al. 2007) and VDDT/TELSA (Beukema and Kurz 1998). Evaluation of vegetation age and size structure change for these models amounts to verification that the predicted forest age changes with each time step in the simulation and/or that the predicted vegetation structural state changes with the model time step in accordance with the probabilities established when the model was calibrated.

Evaluation of Modeled Disturbances.—Evaluation of modeled exogenous disturbances by harvest, weather, fire, insects, or disease in landscape-scale vegetation models is primarily a verification process. For example, harvest is a disturbance that in reality and within a landscape model is controlled by managers. Model verification should be conducted to determine (1) that the timing, location, and cumulative spatial patterning of modeled harvest operations is consistent with those prescribed in the simulation and (2) species composition and age structure of the modeled landscape are adjusted to reflect the anticipated post-harvest state of the forest site. This is simply verification that modeled harvest events operate as intended. Prescribed fire can be thought of as a similar anthropogenic disturbance that requires the same type of model verification.

Modeled patterns of disturbance by weather, wildfire, insects, or disease are usually developed either by analyzing and quantifying observed patterns based on past disturbances (e.g., records of wildfires or insect damage) or by speculation about future patterns of disturbance. In either case, model verification should be conducted to ensure that the predicted patterns and extent of disturbances implemented in the landscape model are consistent with the observed historical data and/or with speculative scenarios. Landscape models often use stochastic methods to predict the timing and location of disturbances due to weather, wildfire, insects, or disease. Thus, model verification should include examination of variability in the location and timing of disturbance events across multiple runs of a given scenario.

Although opportunities for quantitative validation of predictions of disturbance events in landscape-scale vegetation models are limited, there is a recent example. Yang et al. (2007) used historical data on the location of fire ignitions in the Missouri Ozarks from 1970 to 2002 to develop a probability of map of wildfire ignition risk for a 130,000 ha study area. The wildfires in that region are primarily human-caused, and topography, road locations, land ownership, and proximity to communities were shown to be significant predictors of ignition risk. They then used Monte Carlo methods to repeatedly simulate fire spread across the landscape based on the ignition probability and modeled rates of fire spread in response to vegetation type, topography, prevailing wind direction, and locations of fire breaks. The resulting burn probability map summarized the likelihood that any given point on the landscape would burn. This provided for each site a spatially explicit fire probability that is applicable for landscape change simulations. Quantitative validation of the fire risk model was possible by examining the mapped patterns of actual fires in subsequent years (2003–2004) and using categorical analysis to determine if the sites that really burned occurred disproportionately at locations that had a greater than average predicted probability of burning (Yang et al. 2008).

Evaluation of Changes in Species Composition.—Predicting changes in species composition is the most difficult aspect of modeling landscape-scale, long-term vegetation change and of validating those models. In most forest ecosystems the species composition of regeneration varies greatly from site to site (even within a single stand) and is known to be affected by the species composition of the prior stand; the size and species composition of the advance reproduction; the site productivity; the ecological land type; the type, size, frequency, and intensity of the disturbance event(s) that precipitated the regeneration event; and the type, size, frequency, and intensity of disturbances that follow the regeneration event (e.g., drought or surface fire). After a forest is regenerated, tree species composition can continue to change rapidly as the forest progresses through the stand establishment and stem exclusion stages of development; this may last for three decades or more (Oliver and Larson 1996, Johnson et al. 2002). Due to a lack of long-term, site-specific inventory data describing patterns of forest regeneration, a quantitative comparison of observed and modeled long-term changes in tree (or herbaceous) species composition is rarely possible. Short-term comparisons of observed and modeled changes in species composition (e.g., for two to five decades) are possible but rarely sufficient because they usually do not adequately capture successional shifts in species composition (e.g., the gradual shift from oak [*Quercus* spp.] dominated forests to forests dominated by maples [*Acer* spp.] and other mesic species across much of the eastern United States).

Often the most useful verification of predicted long-term changes in tree species composition is the evaluation of "reasonableness" as judged by local experts. In this process a landscape model of vegetation change is applied to a real landscape or a theoretical test landscape while implementing a wide range of disturbance scenarios. Then the predicted changes in species composition over time are vetted by experts who judge the patterns of species change over time to be reasonable or otherwise. Because of the large number of variables involved in landscape modeling, this procedure is facilitated by including modeled scenarios that implement a single disturbance agent while holding other factors constant (e.g., separate scenarios for no disturbance, low intensity fire, high intensity fire, even-aged management, and uneven-aged management). This is a form of sensitivity analysis. Depending on the number of modeled tree species and the complexity of the modeled ecosystems, experts may need to focus on a few key indicator species or scenarios.

An example is the evaluation of tree species composition modeled using LANDIS in southern Indiana. In that region oaks dominate forest overstories on the majority of acres (Woodall et al. 2005). However, five decades of inventory data show that under the current management regime (characterized by low fire frequency and small harvest openings) maples and other mesic and shade-tolerant tree species are increasing in abundance and gradually displacing oaks via greater regeneration success (Shifley and Woodall 2007). Silvics information (Burns and Honkala 1990) and stand-scale regeneration studies indicate that (1) red oak species (*Quercus* section *Lobatae*) are faster growing and shorter-lived than the white oaks (*Quercus* section *Quercus*); (2) oaks are generally more fire tolerant than their mesic competitors, particularly with repeated fires; (3) white oaks tend to be more tolerant of fire than red oaks; and (4) harvesting provides opportunities for some oaks to successfully regenerate via sprouting (Johnson et al. 2002). Three management scenarios applied to the 81,000 ha Hoosier National Forest in southern Indiana provide a qualitative means to examine and evaluate the reasonableness of predicted long-term changes in species composition (Rittenhouse 2008) (Fig. 16-3).

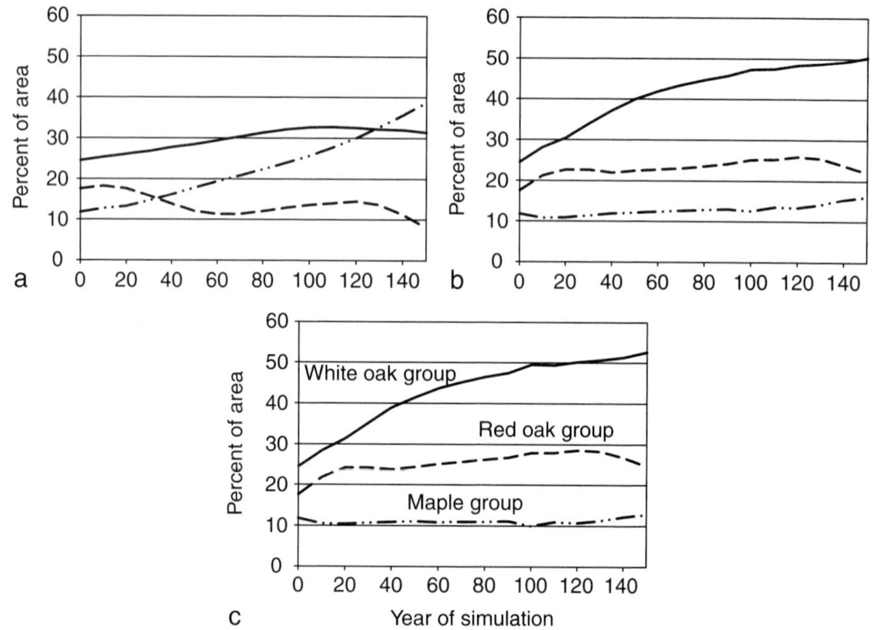

FIG. 16-3

Projected change in dominant tree species composition on each site for three management alternatives on the 81,000 ha Hoosier National Forest. This figure shows projected trends over 150 years for three major species groups: red oaks (*Quercus rubra* L., *Q. Velutina* Lam., *Q. coccinea* Muenchh.), white oaks (*Q. alba*, L., *Q. Prinus* L.), and maple (*Acer saccharum* Marsh., *A. rubrum* L.). With the exception of hickories (*Carya* spp.), the nine remaining species groups were typically dominant on less than 5% of the landscape. Panels ordered in increasing intensity of disturbance: (A) no harvest or prescribed fire, (B) group selection harvest on 3.9% of area per decade and prescribed fire on 3.6% of area per decade, (C) shelterwood or clearcut harvest on 4.3% of area per decade and prescribed fire on 36.7% of area per decade (From Rittenhouse 2008).

EVALUATION OF WILDLIFE HABITAT MODELS

An important consideration in any evaluation of wildlife habitat models is the currency (or metric) used to represent habitat quality because this currency will help decide which type of data to collect or which existing data may be used to validate the model. The most basic measure of habitat suitability is species presence or absence. More complex measures of habitat suitability include species abundance or density (Niemuth et al., this volume), the amount of time spent within different habitat types, vital rates (survival or fecundity), and population viability (Beissinger et al., this volume). Careful consideration of what "suitability" represents is essential in model building, but equally necessary when evaluating model performance. The type of empirical data used for validation should correspond to the objective of the suitability model. For example, if the model is developed to predict breeding habitat suitability, then empirical data used for validation should provide evidence of breeding and ideally some measure of breeding success (e.g., nest success, number of young fledged, season-long or annual fecundity).

Wildlife habitat suitability models are either deductive or inductive (Ottaviani et al. 2004), and the difference between the two is important because some validation approaches are appropriate for only one type. Deductive models include those relying on expert knowledge and published literature (e.g., Dijak and Rittenhouse, this volume). In contrast, inductive models are statistical models developed from wildlife data that are associated with habitat features (e.g., Niemuth et al., this volume). The latter are commonly developed from survey data or radiotelemetry (e.g., resource selection functions; Manly et al. 2002). Validation of deductive models generally involves the use of independent field data, such as animal location, density, or other demographic components (e.g., nest success) to assess the utility of the model. Validation then consists of comparing expected (or predicted) values to observed (or reference) values and quantifying the agreement between the two as a form of accuracy assessment (Rykiel 1996). This comparison is made in different ways depending on several factors, including the type of model and data used in the validation procedure, and whether or not the model output and reference maps are spatially explicit. Validation of inductive models also often uses independent data (e.g., Mladenoff et al. 1999, Luck 2002), but frequently model evaluation includes traditional statistical evaluations, such as resampling (e.g., jackknife and bootstrapping) and data-partitioning (e.g., *k*-fold cross-validation) procedures (Boyce et al. 2002) on the original observations. We view the use of resampling and data-partitioning tools as useful first approximations of model evaluation; however, the use of independent data allows for a richer investigation of model performance and utility issues (e.g., Is the model over-fit? Is performance of the model overrated?).

The most common type of data used to validate wildlife habitat suitability models is presence-absence data. Presence-absence models use data from known occurrences or observations (site is used) and known absences (site is unused). Sites classified as unused must be confirmed via sampling. Otherwise, used locations represent a sample from a distribution of available locations on the landscape. In this sense, presence-only models are a subset of presence-available models because the data contain known occurrences; absence is neither confirmed nor addressed in the model. Several papers have summarized validation procedures for presence-absence models (Fielding and Bell 1997, Boyce et al. 2002, Hirzel et al. 2006) including Receiver Operating Characteristic (ROC) curves and Area Under the Curve (AUC) (Zweig and Campbell 1993, Cumming 2000, Pearce and Ferrier 2000), and confusion matrix-derived measures (Baldessarini et al. 1983) such as the Kappa statistic (Cohen 1960, Fielding and Bell 1997). For presence-available models, Boyce et al. (2002) recommended a *k*-fold cross-validation method and used Spearman-rank correlation to compare observed values (presence only) to categories (bins) of predicted values. The Boyce et al. (2002) approach is also applicable to presence-absence models. All the statistical approaches for validation of presence-absence and presence-available models are applicable to suitability models. When presence and absence data are available, one also has the opportunity to evaluate both omission error (species present, yet model predicts unsuitable habitat) and commission error (species absent, yet model predicts suitable habitat; Ottaviani et al. 2004). We believe that measures such as ROC curves (Fielding and Bell 1997, Pearce and Ferrier 2000) are useful, in part, because they overcome issues with choosing an arbitrary probability threshold when evaluating model performance (although see Termansen et al. 2006).

Roloff and Kernohan (1999) reviewed validation studies for published Habitat Suitability Index (HSI) models and provided a "Checklist for study design and validation considerations for evaluating HSI model performance." They found that the most common deficiencies in HSI validation studies were (1) inadequate consideration of input data variability and how that variability affected interpretation of the final HSI output (Bender et al. 1996); (2) application of the models to inappropriate spatial scales; (3) sampling too narrow a range of habitat conditions; and (4) collection of population data over too short a time frame to reflect variation in population size, density, or demographic rates. Subsequent studies have addressed some of the issues summarized by Roloff and Kernohan (1999). For example, Burgman et al. (2001) used a fuzzy numbers approach for establishing reliability estimates (i.e., confidence intervals) of HSIs. Larson et al. (2004) used a similar approach to create upper and lower limits of ovenbird (*Seiurus aurocapillus*) habitat suitability.

Although many validation procedures exist, relatively few specifically address the use of animal data to assess a suitability model's predictive ability. In the next two sections we highlight general issues and practical approaches for validation of both deductive and inductive wildlife habitat suitability models. Our purpose is to address issues of validation with the explicit goal of quantifying the predictive ability of habitat models using animal data. We focus our discussion around HSI models and not animal counts or viability per se. However, many of the general concepts apply to either objective. Often, we are ultimately interested in the same diagnostics, such as classification rates and the difference between predicted and observed values within the landscape.

General Issues with Validation of Habitat Suitability Models

Here we discuss four specific issues related to validation of habitat suitability models: (1) model uncertainty (also see Millspaugh et al., this volume), (2) types of empirical data used in validation, (3) multiscale validation, and (4) autocorrelation (both spatial and temporal). The increased use of habitat suitability models within large landscapes for policy, conservation, and management decisions (e.g., state and federal land planning) makes a discussion of these issues relevant. Although this is not a complete list of HSI model validation issues (e.g., no discussion of nonconstant detection probabilities [MacKenzie et al. 2006]), it provides an essential starting point that can be expanded with the guidance of other literature that addresses these issues in part or on whole (Scott et al. 2002, Williams et al. 2002, MacKenzie et al. 2006).

Model Uncertainty.—Habitat suitability models include numerous sources of model uncertainty. Regan et al. (2002) defined two main categories of uncertainty: linguistic uncertainty and epistemic uncertainty. Linguistic uncertainty arises from the vagueness, ambiguity, and context dependence of the natural and scientific language used in developing, describing, and applying models. Epistemic uncertainty arises from uncertainty regarding a determinate fact. For example, habitat suitability models often include numerous habitat variables, and each variable is subject to location or assignment error in GIS layers, bias due to study design or data collection methods, and natural variation in biotic or environmental conditions. Rigorous attention to study design and error assessments should be conducted prior to model development.

Model uncertainty also arises from subjective interpretation of wildlife-habitat relationships and the decisions made in developing models (Ray and Burgman 2006). For example, deductive habitat suitability models (e.g., traditional HSI models) use expert knowledge to specify which habitat variables are important to a species, the relative importance of different variables, and how to relate those variables to habitat suitability values (Larson et al. 2003; Dijak and Rittenhouse, this volume). Given that the assumed or known form of the suitability relationship may be logical (presence-absence of feature), linear, or nonlinear (e.g., sigmoid or threshold response), and the HSI equation may use geometric, arithmetic, or logical relationships to calculate the composite HSI value, many different HSI models may be developed from the same set of variables. Ray and Burgman (2006) recommend using "bounded" habitat suitability maps (i.e., maps developed under alternative HSI models) to evaluate subjective uncertainty. This is a variation of sensitivity analysis.

Alternatively, model verification and calibration, conducted prior to validation, may be used to ensure that the model form is correct and all relevant parameters are included. For example, Mitchell et al. (2002) determined the relative contribution of each habitat suitability component (e.g., input data as well as individual suitability indices) to the final HSI model using sensitivity and elasticity analyses (Caswell 1978, Stearns 1992). He quantified how the variability in each component could affect the final HSI values by weighting each individual suitability parameter in the HSI equation. Unequal sensitivity and elasticity values among components suggest variation in model output is due to a subset of model components. In that case, the model may be reduced to a subset of components without substantially affecting model predictions. Beven (1993), Oreskes et al. (1994), Rykiel (1996), Beutel et al. (1999), Morrison et al. (1998), Johnson (2001), and Scott et al. (2002) provide methodological summaries or reviews that address model uncertainty, verification, and calibration.

Data for Model Evaluation.—The second main issue relates to the data type used in model validation. An important consideration in model evaluation is the currency (metric) used to represent habitat quality because this currency will help decide which type of data to collect or which existing data may be used to validate the model. Buskirk and Millspaugh (2006) discussed alternative currencies used in resource selection studies; these are defined as measures of investments by an animal in searching, finding, or using a resource. The currencies include time spent, distance traveled, energy expended, and predation risks incurred. In addition, they defined event sites as those places that animals use intensively or where important life functions occur. Each of these currencies is applicable to habitat model validation using animal data, and they complement more traditional measures of habitat quality.

More complex measures of habitat suitability include species abundance or density, vital rates (survival or fecundity), and population viability (Akçakaya and Brook, this volume; Beissinger et al., this volume; Larson et al., this volume). Van Horne (1983) suggested that habitat quality should be the product of density, survival, and expectation of future offspring (also see Johnson 2007). The data needed to meet this definition of habitat quality are often prohibitive. However, these metrics are the most meaningful and useful, given issues with animal location data (Battin 2004, Rittenhouse 2008) (Fig. 16-4). These metrics must be used when demographics are being modeled. Despite the desirability of demographic data, they can be problematic for validating habitat models and, for this reason, are often not used. For example, other factors, such as territoriality, may limit access to high-quality habitat, resulting in some individuals using low-quality habitat. The existence of demographic phenomena affecting the relationship between populations and habitat (e.g., source-sink) also should be considered to ensure inference about the predictive capability of the model is correct (Conroy and Moore 2002). Thus, it is critical that context be considered. For example, a species below carrying capacity might respond differently to available habitats. In such cases, habitat suitability might be better defined as some demographic parameter.

Spatial Scale and Validation.—The third issue with validation of wildlife suitability models is that validation should address the spatial scale of model development and of the intended application. The level of accuracy or precision should be specified *a priori* because agreement depends in part on the cell size (i.e., landscape resolution; Pontius et al. 2004). In general, maps with coarse resolution have better agreement with observed data than maps with fine resolution because the coarse resolution data are an average or a composite across a relatively large area. This reduces the variation in the observed values. In other words, the focus shifts from spatially explicit prediction at fine resolutions to composition at coarser resolutions. Advancements in computing technology

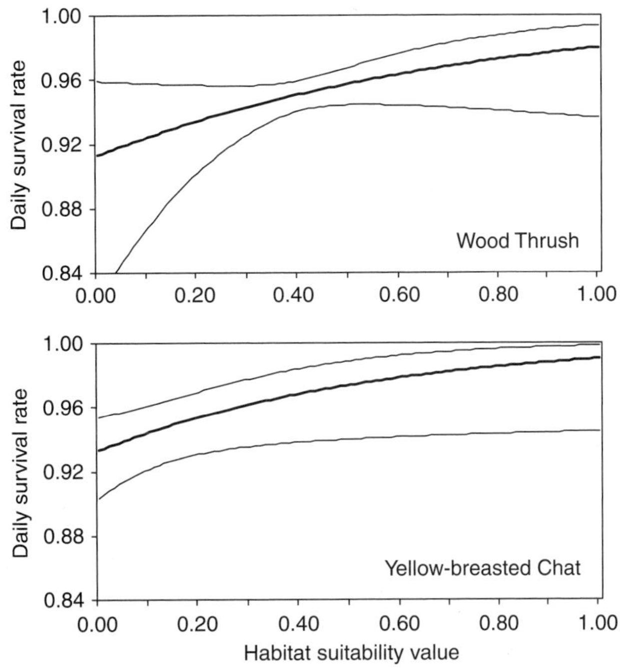

FIG. 16-4

Daily survival rates of wood thrush and yellow-breasted chats in relation to habitat suitability values in Missouri Ozark Forest Ecosystem Project, south-central Missouri, 1991–2002. Thin lines represent Wald 95% confidence limits for the logistic-exposure model (Rittenhouse 2008).

facilitate application of wildlife suitability models to large spatial extents at high resolutions (e.g., small cell sizes), which means that HSI models incorporate multiple spatial scales. Empirical data and the suitability measures developed from them may represent habitat associations at similar or different spatial scales than intended by the model. The most common spatial scale used in wildlife suitability models is the animal's home range size, which corresponds to Johnson's (1980) second-order selection. However, habitat variables at any scale may influence selection and affect the suitability relationship. The correlation of habitat variables within and among scales results in cross-scale correlation; Battin and Lawler (2006) present methods for identifying and incorporating cross-scale correlation in analyses.

Spatial and Temporal Autocorrelation.—The fourth issue with habitat suitability model validation is treatment of spatial and temporal autocorrelation. Spatial autocorrelation results when the distribution of animals or environmental variables in a cell is not independent of surrounding cells (e.g., attraction of con-specifics, dependence in the suitability value in neighboring cells). Spatial autocorrelation may occur in wildlife models with suitability relationships based on patch size (area) functions or the composition of habitat computed within a moving window, because the suitability value is determined in part from a cell's proximity to other cells with similar resource attributes. Spatial autocorrelation may also occur in the animal data used for validation. For example, kernel methods for estimating the distribution and intensity of use by animals inherently are spatially autocorrelated (Marzluff et al. 2004). The concern with spatial autocorrelation is that it may contribute to either stronger agreement or disagreement when comparing the observed map to the predicted map. Without explicitly addressing spatial autocorrelation, wildlife suitability models may chronically under- or over-perform in certain regions of the map. Validation techniques that retain the spatial context of prediction errors (e.g., as maps of prediction errors) rather than providing a global measure of error may facilitate identification of spatial autocorrelation.

Validation of Suitability Models Using Animal Location Data

In this section we focus on approaches that use animal data to validate habitat suitability models, which predict habitat quality and not demographic parameters. These approaches address some of the validation issues discussed in the preceding section, but due to limitations in available animal data, no single approach simultaneously addresses all the issues (Van Horne 1983, Battin 2004). The choice of validation procedure varies by the form of the habitat suitability model and the available animal data, and it should be matched to the intended model use. We primarily discuss procedures that compare the predicted distribution of suitability values to observed locations of individual animals or of distributions (populations) of animals, which may be estimated from radio-telemetry relocations, surveys (e.g., point counts, distance sampling), or territory spot mapping. The features of distributions that are of interest when validating suitability models include the frequency of values with the same classification, the magnitude of the difference between predicted and observed values for a given cell, the cell location, the type of error (agreement or disagreement), and the source of error (e.g., issues of model uncertainty, animal data, spatial scale, autocorrelation).

The first validation approach uses compositional analysis (Aitchinson 1986). Compositional analysis has been used in studies of habitat selection to determine if the proportion of each habitat type within the home range (area used) differs from the proportional occurrence of habitat types at a larger scale (area available—typically the entire study area; Aebischer et al. 1993). A logical extension of compositional analysis for validation is to consider the predicted map as defining "available" habitat and testing whether the proportion of suitability index values within an observed animal's home range (i.e., "used") differ from what is available. This requires categorizing (or binning) suitability values and treating the categorical suitability data as synonymous with a categorical map of habitat types. In addition, the working assumption is that animals should use areas with higher habitat suitability values than is the mean of all available habitat. For example, Ottaviani et al. (2004) validated habitat suitability models for 113 species of terrestrial vertebrates and 82 species of birds using compositional analysis. They compared the mean, covariance structure, and proportion of habitat suitability classes within polygons of used sites to polygons of similar size selected at random from the study area. Differences in modeled habitat suitability class rankings from those for sites that were actually used may reveal which landscape conditions are contributing to agreement or lack thereof. Ottaviani et al. (2004) assumed animal use within a polygon was equal to the raw proportion of habitat suitability classes within the polygon, which means that nonrandom use of habitat suitability classes within polygons was not considered. Millspaugh et al. (2006) suggested weighting the raw proportion of habitat (suitability classes) by the amount of animal use, estimated from empirical data. In this way, both the proportion of each habitat suitability class and the amount of use of that class are considered in the analysis (Millspaugh et al. 2006).

While compositional analysis can indicate that differences exist between observed and predicted values, it does not necessarily provide a spatial context regarding those differences. In other words, do the maps look similar? Pontius et al. (2004) used multiple resolution analysis to answer this question. They partitioned sources of error between observed and predicted values based on cell location and on the quantity of cells with particular suitability values. The Pontius et al. (2004) approach provides a good basis for understanding model performance in terms of the quantity and location of suitability values, and the influence of spatial resolution on model prediction ability. Couto (2003) presented approaches for comparing maps with different classification schemes, which might occur when comparing habitat suitability maps to animal distributions. Cell values in "hard maps" are discrete and mutually exclusive, meaning that a cell has only one value. In contrast, cells in "fuzzy maps" have mixed or uncertain membership, meaning that a cell could have more than one value. Uncertainty in cell membership could arise from model uncertainty, autocorrelation, or differences in cell resolution between predicted and observed maps (i.e., spatial scale). Couto (2003) described measurements based on fuzzy set theory that assessed the frequency, magnitude, source, and nature of errors, for hard maps and fuzzy reference data.

Roloff et al. (2001) validated a spatially explicit habitat effectiveness model for elk (*Cervus elaphus*) in South Dakota using telemetry data. The model scored suitability based on forage quality, quantity, and availability (Fig. 16-5). Using the volume of intersection index (Millspaugh et al. 2004), which compares the surface fit of two utilization distributions (i.e., three-dimensional surfaces), Roloff et al. (2001) compared the suitability model output with the total distribution of elk movements (Fig. 16-6). This global measure of agreement between two utilization distributions offers an overall assessment of fit, but follow-up analyses are required to determine where the model is predicting poorly. We view this approach as appropriate when comparing the global fit of two distributions and when there is interest in determining the relative fit (e.g., does the model perform better or worse during some seasons than others, or does the model perform better for one group of animals versus another group?).

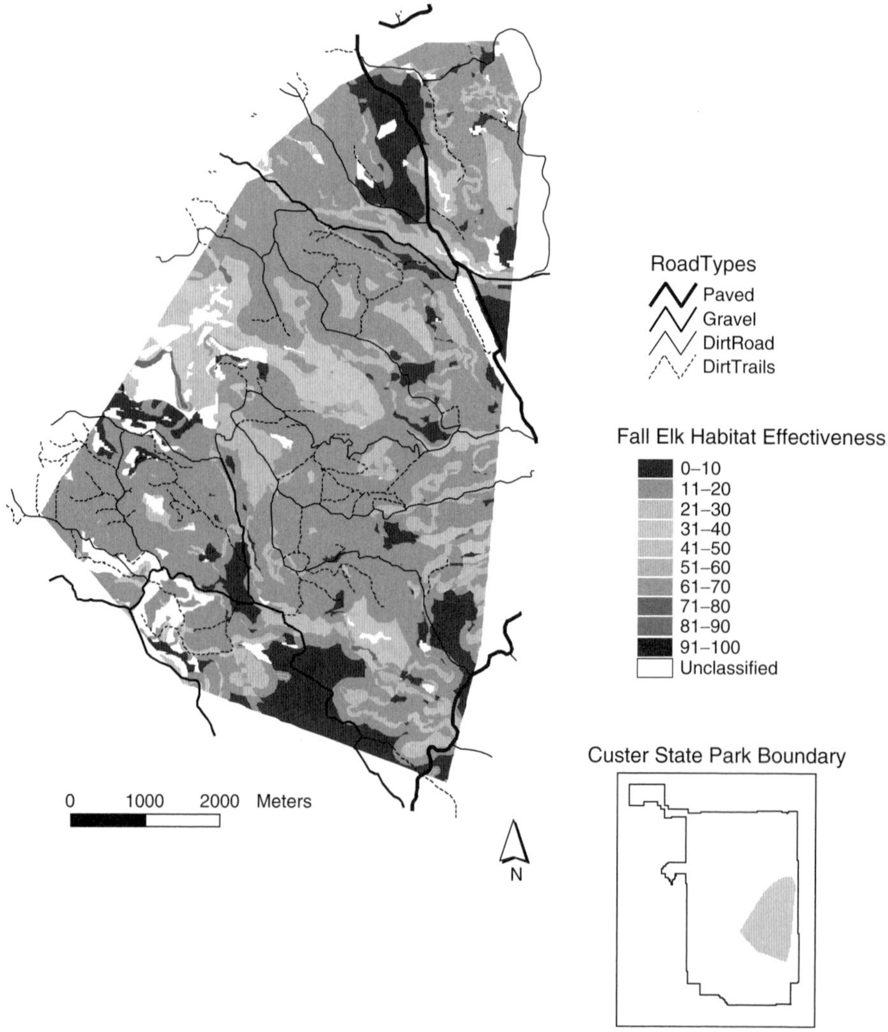

FIG. 16-5 (See color insert.)

Fall habitat quality scores for an elk subherd in Custer State Park, South Dakota, USA (from Roloff et al. 2001).

As with vegetation models, it is useful to simply summarize basic information (e.g., frequency of correct observations) about model performance using an independent data set. Fielding and Bell (1997) proposed several useful calculations to assess model performance when comparing animal observation data with suitability models. First, they constructed a table that identifies the following: number of sites where the species was predicted to occur and it did; number of sites where the species was predicted to occur and it did not; the number of sites where the

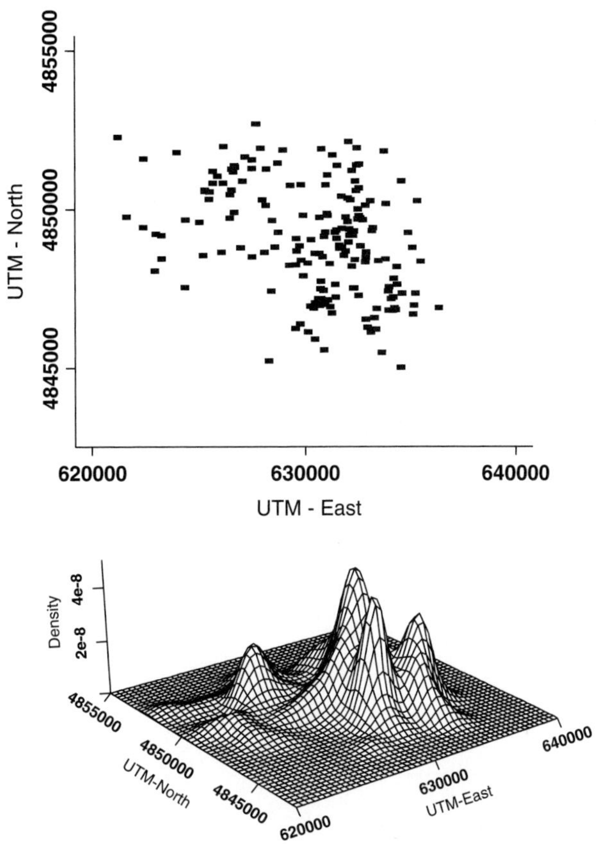

FIG. 16-6

A sample utilization distribution. The upper panel shows the spatial distribution of radiotelemetry observations. The lower panel shows a three-dimensional utilization distribution where the surface height at location x,y represents the intensity of animal use at that point relative to all other locaions (From Millspaugh et al. 2000).

species was predicted to be absent and it was; and finally, the number of sites where the species was predicted to be absent and it was not absent (Table 16-1). From that information, Fielding and Bell (1997) recommended the calculation of several metrics useful for determining model performance (Table 16-2). Simple metrics such as these greatly aid model evaluation and can be used with other measures. For example, an ROC curve (Fielding and Bell 1997, Pearce and Ferrier 2000) is created by plotting sensitivity against 1 – specificity (Table 16-2) across threshold values. The resulting curve gives a measure of model performance (Fielding and Bell 1997, Pearce and Ferrier 2000). We advocate the use of ROC curves; however, it should be noted that the use of ROC curves has recently been criticized in validating species occurrence data (Termansen et al. 2006).

Table 16-1 A Classification Table Used to Summarize the Number of Sites with Observed and Predicted Occurrences of a Species (After Fielding and Bell 1997, Luck 2002)

	Predicted Occurrence	
Observed Occurrence	**Present**	**Absent**
Present	a	c
Absent	b	d

a = number of sites where a species was predicted and observed; b = number of sites where a species was predicted to be present but was not observed; c = number of sites where a species was not predicted to occur yet was observed; and d = the number of sites where the species was predicted to be absent and it was not observed.

Table 16-2 Diagnostic Metrics for Evaluating Model Performance (After Fielding and Bell 1997, Luck 2002) n = the number of cites

Metric	Calculation
Correct classification rate[a]	(a+d)/n
Kappa[b]	[a+d)–((a+c)(a+b)+(b+d)(c+d))/n]/ [n –((a+c)(a+b)+(b+d)(c+d))/n]
Negative predictive power[c]	d/(c+d)
Positive predictive power[d]	a/(a+b)
Prevalence[e]	(a+c)/n
Sensitivity[f]	a/(a+c)
Specificity[g]	(d/(b+d)

Variables a, b, c, and d are as defined in Table 16-1.
[a]measure of the number of sites correctly classified.
[b]measure of the improvement to classification over a null (chance) model. Values <0.4 are poor; 0.4–0.75 are good; and >0.75 is excellent (Landis and Koch 1977).
[c]proportion of sites where the species was predicted to be absent yet was present.
[d]proportion of sites where occurrence was predicted to occur and the species did occur.
[e]proportion of prevalence cases.
[f]the true positive rate.
[g]1 – specificity = false positive rate.

FUTURE DIRECTIONS

The wildlife suitability models discussed in this book are intended to be used for planning over large geographic areas. Such broad applications are useful for addressing complex natural resource issues. Landscape-scale decision support models allow us to investigate these large-scale issues in cases where the collection of field data is logistically and economically prohibitive. The large spatial scales addressed by landscape models are beneficial to planning efforts, but it makes the models hard to validate. Collection of independent data at the necessary temporal and spatial scales is costly and difficult. However, when viewed in an adaptive management framework, continued model refinement and evaluation becomes tractable, reduces uncertainty, and facilitates resource management.

We have discussed models for wildlife habitat in the context of forest vegetation management. Over the next century, changes in land use will also greatly affect habitat suitability for many species at many locations. Increases in the area of land devoted to primary homes, second homes, businesses, and transportation can greatly affect habitat suitability as can rural land use shifts into and out of agricultural production. Efforts are underway to link landscape-scale models of vegetation and wildlife to land use change (e.g., Syphard et al. 2007). Effects of global climate change on forest vegetation and avian species are also under investigation (Iverson et al. 2005, Matthews et al. 2007, Prasad et al. 2007). These factors, and probably other macro effects, will be gradually incorporated into future landscape-scale habitat modeling. That will provide new analysis opportunities and compound difficulties associated with model validation.

Management prescriptions at any given time are made using the best available science and the best available data. However, implementation and monitoring of management prescriptions provides new opportunities to learn more about the system, and prescriptions can change over time as better information becomes available (Millspaugh et al., this volume). Models then serve the dual purposes of (1) quantifying what we know (or think we know) about the system and (2) providing a framework for evaluating key uncertainties in our understanding of the system. This role for models is not new (Williams et al. 2002), but is sorely needed in the future. The value of continuous landscape-scale monitoring, within an adaptive management framework of model performance, cannot be overemphasized (Millspaugh et al., this volume).

Experimental approaches offer another solution to validate models and test our understanding of system processes. In the case of wildlife habitat suitability models, experimental manipulation of vegetation conditions offers a useful opportunity to test the strength of habitat suitability relationships. Projects such as the Missouri Forest Ecosystem Project (MOFEP) are unique in their temporal and spatial scale of experimentation (Brookshire and Shifley 1997, Shifley and Brookshire 2000, Shifley and Kabrick 2002). MOFEP is a large-scale, long-term experiment designed to determine the effects of even-aged, uneven-aged, and no-harvest forest management on biotic and abiotic ecosystem attributes at the landscape scale (Brookshire and Shifley 1997). There are nine sites (three replicates of each treatment) ranging from 312 to 514 ha in size. These are large enough to examine landscape-scale effects of vegetation treatments on wildlife species of management concern. The role of experimentation, even at smaller temporal and spatial scales, would be useful in evaluating suitability relationships.

Much is written and discussed in this book about multispecies models (see Flather et al., this volume; Noon et al., this volume). While we agree with the potential value in multispecies models, validation of these models will present new difficulties. Although it might be possible to survey multiple species at the same time (e.g., birds), there are limitations to multispecies surveys. Further, it will be necessary to develop techniques that are suitable for multiple species validation, which are currently lacking in the literature. However, given the increased emphasis on multispecies assessments, we encourage investigation of appropriate validation procedures.

SUMMARY

Models that predict vegetation and wildlife dynamics at large spatial scales, like all models, are abstractions of reality. Under the best circumstances, models capture the most important features and processes of the real system. However, models are imperfect representations of reality, and every model is wrong. Validation is a critical, yet often neglected component of the modeling process. The key to evaluating a model is determining whether a model is useful for its intended purpose. Validation of wildlife models is universally hampered by the type, quantity, and spatial extent of observations of habitat use that can be quantitatively and qualitatively compared with model predictions. Our ability to conceive and design validation procedures far outstrips our ability to implement those procedures based on field data of habitat use and population dynamics. Fortunately, the model validation process helps resolve this deficiency by identifying model shortcomings, data shortcomings, and opportunities for field research or monitoring that can efficiently address those shortcomings. The essence of model validation and verification is understanding and articulating the strengths, weaknesses, and utility of a model for its intended purpose and relative to alternative methodologies. The principal stages of landscape model validation include (1) vetting the explicit and implicit assumptions of the landscape model; (2) verifying the computer code that implements the model, applies the mathematical equations or algorithms, and handles the bookkeeping; and (3) validating model predictions using real (and preferably independent) data that are invariably limited in scope and spatial extent. For models of sufficient utility to be maintained and improved, these three stages are repeated at periodic intervals, typically in an adaptive management framework. In addition, model evaluation must consider pragmatic aspects such as ease of use, data requirements, the user interface, computational demands, and ease of modification or adaptation. The models presented in this book span a range of intended uses, spatial scales and points of focus. There are vegetation models that predict forest vegetation dynamics for trees, for stands, or for large landscapes. Similarly, there are wildlife models that estimate generic habitat quality and those that predict species population dynamics over time. This chapter addresses universal considerations for evaluation of such models and then presents examples of alternative methodologies from the literature. Despite considerable previous research on model validation, the quantitative validation of landscape model predictions still presents significant practical and technical challenges.

ACKNOWLEDGMENTS

We thank the U.S. Forest Service and the University of Missouri for their support of our research.

LITERATURE CITED

Aebischer, N. J., P. A. Roberston, and R. E. Kenward. 1993. Compositional analysis of habitat use from animal radio-tracking data. *Ecology* 74:1313-1325.

Aitchinson, J. 1986. *The statistical analysis of compositional data*. Chapman and Hall, London, UK.

Akçakaya, H. R. 2001. Linking population-level risk assessment with landscape and habitat models. *Science of the Total Environment* 274:283-291.

Akçakaya, H. R. 2002. *RAMAS GIS: Linking landscape data with population viability analysis (version 4.0t)*. Applied Biomathematics, Setauket, New York, New York, USA.

Akçakaya, H. R., J. Franklin, A. D. Syphard, and J. R. Stephenson. 2005. Viability of Bell's sage sparrow (*Amphispiza belli* ssp. *belli*): Altered fire regimes. *Ecological Applications* 15:521-531.

Akçakaya, H. R., V. C. Radeloff, D. J. Mladenoff, and H. S. He. 2004. Integrating landscape and metapopulation modeling approaches: Viability of sharp-tailed grouse in a dynamic landscape. *Conservation Biology* 19:526-537.

Baldessarini, R. J., S. Finklestein, and G. W. Arana. 1983. The predictive power of diagnostic tests and the effect of prevalence of illness. *Archives of General Psychiatry* 40:569-573.

Barrett, T. M. 2001. *Models of vegetation change for landscape planning: A comparison of FETM, LANDSUM, SIMPPLLE, and VDDT*. U.S. Forest Service, Rocky Mountain Forest and Range Experiment Station, General Technical Report RMRS-GTR-76-WWW, Ogden, Utah, USA.

Battin, J. 2004. When good animals love bad habitats: Ecological traps and the conservation of animal populations. *Conservation Biology* 18:1482-1491.

Battin, J., and J. J. Lawler. 2006. Cross-scale correlations and the design and analysis of avian habitat selection studies. *Condor* 108:59-70.

Bender, L. C., G. J. Roloff, and J. B. Haufler. 1996. Evaluating confidence intervals for habitat suitability models. *Wildlife Society Bulletin* 24:347-352.

Beukema, S. J., and W. A. Kurz. 1998. *Vegetation dynamics development tool: User's guide version 3.0*. ESSA Technologies Ltd., Vancouver, British Columbia, Canada.

Beutel, T. S., R. J. S. Beeton, and G. S. Baxter. 1999. Building better wildlife-habitat models. *Ecography* 22:219-223.

Beven, K. 1993. Prophecy, reality and uncertainty in distributed hydrological modelling. *Advances in Water Resources* 16:41-51.

Box, G. E. P. 1979. Robustness in scientific model building. Pages 201-236 *in* R. L. Launer, G. N. Wilkinson, editors. *Robustness in statistics*. Academic Press, New York, New York, USA.

Boyce, M. S., P. R. Vernier, S. E. Nielsen, and F. K. A. Schmiegelow. 2002. Evaluating resource selection functions. *Ecological Modelling* 157:281-300.

Brand, G. J., and M. R. Holdaway. 1989. Assessing the accuracy of TWIGS and STEMS85 volume predictions: A new approach. *Northern Journal of Applied Forestry* 6:109–114.

Brookshire, B. L., and S. R. Shifley, editors. 1997. *Proceedings of the Missouri Ozark Forest Ecosystem Project symposium: An experimental approach to landscape research*. U.S. Forest Service, North Central Research Station, General Technical Report NC-193, St. Paul, Minnesota, USA.

Buchman, R. G., and S. R. Shifley. 1982. Guide to evaluating forest growth projection systems. *Journal of Forestry* 81:232–234.

Burgman, M. A., D. R. Breininger, B. W. Duncan, and S. Ferson. 2001. Setting reliability bounds on suitability indices. *Ecological Applications* 11:70–78.

Burns, R. M., and B. H. Honkala. 1990. *Silvics of North America*. Volume 1 and 2. U.S. Forest Service, Agriculture Handbook 654, Washington, D.C., USA.

Buskirk, S. W., and J. J. Millspaugh. 2006. Metrics of use and availability in studies of resource selection. *Journal of Wildlife Management* 70:358–366.

Cartwright, N. 1983. *How the laws of physics lie*. Clarendon Press, Oxford, United Kingdom.

Caswell, H. 1978. A general formula for the sensitivity of population growth rate to changes in life history parameters. *Theoretical Population Biology* 14:215–230.

Chew, J. D. 1995. *Development of a system for simulating vegetative patterns and processes at landscape scales*. Dissertation, University of Montana, Missoula, USA.

Chew, J. D., C. Stalling, K. Moeller, E. Bella, R. Schreiner, R. Ahl, and T. Jones. 2007. User guide for SIMPPLLE, Version 2.5 (Draft). U.S. Forest Service, General Technical Report RMRS- xxx. Published on line. http://www.fs.fed.us/rm/missoula/4151/SIMPPLLE/files/publications/SIMPPLLE-MANUAL-2-5.pdf

Cohen, J. 1960. A coefficient of agreement for nominal scales. *Educational and Psychological Measurement* 20:37–46.

Conroy, M. J., and C. T. Moore. 2002. Wildlife habitat modeling in an adaptive framework: The role of alternative models. Pages 205–218 *in* J. M. Scott, P. J. Heglund, M. L. Morrison, J. B. Haufler, M. G. Raphael, W. A. Wall, and F. B. Samson, editors. *Predicting species occurrences: Issues of accuracy and scale.* Island Press, Washington, D.C., USA.

Couto, P. 2003. Assessing the accuracy of spatial simulation models. *Ecological Modelling* 167:181–198.

Cumming, G. S. 2000. Using between-model comparisons to fine-tune linear models of species ranges. *Journal of Biogeography* 27:441–455.

Dixon, G. 1998. *Evaluating stand density management alternatives using the Forest Vegetation Simulator.* U.S. Forest Service, Forest Management Service Center, DRAFT paper, Fort Collins, Colorado, USA.

Dixon, G. E. compiler. 2003. *Essential FVS: A user's guide to the Forest Vegetation Simulator.* U.S. Forest Service, Forest Management Service Center, Internal Report, Fort Collins, Colorado, USA.

Ek, Alan R., S. R. Shifley, T. E. Burk, editors. 1998. *Forest growth modeling and prediction* (Volumes 1 & 2). U.S. Forest Service, North Central Forest Experimental Station, Genral Technical Report NC-120, St.Paul, Minnesota, USA.

Fielding, A. H., and J. F. Bell. 1997. A review of methods for assessment of prediction errors in conservation presence/absence models. *Environmental Conservation* 24:31–46.

Freeman, S. N., D. G. Noble, S. E. Newson, and S. R. Baillie. 2007. Modelling population changes using data from different surveys: The common birds census and the breeding bird survey. *Bird Study* 54:61–72.

Gentil, S., and G. Blake. 1981. Validation of complex ecosystem models. *Ecological Modelling* 14:21–38.

Gustafson, E. J., and T. R. Crow. 1999. HARVEST: Linking timber harvesting strategies to landscape patterns. Pages 309–322 *in* D. J. Mladenoff and W. L. Baker, editors. *Spatial modeling of forest landscape change: Approaches and applications.* Cambridge University Press, Cambridge, United Kingdom.

Gustafson, E. J., N. L. Murphy, and T. R. Crow. 2001. Using a GIS model to assess terrestrial salamander response to alternative forest management plans. *Journal of Environmental Management* 63:281-292.

Gustafson, E. J., and L. V. Rasmussen. 2005. *HARVEST for Windows user's guide: Version 6.1.* U.S. Forest Service, North Central Research Station, St. Paul, Minnesota, USA. <http://www.ncrs.fs.fed.us/4153/Harvest/v61/download/Harv61%20User.pdf>.

Hamming, R. W. 1975. How do you know the simulation is relevant? *Simulation* 25:163-167.

He, H. S., W. Li, B. R. Sturtevant, J. Yang, Z. B. Shang, E. J. Gustafson, and D. J. Mladenoff. 2005. *LANDIS 4.0 user's guide. LANDIS: A spatially explicit model of forest landscape disturbance, management, and succession.* U.S. Forest Service, North Central Research Station, General Technical Report NC-263, St. Paul, Minnesota, USA.

He, H. S., D. J. Mladenoff, and J. Boeder. 1999. An object-oriented forest landscape model and its representation of tree species. *Ecological Modelling* 119:1-19.

Hirzel, A. H., G. Le-Lay, V. Helfer, C. Randin, and A. Guisan. 2006. Evaluating the ability of habitat suitability models to predict species presences. *Ecological Modelling* 199:142-152.

Holling, C. S., editor. 1978. *Adaptive environmental assessment and management.* John Wiley and Sons, New York, New York, USA.

Holthausen, R. S., M. J. Wisdom, J. Pierce, D. K. Edwards, and M. M. Rowland. 1994. *Using expert opinion to evaluate a habitat effectiveness model for elk in western Oregon and Washington.* U.S. Forest Service, Pacific Northwest Research Station, Research paper PNW-RP-479, Portland, Oregon, USA.

Hurley, J. F. 1986. Summary: Development, testing, and application of wildlife-habitat models—The manager's viewpoint. Pages 151-153 *in* J. Verner, M. L. Morrison, and C. J. Ralph, editors. *Wildlife 2000: Modeling habitat relationships of terrestrial vertebrates.* University of Wisconsin Press, Madison, USA.

Husch, B., T. W. Beers, and J. A. Kershaw. 2003. *Forest mensuration.* John Wiley and Sons, New York, New York, USA.

Iverson, L. R., A. M. Prasad, and M. W. Schwartz. 2005. Predicting potential changes in suitable habitat and distribution by 2100 for tree species of the eastern United States. *Journal of Agricultural Meteorology* 61:29-37.

Johnson, D. H. 1980. The comparison of usage and availability measurements for evaluating resource preference. *Ecology* 61:65-71.

Johnson, D. H. 2001. Validating and evaluating models. Pages 105-122 *in* T. M. Shenk, A. B. Franklin, editors. *Modeling in natural resource management: Development, interpretation, and application.* Island Press, Washington, D.C., USA.

Johnson, M. D. 2007. Measuring habitat quality: A review. *Condor* 109:489-504.

Johnson, P. S., S. R. Shifley, and R. Rogers. 2002. *The ecology and silviculture of oaks.* CABI Publishing, New York, New York, USA.

Keane, R. E., D. Long, D. Basford, and B. A. Levesque. 1997. Simulating vegetation dynamics across multiple scales to assess alternative management strategies. Pages 310-315 *in Conference Proceedings—GIS 97, 11th Annual Symposium on Geographic Information Systems. Integrating Spatial Information Technologies for Tomorrow.* GIS World, Vancouver, British Columbia, Canada.

Keane, R. E., R. Parsons, and P. Hessburg. 2002. Estimating historical range and variation of landscape patch dynamics: Limitations of the simulation approach. *Ecological Modelling* 151:29-49.

Larson, M. A., W. D. Dijak, F. R. Thompson, III, and J. J. Millspaugh. 2003. *Landscape-level habitat suitability models for twelve species in southern Missouri.* U.S. Forest Service, North Central Research Station, General Technical Report NC-233, St. Paul, Minnesota, USA.

Larson, M. A., F. R. Thompson, III, J. J. Millspaugh, W. D. Dijak, and S. R. Shifley. 2004. Linking population viability, habitat suitability, and landscape simulation models for conservation planning. *Ecological Modelling* 180:103-118.

Leary, R. A. 1996. Testing models of unthinned red pine plantation dynamics using a modified Bakuzis matrix of stand properties. *Ecological Modelling* 98:35-46.

Luck, G. W. 2002. The habitat requirements of the rufous treecreeper (*Climacteris rufa*). 2. Validating predictive habitat models. *Biological Conservation* 105:395–403.

MacKenzie, D. I., J. D. Nichols, J. A. Royle, K. H. Pollock, L. L. Bailey, and J. E. Hines. 2006. *Occupancy estimation and modeling: Inferring patterns and dynamics of species occurrence*. Elsevier, San Diego, California, USA.

Mankin, J. B., R. V. O'Neill, H. H. Shugart, and B. W. Rust. 1979. The importance of validation in ecosystem analysis. Pages 309–317 *in* H. H, Shugart, R. V. O'Neill, editors. *Systems ecology.* Dowden, Hutchinson and Ross, Stroudsburg, Pennsylvania, USA.

Manly, B. F. J., L. L. McDonald, D. L. Thomas, T. L. McDonald, and W. P. Erickson. 2002. *Resource selection by animals: Statistical design and analysis for field studies*. Second edition. Kluwer Academic, Dordrecht, The Netherlands.

Marcot, B. G., M. G. Raphael, and K. H. Berry. 1983. Monitoring wildlife habitat and validation of wildlife-habitat relationships models. *Transactions of the North American Wildlife and Natural Resources Conference* 48:315–329.

Marzluff, J. M., J. J. Millspaugh, K. R. Ceder, C. D. Oliver, J. Withey, J. B. McCarter, C. L. Mason, and J. Comnick. 2002. Modeling changes in wildlife habitat and timber revenues in response to timber management. *Forest Science* 48:191–202.

Marzluff, J. M., J. J. Millspaugh, P. Hurvitz, and M. S. Handcock. 2004. Resource utilization of an avian nest predator: Relating resources to a probabilistic measure of animal use. *Ecology* 85:1411–1427.

Matthews, S. N., L. R. Iverson, A. M. Prasad, and M. P. Peters. 2007. *Climate change bird atlas: A spatial database of 147 bird species of the eastern United States*. U.S. Forest Service, Northern Research Station, NRS-4151, Delaware, Ohio, USA. <http://www.nrs.fs.fed.us/atlas/bird>.

Mayer, D. G., and D. G. Butler. 1993. Statistical validation. *Ecological Modelling* 68:21–32.

Miller, D. R., G. Butler, and L. Bramall. 1976. Validation of ecological system models. *Journal of Environmental Management* 4:383–401.

Millspaugh, J. J., K. J. Raedeke, G. C. Brundige, and R. A. Gitzen. 2000. Elk and hunter space use sharing in the Southern Black Hills, South Dakota. *Journal of Wildlife Management* 64:994–1003.

Millspaugh, J. J., R. A. Gitzen, B. J. Kernohan, M. Larson, and C. Clay. 2004. Comparability of three analytical techniques to assess joint space use. *Wildlife Society Bulletin* 32:148–157.

Millspaugh, J. J., R. M. Nielson, L. McDonald, J. M. Marzluff, R. A. Gitzen, C. D. Rittenhouse, M. W. Hubbard, and S. L. Sheriff. 2006. Analysis of resource selection using utilization distributions. *Journal of Wildlife Management* 70:384–395.

Miner, C. L., N. R. Walters, and M. L. Belli. 1988. *A guide to the TWIGS program for the North Central United States*. U.S. Forest Service, North Central Forest Experiment Station, General Technical Report NC-125, St. Paul, Minnesota, USA.

Mitchell, M. S., J. W. Zimmerman, and R. A. Powell. 2002. Test of a habitat suitability index for black bears in the southern Appalachians. *Wildlife Society Bulletin* 30:794–808.

Mladenoff, D. J. 2004. LANDIS and forest landscape models. *Ecological Modelling* 180:7–19.

Mladenoff, D. J., T. A. Sickley, and A. P. Wydeven. 1999. Predicting gray wolf landscape recolonization: Logistic regression models vs. new field data. *Ecological Applications* 9:37–44.

Mladenoff, D. J., and H. S. He. 1999. Design and behavior of LANDIS, an object-oriented model of forest landscape disturbance and succession. Pages 163–185 *in* D. J. Mladenoff, and W. L. Baker, editors. *Advances in spatial modeling of forest landscape change: Approaches and applications.* Cambridge University Press, Cambridge, United Kingdom.

Morrison, M. L., and L. S. Hall. 2002. Standard terminology: Toward a common language to advance ecological understanding and application. Pages 43–52 *in* J. M. Scott, P. J. Heglund, M. L. Morrison, J. B. Haufler, M. G. Raphael, W. A. Wall, and F. B. Samson, editors. *Predicting species occurrences: Issues of accuracy and scale.* Island Press, Washington, D.C., USA.

Morrison, M. L., B. G. Marcot, and R. W. Mannan. 1998. *Wildlife-habitat relationships: Concepts and applications*. Second edition. University of Wisconsin Press, Madison, USA.

Mowrer, H. T., K. Barber, J. Campbell, N. Crookston, C. Dahms, J. Day, J. Laacke, J. Merzenich, S. Mighton, M. Rauscher, R. Sojda, J. Thompson, P. Trenchi, and M. Twery. 1997. *Decision support systems for ecosystem management: An evaluation of existing systems*. U.S. Forest Service, Rocky Mountain Forest and Range Experiment Station, General Technical Report RM-GTR-296, Fort Collins, Colorado, USA.

Newson, S. E., R. J. W. Woodburn, D. G. Noble, S. R. Baillie, and R. D. Gregory. 2005. Evaluating the breeding bird survey for producing national population size and density estimates. *Bird Study* 52:42–54.

Oliver, C. D., and B. C. Larson. 1996. *Forest stand dynamics*. Update edition. John Wiley and Sons, New York, New York, USA.

Oreskes, N., K. Shrader-Frechette, and K. Belitz. 1994. Verification, validation, and confirmation of numerical models in the earth sciences. *Science* 263:641–646.

Ottaviani, D., G. J. Lasinio, and L. Boitani. 2004. Two statistical methods to validate habitat suitability models using presence-only data. *Ecological Modelling* 179:417–443.

Pearce, J., and S. Ferrier. 2000. Evaluating the predictive performance of habitat models developed using logistic regression. *Ecological Modelling* 133:225–245.

Pontius, R. G. Jr., D. Huffaker, and K. Derman. 2004. Useful techniques of validation for spacially explicit land-change models. *Ecological Modelling* 179:445–461.

Prasad, A. M., L. R. Iverson, S. Matthews, and M. Peters. 2007. *Climate change tree atlas: A spatial database of 134 forest tree species of the eastern United States*. U.S. Forest Service, Northern Research Station, Delaware, Ohio, USA. <http://www.nrs.fs.fed.us/atlas/tree>.

Ray, N., and M. A. Burgman. 2006. Subjective uncertainties in habitat suitability maps. *Ecological Modelling* 195:172–186.

Regan, H. M., M. Colyvan, and M. A. Burgman. 2002. A taxonomy and treatment of uncertainty for ecology and conservation biology. *Ecological Applications* 12:618–628.

Reynolds, M. R. 1984. Estimating the error in model predictions. *Forest Science* 30:454–469.

Rittenhouse, C. D. 2008. Wildlife response to spatial and temporal changes in forest habitat. Dissertation, University of Missouri, Columbia, USA.

Roloff, G. J., and B. J. Kernohan. 1999. Evaluating the reliability of habitat suitability index models. *Wildlife Society Bulletin* 27:973–985.

Roloff, G. J., J. J. Millspaugh, R. A. Gitzen, and G. C. Brundige. 2001. Verification of a spatially explicit habitat model for Rocky Mountain elk. *Journal of Wildlife Management* 65:899–914.

Root, K. V., H. R. Akçakaya, and L. Ginzburg. 2003. A multispecies approach to ecological valuation and conservation. *Conservation Biology* 17:196–206.

Rykiel, E. J., Jr. 1996. Testing ecological models: The meaning of validation. *Ecological Modelling* 90:229–244.

Scheller, R. M., J. B. Domingo, B. R. Sturtevant, J. S. Williams, A. Rudy, E. J. Gustafson, and D. J. Mladenoff. 2007. Design, development, and application of LANDIS-II, a spatial landscape simulation model with flexible temporal and spatial resolution. *Ecological Modelling* 201:409–419.

Scott, J. M., P. J. Heglund, M. L. Morrison, J. B. Haufler, M. G. Raphael, W. A. Wall, and F. B. Samson, editors. 2002. *Predicting species occurrences: Issues of accuracy and scale.* Island Press, Washington, D.C., USA.

Shifley, S. R., and B. L. Brookshire, editors. 2000. *Missouri Ozark Forest Ecosystem Project: Site history, soils, landforms, woody and herbaceous vegetation, down wood, and inventory methods for the landscape experiment*. U.S. Forest Service, North Central Research Station, General Technical Report NC-208, St. Paul, Minnesota, USA.

Shifley, S. R., and J. M. Kabrick, editors. 2002. *Proceedings of the 2nd Missouri Ozark Forest Ecosystem Project symposium: Post treatment results of the landscape experiment*. U.S. Forest Service, North Central Research Station, General Technical Report NC-227, St. Paul, Minnesota, USA.

Shifley, S. R., F. R. Thompson, III, W. D. Dijak, M. A. Larson, and J. J. Millspaugh. 2006. Simulated effects of forest management alternatives on landscape structure and habitat suitability in the Midwestern United States. *Forest Ecology and Management* 229:361–377.

Shifley, S. R., and C. W. Woodall. 2007. The past, present, and future of Indiana's oak forests. Pages 295–304 *in* D. S. Buckley and W. K. Clatterbuck, editors. *Proceedings of the 15th central hardwood forest conference*. U.S. Forest Service, Southern Research Station, e-General Technical Report SRS-101, Asheville, North Carolina, USA.

Somershoe, S. G., D. J. Twedt, and B. Reid. 2006. Combining breeding bird survey and distance sampling to estimate density of migrant and breeding birds. *Condor* 108:691–699.

Starfield, A. M. 1997. A pragmatic approach to modeling for wildlife management. *Journal of Wildlife Management* 61:261–270.

Starfield, A. M., and A. L. Bleloch. 1991. *Building models for conservation and wildlife management*. Second edition. Burgess International Group, Edina, Minnesota, USA.

Stearns, S. C. 1992. *The evolution of life histories*. Oxford University, Oxford, United Kingdom.

Syphard, A. D., K. C. Clarke, and J. Franklin. 2007. Simulating fire frequency and urban growth in southern California coastal shrublands, USA. *Landscape Ecology* 22:431–445.

Termansen, M., C. J. McClean, and C. D. Preston. 2006. The use of genetic algorithms and Bayesian classification to model species distributions. *Ecological Modelling* 192:410–424.

Thompson, W. L. editor. 2004. *Sampling rare or elusive species: Concepts, designs, and techniques for estimating population parameters.* Island Press, Washington, D.C., USA.

Twery, M. J., P. D. Knopp, S. A. Thomasma, H. M. Rauscher, D. E. Nute, W. D. Potter, F. Maier, J. Wang, M. Dass, H. Uchiyarna, A. Glende, and R. E. Hoffman. 2005. NED-2: A decision support system for integrated forest ecosystem management. *Computers and Electronics in Agriculture* 39:24–43.

U.S. Forest Service. 2008*a*. Forest inventory and analysis national program. USFS webpage. <http://www.fia.fs.fed.us/>. Accessed 3 January 2008.

U.S. Forest Service. 2008*b*. Forest vegetation system. USFS webpage. <http://www.fs.fed.us/fmsc/fvs/>. Accessed 12 January 2008.

University of Washington. 2008. Landscape management system. UW webpage. <http://lms.cfr.washington.edu/index.php>. Accessed 12 January 2008.

Van Horne, B. 1983. Density as a misleading indicator of habitat quality. *Journal of Wildlife Management* 47:893–901.

Van Horne, B. 2002. Approaches to habitat modeling: The tensions between pattern and process and between specificity and generality. Pages 63–72 *in* J. M. Scott, P. J. Heglund, M. L. Morrison, J. B. Haufler, M. G. Raphael, W. A. Wall, and F. B. Samson, editors. *Predicting species occurrences: Issues of accuracy and scale.* Island Press, Washington, D.C., USA.

Vanclay, J. K. 1994. *Modelling forest growth and yield: Applications to mixed tropical forests*. CAB International, Wallingford, United Kingdom.

Vanclay, J. K., and J. P. Skovsgaard. 1997. Evaluating forest growth models. *Ecological Modelling* 98:1–12.

Verner, J., M. L. Morrison, and C. J. Ralph, editors. 1986. *Wildlife 2000: Modeling habitat relationships of terrestrial vertebrates.* University of Wisconsin Press, Madison, USA.

Williams, B. K., J. D. Nichols, and M. J. Conroy. 2002. *Analysis and management of animal populations: Modeling, estimation, and decision making*. Academic Press, San Diego, California, USA.

Woeck, B. N. 2003. *Demographics and space use of mule deer and white-tailed deer in Custer State Park, South Dakota*. Thesis, University of Missouri, Columbia, USA.

Woodall, C., D. Johnson, J. Gallion, C. Perry, B. Butler, R. Piva, E. Jepsen, D. Nowak, and P. Marshall. 2005. *Indiana's forests 1999-2003 (Part A)*. U.S. Forest Service, North Central Research Station, Resource Bulletin NC-253A, St. Paul, Minnesota, USA.

Yang, J., He, H., Shifley, S. R., and Gustafson, E. J. 2007. Spatial patterns of modern period human-caused fire occurrence in the Missouri Ozark Highlands. *Forest Science* 53:1–15.

Yang, J., H. He, and S. R. Shifley. 2008. Spatial controls of occurrence and spread of wildfires in the central hardwood forests of the Missouri Ozark Highlands. *Ecological Applications*, *In press*.

Zweig, M. H., and G. Campbell. 1993. Receiver-operating characteristic (ROC) plots: A fundamental evaluation tool in clinical medicine. *Clinical Chemistry* 39:561–577.

CHAPTER

Methods for Determining Viability of Wildlife Populations in Large Landscapes

17

H. Resit Akçakaya and Barry W. Brook

The viability of a species in a given geographic region is often expressed with measures such as risk of extinction or decline, expected time to extinction, or chance of recovery. Population viability analysis (PVA) uses a variety of models to predict such measures of viability based on demographic data (such as censuses, mark-recapture studies, surveys and observations of reproduction and dispersal events, presence-absence data) and habitat data. Whether a species can persist in a particular landscape, and at the population size the landscape can support, is dependent on the demography of the species and on the spatial distribution, quality, amount, and temporal dynamics of its habitat.

Habitat by itself does not determine whether a species will persist, decline, or recover. There are many situations in which a species may not be viable despite plentiful habitat. One common cause of this disconnection is the effect of threats unrelated to habitat, such as harvest by humans, diseases, and exotic species that are competitors, predators, or parasites. Another common reason is the interaction between the spatial distribution of the habitat and the ecology of the species. Even if there is sufficient habitat, if it is fragmented into patches too small to support populations or too isolated to allow recolonization, the species will not be viable.

Similarly, demography by itself cannot determine how much habitat is required to ensure the viability of the species and where it should be located. Equivalent demographic characteristics (such as survival rates and fecundities) may result in a viable population in one landscape but a declining or extinction-prone population in the other. Depending on the quality and the level of fragmentation of the habitat, population sizes in the latter landscape may be too small for the species to escape the perils of demographic stochasticity, genetic problems, and Allee effects. At small population sizes, these chance effects tend to be the predominant determinant of a species' persistence (Gilpin and Soulé 1986). Even the same demography and the same area of habitat can result in different population dynamics and different chances of persistence, depending on the spatial configuration and heterogeneity of the habitat and intervening matrix, as well as temporal changes in the habitat (including frequency and size distribution of disturbances). Therefore, practical conservation planning in large landscapes must take demography, ecology, and habitat into account, and rely on measures that can integrate the effects of these different factors. Viability is one such measure, and PVA models provide the integrative tool.

Population viability analyses focus on specific types of questions that are relevant to the conservation of wildlife species in large landscapes (Beissinger et al., this volume). Such questions include the probability of persistence of a species in a particular landscape; the cumulative or individual effects of past or future impacts on the species' risk of extinction or decline; and the potential for alternative management and conservation actions to increase the species' chances of survival and/or recovery in the landscape.

Because of its ability to link habitat and demography, PVA can use simultaneously different types of information (see next section) and can consider cumulative and synergistic effects of different types of impacts. Although certain threat factors primarily affect a species' habitat (such as urban sprawl or logging) and others affect its demography (such as harvest or introduced predators), many factors have effects on both habitat and demography (such as pollution, climate change, roads, suppression or increase of fire, invasive species).

The ability to link habitat and demography also allows the user of PVA to compare different types of conservation actions. For example, if a given area of habitat does not support a viable population, potential conservation measures include those that are explicitly habitat-related, such as improving habitat quality, designating a larger area as critical habitat, and changing the spatial configuration of the habitat, as well as a number of measures not directly related to the area and configuration of habitat. These include re-introductions, translocations, regulation of harvest in adjoining areas (which may otherwise act as sink populations because of high exploitation rates), removal or suppression of exotic species, increasing connectivity, and precautions against catastrophic events such as disease epidemics. Methods such as PVA that integrate demographic and habitat information can assess the combined effects of all such conservation measures in terms of the increased viability of the species.

RELATIONSHIP OF PVA TO OTHER METHODS

There are several other quantitative methods and models used for planning wildlife conservation in large landscapes. Population viability analysis is not an alternative to these other methods, but a complementary approach that often uses their output to answer a range of the specific questions discussed previously that other methods cannot address as efficiently (Brook et al. 2000). Indeed, the information these methods provide helps to determine both the complexity of the PVA that can be developed and the types of questions it can be used to address. The relevance of these methods to PVA is briefly discussed in the following paragraph.

Geographic information system (GIS) methods underlie the development and analysis of many of the spatially explicit PVA models. Habitat models predict the distribution of suitable habitat (Dijak and Rittenhouse, this volume; Larson et al., this volume; Niemuth et al., this volume; Fitzgerald et al., this volume), which can determine the spatial structure of the PVA models, as well as characteristics of their spatial units (e.g., carrying capacity of populations or quality of territories). Statistical analyses of population trends allow calibration of demographic models or, if independent of demographic data, used in model construction for the validation of demographic models. Landscape-level measures such as connectivity and fragmentation provide information for the spatial structure and dispersal rates for PVA models. Landscape models (such as LANDIS [He, this volume] and LMS [Oliver et al., this volume]) predict the temporal changes in the landscape, which are then integrated into PVA predictions through simulation of dynamic metapopulation structures (Bekessy et al., this volume; see examples later). Reserve selection methods (Flather et al., this volume) find configuration of protected areas that maximize species representation, and are moving toward considering species viability as well.

ISSUES FOR LARGE-SCALE PVA APPLICATIONS

For wildlife populations in large landscapes, viability analysis requires careful consideration of the issues of spatial scaling and heterogeneity, habitat structure and connectedness, and temporal dynamics of the landscape.

Spatial Extent

The determination of the spatial extent (i.e., geographical range size and boundaries) of the modeled populations is an important issue. For example, when PVA is used to assess the potential impact of a threat to the viability of a wildlife population, and the threat affects only a small part of the landscape, then there might be several alternative ways of selecting the "assessment population," which is the population for which the increased risk due to the specific threat is estimated. Further, the relative importance of such threats to conservation management of the species as a whole can be understood only if its scope is properly contextualized (e.g., the risk of a fire that causes 90% mortality but affects only 10% of a species' range may not be worth managing, whereas a disease that kills 20% of individuals but has a range-wide impact may be). As such, the assessment population may be different from the biological population or metapopulation, and its spatial extent is often more a social (thus, regulatory) issue than a scientific issue. If the assessment population is limited to the affected area, the results will be more sensitive to the potential impacts but less relevant for the overall viability of the population. In such cases, it may be better to assess impacts at multiple spatial extents. Thus, different scales and extents may be needed for different purposes.

Spatial Resolution

In our context, resolution refers to the size of the smallest spatial unit of a model. For PVA, the relevant measure of resolution is related to the spatial unit of the demographic entity being modeled. The demographic entity can be an individual or a population. Thus, for individual-based models, the resolution can be a territory or home range; for population-based models, it can be the smallest habitat patch that supports a subpopulation. For grid-based landscape models, it may be some arbitrarily designated unit that can be defined conveniently within a GIS based on remote sensing.

Most spatially explicit and individual-based PVA models for large landscapes represent the landscape as discrete home ranges or territories of the modeled species. Each territory is often assumed to support one breeding pair. This is often a reasonable assumption, but in many cases the territories do not all have similar sizes or shapes, are often not distributed uniformly in the landscape, and their distribution, sizes, and shapes can change frequently depending on the population size, composition, the level of resources, and the degree of competition. In such cases, models based on a spatially uniform or temporally fixed pattern of territories may miss important structural or dynamic characteristics of the landscape that affect viability.

Most spatially explicit and population-based PVA models for large landscapes represent the landscape as discrete habitable patches within a surrounding matrix that may allow dispersal but does not support populations. Each discrete habitable patch is assumed to support one population (sometimes called a subpopulation of a metapopulation). This is often a necessary simplification in situations in which it is possible to identify habitat patches on the basis of geophysical and ecological landscape metrics, but where detailed point-based presence-absence data are unavailable. When such data are available, they can be used to generate a habitat map using niche modeling (also called species distribution modeling), and the habitat map in turn can be used to identify populations. A biological population can be defined as a group of regularly interbreeding (i.e., panmictic) individuals. One approach to delineating a population is to rephrase the question as: How far apart must two individuals be in order to be considered in different populations? The answer depends on the characteristics of the landscape, as well as a spatial measure (e.g., "Neighborhood distance" in Akçakaya 2005) related to the possibility of interbreeding (for example, the frequency distribution of movement distances of the species or the size of its home range). This approach, combined with modeling and prediction of suitable habitat, is used in habitat-based metapopulation models to delineate populations (Akçakaya 2000, 2005).

There are two other related measures of resolution. One is the size of the smallest unit for applying conservation or management actions. The other is the cell size of the landscape maps (e.g., vegetation maps and satellite imagery) that often form the basis for inferring the distribution of habitat in a PVA model. Although these separate measures of resolution are often different from each other and from the model resolution as discussed previously, sometimes by orders of magnitude, they nevertheless should be consistent with the resolution of the PVA model itself, as defined earlier. For example, if the home range of a species is in the order of 1 ha, habitat maps used as a basis for an individual-based and a population-based model should have cell sizes of much smaller than 1 ha, and smaller than but close to 1 ha, respectively.

Connectivity

Connectivity is the degree to which organisms can move through the landscape. Common measures of connectivity include dispersal rate (proportion of individuals moving from one habitat patch to another) and dispersal probability (probability of an individual moving from one patch or cell to another). Defining connectivity in the context of PVA requires first establishing the model's spatial resolution (the spatial unit of the demographic entity being modeled; see previous discussion). Incorporation of connectivity into PVA models depends on model type. In many models, dispersal rates or probabilities are distance-dependent, age- and/or sex-specific.

Connectivity of wildlife populations in large landscapes is often vital to their persistence in the landscape. However, defining connectivity as a conservation and management goal by itself may be counterproductive. Although increased dispersal often increases viability, this is not universal (see Stacey et al. 1997, Beier and Noss 1998, Lecomte et al. 2004). Even when it does, increasing dispersal by building or maintaining habitat corridors may not be the best option. Whether or not such conservation actions are useful depends on many factors, including the behavior of the species in corridors, relationship between dispersal and viability, and cost of, and alternatives to, corridors (Akçakaya et al. 2007).

Spatial Heterogeneity

The larger the landscape, the more heterogeneous it is likely to be. For PVA models, spatial heterogeneity refers to the variability in habitat quality and other environmental factors across the landscape. This spatial variability is expressed in PVA models as differences in population parameters or occupancy rates in different parts of the landscape—for example, in different populations. The parameters most commonly used to express this variability include population size or density, carrying capacity and vital rates (survival and fecundity), and foraging probability and home range size in individual-based models.

Other aspects of demography also are likely to change across space. For example, catastrophe probabilities and associated effects, dispersal rates, variability in vital rates and other parameters, temporal trends in carrying capacity and other parameters, and type and strength of density dependence (including Allee effects) are among the parameters that may vary among populations as a result of differences in the habitats and other environmental factors they experience.

There are two important aspects of spatial variability that result from its interaction with temporal variability. One is the temporal change in spatial patterns of habitat quality, discussed in the next section. The other is the spatial correlation in temporal variability, which is often expressed as correlation of vital rates or other stochastically varying parameters among the different populations in the landscape. Spatially correlated fluctuations in population parameters are brought about by environmental factors that are often correlated even at relatively large distances (e.g., large-scale climatic factors such as rainfall, temperature, flow rate), and in turn result in more synchronized declines and extinctions among populations (Akçakaya and Ginzburg 1991, Burgman et al. 1993, LaHaye et al. 1994). In a large landscape, models based on an assumption of completely correlated fluctuations among populations are likely to be invalid, at least for those populations that are far away from each other. On the other extreme, models based on an assumption of uncorrelated fluctuations among populations are likely to be invalid for those that are close by, and underestimate the temporal variability of population size and therefore the extinction risk of the metapopulation. Estimating correlations correctly is particularly important in large landscapes because neither of the two simplifying assumptions commonly made in models is likely to be valid.

Temporal Dynamics of the Landscape

An important factor governing the viability of species in large landscapes is the effect of landscape dynamics on the temporal variability of the habitat and hence on the dynamics of the wildlife populations. Landscapes change according to seasons, climatic fluctuations (e.g., droughts, El Ninô events), disturbances (e.g., fire and windthrow) and succession, as well as human impacts (e.g., urban sprawl, global climate change, and agricultural expansion). For a wildlife species in such a dynamic landscape, these changes are manifested as monotonic trends or oscillatory changes in both population characteristics (e.g., carrying capacity and average fecundity) and in metapopulation attributes (e.g., number and sizes of populations and distances among them).

Viability of species in these dynamic landscapes depends on the interaction between landscape change and the species' ecology (i.e., its ability to disperse between and function within the habitat patches). Some PVA models incorporate aspects of a dynamic landscape in a static spatial structure—for example, using a metapopulation model with fixed number of populations whose carrying capacity changes over time (Pulliam et al. 1992, Lindenmayer and Possingham 1996, Stelter et al. 1997, Akçakaya and Raphael 1998, Johst et al. 2002, Keith 2004). Such changes can be deterministic (for example, to simulate forest growth) or stochastic (to simulate the effects of random disturbances such as fires) or both (e.g., using a deterministic function of time since a stochastic disturbance event).

Other PVA models also incorporate a dynamic spatial structure that arises from habitat patches splitting, merging, appearing, and disappearing as the species' habitat changes (Akçakaya 2001, Akçakaya and Root 2003). This approach has been applied to landscape changes brought about by timber harvest, succession, and natural disturbances (Akçakaya et al. 2004, 2005; Wintle et. al 2005; Bekessy et al., this volume).

One temporal process in particular—climate change—has encouraged the development of modeling approaches to determine temporal range shifts (e.g., Keith et al. 2008). There are now many documented examples of a species' bioclimatic tolerance limits shifting toward higher latitudes and upward in elevation as the planet warms (Araujo and Guisan 2006, LaSorte and Thompson 2007). There are also instances of the spread of woody vegetated habitats of threatened species caused by the enrichment of atmospheric CO_2 or altered fire regimes (Bond et al. 2005, Bowman et al. 2006). Given the global extent of climate change, it is likely that the incorporation of such spatio-temporal changes will become routine in future PVA modeling.

TYPES OF MODELS AND EXAMPLES OF THEIR APPLICATIONS

What model should be used to determine viability of species in large landscapes? The answer very much depends on the type of data that are available for model parameterization, the ecology of the species under consideration, the nature of the conservation management question being asked, and the available expertise (Dunning et al. 1995, White 2000). In this section, we describe four commonly used approaches and provide a selection of examples to illustrate their application. A brief summary of the model types and further examples are given in Table 17-1. At the end of this section, we also briefly consider issues of model selection and multimodel inference.

Occupancy Models

We often know very little about the ecology and demography of threatened species. By virtue of their rarity, they are difficult and expensive to study, especially across large landscapes, and long-term monitoring data are usually not available. That said, one form of information that is often routinely collected, or can be

Table 17-1 Types of Population Viability Models and Examples of Their Use

Model Type	Best For	Examples
Occupancy models	Equilibrium metapopulations High rate of local extinction and recolonization Limited demographic data Small, short-lived organisms (e.g., invertebrates) Large number of patches	Granville Fritillary (Hanski et al. 1996) Arboreal forest marsupials (Lindenmayer et al. 1999)
Lattice (grid-based) models	Relatively uniform or undisturbed landscapes Landscapes with continuous environmental gradients Locally abundant organisms	*Aster kantoensis* (Shimada and Ishihama 2000) Asian Water Buffalo (Brook and Bradshaw 2006*a*) Vegetation patterns (Molofsky and Bever 2004)
Demographically structured metapopulation models	Declining populations Locally abundant organisms Vertebrates and plants Large or dynamic landscapes Sufficient demographic data	Sargent's Cherry Palm (Maschinski and Duquesnel 2006) Woodland Brown Butterfly (Kindvall and Bergman 2004) Chinook Salmon (Ruckelshaus et al. 2004) Tree Frog (Pellet et al. 2006) Eastern Indigo Snake (Breininger et al. 2004) Bell's Sage Sparrow (Akçakaya et al. 2005) Brown Creeper (Wintle et al. 2005) Magpie Geese (Brook and Whitehead 2005) Carpentarian Rock Rat (Bowman et al. 2006) Ocelot (Haines et al. 2006)
Individual-based models	Very small populations Abundant demographic and behavioral data Large-bodied, territorial species Modeling impact of genetic threats Determining emergent behaviors	*Micoures demerarae* (Brito and Fernandez 2000) Northern Spotted Owl (Lamberson et al. 1996) Leadbeater's Possum (Lindenmayer and Lacy 1995) *Panthera gombaszoegensis* (O'Regan et al. 2002) Community assembly (Hraber and Milne 1997)

acquired at an acceptable cost, is presence-absence data used to assess site occupancy and distributional ranges. If discrete habitat patches can also be defined, then simple occupancy models can be used to provide a crude assessment of population viability. Occupancy models, which consider whether a species is either present or absent from a patch, were first described within the theoretical context of simple metapopulations (Levins 1970). Although early work considered only the turnover of patch tenancy within a patch network, while ignoring the distance between patches and the size of the populations that these patches could support, more recent advances have relaxed these restrictive assumptions.

Certainly, the most widely adopted occupancy model to be used for practical conservation problems is Hanski's (1994) incidence function model (IFM). This model can be fitted to presence-absence data, while taking explicit account of the location and size of patches to estimate turnover (a dynamic equilibrium between extinction and recolonization is an underlying assumption), and hence long-term viability. The best-known and exemplar application of the IFM is to the Glanville fritillary butterfly (*Melitaea cinxia*), found scattered throughout the grassy meadows of southwest Finland. Extensive metapopulation survey data on this species, covering hundreds of subpopulations over more than a decade, have allowed extensive refinement and testing of the model and associated theory in spatial ecology (Ovaskainen and Hanski 2003).

There are some clear advantages to using the IFM and related stochastic patch occupancy models, including (1) they are relatively easy to parameterize given quite minimal "snapshot" data taken at a landscape-scale; (2) more sophisticated state transition models can be used to incorporate additional information about measured levels of patch turnover, if available (Sjögren-Gulve and Ray 1996), and GIS data on landscape structure and habitat quality can also be integrated into the basic IFM, if available (Moilanen and Hanski 1998); (3) conversely, Bayesian approaches have been developed to augment model fitting and take account of uncertainty better when required information on the study metapopulation is unavailable (O'Hara et al. 2002); and (4) user-friendly generic software tools such as META-X (Grimm et al. 2004) and SPOMSIM (Moilanen 2004) are available to perform the necessary calculations; these also allow the user to implement optimization algorithms (e.g., for reserve design) and undertake scenario modeling (e.g., patch deletion to simulate habitat loss).

Equally, there are obvious limits to inference that can be made with occupancy models (Baguette 2004). For instance, they ignore local population dynamics and assume a disconnect between the time scale of dispersal events and the rate of population growth or decline within patches considered effectively instantaneous, and therefore not amenable to management. Both assumptions may be unrealistic for large, long-lived species with large distributional ranges. Empirical tests of IFM predictions using real-world data have given mixed results, working quite well for some mammal species but poorly for others (Lindenmayer et al. 1999). Even in the case of the well-studied Glanville fritillary, uncertainties in parameter estimation precluded a precise estimation of extinction risks, although the rank order of management scenarios was robustly predicted (Drechsler et al. 2003).

Another limitation of IFM is the equilibrium assumption. Many species of conservation concern are declining for various reasons, which invalidates this assumption. In more detailed models (discussed later), such declines are modeled in various ways—for example, by a gradually declining carrying capacity or by a density-independent decline due to low survival or fecundity. In addition, habitat loss is often accompanied by habitat fragmentation in which the number of patches increases as they split and become smaller. Such changes are incorporated in the more detailed models, but not in IFM, which limits their applicability to threatened species (Pellet et al. 2006).

Lattice (Grid-Based) Models

As described previously, most spatially explicit population dynamics models divide the landscape into a number of discrete units, such as patches. Each patch is then considered to either support a resident subpopulation, or to be vacant. Alternatively, each patch can be modeled as a distinct and structured demographic unit, which may or may not be linked to other patches by dispersal (see next section). Lattice models, however, are not concerned with patches (Nakamaru 2006). Landscapes are instead treated as continuous, gridded surfaces, and the species of interest has a defined probability of moving between adjacent cells within a fixed time step. In lattice models, the entire landscape is considered to be usable, with the relative quality of each cell defined by a Habitat Suitability Index function based on environmental variables. This method considers explicitly the spatial and temporal variation of environments and the influence that this heterogeneity exerts on organisms using these landscapes. Moreover, the raster-based format used by GIS provides a convenient way to integrate the different forms of information required for parameterizing lattice models (e.g., geophysical landscape attributes and vegetation pattern inferred from satellite imagery; With 1997).

Population dynamics can be simulated within each grid cell, although usually in a simplified (nonstructured) form, such as via a density-dependent scalar growth function. For instance, a lattice model of feral ungulates in Kakadu National Park, northern Australia, used a theta-logistic function to model densities within 10 km × 10 km grid cells (Brook and Bradshaw 2006*a*). The model was habitat driven in that animal densities were ultimately linked to the habitat quality of each cell (e.g., vegetation type, distance to water), which determined carrying capacity. The proportion of animals that dispersed to adjacent cells depended on the density within the originating cell. This model, when linked to a previously determined functional response curve developed for buffalo (*Bubalus bubalis*) and pigs (*Sus scrofa*), was used to determine the optimal spatial configuration of helicopter-based culling. By targeting individual grid cells or

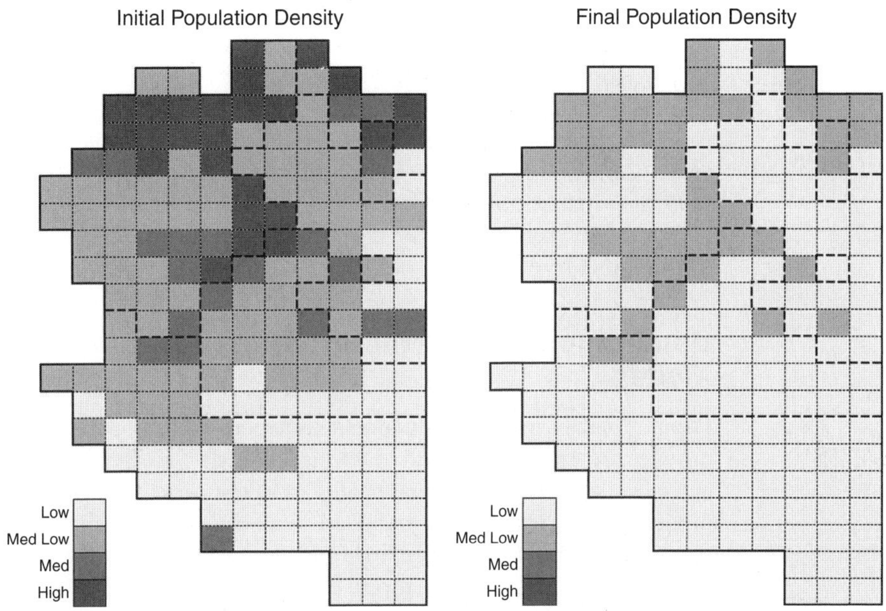

FIG. 17-1

Population density of feral buffalo (*Bubalus bubalis*) in Kakadu National Park, northern Australia, predicted by a lattice population model with 10 × 10 km grid cells. The left map shows densities prior to management, and the right map shows densities after a 10-year culling regime. The dashed lines delineate the land management zones, each subjected to different culling intensities.

clusters of cells, researchers were able to determine how to make the best use of limited logistical resources to minimize damage in the most environmentally and culturally sensitive areas of this large national park (Fig. 17-1).

Lattice models, by making use of a variety of data layers, can be arbitrarily complex. This modeling is useful because such models are naturally tailored to suit the type of data available, but it tends to limit the generalities that can be extracted from such computations; their results tend to be case specific. A more abstract but generalizable approach is to use cellular automata, in which just a few simple rules govern the interactions of individual cells. The fine-scale effects of habitat fragmentation on extinction risk can be examined (Oborny et al. 2005), and complex patterns that resemble real-world landscape features (e.g., banded vegetation in semi-arid savannas) often emerge from such cellular automata models (Rietkerk et al. 2004). Their consistent properties can be used to develop rules of thumb for managing habitat across large landscapes when few species-habitat data are available (With 1997). The challenge is to verify whether the generating mechanisms apply in real-world situations (Molofsky and Bever 2004).

Unlike metapopulation models that are based on discrete habitat patches, the resolution of the grid-based models is not based on the structure of the landscape. However, their resolution should be consistent with the biology of the species. Because population dynamics are simulated within each grid cell, there is an implicit assumption of the applicability of population processes at that scale. Hence, cells that are too small may not be suitable because the "population" in each cell may not have the properties of a biological population, and cells that are too large may not be appropriate because of the heterogeneity of the habitat within each cell.

Demographically Structured Metapopulation Models

Populations in which individuals differ in their contributions to population growth are structured, and these individuals can be classified by state (e.g., age, size, sex, developmental stage). When information is available on variation in vital rates, a tool commonly used to assess spatiotemporal changes in populations in large landscapes is matrix projection models (Caswell 2001). In general, structured models give a more detailed portrayal of metapopulation change through time than occupancy or time-series methods, and they can contribute to more targeted management questions because they identify the vital rates and specific states with the greatest influence on the population rate of change.

The probability of extinction and related viability metrics (e.g., risk of decline, probability of patch occupancy, population abundance) are usually estimated by Monte Carlo computer simulation. These models can take explicit account of demographic parameters (survival and fecundity), density dependence, environmental fluctuations, changes in the status of habitat, and the impact of infrequent, catastrophic events such as fire, cyclones, or disease epidemics (Akçakaya et al. 1999). In fragmented populations or species distributed over large landscapes, spatial correlation of environmental variation and dispersal are incorporated as a metapopulation model, with stage-structured patch-based populations being connected by movement rates defined according to a proportion of population size and distance between patches.

An issue prominent on the agenda of managers of wildlife populations in large landscapes is that of sustainable harvest. That is, what levels of off-take can a given wildlife population support, and importantly, what role do spatial refugia (e.g., reserves that are closed to hunting) play in offsetting harvest in other parts of a species' distributional range? Brook and Whitehead (2005) used a spatially structured demographic model to address this issue for magpie geese (*Anseranas semipalmata*), a species that was once found in abundance throughout eastern Australia, but is now restricted to the tropical wetlands of northern Australia and New Guinea. This model considered spatial and temporal variation in both habitat suitability and off-take. They showed that current levels of indigenous and recreational harvests could be supported over the long term in model systems that incorporated dispersal and spatial structure. Yet, if the same population size of geese were assumed to be distributed and harvested homogenously across the landscape, the species could not compensate sufficiently and was predicted to decline to extinction within decades.

As with most metapopulation models, the magpie goose model treated habitat patches (wetlands) as fixed, clearly defined units. However, in reality the definition of what constitutes habitat versus nonhabitat is often not clear cut. In such instances, habitat suitability models based on observed occurrences of species can facilitate definition of metapopulation spatial structure. One of the earlier examples of a metapopulation model in which patches were defined empirically, based on the distribution of suitable habitat, involved the California Gnatcatcher (*Polioptila c. californica*), a threatened species dependent on coastal sage scrub vegetation. The spatial structure of the metapopulation was based on a habitat suitability map (Fig. 17-2), which was predicted by a statistical habitat model (Akçakaya and Atwood 1997). This was based on sighting locations of observed pairs and maps of variables related to the species' habitat preferences, linked using a logistic generalized linear model. The RAMAS GIS software package (Akçakaya 2005) automates many of the steps required to achieve integration of landscape/vegetation maps, Habitat Suitability Index models, and metapopulation matrix projection models.

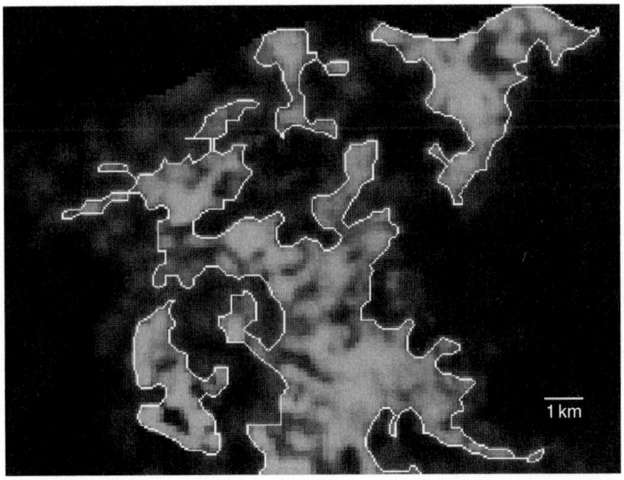

FIG. 17-2

Detail of the spatial structure of the habitat-based metapopulation model of the California Gnatcatcher. Shades of gray represent habitat suitability (the lighter the color, the higher the suitability), the white outlines are the outer borders of patches identified by the model based on a neighborhood distance parameter (Akçakaya and Atwood 1997). Cell (pixel) size of the underlying map is 100 meters. Each patch represents one population of the metapopulation model. The smallest patch in this section of the study area consists of 245 cells.

Recent work has demonstrated the value of incorporating the temporal dynamics of spatial data derived from landscape-scale vegetation models more directly into demographically structured metapopulation models of threatened species. Sharp-tailed grouse (*Tympanuchus phasianellus*) has declined in parts of its range due to loss of its steppe-grassland habitat. Its remaining habitat in Wisconsin is both fragmented and dynamic (i.e., the number, size, and distribution of patches change over time). The viability of this species was analyzed using a metapopulation model with dynamic spatial structure that was based on the predictions of a forest landscape model, which simulated landscape dynamics brought about by processes such as succession, disturbances, and silviculture (Akçakaya et al. 2004). The landscape component of the model predicted forest landscape dynamics in the form of a time series of raster maps. These maps were combined into a time series of patch structures, which formed the dynamic spatial structure of the metapopulation component. The results showed that the viability of this species was sensitive to landscape dynamics and demographic variables such as fecundity and mortality. Ignoring the landscape dynamics gave overly optimistic results, and results based only on landscape dynamics (ignoring demography) lead to a different ranking of the management options than the ranking based on the more realistic model incorporating both landscape and demographic dynamics (Fig. 17-3).

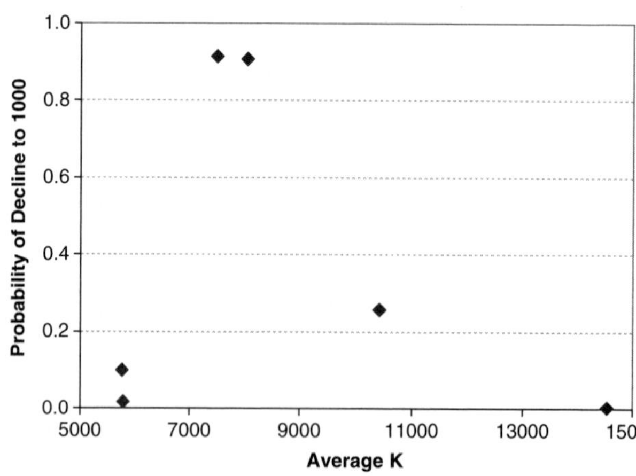

FIG. 17-3

Relationship between the amount of suitable habitat for the Sharp-tailed Grouse in the Wisconsin Pine Barrens region, USA (measured as total average carrying capacity) and viability (measured as risk of decline to 1000 individuals) for seven scenarios of forest management simulated by Akçakaya et al. (2004). Because of the interaction between landscape dynamics and the population dynamics of the species, the amount of habitat does not correctly predict the viability of the species.

Individual-Based Models

Individual-based models (IBMs) simulate the fate of each individual in a population, rather than cohorts, and like matrix-models, IBMs usually incorporate demographic and environmental stochasticity, habitat quality, and density dependence. The family of models IBMs entail has been extensively reviewed by Grimm (1999). The major advantage of IBMs is that individual heterogeneity is modeled explicitly. For instance, individuals may differ in the vital rates because of body condition, breeding status, the habitat quality within their exclusive home range, or because of their pedigree of inbreeding. These attributes reflect the relative contribution that each individual makes toward maintaining population viability, and can therefore be important to represent (White 2000). The caveats to developing IBMs are in the detailed data they demand, and the computational constraints of simulating large population sizes.

The IBM approach in population viability analysis is most frequently used for modeling small populations on the brink of extinction, which may be suffering from demographic failure, habitat loss, or inbreeding depression (e.g., Brito and Fernandez 2000, Lacy 2000). One interesting application of IBMs to populations in large landscapes was the simulation of an extinct predatory cat, *Panthera gombaszoegensis*, based on life history data of modern jaguars (*Panthera onca*). O'Regan et al. (2002) used the IBM software package VORTEX (Lacy 2000) to examine the size of refugial tracts of habitat that remained during repeated periods of intense glaciation in Pleistocene Europe that would have been sufficient to support viable and sufficiently interconnected populations.

An IBM approach more amenable to modeling populations in large landscapes is complex adaptive systems (CAS) modeling (Railsback 2001), which focuses on how the properties of aggregations of individuals can be determined by the characteristics and behavior of the individuals (e.g., habitat fidelity, food preferences, dispersal propensity, predator avoidance strategies). In essence, CAS approaches can translate ecological trends from individuals to landscape or population dynamics. This special class of IBMs is considered promising because unlike standard IBMs, no system-level responses are forced on the CAS (Elliot and Kiel 2002). Instead, ensembles of individuals within a spatially distributed population process information about their environment and modify their behavior according to a goal (Hraber and Milne 1997). The attractiveness of CAS is that it offers an alternative to parametric estimation approaches, which are often hampered by nonlinearity in data, unknown distributions, and problems associated with null data (e.g., determining why species do not forage in apparently suitable habitats). Complex adaptive system models are characterized by emergent responses (the ability to predict a wide range of realistic system-level attributes from a model in which individuals follow simple decision rules and regulatory responses) and can incorporate detailed spatial information and landscape-scale attributes linked to GIS. However, to date, the contribution of CAS to population management at the landscape-scale has been relatively minor, due at least in part to a lack of a validated "top-down" theoretical framework.

Selecting and Parameterizing a Model

Typically, the process of undertaking a population viability analysis goes through the following stages: (1) the identification of threatened species; (2) data collection (or collation) and analyses; (3) model formulation and construction; (4) simulations of the future fate of a population based on a range of potential threats (often projecting 50–100 years); and (5) the suggestion of possible remedial actions. Sensitivity analyses and the outcomes of various "what if?" scenarios are examined to evaluate different management options. Sensitivity analyses are also used to determine which parameters most strongly influence model predictions, and can be used to focus researchers' efforts on improving estimates of the most important variables (Caswell 2001).

The recent literature on viability analysis has emphasized the need to consider multiple working hypotheses and hence, more than one model (Burnham and Anderson 2002, Wintle et al. 2003). This operates at two levels:

1. When estimating the basic ecological attributes of a species, such as survival and reproductive rates or habitat suitability, one commonly uses a likelihood-based statistical model (e.g., generalized linear mixed effects model). It is critical at this stage to use robust methods of model selection that provide a measure of the strength of evidence for supporting models, such as information theoretic or iterative cross-validation methods to weight alternative models for multimodel inference. Alternatively, Bayesian statistical methods can be used to weight alternative model structures according to their posterior probabilities. The philosophical motivation for applying any of these weighting methods is that parameter estimates should be based on weighted estimates derived from all plausible models, using model weights to control for model selection uncertainty (Burnham and Anderson 2002).

2. When one is deciding on the choice of predictive models, the amount and type of available data provide a strong guide as to which method should be used. In general terms, simple approaches with few parameters, such as occupancy models, are easier to parameterize than more complex simulations, but less likely to fully encompass the range of potential factors that may influence the dynamics of a population. In instances in which there are sufficient data to parameterize a complex structure model, it may still be worthwhile also developing simpler approaches simultaneously. This acts as a check on the influence of model choice on predictions. It also makes transparent the structural generality/realism and parameter bias/precision trade-offs that are inevitable when deciding whether to use simple or complex models.

Irrespective of the model selection method used, it is important that parameter estimation is not divorced from biological intuition. For instance, some statistical data analysis techniques may produce estimates that are statistically valid but biologically implausible (such as obtaining negative multiple regression coefficients when estimating stage-specific fecundities; Akçakaya et al. 1999). In other cases the estimation of parameters will be prone to a lack of statistical power, and real trends may be overlooked. Because it is impossible to specify a general protocol to cover all such contingencies, the decision of when to reject or re-evaluate such estimates should be governed primarily by implicit knowledge of the particular species' biology.

Recent developments in wildlife population modeling have highlighted the value in combining the parameter estimation, model formulation, and population project stages. For instance, recent innovations in the development of Bayesian analysis using Markov Chain Monte Carlo (MCMC) allow the fitting of complex data to model, and include key aspects of individual and spatial heterogeneity (Clark et al. 2005). These hierarchical population dynamics models represent population structure as fixed stages and spatial strata, with differing levels of variability among individuals and groups.

Measurement errors and biases are inevitably incorporated into parameter estimates through imperfect data collection in the field (White 2000). For example, the commonly used Cormack-Jolly-Seber method of survival analysis associated with the mark-recapture techniques carries a set of implicit assumptions that, if violated, will lead to inexact parameter estimates (Lebreton et al. 1993). Due to imperfect data collection techniques, such measurement errors are usually impossible to eliminate. However, if there is a consistent bias when one technique of estimation is compared to another independent estimate, then the systematic bias can be calculated, and the value of the parameter(s) or counts can be adjusted accordingly (Akçakaya et al. 1999). Hierarchical Bayesian analysis is particularly robust to incomplete data.

FUTURE DIRECTIONS

There are many recent developments pertaining to PVA methods, especially on data analysis and estimation of demographic parameters (Holmes 2004, Dunham et al. 2006); model selection and multimodel inference (see previous section); prediction of habitat from occurrence data (Elith et al. 2006); detection and modeling of density dependence (Brook and Bradshaw 2006*b*); and estimating natural variability by removing variance due to measurement error and sampling variability (Dennis et al. 2006). We expect this trend to continuc, with new quantitative methods providing more accurate and less biased estimation of parameters for various components of PVA, including habitat modeling, demographic modeling, and interactions among populations.

A major future direction is related to the incorporation of habitat and landscape dynamics discussed previously. Natural landscape dynamics and disturbance regimes interact with human land-use and large-scale human impacts to shape the landscape patterns that determine the spatial structure and dynamics of metapopulations. Ecological research in large landscapes that ignores the human element, or that includes it only as an implicit "extrinsic factor," is becoming increasingly regarded as unrealistic and artificial (Nyhus et al. 2002). The dynamics of coupled human and natural ecological systems are essential to understanding the ecology of populations and communities in large landscapes. In the future, we expect the development of much more explicit links between habitat-based PVA models and anthropogenic drivers of landscape change, such as climate change (including shifting bioclimatic envelopes; Keith et al. 2008) and human land use. These drivers in turn could be linked to models of human population dynamics and socioeconomics (Nyhus et al. 2002).

Closely related to the preceding developments are the challenges posed by compounded uncertainties, especially when the uncertain outputs of one model are used as inputs to the next model. In addition to new approaches for dealing with uncertainties, these challenges will also require new studies of validation, in which the available data are divided, with one set used for model development and the second set used for comparing with model predictions. This separation can be spatial or temporal. Spatial separation of populations in a large landscape can provide the opportunity to validate the generality and applicability of model predictions in the absence of long-term monitoring data. In one of the few validation studies involving PVA, Brook et al. (2000) validated PVA predictions of abundance and risks of decline, by temporally separating the available data, estimating the parameters from the first half of each data set and using the second half to evaluate model performance. They found that predictions were accurate; the risk of population decline closely matched observed outcomes; there was no significant bias; and population size projections did not differ significantly or importantly from reality. Further, the predictions of five software packages they tested were highly concordant. However, this study was based on a limited number of well-studied species; larger studies that attempt to validate PVA predictions would evaluate the reliability of models for different types of questions and available data and provide insights in terms of best modeling practices. Theoretical studies do suggest, however, that the reliability of population viability predictions degrades considerably as projection time frame exceeds a few generations, especially in highly variable systems. Nevertheless, the relative rankings of alternative management scenarios may be robust even when absolute predictions of risk are uncertain.

SUMMARY

We reviewed methods of population viability analysis (PVA) as applied to wildlife populations in large landscapes. For these populations, viability analysis requires careful consideration of the issues of spatial heterogeneity and scaling of ecological processes, habitat connectedness (including "permeability of the intervening matrix"), and temporal dynamics of the landscape. Spatially structured models used for large-scale PVA include occupancy models, grid-based lattice models, demographically structured metapopulation models, and individual-based models. We discussed the assumptions and limitations of model types and the context within which each is more appropriate. Population viability analyses in large landscapes often require the definition of distinct subpopulations, which in turn depend critically on the spatial scale of, and barriers to, dispersal in relation to the distribution of suitable habitat. Another important factor is the effect of landscape dynamics on the temporal variability of the habitat, and hence on the dynamics of the wildlife populations. Viability of species in dynamic landscapes depends on the interaction between landscape change (the pattern, scale, rate, and direction of landscape changes in size, structure, and quality) and the species' ecology (its ability to disperse between and grow in the habitat patches or make use of the matrix). Spatial separation of populations can also provide the opportunity to validate the generality and applicability of model predictions in the absence of long-term monitoring data.

LITERATURE CITED

Akçakaya, H. R. 2000. Viability analyses with habitat-based metapopulation models. *Population Ecology* 42:45-53.

Akçakaya, H. R. 2001. Linking population-level risk assessment with landscape and habitat models. *Science of the Total Environment* 274:283-291.

Akçakaya, H. R. 2005. *RAMAS GIS: Linking spatial data with population viability analysis. Version 5.0*. Applied Biomathematics, Setauket, New York, USA.

Akçakaya, H. R., and J. L. Atwood. 1997. A habitat-based metapopulation model of the California gnatcatcher. *Conservation Biology* 11:422-434.

Akçakaya, H. R, M. A. Burgman, and L. R. Ginzburg. 1999. *Applied population ecology*. Sinauer Associates, Sunderland, Massachusetts, USA.

Akçakaya, H. R., J. Franklin, A. D. Syphard, and J. R. Stephenson. 2005. Viability of Bell's sage sparrow (*Amphispiza belli* ssp. *belli*): Altered fire regimes. *Ecological Applications* 15:521-531.

Akçakaya, H. R., and L. R. Ginzburg. 1991. Ecological risk analysis for single and multiple populations. Pages 73-87 *in* A. Seitz and V. Loeschcke, editors. *Species conservation: A population-biological approach*. Birkhauser Verlag, Basel, Germany.

Akçakaya, H. R., G. Mills, and C. P. Doncaster. 2007. The role of metapopulations in conservation. Pages 64-84 *in* D. W. Macdonald and K. Service, editors. *Key topics in conservation biology*. Blackwell Publishing, Oxford, United Kingdom.

Akçakaya, H. R., V. C. Radeloff, D. J. Mladenoff, and H. S. He. 2004. Integrating landscape and metapopulation modeling approaches: Viability of the sharp-tailed grouse in a dynamic landscape. *Conservation Biology* 18:526-537.

Akçakaya, H. R., and M. G. Raphael. 1998. Assessing human impact despite uncertainty: Viability of the northern spotted owl metapopulation in the northwestern USA. *Biodiversity and Conservation* 7:875-894.

Akçakaya, H. R., and W. T. Root. 2003. *RAMAS landscape: Integrating metapopulation viability with LANDIS forest dynamics model (version 1.0)*. Applied Biomathematics, Setauket, New York, USA.

Araújo, M. B., and A. Guisan. 2006. Five (or so) challenges for species distribution modelling. *Journal of Biogeography* 33:1677-1688.

Baguette, M. 2004. The classical metapopulation theory and the real, natural world: A critical appraisal. *Basic and Applied Ecology* 5:213-224.

Beier, P., and R. F. Noss. 1998. Do habitat corridors provide connectivity? *Conservation Biology* 12:1241-1252.

Bond, W. J., F. I. Woodward, and G. F. Midgley. 2005. The global distribution of ecosystems in a world without fire. *New Phytologist* 165:525-538.

Bowman, D. M. J. S., D. L. McIntyre, and B. W. Brook. 2006. Is the Carpentarian rock-rat (*Zyzomys palatalis*) critically endangered? *Pacific Conservation Biology* 12:134-139.

Breininger, D. R., M. L. Legare, and R. B. Smith. 2004. Eastern indigo snakes (*Drymarchon couperi*) in Florida: Influence of edge effects on population viability. Pages 299-311 *in* H. R. Akçakaya, M. A. Burgman, O. Kindvall, C. C. Wood, P. Sjörgren-Gulve, J. S. Hatfield, and M. McCarthy, editors. *Species conservation and management: Case studies.* Oxford University Press, New York, New York, USA.

Brito, D., and F. A. S. Fernandez. 2000. Metapopulation viability of the marsupial *Micoureus demerarae* in small Atlantic forest fragments in south-eastern Brazil. *Animal Conservation* 3:201-209.

Brook, B. W., and C. J. A. Bradshaw. 2006*a*. *Spatio-Temporal Animal Reduction (STAR) model for feral species in Kakadu National Park*. Report to KNP Board of Management. Darwin, Northern Territory, Australia.

Brook, B. W., and C. J. A. Bradshaw. 2006*b*. Strength of evidence for density dependence in abundance time series of 1198 species. *Ecology* 87:1445-1451.

Brook, B. W., J. J. O'Grady, A. P. Chapman, M. A. Burgman, H. R. Akçakaya, and R. Frankham. 2000. Predictive accuracy of population viability analysis in conservation biology. *Nature* 404:385-387.

Brook, B. W., and P. J. Whitehead. 2005. Sustainable harvest regimes for magpie geese (*Anseranas semipalmata*) under spatial and temporal heterogeneity. *Wildlife Research* 32:459-464.

Burgman, M. A., S. Ferson, and H. R. Akçakaya. 1993. *Risk assessment in conservation biology.* Chapman and Hall, London, United Kingdom.

Burnham, K. P., and D. R. Anderson. 2002. *Model selection and multimodel inference: A practical information-theoretic approach*. Second edition. Springer, New York, New York, USA.

Caswell, H. 2001. *Matrix population models: Construction, analysis, and interpretation*. Second edition. Sinauer Associates, Sunderland, Massachusetts, USA.

Clark, J. S., G. A. Ferraz, N. Oguge, H. Hays, and J. DiCostanzo. 2005. Hierarchical Bayes for structured, variable populations: From recapture data to life-history prediction. *Ecology* 86:2232-2244.

Dennis, B., J. M. Ponciano, S. R. Lele, M. L. Taper, and D. F. Staples. 2006. Estimating density dependence, process noise, and observation error. *Ecological Monographs* 76:323-341.

Drechsler, M., K. Frank, I. Hanski, R. B. O'Hara, and C. Wissel. 2003. Ranking metapopulation extinction risk: From patterns in data to conservation management decisions. *Ecological Applications* 13:990-998.

Dunham, A. E., H. R. Akçakaya, and T. S. Bridges. 2006. Using scalar models for precautionary assessments of threatened species. *Conservation Biology* 20:1499-1506.

Dunning, J. B., Jr., D. J. Stewart, B. J. Danielson, B. R. Noon, T. L. Root, R.H Lamberson, and E. E. Stevens. 1995. Spatially explicit population-models: Current forms and future uses. *Ecological Applications* 5:3-11.

Elith, J., C. H. Graham, R. P. Anderson, M. Dudík, S. Ferrier, A. Guisan, R. J. Hijmans, F. Huettmann, J. R. Leathwick, A. Lehmann, J. Li, L. G. Lohmann, B. A. Loiselle, G. Manion, C. Moritz, M. Nakamura, Y. Nakazawa, J. Overton, A. T. Peterson, S. J. Phillips, K. S. Richardson, R. Scachetti-Pereira, R. E. Schapire, J. Soberón, S. Williams, M. S. Wisz, and N. E. Zimmermann. 2006. Novel methods improve prediction of species' distributions from occurrence data. *Ecography* 29:129-151.

Elliott, E., and L. D. Kiel. 2002. Exploring cooperation and competition using agent-based modeling. *Proceedings of the National Academy of Sciences* 99:7193-7194.

Gilpin, M. E., and M. E. Soulé. 1986. Minimum viable populations: Processes of species extinction. Pages 19–34 *in* M. E. Soulé, editor. *Conservation biology: The science of scarcity and diversity.* Sinauer, Sunderland, Massachusetts, USA.

Grimm, V. 1999. Ten years of individual-based modelling in ecology: What have we learned and what could we learn in the future? *Ecological Modelling* 115:129–148.

Grimm, V., H. Lorek, J. Finke, F. Koester, M. Malachinski, M. Sonnenschein, A. Moilanen, I. Storch, A. Singer, C. Wissel, and K. Frank. 2004. META-X: Generic software for metapopulation viability analysis. *Biodiversity and Conservation* 13:165–188.

Haines, A. M., M. E. Tewes, L. L. Laack, J. S. Horne, and J. H. Young. 2006. A habitat-based population viability analysis for ocelots (*Leopardus pardalis*) in the United States. *Biological Conservation* 132:424–436.

Hanski, I. 1994. A practical model of metapopulation dynamics. *Journal of Animal Ecology* 63:151–162.

Hanski, I., A. Moilanen, T. Pakkala, and M. Kuussaari. 1996. The quantitative incidence function model and persistence of an endangered butterfly metapopulation. *Conservation Biology* 10:578–590.

Holmes, E. E. 2004. Beyond theory to application and evaluation: Diffusion approximations for population viability analysis. *Ecological Applications* 14:1272–1293.

Hraber, P. T., and B. T. Milne. 1997. Community assembly in a model ecosystem. *Ecological Modelling* 103:267–285.

Johst, K., R. Brandl, and S. Eber. 2002. Metapopulation persistence in dynamic landscapes: The role of dispersal distance. *Oikos* 98:263–270.

Keith, D. 2004. Australian heath shrub *(Epacris barbata)*: Viability under management options for fire and disease epidemics. Pages 90–102 *in* H. R. Akçakaya, M. A. Burgman, O. Kindvall, C. C. Wood, P. Sjögren-Gulve, J. S. Hatfield, and M. A. McCarthy, editors. *Species conservation and management: Case studies.* Oxford University Press, United Kingdom.

Keith, D. A., H. R. Akcakaya, W. Thuiler, G. F. Midgley, R. G. Pearson, S. J. Phillips, H. M. Regan, M. B. Araujo, and T. G. Rebelo. 2008. Predicting extinction risks under climate change: Coupling stochastic population models with dynamic bioclimatic habitat models. *Biology Letters* (in press).

Kindvall, O., and K.-O. Bergman. 2004. Woodland brown butterfly *Lopinga achine* in Sweden: Viability in a dynamic landscape maintained by grazing. Pages 171–178 *in* H. R. Akçakaya, M. A. Burgman, O. Kindvall, C. C. Wood, P. Sjögren-Gulve, J.S Hatfield, and M. A. McCarthy, editors. *Species conservation and management: Case studies.* Oxford University Press, United Kingdom.

Lacy, R. C. 2000. Considering threats to the viability of small populations using individual-based models. *Ecological Bulletins* 48:39–51.

LaHaye, W. S., R. J. Gutiérrez, and H. R. Akçakaya. 1994. Spotted owl metapopulation dynamics in southern California. *Journal of Animal Ecology* 63:775–785.

Lamberson, R. H., R. McKelvey, B. R. Noon, and C. Voss. 1996. A dynamic analysis of northern spotted owl viability in a fragmented forest landscape. *Conservation Biology* 6:505–512.

LaSorte, F. A., and F. R. Thompson, III. 2007. Poleward shifts in winter ranges of North American birds. *Ecology* 88:1803–1812.

Lebreton, J.-D., R. Pradel, and J. Clobert. 1993. The statistical analysis of survival in animal populations. *Trends in Ecology and Evolution* 8:91–95.

Lecomte, J., K. Boudjemadi, F. Sarrazin, K. Cally, and J. Clobert. 2004. Connectivity and homogenisation of population sizes: An experimental approach in *Lacerta vivipara*. *Journal of Animal Ecology* 73:179–189.

Levins, R. 1970. Extinction. Pages 77–100 *in* M. Gerstenhaber, editor. *Some mathematical problems in biology.* American Mathematical Society, Providence, Rhode Island, USA.

Lindenmayer, D. B., and R. C. Lacy. 1995. Metapopulation viability of Leadbeater's Possum, *Gymnobelideus leadbeateri*, in fragmented old-growth forests. *Ecological Applications* 5:164–182.

Lindenmayer, D. B., M. A. McCarthy, and M. L. Pope. 1999. Arboreal marsupial incidence in eucalypt patches in south-eastern Australia: A test of Hanski's incidence function metapopulation model for patch occupancy. *Oikos* 84:99–109.

Lindenmayer, D. B., and H. P. Possingham. 1996. Ranking conservation and timber management options for Leadbeater's Possum in southeastern Australia using population viability analysis. *Conservation Biology* 10:235–251.

Maschinski, J., and J. Duquesnel. 2006. Successful reintroductions of the endangered long-lived Sargent's cherry palm, *Pseudophoenix sargentii*, in the Florida Keys. *Biological Conservation* 134:122–129.

Moilanen, A. 2004. SPOMSIM: Software for stochastic patch occupancy models of metapopulation dynamics. *Ecological Modelling* 179:533–550.

Moilanen, A., and I. Hanski. 1998. Metapopulation dynamics: Effects of habitat quality and landscape structure. *Ecology* 79:2503–2515.

Molofsky, J., and J. D. Bever. 2004. A new kind of ecology? *Bioscience* 54:440–446.

Nakamaru, M. 2006. Lattice models in ecology and social sciences. *Ecological Research* 21:364–369.

Nyhus, P. J., F. R. Westley, R. C. Lacy, and P. S. Miller. 2002. A role for natural resource social science in biodiversity risk assessment. *Society and Natural Resources* 15:923–932.

Oborny, B., G. Meszéna, and G. Szabó. 2005. Dynamics of populations on the verge of extinction. *Oikos* 109:291–296.

O'Hara, R. B., E. Arjas, H. Toivonen, and I. Hanski. 2002. Bayesian analysis of metapopulation data. *Ecology* 83:2408–2415.

O'Regan, H. J., A. Turner, and D. M. Wilkinson. 2002. European Quaternary refugia: A factor in large carnivore extinction? *Journal of Quaternary Science* 17:789–795.

Ovaskainen, O., and I. Hanski. 2003. Extinction threshold in metapopulation models. *Annales Zoologici Fennici* 40:81–97.

Pellet, J., G. Maze, and N. Perrin. 2006. The contribution of patch topology and demographic parameters to population viability analysis predictions: The case of the European tree frog. *Population Ecology* 48:353–361.

Pulliam, H. R., J. B. Dunning, Jr., and J. Liu. 1992. Population dynamics in complex landscapes: A case study. *Ecological Applications* 2:165–177.

Railsback, S. F. 2001. Concepts from complex adaptive systems as a framework for individual-based modelling. *Ecological Modelling* 139:47–62.

Rietkerk, M., S. C. Dekker, P. C. de Ruiter, and J. van de Koppel. 2004. Self-organized patchiness and catastrophic shifts in ecosystems. *Science* 305:1926–1929.

Ruckelshaus, M., P. McElhany, M. McClure, and S. Heppell. 2004. Chinook Salmon *(Oncorhynchus tsawytscha)* in Puget Sound: Effects of spatially correlated catastrophes on persistence. Pages 208–218 *in* H. R. Akçakaya, M. A. Burgman, O. Kindvall, C. C. Wood, P. Sjögren-Gulve, J. S. Hatfield, and M. A. McCarthy, editors. *Species conservation and management: Case studies.* Oxford University Press, United Kingdom.

Shimada, M., and F. Ishihama. 2000. Asynchronization of local population dynamics and persistence of a metapopulation: A lesson from an endangered composite plant, *Aster kantoensis. Population Ecology* 42:63–72.

Sjögren-Gulve, P., and C. Ray. 1996. Using logistic regression to model metapopulation dynamics: Large-scale forestry extirpates the pool frog. Pages 111–137 *in* D. R. McCullough, editor. *Metapopulations and wildlife conservation.* Island Press, Washington, D.C., USA.

Stacey, P. B., V. A. Johnson, and M. L. Taper. 1997. Migration within metapopulations: The impact upon local population dynamics. Pages 267–291 *in* I. Hanski and M. E. Gilpin, editors. *Metapopulation biology: Ecology, genetics, and evolution.* Academic Press, San Diego, California, USA.

Stelter, C., M. Reich, V. Grimm, and C. Wissel. 1997. Modelling persistence in dynamic landscapes: Lessons from a metapopulation of the grasshopper *Bryodema tuberculata. Journal of Animal Ecology* 66:508–518.

White, G. C. 2000. Population viability analysis: Data requirements and essential analyses. Pages 288–331 *in* L. Boitani and T. K. Fuller, editors. *Research techniques in animal ecology: Controversies and consequences.* Columbia University Press, New York, New York, USA.

Wintle, B. A., S. A. Bekessy, L. A. Venier, J. L. Pearce, and R. A. Chisholm. 2005. Utility of dynamic-landscape metapopulation models for sustainable forest management. *Conservation Biology* 19:1930–1943.

Wintle, B.A, M. A. McCarthy, C. T. Volinsky, and R. P. Kavanagh. 2003. The use of Bayesian model averaging to better represent uncertainty in ecological models. *Conservation Biology* 17:1579–1590.

With, K. A. 1997. The application of neutral landscape models in conservation biology. *Conservation Biology* 11:1069–1080.

CHAPTER

Dynamic Landscape Metapopulation Models and Sustainable Forest Management

18

Sarah Bekessy, Brendan Wintle, Ascelin Gordon, Ryan Chisholm, Lisa Venier, and Jennie Pearce

Sustainable forest management is a widely held international goal (Mulder et al. 1999, Smith and Zollner 2005) and in many cases a legislated mandate (Commonwealth of Australia 2001). One component of sustainable forest management involves assessing the impact of management on biodiversity, which is frequently carried out using individual species as indicators (Mulder et al. 1999). However, the choice of indicators and determination of how they should be monitored is far from resolved. The urgency with which meaningful, practical, and immediate assessments of sustainability must be developed is highlighted by national and international sustainable forest management certification systems that are expanding rapidly and influencing market demand (Kanowski et al. 1999) and the rate of timber harvesting in forests is continuing at substantial levels (Canadian Council of Forest Ministers 2000).

The use of indicators is often associated with long-term and broad-scale monitoring of populations to assess population trends and inform management in an adaptive manner (Walters and Holling 1990, Johnson 1999, Elzinger et al. 2001). However, there is increasing evidence that long-term monitoring alone, especially at broad scales, is unable to provide useful information to address the most immediate concerns of forest management for sustainability (Temple and Wiens 1989, Ralph et al. 1995). This is partly because monitoring programs require long time frames to identify change (Green and Hirons 1991). It is often difficult to use broad-scale monitoring data to establish whether real and important changes are occurring in populations (Temple and Wiens 1989), especially at scales necessary to inform sustainable forest management. Broad-scale programs are suited to detecting changes over long time periods and over large areas.

Long-term, broad-scale monitoring has also been criticized because it is difficult to assign a cause to changes that are measured. Morrison (1986) suggested that species respond to change, but not uniquely to specific changes; therefore, it becomes difficult to identify the responsible mechanisms unless the data are collected within carefully designed experiments. Population trends, in and of themselves, do not help identify cause-and-effect relationships and therefore do not help identify management options (Ralph et al. 1995). The general lack of controls in large-scale monitoring programs is also problematic because of the pervasive impacts of large-scale disturbances, such as climate change. Information about underlying causes of change are more likely to be obtained through intensive local-scale studies (Venier and Pearce 2004). The major drawback of such studies is that they are not as likely to provide as widely applicable results as broader scale studies.

Alternative indices of sustainability such as vegetation structural indicators have been proposed (Lindenmayer et al. 2000) on the basis that they do not suffer from the specific concerns identified for taxon-based indicators. Structural indicators are appealing because they are relatively easy to measure, and the structural consequences of forest management are easier to predict than individual species and demographic responses. Structural complexity maintenance, forest connectivity, and spatial heterogeneity in stand structure are important aspects of forest management and should be included as criteria for sustainable forest management. Such criteria would be relatively easy to monitor. However, the link between species persistence and structural metrics is seldom clear, and structural indices may not be a good indicator of species richness or persistence (Abensperg-Traun et al. 1997).

An important aspect of sustainable forest management is maintaining viable populations of associated organisms (Noss 1990, Poiani et al. 2000, Smith and Zollner 2005), and measuring structural metrics does not ensure that management achieves this outcome. The measurement of structural metrics alone does not assist managers in balancing the ecological, economic, and social values of a forest because it does not answer the question of how much connectivity, heterogeneity, and complexity is enough to ensure species persistence and the maintenance of biodiversity in a region. The question of how much of these structural attributes in a landscape is enough can only be properly answered in the long term through carefully designed biodiversity monitoring strategies within an adaptive management context. Even so, such a system provides no immediate guidance and no forecasting of the likely consequences of current actions and management alternatives. Options for exploring alternatives are required in the short term, even if a comprehensive and reliable monitoring system is in place.

Recently, there has been an increasing emphasis on exploring indicator responses through habitat models (Mulder et al. 1999) where habitat relationships are assessed based on small-scale studies and expert opinion (e.g. Yahner 1992, Petranka et al. 1994). Predictions of the future distribution of habitat together with known habitat occupancy rates provide an approximation of the future abundance of a species under alternative management approaches (e.g., Gustafson et al. 2001). Such approaches are more appealing than simple structural indicators and abundance measurements because they make explicit use of available knowledge about habitat requirements to make predictions about the local persistence of a species. Often these requirements incorporate forest structural attributes, but in a way that is more biologically meaningful than simply measuring the structural attributes themselves. However, some of these methods do not consider environmental and demographic stochasticity explicitly. Further, some do not account for the spatial attributes of a species' biology that may occur at broader scales such as dispersal dynamics and Allee effects. Some of these methods are also constrained by inadequate estimates of detection probability and the ability to confidently determine absence (Wintle et al. 2005*b*). Consequently, these methods may be unable to capture the potential landscape-scale effects of forest management activities on habitat composition and configuration and temporal fluctuations in habitat occupancy that in turn affect population persistence (Andren 1994, McGarigal and McComb 1995, DesRochers and Hannon 1997, Schmiegelow et al. 1997).

Metapopulation models can be used to address these additional concerns. They incorporate the dynamic consequences of dispersal among local populations and the conditions that lead to regional persistence of a species (Hanksi 1998). They provide a mathematical representation of the demographics within populations and dispersal between populations, and allow predictions of population size over time. Metapopulation models have been used widely in endangered species management (Boyce 1992, Akçakaya et al. 1995), but not in more general management problems, though the potential for such an application has been recognized (Burgman and Possingham 2000). The predictions of metapopulation models for a range of indicator species may provide a useful means of evaluating the sustainability of current and alternative forest management activities and predicting ecological changes.

Metapopulation models have been used in conservation planning under the umbrella of population viability analysis (PVA; Akçakaya et al. 1995; Lindenmayer and Possingham 1996; Burgman and Lindenmayer 1998; Akçakaya and Brook, this volume; Beissinger et al., this volume). Population viability analysis has been described as any systematic attempt to understand the processes that make a population vulnerable to decline or extinction (Gilpin and Soulé 1986, Burgman and Lindenmayer 1998) and may be used to assess the likelihood that a population will persist for some arbitrarily chosen time into the future (Shaffer 1990, Boyce 1992, Smith and Person 2008). It is an interactive process of model construction, parameterization, sensitivity analysis, and validation (Akçakaya 2000, Burgman and Possingham 2000). While there is considerable uncertainty associated with using population viability models to predict actual risks of extinction (Taylor 1995, McCarthy 1996, Beissinger and Westphal 1998, Fieberg and Ellner 2000), PVA models appear to be useful for predicting changes in population size and ranking the severity of the effect of different management strategies (Boyce 1992, Beissinger and Westphal 1998, McCarthy et al. 2003). The models allow the available data and information to be integrated in a manner that is comprehensive, explicit, and repeatable, which then allows a transparent assessment of the consequences of different management strategies (McCarthy et al. 2004).

The development of a species metapopulation structure involves the identification of habitat requirements including the finer scale dependencies derived from habitat studies (e.g., Yahner 1992, Petranka et al. 1994, Smith et al. 2004, Smith and Person 2008) and the ecosystem stresses it responds to. Metapopulation models may predict change in habitat attributes or structural indicators in a way that directly addresses the impact of such changes on populations of forest-dependent species in terms of their probability of decline or loss. Moreover, it provides a way to incorporate available information about specific spatial and demographic requirements of species. Demographic models also allow the comparison of management scenarios that do not explicitly change forest structure or habitat quality indices but may impact on biodiversity, such as hunting or the application of herbicides and pesticides.

A modeling approach also addresses the need for more immediate information to make informed management decisions. The nature of the system can be hypothesized, model predictions can be generated, and the impacts of management can be measured using metrics such as minimum expected population size. Models can provide the capacity to compare alternative management options using the best information available and to quantify the uncertainty in what we know.

Constructing metapopulation models under a range of management scenarios requires a dynamic landscape model (Burgman et al. 1993; Holt et al. 1995; Mladenoff and Baker 1999; He, this volume) to characterize future changes in landscape vegetation composition and structure resulting from each management scenario. Dynamic landscape models predict the vegetation composition and structure of future landscapes by incorporating the effects of deterministic and stochastic disturbance (such as timber harvesting and fire) and succession. Successional processes may be described on the basis of establishment and persistence probabilities for individual vegetation species or vegetation types (Mladenoff and He 1999). Linkages between dynamic landscape models and metapopulation models are very recent and pose a number of challenges including software and computing challenges. Previous studies manually linked outputs from dynamic landscape models with habitat suitability models or population models (Larson et al. 2004, Shifley et al. 2006). More sophisticated packages devoted to these approaches are emerging. The case studies described here and one other study (Akçakaya et al. 2004) have used the dynamic landscape metapopulation (DLMP) modeling software package RAMAS Landscape (Akçakaya et al. 2003).

Further challenges to sustainability assessments are brought about by the multitude of scales at which forest management takes place. Cumulative effects of forest management on ecosystem composition and function arise from activities at the level of prescription (usually stands and management units) to the level of resource allocation (usually regions, license areas, and provinces). Assessment of sustainable management should therefore encompass the influences of both prescription- and allocation-level decisions. Dynamic landscape metapopulation models have the flexibility to be developed at multiple scales and to incorporate influences at multiple scales.

INDICATOR SPECIES

Monitoring a few indicator species is an intuitively appealing method of measuring the ecological sustainability of forest management because it is impossible to measure and monitor the effects of forest management on all species or environmental conditions of interest (Landres et al. 1988; Noon et al., this volume). Lindenmayer et al. (2000) define seven types of indicator species: (1) species whose presence indicates the presence of a set of other species; (2) keystone species (*sensu* Terborgh 1986) whose addition to or loss from an ecosystem leads to major changes; (3) species whose presence indicates human-created abiotic conditions such as air or water pollution (Spellerberg 1994); (4) dominant species that provide much of the biomass or number of individuals in an area; (5) species that indicate particular environmental conditions such as certain soil or rock types; (6) species thought to be sensitive to, and therefore serve as an early warning of, environmental change (also called indicator species); and (7) management indicator species, which reflect the effects of a disturbance regime or the efficacy of particular efforts to mitigate disturbance (Milledge et al. 1991).

These seven types of indicator species can be effectively classified into three classes of indicators: (1) biodiversity, (2) environmental, and (3) ecological. Biodiversity indicators indicate the presence of a set of other species (Noss 1990, Gaston and Blackburn 1995, Flather et al. 1997) and therefore provide a descriptive function. Environmental indicators are also descriptive in that they indicate changes in the state of the abiotic environment directly. Ecological indicators demonstrate the effects of environmental change on the biotic systems including species, communities, and ecosystems (Meffe and Carroll 1994), which provides an indication of change in the functioning of the system. Biological indicators of sustainable forest management are ecological indicators in that they must provide information on the effects of forest management on the functioning of the forest ecosystem to be useful. They can be keystone species, dominant species, sensitive species, or species that reflect the ecological effects of a disturbance regime. To be most effective, some must target anticipated stresses that are known to result from current or potential forest management approaches (Mulder et al. 1999, Venier et al. 2007). Examples of such stresses might include the truncation of older forest tree cohorts (e.g., McRae et al. 2001) or reduction in coarse woody debris in the form of snags or fallen logs.

The choice of a wide range of indicator species that target a range of potential ecosystem stresses would increase the likelihood that changes in ecosystem process resulting from forest management would be detected in monitoring systems. Indicators should also be chosen that represent a full range of spatial scales from local to regional, and a range of life history characteristics to capture as much of the ecological spectrum as possible. The rarity and detectability of an indicator species is likely to influence its effectiveness as an indicator of sustainable forest management. A characteristic of rare species is that suitable habitat may remain unoccupied for long periods of time. This creates problems for model-based assessments of sustainability due to difficulties in estimating initial population size, identifying habitat requirements, and describing metapopulation structure. We recommend the use of relatively common, widespread species, as they will serve as a better index of ecosystem condition over a greater proportion of the region and not just in the areas in which they exist.

The principle behind the use of indicator species implies a shift toward an ecosystem approach to management and monitoring. In our framework, indicator species are chosen because they reflect the ecosystem conditions necessary for their persistence. A change in the status of indicator species indicates a change in the state of the system. Likewise, no change in indicator species infers that an ecosystem is healthy. However, this second assertion will hold only if a sufficient number of indicator species are chosen on the basis that they target a variety of different ecosystem stressors predicted to arise from management activities. Using multiple species may also limit the problem of regional ecological differences weakening the effectiveness of indicator species (Smith et al. 2005).

Although the use of indicator species is attractive and could be a valuable management tool (Roberge and Angelstam 2004), there have been several major criticisms of current approaches to using biological indicators to inform sustainable forest management. These include the long time frames required to produce useful information, the lack of cause-and-effect linkages between management and indicator responses (Andelman and Fagan 2000, Smith et al. 2005), and the lack of transparency in the process. The indicator species approach is fraught with the difficulty of defining threatening processes, the species most sensitive to each process, and the manner in which species are affected by each process (Lindenmayer et al. 2002). These difficulties are compounded by complicated interactions between threatening processes and biases in biological data toward well-known vertebrates (Lindenmayer et al. 2002). The indicator species approach, like other taxon-based surrogate schemes, is based on the implicit assumption of nestedness among species; that is, the response of a certain species is assumed to be representative of the response of a broader range of species. Surrogate approaches that rely on single species, or aggregations of a few species, have been criticized because qualitatively similar species may have substantially different responses to environmental change (Lindenmayer et al. 2002; Noon et al., this volume). In a study by Andelman and Fagan (2000), taxon-based surrogate schemes were shown to be no more effective than species selected at random for capturing species or protecting habitat. Moreover, combining multiple assessments of impacts across species remains a significant challenge (Wintle et al. 2005*a*; Noon et al., this volume).

Effective use of indicator species approaches requires an understanding of the relationships between the response of surrogate species and the response of broader biodiversity to management actions (Bekessy et al. 2008). Indicator species can be used to guide habitat-based approaches to biodiversity planning and can be incorporated into population process-based assessments. If an indicator species approach is adopted, the consequences of given management scenarios for individual species persistence need to be understood. Of course, species can also be the focus of modeling or planning because of concern or interest in a particular species, and not because they are to be used as indicators. In this chapter, we propose the development of DLMP models of biological indicators for assessing the sustainability of forest management and guiding forest management decisions at various scales as a means of addressing current criticisms. We define ecological sustainability here as the maintenance of forest-dependent species within the managed forest estate. We review the practical advantages and problems of these methods based on our experiences gained through case studies and provide recommendations for appropriate implementation of the method in an adaptive management setting.

CONSTRUCTING AND INTERPRETING DLMP MODELS FOR INDICATOR SPECIES

Dynamic landscape metapopulation models integrate information on forest succession, natural disturbance regimes, and forest management actions to provide a spatial representation of the landscape and how this landscape changes through time. These models integrate dynamic landscape models with models of the population dynamics of the indicator species, representing the response of the species to this spatially and temporally variable environment. This holistic modeling approach allows prediction of future population sizes of the indicator species under a range of forest management scenarios. Thus, a DLMP model is an integration of modeling techniques currently applied in forest management and conservation planning including habitat modeling, landscape modeling, and metapopulation modeling. The landscape model and the population model are linked via a habitat model, which identifies areas in the landscape that may be suitable for occupation by indicator species. The habitat model provides information for the metapopulation model regarding the location and quality of habitat "patches" and the number of individuals of an indicator species that each patch is likely to support. The landscape dynamics model describes how habitat availability changes through time due to succession and disturbance.

Developing the Model

A DLMP model is developed in five steps (Wintle et al. 2005*a*):

1. Build a habitat suitability map relating species presence and abundance to environmental variables. A habitat model may take various forms and is simply a description of how a species' presence or abundance is related to the landscape. It may be defined by a regression model (e.g., Pereira and Itami 1991; Buckland and Elston 1993; Ford et al. 2004; Niemuth et al., this volume), a classification tree (e.g., Hastie et al. 2001), a Habitat Suitability Index (e.g., Rand and Newman 1998; Burgman et al. 2001; Dijak and Rittenhouse, this volume), a machine learning algorithm, or any other statistical function. This model is extrapolated across the landscape using a geographical information system (GIS; e.g., Menzel et al. 2006), to produce a continuous map of habitat suitability.

2. Develop a population dynamic model for indicator species describing demographic attributes such as age or stage-specific birth and death rates through time. A population model allows predictions of population size over time by modeling the demographic attributes of a species (e.g., Smith and Person 2008). The structure of the population is specified in terms of survival, fecundity, and mortality among juvenile and adult life stages. Demographic stochasticity is included by specifying each parameter as a mean value with a standard deviation.

3. Develop a metapopulation model by linking the population dynamic model to the habitat suitability model to reflect spatial dynamics across time and space. The landscape model and the population model are linked via the habitat model. Patches of contiguous habitat are defined as populations, with carrying capacities. Discrete habitat patches are identified using estimates of species' range movements and a threshold of habitat suitability below which cells would be considered unsuitable and therefore unoccupied. Dispersal rates between populations are specified by the user and describe the degree of interaction between populations. Lindenmayer et al. (1995) provide a review of metapopulation modeling methods.

4. Develop a forest dynamic model to describe how forest composition and structure are expected to change over time given natural and anthropogenic disturbance regimes. A sucession model describes the tree species composition of the landscape and how this composition changes through time, based on species life-history attributes, site conditions, disturbance regimes, and management. Life-history characteristics considered include longevity, age at sexual maturity, shade and fire tolerance, and seed dispersal distance of each tree species. Disturbance regimes include natural processes, such as fire and windthrow, and anthropogenic processes such as timber harvesting and prescribed burning and the extent to which anthropogenic disturbance influences the likelihood of natural disturbance.

5. Link the dynamic forest model to the metapopulation model. The resulting model then provides a spatially and temporally explicit representation of habitat and population dynamics. The succession model is linked to the habitat model to describe habitat availability through time. Changes in fecundity and survivorship, as well as presence or absence of the species, are developed to reflect changing landscape conditions (e.g., survivorship can be set lower in areas buffering harvesting activities). The metapopulation model uses this information to describe population sizes through time.

DLMP Model Software

The five steps outlined in the preceding section are implemented in a new software package called RAMAS Landscape (Akçakaya et al. 2003). RAMAS Landscape is the only standalone software package that is currently designed to implement DLMP models (Akçakaya et al. 2004), and it does this by linking the dynamic landscape modeling package LANDIS 3.7 (Mladenoff and He 1999) with the metapopulation modeling package RAMAS GIS 4 (Akçakaya and Root 2002). The software allows managers to integrate ideas about species habitat, population dynamics, landscape dynamics, and management.

The RAMAS GIS module of RAMAS Landscape simulates species metapopulation dynamics over time. The RAMAS GIS module is composed of various submodules designed to identify the metapopulation patch structure; specify the population model parameters, catastrophes, and management actions; and implement Monte Carlo simulations to evaluate predictive uncertainty. The user must specify the structure of the population in terms of survival, fecundity, and mortality rates among juvenile and adult life stages. Simulations are stochastic in that population parameter estimates, and catastrophic events are specified from a distribution of possible values with the mean and standard deviation of distributions specified by the user.

The LANDIS module simulates forest change by modeling tree species in 10-year age classes (He, this volume). It models succession based on interactions among species life-history attributes, site conditions, disturbance regimes, and management, all of which are set by the user. Life-history characteristics include longevity, age at sexual maturity, shade and fire tolerance, and seed dispersal distance of each tree species. Any number of tree species can be included in the model. Site conditions are encapsulated by "land types," which can be derived from climatic, physiographic, and edaphic properties. The LANDIS model incorporates natural processes (fire, windthrow, succession, and seed dispersal) and anthropogenic processes (e.g., timber harvesting and prescribed burning). It allows many different silvicultural treatments such as thinning, selection, gap harvesting, and clearcut harvesting to be modeled.

A detailed discussion of the theory, design, and implementation of RAMAS Landscape, RAMAS GIS, and the LANDIS model are provided elsewhere (He et al. 1996; Mladenoff and He 1999; Akçakaya and Root 2002; Akçakaya et al. 2003; Akçakaya and Brook, this volume; He, this volume).

Interpreting Model Predictions

The DLMP model estimates the expected metapopulation size at each time step of the simulation and presents this as a population trajectory, with time on the x-axis and population size on the y-axis (e.g., Fig. 18-3B). A population trajectory illustrates fluctuations in population size over time in response to environmental changes and demographic processes. Uncertainty in metapopulation model predictions is characterized by running the model many times and generating predicted population trajectories for each run of the model to form a distribution of predictions. The shape and spread of the predictive distribution defines the magnitude and type of uncertainty inherent in model predictions.

The response of indicator species population size to alternative management scenarios may be defined in terms of the risk of decline, measured by change in the expected minimum population size (EMP; McCarthy and Thompson 2001) between a reference state (e.g., a "natural" or base model with no anthropogenic disturbance) and the particular scenario being evaluated. The EMP is defined as the mean of the predicted minimum population sizes from all simulations of a given model and provides a representation of the lowest population size expected over the duration of a simulation under each management scenario. The EMP has been recommended as a suitable single metric for comparing population trajectories that is easily interpreted and more meaningful than other metrics such as mean population size or quasiextinction probability (McCarthy and Thompson 2001). The change in EMP can be calculated as

$$S_i = \frac{EMP_i - EMP_b}{EMP_b} \times 100 \quad (1)$$

where S_i is sensitivity of model i (the model being investigated), EMP_i is the expected minimum population size of the model i, and EMP_b is the expected minimum population size of the base model. Sensitivity calculated in this way provides an indication of both the magnitude and direction (positive or negative) of the change in EMP.

Results may also be graphically represented as risk curves. These describe the probability that the population will decline below a given threshold value over the course of the simulation. They are constructed by plotting simulation results, such as the minimum population size observed in a replication, as a cumulative probability function of population size (e.g., Fig. 18-1). Management scenarios may be compared in terms of the added risk of the species declining below a particular population size under each scenario relative to some reference state such as a "no timber harvesting" scenario.

The combination of predicted population trajectories, risk curves, expected minimum population size and sensitivity analysis provides a range of options for interpreting the predictions of a DLMP model and ranking management options. A particularly useful method for ranking management options is by comparing their EMPs and by assessing the sensitivity of each option compared with a

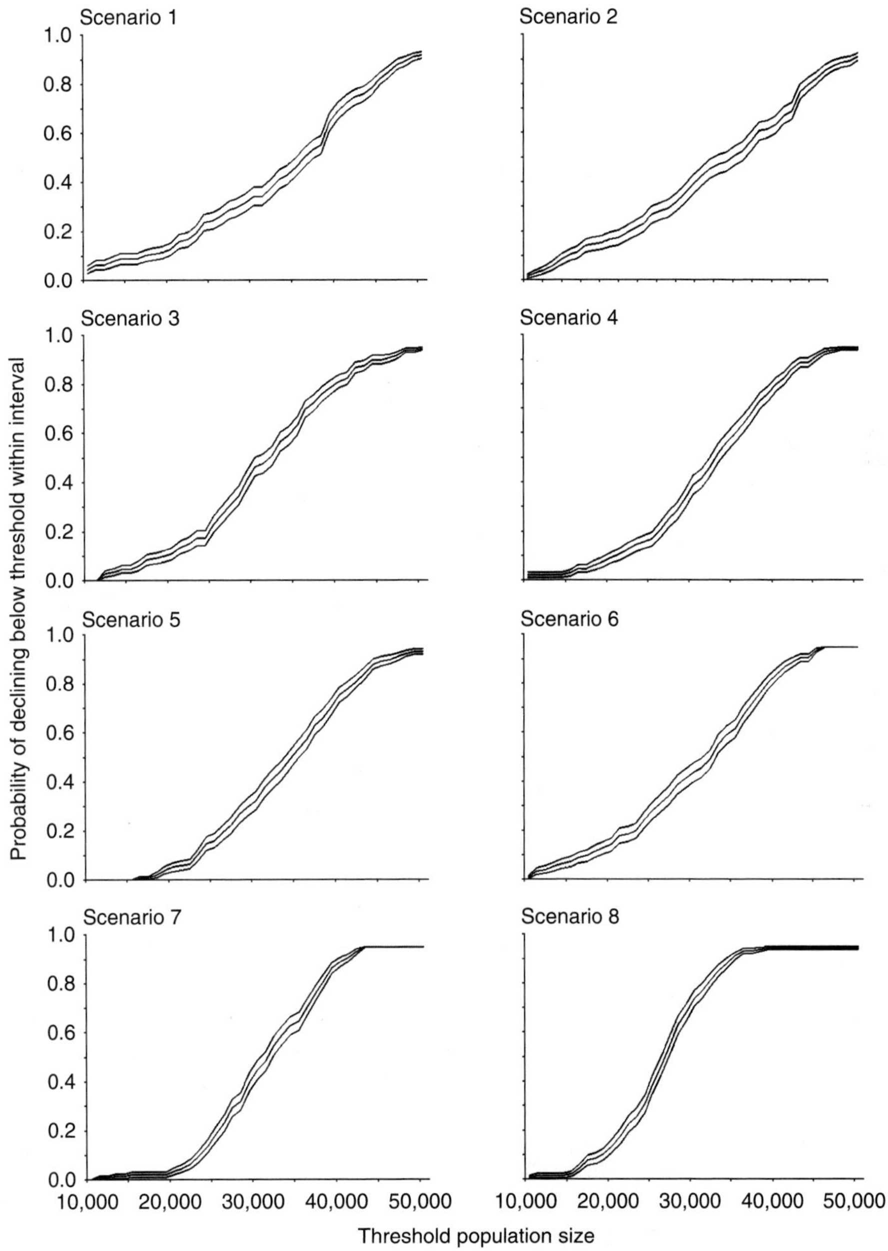

FIG. 18-1

Interval extinction risk curves for the brown creeper population model for eight scenarios (see Table 18-1). On each graph, the middle line shows the estimated probability of declining below the threshold value, while the upper and lower lines show one standard error from this estimate (from Wintle et al. 2005*a, b*). The shift in the risk curve for each scenario, relative to the base scenario, represents the increased risk of smaller population sizes resulting from each scenario.

reference state such as a "no timber harvesting" management option (McCarthy and Thompson 2001). While quantitative comparisons of the impacts of various options are appealing, caution is recommended due to the multitude of uncertainties inherent in predictions (Beissinger and Westphal 1998, McCarthy et al. 2003). For the same reason, ranking of management options is preferred to interpretation of absolute predictions.

SUMMARY OF CASE STUDIES

Study Area

We conducted case studies in north central Ontario, Canada, in a 150 km^2 section of the White River management area (Fig. 18-2), which has been actively managed for timber production for approximately 35 years. The northeast corner of Pukaskwa National Park was also included in the study area. As of 1972, approximately 83% of the research area was covered with mature closed-canopy forest, of which 43% was dominated by conifer forest, 33% by deciduous forest, and 25% dominated by mixed forest. The main tree species in the study area were jack pine (*Pinus banksiana* Lamb.), black spruce (*Picea mariana* Mill.), trembling

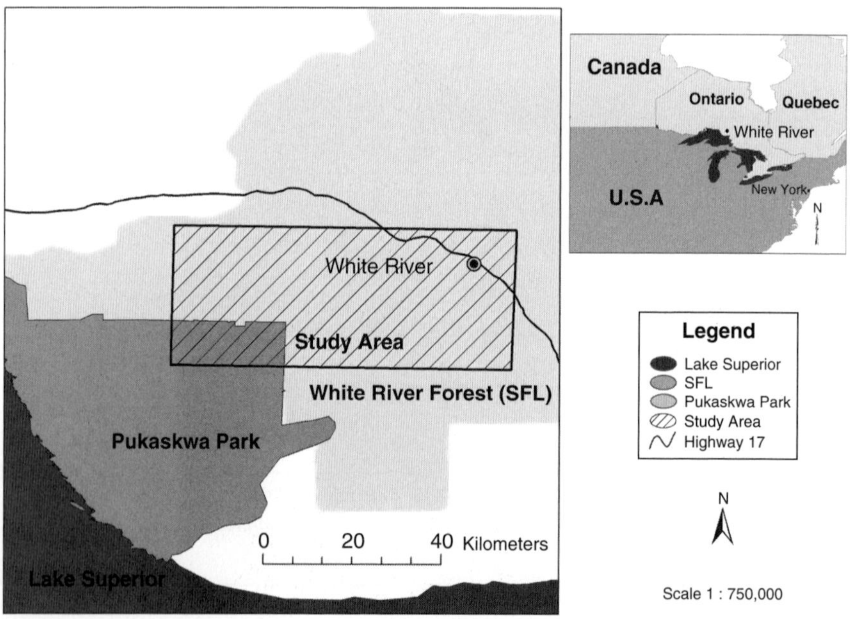

FIG. 18-2

Map of the study area in north central Ontario, Canada.

aspen (*Populus tremuloides* Michx.), balsam fir (*Abies balsamifera* [L.] Mill.), and white birch (*Betula papyrifera* Marsh.), with lesser amounts of white spruce (*Picea glauca* [Moench] A. Voss), eastern white cedar (*Thuja occidentalis* L.), and tamarack (*Larix laricina* Koch). Harvesting activities within the White River management area have concentrated on harvesting mature jack pine or jack pine mixed-wood stands. As of 1998, approximately 21% of the forested portion of the research area had been harvested and replanted principally to jack pine (although sometimes to black spruce, or natural regeneration).

Species

We chose three species for this exercise: the brown creeper (*Certhia americana*), the red-backed salamander (*Plethodon cinereus*), and the red-backed vole (*Clethrionomys gapperi*). We chose these species as potential indicators for three reasons. First, they are dependent on components of the forest that may be affected by forest management. Second, they have very different life histories, allowing us to explore how easily different aspects of a species' biology may be incorporated in DLMP models. Third, each species uses habitat at different scales.

The brown creeper is a monogamous, territorial species that is dependent on snags and old trees for nesting and foraging (Hejl et al. 2002). It is known to be sensitive to timber harvesting that degrades these old growth characteristics (Hobson and Schieck 1999). The red-backed salamander is highly sensitive to environmental change or disturbance due to a strong reliance on their environment for temperature and moisture regulation (Welsh and Droege 2001). They occupy small activity ranges, are quite long-lived, and breed biennially. The red-backed vole is dependent on old, moist forest sites with woody debris and is potentially sensitive to timber management practices that may alter understory conditions (Thompson et al. 2003). They undergo large population fluctuations, are short-lived, and are polygynous breeders.

Details on parameter estimates are provided in Pearce and Venier (2004*a*, 2005, in prep) and Wintle et al. (2005*a*), Venier and Pearce (2005, 2007), and Gordon et al. (in prep). All three species are considered relatively common in the region. Hence, we evaluate their relative abundance and changes in EMP under different management scenarios, rather than their risk of extinction.

Modeling Approach and Scenarios Evaluated

We developed an integrated DLMP model for the three species using RAMAS Landscape. We developed habitat models to describe the presence-absence (brown creeper) and abundance (red-backed salamander, red-backed vole) as a function of forest age, forest composition, microclimate, and elevation. Using these parameters, we developed metapopulation models for each species based on information from biologists and the literature. The succession model was based on Forest Resource Inventory (FRI) maps interpreted from aerial photographs, and tree life history parameters provided by Farrar (1995). The model for each species was used to rank the sustainability of eight forest management scenarios in terms of their impact on the abundance of each species during two timber rotations spread over 160 years (brown creeper) or 100 years (red-backed salamander and red-backed vole). The eight alternative forest management scenarios modeled ranged in intensity from "no timber harvesting and a natural fire regime" to "intensive timber harvesting with salvage logging after fire" (Table 18-1). For the brown creeper, fifty landscape realisations and three population replicates per landscape realisation were conducted for each scenario. Fifty landscape replicates and three population replicates were

Table 18-1 Details of the Four Forest Management Approaches Modeled. Two Fire Regimes were Modeled. The First was a Natural Fire Regime (Scenarios 1, 3, 5, and 7; Fire Size Return Time Parameters were Set to Match Current Estimates), and the Second Assumed Fire Suppression (Scenarios 2, 4, 6, and 8; Fire Return Times were Set to 320 Years for Jack Pine Dominated Forest, and 700 Years for Mixed Forests. The Fire Size Distribution was Set to a Mean of 8,000 ha, an Upper and Lower Limits of 10,000 and 6,000, Respectively) (From Venier et al. 2007)

Scenarios	Harvesting Regime
1 and 2	No timber harvesting
3 and 4	Harvesting according to Natural Disturbance Emulation guidelines (NDE; OMNR 2001). Under these guidelines, 20% of the harvested area in the region is allowed to regenerate naturally to mixed forest, with the remainder replanted to jack pine. Ten percent of the stands nominated for harvesting are retained in one-hectare blocks as wildlife habitat and are not harvested. Replanted areas remain as jack pine for the length of the simulation. The total area harvested is approximately 18,000 ha in each of two rotations. The first rotation starts at the beginning of the simulation. All 18,000 ha are harvested within the first 20 years (salamander and vole) or the first 40 years (creeper) of the simulation. Harvesting in the second rotation is completed between the 90th and 100th year (salamander and vole) and the 90th and 130th year (creeper) of the simulation. Other prescriptions within the NDE standards and guidelines (OMNR 2001) were not modeled due to a lack of data.
5 and 6	Similar to scenario 3, but involves an increase in the intensity of silviculture. The timing of harvesting events is the same as in scenario 3. All areas nominated for harvesting are clearcut and replanted to jack pine. All replanted areas remain as jack pine for the duration of the simulation.
7 and 8	Similar to scenario 4, though the total harvested area effectively increases, as areas burned by wild fire are then salvage logged. Harvested and burnt areas are replanted with jack pine. No salvage logging occurs in Pukaskwa National Park.

conducted for each scenario. This particular ratio of landscape replicates to population replicates was established using the results of an investigation into the relative contributions of landscape and population stochasticity on DLMP predictions. For the red-backed salamander and red-backed vole models we refined our analysis and developed software that iteratively calculated the number of population replicates per landscape realisation (the Repeater package [Chisholm and Wintle 2007]; see below). Full details of model development, management scenarios, and uncertainty analysis are provided elsewhere (Pearce and Venier 2004, Wintle et al. 2005*a*,*b*, Venier et al. 2007).

Sensitivity analysis was conducted to determine if the results of the model were sensitive to estimates of parameter values and other key assumptions. S_i, the change in EMP relative to Scenario 1, was used to compare the sustainability of the various management scenarios.

Key Findings

The results of this case study illustrate that, under the assumptions made in the models, the current style of forest management (most closely resembling scenario 4) is expected to result in a 9% to 25% decrease in the expected minimum population size of the species modeled over the next 100/160 years compared to the option of no timber harvesting (Table 18-2). The threat of local extinction is close to zero for all species under all scenarios. The differences between natural disturbance emulation (scenarios 3 and 4) and more intensive styles of logging (scenarios 5 and 6) were mixed, with the brown creeper showing greatest sensitivity to intensive logging (Table 18-2). Salvage logging led to at least a 15% increase in the area harvested and had a substantial impact on all species modeled (Tables 18-1 and 18-2). Fire was also an important variable for all species, with scenarios including large, infrequent fires increasing the risk of decline in many cases (Table 18-2). Each model incorporated our current knowledge of landscape succession, disturbance regimes, and indicator species biology.

As such, these models provided a synthesis of our current knowledge base and identified information needs, and allowed us to explore the impact of model uncertainties on predicted outcomes of forest management. Therefore, this approach provided a transparent and explicit statement of the predicted cost of management actions in terms of predicted population change, within stated bounds of certainty. The decisions about whether such costs are unacceptable are inevitably value-based, but this method provides a means to describe the risks more clearly. For example, model results indicate that under a fire suppression regime, the additional cost of salvage logging is between a further 1% to 11% decrease in expected minimum population size over the next 100/160 years (Table 18-2). Model results also help to guide future research. For example, all three species' models were sensitive to the specification of density dependence, highlighting that this parameter is a priority for future research, and variations in this parameter should be considered when comparing scenarios.

Table 18-2 Summary of the Population Decline and Carrying Capacity of All Study Species Due to Anthropogenic Disturbance. The Values Presented Represent the Percentage Decline in Expected Minimum Population Size Relative to Scenario 1 (No Anthropogenic Disturbance) and the Minimum Carrying Capacity as the Percentage of the Original Carrying Capacity

		Scenario							
		1	2	3	4	5	6	7	8
Brown creeper (*Certhia americana*)*	% Decline in EMP		8.51	15.20	24.50	23.27	23.99	21.24	31.64
	Minimum K	99.33	75.46	89.01	77.59	90.56	73.56	76.68	51.96
Red-backed salamander (*Plethodon cinereus*)*	% Decline in EMP		5.9	7.0	9.2	5.9	17.0	27.0	28.4
	Minimum K	100.00	100.00	92.3	90.6	93.5	95.7	72.4	73.7
Red-backed vole (*Clethrionomys gapperi*)#	% Decline in EMP		0.5	15.1	16.8	15.5	18.1	20.3	19.0
	Minimum K	96.4	95.2	90.7	88.5	90.2	88.9	86.3	83.6

**160 year simulation; #100 year simulation*

The relative insensitivity of the red-backed salamander to harvesting scenarios (Table 18-2) was unexpected, as terrestrial salamanders have been widely recommended as bioindicators (e.g., deMaynadier and Hunter 1995, Welsh and Droege 2001). Very few population parameters were sensitive to misspecification within this model, suggesting that sufficient connected habitat is available on the landscape to maintain salamander populations, irrespective of forest management actions. However, before accepting these findings, the habitat model needs to be validated. Key microhabitat features expected to be relevant to red-backed salamanders were not included in the habitat model. Although this model was based on data collected within the study area, and represented our best understanding of salamander distribution there, this habitat model had poor predictive performance, suggesting that it did not adequately capture environmental features important to salamanders. Many potentially important predictors were not available in mapped form; others that were available and considered important within other parts of the salamander's range were not significant within the study area. It is likely that including these key habitat attributes would increase the sensitivity of the model to timber harvesting, as refining the distribution of the salamander is expected to reduce the amount of habitat available.

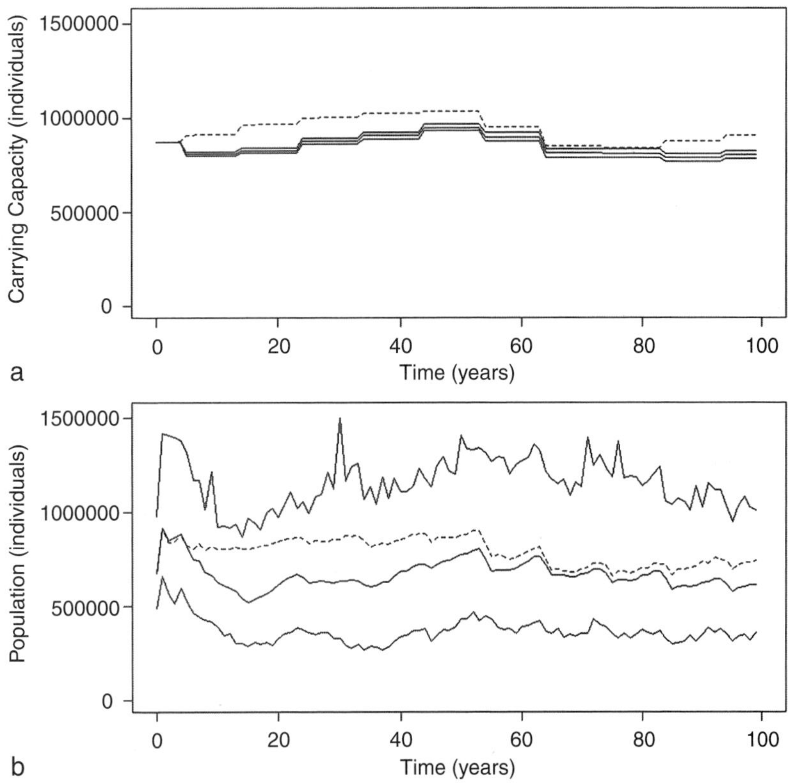

FIG. 18-3

(A) Carrying capacity and (B) population trajectory for the red-backed vole under scenario 3 (natural disturbance emulation with small, frequent fires). The middle line shows the mean carrying capacity or population size, while the upper and lower lines show one standard deviation from the mean. The dashed line shows the mean value of scenario 1.

Although the red-backed salamander model may be more sensitive to habitat availability than any other feature, DLMP models have the capacity to provide greater information on species decline than would be obtained from habitat supply models on their own (Akçakaya et al. 2003). For example, the population trajectory for the red-backed vole under scenario 3 (Fig. 18-3B) followed a substantially different pattern to the predicted habitat availability (expressed in terms of carrying capacity, Fig. 18-3A). This is most likely related to the tendency of populations of red-backed voles to fluctuate in response to disturbance (Fryxell et al. 1998). Short-term loss of habitat is followed by rapid recovery due to the high population growth rate of the species. In this case, predictions based purely on habitat would be optimistic, as including demographic considerations led to a greater estimated risk of decline. Other spatial factors, such as dispersal and connectivity, also affect habitat use and make the predictions of habitat

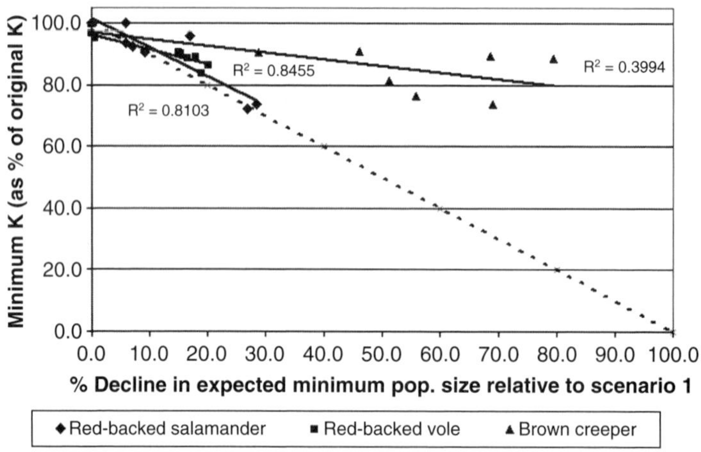

FIG. 18-4

Regressions of decline in expected minimum population size relative to scenario 1 and the minimum carrying capacity as a percentage of the original carrying capacity. Data are presented for three indicator species: the red-backed vole, brown creeper, and red-backed salamander. The dashed line indicates a perfect correlation for comparison.

supply models overly optimistic. For example, if habitat connecting two populations is harvested, it may have a greater impact than simply reducing the total extent of habitat.

Table 18-2 presents a summary of the predicted impact of the eight management scenarios on the three case study species in terms of habitat availability and EMP, and Fig. 18-4 presents the regression between these predictions. Regressions varied substantially among species, with the predictions of the population model for the red-backed salamander most closely following habitat supply and the predictions of the brown creeper least well correlated (Fig. 18-4). Importantly, the ranking of management scenarios based on habitat supply was considerably different for all species when compared with the ranking based on the population models. This result provides some evidence to suggest that habitat supply maps may not be adequate to describe the response of all species to alternative management scenarios.

How Feasible and Realistic Are DLMP Models?

The modeling methods were generally straightforward to apply, especially with the advent of the commercially supported software, RAMAS Landscape. DLMP models can be implemented, however, by manually linking outputs from independent landscape models and population models. The reliability of DLMP model predictions depends on the realism of landscape, habitat, and metapopulation models and assumptions. These, in turn, rely on the quality of available data and limits to understanding of ecological processes and stressors acting on indicator species populations. Model development therefore requires biologists and foresters to work together to parameterize the model and ensure realism within the bounds of data availability. For a similar study conducted in production forest in Tasmania, Australia, researchers developed software and found the programming and debugging elements debilitating for routine applications. With RAMAS Landscape, some data preparation is required outside the package (i.e., GIS data) although it is predominantly standalone and relatively straightforward to use.

The DLMP model adequately met our goal of considering the ecology of the whole forest system when assessing sustainability of forest management. We were able to build a succession model, represent the stochasticity in natural disturbance patterns, and consider indicator species persistence and viability within this system. Using this model, we were then able to consider the additional impact of forest management actions on population persistence, and rank management actions in terms of their impact on the indicator species. Although no forest management actions in the case studies predicted the local extinction of the indicator species, several forest management actions substantially reduced the population of the indicator species relative to a natural disturbance regime. For widespread and abundant indicator species, this is the type of reaction we expect. Thus indicator species demonstrate that the management action under consideration has a large negative impact on the forest system relative to natural disturbance, and is therefore potentially unsustainable. Decisions regarding whether this level of ecological unsustainability warrants a change in management is ultimately a value-based decision, taking into account other social, economic, and ecological values.

The greatest asset of this approach is that DLMP models allow us to incorporate environmental variability into the model, providing a distribution of predictions, rather than a single value. This is a significant advancement over existing sustainability assessment methods such as trend monitoring and habitat supply analysis. We were also able to consider uncertainty in our estimates of metapopulation model parameters, and thus identify areas of the model requiring further study. Parameterization of the metapopulation model is often done with "expert guesses" due to a lack of detailed life history information for most species. Sensitivity analysis allowed us to question and explore these model assumptions and measure how these estimates impacted model predictions.

Do DLMP Models Provide More Information Than Habitat Supply Models?

The relationship between species persistence and forest habitat structure and complexity is an important issue. If model results are simply an index of habitat availability, then this supports the use of habitat supply maps for forest management purposes, rather than a DLMP that includes demographic considerations. Demographic information is expected to be of most importance when habitat patches are isolated through habitat management, habitat becomes limiting through forest management, the indicator species is nonterritorial and polygamous, or when nonhabitat-related influences such as hunting or high periodic mortality rates from environmental stochasticity are important.

One method to assess the importance of habitat availability versus demographic information is to compare the population trajectory with the trajectory of habitat supply. If model results were simply an index of habitat availability, then the population trajectories would be expected to follow the pattern of change in carrying capacity, which is a function of habitat quality and quantity. As already noted, the trajectories for the red-backed vole differed markedly (Fig. 18-3). The lack of a strong correlation between the predictions based on habitat supply and population models for some species (Fig. 18-4) and the difference in ranking of management scenarios based on the two measures (Table 18-2) suggest that DLMP models do provide more information than habitat supply models.

While we argue that the inclusion of spatial metapopulation dynamics adds important elements to the interpretation to sustainability assessments for forest management, we accept that the information required to undertake these studies will only be readily available for a small number of species without substantial investment in data collection. Furthermore, the trade-off between realism and simplicity needs to be carefully examined with respect to the availability of data, as more complex models are not automatically more informative (Beissinger and Westphal 1998; Millspaugh et al., this volume). Attempts to include more details than can be justified by the quality of the available data may result in decreased predictive power (Ginzburg and Jensen 2004).

Limitations of the DLMP Software, RAMAS Landscape

Given the extent to which the habitat map determines model outcomes (including patch structure, population abundance, response to management scenarios, etc.), the results may be highly dependent on our ability to map habitat supply adequately. Our ability to model the habitat relationships of the case study species was unknown but may be low. This is primarily due to a lack of research within the study area defining habitat relationships, the paucity of mapped predictor variables, and the coarse resolution of many of the mapped variables that are available. A primary concern with the RAMAS Landscape package is the practical difficulty associated with incorporating sensitivity analysis on the spatial attributes, such as the habitat supply map, the succession model parameters (e.g., tree species establishment probabilities), and the natural disturbance model parameters (e.g., fire size and frequency). Currently, a full investigation of these forms of uncertainty would involve a long and tedious process of manually simulating stochasticity in such parameters.

This issue is a significant limitation in the current version of RAMAS Landscape. In particular, fire and succession regimes have a stochastic element, but these sources of prediction uncertainty are not linked with the representation of uncertainty in the metapopulation model. The model currently uses a single realization of the landscape for calculating species persistence, with multiple realizations of the species response to this single landscape examined to rank management scenarios. In our examples, fire regimes could not be held constant between scenarios because of the interaction between harvesting history and fire. For example, if an area was burned early in the simulation, it was unavailable for harvesting. Similarly, recently harvested areas were less likely to be burned by wildfire. While this is a realistic basis on which to model fire and harvesting, it limits the generality of results based on a single run of the landscape dynamics model. To overcome this we built our own software (the Repeater package, Chisholm and Wintle 2007) to automate the process of running the metapopulation model over multiple landscape realizations. This software enables the magnitudes of landscape- and demographic-induced variance in model outcomes to be separated, and iteratively calculates the optimal number of metapopulation realizations per landscape to minimize the combined landscape- and demographic-induced variance. This enabled assessment of the impact of the stochastic landscape simulation in our DLMP model. The Repeater software is freely available from <http://www.esapubs.org/archive/appl/A017/013/suppl-1.htm>.

RAMAS Landscape has a number of other limitations that impinge on its versatility in forest management settings. First, LANDIS was developed in the United States to be generalizable to a range of landscape settings. While the way in which the landscape is described meets this criterion of generality, it does not easily allow for incorporation of planning maps used in a specific area. For example, in northern Ontario, FRI maps, derived from interpretation of aerial photographs, are used for planning to describe the vegetation composition and structure of the landscape. Vegetation types are described in terms of the proportion of each tree species present. However, LANDIS cannot use this information. LANDIS describes vegetation types in terms of species presence or absence on each pixel, and dominance is assigned based on the relative age of the trees present. Although FRI information can be transformed into presence-absence form, significant information is lost. The realism of RAMAS Landscape would be enhanced by allowing base maps of vegetation type to be imported directly into LANDIS, and the vegetation types defined automatically based on these maps. Currently, all vegetation types must be specified manually, which is quite tedious.

The second problem we encountered was the inability to model more than 500 populations of the indicator species over the life of the simulation. This limitation meant that we needed to reduce the number of years that could be simulated or limit the spatial extent of the study. This assumption may be realistic for rare and endangered species for which RAMAS GIS was originally developed, but was not a realistic expectation for a DLMP model used to assess forest management over large areas using common and widespread species. Applied Biomathematics should address this concern if it is to continue promoting RAMAS Landscape for use in forest management.

CONCLUSIONS

The modeling approach we present in this chapter is proposed as a fundamental component of sustainable forest management. Dynamic landscape metapopulation models allow forest managers to explore aspects of the ecological sustainability of management actions before they take place. Management decisions are therefore made on the basis of anticipated impacts rather than as a reactive measure following environmental harm.

The DLMP model also helps to focus monitoring efforts by identifying important knowledge gaps. These gaps may be in terms of both species biology and ecosystem functioning. Construction of the DLMP model also highlights both environmental and model uncertainties, and incorporates them directly into the decision-making process. The DLMP model clarifies the causal linkages between management actions and indicator response.

The DLMP model may be used to assess the sustainability of forest management in cases where forest planning is done spatially or aspatially. It is most effective when used as part of an adaptive management system. Both models and strategic monitoring are used to iteratively design and evaluate forest management actions that minimize ecological harm, while maximizing social and economic gain from forest resources.

The models must be a component of an adaptive management system in which the results of monitoring are used iteratively to refine model parameters and predictions. The aim of using DLMP models is to stay a step ahead of environmental harm by prospectively assessing the sustainability of management options (Mulder et al. 1999). Data obtained by the monitoring system are used to improve models, which are in turn used to focus monitoring programs by describing the causal relationship between population processes and environmental stressors (Mulder et al. 1999).

SUMMARY

Sustainable forest management is a widely held international goal and in many cases a legislated mandate. Reliable, practical, and affordable means of assessing the sustainability of forest management remain elusive. Monitoring of biological indicators is an important element, but sufficiently powerful monitoring strategies are expensive and monitoring alone may not provide answers in time to avoid irreversible environmental or ecological damage.

We proposed a model-based approach to assessing sustainability using indicator species of ecosystem condition (as distinct from indicators of biodiversity or species richness) to provide timely feedback to managers about the sustainability of current and alternative forest management options, and to support the development of better-targeted and more relevant monitoring systems. Dynamic landscape metapopulation (DLMP) models integrate spatial models of forest change (also known as landscape dynamic models or forest succession models) with metapopulation models, which describe demographic and biological attributes of species and the dynamic consequences of dispersal and habitat change. We reviewed some of the benefits and criticisms of the indicator species approach and the advantages and problems associated with using DLMP models of indicator species to evaluate the sustainability of forest management options. We drew on results of a case study in northern Ontario, Canada, that utilized three indicator species to explore the sustainability of competing forest management scenarios. We compared those results with other recent studies undertaken in Australia and the United States that explored the utility of DLMP models in forest planning. Based on case study results, DLMP models of indicator species appear to be useful for assessing and ranking the sustainability of management options, quantifying the stresses placed on ecosystems by particular management activities, and targeting future research and data collection. Dynamic landscape metapopulation models have the potential to play an important role in assessments of sustainability, and we propose that such models should be considered a fundamental adaptive management tool. Such models will complement monitoring studies by providing a context for interpreting observed population fluctuations, identifying sensitive parameters and biologically important effect sizes, thereby supporting ecologically meaningful and cost-effective monitoring systems.

LITERATURE CITED

Abensperg-Traun, M., G. Arnold, D. Steven, G. Smith, L. Atkins, J. Viveen, and M. Gutter. 1997. Biodiversity indicators in contrasting vegetation types: A case study from western Australia. *Memoirs of the Museum of Victoria* 56:637–641.

Akçakaya, H. R. 2000. Population viability analyses with demographically and spatially structured models. *Ecological Bulletins* 48:23–38.

Akçakaya, H. R., M. A. McCarthy, and J. L. Pearce. 1995. Linking landscape data with population viability analysis: Management options for the helmeted honeyeater *Lichenostomus melanops cassidix*. *Biological Conservation* 73:169–176.

Akçakaya, H. R., D. J. Mladenoff, and H. S. He. 2003. *RAMAS Landscape: Integrating metapopulation viability with LANDIS forest dynamics model. User manual for version 1.0*. Applied Biomathematics, Setauket, New York, USA.

Akçakaya, H. R., V. C. Radeloff, D. R. Mladenoff, and H. S. He. 2004. Integrating landscape and metapopulation modelling approaches: Viability of the sharp-tailed grouse in a dynamic landscape. *Conservation Biology* 18:526–537.

Akçakaya, H. R., and W. Root. 2002. *RAMAS GIS: Linking spatial data with population viability analysis (version 4.0)*. Applied Biomathematics, Setauket, New York, USA.

Andelman, S. J., and W. F. Fagan. 2000. Umbrellas and flagships: Efficient conservation surrogates or expensive mistakes? *Proceedings of the National Academy of Sciences* 97:5954–5959.

Andren, H. 1994. Effects of habitat fragmentation on birds and mammals in landscapes with different proportions of suitable habitats: A review. *Oikos* 71:355–366.

Beissinger, S. R., and M. I. Westphal. 1998. On the use of demographic models of population viability in endangered species management. *Journal of Wildlife Management* 62:821–841.

Bekessy, S. A., J. Banks, M. A. McCarthy, J. Dorrough, K. Lowe, and B. A. Wintle. 2008. Tools for biodiversity conservation planning in rural landscapes. *Journal of Environmental Management: In press.*

Boyce, M. S. 1992. Population viability analysis. *Annual Review of Ecology and Systematics* 23:481–506.

Buckland, S. T., and D. A. Elston. 1993. Empirical models for the spatial distribution of wildlife. *Journal of Applied Ecology* 30:478–495.

Burgman, M. A., D. R. Breininger, B. W. Duncan, and S. Ferson. 2001. Setting reliability bounds on habitat suitability indices. *Ecological Applications* 11:70–78.

Burgman, M. A., S. Ferson, and H. R. Akçakaya. 1993. *Risk assessment in conservation biology.* Chapman and Hall, London, United Kingdom.

Burgman, M. A., and D. L. Lindenmayer. 1998. *Conservation biology for the Australian environment.* Surrey Beatty and Sons, Chipping Norton, Sydney, Australia.

Burgman, M. A., and H. Possingham. 2000. Population viability analysis for conservation: The good, the bad and the undescribed. Pages 98–112 *in* A. G. Young and G. M. Clarke, editors. *Genetics, demography and viability of fragmented populations.* Cambridge University Press, Cambridge, United Kingdom.

Chisholm, R. A., and B. A. Wintle. 2007. Landscape stochasticity in population viability analysis. *Ecological Applications* 17:317–322.

Commonwealth of Australia. 2001. Regional Forest Agreements Bill. <http://www.aph.gov.au/library/pubs/bd/2001-02/02bd061.htm>. Accessed January 2004.

deMaynadier, P. G., and M. L. Hunter, Jr., 1995. The relationship between forest management and amphibian ecology: A review of the North American literature. *Environmental Review* 3:230–261.

DesRochers, A., and S. J. Hannon. 1997. Gap crossing decisions by forest songbirds during the post-fledging period. *Conservation Biology* 11:1204–1210.

Elzinger, C. L., D. W. Salzer, J. W. Willoughby, and J. P. Gibbs. 2001. *Monitoring plant and animal populations.* Blackwell Science, Massachusetts.

Farrar, J. L. 1995. *Trees in Canada.* Canadian Forest Service, Saint-Laurent, Quebec, Canada.

Fieberg, J., and S. P. Ellner. 2000. When is it meaningful to estimate an extinction probability? *Ecology* 81:2040–2047.

Flather, C. H., K. R. Wilson, D. J. Dean, and W. C. McComb. 1997. Identifying gaps in conservation networks: Of indicators and uncertainty in geographic-based analyses. *Ecological Applications* 7:531–542.

Ford, W. M., S. L. Stephenson, J. M. Menzel, D. R. Black, and J. W. Edwards. 2004. Habitat characteristics of the endangered Virginia northern flying squirrel (*Glaucomys sabrinus fuscus*) in the central Appalachian Mountains. *American Midland Naturalist* 152:430–438.

Fryxell, J. M., J. B. Falls, E. A. Falls, and R. J. Brooks. 1998. Long-term dynamics of small-mammal populations in Ontario. *Ecology* 79:213–225.

Gaston, K. J., and T. M. Blackburn. 1995. Birds, body size and the threat of extinction. *Philosophical Transactions of the Royal Society of London Series B: Biological Sciences* 347:205–212.

Gilpin, M. E., and M. E. Soulé. 1986. Minimum viable populations: Processes of species extinctions. Pages 19–34 *in* M. E. Soulé, editor. *Conservation biology: The science of scarcity and diversity.* Sinauer Associates, Sunderland, Massachusetts, USA.

Ginzburg, L. R., and C. X. J. Jensen. 2004. Rules of thumb for judging ecological theories. *Trends in Ecology and Evolution* 19:121–126.

Green, R. E., and G. J. M. Hirons. 1991. The relevance of population studies to the conservation of threatened birds. *In* C. M. Perrins, J. D. Leberton, and G. J. M. Hirons, editors. *Bird population studies: Their relevance to conservation and management.* Oxford University Press, Oxford, United Kingdom.

Gustafson, E. J., N. L. Murphy, and T. R. Crow. 2001. Using a GIS model to assess terrestrial salamander response to alternative forest management plans. *Journal of Environmental Management* 63:281-292.

Hanksi, I. 1998. Metapopulation dynamics. *Nature* 396:41-49.

Hastie, T., R. Tibshirani, and J. Friedman. 2001. *The elements of statistical learning: Data mining, inference and prediction*. Springer, New York, New York, USA.

He, H. S., D. J. Mladenoff, and J. Boeder. 1996. *LANDIS, a spatially explicit model of forest landscape disturbance, management and succession. LANDIS 2.0 user guide*. University of Wisconsin, Madison, USA.

Hejl, S. J., K. R. Newlon, M. E. McFadzen, J. S. Young, and C. K. Ghalambor. 2002. Brown Creeper (*Cethis americana*). Account 669 in A. Poole and F. Gill. *The birds of North America.* The Academy of Natural Sciences, Philadelphia, Pennsylvania, and The American Ornithologists' Union, Washington, D. C., USA.

Hobson, K. A., and J. Schieck. 1999. Changes in bird communities in boreal mixedwood forest: Harvest and wildfire effects over 30 years. *Ecological Applications* 9:849-863.

Holt, R. D., G. R. Robinson, and M. S. Gaines. 1995. Vegetation dynamics in an experimentally fragmented landscape. *Ecology* 76:1610-1624.

Johnson, B. L. 1999. Introduction to the special feature: Adaptive management—Scientifically sound, socially challenged? *Conservation Ecology* 3:10.

Kanowski, P., D. Sinclair, and P. Freeman. 1999. *International approaches to forest management certification and labelling of forest products: A review.* Canberra, Agriculture, Fisheries and Forestry, Australia.

Landres, P. B., J. Verner, and J. W. Thomas. 1988. Ecological uses of vertebrate indicator species: A critique. *Conservation Biology* 2:316-327.

Larson, M. A., F. R. Thompson, III, J. J. Millspaugh, W. D. Dijak, and S. R. Shifley. 2004. Linking population viability, habitat suitability, and landscape simulation models for conservation planning. *Ecological Modelling* 180:103-118.

Lindenmayer, D. B., M. A. Burgman, H. R. Akçakaya, R. C. Lacy, and H. P. Possingham. 1995. A review of the generic computer programs ALEX, RAMAS/space and VORTEX for modelling the viability of wildlife metapopulations. *Ecological Modelling* 82:161-174.

Lindenmayer, D. B., A. D. Manning, P. L. Smith, H. P. Possingham, J. Fischer, I. Oliver, and M. A. McCarthy. 2002. The focal-species approach and landscape restoration: A critique. *Conservation Biology* 16:338-345.

Lindenmayer, D. B., C. R. Margules, and D. B. Botkin. 2000. Indicators of biodiversity for ecologically sustainable forest management. *Conservation Biology* 14:941-950.

Lindenmayer, D. W., and H. Possingham. 1996. Ranking conservation and timber management options for Leadbeater's Possum in southeastern Australia using population viability analysis. *Conservation Biology* 10:235-251.

McCarthy, M. A. 1996. Extinction dynamics of the helmeted honeyeater: Effects of demography, stochasticity, inbreeding and spatial structure. *Ecological Modelling* 85:151-163.

McCarthy, M. A., S. A. Andelman, and H. P. Possingham. 2003. Reliability of relative predictions in population viability analysis. *Conservation Biology* 17:982-989.

McCarthy, M. A., K. M. Parris, R. van der Ree, M. J. McDonnell, M. A. Burgman, N. S. G. Williams, N. McLean, M. Harper, R. Meyer, A. Hahs, and T. Coates. 2004. The habitat hectares approach to vegetation assessment: An evaluation and suggestions for improvement. *Ecological Management and Restoration* 5:21-27.

McCarthy, M. A., and C. Thompson. 2001. Expected minimum population size as a measure of threat. *Animal Conservation* 4:351-355.

McGarigal, K., and W. C. McComb. 1995. Relationships between landscape structure and breeding birds in the Oregon Coast Range. *Ecological Monographs* 65:235-260.

McRae, D. J., L. C. Duchesne, B. Freedman, T. J. Lynham, and S. Woodley. 2001. Comparisons between wildfire and forest harvesting and their implications in forest management. *Environmental Reviews* 9:223-260.

Meffe, G. K., and C. R. Carroll. 1994. *Principles of conservation biology.* Sinauer Associates, Sunderland, Massachusetts, USA.

Menzel, J. M., W. M. Ford, J. W. Edwards, and L. J. Ceperley. 2006. *A habitat model for the Virginia northern flying squirrel* (Glaucomys sabrinus fuscus) *in the central Appalachian Mountains.* U. S. Forest Service, Northeastern Research Station, Research Paper NE-729, Newton Square, Pennsylvania, USA.

Milledge, D. R., C. L. Palmer, and J. L. Nelson. 1991. Barometers of change: The distribution of large owls and gliders in Mountain Ash forests of the Victorian central highlands and their potential as management indicators. Pages 55-65 *in* D. Lunney, editor. *Conservation of Australia's forest fauna.* Royal Zoological Society of New South Wales, Sydney, Australia.

Mladenoff, D. J., and W. L. Baker. 1999. *Advances in spatial modeling of forest landscape change: Approaches and applications.* Cambridge University Press, Cambridge, United Kingdom.

Mladenoff, D. J., and H. S. He. 1999. Design and behavior of LANDIS, an object-oriented model of forest landscape disturbance and succession. Pages 163-185 *in* D. J. Mladenoff and W. L. Baker, editors. *Advances in spatial modeling of forest landscape change: Approaches and applications.* Cambridge University Press, Cambridge, United Kingdom.

Morrison, M. L., M. F. Dedon, M. G. Raphael, and M. P. Yoder-Williams. 1986. Snag requirements of cavity-nesting birds: Are USDA Forest Service guidelines being met? *Western Journal of Applied Forestry* 1:38-40.

Mulder, S., D. Evans, T. Hamer, P. Henson, S. Miller, S. K. Nelson, D. Roby, and M. Stapanian. 1999. *Marbled murrelet effectiveness monitoring plan for the Northwest Forest Plan.* U.S. Forest Service, Pacific Northwest Research Station, General Technical Report PNW-GTR-439, Portland, Oregon, USA.

Noss, R. F. 1990. Indicators for monitoring biodiversity: A hierarchical approach. *Conservation Biology* 4:355-364.

Ontario Ministry of Natural Resources. 2001. *Forest management guide for natural disturbance pattern emulation. Version 3.1.* OMNR, Queens's printer for Ontario, Toronto, Canada.

Pearce, J. L., and L. A. Venier. An operational methodology for measuring and analyzing bio-indicators to support sustainable forest management. *Final Report to Living Legacy Trust: Project Number*: 07-004, *in press.*

Pereira, J. M. C., and R. M. Itami. 1991. GIS-based habitat modeling using logistic multiple regression: A study of the Mt. Graham red squirrel. *Photogrammetric Engineering and Remote Sensing* 57:1475-1486.

Petranka, J. W., M. P. Brannon, M. E. Hopey, and C. K. Smith. 1994. Effects of timber harvesting on low elevation populations of southern Appalachian salamanders. *Forest Ecology and Management* 67:135-147.

Poiani, K. A., B. D. Richter, M. G. Anderson, and H. E. Richter. 2000. Biodiversity conservation at multiple scales: Functional sites, landscapes and networks. *Bioscience* 50:133-146.

Ralph, C. J., S. Droege, and J. R. Sauer. 1995. *Monitoring bird populations with point counts.* U.S. Forest Service, Pacific Southwest Research Station, General Technical Report PSW-GTR-149, Albany, California, USA.

Rand, G. M., and J. J. Newman. 1998. The applicability of habitat evaluation methodologies in ecological risk assessment. *Human and Ecological Risk Assessment* 4:905-929.

Roberge, J., and P. Angelstam. 2004. Usefulness of the umbrella species concept as a conservation tool. *Conservation Biology* 18:76-85.

Schmiegelow, F. K. A., C. S. Machtans, and S. J. Hannon. 1997. Are boreal birds resilient to forest fragmentation? An experimental study of short-term community responses. *Ecology* 78:1914–1932.

Shaffer, M. L. 1990. Population viability analysis. *Conservation Biology* 4:39–40.

Shifley, S. R., F. R. Thompson, III, W. D. Dijak, M. A. Larson, and J. J. Millspaugh. 2006. Simulated effects of forest management alternatives on landscape structure and habitat suitability in the Midwestern United States. *Forest Ecology and Management* 229:361–377.

Smith, W. P., S. M. Gende, and V. Nichols. 2005. The northern flying squirrel as an indicator species of temperate rain forest: Test of an hypothesis. *Ecological Applications* 15:689–700.

Smith, W. P., J. V. Nichols, and S. M. Gende. 2004. Ecological correlates of flying squirrel microhabitat use and density in temperate rain forests of southeastern Alaska. *Journal of Mammalogy* 85:663–674.

Smith, W. P., and D. K. Person. Estimated persistence of northern flying squirrel populations in old-growth rain forest fragments. *Biological Conservation: In press.*

Smith, W. P., and P. A. Zollner. 2005. Sustainable management of wildlife habitat and risk of extinction. *Biological Conservation* 125:287–295.

Spellerberg, I. F. 1994. *Monitoring ecological change*. Second edition. Cambridge University Press, Cambridge, United Kingdom.

Taylor, B. L. 1995. The reliability of using population viability analysis for risk classification of species. *Conservation Biology* 16:551–558.

Temple, S. A., and J. A. Wiens. 1989. Bird populations and environmental changes: Can birds be bioindicators? *American Birds* 43:260–270.

Terborgh, J. 1986. Keystone plant resources in the tropical forest. Pages 330–344 *in* M. E. Soulé, editor. *Conservation biology: The science of scarcity and diversity.* Sinauer Associates, Sunderland, Massachusetts, USA.

Thompson, I. D., J. A. Baker, and M. Ter-Mikaelian. 2003. A review of the long-term effects of post-harvest silviculture on vertebrate wildlife, and predictive models, with an emphasis on boreal forests in Ontario, Canada. *Forest Ecology and Management* 177:441–469.

Venier, L. A., and J. L. Pearce. 2005. Boreal bird community response to jack pine forest succession. *Forest Ecology and Management* 217:19–36.

Venier, L. A., and J. L. Pearce. 2007. Boreal forest landbirds in relation to forest composition, structure, and landscape: implications for forest management. *Canadian Journal of Forest Research* 37:1214–1226.

Venier, L. A., J. L. Pearce, B. A. Wintle, and S. A. Bekessy. 2007. Future forests and indicator-species population models. *Forestry Chronicle* 83:36–40.

Walters, C., and C. S. Holling. 1990. Large-scale management experiments and learning by doing. *Ecology* 71:2060–2068.

Welsh, H. H., and S. Droege. 2001. A case for using plethodontid salamanders (Family Plethodontidae) for monitoring biodiversity and ecosystem integrity on North American forestlands. *Conservation Biology* 15:558–569.

Wintle, B. A., S. A. Bekessy, L. A. Venier, J. L. Pearce, and R. A. Chisholm. 2005*a*. Utility of dynamic-landscape metapopulation models for sustainable forest management. *Conservation Biology* 19:1930–1943.

Wintle, B. A., R. P. Kavanagh, M. A. McCarthy, and M. A. Burgman. 2005*b*. The magnitude and management consequences of false negative observation error in surveys of arboreal marsupials and large forest owls. *Journal of Wildlife Management* 69:905–917.

Yahner, R. H. 1992. Dynamics of a small mammal community in a fragmented forest. *American Midland Naturalist* 127:381–391.

CHAPTER

Habitat Networks for Terrestrial Wildlife: Concepts and Case Studies

19

Mary M. Rowland and Michael J. Wisdom

Species of conservation concern, which we define as species with rare or declining populations or habitats, often number in the hundreds or even thousands within a given ecosystem. Moreover, these species typically span a wide spectrum of taxa and are associated with a broad set of ecological characteristics and diverse management challenges. Management designed to fully meet the needs of large numbers of species is by definition impossible: Each species occupies its own niche, and explicitly addressing each of these multidimensional niches would far exceed resources available to managers (Noss and Cooperrider 1994). The management challenge is thus how the many dimensions of multispecies requirements can be reduced to a workable number for practical management application and yet be sufficiently robust to represent the broad, ecological needs of the comprehensive set of species that management must address based on current policies and regulations.

Further complicating this management challenge is the need to address species' requirements in space and time. These requirements vary by activity, season, and life history, and proper arrangement of resources to fulfill these needs within a space compatible with daily and seasonal movements is essential. Moreover, maintenance of desired conditions over time is challenged by pervasive disturbances such as wildfire, exotic species invasions, and human impacts, many of which interact synergistically in ways unpredictable and little understood.

One modeling approach that addresses the spatial and temporal requirements of single or multiple species is the use of habitat networks. We define a habitat network as a spatially explicit portrayal of environmental conditions across large landscapes that can be used to understand the status and trends of species of conservation concern, particularly in relation to how species' needs are met through management of habitat abundance and distribution. Habitat networks are specifically designed to account for and summarize spatial information across landscapes compatible in size and arrangement with the targeted species' activities and movements (Hobbs 2002).

Various alternative definitions have been used for habitat networks, resulting in contrasting applications and interpretations. For example, habitat networks have been defined as "core areas connected by corridors and shielded by buffer zones" (referred to as "ecological networks" by Bani et al. [2002]), "habitat corridors and stepping stones to maintain genetic connectivity between populations" (von Haaren and Reich 2006), "nodes associated with hospitable habitat patches, and links, associated with corridors, for spatial connectivity to support viable metapopulations" (Nikolakaki and Dunnett 2005), and "an interconnected set of habitat elements that together allow for movement of biota and enhance survival probabilities" (Hobbs 2002).

Opdam (2002) defined habitat networks based on the "functional cohesion" among habitat patches in relation to dispersal and other movements, rather than the physical connectedness of patches. Schulte et al. (2006) grouped networks with patchworks and gradients as one class of conservation concepts, "landscape configuration," and described the interrelationships between networks and other theories related to biodiversity conservation. Most definitions of habitat networks, including ours, share two key characteristics: (1) identification of suitable habitat patches and connections among habitat blocks at a scale compatible with species' movements; and (2) evaluation of the entire landscape in relation to meeting species' needs, rather than a limited subset of landscapes such as bioreserves (Haufler 1999).

Habitat networks provide several potential benefits, including (1) conditions for large numbers of species of conservation concern can be addressed efficiently across space and time; (2) a wide variety of habitat characteristics can be holistically integrated; and (3) ecological characterizations provided as part of the network do not dictate a particular form of management, but rather provide the basis for development of a variety of follow-up on management strategies and options.

With these benefits in mind, in this chapter we describe two case examples of habitat networks in conservation planning. Our objectives are to (1) describe the conceptual basis of habitat networks; (2) illustrate practical methods for characterizing habitat networks for species of conservation concern; (3) discuss how network analyses can be interpreted for management; and (4) identify additional knowledge needed for the improved use of networks.

CONCEPTS OF HABITAT NETWORKS

The conceptual basis for habitat networks stems primarily from conservation theories of island biogeography (MacArthur and Wilson 1967) and metapopulation dynamics (Levins 1969, Hanski and Gilpin 1991), which are the foundations of conservation biology (Noss 1983, Noss and Harris 1986) and landscape ecology (Forman and Godron 1986). As applied to species management, these disciplines share the central tenet of seeking to understand the spatial structure of habitats and its influence on population dynamics. Knowledge of this spatial structure is essential for understanding, and managing for, population persistence. A spatial structure composed of large, relatively unfragmented, and well-connected habitats increases the probability of persistence. Small, fragmented, and isolated habitats decrease that probability.

While such generalizations are logical, understanding the landscape context of habitat—how habitat abundance, patch size, quality, configuration, and connectivity affect persistence of individual species in time and space—is one of the most complicated and challenging aspects of species- and community-level research and management (Hobbs 2002, Bennett 2003). In essence, understanding these spatial characteristics of habitat and their effects on populations is the foundation for habitat networks and their effective application in management (Opdam 2002). Further complicating this challenge is the dynamic nature of habitats, which can change dramatically over time in response to a variety of disturbance regimes.

Although many conceptual and theoretical approaches to habitat networks have been developed (see Hobbs [2002] and Opdam [2002] for review), published examples of practical or "operational" management applications are limited (Hobbs 2002, Schulte et al. 2006). Nonetheless, habitat networks and related conservation concepts (e.g., emphasis areas, patchworks, coarse-filter strategies) have been widely proposed for conservation planning and management (Noss and Cooperrider 1994, Haufler 1999, Hobbs 2002, Opdam 2002, Schulte et al. 2006). Regardless of the specific approach, information considered in designing habitat networks typically includes estimates of abundance, quality, configuration, and connectivity of habitats. (We adhere to the definition of habitat by Hall et al. [1997:3] as "the resources and conditions present in an area that produce occupancy—including survival and reproduction—by a given organism.")

A fundamental premise of habitat networks is that habitat either is naturally fragmented or has become fragmented, and thus some configuration of habitat patches and linkages is necessary to support populations of the species of interest (Vos et al. 2001, Opdam 2002). In a habitat network, contiguous blocks of habitat are defined as habitat patches or core areas and are surrounded by a matrix of nonhabitat or less suitable habitat (e.g., Opdam 2002, Nikolakaki and Dunnett 2005). Linkages or corridors that connect patches also may be explicitly identified. For example, Bani et al. (2002) identified and mapped corridors for avian and carnivore focal species in woodland habitats in a densely populated area in northern Italy by developing an index of "matrix resistance." The lines of lowest resistance represented linkages between core areas of habitat and were located in paths of 30 $\times$ 30 m cells of the "best available land cover" (Bani et al. 2002).

In our case studies, we further emphasized evaluation of resistance and resiliency of habitats—that is, the degree to which habitats can resist or recover from disturbance. Estimates of population size for local populations and corresponding metapopulations are sometimes considered in network design when spatially explicit demographic, movement, or dispersal data are available (e.g., Bani et al. 2002, Opdam 2002, Gutiérrez 2005, Nikolakaki and Dunnett 2005). However, such data are unavailable for most species of conservation concern (Baguette and Van Dyck 2007), making the network characterization process largely habitat driven and based on more general, but incomplete, knowledge of how species respond to the spatial structure of habitat. Consequently, the challenge is how best to incorporate this incomplete knowledge in designing networks that support persistent populations.

Steps in Characterizing Habitat Networks

Although each habitat network portrays a unique characterization of environmental conditions, based on objectives of the network and targeted taxa, a basic sequence of steps is applicable in developing most networks (Fig. 19-1; see also Opdam [2002]: Fig. 21.3). We present these steps sequentially; however, some may be undertaken simultaneously or in different order (e.g., species selection, determination of spatial scale). The most critical step in designing a habitat network is the first: developing a well-defined set of objectives or conservation aims (Opdam 2002). Conservation of biodiversity in the planning area is a common network objective (e.g., Opdam 2002, Schulte et al. 2006), but more focused objectives may include protecting particular rare or sensitive species within the planning area (Wiersma and Urban 2005), or identifying blocks of contiguous habitat that are suitable for restoration for species groups (Wisdom et al. 2005*b*).

Depending on the objectives of the network, it may be developed for individual species (Baguette et al. 2000, Nikolakaki and Dunnett 2005), surrogate species, or groups of species with similar environmental requirements or responses to habitat change (Vos et al. 2001; Bani et al. 2002; Wisdom et al. 2002, 2005*b*). If surrogate species or species groups are used to represent the needs of a larger suite of species in a network, a rigorous, peer-reviewed process is needed to establish the surrogates or groups that are assumed to represent the full set of species for which the network is targeted. This process of selecting and using surrogate species or groups of species has been described conceptually and operationally by Wiens et al. (2008). The case example used by Wiens et al. (2008) to illustrate this process drew in part on the data sources and research from the Interior Columbia Basin (Wisdom et al. 2000) that form the basis for our first case example (Wisdom et al. 2002), described later.

Selection of the spatial scale and extent of the network also is important (Fig. 19-1). Ideally, this choice will be dictated by the life history and distribution of the targeted species in the planning area, but in reality the spatial scale of the network is often driven by the resolution and affordability of available spatial data layers. Some trade-offs are necessary, as the selected scale must not only be appropriate in terms of species' ecology, but also match administrative scales used in conservation planning and management.

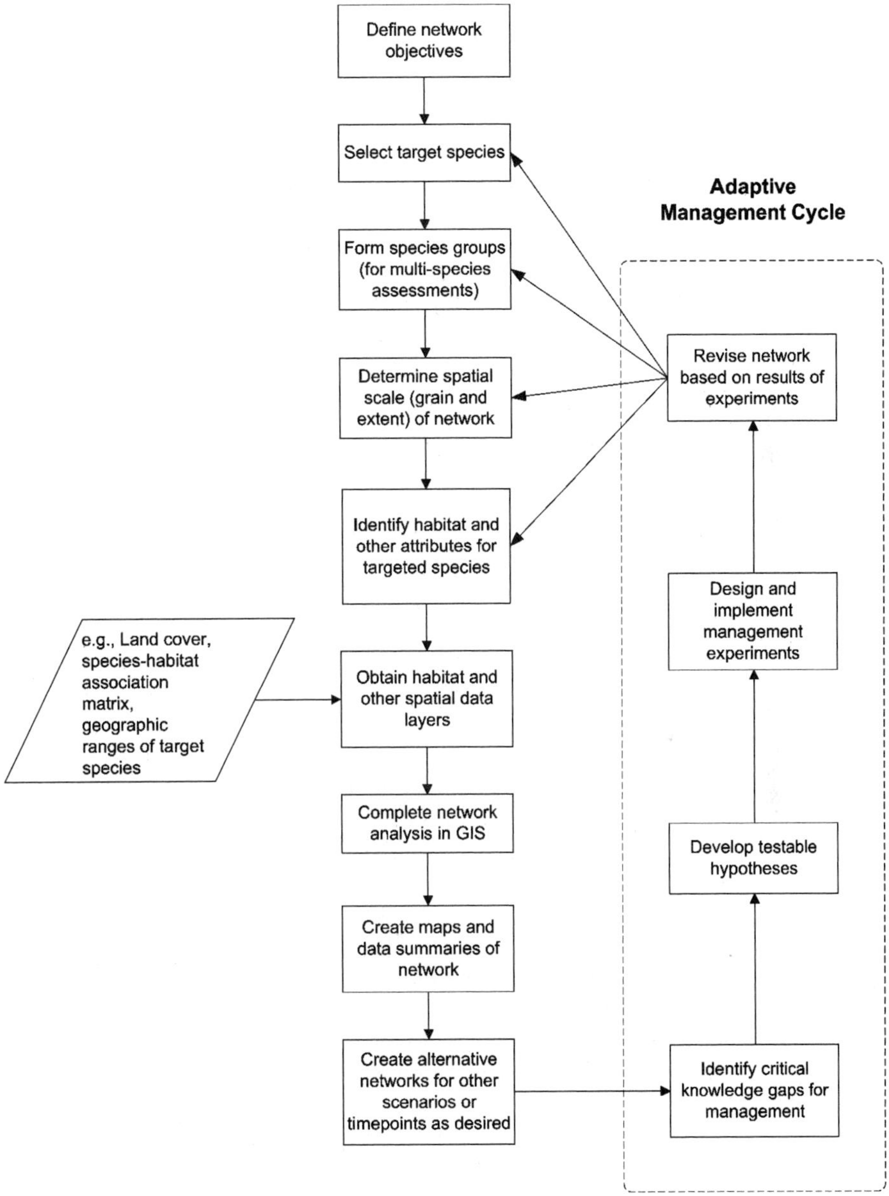

FIG. 19-1

Steps in development of a habitat network, including an adaptive management cycle to integrate results of follow-on research in network design.

Classification of habitat for species in the network can be based on a variety of sources, such as existing species-habitat matrices (e.g., Mayer and Laudenslayer 1988), literature review, or expert opinion (Beck and Suring, this volume; Larson et al., this volume). Alternatively, species-habitat associations can be developed explicitly for the network through field studies. For example, León-Cortés et al. (2004) determined habitat associations for a unique butterfly species, *Baronia brevicornis*, in southern Mexico by walking >1,300 transects while developing a habitat network for conservation of this species. Regardless of origin, the specificity of habitat as defined for the network will strongly influence measures of habitat abundance and connectivity (Hobbs 2002). Information beyond habitat may be included in the network, such as key ecological processes that affect target species, effects of human disturbance, or population density.

Habitat Networks as Wildlife Habitat Models

Habitat networks have been variously defined, but all definitions support the concept of habitat networks as models: abstractions or simplifications of the real world (Nichols 2001). We can never completely identify or accurately measure the comprehensive suite of environmental conditions that constitute habitat, or habitat linkages, for a species. We can, however, with varying levels of certainty, measure and map habitat components that are consistently associated with population status or trends for targeted species of concern, such as amount of interior old-growth forest for northern spotted owls (*Strix occidentalis caurina;* Franklin et al. 2000; Hicks et al., this volume). For many species, especially birds and mammals, these components typically include vegetation structure and composition and the quantity, quality, and configuration of these in the landscape, which can be spatially depicted in a geographic information system (GIS).

A prerequisite for assessing the utility of any model is a clear statement of the model's objectives (Millspaugh et al., this volume). In the creation of habitat networks, very different model structures and inputs may be realized, depending on network objectives. For example, consider two contrasting objectives: conservation of all native biota within a defined landscape versus habitat restoration for a particular species group. In the first example, habitat patches for the network would be selected from a broad cross-section of ecosystem conditions to encompass the greatest biodiversity. Model inputs might include measures of species richness, land ownership, and land cover. By contrast, in the second example existing and potential habitat for species in the group would be mapped, and areas with high restoration potential would be emphasized. Model inputs in this case might include population and habitat distribution data for species in the group, estimated restoration potential, and risk of habitat loss. Careful consideration of the resulting habitat network and its utility in meeting its prescribed objectives is imperative: "What are my conservation objectives? Will the habitat network as designed help meet them?"

Habitat networks might not be a pragmatic or effective tool in meeting all conservation objectives, such as habitat restoration for very rare species with limited known distributions, in which case all habitat patches would be identified and targeted for maintenance or restoration, regardless of their spatial relationship. Alternatively, species for which habitat is poorly defined, especially species with no known strong alliance with vegetation composition or structure, might not be suitable candidates for development of a habitat network, especially at landscape levels. In general, however, species that occupy large landscapes and for which spatial population structure and distribution of habitats are important will likely benefit from a network approach.

Spatial and Temporal Basis of Habitat Networks

Habitat networks exemplify spatial relationships in wildlife ecology; habitat patches are not only defined and located, but also mapped in relation to each other. Mapping habitat networks in a GIS thus allows for "spatial depictions of theoretical constructs," such as core habitat and linkages (O'Neil et al. 2005:418). Habitat networks are most appropriately applied across large landscapes, such as multiple watersheds or subbasins, or even ecoregions, for two reasons. First, these large spatial extents typically encompass the seasonal or year-round ranges of individuals or populations of many wide-ranging species. Second, the data layers commonly available to construct networks often lack the resolution to accurately depict fine-scale habitat features (Opdam 2002). Thus, habitat networks are typically characterized by coarse-scale features (e.g., canopy cover of dominant vegetation or topographically derived variables), rather than fine-scale features (e.g., site-specific forage resources or seeps, springs, and caves).

Another consideration in development of a habitat network in GIS is data type (Roloff et al., this volume). Ideally, the network should be developed from primary base data layers (e.g., tree density by size class), rather than derived or interpreted attributes, such as existing vegetation classes (O'Neil et al. 2005). Thus, if habitat is redefined for some targeted species through the development of new habitat relationship models, the base layers may still be used to map habitat in the new network without re-creating the entire system.

Habitat networks are typically developed to represent current environmental conditions (e.g., Baguette et al. 2000, Bani et al. 2002). However, networks can also be used to project future conditions or conditions under alternative management scenarios (Verboom et al. 2001, Opdam 2002). For example, a habitat network was designed for red deer (*Cervus elaphus*) in northwestern Europe that identified areas not currently occupied, but that could support viable populations in the future (Bruinderink et al. 2003). Effects of climate change on future spatial patterns of habitat and metapopulations will require dynamic network models that portray a range of potential outcomes (Opdam and Wascher 2004). Alternatively, a habitat network can reflect changes from historical to current conditions (e.g., Wisdom et al. [2002] and "Case Studies" below). Ultimately, the objectives of the network will dictate its spatial and temporal scale.

CASE STUDIES

We present two examples of habitat networks. Both were developed for use in broad-scale land management and conservation planning in the western United States, and evaluated habitat conditions for groups of terrestrial vertebrates of conservation concern across multiple land ownerships and state boundaries. All vertebrates selected for analysis were wide-ranging and not reliant on fine-scale habitat features (e.g., riparian corridors), and thus were suitable for assessment across large landscapes. In the first example, habitat networks in the Interior Columbia Basin were characterized by measures of habitat abundance coupled with measures of habitat resiliency and quality. In the second, composite habitat conditions in the Great Basin were based on estimates of habitat abundance and risk of habitat loss.

Many other landscape and ecological characteristics beyond measures of habitat abundance and quality can be incorporated in habitat networks, including dispersal rates, predicted population persistence, and connectivity (Bani et al. 2002, Opdam 2002). Studies in which dispersal and movement behavior of multiple species have been explicitly considered in the use of habitat networks include work with the marshland bird networks in The Netherlands (Verboom et al. 2001), disturbance-sensitive mammals in the Yukon, Canada (Wiersma and Urban 2005), woodland birds and mammalian carnivores in Italy (Bani et al. 2002), and butterfly species networks in southern Belgium (Baguette et al. 2000).

Our case examples address problems and conditions commonly faced by managers charged with maintenance and recovery of habitats and populations of large numbers of species of conservation concern. First, landscapes in our examples are vast, encompassing millions of hectares; consequently, the available spatial data were coarse in resolution and limited in numbers and types of habitat variables represented. Second, the number of species to be addressed (40 and 91 for the two examples) was too great to allow development of networks for individual species, thus necessitating the use of species groups in network designs. And third, knowledge of the species' habitat requirements was highly variable and incomplete, with little spatially explicit demographic or movement data available for many species, thus requiring use of more general knowledge of species' associations with dominant existing vegetation cover types and the abundance and arrangement of these cover types in space and time. While these issues constrained the potential approaches to characterize habitat networks in the two case examples, the networks designed for each case example supported the management goal to characterize broad-scale habitat conditions for a comprehensive set of targeted species.

Use of surrogate species or groups of species, in particular, has been criticized as not reflecting the needs of the full suite of species that the surrogate or group is intended to represent. This criticism was addressed in detail by Wiens et al. (2008), who described conditions for which surrogate species and species grouping methods were not only helpful but necessary for effective management. These conditions included (1) a large number of species to be addressed (e.g., >50 species), such that individual species management is infeasible; (2) a management area intermediate in size, between continental areas (too large) and local assemblages of patches (too small); and (3) sufficient knowledge of taxa requirements and associated spatial data, allowing habitat conditions to be mapped and assessed and application of rigorous, quantitative methods to select surrogates or groups from the full set of species.

Interior Columbia Basin

Background.—The Interior Columbia Basin Ecosystem Management Project (ICBEMP) was a cooperative endeavor between the U.S. Forest Service (USFS) and Bureau of Land Management (BLM). The project's aim was to develop an ecosystem-based strategy for USFS- and BLM-managed lands across the vast expanse (58 million ha) of the Interior Columbia Basin (hereafter referred to as "Columbia Basin;" Fig. 19-2) (U.S. Forest Service 1996; see also Gravenmier et al. [1997] and Hann et al. [1997] for further description of the Columbia Basin and ICBEMP). The science assessment area of the ICBEMP extended from northwest Washington to Wyoming, with public lands composing 53%. This monumental multiyear and multiscale effort was undertaken to develop an ecosystem-wide management plan to supersede >50 existing federal land management plans in place at the onset of the project in 1994 (U.S. Forest Service 1996).

As part of science assessments for the ICBEMP, a coarse-scale evaluation of habitat conditions for 91 species of upland terrestrial vertebrates of concern was conducted (Wisdom et al. 2000). Habitats for individual species, as well as for species groups and "families" of groups, were analyzed at multiple spatial scales using a hierarchical classification system to assign species to groups and groups to families (Wisdom et al. 2000). Habitat trends were assessed by comparing current (mid-1990s) with historical (circa 1850–1890) conditions. The habitat network analysis described here was conducted as part of the ICBEMP.

Of the 91 species for which conditions were assessed, 44 species, composing five families and 19 groups, were selected for the habitat network analysis. The species in these five families were characterized by declining habitat conditions, range contractions, and relatively narrow habitat requirements; the families included species associated with old forests, forest stand initiation, sagebrush (*Artemisia* spp.), and grasslands. The primary goal in creating the habitat networks was to characterize broad-scale conditions that reflected composite differences among species in the quantity, quality, and connectivity of habitat (Wisdom et al. 2002:3).

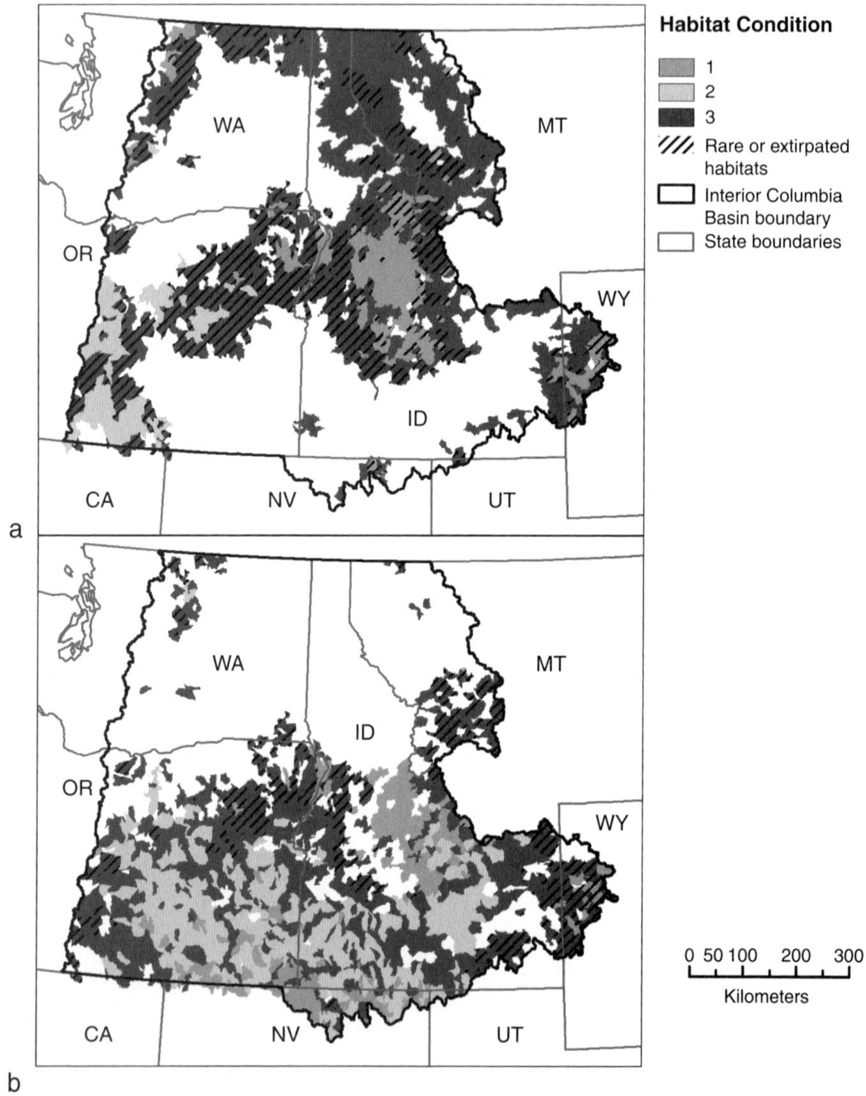

FIG. 19-2 (See color insert.)

Habitat network for Family 1 (old forest, low elevation; A) and Family 11 (sagebrush; B) in the Interior Columbia Basin, USA. See text for explanation of habitat condition classes and rare or extirpated habitats. Blank areas are watersheds that contain no public lands or are outside the range of the family. Adapted from Wisdom et al. (2002).

Assumptions of the habitat network analysis included (1) local (i.e., small-scale) assessments would be conducted to complement the broad-scale characterization provided by the networks; (2) suitable habitats were correctly identified for the species of concern; and (3) the broad-scale approach provided by the habitat networks would assist in conservation planning over the entire Columbia Basin, not only for the 44 species evaluated, but also for other species of concern whose habitats overlapped those of the selected species.

Methods.—Coarse-scale (1-km^2 pixels) measures of habitat conditions for the 44 selected terrestrial vertebrates were evaluated by using two variables: (1) habitat abundance and (2) disturbance departure and fragmentation, which reflects habitat quality and resiliency. Habitat was mapped for each species with a comprehensive species-habitat association matrix developed explicitly for the ICBEMP (Wisdom et al. 2000). Digital maps of historical and existing vegetation cover types and structural stages in the Columbia Basin were derived from a vegetation succession model developed for the ICBEMP (Keane et al. 1996, Hann et al. 1997). Species experts then used the >150 cover type-structural stage combinations, such as old multistory western larch (*Larix occidentalis*), to assign habitat for each species in the matrix. Habitat was then mapped in a GIS for each species within its geographic range in the Columbia Basin. See Wisdom et al. (2000) for additional methods of identifying, quantifying, and mapping habitat for the species.

To measure habitat abundance, habitat at the watershed level (5th hydrologic unit code; Gravenmier et al. 1997) was mapped and summarized for each of the 19 groups to which the 44 species were assigned. Next, mean abundance (in hectares) of habitat among all groups within a family was calculated for each watershed (n = 2,562 watersheds) in the Columbia Basin. Watersheds for each family were then ranked from highest to lowest, based on mean habitat abundance, and assigned to one of three classes: (1) Class A, which included all watersheds in the top two quartiles; (2) Class B, watersheds in the next lowest (third) quartile; and (3) Class C, watersheds in the lowest quartile of habitat abundance.

The second variable, the disturbance departure and fragmentation index (hereafter referred to as disturbance departure), reflects composite effects of changes from the natural or native system at multiple scales (Hann et al. 2003). The variable represents several broad-scale processes related to habitat quality and resiliency, such as changes in vegetation patch size, composition, and arrangement; frequency and intensity of fire; composition of native versus nonnative vegetation; and human disturbance. The index was derived from three primary, coarse-scale input variables: landscape management pattern, landscape vegetation pattern, and potential vegetation group pattern (Hann et al. 2003). These three variables were selected as those most useful for "accurately representing the major patterns and effects of human activities and management on the quality and resiliency of wildland landscapes in the Basin" (Hann et al. 2003:5). The disturbance departure variable was derived as four classes—low, moderate, high, and very high—with the last representing the greatest deviation from historical conditions. Each watershed in the Columbia Basin was assigned to one of these four classes.

The three classes of habitat abundance were then combined with the four classes of disturbance departure to create three habitat condition classes for each family, at the watershed level, during the current time period:

1. *Condition 1:* Watersheds with low disturbance departure and any class of habitat abundance. Such watersheds were considered very resilient and to have changed little in habitat abundance or quality since historical times.

2. *Condition 2:* Watersheds with moderate disturbance departure and habitat abundance in Class A. Conditions in these sites reflect moderate resiliency and some degradation in quality, but relatively abundant habitat.

3. *Condition 3:* All other watersheds not classified as Condition 1 or 2. Watersheds in this class typically contained degraded and uncommon, rare (present but <1% of the watershed), or extirpated habitats.

Each watershed was assigned to a condition class for a family if the current geographic range of any species in the family overlapped the watershed and the watershed contained habitat for that species, either historically or currently. Resulting habitat condition classes were mapped across the Columbia Basin for each of the five families for the current time period, with the exclusion of 461 (18%) watersheds that contained no public lands.

Last, watersheds were highlighted in which habitat for a family was present historically but either had been extirpated or was now rare. Although no formal connectivity analysis was conducted, spatial gaps in connectivity that could be addressed through habitat restoration and conservation planning were identified by this analysis. Spatial gaps in connectivity were characterized as watersheds in Condition 3 with rare or extirpated habitats that were adjacent to watersheds in Condition 1 or 2. Such Condition 3 watersheds represented areas where the greatest declines in habitat abundance and quality had occurred, and where increasing connectivity through restoration would be most beneficial. Those situations were noted in terms of the geographic areas in which these types of habitat "gaps" were present (Wisdom et al. 2002).

Results.—Here we focus on results for two contrasting families, Family 1 (old forest, low elevation) and Family 11 (sagebrush; Table 19-1). Habitat for Family 1 species was broadly distributed across forested areas of the Columbia Basin (Fig. 19-2A). Likewise, the sagebrush habitats of Family 11 were found throughout the Columbia Basin, primarily in lower-elevation rangelands but especially in eastern Oregon and central and southern Idaho (Fig. 19-2B).

Condition 3 was dominant among watersheds for both families, especially Family 1, indicating substantial declines in amount, quality, and resiliency of habitats for species in these families (Table 19-2, Fig. 19-2). Moreover, habitat

Table 19-1 Vertebrate Species of Conservation Focus from Families 1 and 11, Selected for Characterization of Habitat Conditions in the Interior Columbia Basin (Adapted from Wisdom et al. [2000, 2002])

Family	Group	Common Name	Scientific Name
1 (old forest, low elevation)[a]	1	White-headed woodpecker	*Picoides albolarvatus*
	1	White-breasted nuthatch	*Sitta carolinensis*
	1	Pygmy nuthatch	*Sitta pygmaea*
	2	Lewis' woodpecker	*Melanerpes lewis*
	3	Western gray squirrel	*Sciurus griseus*
11 (sagebrush)[b]	33	Greater sage-grouse	*Centrocercus urophasianus*
	33	Sage thrasher	*Oreoscoptes montanus*
	33	Brewer's sparrow	*Spizella breweri*
	33	Sage sparrow	*Amphispiza belli*
	33	Lark bunting	*Calamospiza melanocorys*
	33	Pygmy rabbit	*Brachylagus idahoensis*
	33	Sagebrush vole	*Lemmiscus curtatus*
	34	Black-throated sparrow	*Amphispiza bilineata*
	34	Kit fox	*Vulpes macrotis*
	35	Loggerhead shrike	*Lanius ludovicianus*

[a]*Habitats consist primarily of lower montane forests in late-seral condition.*
[b]*Habitats consist primarily of sagebrush communities.*

has been extirpated in many of the watersheds in this condition class; for example, more than one third of the watersheds for Family 1 in Condition 3 no longer support habitat for species in this family (Table 19-2). Watersheds with rare or extirpated habitat were widely distributed across the current range of species in both families (Table 19-2, Fig. 19-2).

Families 1 and 11 also had the lowest percentage of watersheds in Condition 1 among all five families, evidence of the paucity of habitats resembling historical conditions for these species. A large majority of watersheds in Condition 1 (64–81%) for both families was found in protected areas such as national parks, wilderness, or roadless areas (Wisdom et al. 2002). Condition 2, representing moderate disturbance departure but relatively abundant habitat, was rare (6%) for Family 1. By contrast, Family 11 had the greatest percentage (25%) of watersheds in Condition 2 among all families (Table 19-2).

Table 19-2 Watersheds by Habitat Condition Class and those Containing Extirpated or Rare Habitats for Two Families of Vertebrate Species of Conservation Focus in the Interior Columbia Basin (See Text and Wisdom et al. [2002] for Details)

			Percentage of Watersheds		
Family	***n***	**Habitat Condition Class**	**By Habitat Condition**	**With Extirpated Habitats**	**With Rare Habitats**
1	1,248	1	14	0	4
		2	6	0	0
		3	80	30	11
		All	100	30	15
11	1,229	1	15	0	<1
		2	25	0	0
		3	59	15	5
		All	100[a]	15	5

[a]*Discrepancies between sums in columns are due to rounding.*

Great Basin

Background.—Habitats for species associated with the sagebrush ecosystem have undergone dramatic declines in extent and quality since European settlement (Knick et al. 2003, Wisdom et al. 2005*a*, Chambers et al. 2007). Causes of these changes are diverse, and include intensive livestock grazing, energy extraction, invasion of exotic species such as cheatgrass (*Bromus tectorum*), encroachment of pinyon-juniper woodlands (*Pinus* spp. - *Juniperus* spp.), and altered fire regimes (Knick et al. 2003, Wisdom et al. 2005*a*). These alterations have prompted resource managers to develop and apply innovative approaches to conserve and restore habitats for sagebrush-associated species (e.g., Bureau of Land Management 1999).

In response, we conducted a regional assessment of habitat threats for vertebrate species of concern in the Great Basin Ecoregion, which encompasses most of Nevada and portions of eastern California and western Utah (Fig. 19-3; Nachlinger et al. 2001). This region not only harbors some of the most extensive remaining expanses of sagebrush in the United States, but also has experienced unprecedented losses of sagebrush from catastrophic wildfires (Nachlinger et al. 2001, Rowland and Wisdom 2005, Chambers et al. 2007). The BLM, which manages the majority (52%) of sagebrush nationwide, solicited and funded the Great Basin assessment to help meet its goal to complete broad-scale assessments of

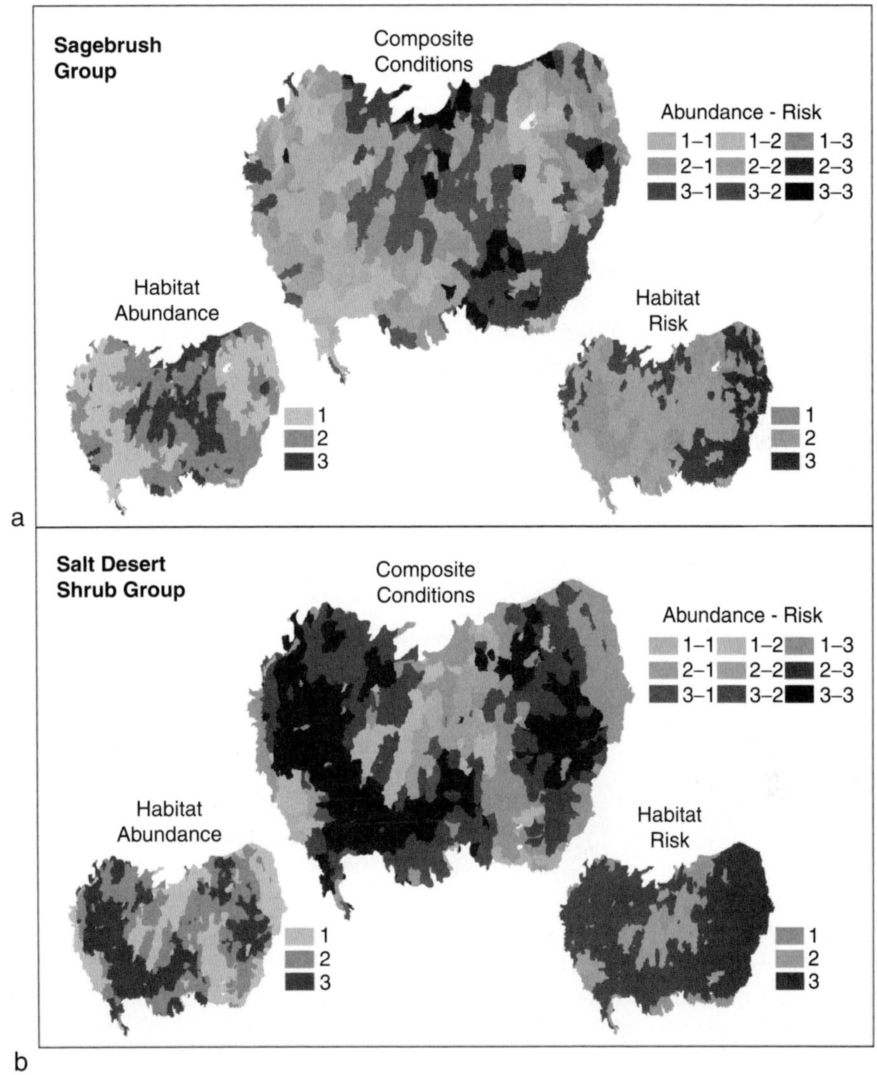

FIG. 19-3 (See color insert.)

Habitat abundance, habitat risk, and composite habitat conditions (all combinations of habitat abundance and risk) for the sagebrush (A) and salt desert shrub (B) groups of species in watersheds of the Great Basin Ecoregion, USA. See text for explanations of habitat abundance and risk classes. For composite conditions, the first number represents the abundance class; and the second number, the risk class (e.g., 1-3 is low habitat abundance and moderate-high risk). Mean size of the 367 watersheds that occur entirely within the ecoregion was 66,000 ha; n = 521 for all watersheds intersecting the ecoregion. Adapted from Wisdom et al. (2005*b*).

habitat conditions in all ecoregions of the sagebrush ecosystem (Wisdom et al. 2005*a*). To initiate this project, a series of protocols was developed for regional assessment of habitats in sagebrush ecosystems (Wisdom et al. 2005*a*). These protocols include selection of species of conservation concern, assignment of species to groups, and estimation of habitats at risk for individual species and species groups.

Primary goals of the Great Basin assessment were to (1) evaluate habitat conditions and threats for selected species of concern; (2) demonstrate application of the newly developed protocols in the Great Basin; and (3) describe the application of results for land management and conservation planning. Secondary goals related to the use of species groups were to (1) reveal regional patterns of habitat conditions and (2) characterize habitat conditions at the watershed level for land management planning.

Methods.—Forty vertebrates of concern, including 13 mammals, 17 birds, and 10 herptiles, were selected for analysis (Table 19-3). Criteria for selection

Table 19-3 Vertebrate Species of Conservation Concern Selected for Assessment in the Great Basin Ecoregion (From Wisdom et al. 2005*b*)

Group	Common name	Scientific name
Sagebrush	Greater sage-grouse	*Centrocercus urophasianus*
	Sage thrasher	*Oreoscoptes montanus*
	Sage sparrow	*Amphispiza belli*
	Vesper sparrow	*Pooecetes gramineus*
	Brewer's sparrow	*Spizella breweri*
	Wyoming ground squirrel	*Spermophilus elegans nevadensis*
	Pygmy rabbit	*Brachylagus idahoensis*
Salt desert shrub	Great Basin collared lizard	*Crotaphytus bicinctores*
	Long-nosed leopard lizard	*Gambelia wislizenii*
	Desert horned lizard	*Phrynosoma platyrhinos*
	Desert spiny lizard	*Sceloporus magister*
	Long-nosed snake	*Rhinocheilus lecontei*
	Groundsnake	*Sonora semiannulata*
	Merriam's kangaroo rat	*Dipodomys merriami*
	Chisel-toothed kangaroo rat	*Dipodomys microps*

continues

Table 19-3 Vertebrate Species of Conservation Concern Selected for Assessment in the Great Basin Ecoregion (From Wisdom et al. 2005*b*) *cont...*

Group	Common name	Scientific name
Sagebrush-woodland	Gray flycatcher	*Empidonax wrightii*
	Green-tailed towhee	*Pipilo chlorurus*
	Merriam's shrew	*Sorex merriami*
	Sagebrush vole	*Lemmiscus curtatus*
	White-tailed jackrabbit	*Lepus townsendii*
Shrubland	Common sagebrush lizard	*Sceloporus graciosus*
	Northern harrier	*Circus cyaneus*
	Prairie falcon	*Falco mexicanus*
	Short-eared owl	*Asio flammeus*
	Western burrowing owl	*Athene cunicularia hypugaea*
	Loggerhead shrike	*Lanius ludovicianus*
	Black-throated sparrow	*Amphispiza bilineata*
	Kit fox	*Vulpes macrotis*
	Pronghorn	*Antilocapra americana*
	Ord's kangaroo rat	*Dipodomys ordii*
	Dark kangaroo mouse	*Microdipodops megacephalus*
	Little pocket mouse	*Perognathus longimembris*
	Northern grasshopper mouse	*Onychomys leucogaster*
Generalist	Great Basin spadefoot	*Spea intermontana*
	Nightsnake	*Hypsiglena torquata*
	Striped whipsnake	*Masticophis taeniatus*
	Ferruginous hawk	*Buteo regalis*
	Swainson's hawk	*Buteo swainsoni*
	Lark sparrow	*Chondestes grammacus*
	Brewer's blackbird	*Euphagus cyanocephalus*

included association with sagebrush habitats and with habitat features that can be accurately mapped with coarse-scale data, a geographic range encompassing >5% of the study area (or about 1.5 million ha), and risk status (determined from state-level ranks obtained from NatureServe [2005]) (Wisdom et al. 2005*a*). For example, rock wrens (*Salpinctes obsoletus*) and rock squirrels (*Spermophilus variegatus*) were dropped from the list due to their strong affinity for rock outcrops, which could not be feasibly mapped at the spatial extent of our study area. The 40 species selected represented a diverse group of widely distributed sagebrush-associated species (Table 19-3).

We quantified habitat for each species within its geographic range in the ecoregion using a species-habitat association matrix as follows. Existing vegetation in the study area was mapped with a land cover classification of 90-m resolution developed for regional assessment of sagebrush habitats in the western United States (Comer et al. 2002), but that incorporated all existing vegetation types. This coverage included 47 land cover types (e.g., mountain big sagebrush [*Artemisia vaseyana*]) in the Great Basin. We developed a habitat association matrix for the 40 species of concern with this land cover layer, based on existing species-habitat databases (e.g., Maser et al. 1984) and consultation with species experts. Ideally, habitat would have been identified for the network by conducting field studies to document occurrence or abundance of the species of interest in various cover types within the Great Basin, but the immense size of the study area and number of species in the assessment precluded such data collection.

Each of the 40 species was then assigned to one of five groups—sagebrush, shrubland, salt desert shrub, sagebrush-woodland, and generalist—based on similarities in habitat associations and habitat abundance among species in each group. Current habitat conditions for each group were evaluated by watersheds, due to the increasing prevalence and preference of this spatial extent for research and management in sagebrush ecosystems of the western United States (Bureau of Land Management 1999, Wisdom et al. 2005*b*).

For each watershed and species group, habitat abundance and risk of habitat loss were estimated and mapped, and then combined to estimate composite habitat conditions. To quantify habitat abundance for species groups, the amount and percentage of habitat for each species were first calculated in each watershed within the species' range in the study area. Next, the mean percent habitat across all species within a group was calculated at the watershed level. Last, habitat abundance was classified, by group, in each watershed as follows: (1) low: mean habitat <25%; (2) moderate: mean habitat 25–50%; and (3) high: mean habitat >50%.

To estimate risk of habitat loss, a rule-based model of risk of displacement of native vegetation by cheatgrass was developed and applied (Suring et al. 2005). Model output was classified as none, low, moderate, or high risk for each 90-m pixel in the study area. At the watershed level, the percentage of each species' habitat within the four risk categories was calculated. The mean percent habitat, by risk category, was then calculated among all species in each group. Last, watersheds were classified as follows: (1) none-low: habitat in the none and low-risk categories combined >50%; (2) low-moderate: habitat in the low- and moderate-risk categories combined >50%; and (3) moderate-high: habitat in the moderate- and high-risk categories combined >50%. Finally, the three habitat abundance classes were combined with the three risk classes, yielding nine possible combinations for assignment of habitat condition at the watershed level.

Results.—Here we present a subset of the results of the Great Basin assessment, concentrating on the contrasting patterns for the sagebrush and salt desert shrub species groups. Across the Great Basin, habitat abundance for the sagebrush group was dominated by the moderate class, or watersheds with mean habitat from 25–50% of the watershed area (Table 19-4, Fig. 19-3A). This group also had the lowest percentage (22%) of watersheds in the high abundance class among all groups, indicating that relatively few watersheds in the Great Basin are currently dominated by sagebrush habitats. In contrast to this pattern, results for the salt desert shrub group indicated an even distribution of watersheds among the three classes of habitat abundance (Table 19-4, Fig. 19-3B). Watersheds with the most habitat for species in the sagebrush group were in the mountains of the ecoregion's center and along its northern edge; habitat for salt desert shrub species was most abundant in the western and eastern portions of the ecoregion, with less habitat in the central area (Fig. 19-3).

Table 19-4 Percentage of Watersheds in the Great Basin by All Combinations of Habitat Abundance and Risk for Two Sample Species Groups (Adapted from Wisdom et al. 2005*b*)

			Risk of Habitat Displacement by Cheatgrass			
Species Group	***n***	**Habitat Abundance**	**None-Low**	**Low-Moderate**	**Moderate-High**	**All Risk Classes Combined**
Sagebrush	168	Low	14	10	9	32[a]
	236	Moderate	7	21	18	46
	115	High	2	14	7	22
	519	Total	24	44	32	100
Salt desert shrub	180	Low	6	8	21	35
	156	Moderate	4	4	23	31
	171	High	5	3	26	34
	507	Total	14	15	71	100

[a]*Discrepancies between sums and numbers in rows and columns are due to rounding.*

Patterns of habitat risk contrasted sharply between the two groups, both quantitatively and spatially. Watersheds for the sagebrush group were somewhat equally divided among the three risk classes, although the none-low risk class was least common (24%). However, moderate-high risk was clearly the dominant class for species in the salt desert shrub group (Table 19-4, Fig. 19-3B). For the sagebrush group, lower-risk habitat was distributed throughout the ecoregion but especially scarce in the central portion (Fig. 19-3A). Higher-risk habitats were found along the eastern and northern perimeter of the study area. Spatial patterns of habitat risk for the salt desert shrub group were markedly different from those for the sagebrush group, with high-risk habitat blanketing most of the ecoregion, with the exception of the central core (Fig. 19-3B).

Examination of composite conditions revealed that watersheds in the "best condition," i.e., those with abundant habitat at low risk (abundance-risk class 3-1), were very rare for both groups (Table 19-4, Fig. 19-3). The most common composite condition was that of moderate habitat abundance with low-moderate risk (class 2-2; 21%) for the sagebrush species group, and high habitat abundance with moderate-high risk (class 3-3; 26%) for the salt desert shrub group (Table 19-4, Fig. 19-3).

Discussion.—The two case studies of habitat networks share several traits, including (1) use of species groups, (2) incorporation of past disturbance or future risk, and (3) "wall-to-wall" characterization of regional habitat conditions across all land ownerships. In both the Columbia Basin and Great Basin, separate networks were developed for groups of species that contrasted in their habitat associations and past levels of habitat loss (Columbia Basin) or predicted risk of habitat loss (Great Basin). Other authors also have described network approaches for species groups; for example, Vos et al. (2001) grouped species by "ecological profile" using individual area requirements and dispersal distance to reflect metapopulation response to landscape change.

Use of disturbance departure in the Columbia Basin and risk of habitat loss from displacement by cheatgrass in the Great Basin provided further discrimination between watersheds with similar amounts of habitat but often dramatically different risk. Such an approach goes beyond simple identification of habitat patches within a network, and parallels that of McIntyre and Hobbs (1999), who used a continuum of habitat loss factors to characterize habitat in the matrix. Similarly, Frank (2004) described "strong" habitat patches, based on distance to neighboring patches that were able to withstand negative effects of environmental stochasticity.

The two case examples demonstrate spatial characterization of regional-level habitat conditions across multiple landscapes and ownerships, providing a springboard for more small-scale evaluations to determine what specific locales within watersheds warrant management action and what actions are feasible. This contrasts with habitat networks that identify individual habitat patches or core areas and corridors between patches (e.g., Bani et al. 2002, Nikolakaki and Dunnett 2005). Portions of our networks, however, can be interpreted more traditionally. For example, condition 3 watersheds in the Columbia Basin are most likely dominated by matrix habitats, with few functioning "core" habitat patches remaining. Well-informed management of the "semi-natural matrix" that constitutes most of the land area in the United States (Noss and Cooperrider 1994) may be the most prudent approach to biodiversity conservation.

POLICY AND MANAGEMENT IMPLICATIONS

Habitat networks depict the spatial structure of habitats and linkages between them, providing a comprehensive assessment of how single or multiple species use, or respond to, the spatial structure of their environment (Hobbs 2002, Opdam 2002). Thus, habitat networks confer multiple potential benefits to conservation planning and land management. First, different networks can be established and managed for different species or species groups, and potential differences and trade-offs in management strategies within and among groups can be assessed and reconciled. Second, network characterizations, by design, directly inform management actions to maintain, restore, or improve conditions for targeted species. Management strategies based on networks can be designed and implemented over several spatial and temporal scales, allowing priorities to be established in both time and space in relation to associated disturbance regimes. Finally, management strategies based on networks can be assessed and adjusted in relation to trade-offs between strategies developed for individual species, such as recovery plans for federally threatened or endangered species versus other policy or resource objectives (e.g., timber production, grazing, or recreation).

In the contrast to these benefits, management use of a habitat network may confer a sense of false confidence if information on which the network is based is insufficient to warrant its use. Knowledge of species' requirements is variable, and different species may respond differently to management at a given scale. Consequently, use of a network designed for multiple species, developed and implemented at a fixed scale, and with variable knowledge of requirements among the species represented in the network is likely to confer greater benefits to some species than others, and might not fully depict spatial patterns of importance for some species. These potential problems illustrate the necessity of an adaptive management approach when characterizing and implementing habitat networks (see "Implementing Habitat Networks Through Adaptive Management").

Management Integration Within and Among Species Groups

The use of habitat networks for species groups does not limit the degree to which conditions for individual species can be assessed and managed. On the contrary, the concept of networks recognizes the inherent limits of time and resources to assess and manage conditions for individual species beyond a few special cases, and allows managers to take a more holistic approach when managing large landscapes for multiple species of conservation concern. Incorporating species groups in network design and comparing outcomes of group-based networks with those developed for individual target species can reveal how well strategies designed for groups support goals for individual species.

For large landscapes, different networks must be developed for different suites of species to accommodate the intrinsic biodiversity at this scale. That is, different species have different habitat associations, geographic ranges, and areas over which they conduct daily and seasonal activities in relation to the amount, distribution, configuration, and connectivity of habitats. Such differences mandate the use of multiple habitat networks, with each individual network designed to reflect conditions for species with similar habitat requirements and responses to habitat change.

Direct Links to Management

Spatial information from habitat networks is intentionally derived to link directly with management for targeted species, and can help guide prioritization of management activities in space and time. In our case studies, past land management has profoundly affected the quantity and quality of habitats for species of concern. For example, in the Columbia Basin network, the dominant condition, i.e., Condition 3, for Families 1 and 11 represents suboptimal habitats, typically low-elevation sites that have been intensively managed for livestock, timber production, and other commodity uses. The preponderance of watersheds in Condition 3 exemplifies the urgent need to actively protect remnant habitats that still function effectively (i.e., those in Condition 1) and restore degraded or diminished habitats to improve connectivity among watersheds.

Often, initial development of a habitat network for management requires subsequent modification to address new or evolving management objectives. One example is the ongoing National Forest Plan revision process for three forests in eastern Washington state (Colville, Okanogan, and Wenatchee National Forests) (U.S. Forest Service 2006). To incorporate multispecies conservation strategies in the plans, the biologists modified the habitat network framework developed by Wisdom et al. (2002). To do this, the biologists selected focal species for analysis, including white-headed woodpecker (*Picoides albolarvatus*) and tiger salamander (*Ambystoma tigrinum*), and developed and applied Bayesian Belief Network models (Marcot et. al. 2001) for each species to estimate habitat suitability at the watershed level. Habitat suitability scores then were combined in a GIS with other attributes (e.g., amount of source habitat relative to historic median, land ownership pattern) to assign a habitat condition class to each watershed.

The resulting habitat condition classes then guided selection of a management strategy (e.g., restoration, protection, connectivity) for each focal species in each watershed. Overlays of networks for the focal species will guide prioritization of management through the application of conservation strategies to benefit multiple species. For example, one strategy is to reduce road construction and access in areas occupied by several focal species that are sensitive to effects of roads. The conservation strategies developed for the watersheds were created through an interdisciplinary process in order to address multiple resource objectives.

Reconciling Network Strategies in Relation to other Objectives

A variety of laws, policies, and regulations guide resource management on state, federal, private, and tribal lands in the United States and elsewhere. Some of these directives clarify the need to maintain populations of all native biota in their native environments. For example, the U.S. Endangered Species Act dictates that no species will be managed so as to cause its designation as threatened or endangered. Similarly, regulations supporting implementation of the National Forest Management Act call for sustaining native ecological systems by providing conditions to support diversity of native plants and animals in the planning area (U.S. Government 2005). These regulations thus provide a clear basis for development and implementation of holistic approaches for species management, such as habitat networks.

Other laws, regulations, and policies, however, provide direction for resource objectives beyond maintenance of native biota. The U.S. Multiple-Use Sustained-Yield Act, for example, requires National Forests to be managed for a wide variety of commodities and uses, including timber, livestock, mining, water, and recreation. The resultant challenge is to ensure compatibility between management to maintain populations of native species and management for other, potentially competing resources.

One method to reconcile species management with other resource objectives is to map conditions for each set of resources, including a habitat network for species of concern, and use results to identify compatibilities and conflicts in strategies among all featured resources. Trade-offs among resource objectives can then be explicitly considered, and all resource objectives integrated.

In the Great Basin, watersheds with abundant habitat and low risk of habitat loss are likely to represent “habitat strongholds” for the associated species. Under these conditions, other resource uses such as grazing and mining operations may be compatible with habitat maintenance. By contrast, watersheds of low or moderate habitat abundance but high risk may warrant subordination of livestock grazing, energy extraction, and other conflicting uses to habitat preservation for targeted species of conservation concern.

These examples illustrate the degree to which spatial information about species of conservation concern can be represented by habitat networks for setting joint policies and strategies to concurrently meet a variety of management objectives among multiple stakeholders. In particular, the use of habitat networks to evaluate trade-offs among competing resource objectives, and to ultimately reconcile potential conflicts among these objectives, can be an essential component of multiple use management within and among all land ownerships in relation to a diverse array of laws and policies.

FUTURE DIRECTIONS

Knowledge Gaps

Concepts and uses of habitat networks will evolve as knowledge about networks is gained and applied. Currently, information needed to fully implement effective networks is limited for most species (Hobbs 2002). Fundamental knowledge to develop habitat networks is lacking for four key topics: (1) environmental requirements of individual species, in terms of number, distribution, and configuration of habitat patches; (2) size of and distance between habitat patches to best facilitate dispersal and other inter-patch movements; (3) resistance and resiliency of species' habitats given prevalent disturbance regimes; and (4) integration of the above information for multiple species. All four topics encompass issues of spatial and temporal scale, specifically how habitats and associated populations of species will be evaluated and maintained at the appropriate spatial and temporal extents (appropriate geographic area and time period for managing the species) and at the appropriate spatial and temporal grain (i.e., the resolution at which habitats and populations are measured and how often the measurements are taken). Traditionally, habitat networks have emphasized the first two topics, but the third and fourth are equally relevant for holistic conservation planning. Consequently, new research is required to address knowledge gaps about habitat networks and their application (see Hobbs [2002] and Opdam [2002] for additional review of knowledge gaps related to habitat networks).

For example, little is known about the environmental requirements of most species of conservation concern beyond birds and large mammals (Bonnett et al. 2002, Clark and May 2002), particularly in relation to the spatial configuration of habitat patches and their colonization potential (topic 1). Research needed to address this topic would focus on species occurrence and persistence in habitat patches that vary in size, configuration, vegetation structure and composition, landscape setting, and other environmental factors. Similarly, for most species, little is known about habitat connectivity and flow of individuals among habitat patches (Opdam 2002, Schulte et al. 2006) (topic 2). Species-level (i.e., fine-filter) research needed to address this topic would include studies of metapopulation dynamics, dispersal rates, movement behavior, use of corridors or other linkages, and optimal inter-patch distances.

Habitat networks are unlikely to remain stable over time. Disturbance agents such as wildfire, floods, drought, invasive species, and myriad human activities must be considered when planning for maintenance of habitat networks. It follows that the concepts of habitat resistance and resiliency should be addressed when planning and developing habitat networks (topic 3). If habitats lack resistance and resiliency, then larger habitat patches must be conserved to buffer against environmental stochasticity. Unfortunately, knowledge of habitat resistance and resiliency is limited, particularly in the face of global climate change and the increasing human footprint affecting habitats worldwide (Sanderson et al. 2002). Moreover, knowledge is lacking about species response time to habitat changes, or effects of shifts in habitat configuration on metapopulation dynamics (Opdam 2002, Frank 2004).

The potential for holistic integration of multiple species of conservation concern in habitat networks is seldom broached. Concepts of habitat networks initially addressed individual species and how they used geographically isolated habitat patches in fragmented landscapes. The metapopulation dynamics exhibited by these species provided a basis for designing habitat networks that were sufficiently connected to allow for interaction among individuals from different local populations. Applying networks for individual species, however, is an impractical approach to meet overarching goals of biodiversity conservation (Noss and Cooperrider 1994, Opdam 2002). The challenge now lies with understanding how metapopulation dynamics of many species of conservation concern can be considered comprehensively. This challenge is probably impossible to fully meet, given the multitude of species that typically deserve management attention in a single large landscape.

Integration of a comprehensive set of species' needs in network design and management can be approached with both top-down and bottom-up methods (Gripenberg and Roslin 2007). Top-down methods of research would attempt to elucidate spatial structure of populations of multiple species with similar environmental requirements, space use, and movements, without conducting detailed research about metapopulation characteristics of each species. By contrast, bottom-up methods would focus on how selected species of concern use habitats, relying on traditional metapopulation research techniques. Because this research is extremely costly and time-consuming, species selected for such detailed studies would ideally represent a larger set of species of conservation concern, using concepts such as focal species (Lambeck 1997).

Implementing Habitat Networks Through Adaptive Management

Habitat networks and associated conservation planning serve as regional hypotheses that can be tested and evaluated through landscape-level management experiments under the auspices of adaptive resource management (Fig. 19-1; Walters 1986, Kendall 2001). Under the paradigm of adaptive management, research and management collaborate to identify knowledge gaps, especially those affecting key economic, social, political, and ecological issues. The collaborators then develop testable management hypotheses, followed by management experiments and implementation of results in land management. This cycle is repeated as necessary to identify additional knowledge gaps and improve outcomes of management. (See Haddad et al. [2003] for an excellent example of a broad-scale, manipulative experiment to evaluate corridor use by different taxa.)

If validation research, ideally applied through adaptive management, is not conducted to address primary sources of uncertainty associated with habitat networks, the utility and credibility of networks may be questioned. Consequently, such research is integral to fostering use of habitat networks in land management and conservation planning. A benefit of using an adaptive management framework is that managers have the freedom to design and implement habitat networks despite uncertainty about the networks and their efficacy in maintaining ecosystem diversity or protection for individual species.

Schulte et al. (2006) found limited and inconsistent application of habitat networks and other related concepts in planning for biodiversity. Despite obstacles to implementation of networks, however, land managers can learn much by incorporating habitat networks as part of standard operations for conservation planning and management. For example, under the 2008 final rule describing the future land management planning framework of the Forest Service, an environmental management system (EMS) will be required for land management planning throughout the agency (U.S. Government 2008). The EMS will serve as a framework for adaptive management in the Forest planning process; habitat networks offer one tool for assessing environmental conditions for multiple species on National Forest lands under such a framework.

Management use of habitat networks will be challenging, given existing knowledge gaps, limited resources, and diverse management objectives. However, the alternative—ignoring spatial structure of habitats—will ultimately lead to failure. That as context, we advocate increasing emphasis on management application of, and research focus on, habitat networks.

SUMMARY

Powerful analysis tools now enable ecologists to characterize landscapes at a variety of spatial scales directly applicable to management. One such application is the use of habitat networks to characterize and manage large landscapes for habitats and species of interest. A habitat network is a spatially explicit portrayal of environmental conditions across large areas that can be used to understand the status and trends of wildlife species, particularly in relation to how species needs are met through management of habitat abundance and distribution. In this chapter, we discussed the concepts of habitat networks and provided case examples for two areas in the western United States: the Interior Columbia Basin and the Great Basin. To address the need for broad-scale, comprehensive planning within the Interior Columbia Basin, we developed a habitat network for five groups of terrestrial vertebrates. We used habitat abundance and habitat quality to describe watersheds in the 58 million-ha basin. Watersheds were assigned to one of three habitat condition classes for each species group. In a similar analysis, we mapped habitats for groups of sagebrush-associated vertebrates in the Great Basin. Here, we characterized watersheds for each species group by (1) habitat abundance, (2) habitat at risk of displacement by cheatgrass, and (3) the composite conditions of habitat abundance and risk. Mapping habitat networks can foster efficient conservation planning at regional levels by guiding the spatial prioritization of limited resources for habitat conservation and restoration. The methods we described can be augmented with additional spatial models that incorporate other landscape and ecological characteristics, such as habitat or population connectivity and land protection status. We discussed implications of mapping habitat networks in the context of current management and policies related to wildlife habitats, as well as future needs for characterizing wildlife habitats across large landscapes.

ACKNOWLEDGMENTS

W. J. Hann, M. A. Hemstrom, R. S. Holthausen, and B. C. Wales were instrumental in development and application of habitat networks for the Interior Columbia Basin project. L. H. Suring, C. W. Meinke, K. Preston, L. Schueck, and S. Knick collaborated closely with us on the Great Basin ecoregional assessment, including development of the watershed-level analyses of habitat conditions. We thank W. Gaines and B. C. Wales for providing information on the northeastern Washington National Forests and their use of habitat networks. J. S. Boyd and A. Youngblood assisted with preparation of graphics and spatial analysis, and J. B. Haufler provided relevant literature. Last, we thank an anonymous reviewer and the editors for their insightful comments. Funding for the work reported here was provided by the U.S. Forest Service, Bureau of Land Management, and U.S. Geological Survey.

LITERATURE CITED

Baguette, M., S. Petit, and F. Quéva. 2000. Population spatial structure and migration of three butterfly species within the same habitat network: Consequences for conservation. *Journal of Applied Ecology* 37:100–108.

Baguette, M., and H. Van Dyck. 2007. Landscape connectivity and animal behavior: Functional grain as a key determinant for dispersal. *Landscape Ecology* 22:1117–1129.

Bani, L., M. Baietto, L. Bottoni, and R. Massa. 2002. The use of focal species in designing a habitat network for a lowland area of Lombardy, Italy. *Conservation Biology* 16:826–831.

Bennet, A. F. 2003. *Linkages in the landscape: The role of corridors and connectivity in wildlife conservation.* IUCN Forest Conservation Programme, Conserving Forest Ecosystems Series

Number 1. International Union for Conservation of Nature and Natural Resources, Gland, Switzerland.

Bonnet, X., R. Shine, and O. Lourdais. 2002. Taxonomic chauvinism. *Trends in Ecology and Evolution* 17:1-3.

Bruinderink, G. G., T. Van Der Sluis, D. Lammertsma, P. Opdam, and R. Pouwels. 2003. Designing a coherent ecological network for large mammals in northwestern Europe. *Conservation Biology* 17:549-557.

Bureau of Land Management. 1999. *The Great Basin Restoration Initiative: Out of ashes, an opportunity.* U.S. Bureau of Land Management, National Office of Fire and Aviation, Boise, Idaho, USA.

Chambers, J. C., B. A. Roundy, R. R. Blank, S. E. Meyer, and A. Whittaker. 2007. What makes Great Basin sagebrush ecosystems invasible by *Bromus tectorum*? *Ecological Monographs* 77: 117-145.

Clark, J. A., and R. M. May. 2002. How biased are we? *Conservation in Practice* 3:28-29.

Comer, P., J. Kagan, M. Heiner, and C. Tobalske. 2002. *Current distribution of sagebrush and associated vegetation in the western United States (excluding NM and AZ). Digital Map 1:200,000 scale.* U.S. Geological Survey, Forest and Rangeland Ecosystems Science Center, Boise, Idaho, and The Nature Conservancy, Boulder, Colorado, USA.

Forman, R. T. T., and M. Godron. 1986. *Landscape ecology.* John Wiley and Sons, New York, New York, USA.

Frank, K. 2004. Ecologically differentiated rules of thumb for habitat network design—Lessons from a formula. *Biodiversity and Conservation* 13:189-206.

Franklin, A. B., D. R. Anderson, R. J. Gutiérrez, and K. B. Burnham. 2000. Climate, habitat quality, and fitness in northern spotted owl populations in northwestern California. *Ecological Monographs* 70:539-590.

Gravenmier, R. A., A. Wilson, and J. Steffenson. 1997. Information system development and documentation. Pages 2012-2066 *in* T. M. Quigley and S. J. Arbelbide, technical editors. *An assessment of ecosystem components in the interior Columbia basin and portions of the Klamath and Great Basins.* U.S. Forest Service, Pacific Northwest Research Station, General Technical Report PNW-GTR-405, Portland, Oregon, USA.

Gripenberg, S., and T. Roslin. 2007. Up or down in space? Uniting the bottom-up versus top-down paradigm and spatial ecology. *Oikos* 116:181-188.

Gutiérrez, D. 2005. Effectiveness of existing reserves in the long-term protection of a regionally rare butterfly. *Conservation Biology* 19:1586-1597.

Haddad, N. M., D. R. Bowne, A. Cunningham, B. J. Danielson, D. J. Levey, S. Sargent, and T. Spira. 2003. Corridor use by diverse taxa. *Ecology* 84:609-615.

Hall, L. S., P. R. Krausman, and M. L. Morrison. 1997. The habitat concept and a plea for standard terminology. *Wildlife Society Bulletin* 25:173-182.

Hann, W. J., J. L. Jones, M. G. Karl, P. F. Hessburg, R. E. Keane, D. G. Long, J. P. Menakis, C. H. McNicoll, S. G. Leonard, R. A. Gravenmier, and B. G. Smith. 1997. Landscape dynamics of the basin. Pages 337-1055 *in* T. M. Quigley and S. J. Arbelbide, technical editors. *An assessment of ecosystem components in the interior Columbia basin and portions of the Klamath and Great Basins.* U.S. Forest Service, Pacific Northwest Research Station, General Technical Report PNW-GTR-405, Portland, Oregon, USA.

Hann, W. J., M. J. Wisdom, and M. M. Rowland. 2003. *Disturbance and fragmentation of natural systems in the Interior Columbia Basin.* U.S. Forest Service, Pacific Northwest Research Station, Research Paper PNW-RP-545, Portland, Oregon, USA.

Hanski, I., and M. Gilpin. 1991. Metapopulation dynamics: Brief history and conceptual domain. *Biological Journal of the Linnaean Society* 42:3-16.

Haufler, J. 1999. Strategies for conserving terrestrial biological diversity. Pages 17-30 *in* R. K. Baydack, H. Campa, III, and J. B. Haufler, editors. *Practical approaches to the conservation of biological diversity. Part I: Conserving biological diversity—principles and perspectives.* Island Press, Washington, D.C., USA.

Hobbs, R. J. 2002. Habitat networks and biological conservation. Pages 150–170 *in* K. J. Gutzwiller, editor. *Applying landscape ecology in biological conservation.* Springer, New York, New York, USA.

Keane, R. E., D. G. Long, J. P. Menakis, W. J. Hann, and C. D. Bevins. 1996. *Simulating coarse-scale vegetation dynamics using the Columbia River basin succession model-CRBSUM.* U.S. Forest Service, Intermountain Research Station, General Technical Report INT-GTR-340, Ogden, Utah, USA.

Kendall, W. H. 2001. Using models to facilitate complex decisions. Pages 147–170 *in* T. M. Shenk and A. B. Franklin, editors. *Modeling in natural resource management: Development, interpretation, and application.* Island Press, Washington, D.C., USA.

Knick, S. T., D. S. Dobkin, J. T. Rotenberry, M. A. Schroeder, W. M. Vander Haegen, and C. van Riper, III. 2003. Teetering on the edge or too late? Conservation and research issues for avifauna of sagebrush habitats. *Condor* 105:611–634.

Lambeck, R. J. 1997. Focal species: A multi-species umbrella for nature conservation. *Conservation Biology* 11:849–856.

León-Cortés, J. L., F. Pérez-Espinoza, L. Marín, and A. Molina-Martínez. 2004. Complex habitat requirements and conservation needs of the only extant Baroniinae swallowtail butterfly. *Animal Conservation* 7:241–250.

Levins, R. 1969. Some demographic and genetic consequences of environmental heterogeneity for biological control. *Bulletin of the Entomological Society of America* 15:237–240.

MacArthur, R. H., and E. O. Wilson. 1967. *The theory of island biogeography.* Princeton University Press, Princeton, New Jersey, USA.

Marcot, B. G., R. S. Holthausen, M. G. Raphael, M. M. Rowland, and M. J. Wisdom. 2001. Using Bayesian belief networks to evaluate fish and wildlife population viability under land management alternatives from an environmental impact statement. *Forest Ecology and Management* 153:29–42.

Maser, C., J. W. Thomas, and R. G. Anderson. 1984. *Wildlife habitats in managed rangelands—The Great Basin of southeastern Oregon.* The relationship of terrestrial vertebrates to plant communities and structural conditions, Part 1 (text) and Part 2 (appendices). U.S. Forest Service, Pacific Northwest Forest and Range Experiment Station, General Technical Report PNW-GTR-172, Portland, Oregon, USA.

Mayer, K. E., and W. F. Laudenslayer, Jr., editors. 1988. *A guide to wildlife habitats of California.* State of California, Resources Agency, Department of Fish and Game, Sacramento, California, USA.

McIntyre, S., and R. Hobbs. 1999. A framework for conceptualizing human effects on landscapes and its relevance to management and research models. *Conservation Biology* 13:1282–1292.

Nachlinger, J., K. Sochi, P. Comer, G. Kittel, and D. Dorfman. 2001. *Great Basin: An ecoregion-based conservation blueprint.* The Nature Conservancy, Reno, Nevada, USA.

NatureServe. 2005. *NatureServe Explorer: An online encyclopedia of life. Version 4.4.* NatureServe, Arlington, Virginia. <http://www.natureserve.org/explorer>.

Nichols, J. D. 2001. Using models in the conduct of science and management of natural resources. Pages 11–34 *in* T. M. Shenk and A. B. Franklin, editors. *Modeling in natural resource management: Development, interpretation, and application.* Island Press, Washington, D.C., USA.

Nikolakaki, P., and N. Dunnett. 2005. The use of spatial concepts as a basis for designing a viable-habitat network: Conserving redstart (*Phoenicurus phoenicurus*) populations in Sherwood Forest, England. *Journal for Nature Conservation* 13:31–48.

Noss, R. F. 1983. A regional landscape approach to maintain biodiversity. *BioScience* 33:700–706.

Noss, R. F., and A. Y. Cooperrider. 1994. *Saving nature's legacy.* Island Press, Washington, D.C., USA.

Noss, R. F., and L. D. Harris. 1986. Nodes, networks, and MUMs: Preserving diversity at all scales. *Environmental Management* 10:299–309.

O'Neil, T. A., P. Bettinger, B. G. Marcot, B. W. Luscombe, T. Koeln, H. J. Bruner, C. Barrett, J. A. Pollock, and S. Bernatas. 2005. Application of spatial technologies in wildlife biology. Pages 418-447 *in* C. E. Braun, editor. *Techniques for wildlife investigations and management.* Sixth edition. The Wildlife Society, Bethesda, Maryland, USA.

Opdam, P. 2002. Assessing the conservation potential of habitat networks. Pages 381-404 *in* K. J. Gutzwiller, editor. *Applying landscape ecology in biological conservation.* Springer, New York, New York, USA.

Opdam, P., and D. Wascher. 2004. Climate change meets habitat fragmentation: Linking landscape and biogeographical scale levels in research and conservation. *Biological Conservation* 117:285-297.

Rowland, M. M., and M. J. Wisdom. 2005. The Great Basin at risk. Pages 83-93 *in* M. J. Wisdom, M. M. Rowland, and L. H. Suring, editors. *Habitat threats in the sagebrush ecosystem: Methods of regional assessment and applications in the Great Basin.* Alliance Communications Group, Lawrence, Kansas, USA.

Sanderson, E. W., M. Jaiteh, M. A. Levy, K. H. Redford, A. V. Wannebo, and G. Woolmer. 2002. The human footprint and the last of the wild. *BioScience* 52:891-904.

Schulte, L. A., R. J. Mitchell, M. L. Hunter, Jr., J. F. Franklin, R. K. McIntyre, and B. J. Palik. 2006. Evaluating the conceptual tools for forest biodiversity conservation and their implementation in the U.S. *Forest Ecology and Management* 232:1-11.

Suring, L. H., M. J. Wisdom, R. J. Tausch, R. F. Miller, M. M. Rowland, L. Schueck, and C. W. Meinke. 2005. Modeling threats to sagebrush and other shrubland communities. Pages 114-149 *in* M. J. Wisdom, M. M. Rowland, and L. H. Suring, editors. *Habitat threats in the sagebrush ecosystem: Methods of regional assessment and applications in the Great Basin.* Alliance Communications Group, Lawrence, Kansas, USA.

U.S. Forest Service. 1996. *Status of the Interior Columbia Basin: Summary of scientific findings.* U.S. Forest Service, Pacific Northwest Research Station, General Technical Report PNW-GTR-385, Portland, Oregon, USA.

U.S. Forest Service. 2006. *Terrestrial species assessments: R6 Forest Plan revisions.* U.S. Forest Service, Pacific Northwest Research Station, Portland, Oregon, USA.

U.S. Government. 2005. Rules and regulations, Department of Agriculture, Forest Service, National Forest System land management planning. *Federal Register* 70:1023-1061.

U.S. Government. 2008. Department of Agriculture, Forest Service. 36 CFR Part 219. National Forest System land management planning; final rule. *Federal Register* 73:21468-21512.

Verboom, J., R. Foppen, P. Chardon, P. Opdam, and P. Luttikhuizen. 2001. Introducing the key patch approach for habitat networks with persistent populations: An example for marshland birds. *Biological Conservation* 100:89-101.

von Haaren, C., and M. Reich. 2006. The German way to greenways and habitat networks. *Landscape and Urban Planning* 76:7-22.

Vos, C. C., J. Verboom, P. F. M. Opdam, and C. J. F. Ter Braak. 2001. Toward ecologically scaled landscape indices. *American Naturalist* 157:24-41.

Walters, C. 1986. *Adaptive resource management of renewable resources.* Macmillan, New York, New York, USA.

Wiens, J. A., G. D. Hayward, R. S. Holthausen, and M. J. Wisdom. 2008. Using surrogate species and groups for conservation planning and management. *BioScience* 58:241-252.

Wiersma, Y. F., and D. L. Urban. 2005. Beta diversity and nature reserve design in the Yukon, Canada. *Conservation Biology* 19:1262-1272.

Wisdom, M. J., R. S. Holthausen, B. C. Wales, C. D. Hargis, V. A. Saab, D. C. Lee, W. J. Hann, T. D. Rich, M. M. Rowland, W. J. Murphy, and M. R. Eames. 2000. *Source habitats for terrestrial vertebrates of focus in the Interior Columbia Basin: Broad-scale trends and management implications.* U.S. Forest Service, Pacific Northwest Research Station, General Technical Report PNW-GTR-485, Portland, Oregon, USA.

Wisdom, M. J., M. M. Rowland, and L. H. Suring, editors. 2005*a*. *Habitat threats in the sagebrush ecosystem: Methods of regional assessment and applications in the Great Basin*. Alliance Communications Group, Lawrence, Kansas, USA.

Wisdom, M. J., M. M. Rowland, L. H. Suring, L. Schueck, C. A. Meinke, S. T. Knick, and B. C. Wales. 2005*b*. Habitats for groups of species. Pages 205–231 *in* M. J. Wisdom, M. M. Rowland, and L. H. Suring, editors. *Habitat threats in the sagebrush ecosystem: Methods of regional assessment and applications in the Great Basin*. Alliance Communications Group, Lawrence, Kansas, USA.

Wisdom, M. J., B. C. Wales, R. S. Holthausen, M. A. Hemstrom, and M. M. Rowland. 2002. A habitat network for terrestrial wildlife in the Interior Columbia Basin. *Northwest Science* 76:1-14.

CHAPTER

Landscape-Level Planning for Conservation of Wetland Birds in the U.S. Prairie Pothole Region

20

Neal D. Niemuth, Ronald E. Reynolds, Diane A. Granfors, Rex R. Johnson, Brian Wangler, and Michael E. Estey

There are many ongoing, extensive, and well-defined wildlife conservation issues in the Prairie Pothole Region (PPR) of North America. Substantial programs have been developed to address these conservation issues in the United States portion of the PPR. These programs require biologically sound and scientifically rigorous tools to provide programmatic accountability as well as guidance for conservation actions. Consequently, conservation scientists have developed an integrated process of model development and application that encompasses biology, social issues, ecological threats, program delivery, landscape management, and public policy. We describe a portion of that process as it is applied to conservation of wetland-dependent birds such as waterfowl, waterbirds, and shorebirds, and we have illustrated the process using examples of spatial models for five species of upland-nesting ducks (*Anas* spp.), sora (*Porzana carolina*), and marbled godwit (*Limosa fedoa*). Our emphasis is on conservation as a process, and we have included considerable background information that we feel is necessary to convey the importance of context and program delivery to developing effective models for conservation at broad spatial scales.

THE PRAIRIE POTHOLE REGION

The PPR is located in the north central part of North America where areas of high wetland density intersect with grasslands of the northern Great Plains (Fig. 20-1). "Pothole" basins in the PPR are of glacial origin and contain a variety of wetland types ranging from wet meadows and shallow-water ponds to saline lakes, marshes, and fens (Cowardin et al. 1979, Kantrud et al. 1989). Most wetlands in

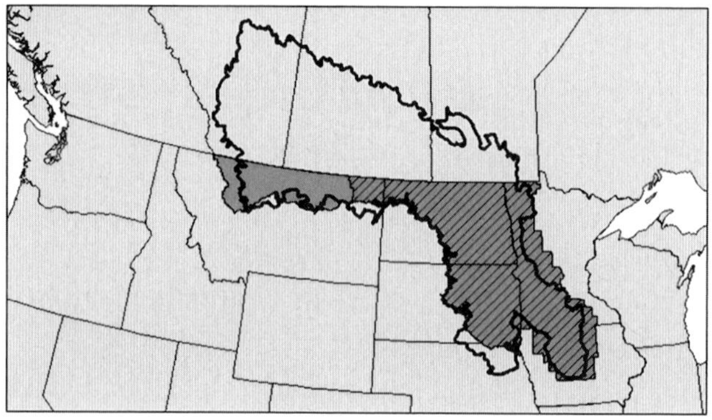

FIG. 20-1

Location of the Prairie Potholes Bird Conservation Region (black outline), which approximates the Prairie Pothole Region of North America. The U.S. Prairie Pothole Joint Venture is shown in dark gray; cross-hatching identifies the primary working area within the PPJV.

the PPR are < 0.5 ha in size, and wetland density exceeds 40 per km^2 in some areas (Kantrud et al. 1989). The numerous wetlands of the PPR help make it the most productive area for waterfowl in North America, hosting >50% of the continent's surveyed populations for 8 of 12 waterfowl species found in the region (Batt et al. 1989). The North American Waterfowl Management Plan (NAWMP; North American Waterfowl Management Plan Committee 1986) identified the PPR as the continent's top priority for waterfowl conservation and emphasized the need for innovative, landscape-level conservation strategies.

The myriad wetlands also make the PPR valuable to a host of other wetland-dependent species, especially waterbirds. Although population data for waterbirds are imperfect, the PPR appears to harbor ~70% of the continental population of Franklin's gull (*Larus pipixcan*); >50% of the continental population of pied-billed grebe (*Podilymbus podiceps*), American bittern (*Botaurus lentiginosus*), sora, American coot (*Fulica americana*), and black tern (*Chlidonia niger*); and ~30% of the continental population of American white pelican (*Pelecanus erythrorhynchos*) and California gull (*Larus californicus*; Beyersbergen et al. 2004). Grasslands in the PPR complement wetlands, as many species of wetland birds nest in surrounding grasslands, and nesting success of many species of grassland-nesting birds increases with amount of grass in the landscape (Greenwood et al. 1995, Reynolds et al. 2001, Herkert et al. 2003, Stephens et al. 2005).

CONSERVATION EFFORTS

Loss and degradation of wetland and upland habitats are the primary conservation issues affecting wetland birds in the PPR (Beyersbergen et al. 2004). Settlement by Europeans greatly transformed the PPR, largely through conversion of native grasslands and wetlands to agricultural fields. As a consequence, populations of most species of wetland birds have declined from historic levels, and habitat is considered the limiting factor for populations of most wetland bird species in the region. Accordingly, primary conservation treatments in the PPR are protection of existing wetlands and grasslands through purchase of conservation easements and restoration of degraded or converted wetlands and grasslands.

Conservation efforts in the PPR began in earnest in 1934, with passage of the Migratory Bird Hunting Stamp ("Duck Stamp") Act. Money from Duck Stamps went into the Migratory Bird Conservation Fund (MBCF), which was used to buy National Wildlife Refuges, and, since 1958, Waterfowl Production Areas in the form of fee-title acquisitions and purchase of wetland and grassland easements. Since 1989, when Congress passed the North American Wetlands Conservation Act (NAWCA) to provide funding to implement the NAWMP, the bulk of conservation efforts for permanent habitat in the PPR have been funded by the MBCF, NAWCA, and nonfederal match for NAWCA funds. Presently, approximately $13 million is spent annually on waterfowl conservation in the U.S. PPR, with about $11 million coming from the MBCF and the remainder from NAWCA. Ducks Unlimited and state wildlife agencies are the primary providers of matching funds for NAWCA grants.

Identification of priority conservation areas within the PPR is important given limited conservation funds and the large area, diversity of landcover, and variation in bird distribution and density within the region. Early conservation efforts in the PPR were largely opportunistic and focused on waterfowl. Following establishment of the NAWMP, the Prairie Pothole Joint Venture (PPJV) was formed as a regional, cooperative entity to coordinate waterfowl management by member groups and agencies in the U.S. PPR. Two Habitat and Population Evaluation Team (HAPET) offices were created in 1989 to provide strategic guidance for conservation in the PPJV; the modeling and conservation planning process we describe is the approach taken by these offices.

APPROACH TO CONSERVATION PLANNING

Realities of Conservation

Landscape-level conservation planning for wetland birds is complicated by a variety of social, ecological, and programmatic issues. The PPR landscape and economy are dominated by agriculture, which greatly influences conservation needs, opportunities, and implementation. Wetlands are viewed by some landowners as an impediment to farming, and these individuals may have little appreciation for, or even an active dislike of, some wildlife, wetlands, and conservation programs (Leitch 1989). On the other hand, grasslands and wetlands provide forage and water for cattle, and many ranchers in the region are supportive of efforts to maintain these habitats and are willing cooperators in conservation programs. Therefore, maintaining grasslands in the landscape has direct benefits for wetland conservation as well as the ranching community. However, government subsidies create financial incentives to raise commodity crops (i.e., corn, soybeans) instead of cattle, which have led to plowing of grasslands and draining or degradation of wetlands.

Land use and conservation issues vary across the PPR. Some physiographic regions such as the Agassiz Lake Plain along the Minnesota/North Dakota border are heavily cultivated and have virtually no wetlands or grasslands remaining. Landscapes such as these require extensive habitat restoration, although high land prices and competing land uses limit what can be accomplished. Other parts of the PPR have considerable areas of intact wetlands and/or grasslands that are used for cattle ranching or operations that include both cattle and crop agriculture. These landscapes are more conducive to conservation of existing habitat, which is invariably cheaper than habitat restoration; however, some conservation programs have funding dedicated to habitat restoration. Therefore, all landscapes and treatments are considered when planning for conservation action. The HAPET approach is to identify the best conservation treatment for every location based on landscape characteristics and the best location for every appropriate conservation treatment.

Water conditions in the PPR vary greatly over time and space (Fig. 20-2), which influences density and distribution of waterfowl (Stewart and Kantrud 1973, Brewster et al. 1976) and other waterbirds (Alisauskas and Arnold 1994, Peterjohn and Sauer 1997, Niemuth and Solberg 2003; Fig. 20-2). Consequently, areas that may experience high use by wetland birds one year may be completely unsuitable a few years later simply because of lack of water. Finally, limited funding also constrains conservation efforts in the PPR. Even though the PPJV's efforts have been expanded to include all priority migratory bird species that routinely inhabit the region, expanded commitments have not been met with corresponding new funding and the vast majority of funding historically and presently comes from and is directed toward waterfowl.

Model Development and Integration

Most conservation efforts in the PPR focus on protection and restoration of grasslands and wetlands, and there is great potential for providing benefits for multiple species. We developed standalone, single-species models because of the targeted nature of funding and because diversity metrics are often inappropriate as a response variable in models for conservation planning (Conroy and Noon 1996, Villard et al. 1998, Goldstein 1999). This approach allows targeting of locations and treatments to address different needs (e.g., preservation, restoration, or enhancement of wildlife habitat) for any focal species, combination of species, or program. This approach also allows for rapid response to requests for specific decision support (see Johnson et al., this volume), is conducive to adaptive changes in models as new information becomes available, and maximizes ability

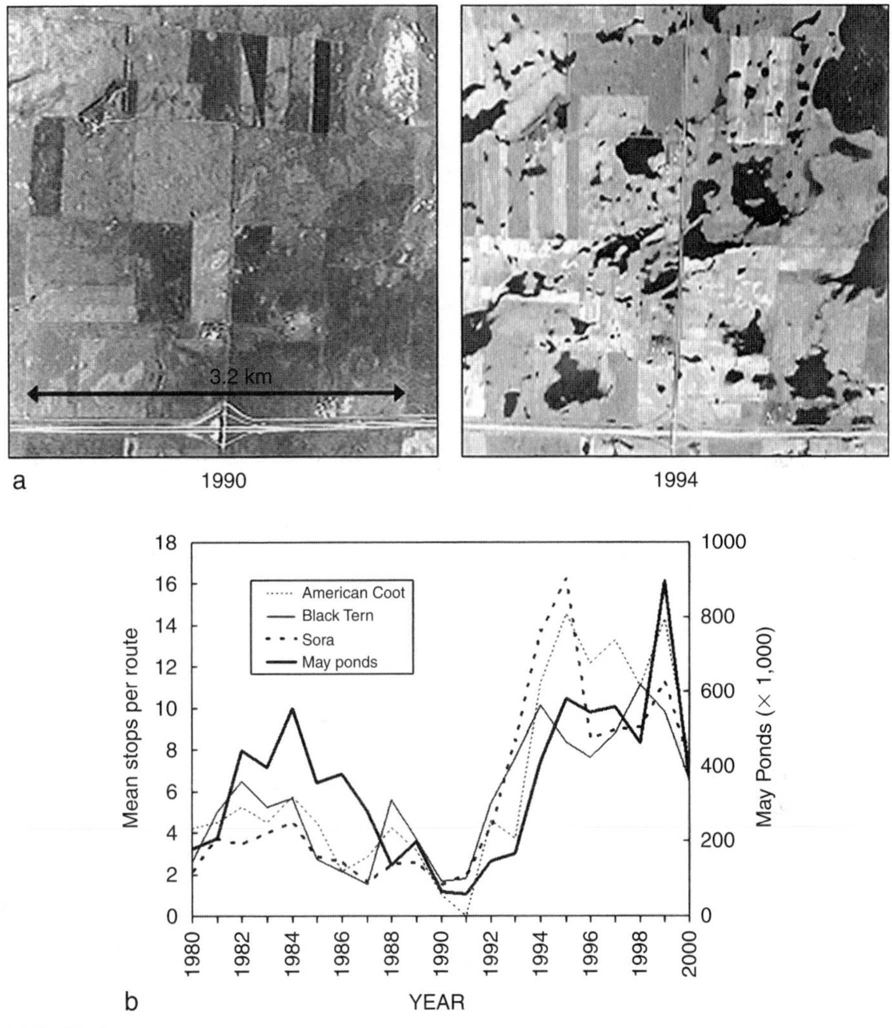

FIG. 20-2

(A) Aerial videograph of Four-Square-Mile Survey Plot 182 in central North Dakota, USA, in 1990 and 1994; note highway interchange at bottom of videographs. (B) Variation in estimated number of May ponds and number of Breeding Bird Survey stops on which waterbirds were detected in north-central North Dakota, 1980–2000 (Niemuth and Solberg 2003).

to integrate programs and species while maintaining biological integrity of models used as conservation planning tools. Spatial tools may vary depending on location and are developed to meet specific needs of partners and programs. We believe the philosophy of separate planning and integrated action allows maximum flexibility while maintaining biological integrity of models. This approach also preserves the unique priorities and objectives of conservation partners, as opposed to an approach where partners may be pressured to work only in areas of overlap identified for multiple species. Promoting conservation actions only in areas of maximum species overlap (i.e., a local species richness approach) can lead to "conservation mediocrity" and is often inconsistent with the greater need to conserve biodiversity at the continental and ecosystem scales.

Spatial models developed for conservation planning in the PPR incorporate different types of biological responses; unfortunately, costs of acquiring data used to parameterize models typically increase with the usefulness or completeness of the response (Fig. 20-3A). The type of model developed for conservation

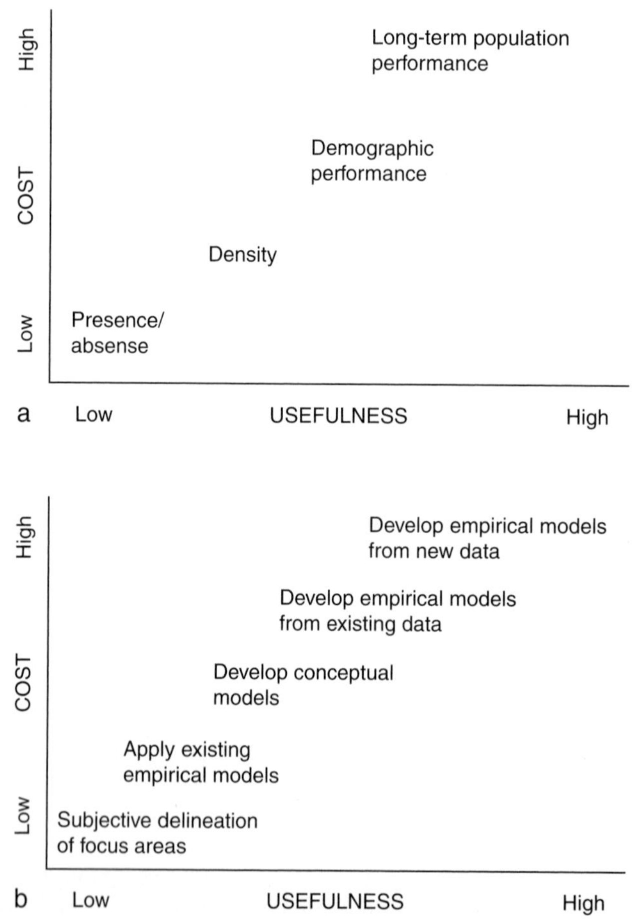

FIG. 20-3

General relationships between usefulness and cost for (A) various levels of biological response used in spatial models and (B) methods of developing spatial models.

planning in the PPR depends primarily on the question being asked and resources available for data collection and model development (Fig. 20-3B). We use presence-absence data and logistic regression for identifying potential habitat when species occur in low numbers or when more comprehensive data are not available. When count data are available, we use linear or Poisson regression to model densities of birds across the landscape with the understanding that high density does not necessarily indicate habitat quality. However, if habitat quality is defined as the product of density, survival, and reproduction vital rates (Van Horne 1983), models predicting density are valuable as a component of that definition, particularly when used in combination with additional information that provides some indication of population response such as survival or nesting success. In addition, models of bird abundance help ensure that assessments of risk or expenditure of limited conservation resources consider many, rather than few, birds. Our goal is to create biologically and scientifically sound empirical models, developed at appropriate scales and meeting specific purposes. Finally, we ensure that our models are of sufficiently fine spatial and temporal resolution that they can provide guidance at the scale at which conservation actions take place, and we acknowledge the assumptions and limitations inherent to modeling (see Johnson et al., this volume). In the absence of data suitable for developing empirical models, we have found considerable value in using conceptual models to assess landscapes (Niemuth et al. 2005) and aid in the identification of assumptions and knowledge gaps (also see Dijak and Rittenhouse, this volume; and Fitzgerald et al., this volume for applications of habitat suitability models). In this chapter, we illustrate the development of empirical models for waterfowl and marbled godwit using field data collected through our offices; the development of empirical models for sora using Breeding Bird Survey (BBS) data; and the development of a conceptual model for marbled godwit. These models follow a generalized hierarchy of usefulness and cost (Fig. 20-3B).

Data Availability and Quality

Reliable wetland and landcover data are the foundation of all these modeling efforts. We used the National Wetlands Inventory (NWI) digital database, which is based on the Cowardin et al. (1979) wetland classification system, and provided finer thematic and spatial resolution for wetlands than was possible using available satellite imagery. National Wetlands Inventory data for our study region are based on aerial photographs collected in the late 1970s and early 1980s; a 2005 evaluation of wetland loss in North Dakota and South Dakota indicates that $<3\%$ of wetland basins identified by NWI showed indications of new surface drainage since aerial photography was collected (C. R. Loesch, U.S. Fish and Wildlife Service, unpublished data). Aerial photography used by the NWI was collected during periods of average precipitation. Some very small wetlands identified by the NWI were delineated by a single point on a map, to which we assigned a buffer of 7.6 m, creating a polygon with an area of 0.015 ha. The NWI delineated different water regime and vegetation zones, when present, within large wetlands. In these cases we created wetland basin polygons identified by the most permanent water regime within each basin (Cowardin et al. 1995, Johnson and Higgins 1997). Our final classes were temporary, seasonal, semipermanent, permanent, and riverine wetland basin polygons, which we used in the models. National Wetland Inventory data were not available for all of Montana, which is why our primary working area did not include all of the PPR in Montana.

For the sora and marbled godwit models, we used landcover data derived from Thematic Mapper satellite images (30 m resolution) acquired from May 1992 through September 1996. Individual images were classified, upland landcover classes were resampled to 2.02 ha minimum mapping unit, and NWI basin data were integrated into the grid with a 0.09 ha minimum size of individual wetland basins (Table 20-1). User accuracy for all images exceeded 80% (U.S. Fish and Wildlife Service, unpublished data). Biological response (i.e., bird) data were collected at scales and times appropriate to the questions being addressed in models and are discussed in the following sections.

Conservation Planning for Ducks

An early priority of the PPJV was development of tools to identify priority areas for waterfowl conservation efforts. In this chapter, we summarize the waterfowl modeling work of Reynolds et al. (2006) and compare their results to models developed for sora and marbled godwit. Modeling efforts focused on five species of upland-nesting ducks because of their high numbers in the PPR and importance to continental harvest: blue-winged teal (*Anas discors*), gadwall (*A. strepera*), mallard (*A. platyrhynchos*), Northern pintail (*A. acuta*), and Northern shoveler (*A. clypeata*). For a complete description of field methodology and model development as it was applied in North Dakota, South Dakota, and northeastern Montana, see Cowardin et al. (1995), Reynolds et al. (1996), and Reynolds et al. (2006); data collection and model development were similar in the U.S. Fish and Wildlife Service (USFWS) Region 3 (Minnesota and Iowa) portion of the PPR (R. R. Johnson, unpublished data).

Conservation Planning for Waterbirds and Shorebirds

Relatively little information exists regarding relationships between nonwaterfowl birds and landscapes (Scott et al. 1993, Flather and Sauer 1996), and development of spatial planning tools for nonwaterfowl birds in the PPR lags behind development of spatial planning tools for ducks. Breeding biology, species status, and available data strongly influence the approach taken for model

Table 20-1 Candidate Predictor Variables Used to Model Number of Soras Detected at Breeding Bird Survey Stops in North Dakota, USA. All Landcover Variables were Calculated from Variably Sized Buffers Around BBS Stops

Landscape Variable	Description
Temporary (%)	Percent of wetland area within the buffer composed of temporary wetland basins derived from NWI data.
Seasonal (%)	Percent of wetland area within the buffer composed of seasonal wetland basins derived from NWI data.
Semipermanent (%)	Percent of wetland area within the buffer composed of semipermanent wetland basins derived from NWI data.
Wetland variety (n)	Number of different wetland water regimes (temporary, seasonal, etc.) within moving window.
Wetland number (n)	Number of wetland basins within moving window.
Wetness (%)	Percent of area of seasonal and semipermanent wetland basins containing water, interpolated from >23,000 basins videographed during 1995 waterfowl surveys (see Reynolds et al. 2006) in North Dakota, South Dakota, and eastern Montana.
Undisturbed Grass (%)	Percent of buffer composed of mix of cool-season grass and forb species planted on previously cropped land; generally undisturbed but may be hayed or grazed intermittently. Includes CRP plantings and dense nesting cover on waterfowl production areas.
Forest (%)	Percent of buffer composed of forest cover within each buffer.
Northing	Universal Transverse Mercator coordinate indicating north-south position. Also included as quadratic term.
Easting	Universal Transverse Mercator coordinate indicating east-west position. Also included as quadratic term.
Observer	Identifier for each observer, coded as 0/1 binary variable.
Stop Number	Number (1–50) of stop within each route.

development and conservation of traditional waterbird species in the PPR. For example, American white pelican and Franklin's gull are highly colonial and typically nest in the same few locations each year. Conservation of habitat for these species is simplified, as their nesting sites are generally known and, in most cases, already protected. Species such as sora and black tern are more broadly distributed across the landscape, and their distribution and density can vary greatly among years. Therefore, we have adopted a landscape approach to modeling and conservation of species such as these. Because population size of most waterbird species in the PPR is poorly known, conservation scientists have not set numeric population goals but instead strive for "no net loss" of populations through conservation of existing habitat (Beyersbergen et al. 2004). Populations and ranges of several species of shorebirds breeding in the PPR also are shrinking as wetland and grassland habitats are lost; for marbled godwit, planners have set a tentative population target of a 35% increase (commensurate with past habitat loss) over a present estimate of 168,000 in the Great Plains subpopulation (Brown et al. 2000).

We used stop-level BBS data in conjunction with landcover information to model correlates of the number of soras detected at BBS stops in the PPR portion of North Dakota in 1995; we did not model soras in South Dakota due to limited BBS coverage in that state. Because seasonal timing of the BBS (Bystrak 1981) was suspected of being suboptimal for detecting species such as marbled godwit, we used data from a regional survey of breeding shorebirds instituted by the Bismarck HAPET office to develop a similar landscape-level model predicting presence of marbled godwits. Empirical models presented in this chapter that predict density and distribution of sora and marbled godwit are examples based on one year of data; in practice, models from multiple years are used to reduce variation caused by changes in moisture and wetland conditions among years. Spatial and temporal variation in bird numbers reinforces the importance of long-term data sets collected over broad spatial extents to effective conservation planning in the PPJV.

MODELING SPECIES ABUNDANCE AND DISTRIBUTION

Waterfowl Models

Following methodology detailed in Reynolds et al. (2006), we sampled waterfowl annually from 1987–1998 on 626 10.4 km^2 primary sampling blocks that were stratified based on the area of land that the USFWS owned or had under easement in the surrounding 93.2 km^2 township (Cowardin et al. 1995). We randomly selected approximately 4,435 wetland basins from within the primary sampling blocks that we visited twice each year, once from 1 May–15 May and again from 20 May–5 June; we matched data with peak occurrence of each species (Reynolds et al. 2006; R. R. Johnson et al., unpublished data). Surveyors estimated the percentage of surface area of each wetland basin covered by water by comparing the extent of water observed in wetland basins to mapped NWI wetland boundaries overlaid on aerial photographs. We used these ground data to develop models that incorporated both temporal and spatial variation in wetland condition. We did not include riverine wetlands, which composed <0.03% of wetlands in the study area, in the survey design, but modeled waterfowl presence on riverine wetlands using pair-ratio models from surveys conducted 1983–1986 (Reynolds et al. 2006).

Statistical analysis.—Given the number and distribution of waterfowl observed on wetland basins during surveys, we used linear regression in Montana and the Dakotas (Reynolds et al. 2006) and Poisson regression in Minnesota and Iowa (R. R. Johnson, unpublished data) to relate numbers of duck pairs to wetland and spatial variables. Because models were developed to be applied to approximately 3.3 million wetland basins in the PPJV area, we only considered predictor variables in model development that had been measured remotely for all wetlands. We developed models for each of the five priority species based on the nonlinear relationship identified by Cowardin et al. (1988) between duck pairs and wetland size for the four classes of wetland basins (temporary, seasonal, semipermanent, permanent) considered in analysis. We included Universal Transverse Mercator (UTM) coordinates in analysis because ducks were not distributed evenly throughout the PPR (Stewart and Kantrud 1973). We used backward stepwise procedures to fit each model, deleting terms with $P > 0.05$ in each step (Reynolds et al. 2006).

Model application.—Because ducks use nesting cover away from core wetlands used for feeding and resting (Duebbert et al. 1983), we used published home range characteristics to model potential accessibility of land units surrounding wetlands to female ducks on a species-specific basis (Reynolds et al. 2006). We derived potential accessibility to breeding hens in the region by summing the number of breeding pairs predicted to have access to 390 × 390 m (15.2 ha) land units for the five target species using the ArcInfo GRID module (Environmental Systems Research Institute, Redlands, California). Relationships between ducks and wetlands were described for individual wetland basins, which varied in size, so that portion of the waterfowl model did not have an explicit measure of scale. However, proximity zones used to determine accessibility of land units to breeding hens varied among species and ranged from 1.2 to 4.0 km (Reynolds et al. 2006).

Sora Models

We obtained 1995 data for 27 BBS routes within the PPR portion of North Dakota from the U.S. Geological Survey, Patuxent Wildlife Research Center, Laurel, Maryland. Each 40 km BBS route contained 50 stops, or survey points, 0.81 km apart; details of route placement and sampling were described by Bystrak (1981). We acquired digitized survey routes from the National Atlas of the United States (<http://nationalatlas.gov>) as an ArcView shapefile (Environmental Systems Research Institute, Redlands, California). We calculated locations for 50 stops on each BBS route by creating a point at the start node of each digitized route and every 0.81 km thereafter to the end of the route.

Because many bird species are influenced by the landscape beyond the area included by traditional bird survey methods (e.g., point-count circles; Howell et al. 2000), we sampled habitat at three scales using circular moving window analysis, which summarizes data within a "window" of a selected size around each 30 × 30 m cell in a raster GIS data layer. The area within each moving window was 48, 191, and 452 ha, respectively, for circles with radii approximating 400, 800, and 1,200 m.

Analyzing BBS data at the stop level allows inferences to be made at a much finer spatial resolution than using BBS data at the route level. However, developing predictive models from stop-level BBS data is complicated by the presence of spatial autocorrelation, which can lead to overestimation of the precision of parameter estimates (Legendre 1993) and obscure ecological patterns (Carroll and Pearson 2000). We addressed several forms of spatial structure and nuisance factors in stop-level BBS data. First, BBS stops are nested within routes, and varying ability of observers (see Sauer et al. 1994) on different routes may result in spatial patterns in detection. Therefore, we included observer identity as a dummy variable (Table 20-1) to incorporate differences in observer ability in our models. Second, detection of some species of birds varies substantially during the daily survey period (Robbins 1981), which begins 0.5 hour before sunrise and typically lasts 4 to 4.5 hours (Bystrak 1981). Thus, birds that are most vocal early in the day are more likely to be found on stops at the beginning of a route than at stops toward the end of a route. We included stop numbers to provide an index to time relative to sunrise (Table 20-1), which enabled incorporation of time-related differences in detection in predictive models. Third, bird distribution across large geographic extents may follow gradients as a consequence of trends in climate and landcover (see O'Connor et al. 1999). Consequently, adjacent stops were more likely to have similar landcover and avifauna than stops farther apart. We included easting and northing UTM coordinates as linear and quadratic terms (Table 20-1) to model broad-scale gradients in bird distribution as trend surface variables (Legendre 1993).

Statistical analysis.—We found that the number and distribution of soras detected on BBS stops in North Dakota in 1995 followed a Poisson distribution, so we used Poisson regression to model the number of soras detected at BBS stops as a function of predictor variables (Table 20-1). We developed a set of candidate models at each of the three scales and then used information-theoretic methods to evaluate how well models were supported by the data at each scale (Burnham and Anderson 1998). Based on previous studies, we assumed that soras would be positively associated with annual precipitation and local water availability; complexes of temporary, seasonal, and semipermanent wetlands; and dense grasslands surrounding wetlands (Kantrud and Stewart 1984, Fairbairn and Dinsmore 2001, Naugle et al. 2001, Niemuth and Solberg 2003). In an attempt to develop a parsimonious model and avoid spurious correlations, we only evaluated main effects of linear relationships, with the exception of the trend surface variables described previously. We assessed models for overdispersion based on goodness-of-fit of the global model, using Akaike's Information Criterion corrected for overdispersion and small sample size ($QAIC_c$) for model selection and adjusting variance estimates as appropriate (Burnham and Anderson 1998). We considered all models with AIC differences

(Burnham and Anderson 1998) ≤4.0; for purposes of this example, we did not average models but used the model that was best supported by the data. We used 80% of the data for model building and 20% for validation. We used Number Cruncher Statistical System (Hintze 2004) for statistical analysis and program PASSAGE (Rosenberg 2003) to assess autocorrelation.

We evaluated spatial dependencies in the data and the ability of models to account for spatial dependencies by creating Moran's *I* correlograms, which evaluate spatial dependence at increasing distances between points (Moran 1950, Legendre and Legendre 1998). Values of Moran's *I* range from −1 to 1 indicating greater levels of negative and positive spatial autocorrelation, respectively. We created correlograms for the amount of seasonal and semipermanent wetland in the landscape, number of soras detected at BBS stops, and residuals from models that incorporated observer effect, stop number, and stop location.

Model application.—We created maps showing predicted number of birds throughout the study region by incorporating GIS layers for habitat and location into the Poisson regression equation for the final model. Because the maximum number of soras detected at any BBS stop in the study area in 1995 was six, we capped predicted values at six individuals. Model output consisted of GIS cells representing the number of individuals predicted to be present at a BBS stop, which we reclassified into 60 categories ranging from 0 to 6 at intervals of 0.1. We then resampled resolution of GIS cells to an area equaling the 125 m effective detection distance assumed by Rosenberg and Blancher (2005).

Marbled Godwit Statistical Models

Prior to instituting the breeding shorebird survey, the Bismarck HAPET office spent two years assessing roadside bias, daily timing of surveys, and seasonal timing of surveys; based on these evaluations, roadside surveys were adopted. Survey routes were 40 km long and were randomly located within physiographic strata. Surveys were similar to the BBS in that stops were 0.8 km apart, were surveyed for three minutes, and included birds within a 400 m radius, but differed from the BBS in that surveys started at sunrise, routes were sampled once in early May and once in early June, and only breeding shorebirds were recorded.

Statistical analysis.—Methodology and landscape data were similar to that used to develop the sora model except that observer effects, stop number, and annual wetness were not included in models, as marbled godwit numbers do not appear to fluctuate with water conditions and we did not expect marbled godwit detection to be strongly influenced by time of day or observer ability. Given the low numbers of stops on which marbled godwits were detected and the low numbers of marbled godwits detected, we modeled probability of detecting marbled godwit using logistic regression, which models a binary response (detection/nondetection per stop, in this case); we developed models separately for North Dakota and South Dakota. We evaluated goodness of fit (Hosmer and Lemeshow 2000) of the global model and receiver operating characteristics (ROC) plots to indicate how models performed on data with which they were built, with the caveat that absolute use or nonuse at stops was not known. Receiver operating characteristics scores range from 0 to 1 and indicate the ability of a model to discriminate between two groups; a score of 0.5 indicates random performance and higher values indicate better discrimination (Hosmer and Lemeshow 2000).

Model application and comparison of models.—We applied the marbled godwit model using techniques similar to those described for soras. Model output consisted of GIS cells representing the relative probability of a marbled godwit being detected at a shorebird survey stop. Finally, we examined similarity in areas identified as high priority/density by the marbled godwit, sora, and waterfowl models. We calculated the correlation between GIS grid layers of the North Dakota portion of the PPR for all pair-wise combinations of the three models.

Marbled Godwit Conceptual Model

In the absence of data suitable for developing statistical models, conceptual models can provide guidance for conservation efforts (Fig. 20-3B). To identify important breeding sites for marbled godwits in Minnesota, the USFWS Region 3 HAPET met with marbled godwit experts from state and federal agencies in Minnesota. By leading the group on a tour of habitats ranging from suitable to unsuitable, essential elements of breeding godwit habitat patches and landscapes were identified. We formalized these concepts into rules (Table 20-2), and applied the rules to elevation, NWI, and classified landcover data.

Table 20-2 Parameters Used in a Conceptual Model to Predict Marbled Godwit Habitat Quality in Minnesota, USA

Scale	Characteristic	Criteria
Patch	Size	$\geq$ 130 ha: $\geq$ 400 m wide – required $\geq$ 800 m wide – better
	Wetlands	$\geq$ 1.6 ha of temporary or saturated wetlands per 130 ha patch; predicted quality did not increase with additional wetlands
	Trees	>100 m between patch and trees
Landscape	Percent grass (3.2 km radius)	10–30% – required >30% – better
	Topography (535 m radius)	$\geq$4% average slope – poor $\leq$3% average slope – better

RESULTS AND CHARACTERISTICS OF MODELS

Waterfowl Models

Waterfowl response to wetland basins varied both spatially and temporally among species and water regimes as described in Reynolds et al. (2006) with similar responses in Minnesota and Iowa (R. R. Johnson, unpublished data). In general, the number of duck pairs per unit of wet area increased from south to north and from east to west for all wetland classes; number of pairs increased nonlinearly, with higher densities on smaller wetlands (Reynolds et al. 2006). Cross-validation indicated that these models performed substantially better than models that did not account for spatial variation or nonlinearity. Models predicting the number of pairs on individual wetlands had R^2 values of approximately 0.30; this value increased to 0.88 when predicting total number of pairs occupying wetlands on landscapes of 41.6 km^2 (Reynolds et al. 2006). When we applied regression coefficients to the corresponding GIS layers, summed accessibility, and displayed results, the model resembled a radar image of a thunderstorm weather system crossing the region and was called the "Thunderstorm Map" (Fig. 20-4).

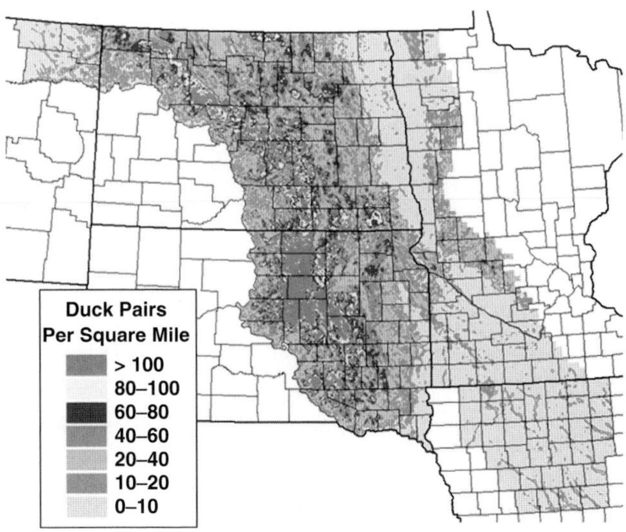

FIG. 20-4 (See color insert.)

Results of applying duck pair-wetland regression models to wetland basins, pair ratio models to riverine wetlands, and accessibility models to 390 × 390 m (15.2 ha) land units to the primary working area within the Prairie Pothole Region of the United States. This figure (referred to as the "Thunderstorm Map") shows which land units would be accessible to different densities of nesting hens and, thus, where grassland conservation efforts would provide the greatest benefits. Results are presented as pairs per km^2 for proximity zones around each land unit, where the area of proximity for each of five upland-nesting duck species was the approximate distance hens have been known to travel from core wetlands to nesting cover (derived from Reynolds et al. 2006; and R. R. Johnson, unpublished data).

Even though the model was developed using wetland data, the results (potential number of duck pairs in a community of wetlands with access to land units) are used to target uplands with potential for access by high numbers of nesting hens.

Sora Model

The area of seasonal and semipermanent wetlands within 800 m of BBS stops showed strong positive spatial autocorrelation (Fig. 20-5A), as did the number of soras (Fig. 20-5B). The number of birds detected per stop ranged from 0 to 6, with 365 soras detected at 238 of the 1,080 stops in the model-building data set. In addition to being influenced by observer ability, time of day, and location, the number of individuals detected was positively associated with amount of water in wetland basins; area of temporary, seasonal, and semipermanent

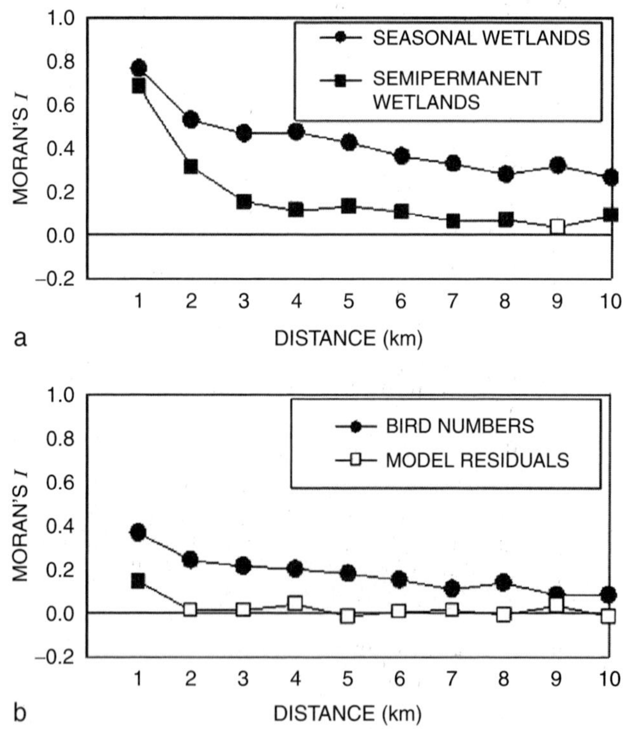

FIG. 20-5

(A) Moran's *I* correlograms for amount of seasonal and semipermanent wetlands within 800 m of 1,350 North Dakota Breeding Bird Survey stops. (B) Moran's *I* correlograms for number of soras detected at Breeding Bird Survey stops in 1995 (circles) and residuals from model including habitat, trend surface, observer, and time of day variables (squares). Filled symbols denote statistically significant ($P < 0.05$) positive spatial autocorrelation.

wetlands; area of undisturbed grass; number of wetland basins; and variety of water regimes in the surrounding landscape (Equation 1; nuisance factors such as observer effect and time of day not presented). There was little model uncertainty, as the best model had an AIC weight of 0.85, and the only other competing model with an AIC difference ≤ 4.0 had an AIC weight of 0.15. The rank of candidate models was similar but with consistent differences in AIC values, among scales, with lowest AIC values at the 800 m scale.

$$\begin{aligned}\text{Soras} = \text{Exp}(-38.18 - (5.76\text{E-6} * \text{East}) + (5.78\text{E-6} * \text{North})\\ +(0.1 * \text{Temporary}) + (0.09 * \text{Seasonal}) + (0.05 * \text{Semipermanent})\\ +(0.17 * \text{Wetland variety}) + (0.013 * \text{Wetland number}) + (0.11 * \text{Wetness})\\ +(0.008 * \text{Undisturbed Grass}) - (0.007 * \text{Forest}))\end{aligned} \qquad (20\text{-}1)$$

The final model fit moderately well ($R^2 = 0.33$); as expected, predicted numbers were significantly ($P < 0.0001$) correlated with actual number of birds detected in the validation portion of the data, although the correlation coefficient was low (0.37). Inclusion of trend surface, observer effect, moisture, and time-of-day terms substantially improved model fit and reduced positive spatial autocorrelation in residuals (Fig. 20-5B). Spatial patterns in density are readily discernible on the map showing estimated number of individuals (Fig. 20-6).

Marbled Godwit Models and Comparison of Statistical Models

Participants observed marbled godwits at 144 (11.5%) of 1,250 stops along 25 survey routes in North Dakota and 32 (3.8%) of 850 stops along 19 survey routes in South Dakota. Some model uncertainty existed, particularly in South

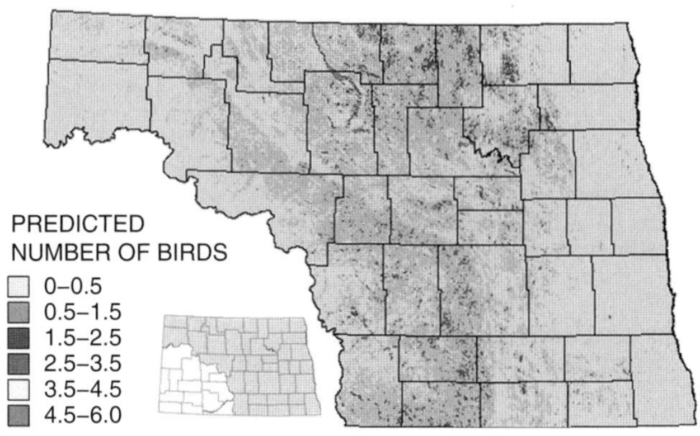

FIG. 20-6 (See color insert.)

Number of soras predicted to be present per 5-ha unit in the Prairie Potholes Bird Conservation Region portion of North Dakota, USA, in 1995 as a function of landscape-level spatial model. Low numbers relative to waterfowl model (Fig. 20-4) are due in part to different spatial scales used in model output.

Dakota, although competing models within states were similar (Table 20-3). Even though marbled godwits were more frequently observed in North Dakota, parameter estimates from final models indicated that marbled godwits showed similar responses to landscape characteristics in analyses for both states (Table 20-4); the lack of discontinuities along the North Dakota/South Dakota state line (Fig. 20-7A) reinforces the similarity in results. Final models performed well, with ROC scores of 0.74 and 0.82, for North Dakota and South Dakota, indicating acceptable and excellent discrimination, respectively (Hosmer and Lemeshow 2000). Models performed best using variables sampled with an 800-m window.

Areas of high predicted occurrence of marbled godwits in the PPR of North Dakota and South Dakota generally coincided with areas identified as having high potential waterfowl density (Fig. 20-7). Predicted presence of marbled godwits was positively correlated with predicted waterfowl accessibility ($r = 0.55$)

Table 20-3 State, AIC differences (Δ_i), variables included in model, number of parameters (K), and AIC weights (w_i) for logistic regression models predicting detection of marbled godwits in North Dakota and South Dakota

State	(Δ_i)	Variables in model	K	(w_i)
ND	0.0	East, north, grassland, temporary, seasonal, semipermanent	7	0.58
ND	2.0	East, north, grassland, temporary, seasonal, semipermanent, forest	8	0.21
ND	2.0	East, north, grassland, temporary, seasonal, semipermanent, number of wetlands	8	0.21
SD	0.0	East, north, grassland, temporary, seasonal, semipermanent, forest	8	0.44
SD	1.0	East, north, grassland, temporary, seasonal, semipermanent, forest, variety of wetlands	9	0.27
SD	2.0	East, north, grassland, temporary, seasonal, semipermanent, forest, number of wetlands	9	0.16
SD	2.5	East, north, grassland, temporary, seasonal, semipermanent	7	0.13

Table 20-4 Parameter Estimates for Landscape-Level Logistic Regression Models Predicting Detection of Marbled Godwits in North Dakota and South Dakota. Variable Labels Follow those of Table 20-1

State	Intercept	East	North	Grassland	Temporary	Seasonal	Semiperm.	Forest
ND	41.3	-8.6E-6	-7.3E-6	0.006	0.23	0.09	0.07	
SD	-14.1	-1.1E-5	3.7E-6	0.007	0.08	0.09	0.07	-4.3

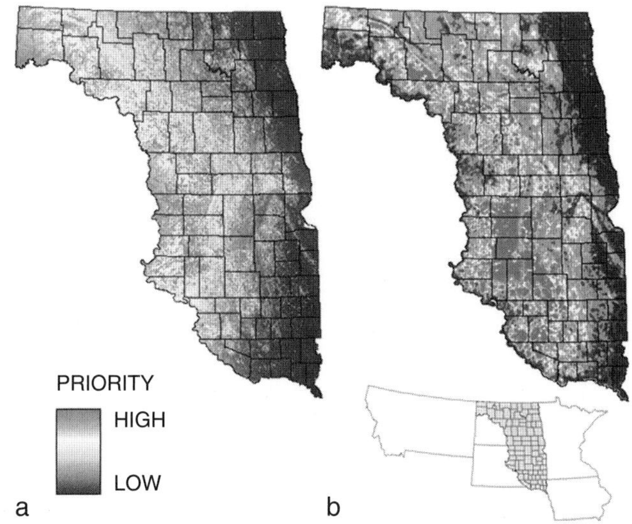

FIG. 20-7 (See color insert.)

(A) Predicted distribution of marbled godwit in the Prairie Pothole Region of North Dakota and South Dakota, USA, in 2004, where landscapes with high probability of detecting marbled godwits are given high priority. (B) Breeding duck "Thunderstorm Map" from Fig. 20-5 for the same geographic region and using same color scheme as (A); priority in this example was defined by number of five species of waterfowl predicted to have access to 390 × 390 m land units. Southern South Dakota is outside the range of marbled godwit; hence, predicted marbled godwit presence is low even though predicted duck numbers are high.

and modeled sora density in North Dakota ($r = 0.40$); predicted densities of sora and waterfowl also were positively correlated ($r = 0.42$; $P < 0.001$ in all cases).

The conceptual model developed for marbled godwit conservation in Minnesota identified areas most likely to support breeding marbled godwits and areas with varying potential for habitat restoration (Fig. 20-8A). Although the model was knowledge (versus data) driven, independent surveys and ancillary data correlated well with the model output (Fig. 20-8B).

APPLICATIONS TO CONSERVATION

The models we have described, particularly the waterfowl model and its derivatives, are the primary tools used to guide the annual expenditure of millions of dollars for acquisition of conservation easements within the PPJV. The primary criterion for acquiring a grassland easement in the PPJV is that the property to be acquired falls within a zone having $\geq$25 pairs of waterfowl per square mile (2.6 km^2) as identified by the Thunderstorm Map. In addition, the Thunderstorm Map is used to identify areas with high potential to attract nesting pairs but little

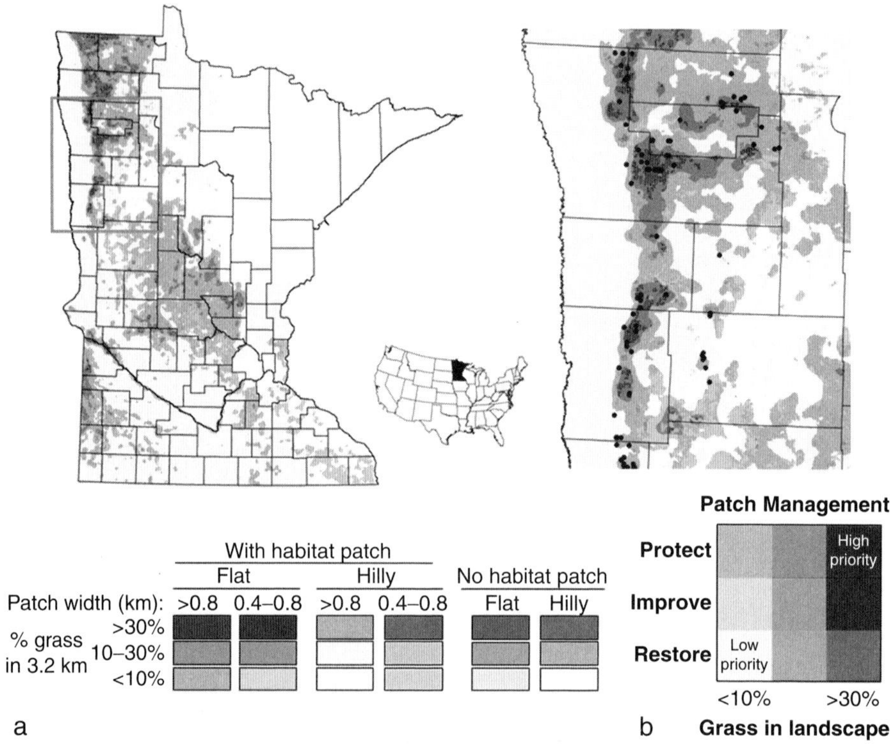

FIG. 20-8 (See color insert.)

(A) Conceptual model for marbled godwit habitat quality in Minnesota, USA, driven by expert knowledge. Dark red areas indicate the full complement of requirements are met and should be protected; purple areas would require restorations to meet species needs (i.e., patch development: increase grass patch size, wetland restoration within patch, etc.). Yellow and colorless areas should not be targeted for godwit management because of the godwit's preference for low-relief areas. (B) Portion of conceptual marbled godwit model in northwestern Minnesota showing location (◙) of marbled godwits observed along *ad hoc* roadside survey.

grassland cover to target for upland treatments such as grassland restoration. Finally, the Thunderstorm Map has been used by the U.S. Government Accountability Office (GAO; 2007) to evaluate the cost effectiveness of easements acquired by the U.S. Fish and Wildlife Service. As mentioned previously, models predicting density are particularly valuable when used in combination with additional information that provides an indication of population response such as survival or nesting success. Because waterfowl nesting success increases with the amount of grass in the landscape (Greenwood et al. 1995, Reynolds et al. 2001, Stephens et al. 2005), typical conservation treatments include acquisition of easements on grasslands in areas with high potential waterfowl accessibility or restoration of grasslands in areas with high potential duck numbers but little grassland.

Conservation models such as those we have described are not only useful for identifying areas for conservation action, but also allow assessment of how past conservation actions have benefited other species, identification of areas of overlapping priority, assessment of past and ongoing habitat loss, and estimating the relative impacts of wetland versus upland habitat change. As Fig. 20-7 and the correlations demonstrate, priority landscapes for waterfowl, sora, and marbled godwit in our region are similar, and it is likely that sora and marbled godwit have benefited substantially from waterfowl conservation efforts. Waterfowl production areas and conservation easements protect >2.7 million acres of wetlands and grasslands in the PPJV (Beyersbergen et al. 2004). These wetland and grassland complexes, protected primarily through waterfowl conservation dollars, have conserved substantial amounts of habitat for many species of non-waterfowl birds (Fig. 20-9; Naugle et al. 2001). Of course, species vary in their

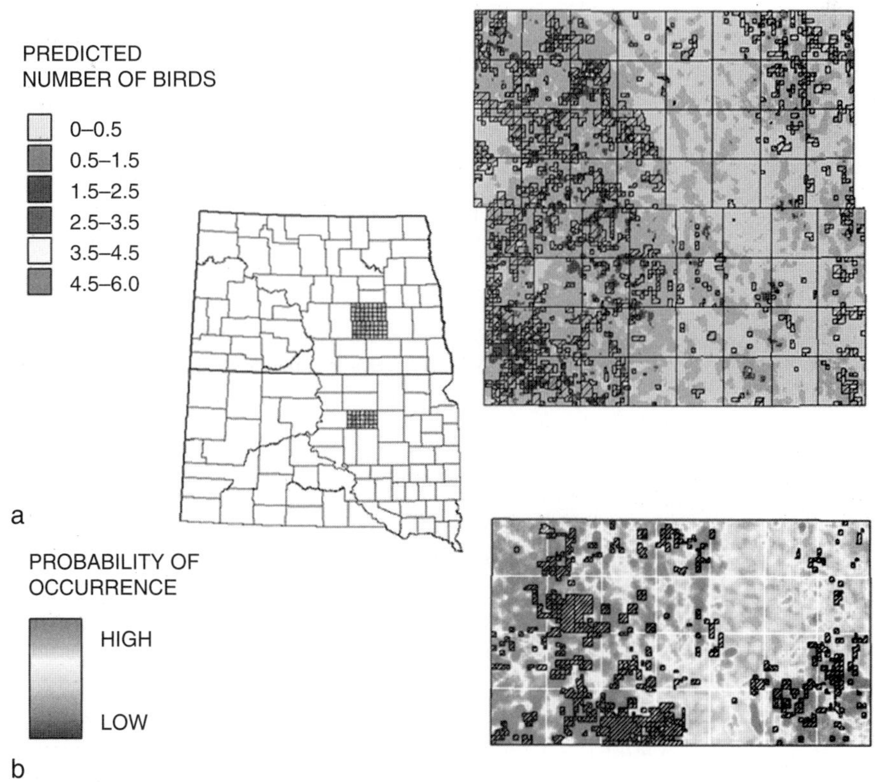

FIG. 20-9 (See color insert.)

(A) Waterfowl Production Areas and wetland easements (black hatching) overlaid on map of predicted number of soras in Stutsman County, North Dakota. (B) Waterfowl Production Areas and grassland easements (black hatching) overlaid on map of predicted probability of occurrence of marbled godwit in Faulk County, South Dakota. Gridlines in both maps indicate civil townships at 10 km intervals.

habitat requirements and not all species will show as much overlap as the species we have presented here. For the most part, conservation needs in the PPR are simple: Preserve and enhance wetlands and grassland where they exist and restore wetlands and grasslands where they have been converted to other uses. But that simplicity quickly disappears when multiple management treatment types (e.g., wetland protection, riparian restoration, etc.) with varying costs and opportunities are considered for multiple species over a broad geographic region. Spatial models enable conservation scientists to assess these factors and identify gaps in coverage as well as areas of overlap, with the goal of maximizing the benefits of conservation programs in the PPR. Spatially and biologically explicit models also allow critical examination of costs and benefits of proposed policies and programs.

Policy and Management Implications

Despite the relative simplicity of conservation needs in the PPR, current conservation efforts appear insufficient to maintain present levels of habitat. Spatial models such as we have described are useful in that they help prioritize and target areas for conservation, thus increasing efficiency of conservation efforts. But conservation in the PPR is not limited by lack of planning tools such as the spatial models we have described or the innumerable planning documents that are the devil's spawn of various conservation initiatives. Conservation is limited by funding to conserve and restore habitat. Wetland-dependent birds in the PPR have benefited greatly from agricultural programs such as the Conservation Reserve Program (CRP) and the "Swampbuster" provision of the 1985 U.S. Food Security Act (Johnson et al. 1996, Reynolds et al. 2001, Beyersbergen et al. 2004, Niemuth et al. 2006). In the absence of substantial increases in conservation funding, continuation and expansion of wildlife-friendly agricultural policy count as the single factor that can provide the greatest benefit to wetland bird populations in the PPR. Limited resources and competing demands for land will require strategic application of such programs to increase benefits to wildlife. Spatial models are ideal for such applications (Reynolds et al. 2006) and are being used to target enrollment of lands in the CRP across the PPR (Farm Service Agency 2006) and the Wetland Reserve Program in Minnesota (R. R. Johnson, personal observation). Spatial models also will be useful for strategic targeting of landscapes for wildlife benefits in other programs, such as enrollment of wetlands and grasslands in future carbon sequestration programs.

Wildlife management in the PPR is complicated by several factors. First, because the majority of the land in the PPR is privately owned and used for agricultural production, cooperation with private landowners will be essential to continued conservation efforts in the region. Second, many wetland bird species in the region respond to wetland and grassland complexes at a landscape scale (Naugle et al. 2001, Reynolds et al. 2001, Niemuth et al. 2006). Conservation treatment and management actions need to consider landscape context and may involve many landowners, as well as a variety of governmental and nongovernmental agriculture and conservation programs. Finally, even though waterfowl conservation efforts have conserved substantial amounts of habitat used by nongame species, the value of waterfowl conservation to nongame species is sometimes questioned. Additional, nonwaterfowl funding will be necessary to meet the conservation needs of all priority species in the region. These considerations, and others, further emphasize the need for a cooperative approach to conservation in the PPR.

FUTURE DIRECTIONS

Considerable uncertainty exists regarding future directions for wetland bird conservation in the PPR, as several factors indicate that avian distribution, land use, and conservation issues are likely to change in the future. Potential effects of global climate change in the PPR are poorly understood, but likely will influence water conditions and wetland distribution by altering precipitation and evapotranspiration levels (Johnson et al. 2005). Agricultural land use in the region likely will intensify as native prairie continues to be converted to cropland and genetically modified crops are planted to help meet increasing demands for food commodities and biofuels (Higgins et al. 2002, Krapu et al. 2004). The extent and degree of these changes may be shaped by federal farm programs, which greatly influence conservation in the region (Johnson and Igl 1995, Johnson et al. 1996, Reynolds et al. 2001), and are in turn influenced by a host of political and economic factors. Spatial models will be necessary to assess and address the effects of these, and other, changes as they influence conservation delivery in the PPR.

Several factors could substantially improve future modeling efforts and our ability to increase efficiency of conservation planning in the PPJV. High priority should be given to the acquisition of more and better bird data to be used as a response variable in model development. This need is particularly important for nonwaterfowl species, of which many priority species are infrequently detected on existing surveys. Presently, most of the models we use for conservation planning in the PPJV focus on species presence or density; acquiring survival and productivity data would enable better consideration of aspects of avian conservation that presently are poorly addressed, particularly for nongame species. Better understanding of existing data and models is also a priority. For example, the effect of roadside bias on parameter estimates in spatial models developed from BBS data is unknown. Similarly, detection probabilities have not been assessed for any of the data sets we presently use in development of spatial models. Finally, inclusion of upland habitat variables in waterfowl models may be a valuable next step, especially as models are used to make decisions regarding placement of upland treatments such as CRP grasslands.

Expanded species coverage also will be needed. Current planning efforts emphasize focal species and conservation of areas that provide benefits for multiple species (see Johnson et al., this volume). As models for additional species are developed and planning efforts become more refined, planning will better incorporate species whose habitats show little or no overlap with other species. Evaluation and adoption of new modeling techniques (e.g., Elith et al. 2006) may help improve modeling efforts and enable development of useful models for species for which little data are available. However, it is imperative that pursuit of improvements in statistical methodology does not overshadow species biology and the development and use of models for specific applications and treatments. Acquiring finer-grained remotely sensed data would enable incorporation of additional information (e.g., vegetation species composition and structure, distribution of small clumps of trees and shrubs, amount and configuration of emergent vegetation in wetland basins) into models. However, some fine-grained features are ephemeral and influenced by annual precipitation levels, and are therefore of little value for long-term planning. Finally, all of our models are based on assumptions that landcover data are accurate, bird-habitat relationships are adequately modeled, and that conservation treatments adequately address factors presently limiting populations of species of concern. All these assumptions must be assessed and models refined in an adaptive manner, especially as human pressures on wildlife continue to increase.

Given increasing demands on resources, future conservation efforts will require even greater collaboration among federal and state agencies, nongovernmental organizations, and other partners. However, collaboration must go beyond the formation of partnerships and plans. The solution to conservation needs in the PPR lies not with better planning and modeling, but with increased on-the-ground actions that benefit wetland birds.

SUMMARY

The millions of wetlands that define the Prairie Pothole Region (PPR) make it the most productive area for waterfowl in North America. These wetlands are equally important to many other wetland-dependent bird species, particularly shorebirds and waterbirds. In response to ongoing loss of wetlands and grasslands, extensive conservation initiatives, particularly acquisition of conservation easements, have been undertaken to conserve wetland bird habitat in the United States portion of the PPR. In recent years, these conservation efforts have been guided largely by the results from spatial models that evaluate landscapes relative to their accessibility to breeding waterfowl and their potential for waterfowl production. We presented a philosophy of conservation planning and illustrated that philosophy with examples of spatial models that predict density and distribution of sora and marbled godwit, in addition to upland-nesting waterfowl. Given the targeted nature of conservation funds and that diversity metrics are inappropriate as a response variable in models used for conservation planning, we developed species-specific models that can stand alone or be integrated with results of other models. This allows targeting of locations and treatments to address different needs (e.g., preservation, restoration, or enhancement of wildlife habitat) for any focal species, combination of species, or program while maintaining biological integrity of information used in conservation planning tools. We developed models by using National Wetlands Inventory data, landcover data, and estimates of water conditions as predictor variables for species presence-absence or count data. We determined that target bird species in the region were influenced by landscape composition and configuration, wetland class, and amount of water in wetland basins. Priority areas for several wetland-dependent bird species in the region overlap considerably, and many nonwaterfowl species have benefited substantially from waterfowl conservation efforts in the PPR.

ACKNOWLEDGMENTS

We thank the many volunteers and cooperators who collected data during annual waterfowl, shorebird, and breeding bird surveys. We particularly wish to acknowledge participants from U.S. Fish and Wildlife Service Refuge field offices, who support much of our work. We also thank the USFWS National Wetlands Inventory and the Northern Great Plains Office of Ducks Unlimited for helping create digital databases used in our models.

LITERATURE CITED

Alisauskas, R. T., and T. W. Arnold. 1994. American coot. Pages 127–143 *in* T. C. Tacha and C. E. Braun, editors. *Migratory shore and upland game bird management in North America.* International Association of Fish and Wildlife Agencies, Washington, D.C., USA.

Batt, B. D., M. G. Anderson, C. D. Anderson, and F. D. Caswell. 1989. Use of prairie potholes by North American ducks. Pages 204–227 *in* A. van der Valk, editor. *Northern prairie wetlands.* Iowa State University, Ames, USA.

Beyersbergen, G. W., N. D. Niemuth, and M. R. Norton. 2004. *Northern Prairie and Parkland Waterbird Conservation Plan.* Prairie Pothole Joint Venture, Denver, Colorado, USA.

Brewster, W. G., J. M. Gates, and L. D. Flake. 1976. Breeding waterfowl populations and their distribution in South Dakota. *Journal of Wildlife Management* 40:50–59.

Brown, S., C. Hickey, and B. Harrington, editors. 2000. *United States Shorebird Conservation Plan.* Manomet Center for Conservation Sciences, Manomet, Massachusetts, USA.

Burnham, K. P., and D. R. Anderson. 1998. *Model selection and inference: A practical information-theoretic approach.* Springer-Verlag, New York, New York, USA.

Bystrak, D. 1981. The North American Breeding Bird Survey. *Studies in Avian Biology* 6:34–41.

Carroll, S. S., and D. L. Pearson. 2000. Detecting and modeling spatial and temporal dependence in conservation biology. *Conservation Biology* 14:1893–1897.

Conroy, M. J., and B. R. Noon. 1996. Mapping of species richness for conservation of biological diversity: Conceptual and methodological issues. *Ecological Applications* 6:763–773.

Cowardin, L. M., V. Carter, F. C. Golet, and E. T. LaRoe. 1979. *Classification of wetlands and deepwater habitats of the United States*. U.S. Fish and Wildlife Service, FWS/OBS/-79/31, Washington, D.C., USA.

Cowardin, L. M., D. H. Johnson, T. L. Shaffer, and D. W. Sparling. 1988. *Applications of a simulation model to decisions in mallard management*. U.S. Fish and Wildlife Service, Technical Report 17, Washington, D.C., USA.

Cowardin, L. M., T. L. Shaffer, and P. M. Arnold. 1995. *Evaluation of duck habitat and estimation of duck population sizes with a remote-sensing based system*. National Biological Service, Biological Science Report 2, Washington, D.C., USA.

Duebbert, H. F., J. T. Lokemoen, and D. E. Sharp. 1983. Concentrated nesting of mallards and gadwalls on Miller Lake Island, North Dakota. *Journal of Wildlife Management* 47:729–740.

Elith, J., C. H. Graham, R. P. Anderson, M. Dudik, S. Ferrier, A. Guisan, R. J. Hijmans, F. Huettmann, J. R. Leathwick, A. Lehmann, J. Li, L. G. Lohmann, B. A. Loiselle, G. Manion, C. Moritz, M. Nakamura, Y. Nakazawa, J. McC. Overton, A. T. Peterson, S. J. Phillips, K. S. Richardson, R. Scachetti-Pereira, R. E. Schapire, J. Soberón, S. Williams, M. S. Wisz, and N. E. Zimmermann. 2006. Novel methods improve prediction of species' distributions from occurrence data. *Ecography* 29:129–151.

Fairbairn, S. E., and J. J. Dinsmore. 2001. Local and landscape-level influences on wetland bird communities of the Prairie Pothole Region of Iowa, USA. *Wetlands* 21:41–47.

Farm Service Agency. 2006. *Conservation Reserve Program duck nesting habitat initiative fact sheet*. U.S. Farm Service Agency, Washington D.C., USA.

Flather, C. H., and J. R. Sauer. 1996. Using landscape ecology to test hypotheses about large-scale abundance patterns in migratory birds. *Ecology* 77:28–35.

Goldstein, P. Z. 1999. Functional ecosystems and biodiversity buzzwords. *Conservation Biology* 13:247–255.

Greenwood, R. J., A. B. Sargeant, D. H. Johnson, L. M. Cowardin, and T. L. Shaffer. 1995. Factors associated with duck nest success in the Prairie Pothole Region of Canada. *Wildlife Monographs* 128.

Herkert, J. R., D. L. Reinking, D. A. Wiedenfeld, M. Winter, J. L. Zimmerman, W. E. Jensen, E. J. Finck, R. R. Koford, D. H. Wolfe, S. K. Sherrod, M. A. Jenkins, J. Faaborg, and S. K. Robinson. 2003. Effects of prairie fragmentation on the nest success of breeding birds in the midcontinental United States. *Conservation Biology* 17:587–594.

Higgins, K. F., D. E. Naugle, and K. J. Forman. 2002. A case study of changing land use practices in the northern Great Plains, U.S.A.: An uncertain future for waterbird conservation. *Waterbirds 25, special publication* 2:42–50.

Hintze, J. 2004. *NCSS and PASS*. Number Cruncher Statistical Systems, Kayesville, Utah, USA.

Hosmer, D. W., and S. Lemeshow. 2000. *Applied logistic regression*. Second edition. John Wiley & Sons, New York, New York, USA.

Howell, C. A., S. C. Latta, T. M. Donovan, P. A. Porneluzi, G. R. Parks, and J. Faaborg. 2000. Landscape effects mediate breeding bird abundance in Midwestern forests. *Landscape Ecology* 15:547–562.

Johnson, D. H., and L. D. Igl. 1995. Contributions of the Conservation Reserve Program to populations of breeding birds in North Dakota. *Wilson Bulletin* 107:709–718.

Johnson, R. R., and K. F. Higgins. 1997. *Wetland resources of eastern South Dakota*. South Dakota State University, Brookings, South Dakota, USA.

Johnson, R. R., D. E. Naugle, M. E. Estey, and K. F. Higgins. 1996. Characteristics of eastern South Dakota wetlands basins and implications of changes in jurisdictional wetland definitions. *Transactions of the North American Wildlife and Natural Resources Conference* 61:127–136.

Johnson, W. C., B. V. Millett, T. Gilmanov, R. A. Voldseth, G. R. Guntenspergen, and D. E. Naugle. 2005. Vulnerability of northern prairie wetlands to climate change. *BioScience* 55:863–872.

Kantrud, H. A., G. L. Krapu, and G. A. Swanson. 1989. Prairie basin wetlands of the Dakotas: A community profile. *U.S. Fish and Wildlife Service, Biological Report* 85(7.28), Washington, D.C., USA.

Kantrud, H. A., and R. A. Stewart. 1984. Ecological distribution and crude density of breeding birds on prairie wetlands. *Journal of Wildlife Management* 48:426-437.

Krapu, G. L., D. A. Brandt, and R. R. Cox, Jr. 2004. Less waste corn, more land in soybeans, and the switch to genetically modified crops: Trends with important implications to wildlife management. *Wildlife Society Bulletin* 32:127-136.

Legendre, P. 1993. Spatial autocorrelation: Trouble or new paradigm? *Ecology* 74:1659-1673.

Legendre, P., and L. Legendre. 1998. *Numerical ecology.* Second edition. Elsevier Scientific Publishing, Amsterdam, The Netherlands.

Leitch, J. A. 1989. Politicoeconomic overview of prairie potholes. Pages 3-14 *in* A. van der Valk, editor. *Northern Prairie Wetlands.* Iowa State University, Ames, USA.

Moran, P. A. P. 1950. Notes on continuous stochastic phenomena. *Biometrika* 37:17-23.

Naugle, D. E., R. R. Johnson, M. E. Estey, and K. F. Higgins. 2001. A landscape approach to conserving wetland bird habitat in the Prairie Pothole Region of eastern South Dakota. *Wetlands* 21:1-17.

Niemuth, N. D., M. E. Estey, and C. R. Loesch. 2005. Developing spatially explicit habitat models for grassland bird conservation planning in the Prairie Pothole Region of North Dakota. Pages 469-477 *in* C. J. Ralph and T. D. Rich, editors. *Bird conservation implementation and integration in the Americas.* Proceedings of the 3rd International Partners in Flight Conference 2002. U.S. Forest Service, Pacific Southwest Research Station, PSW-GTR-191, Albany, California, USA.

Niemuth, N. D., M. E. Estey, R. E. Reynolds, C. R. Loesch, and W. A. Meeks. 2006. Use of wetlands by spring-migrant shorebirds in agricultural landscapes of North Dakota's Drift Prairie. *Wetlands* 26:30-39.

Niemuth, N. D., and J. W. Solberg. 2003. Response of waterbirds to numbers of wetlands in the Prairie Pothole Region of North Dakota, U.S.A. *Waterbirds* 26:233-238.

North American Waterfowl Management Plan Committee. 1986. *North American Waterfowl Management Plan.* U.S. Fish and Wildlife Service, Washington, D.C., USA.

O'Connor, R. J., M. T. Jones, R. B. Boone, and T. B. Lauber. 1999. Linking continental climate, land use, and land patterns with grassland bird distribution across the conterminous United States. *Studies in Avian Biology* 19:45-59.

Peterjohn, B. G., and J. R. Sauer. 1997. Population trends of black terns from the North American Breeding Bird Survey, 1966-1996. *Colonial Waterbirds* 20:566-573.

Reynolds, R. E., D. R. Cohan, and M. A. Johnson. 1996. Using landscape information approaches to increase duck recruitment in the Prairie Pothole Region. *Transactions of the North American Wildlife and Natural Resources Conference* 61:86-93.

Reynolds, R. E., T. L. Shaffer, C. R. Loesch, and R. R. Cox, Jr. 2006. The Farm Bill and duck production in the Prairie Pothole Region: Increasing the benefits. *Wildlife Society Bulletin* 34:963-974.

Reynolds, R. E., T. L. Shaffer, R. W. Renner, W. E. Newton, and B. D. J. Batt. 2001. Impact of the Conservation Reserve Program on duck recruitment in the U.S. Prairie Pothole Region. *Journal of Wildlife Management* 65:765-780.

Robbins, C. S. 1981. Effect of time of day on bird activity. *Studies in Avian Biology* 6:275-286.

Rosenberg, K. V., and P. J. Blancher. 2005. Setting numerical population objectives for priority landbird species. Pages 57-67 *in* C. J. Ralph and T. D. Rich, editors. *Bird conservation implementation and integration in the Americas.* Proceedings of the 3rd International Partners in Flight Conference 2002. U.S. Forest Service, Pacific Southwest Research Station, General Technical Report PSW-191, Albany, California, USA.

Rosenberg, M. S. 2003. *PASSAGE: Pattern analysis, spatial statistics, and geographic exegesis. Version 1.1.* Department of Biology, Arizona State University, Tempe, USA.

Sauer, J. R., B. G. Peterjohn, and W. A. Link. 1994. Observer differences in the North American Breeding Bird Survey. *Auk* 111:50-62.

Scott, J. M., F. Davis, B. Csuti, R. Noss, B. Butterfield, C. Groves, H. Anderson, S. Caicco, F. D'Erchia, T. C. Edwards, Jr., J. Ulliman, and R. G. Wright. 1993. Gap analysis: A geographic approach to protection of biological diversity. *Wildlife Monographs* 123.

Stephens, S. E., J. J. Rotella, M. S. Lindberg, M. L. Taper, and J. K. Ringelman. 2005. Duck nest survival in the Missouri Coteau of North Dakota: Landscape effects at multiple spatial scales. *Ecological Applications* 15:2137–2149.

Stewart, R. E., and H. A. Kantrud. 1973. Ecological distribution of breeding waterfowl populations in North Dakota. *Journal of Wildlife Management* 37:39–50.

U.S. Government Accountabilty Office. 2007. *Prairie Pothole Region land acquisitions*. A report to the subcommittee on Interior, Environment, and related agencies, Committee on Appropriations, House of Representatives. GAO-07-1093.

Van Horne, B. 1983. Density as a misleading indicator of habitat quality. *Journal of Wildlife Management* 47:893–901.

Villard, M.-A., E. V. Schmidt, and B. A. Maurer. 1998. Contribution of spatial modeling to avian conservation. Pages 49–64 *in* J. M. Marzluff and R. Sallabanks, editors. *Avian conservation: Research and management.* Island Press, Washington, D.C., USA.

CHAPTER

Plum Creek's Central Cascades Habitat Conservation Plan and Modeling for the Northern Spotted Owl

21

Lorin L. Hicks, William O. Vogel, Dale R. Herter, Richard J. Early, and Henning C. Stabins

The northern spotted owl (*Strix occidentalis caurina*) is the largest of the three subspecies of spotted owl inhabiting western North America (Gutiérrez et al. 1995). The species was listed as threatened by the U.S. Fish and Wildlife Service (1990) under the Endangered Species Act (ESA) in 1990 based on three findings by the agency: (1) suitable forest habitat was declining throughout its range; (2) populations showed declining trends; and (3) existing regulatory mechanisms were not adequate to protect the owl.

Following the federal listing of the spotted owl, confusion ensued regarding the regulatory impact of listing on state and private lands within the range of the species (Dietrich 1992). Not only were the numbers of spotted owl sites on nonfederal lands poorly known due to lack of adequate surveys, but the amount and configuration of suitable habitat needed to support each site were the subject of vigorous public debate (Yaffee 1994). This confusion was especially true for private forest landowners who found themselves caught on two separate playing fields as both state and federal agencies sought to impose conflicting regulations protecting the spotted owl on nonfederal lands. Eventually, what emerged in Washington state was a "circles and survey" regulatory strategy whereby landowners were required to conduct protocol surveys in suitable habitat for spotted owls and verify reproductive status for each pair found. Moreover, for each pair located, a circle with a radius approximating the mean annual home range was designated. In the central Cascades, a 2.9 km radius circle totaling 2,695 ha was designated, within which 40% (1,078 ha) must be maintained as suitable spotted owl habitat. The size of the regulatory circle varied, depending on spotted owl home range and habitat use studies conducted in various regional provinces such as the Olympic Peninsula, Cascades, Oregon Coast, or Northern California (U.S. Fish and Wildlife Service 1989, Lemkuhl and Raphael 1993, Bart 1995, Meyer et al. 1998). Within a few years of listing, landowners within the range of the spotted owl found themselves with both expensive survey requirements and significant acres of timber assets encompassed in regulatory circles. Within Washington's Interstate-90 corridor in the Central Cascades, 107 regulatory circles affected thousands of acres of Plum Creek property (Fig. 21-1). It is important to note that these circles moved as spotted pairs moved, requiring constant surveys and monitoring of spotted owl pairs, as well as dispersing offspring forming new pairs, creating yet more regulatory circles. This regulatory scenario not only was problematic for forest managers, but was not a desirable biological approach to managing spotted owls and their habitat across landscapes. Management circles did not incorporate landscape variability such as

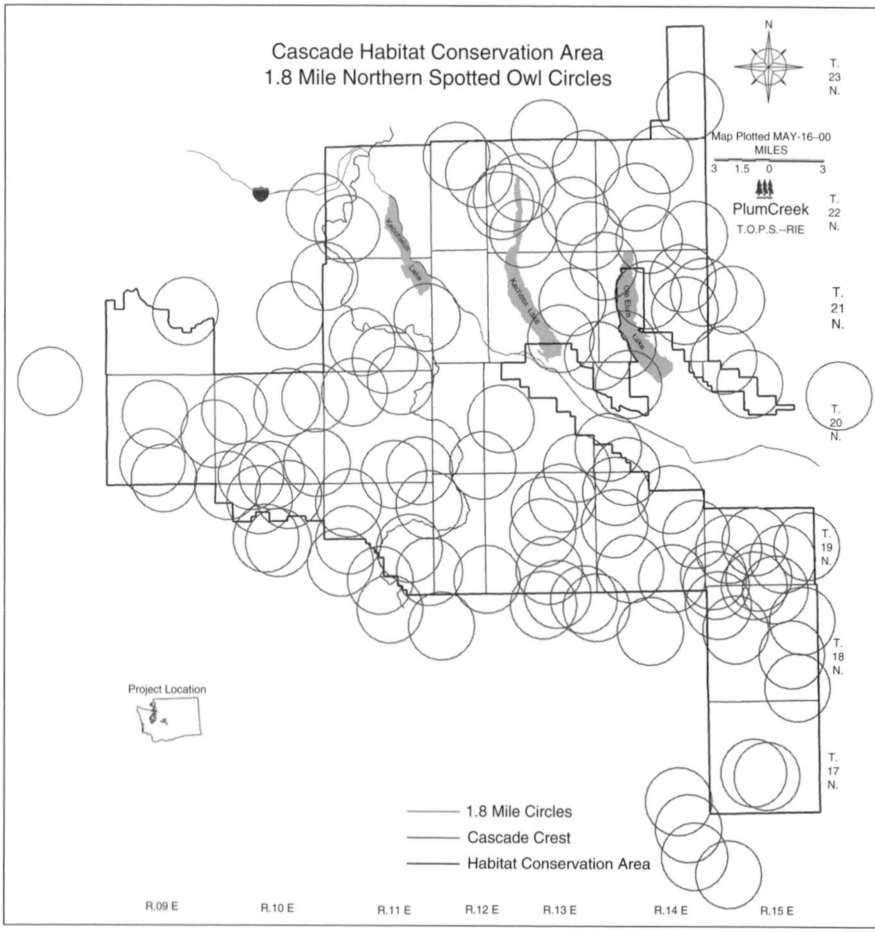

FIG. 21-1

Circles represent 2.9 km radius spotted owl activity centers (nest sites or locations of multiple owl observations) in the 170,000 ha Cascades Habitat Conservation Plan Area, central Cascades Mountain Range, Washington, USA.

topography, water bodies, and habitat juxtaposition, which are important factors affecting spotted owl use. In areas of high spotted owl density, regulatory circles often overlapped, including habitat for multiple pairs that logically would only be used by one pair of nesting owls. Additionally, regulatory circles and habitat within could be removed as spotted owl sites became vacant over time.

In addition to the northern spotted owl, other species in the region were subsequently listed or proposed for listing, such as stocks of Pacific salmon (*Oncorhynchus* spp.), marbled murrelet (*Brachyramphus marmoratus*), bull trout (*Salvelinus confluentus*), and Canada lynx (*Lynx canadensis*). Combined with species that were federally listed prior to the northern spotted owl, such as the grizzly bear (*Ursus arctos horribilis*), gray wolf (*Canis lupus*), bald eagle (*Haliaeetus leucocephalus*), and peregrine falcon (*Falco peregrinus*), wildlife management in the Pacific Northwest became exceedingly contentious and complex. The threat of additional species listings further eroded the regulatory predictability desired by federal, state, and private forest landowners.

In the early 1990s, several conservation strategies were proposed for lands managed by the U.S. Forest Service and Bureau of Land Management (Thomas et al. 1990, Lujan et al. 1992), culminating in the Northwest Forest Plan, which is presently guiding federal land management activities in Oregon, Washington, and northern California (Forest Ecosystem Management Assessment Team 1993). All these strategies sought to avoid the pitfalls of single-species management by addressing multiple species at the landscape scale.

The voluntary habitat conservation planning process, added to the ESA in 1982, provided the opportunity for private forest landowners such as Plum Creek to develop a landscape-scale long-term management plan in cooperation with the federal agencies to address the needs of listed species like the spotted owl, as well as other wildlife species that could be federally listed in the future. Several industrial timber companies in the Pacific Northwest initiated landscape-scale, multispecies conservation plans during the 1990s under either federal or state regulations with varying outcomes (Loehle et al. 2002).

In this chapter, we describe how Plum Creek's Central Cascades Habitat Conservation Plan (HCP) was developed to embrace a multispecies approach to habitat management at the landscape level. We discuss how the HCP incorporated models to both quantify habitat relationships and provide a means to develop alternatives and evaluate impacts of HCP implementation on associated resources. Finally, we offer some insights into HCP effectiveness, based on the first 10 years of plan implementation.

THE HABITAT CONSERVATION PLAN PROCESS

Section 10 of the ESA provides nonfederal land managers with the ability to apply for an incidental take permit when their otherwise lawful management activities may affect listed species in such a way as to harass or harm them, which is considered to be "incidental take." Such incidental take can be the result of habitat modification or destruction. An application for an incidental take permit must be accompanied by a conservation plan which specifies the impacts anticipated to result from that incidental take; the methods the land managers will use to minimize, mitigate, and monitor the incidental take; alternatives considered; and other measures that may be necessary.

These conservation plans are often called Habitat Conservation Plans (or HCPs), as they usually focus on providing habitat for the species in question (Beatley 1994). These plans may address unlisted species in addition to listed species. Habitat Conservation Plans form the basis for agreements between the land manager and the federal agencies (U.S. Fish and Wildlife Service and National Marine Fisheries Service; hereafter referred to as the Services), and are often long-term agreements with assurances on both sides. The HCP also describes the current status of the environment, the status of the various species in question, and the relevant science and stressors surrounding them.

A comparison between the species and habitat conditions that would be expected to result from permit issuance through the HCP and the conditions that would be expected to result from the status quo without permit issuance is normally a part of an HCP as well, including discussion of alternate scenarios. Estimating species and habitat conditions through various action and no-action scenarios over long periods of time can be a challenge. This is especially true for forest ecosystems because the permutations and consequences of management may last for decades. Moreover, there are large uncertainties about outcomes under both the "action" (e.g., HCP) and "no-action" (circles and surveys) scenarios. For instance, if owl sites were to become vacant and regulatory circles removed, less habitat would be available under the "no-action" scenario than if owls had persisted at those sites.

In landscape planning of this nature, various actions of the HCP applicant must be considered in context with actions by adjacent landowners, such as the U.S. Forest Service. Additionally, species considered in the HCP may be affected by factors completely outside the planning area, such as ocean conditions for salmon and marbled murrelet populations. Compliance with the National Environmental Policy Act often requires projections of effects for various management alternatives, thereby necessitating modeling or other projection analyses to understand long-term plans such as an HCP. Because of this uncertainty and lack of complete managerial control, HCPs generally focus on habitat provided by the applicant and avoid population metrics as a measure of plan success.

When processing an application for an incidental take permit, the federal agencies must determine that the action will not jeopardize any listed species nor destroy critical habitat for any listed species. This requirement for the federal action of permit issuance is the same as the requirements for any project that is conducted, funded, or authorized by a federal agency, and is addressed by completion of a Biological Opinion under section 7 of the ESA. In addition, Section 10 specifies a number of specific issuance criteria that must be met in order for such an incidental take permit to be issued. For instance, one of these criteria is that the impacts associated with the taking of the species must be minimized and mitigated to the maximum extent practicable. The federal agencies generally document the fulfillment of the issuance criteria in a document known as a Statement of Findings. When an HCP is properly prepared, it will provide the information needed by the public to understand how the federal agencies may be capable of making their independent determinations that will be contained within the Biological Opinion and Statement of Findings.

Since its inception in 1982, the incidental take permit program under the ESA has grown considerably to include many applicants and a substantial land area. As of 2007, the U.S. Fish and Wildlife Service (USFWS) approved 537 HCPs covering >18.2 million hectares (U.S. Fish and Wildlife Service 2007). These HCPs include plans for single species and plans that seek to address the needs of multiple species within the planning area. Multispecies HCPs have been encouraged by agencies and conservation interests because it is thought to improve the potential for an effective management program and reserve system. Additionally, permittees have incentives to cover as many species as possible in the incidental take permit to protect themselves against future listings. Consequently, including more species in an HCP would seem to serve both interests, providing more certainty for the permittee and increasing overall conservation value. However, multispecies HCPs have been criticized as inadequate and ineffective. Critics have claimed that many HCPs lacked the supporting data to justify conservation measures offered in the plans, and that few HCPs were designed to include adequate monitoring to inform future decisions (Kareiva et al. 1999). For instance, Rahn et al. (2006) cited three deficiencies in 22 multispecies HCP reviewed. Shortcomings that could limit conservation value were too broad and considered species for which there was no localized scientific information. Second, most unconfirmed species did not have specific conservation actions. Finally, the degree of justification was quite variable for included species and the extent the plans offered species-specific conservation actions.

As will be discussed later, the Plum Creek Cascades HCP incorporated a combination of general habitat-related conservation measures (e.g., forest structure stage diversity, talus slopes, ponderosa pine habitat, springs and seeps, etc.) and species-specific conservation actions (e.g., northern spotted owl, marbled murrelet, grizzly bear, and gray wolf). Even though some vertebrate species were not confirmed to be present in the planning area at the time of HCP development, subsequent surveys confirmed presence and that general habitat-related conservation measures were effective. For instance, the Larch Mountain salamander (*Plethodon larselli*) was suspected to occur in the planning area but was not confirmed until the HCP had been implemented for three years. The species was found to occur in areas where talus slope conservation measures were implemented to minimize ground disturbance and maintain conifer cover and moss for micro-site moist climate conditions that this species prefers.

Another shortcoming of HCPs is that they are expensive to develop and have become more difficult to complete in the last 10 years. The Services have made planning funds available, but there is no certainty that an applicant beginning an HCP process will be successful. Many applicants are drawn to HCPs out of necessity in the short term to mitigate an ESA-related conflict (e.g., a land development project) and others by the desire for long-term certainty (e.g., forest landowners with 50-year timber rotations).

The "no surprises" rule (50 CFR 17.22, 17.32) provides that HCP permittees will not be required to provide more money or land for conservation efforts once an HCP has been approved, except under extraordinary circumstances. The "no surprises" policy was installed by the Department of Interior in 1994 to address uncertainty inherent in the HCP process that was leading to very low HCP participation levels. The "no surprises" policy, including the subsequent rule promulgated in 1998, was very effective in providing an incentive for landowners to develop HCPs, particularly multispecies HCPs that include both listed and unlisted species (Slingerland 1999). Conversely, the "no surprises" rule has been criticized as effectively precluding adaptive management and changes needed to inform and improve HCPs over time (Wilhere 2002, Rahn et al. 2006).

Adaptive management is necessary where substantial risk exists and can also be used where agreement could not otherwise be reached. In the realm of HCPs, adaptive management may mean that a land manager may need to operate differently in the future to accommodate changing or unexpected conditions, or to accommodate the development of new information. Operating differently might involve additional land or water encumbrances, or other expenses. However, because this added conservation was part of and contemplated in the HCP Implementation Agreement, it does not conflict with the "no surprises" rule. Therefore, adaptive management must be carefully considered prior to entering into such an agreement. Modeling is an important component of examining various management scenarios through time and assessing the ecologic, economic, and social ramifications of a conservation plan. The relationship between the "no surprises" policy and adaptive management in the Plum Creek Cascades HCP is explored later in this chapter.

CASCADES HABITAT CONSERVATION PLAN

The Cascades Habitat Conservation Plan area is located in the central Cascade Mountain Range of Washington. The 170,000 ha planning area lies between 100 and 160 km east of the city of Seattle along the Interstate-90 corridor (Fig. 21-2). Because of the "checkerboard" configuration of land ownership, the HCP planning area included 69,000 ha of Plum Creek ownership and about

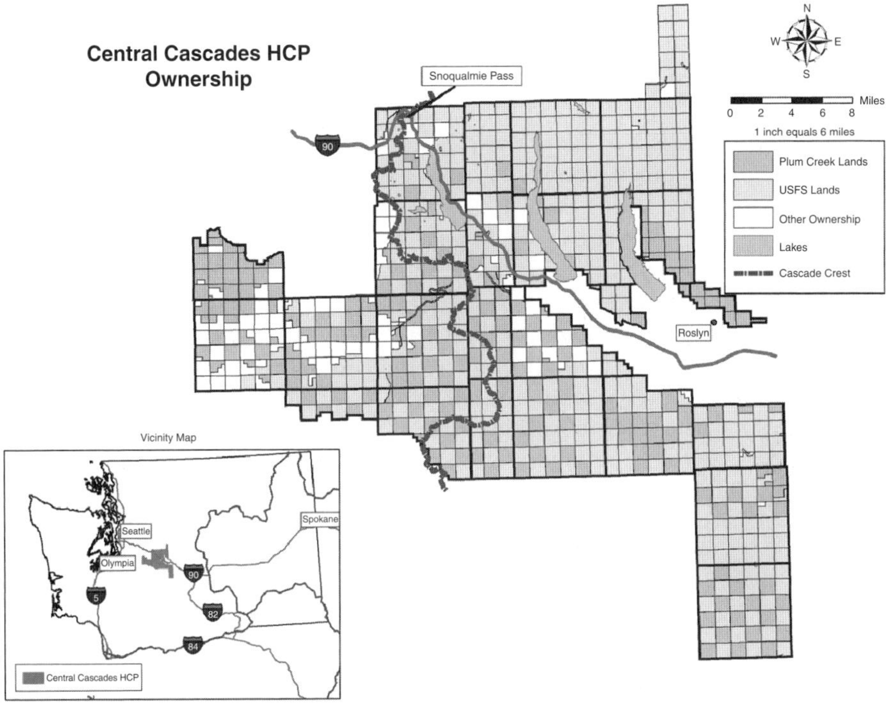

FIG. 21-2 (See color insert.)

Land ownership pattern in the 170,000 ha Cascades Habitat Conservation Plan Area, central Cascades Mountain Range, Washington, USA.

101,000 ha of other ownership intermingled and surrounding the Plum Creek lands (Fig. 21-2). Other major ownerships include the U.S. Forest Service, Washington State Department of Natural Resources, and the City of Tacoma. The checkerboard configuration is a result of land grants provided to private railroad companies by the federal government in the 1800s as an incentive to build rail lines into largely unsettled territories and provide U.S. military transport. The railroad companies were typically granted alternating 2.5 km^2 sections of land for 32 km on both sides of the rail line. Plum Creek purchased these lands from the railroad in 1989.

The HCP planning area straddles the Cascades mountain range crest and includes the varying climatic conditions and resulting forest conditions of two physiographic provinces: the western Washington Cascades and eastern Washington Cascades (Bailey 1988, Thomas et al. 1990). The western portions of the project area are dominated by humid forests composed primarily of Douglas fir (*Pseudotsuga menziesii*) and western hemlock (*Tsuga heterophylla*) at middle and lower elevations and noble fir (*Abies procera*), Pacific silver fir (*A. amabilis*), and mountain hemlock (*T. mertensiana*) at higher elevations.

Westside forests are generally represented by only one or two tree species and greater uniformity in tree size within stands. Climatic conditions in this region are relatively mild with moist winters.

The eastern portions of the planning area are dominated by mixed-conifer (grand fir [*A. grandis*], Douglas fir, western larch [*Larix occidentalis*], ponderosa pine [*Pinus ponderosa*], western white pine [*P. monticola*], lodgepole pine [*P. contorta*]) forests at middle and lower elevations. Forests east of the Cascade crest are typically composed of a greater diversity of tree species (i.e., typically five or more species per stand) and greater diversity of tree sizes within stands than the forests west of the Cascade crest. Higher elevations in the eastern Cascades are dominated by subalpine fir (*A. lasiocarpa*) and Engelmann spruce (*Picea englemannii*). The interior climate east of the Cascade crest produces less precipitation overall, which usually occurs in the form of snow during the winter. Forests in this eastern region are fragmented due to poor soils, high fire frequencies, alpine meadows, and timber harvesting.

Multispecies Approach

Early in the negotiation process between Plum Creek and the Services, fundamental points of agreement were documented in the form of a "term sheet" that facilitated development of the HCP. The term sheet identified objectives and expectations for the HCP such as provisions for spotted owl dispersal habitat, coordination with the Northwest Forest Plan on intermingled U.S. Forest Service land, and attention to issues threatening forest health on Plum Creek land. Based on conversations with the Services, and interaction of a Plum Creek science team made up of staff and outside consulting experts, a more refined list of HCP components emerged, which influenced the selection of models, tools, and techniques to construct the HCP. These components include a riparian strategy, adaptive management, habitat "futuring," and a reserve network designed to complement and augment conservation measures implemented on federal land.

Several alternative management scenarios were developed early in the Cascade HCP process. These included a riparian alternative, whereby most of the habitat for federally listed species would be concentrated along major fish-bearing streams. Another alternative was termed the "dispersal strategy" where Plum Creek's contribution to spotted owl habitat would be restricted to middle-successional forests that could support dispersing juveniles to nesting habitat on public land. Although we used models and analysis techniques described later in this chapter to analyze these alternatives, the advent of the "no surprises" policy described earlier provided the impetus to expand the scope of the Cascades HCP to include multiple species and a wider variety of habitat conservation measures (Slingerland 1999).

Due to the substantial number of species that are currently listed or may be listed in the future on and adjacent to Plum Creek's lands in the planning area, we believed that a multispecies, ecosystem-based conservation plan was the most effective way to plan for ongoing forest management within the laws governing and protecting wildlife species (Noon et al., this volume). By attempting to address the habitat requirements of multiple species in conjunction with planning efforts by federal, state, and other private landowners, the HCP could contribute to proactive conservation. By maintaining healthy populations, these conservation plans may possibly negate the need for formal ESA protection for certain species in the future, protecting Plum Creek from potential ESA-related regulatory constraints.

The Plum Creek Cascades HCP addresses the biological needs of 315 species of fish and wildlife known to occur in the planning area and other vertebrate species across all taxa that depend on similar habitats and are suspected to occur in the planning area. The multispecies approach can help to reduce conflicts over resource management by providing a mechanism for consideration of overall ecosystem health, habitat availability, and the needs of multiple species. To achieve this broad goal, the HCP focuses on ecosystems and habitats rather than species, addresses impacts not only at the site scale but also on an ecosystem scale, and concentrates on potential long-term or future impacts rather than on intermediate or short-term impacts. Further discussion of an ecosystem diversity focus in private land planning is explored by Haufler and Kernohan (this volume).

Because of the intermingled land ownership in the HCP planning area, we designed the HCP to be consistent with the goals and objectives of management efforts on federal lands by the U.S. Forest Service. The multispecies approach in the HCP tiers to the multispecies approach in the Northwest Forest Plan. In concert, these two landscape plans protect forest habitat, provide management options for the protection of stream corridors to enhance conditions for associated aquatic and terrestrial species, and provide forest connectivity among patches of various forest structure stages.

The primary objectives of the HCP are:

1. To comply with the requirements of Section 10 of the ESA, including, to the maximum extent practicable, minimizing and mitigating impacts of any "take" incidental to lawful timber harvest and related forest management activities;

2. To provide Plum Creek with predictability and flexibility to manage its timberlands economically while contributing in a meaningful way to the conservation of the listed species and numerous other unlisted species; and

3. To provide habitat conditions to conserve the ecosystem upon which all species depend in the planning area.

The components of the HCP help to avoid, reduce, or eliminate potentially adverse environmental impacts resulting from Plum Creek's forest management activities. The components include measures to maintain and protect riparian habitat areas, spotted owl habitat management, a watershed protection program, retention of green trees and snags, and protection of special habitats for all vertebrate species of wildlife known or suspected to occur in the project area, in addition to monitoring to ensure effective implementation and guide potential changes.

HABITAT CONSERVATION PLAN MODELING

Forest Stand Classification

A primary focus of the ecosystem management approach in the HCP planning area is to link the biological needs of forest wildlife to the physical and vegetative characteristics of the forest environment. This approach requires:

1. A system to classify the diverse forest stands over a large landscape;
2. An alignment of the biological requirements of resident wildlife species with the forest stand classification system; and
3. A modeling capability to predict the amount and location of the various forest classes in the future as a result of forest growth and management across the intermingled land ownership.

To develop the HCP, we evaluated the planning area using geographic information systems (GIS) to describe attributes including elevation, slope, aspect, annual precipitation, soil type, vegetation, and ownership. A hierarchical ecological classification system identified seven levels and included both abiotic and biotic parameters: Ecoregion, Geologic District, Landtype Association, Landtype, Valley-bottom Type, Landform, and Vegetation Type. Combined with existing timber inventories, we used this information to establish a forest stand structural classification system for wildlife, following Oliver and Larson (1990) and further described in Oliver et al. (1995). Important features of the classification system include the use of common timber inventory parameters that are often already measured by foresters and exist in a format which, due to the intermingled land ownership, was compatible with U.S. Forest Service timber inventory data. We refined the system using a stand visualization program (McGaughey 2002) that constructed visual plots of the structure stages from actual data. From this process, we identified eight forest stand structure stages: stand initiation, shrub/sapling, young forest, pole timber, dispersal forest, mature forest, managed old growth, and old growth.

The structural stages are defined by tree diameter ranges. Stand age is used in part to identify old-growth stands. Classification criteria differed between the westside and eastside Cascade mountain range portions of the HCP project area to account for the climatic and tree species composition differences in the two ecological provinces. We developed two structure stages—dispersal forest and managed old growth—for special conditions and needs of the HCP. Dispersal forest encompasses the "entry level" stand conditions favorable to dispersal of spotted owls by providing minimal conditions for roosting and foraging, allowing owls to move between more favored habitats. Managed old-growth stands contain large diameter trees, are generally <200 years old, and typically have a history of selective timber harvest. Managed old growth was developed from experimentation in structural retention as discussed by Lindenmayer and Franklin (2002) and Franklin et al. (1997). For managed old-growth stands to be included in further calculations of spotted owl habitat, sufficient forest structure had to be retained to maintain conditions favorable to spotted owls as nesting, roosting, and foraging habitat (Hicks 1991). Our specifications for spotted owl foraging/dispersal habitat and nesting, roosting, and foraging habitat used in the HCP were based on over 1,000 vegetation plots collected on locations of radio-tagged spotted owls throughout the planning area from 1988 to 1995 (Hicks and Stabins 1995, Hicks et al. 1995, Herter et al. 2003).

Lifeforms and Wildlife Habitat Analyses

To evaluate, model, and plan for a variety of habitat conditions to benefit the multiple wildlife species found in the HCP project area, we developed a wildlife habitat relationship matrix following Thomas (1979) and Brown (1985). The "lifeform" approach developed and described by Thomas (1979) for the Blue Mountains of Oregon and Washington and by Brown (1985) for western Washington and Oregon were the state-of-the art compendiums for multispecies habitat relationship data at the stand and landscape level when work on the HCP began. Moreover, the format was familiar to our peers and reviewers and was applicable to other ownerships in the HCP planning area. We consulted with local experts to refine and adjust the matrix accounts for the known or suspected occurrences of vertebrate wildlife species across the array of forest types, forest stand structural stages, and special habitats that occur in the planning area, which allowed quantitative assessments of wildlife habitat across the landscape (Lundquist and Hicks 1995). Further analyses using the matrix allowed the comparison of various outcomes from alternative management scenarios or changes to the plan's assumptions on wildlife habitat distribution and amounts over the HCP's 50-year planning period.

We grouped over 300 wildlife species into 16 guilds, or lifeforms, that describe their breeding and feeding strategies (Table 21-1). We assumed that habitat conditions are the primary determinants of the number of wildlife species and numbers of individuals in a given area. We then assigned forest stand structure stages developed for the forest classification system, and special habitat types such as wetlands or talus slopes, to each of the lifeforms. We designated stand structures as nonhabitat, primary habitat, and secondary habitat. Primary habitats were those on which the species relied, while secondary

Table 21-1 Lifeform Descriptions Used in Plum Creek's Cascades Habitat Conservation Plan, Washington, USA

No.	Lifeform Type	Search Area	No. Spp.	Reproduces	Feeds	Habitat
1	Fish	RHA	34	in water	in water	Primary: Water
2	Frogs, salamanders	RHA	10	in water	on the ground, in bushes, and/or in trees	Primary: DF/MF/MOG/OG Secondary: SI/SS/YF/PT
3	Turtles, ducks	RHA	36	on the ground around water	on the ground, and in bushes, trees, and water	Primary: DF/MF/MOG/OG Secondary: SI/SS/YF/PT
4	Falcons, goats	Rocks & talus	17	in cliffs, caves, rimrock, and/or talus	on the ground or in the air	Primary: PT/DF/MF/MOG/OG Secondary: SI/SS/YF
5	Grouse, hares, elk/deer (gray wolf)	0.5 mile window	33	on the ground without specific water, cliff, rimrock, or talus association	on the ground	Forage: SI/SS/YF Cover: PT/DF/MF/MOG/OG
6	Warblers, porcupines	RHA	8	on the ground	in bushes, trees, or in the air	Primary: SI/SS/YF Secondary: PT/DF/MF/MOG/OG
7	Sparrows, blackbirds, thrushes	RHA	19	in bushes	on the ground, in water, or in the air	Primary: SS/YF/PT/DF Secondary: MF/MOG/OG
8	Warblers, flycatchers	HCP	7	in bushes	in trees, bushes, or in the air	Primary: SS/YF/PT/DF Secondary: MF/MOG/OG
9	Waxwings, grosbeaks	RHA	5	primarily in deciduous trees	in trees, bushes, or in the air	Primary: YF/PT/DF Secondary: MF/MOG/OG
10	Squirrels, tanagers, warblers	HCP	12	primarily in conifers	in trees, bushes, or in the air	Primary: PT/DF/MF/MOG/OG Secondary: SS/YF

11	Vireos, hawks	HCP	28	in conifers or deciduous trees	in trees, bushes, on the ground, or in the air	Primary: PT/DF/MF/MOG/OG Secondary: SI/SS/YF
12	Herons, osprey, great horned owl	RHA	6	on very thick branches	on the ground or in the water	Primary: DF/MF/MOG/OG Secondary: PT
13 13a	Woodpeckers: Lewis' woodpecker, white-headed woodpecker, pileated woodpecker	HCP HCP	14	in own holes excavated in trees	in trees, bushes, on the ground, or in the air	Primary: DF/MF/MOG/OG Secondary: YF/PT Primary: MF/MOG/OG Secondary: SI/SS after 10 yrs. YF/PT after 20 yrs. DF every year
14 14a	Bats, owls, bluebirds Flammulated owl, Vaux's swift, fisher	HCP HCP	43	in a hole made by another species or a natural hole	on the ground, in water, or in the air	Primary: DF/MF/MOG/OG Secondary: SI/SS/YF/PT Primary: MF/MOG/OG Secondary: DF
15	Shrews, bears voles	HCP	36	in a burrow underground	on the ground or underground	Young Aged: SI/SS/YF Mid-Aged: PT/DF Late Aged: MF/MOG/OG
16	Kingfishers, otters, beavers	RHA	7	in a burrow underground	in the air or in the water	Primary: DF/MF/MOG/OG Secondary: SI/SS/YF/PT

Search Area: RHA—Riparian Habitat Areas; HCP—Habitat Conservation Plan Area; 0.5 mile window—scanning radius which provides a basis for sampling edge habitat (i.e., the area between forage and cover habitats) in the HCP planning area.
Habitat: Primary—habitats which the Lifeform species rely on; Secondary—habitats also used by the Lifeform species. DF—Dispersal Forest; MF—Mature Forest; MOG—Managed Old Growth; OG—Old Growth; SI—Stand Initiation; SS—Shrub/Sapling; YF—Young Forest.

habitats also were used by the species. Primary habitat was emphasized during evaluations of the HCP to ensure the most important habitats would not be reduced to undesirable levels. Secondary habitat was allotted only half the weight of primary habitat during assessments of total suitable habitat (Table 21-1).

Because the forest stand classification system was tied to a GIS polygonal database, analyses resulted in describing the distribution and amount of wildlife habitat for all 16 lifeforms across the project area. Spatial analyses varied among the lifeforms and were tailored to the life history requirements of the species. In some cases, we used the entire planning area for the analysis and for some lifeforms, a more restricted analysis area focused on habitats such as streams, wetlands, talus, or forest edges (Table 21-1). For instance, lifeform 4 included species that are associated with cliffs, talus, and rocky outcroppings. Species in this lifeform included the Larch Mountain salamander and Townsend's big-eared bat (*Plecotus townsendii*). Primary habitat for these species was pole timber and older forests found in the vicinity of these rock features.

Several methods were used to refine and improve the lifeform matrix during the development and modeling process. First, the lifeform matrix was peer-reviewed by biologists from government and private organizations with local knowledge of wildlife species in the HCP planning area and familiarity with forest wildlife habitat relationship matrices. Second, as a result of peer review comments, we delineated subgroups within the primary cavity-excavator and secondary cavity-user lifeforms (13/13a and 14/14a, Table 21-1) for those species that are more associated with late successional structural stages (e.g., pileated woodpecker [*Dryocopus pileatus*], flammulated owl [*Otus flammeolus*], fisher [*Martes pennanti*]). Finally, we did not count recently harvested areas as habitat for some lifeforms (e.g., cavity-excavators) until 10 or 20 years into the HCP period, since some wildlife conservation measures (e.g., snag and green wildlife tree retention) were not implemented historically and, therefore, may not provide adequate structural elements for these lifeforms at the time of HCP inception (Table 21-1). These modifications to the habitat relationship model provided opportunities for hypothesis testing under the HCP's adaptive management program, discussed later in this chapter.

Wildlife Habitat Futuring

Because the HCP is a 50-year agreement, we used the strategic planning program OPTIONS (Reimer 2007) to model the forest structure stages through time. The OPTIONS model is a state-of-the-art forest estate planning model that plays a central role in the maintenance of the Central Cascades HCP. Inputs to the model include tabular forest inventory information (tree species, diameter, height, etc.), spatial forest inventory information (stand polygon locations), spatial landscape information (sensitive sites, view sheds, etc.), and localized growth-and-yield information. OPTIONS uses both existing forest inventory data

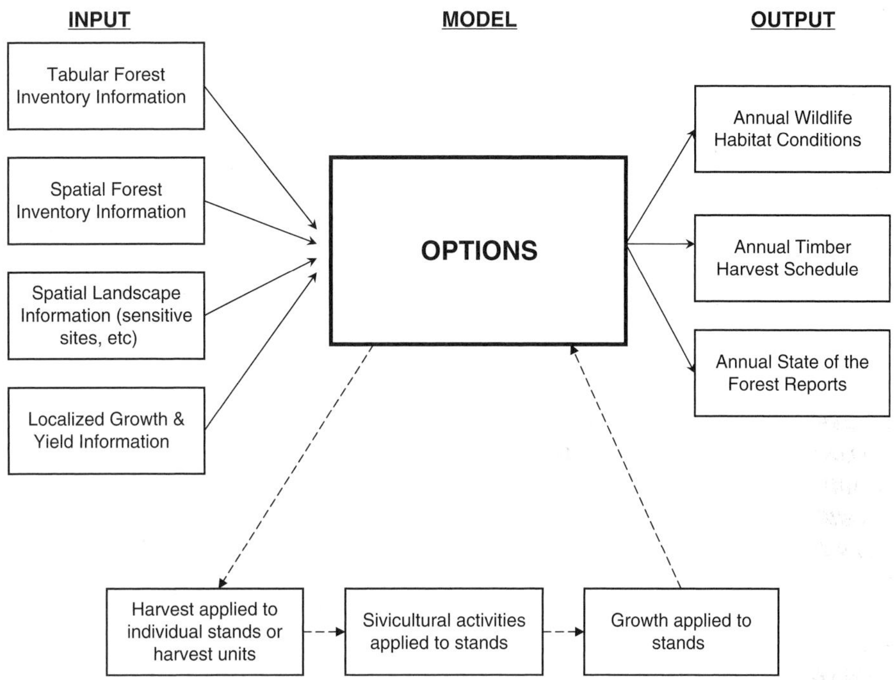

FIG. 21-3

Input and output parameters of the forest growth and modeling software OPTIONS used to project forest structural stage and wildlife habitat amounts and trends over a 50-year planning period in the 170,000 ha Central Cascades Habitat Conservation Plan area, Washington, USA.

and regionally calibrated growth algorithms to project forest stands into the future (Fig. 21-3). With the given inputs, the model provides reliable forecasts of stand attributes such as stand height, diameter, age, canopy stratification, and tree species richness. Management activities are reported for each stand for each year of the planning horizon. Based on the necessary condition and relative importance of various stand-level attributes, habitat suitability and capability for numerous special species, lifeforms, or habitat type can be assessed, forecast, and planned.

Finally, with the OPTIONS model, the abundance and spatial arrangement of individual stands of habitat that influence overall wildlife habitat conditions can be evaluated. Forest-level analysis of stand-level habitat classifications facilitates the assessment of broad-level habitat conditions. Landscape attributes such as fragmentation, isolation, and connectivity can be modeled and forecast so that long-term management prescriptions can be assessed. We used OPTIONS to assess the amounts of primary and total suitable habitat for each lifeform at 10-year intervals over the 50-year life of the HCP, accounting for forest growth and timber management. The modeling results were instrumental in projecting wildlife habitat conditions and the potential impacts of various HCP planning alternatives. The analysis also assessed timber resources, such as long-term supply, growth-and-yield, and timberland management strategies. We obtained stand inventory data from other land owners in the HCP planning area and assessed habitat conditions across the entire 170,000 ha planning area. Assumptions about rates of timber harvest were made for other landowners to inform future habitat projections.

Results of the lifeform habitat modeling during early iterations led to changes in the conservation plan design. Initial modeling results indicated a marked decrease in late successional habitat during the first decades of the plan period under conventional timber harvest strategies. We established harvest deferrals at key locations in the project area for 20 years to maintain the amount of late successional habitat during these low points (see Spotted Owl discussion). In addition, initial OPTIONS output indicated large amounts of the dispersal forest type across the planning area. This led to changes in the timber harvest schedule to direct harvest toward mid-successional stands, allowing for a more balanced diversity of young, middle, and late successional forest structural stages across the planning area.

As stated earlier, a key feature of OPTIONS is its use of input data that are common to most forest inventory systems. Consequently, we modeled both the future of habitat on Plum Creek lands under the HCP for 50 years as well as habitat availability on federal lands under harvest scenarios developed in the Northwest Forest Plan. Inventory data were acquired from the Wenatchee and Mt. Baker-Snoqualmie National Forests and from timber stand classification (phototyping) completed by outside contractors. We cross-checked proposed harvest levels with U.S. Forest Service staff for accuracy. In the 10 years of the HCP operation, it has been evident that harvest levels on U.S. Forest Service land in the HCP planning area have been significantly less than proposed under the Northwest Forest Plan, and less than originally modeled under OPTIONS for habitat supply projections. Thus, the modeling constraints on the U.S. Forest Service lands have been modified. Consequently, U.S. Forest Service lands are currently providing more late-successional habitat than estimated 10 years ago during HCP development.

The modeling output also allowed land managers, scientists, and the public to communicate, understand, and make decisions regarding habitat trade-offs in the HCP. For example, lifeform modeling quantified the future decrease in habitat for early successional and edge-associated species (lifeform 5) such as elk (*Cervus canadensis*) and deer (*Odocoileus* spp.) (Fig. 21-4) and for species reliant on early successional and shrubby habitats in wetland and riparian areas (lifeform 6). The decrease is due to the focus of the HCP on meeting the biological needs of the spotted owl, retention of buffers adjacent to streams for aquatic species protection, and the relatively low levels of timber harvest on federal lands in the Northwest Forest Plan. This example describes how an analysis outcome was useful in understanding and communicating the response of wildlife use to changing habitat conditions across the HCP project area through time.

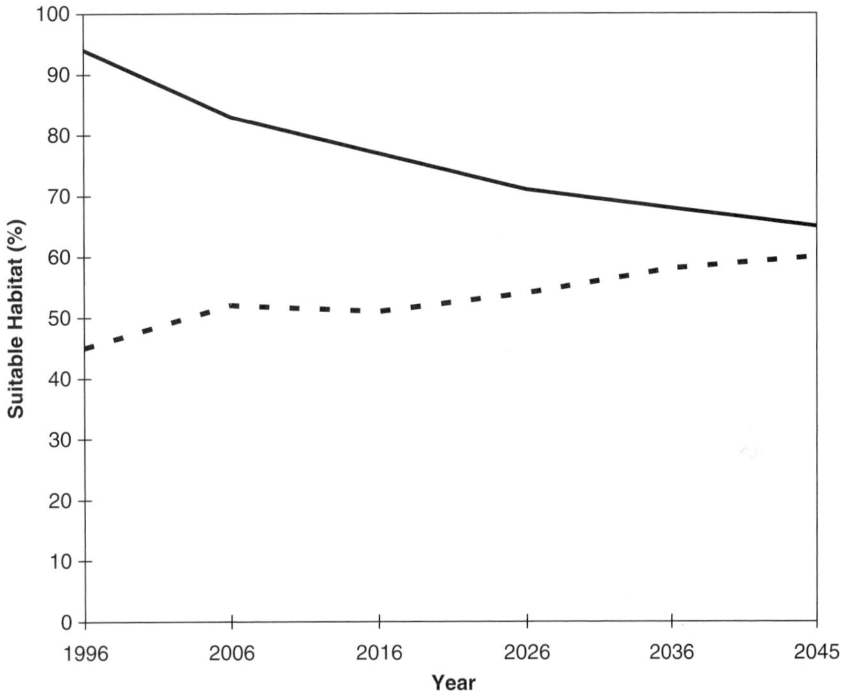

FIG. 21-4

Wildlife habitat futuring over a 50-year planning period identifies trade-offs in the amount and trend of suitable habitat between early successional forest species like deer and elk (solid line) and late successional forest species such as the spotted owl (dashed line) across multiple ownerships in a 170,000 ha Habitat Conservation Plan area in the Central Cascades Mountain Range, Washington, USA.

The wildlife habitat relationship matrix and habitat modeling in the HCP were aids to developing a set of working hypotheses of the expected use patterns of wildlife species across the vegetation types, structural stages, and special habitats in the project area. Although the wildlife habitat matrix did not represent a detailed, site-specific inventory of wildlife occurrence across the diverse habitats of the planning area, the matrix provided information on potential or likely wildlife use patterns based on the available literature. Habitat futuring under various management alternatives also allowed assessment of potential future risks due to fuel loading and fires, as well as susceptibility to pathogens such as the western spruce budworm (*Choristoneura occidentalis*). The habitat futuring exercise allowed land managers, regulators, and the public to evaluate how the proposed HCP and alternate land management plans might affect wildlife species and other aspects of the environment.

Spotted Owls and a Resource Selection Probability Function

Studies of habitat use have indicated that northern spotted owls generally use mature and old-growth forest as much or more than expected, and early seral stage forest less than expected (Forsman 1980; Forsman et al. 1984; Carey et al. 1990, 1992; Sisco 1990; Solis and Gutiérrez 1990; Meyer et al. 1998). A resource selection function (RSF) analysis was carried out in the HCP to better quantify these and other aspects of the use of habitat by spotted owls. Initially, spotted owl activity centers were located in the HCP planning area using standard and extensive survey protocols (Herter and Hicks 2000), likely locating nearly all the regularly occupied owl sites. The selection of a 1.1 km radius circle around these sites as the unit for describing the habitat followed the process described by Irwin and Hicks (1995), with similar methods used later by Meyer et al. (1998), Swindle et al. (1999) and Franklin et al. (2000). This size analysis circle describes the core area of an owl territory and does not analyze the entire area used on an annual basis. Forest habitat designations followed the protocol established by Oliver et al. (1995) and are described further in the HCP (Plum Creek Timber Company 1996).

While information was available to quantify owl nesting habitat, there was a dearth of information to define foraging habitat. Radio telemetry was used to monitor adult nonnesting owls. Daytime roost sites were located and vegetative analyses were conducted surrounding these roost sites. This information helped inform the definition of foraging/dispersal habitat for spotted owls. Separate definitions were developed for nesting habitat and foraging/dispersal habitat, and these each were dependent on the ecological zone (east side or west side of the Cascades mountain range). Definitions included factors available from standard timber inventories including size, density, and species composition of overstory trees.

The RSF model fulfills two important functions in the HCP. The first is to provide a method to assess the likelihood that habitat retained in the plan or projected to grow over the 50-year life of the plan would have a low, medium, or high probability of occupancy by spotted owls. The second was to use the RSF to provide estimates of the spotted owl carrying capacity under different management scenarios, with the carrying capacity defined to be the most reasonable maximum number of spotted owl activity centers that could be accommodated in the planning area with implementation of both the Northwest Forest Plan on federal lands and the HCP on Plum Creek lands. This latter feature of the RSF is an important monitoring component of the HCP because it makes it possible to evaluate the effectiveness of the plan after years of implementation. The RSF was initially developed in 1995 as the HCP was in its formative stage, but was revised in 2001, making use of extensive data then available in the GIS database for the HCP planning area.

The original RSF was used in the evaluation of the HCP. Using logistic regression analysis, we estimated RSFs for an array of 1.1 km "moving windows" across the HCP planning area (Irwin and Hicks 1995) because this distance provided the greatest level of discrimination between occupied and unoccupied sites. Analysis of the RSF values indicated that spotted owls were distributed nonrandomly across the planning area and distributions varied with respect to available nesting habitat, topographic variation, and fire-management activity zones (reflective of ecologically significant moisture and species composition gradients).

At the inception of the HCP in 1997, there were known to be 104 spotted owl activity centers in the HCP planning area, although only a subset of these sites were active in any one year. The revised RSF was estimated from a comparison between 92 sites known to have been actively used by spotted owls, and 51 randomly selected unused sites. Because the original 104 spotted owl sites included many sites where only a single spotted owl used the habitat temporarily, we compared only 92 used sites in which we could document multiyear use by one or more spotted owls. Later the number of unused randomly selected sites was increased to 170 to improve the accuracy of the estimated function. In order to ensure that unused sites really were unused, their circles were not permitted to overlap more than 50% with the 1.1 km radius circles for any sites in the used sample. It is also important to note that not all known activity centers are occupied by spotted owls in any given year. As discussed later, occupancy may be affected by factors other than habitat, such as the presence of predators or competitors of the spotted owl.

The major difficulty with determining the RSF was in the choice of the variables to include in the function. There were over 100 variables available in the GIS, and it was thought from the onset that no more than 10 of these should be used. Looking at all possible combinations of the 100 variables was not realistic because the number of possible models is overwhelming (Burnham and Anderson 1998). Furthermore, we combined the variables into 13 groups (e.g., slope, habitat type, stand structure, riparian/upland land status), where in some cases all the variables in a group should either be in or out of the equation. This made automatic stepwise selection procedures difficult to apply. For these reasons, we decided that an initial screening process was necessary to select biologically meaningful variables. This screening process is described in detail by Hicks et al. (2003*b*) and involved multivariate tests on the 13 groups of variables to see whether there were significant differences between unused and used sites. Following the initial screening, an RSF was fitted separately for each group of variables using logistic regression (Manly et al. 2002). In addition to the 13 groups of variables, 14 other equations with various changes and combinations of variables were fitted by logistic regression (see Hicks et al. 2003*b*). Consistency with the data for each of the 27 equations was assessed using Akaike's Information Criterion (AIC) with small values of this criterion being preferred to large values in terms of the compromise between keeping the number of variables in the equation as small as possible and fitting the data well (Burnham and Anderson 1998).

The logistic regression RSF was converted to a resource selection probability function (RSPF) using a procedure described in Hicks et al. (2003*b*) and Manly et al. (2002). For the Cascades HCP, the estimated RSPF is

$$w^* = \exp(z^*)/\{1 + \exp(z^*)\},$$

where

$$z^* = -6.369 + 0.002550XMF + 0.002531XMOG + 0.002942XOG + 0.006225XRIP + 0.001676XELEV + 0.03286XSI$$

XMF = the number of mature forest acres, XMOG = the number of managed old-growth acres managed as spotted owl habitat, XOG = the number of old-growth acres, XRIP = the total riparian acres, XELEV = the maximum - minimum elevation in meters, and XSI = the minimum 50-year site index in feet.

We used the RSPF to calculate an expected number of spotted owl sites in the HCP project area (Hicks et al. 2003*b*). One of the most useful aspects of the analysis is that it can be extended using the habitat distribution that is expected to exist in the future as a result of timber harvest, forest regrowth, and habitat protection resulting from implementation of the HCP on Plum Creek lands and the Northwest Forest Plan on adjacent checkerboard federal lands. Management actions contemplated under these plans can be modeled in the GIS and then the carrying capacity can be predicted at different times in the future (Fig. 21-5). The carrying capacity of spotted owl sites in the HCP planning area is projected to increase over the 50-year planning period, based on habitat availability, with forest management measures implemented under the HCP and Northwest Forest Plan.

Using the RSPF to predict changes in the carrying capacity of the study area based on forest management actions is an application recommended by Boyce and McDonald (1999), with some controversy resulting (Mysterud and Ims 1999, Boyce et al. 1999). At issue is the question of whether it can be assumed that the RSPF remains constant when the amounts of different types of habitat changes. In general, it can be expected that large changes in the availability of different types of habitat are likely to result in changes in the way that animals select the habitat, and hence in the RSPF. However, with the HCP application, the nature of the habitat in the planning area is not expected to change substantially in the near future (Fig. 21-5). Therefore, the assumption of a constant RSPF seems reasonable.

When the carrying capacity of the area remains constant or increases, this does not mean that the number of animals will also necessarily increase. In fact, we have evidence that the number of spotted owls in the HCP planning area has declined in the last 10 years while the available habitat has changed little. This decrease in spotted owls coincides with an increase in the number of barred owls (*Strix varia*) in the region (Anthony et al. 2006).

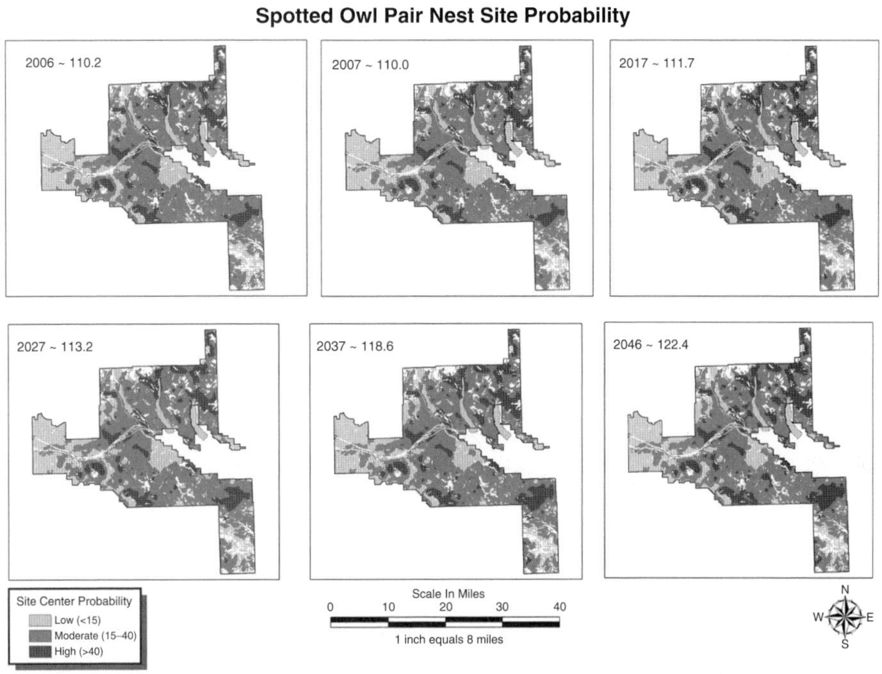

FIG. 21-5 (See color insert.)

Map of estimated values from the resource selection probability function (RSPF) for the Central Cascades Habitat Conservation Plan area, Washington, USA, for the present (2006) and as projected for 2045. The value of the RSPF for every pixel in the geographic information system was evaluated and plotted using a gradient from nonforested areas (white) to areas with the highest probabilities of spotted owl use (maroon). Map headings are the year, followed by the estimated carrying capacity of owl pairs in that year.

It seems likely that predation, interspecific territoriality, and competition for habitat with barred owls is largely responsible for the decline in spotted owl numbers. The RSPF was revised at a time when barred owl sites were approximately equal to spotted owl sites in number (Herter and Hicks 2000), but prior to a consecutive sequence of poor spotted owl reproductive years. While the HCP was written to anticipate some effects of stochastic weather events and barred owl competition, we were nonetheless surprised by the combined effects of these two factors on resident spotted owl populations in the planning area. Conversely, this development points to one of the strengths of the multispecies HCP approach. Without the HCP, habitat in regulatory circles vacated by spotted owls would have been harvested. Under the HCP, Plum Creek continues to maintain vacated spotted owl habitat to the benefit of other species associated with late-successional forests.

ADAPTIVE MANAGEMENT AND MONITORING

Although a significant body of scientific information and expertise was used to develop Plum Creek's Cascades HCP, not all the questions about the long-term effects of HCP implementation on fish and wildlife species and their habitats can be answered with total certainty today. This is particularly pertinent to the use of models in the development of the HCP. However, uncertainty can be addressed by implementation of an adaptive management approach, which incorporates research and monitoring into a responsive program to evaluate the HCP as a "management experiment" that may be modified as necessary to meet objectives.

Adaptive management is a process that can improve management practices incrementally by implementing plans in ways that maximize opportunities to learn from experience. Adaptive management (Holling 1978; Romesburg 1981; MacNab 1983, 1985; Walters 1986; Eberhardt 1988; Thomas et al. 1990; Wilhere 2002) can provide a reliable means for assessing the HCP, producing better ecological knowledge, and developing appropriate modifications to improve forest management. The primary challenge for using an adaptive management approach is to demonstrate simply and clearly why a change in management would be worthwhile.

Plum Creek's Cascades HCP was the first HCP to openly include adaptive management in the sense that the permittees acknowledged that in certain situations they may need to modify their actions so that additional expense, in the form of additional land, water, or other resources, may need to be provided as mitigation. This voluntary "partial set-a-side" of assurances under the "no surprises" policy was a significant step for Plum Creek. Plum Creek's HCP included adaptive management concepts throughout, but explicitly for several topics including whether spotted owl populations were following the model results and involved the use of deferral habitat. Another explicit topical area included whether riparian buffers were functioning adequately regarding aquatic resource values and processes.

The concept of adaptive management was soon incorporated into additional HCPs such as the Washington Department of Natural Resources HCP, Plum Creek's Native Fish HCP, and the Green Diamond Resources (formally Simpson Timber) HCP in Washington. With the development of each subsequent HCP, the concepts behind adaptive management within the "no surprises" framework of regulatory assurances was refined and cultured. Eventually, adaptive management became a common tool in HCPs across the nation.

Not only does adaptive management make sense for a long-term landscape management plan like an HCP, but recent USFWS policy (U.S. Fish and Wildlife Service and National Oceanic and Atmospheric Administration 2000) addressed incorporation of adaptive management into HCPs where significant uncertainty exists. Not all HCPs require an adaptive management approach. However, adaptive management can be essential for HCPs that otherwise would pose a substantial risk as a result of information gaps regarding the species biology, the impacts likely to result from HCP actions, or the response of the species to minimization and mitigation features of the HCP. Adaptive management is also helpful in guiding plan amendments in response to changed circumstances. Through the Cascades HCP, it was determined that adaptive management in HCPs could be stringent with specific monitoring thresholds for initiating changes and active experimentation, but that adaptive management could also include observational monitoring and incorporation of outside research findings. Through these early HCPs, the Services came to understand how adaptive management could be selectively limited in scope or could be constructed with landowner involvement into cooperative and beneficial approaches.

Adaptive management is incorporated into many elements of the Cascades HCP, but most relevant to this discussion is the use of adaptive management in the development of thresholds for initiating corrective action. As stated in the HCP (Plum Creek Timber Company 1996, Sec. 5.4.2.2, pg 328): "A key element of adaptive management is the establishment of testable hypotheses tied to management objectives. Should resultant monitoring determine that biological conditions are outside the 'bounds' estimated in the HCP, Plum Creek and the Services will review assumptions, refine models and modify management to protect public resources." These thresholds "for triggering corrective action must be linked to key elements of the HCP by being related to statistically significant, biologically relevant elements and obtainable through monitoring data collected during the permit period."

To that end, model outputs contribute significantly to the future direction of the HCP. For instance, a variance of more than 20% between the projected and actual habitat condition for any lifeform creates a need to consult with the Services on causative factors and potential remedial actions. Additionally, a finding of actual spotted owl populations (as determined by spotted owl monitoring in the planning area and outside peer review) that are less than 80% of the carrying capacity estimated in the RSPF model will also trigger possible corrective action, such as retention of more habitat deferrals, extension of nesting habitat deferrals for longer time periods, or habitat deferrals may be moved to more effective locations. Key questions regarding the tripping of these threshold triggers are "Is the observed deficiency a result of HCP management?" and "Will modification of the HCP remedy conditions in a substantive way?"

In the 10 years of HCP implementation and management, model outputs have triggered thresholds and caused further evaluation of causative actions. The distribution of the eight stand structure types has deviated more than 20% from that predicted by the OPTIONS model. This occurred as the actual percentage of early successional stand structure types (stand initiation, shrub sapling, and young forest) dropped below that predicted by OPTIONS for the first decade (year 2006) of the plan. It was determined to be a result of HCP management because the deficiency was caused by a reduction in timber harvest activity in the HCP compared to levels initially estimated on Plum Creek lands during HCP development.

However, the biological consequences of less early successional habitat on the landscape than predicted was judged to be minimal, because none of the 315 species covered in the HCP would be jeopardized by less early successional habitat on Plum Creek property. The Services determined that species dependent on early successional habitats in the central Cascades are generally able to use other habitats or have evolved with such scarcity of habitats and compensated by developing good dispersal abilities. Additionally, early successional habitats still occur at near historic levels within the planning area.

As stated earlier, populations of spotted owls within the HCP project area have dropped significantly during the first decade of HCP implementation, based on spotted owl site monitoring in the HCP and demographic monitoring of owl populations mandated under the Northwest Forest Plan. Current estimates of active spotted owl sites within the HCP planning area are approximately 40% of the number when the HCP was implemented in 1997 (Raedeke Associates 2006), comparable to trends observed in the majority of other populations in the region (Anthony et al. 2006). Habitat amounts have remained relatively static within the HCP planning area over the last decade (Hicks et al. 2003*a*). Barred owls have been studied in Washington for over 20 years and were known to be present in the HCP planning areas when surveys were initiated in 1990 (Hamer 1988, Herter and Hicks 2000). However, barred owls have significantly increased within the HCP planning area within the last 10 years and now threaten the persistence of the northern spotted owl through much of its extant range (Anthony et al. 2006). The barred owl threat has extended across all types of landscapes, from industrial forest lands to national parks where no significant timber harvest has occurred in recent decades (Herter and Hicks 2000, Pearson and Livezey 2003, Gremel 2005, Livezey 2007). In this application of model output and trigger thresholds, the answer to the first question—"Is the observed deficiency a result of HCP management?"—is clearly that barred owl range expansion and competition with spotted owls has not been caused by the implementation of the HCP. However, the answer to the second question—"Will modification of the HCP remedy conditions in a substantive way?"—remains a significant and problematic management challenge for agencies and landowners within the HCP area as well as throughout the range of the spotted owl in the Pacific Northwest (Gremel 2005, Livezey 2007). The USFWS Recovery Plan for the spotted owl identifies the barred owl range expansion as a major impediment to the recovery of the spotted owl and describes an aggressive program for barred owl control through direct reduction (U.S. Fish and Wildlife Service 2008).

POLICY AND MANAGEMENT IMPLICATIONS

The Cascades HCP was a major achievement in furthering the concept of multispecies conservation planning at the landscape scale. The significance of the Cascades HCP was enhanced by the fact that it was completed within a checkerboard ownership pattern of public and private land in a controversial area of Washington known as the Interstate-90 corridor (Hicks 1997). Although the HCP did resolve ESA-related land management concerns, other issues such as back-country recreation, aesthetics, and road access were not addressed by the HCP and remained to be resolved in the Interstate-90 Corridor. Following inception of the HCP, the Interstate-90 Land Exchange between Plum Creek and the U.S. Forest Service was completed in 2000 to consolidate checkerboard ownership in the Interstate-90 Corridor and address these and other non-ESA-related public issues.

The Cascades HCP was instrumental in facilitating the land exchange. Plum Creek land could be effectively appraised for its value to the U.S. Forest Service because specific ESA-related constraints could be more accurately quantified. Conversely, U.S. Forest Service land with ESA-related habitat values could be exchanged to Plum Creek and incorporated into the multispecies HCP. The Cascades HCP was concurrently amended to reflect ownership changes resulting from the land exchange. In addition to the Interstate-90 Land Exchange, the HCP provided the framework for other conservation transactions, including conservation land sales to the U.S. Forest Service, and sales to private/public conservation groups, including the Mountains to Sound Greenway, Northwest Ecosystem Alliance, and the Cascade Coalition Partnership.

The science-based foundation of the Cascades HCP provided the platform for other research investigations during the first decade of its existence. These research projects have provided information to assess and refine models used in the HCP such as investigations in use of midsuccessional forests by avian species, which helped evaluate the lifeform modeling and "guilding" of vertebrate species to forest structure stages (Manuwal and Gergen 2001). Additionally, the lifeform guilding and structure stage classification framework used in the HCP provided the format for a research project to evaluate habitat managed under the HCP as biodiversity "hot spots" (Hanson et al. 2006).

The use of models developed by HCP applicants can be helpful in assessing impacts, but blind reliance by U.S. Fish and Wildlife Service staff upon applicant-developed models is not appropriate or in the best interest of the species, trust responsibilities, or the general public. The RSPF model for spotted owls was subjected to such scrutiny. The U.S. Fish and Wildlife Service worked with the Washington Department of Fish and Wildlife to develop a test for the Plum Creek model. Agency biologists were concerned that the model might not adequately account for habitat fragmentation. A test was developed that included equal amounts of habitat; but, under one scenario, the habitat patches were smaller and less contiguous. The agencies wanted to see if the RSPF model would predict the same number of spotted owls for both scenarios. The RSPF model successfully predicted a lower carrying capacity for spotted owls in the more fragmented test landscape (Irwin and Hicks 1995: Appendix 2), which was consistent with what was known about spotted owl habitat use. Also, in conducting its analyses under section 7 of the ESA, the U.S. Fish and Wildlife Service used a variety of analyses developed by its own staff to test or corroborate the RSPF model. The combination of independent analyses in conjunction with the RSPF model allowed the U.S. Fish and Wildlife Service to explore different scenarios and develop its own independent analysis that was vastly better than the analysis that could have been conducted in the absence of the modeling and supportive work undertaken by Plum Creek. The experiences that the U.S. Fish and Wildlife Service obtained from examining the models and participating in the analyses with Plum Creek provided a preparatory education to help the Services deal with future applicants. The Services had a better understanding of informational needs, availability of surrogate information, logistical limitations, and alternate solutions.

Most of the modeling and analytical work to prepare the Cascades HCP was completed between 1993 and 1996. There have been major improvements in the tools and techniques we used to build the HCP that could facilitate a similar effort if attempted today. The data available to explore landscape planning for wildlife have been improved substantially for the Pacific Northwest by the publication of the Wildlife-Habitats Relationships in Oregon and Washington compendium, which now includes digital databases for improved access and timely updates. The early versions of the Stand Visualization software we used to illustrate silvicultural experiments and spotted owl habitat have been significantly improved and now include the ability to display and analyze data at the landscape level. The OPTIONS program has also been improved to include the ability to incorporate more treatments and greater storage of stand-level data in the analysis of growth and harvest scenarios. Finally, the advancement of GIS now allows even more sophisticated analysis of complex biotic and physical landscape data than was available a decade ago. As we have seen with the revision of the RSPF model described earlier, GIS capabilities to analyze and compute data at the landscape level have transcended the limitations of early systems to simply make maps. Although the tools and techniques have improved since we initially developed the HCP, we have not seen other approaches or processes that we would use in their place today.

In 2007, Plum Creek and the Services completed a major 10-year review of the Cascades HCP. During that review, we assessed the quantity and quality of information obtained by monitoring and implementation of management practices mandated by the HCP. Both Plum Creek and the Services concluded that the tools and techniques used to construct and monitor the HCP have been effective and continue to provide useful information to serve both the science and management objectives of the HCP. No major changes to management practices were required as a result of the 10-year review.

However, the most disturbing development to occur during the first decade of the HCP is the reduction of the resident spotted owl population in response to the influx of barred owls. In hindsight, we underestimated the impact of the barred owl as a factor in depressing spotted owl populations within our planning area. Our modeling efforts did not incorporate the barred owl as a factor influencing habitat use by spotted owls. As we learn more about the influence of barred owls on spotted owls, it becomes evident that there are both habitat and behavioral aspects of the issue. Because little was known about barred owl and spotted owl interactions at the time of HCP development, and the dynamic nature of the issue, it is speculative that modeling could have been helpful in anticipating or addressing this complex challenge in the HCP. Given that the increase of barred owls is affecting spotted owl populations throughout the current range, it does seem plausible that the HCP could provide an opportunity for Plum Creek and state and federal agencies to collaboratively investigate biological dynamics between the two species in the planning area and develop adaptive management responses that might have value and applicability over a larger area.

Despite disappointments and challenges regarding the future of spotted owls in our planning area, the other components of the HCP continue to perform at or above expectations. The riparian strategy has been successful in maintaining water quality and habitat diversity within streams; habitat targets for maintaining the eight forest structure stages across ownerships have largely been met; and protection of special habitats such as microclimates on talus slopes, seeps and springs, late-successional ponderosa pine stands, and wildlife trees for cavity-nesting species has been achieved. One advantage to developing a multispecies HCP is that other species can benefit even if challenges develop for some target species.

FUTURE DIRECTIONS

The Cascades HCP recently completed its first decade in the 50-year life of the plan. The use of modeling to evaluate biological relationships, quantify forest harvest and growth effects on habitat, and "future out" the biological consequences of HCP implementation and management has been instrumental to the plan. Monitoring requirements in the HCP and the HCP Implementation Agreement mandate major reviews of the plan at 5- and 10-year intervals. During these reviews, new information regarding forest inventory updates and management activities are brought to the Services. Additionally, all terrestrial and aquatic monitoring data are summarized and reviewed to determine if threshold "triggers" are tripped or if mitigation measures and management activities mandated by the HCP are ineffective. In this way, adaptive management is achieved using information acquired within the HCP and applied to models and analyses developed for the HCP.

The emerging issue of barred owl and spotted owl interactions may form the basis of additional monitoring and modeling in the HCP. This would be completed in conjunction with other agencies and entities, since this is a situation affecting spotted owl recovery throughout the species range.

The HCP was designed to build upon conservation strategies used on federal lands in the Northwest Forest Plan. The HCP has undergone a major amendment to incorporate changing land ownership arising from the Interstate-90 Land Exchange. Subsequent conservation land sales have increased the probability that what is now a single-landowner HCP may become a multilandowner HCP as other forest interests (private or conservation-based) acquire ownership within the HCP project area and assume conservation responsibilities under the plan. If that scenario develops, analytical models such as lifeform/structure stage guilding and the OPTIONS model will likely be refined and applied to the new ownerships to coordinate the continued achievement of the HCP landscape conservation objectives.

SUMMARY

Plum Creek Timber Company is one of the largest private owners of forested habitat occupied by the northern spotted owl *(Strix occidentalis caurina)* in the United States. At the time of federal listing in 1990, 107 spotted owl territories had been located near the company's property intermingled with other lands in the Central Cascade Mountains of Washington. To facilitate the management of the spotted owls and address the economic impacts that the federal listing of the species had created, Plum Creek, the U.S. Fish and Wildlife Service, and the National Marine Fisheries Service (the Services) began the development of a Habitat Conservation Plan (HCP), a process under the Endangered Species Act, which allows private landowners to obtain a federal permit to incidentally "take" a listed species or its habitat under otherwise lawful activities in accordance with an approved plan that specifies actions to mitigate and minimize the anticipated impact on the listed species. The HCP area is 169,510 hectares in size, including 49,239 hectares of Plum Creek property. The HCP, which was approved by the Services in June 1996, is a 50-year plan which addresses the habitat needs of 315 vertebrate species, including the northern spotted owl. Modeling was an important tool to develop several key components of the HCP described in this chapter. The first is a forest stand structure classification system based on spotted owl monitoring and habitat analysis. The second is the OPTIONS™ timberland harvest and planning model used to simulate different management strategies and project spotted owl habitat through the 50-year planning period. Finally, a resource selection function was used to assess the likelihood that habitat retained in the plan, or projected to grow over the life of the plan, would have a low, medium, or high probability of occupancy by spotted owls. Adaptive management was incorporated into the Cascades HCP to address uncertainty by establishing thresholds for initiating corrective action. The tools and techniques used to construct and monitor the HCP have been effective and continue to provide useful information to serve both the science and management objectives of the HCP.

LITERATURE CITED

Anthony, R. G., E. D. Forsman, A. B. Franklin, D. R. Anderson, K. P. Burnham, G. C. White, C. J. Schwarz, J. D. Nichols, J. E. Hines, G. S. Olson, S. H. Ackers, L. S. Andrews, B. L. Biswell, P. C. Carlson, L. V. Diller, K. M. Dugger, K. E. Fehring, T. L. Fleming, R. P. Gerhardt, S. A. Gremel, R. J. Gutiérrez, P. J. Happe, D. R. Herter, J. M. Higley, R. B. Horn, L. L. Irwin, P. J. Loschl, J. A. Reid, and S. G. Sovern. 2006. Status and trends in demography of northern spotted owls, 1985–2003. *Wildlife Monographs* 163:1–48.

Bailey, R. G. 1988. *Ecogeographic analysis: A guide to the ecological division of land for resource management*. U.S. Forest Service, Miscellaneous Publication 1465, Washington, D.C., USA.

Bart, J. 1995. Amount of suitable habitat and viability of northern spotted owls. *Conservation Biology* 9:943–946.

Beatley, T. 1994. *Habitat conservation planning: Endangered species and urban growth*. University of Texas Press, Austin, USA.

Boyce, M. S., and L. L. McDonald. 1999. Relating populations to habitats using resource selection functions. *Trends in Ecology and Evolution* 14:268–272.

Boyce, M. S., L. L. McDonald, and B. F. J. Manly. 1999. Reply. *Trends in Ecology and Evolution* 14:490.

Brown, E. R., editor. 1985. *Management of wildlife and fish habitats in forests of western Oregon and Washington*. U.S. Forest Service, Pacific Northwest Research Station, Report number R6-F&WL-192, Portland, Oregon, USA.

Burnham, K. P., and D. R. Anderson. 1998. *Model selection and inference: A practical information-theoretic approach*. Springer-Verlag. New York, New York, USA.

Carey, A. B., S. P. Horton, and B. L. Biswell. 1992. Northern spotted owls: Influence of prey base and landscape character. *Ecological Monographs* 62:223–250.

Carey, A. B., J. A. Reid, and S. P. Horton. 1990. Spotted owl home range and habitat use in southern Oregon coast ranges. *Journal of Wildlife Management* 54:11–17.

Dietrich, W. 1992. *The final forest: The battle for the last great trees of the Pacific Northwest*. Simon and Schuster. New York, New York, USA.

Eberhardt, L. L. 1988. Testing hypotheses about populations. *Journal of Wildlife Management* 52:50–56.

Forest Ecosystem Management Assessment Team. 1993. *Forest ecosystem management: An ecological, economic, and social assessment*. U.S. Government Printing Office, Washington, D.C., USA.

Forsman, E. D. 1980. *Habitat utilization by spotted owls in the west-central Cascades of Oregon*. Dissertation, Oregon State University, Corvalis, USA.

Forsman, E. D., E. C. Meslow, and H. M. Wight. 1984. Distribution and biology of the spotted owl in Oregon. *Wildlife Monographs* 87:1–64.

Franklin, A. B., D. R. Anderson, R. J. Gutiérrez, and K. P. Burnham. 2000. Climate, habitat quality, and fitness in northern spotted owl populations in northwestern California. *Ecological Monographs* 70:539–590.

Franklin, J. F., D. R. Berg, D. A. Thornburgh, and J. C. Tappeiner. 1997. Alternative silviculture approaches to timber harvesting: Variable retention harvest systems. Pages 111–139 *in* K. A. Kohm and J. F. Franklin, editors. *Creating a forestry for the 21st century: The science of ecosystem management*. Island Press, Washington, D.C., USA.

Gremel, S. 2005. *Factors controlling distribution and demography of northern spotted owls in a reserved landscape*. Thesis, University of Washington, Seattle, USA.

Gutiérrez, R. J., A. B. Franklin, and W. S. Lahaye. 1995. Spotted owl (*Strix occidentalis*). Account 179 *in* A. Poole and F. Gill, editors. *The birds of North America*. The Academy of Natural Sciences, Philadelphia, Pennsylvania, and The American Ornithologists' Union, Washington, D.C., USA.

Hanson, A. J., J. P. Verschuyl, and D. B. McWethy. 2006. *Biodiversity potential in the Pacific and Inland Northwest: Phase II—Applications to industry planning areas*. Landscape Biodiversity Lab, Technical Report number 3, Montana State University, Bozeman, Montana, USA.

Herter, D. R., and L. L. Hicks. 2000. Barred owl and spotted owl populations and habitat in the central Cascade range of Washington. *Journal of Raptor Research* 34:279–286.

Herter, D. R., L. L. Hicks, H. C. Stabins, J. J. Millspaugh, A. J. Stabins, and L. D. Melampy. 2003. Roost site characteristics of northern spotted owls in the nonbreeding season in central Washington. *Forest Science* 48:437–444.

Hicks, L. L. 1991. Plum Creek Timber Company's new forestry experiments: Lessons learned and future directions. *Forest Perspectives* 1:9–10.

Hicks, L. L. 1997. Plum Creek's Cascades Habitat Conservation Plan: A corporate approach to ecosystem management. *Endangered Species Update* 14:35–39.

Hicks, L. L., D. R. Herter, and R. J. Early. 2003*a*. Clines in life history characteristics of the spotted owl in Washington. *Northwestern Naturalist* 84:57–67.

Hicks, L. L., D. R. Herter, and W. Eschholz. 1995. *Spotted owl home range and habitat use in the Plum Creek Habitat Conservation Plan planning area*. Plum Creek Timber Company, L.P. Technical Report number 5, Seattle, Washington, USA.

Hicks, L. L., B. F. J. Manly, D. R. Herter, R. J. Early, and H. C. Stabins. 2003*b*. A resource selection probability function for the northern spotted owl in Plum Creek's central Cascade Habitat Conservation Plan, Washington State. Pages 121–129 *in* S. Huzurbazar, editor. *Resource selection methods and applications. Proceedings of the 1st International Conference on Resource Selection*. Omnipress, Madison, Wisconsin, USA.

Hicks, L. L., and H. C. Stabins. 1995. *Spotted owl habitat descriptions for Plum Creek's Cascades Habitat Conservation Plan*. Plum Creek Timber Company, L.P. Technical Report number 4, Seattle, Washington, USA.

Holling, C. S., editor. 1978. *Adaptive environmental assessment and management*. John Wiley and Sons, New York, New York, USA.

Irwin, L. L., and L. L. Hicks. 1995. *Assessment of methods for estimating potential impacts of timber harvesting on spotted owls in the Plum Creek Cascades Habitat Conservation Plan project area*. Plum Creek Timber Company, Technical Report number 6, Seattle, Washington, USA.

Kareiva, P., S. Andelman, D. Doak, B. Elderd, M. Groom, J. Hoekstra, L. Hood, F. James, J. Lamoreux, G. LeBuhn, C. McCulloch, J. Regetz, L. Savage, M. Ruckelshaus, D. Skelly, H. Wilbur, K. Zamudio, and NCEAS Habitat Conservation Plan Working Group. 1999. *Using science in habitat conservation plans*. National Center for Ecological Analysis and Synthesis, University of California, Santa Barbara, and American Institute of Biological Sciences, Washington, D.C., USA.

Lemkuhl, J. F., and M. G. Raphael. 1993. Habitat pattern around northern spotted owl locations on the Olympic Peninsula, Washington. *Journal of Wildlife Management* 57:302–315.

Lindenmayer, D. B., and J. F. Franklin. 2002. *Conserving forest biodiversity: A comprehensive multiscaled approach*. Island Press, Washington, D.C., USA.

Livezey, K. B. 2007. Barred owl habitat and prey: A review and synthesis of the literature. *Journal of Raptor Research* 41:177–201.

Loehle, C., J. G. MacCracken, D. Runde, and L. Hicks. 2002. Forest management at landscape scales: Solving the problems. *Journal of Forestry* 100:25–33.

Lujan, M., Jr., D. R. Knowles, J. Turner, and M. Plenert. 1992. *Recovery plan for the northern spotted owl—Final draft*. U.S. Fish and Wildlife Service, Portland, Oregon, USA.

Lundquist, R. W., and L. L. Hicks. 1995. *Development of a wildlife habitat relationship matrix for the Plum Creek Cascades Habitat Conservation Plan*. Plum Creek Timber Company, Technical Report number 9, Seattle, Washington, USA.

MacNab, J. 1983. Wildlife management as scientific experimentation. *Wildlife Society Bulletin* 11:397–401.

MacNab, J. 1985. Carrying capacity and related slippery shibboleths. *Wildlife Society Bulletin* 13:403–410.

Manly, B. F. J., L. L. McDonald, D. L. Thomas, T. L. McDonald, and W. P. Erickson. 2002. *Resource selection by animals: Statistical design and analysis for field studies*. Second edition. Kluwer Academic, Dordrecht, The Netherlands.

Manuwal, D. A., and T. Gergen. 2001. *Evaluating forest breeding birds in the Plum Creek Habitat Conservation Plan area*. Wildlife Science Group, College of Forest Resources, University of Washington, Seattle, USA.

McGaughey, R. 2002. *Stand visualization system*. U.S. Forest Service, Pacific Northwest Research Station. <http://forsys.cfr.washington.edu/svs.html>.

Meyer, J. S., L. L. Irwin, and M. S. Boyce. 1998. Influence of habitat abundance and fragmentation on northern spotted owls in western Oregon. *Wildlife Monographs* 139:1-51.

Mysterud, A., and R. A. Ims. 1999. Relating populations to habitat. *Trends in Ecology and Evolution* 14:489-490.

Oliver, C. D., R. Greggs, S. Boyd, and L. L. Hicks. 1995. *Forest stand structural classification system developed for the Plum Creek Cascades Habitat Conservation Plan*. Plum Creek Timber Company, Technical Report number 10, Seattle, Washington, USA.

Oliver, C. D., and B. C. Larson. 1990. *Forest stand dynamics*. McGraw-Hill, New York, New York, USA.

Pearson, R. R., and K. B. Livezey. 2003. Distribution, numbers, and site characteristics of spotted owls and barred owls in the Cascade Mountains of Washington. *Journal of Raptor Research* 37:265-276.

Plum Creek Timber Company. 1996. *Multi-species habitat conservation plan on forestlands owned by Plum Creek Timber Company in the I-90 corridor of the central Cascades mountain range, Washington*. Plum Creek Timber Company, Seattle, Washington, USA.

Raedeke Associates. 2006. *Northern spotted owl demographic surveys in the Cascades Habitat Conservation Plan Area*. Raedeke Associates, 21 November 2006, Seattle, Washington, USA.

Rahn, M. E., H. Doremus, and J. Diffendorfer. 2006. Species coverage in multispecies habitat conservation plans: Where's the science? *BioScience* 56:613-619.

Reimer, D. R. 2007. *OPTIONS*. D. R. Systems, Nanaimo, British Columbia, Canada. <http://www.drsystemsinc.com/contact.html>.

Romesburg, H. C. 1981. Wildlife science: Gaining reliable knowledge. *Journal of Wildlife Management* 45:293-313.

Sisco, C. L. 1990. *Seasonal home range and habitat ecology of spotted owls in northwestern California*. Thesis, Humbolt State University, Arcata, California, USA.

Slingerland, G. R. 1999. *The effect of the "no surprises" policy on habitat conservation planning and the Endangered Species Act*. Thesis, Virginia Polytechnic Institute and State University, Blacksburg, USA.

Solis, D. M. Jr., and R. J. Gutiérrez. 1990. Summer habitat ecology of northern spotted owls in northwestern California. *Condor* 92:739-748.

Swindle, K. A., W. J. Ripple, E. C. Meslow, and D. Schafer. 1999. Old-forest distribution around spotted owl nests in the central Cascades Mountains, Oregon. *Journal of Wildlife Management* 63:1212-1221.

Thomas, J. W. 1979. *Wildlife habitats in managed forests: The Blue Mountains of Oregon and Washington*. U.S. Forest Service, Agricultural Handbook number 553, Washington, D.C., USA.

Thomas, J. W., E. D. Forsman, J. B. Lint, E. C. Meslow, B. R. Noon, and J. Verner. 1990. *A conservation strategy for the northern spotted owl*. Interagency Scientific Committee to Address the Conservation of the Northern Spotted Owl, Portland, Oregon. U.S. Government Printing Office 791-171/20026, Washington, D.C., USA.

U.S. Fish and Wildlife Service. 1989. *The northern spotted owl status review supplement*. USFWS, Portland Oregon, USA.

U.S. Fish and Wildlife Service. 1990. *Status review for the northern spotted owl* (Strix occidentalis caurina). USFWS, Portland, Oregon, USA.

U.S. Fish and Wildlife Service. 2007. USFWS webpage. <http://ecos.fws.gov/conserv_plans/index.jsp>.

U.S. Fish and Wildlife Service. 2008. *Recovery Plan for the Northern Spotted Owl* (Strix occidentalis caurina). USFWS, Portland, Oregon, USA.

U.S. Fish and Wildlife Service and National Oceanic and Atmospheric Administration. 2000. Notice of availability of a final addendum to the handbook for habitat conservation planning and incidental take permitting process. *Federal Register* 65:35242–35257.

Walters, C. J. 1986. *Adaptive management of renewable resources*. Macmillan, New York, New York, USA.

Wilhere, G. F. 2002. Adaptive management in habitat conservation plans. *Conservation Biology* 16:20–29.

Yaffee, S. L. 1994. *The wisdom of the spotted owl: Policy lessons for a new century*. Island Press, Washington, D.C., USA.

CHAPTER

Application of Models to Conservation Planning for Terrestrial Birds in North America

22

Jane A. Fitzgerald, Wayne E. Thogmartin, Randy Dettmers, Tim Jones, Christopher Rustay, Janet M. Ruth, Frank R. Thompson, III, and Tom Will

Bird conservation in the United States is a good example of the use of models in large-scale wildlife conservation planning because of its geographic extent, focus on multiple species, involvement of multiple partners, and use of simple to complex models. We provide some background on the recent development of bird conservation initiatives in the United States and the approaches used for regional conservation assessment and planning. We focus on approaches being used for landscape characterization and assessment, and bird population response modeling.

BIRD CONSERVATION INITIATIVES

Bird conservation planning in the United States is guided by four major partnership-driven initiatives organized around taxonomic groups of birds that differ fundamentally in aspects of their basic biology or role in sport recreation (e.g., waterfowl, landbirds, shorebirds, and waterbirds). Each initiative has assessed the conservation status of each species under its purview based on parameters such as population size, population trend, and vulnerability to external threats. The assessment results have been used to determine which species are most in need of conservation action. Continental or national population goals and recommended conservation actions have been assigned to the species of highest priority. Each of these broad-scale plans has been (or is in the process of being) stepped down to ecoregional scales. The four plans attempt to provide basic guidance on the conservation needs of each species in the respective groups they cover, although the degree to which each succeeds in this effort depends to a large extent on the amount and quality of information currently available. In chronological order of initiative formation, the current guiding documents produced by the partnerships at the national or international scale are the North American Waterfowl Management Plan (NAWMP; NAWMP Plan Committee 2004), the Partners in Flight (PIF) North American Landbird Conservation Plan (Rich et al. 2004), the U.S. Shorebird Conservation Plan (Brown et al. 2001), and Waterbird Conservation for the Americas (Kushlan et al. 2002).

The development of these four national/international bird conservation planning efforts catalyzed the formation of the North American Bird Conservation Initiative (NABCI) in 1999 to facilitate integration and cooperation among the various initiatives. The North American Bird Conservation Initiative also provided more formal links between Canada, the United States, and Mexico. Supplementing the efforts of these taxonomically based planning initiatives are individual species initiatives for which strong constituencies have developed (e.g., Northern Bobwhite Conservation Initiative, North American Grouse Management Strategy). Prior to the development of NABCI, no consistent geographic framework existed in which to integrate the emerging regional conservation plans. As a response to that need, NABCI delineated ecologically distinct regions with similar bird communities, habitats, and resource management issues. These regions range in size from 52,000 to 2.9 million km^2 and are known as Bird Conservation Regions, or BCRs (Fig. 22-1).

With the development and implementation of the NAWMP in the mid-1980s came a recognition that the conservation actions required to restore declining populations needed to be applied at landscape scales and targeted to specific geographies where the biological impacts would be most profound. Regional partnerships of federal and state natural resource management agencies and private conservation organizations formed what are now known as joint ventures (JVs), understanding that their conservation actions needed to be coordinated to produce cumulative, positive, and ecologically relevant impacts. These JVs began to develop biological models (Cowardin and Johnson 1979, Cowardin et al. 1995, Reynolds et al. 1996) to link waterfowl numbers to specific acreage targets. These models have been further refined with the use of geographic information systems (GIS) to create spatially explicit hypotheses predicting where habitat acres could most efficiently and economically achieve target objectives. This process has recently been dubbed "Conservation Design."

United under the NABCI mission "to deliver the full spectrum of bird conservation through regionally based, biologically driven, landscape oriented partnerships," the original waterfowl JVs have now accepted responsibility for implementing conservation objectives for "all birds" (i.e., waterfowl, waterbirds, shorebirds, and landbirds). In addition, new JVs have formed to guide

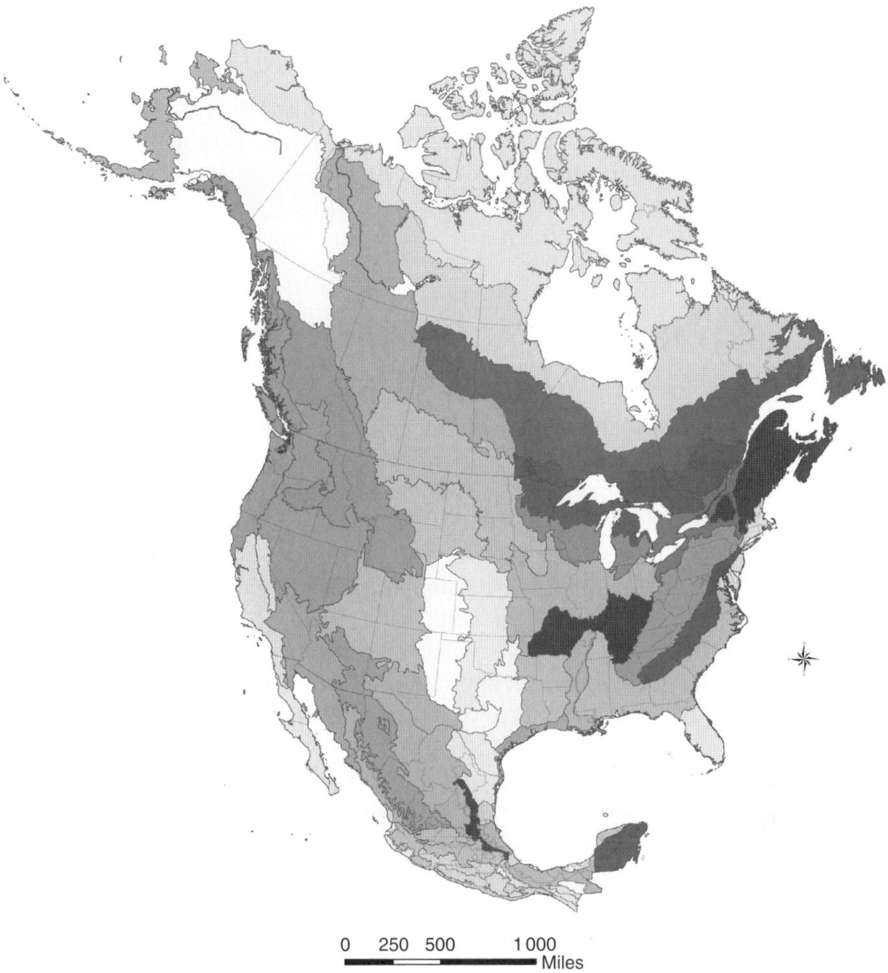

FIG. 22-1

Regions for bird conservation planning in North America (n = 67 Bird Conservation Regions).

conservation efforts in areas where none previously existed. Each of these regional conservation partnerships is encouraged to "step down" national and international population goals for all birds to the BCR scale and to develop spatially explicit habitat objectives needed to reach identified goals.

The Five Elements process of conservation design (Will et al. 2005), as proposed by PIF for implementation of all bird conservation at the ecoregional scale, entails (1) Landscape Characterization and Assessment; (2) Bird Population Response Modeling; (3) Conservation Opportunities Assessment; (4) Optimal Landscape Design; and (5) Monitoring and Evaluation. This chapter addresses newly developing geospatial techniques for assessing landscapes and patterns of landbird distribution and abundance in response to habitat characteristics. These capabilities will allow conservation planners to evaluate the capacity of landscapes and ecoregions to support priority species at desired levels. Although the focus of this chapter is on landbird applications, the same or similar approaches could be applied to other species of conservation concern. We defer discussion of multispecies response modeling, imperative to the Fourth Element of conservation design, to Noon et al. (this volume).

GIS-BASED HABITAT ASSESSMENT AND LANDSCAPE CHARACTERIZATION

A landscape-scale assessment of the current amount and condition of habitat types across an ecoregion, along with a characterization of the ability of those habitat types to support and sustain bird populations, is fundamental to the conservation design process. Habitat assessment and landscape characterization should not only describe the current amounts of different habitat types across an ecoregion but also summarize patch characteristics and landscape configurations that define the ability of a landscape to sustain healthy bird populations. Ultimately, landscape characterization should provide the capacity to assess the relative current and potential contributions of different land parcels to meet conservation objectives most efficiently.

If we are to conduct a habitat assessment, land-cover data that can be consistently applied across biologically appropriate scales (e.g., ecosystems, ecoregions) must be available. The National Land Cover Dataset (NLCD) 2001 (Homer et al. 2007) provides "seamless, consistent" land-cover data at a 30 m^2 cell resolution for the conterminous United States (http://www.epa.gov/mrlc/nlcd.html; though see Thogmartin et al 2004*a*). However, NLCD 2001 is limited in its ability to distinguish between fine classifications within general habitat types and also has issues with accuracy in some regions. The current version of NLCD classified Landsat Thematic Mapper imagery into 21 categories of terrestrial land cover, but in the conterminous United States these include only three classes of forested upland (deciduous, mixed, and evergreen), one class of shrubland, one class of herbaceous upland (grasslands/herbaceous), and two classes of wetland (woody wetland and emergent herbaceous wetlands).

More detailed land-cover data are available from other sources for almost all portions of the United States, but none of these other sources can currently be applied in a consistent fashion across the entire country. Land-cover products from the Gap Analysis Program (GAP; http://gapanalysis.nbii.gov) provide more detailed habitat classifications on a state-by-state basis. In addition, regional GAP analysis projects are now underway, which will provide consistent land cover across major regions of the United States (e.g., Southwest, Southeast, and Northwest Regional Gap Analysis projects).

Another source of habitat information that could be useful for some bird conservation design applications is the Landscape Fire and Resource Management Planning Tools Project (LANDFIRE) data set (Rollins et al. 2003), which provides regionally consistent data across much of the United States for existing vegetation composition and structure, wildland fuel loads, historical vegetation conditions, and historical fire regimes (see http://www.landfire.gov).

Aerial photography is an alternative that provides a source of high-resolution, detailed land-cover data (Paine and Kiser 2003). It provides improved classification of specific habitat types and seral stages along with better definition of patch boundaries compared to land-cover data derived from satellite imagery (e.g., NLCD). However, aerial photography also suffers from human subjectivity in interpreting the photo images, inconsistency between observers, issues with photo availability and quality (they vary with year and season), and higher costs in time and money. Automated land-cover classification as implemented with, for instance, Feature Analyst (Visual Learning Systems, Missoula, MT), offers the potential for some future relief in these matters.

All the sources mentioned here are available for conducting assessments of the amounts of different habitat types across large spatial extents. Which sources are most appropriate will depend on the location and extent of the area of interest, the desired level of detail for discriminating habitat types, availability, and resources in terms of time and money available for conducting a project.

A critical consideration for the proper use of these spatial data is their accuracy. Many spatial data are most accurate at a minimum mapping unit that is coarser than the resolution of the data. For instance, the NLCD 1992 (Vogelmann et al. 2001) possessed a spatial resolution of 30 m^2, but the minimum mapping unit has been suggested to be at least 1 ha, an order of magnitude coarser. At a regional scale relevant to conservation design, such coarseness is generally not prohibitive. A larger issue is the limited and often poorly classified thematic resolution of the spatial data. Most image classification methods, such as classification trees, poorly classify rare land covers (Stehman et al. 2003). Given that many species are of conservation concern because of declines in the abundance of their habitat, such habitat misclassification is particularly problematic given that it makes it difficult if not impossible to correctly assess a species' habitat. In addition, any given land cover class label may not be consistent across map products. Thogmartin et al. (2004*a*) found in the NLCD 1992, for instance, that pasture/hay was confused with herbaceous grassland in the upper midwestern United States. Thogmartin et al. (2004*a*) also reported that emergent herbaceous wetlands were more likely to be mapped in one mapping region as compared to others despite each mapping region occurring in the same ecoregion. Further, it is unknown if the seams that were observed in the NLCD 1992 have been rectified for the NLCD 2001. These sorts of mapping errors can percolate into mapped models of species-habitat associations, yielding misleading conservation decisions.

In addition to an assessment of how much of different habitat types exist within a region of interest, a characterization of landscape attributes can be important. Landscape characterization typically involves calculating metrics describing the size, shape, and configuration of habitat patches as well as the level of spatial heterogeneity within the region. Landscape characterization is important because these patterns are often linked to ecological processes (Gustafson 1998), such as increased amounts of habitat edge or other measures of fragmentation relating to increased predation rates (Andren and Anglestam 1988, Hartley and Hunter 1998). Metrics that should be considered for measurement as part of landscape characterization include those for size and shape of patches (total area, core area, perimeter, width), landscape composition (proportional cover of a given land-cover class, richness and evenness of land-cover classes), configuration of patches in the landscape (contagion, dispersion, isolation), and neighborhood characteristics (distance between similar patches, distance to important features such as water or roads) (Li et al. 2005).

Various computer applications exist to assist in calculating metrics for landscape characterization. Most GIS programs (e.g., ArcGIS 9.2 [Environmental Systems Research Institute, Redlands, CA], GRASS 6.2 [GRASS Development Team 2006], ERDAS Imagine 8.7 [Leica Geosystems GIS and Mapping, Atlanta, GA], IDRISI Andes [Clark Labs, Worcester, MA]) have functions for calculating many of these metrics, although usually on an individual basis. Several software applications specifically designed for calculating these metrics make this process easier and faster. These include FRAGSTATS (McGarigal et al. 2002) and IAN (DeZonia and Mladenoff 2004). In addition, extensions for GIS programs such as ArcView and GRASS exist to enhance capacity of these programs for calculating these metrics.

In addition to land-cover data depicting amount and configuration of different habitat types across ecoregions, other data relating to the physical and climatic aspects of the environment can also be very useful in conservation design applications for birds, especially in developing species-habitat models. The National Elevation Dataset (NED) provides seamless 10 and 30 m digital elevation data across the conterminous United States as well as Alaska and Hawaii (<http://ned.usgs.gov/>). These data can be particularly useful in describing elevational and moisture gradients that are important in defining bird distributions. Other digital data sources relating to physical characteristics include the National Hydrography Dataset (NHD; <http://nhd.usgs.gov/>), which depicts surface water features such as lakes, rivers, and streams; and databases on soil types, the General Soil Map (<http://www.ncgc.nrcs.usda.gov/products/datasets/statsgo>) for the United States (updated in 2006 and providing data at a 1:250,000 scale), and the Soil Survey Geographic Database (<http://www.ncgc.nrcs.usda.gov/products/datasets/ssurgo>), which contains detailed county-level soils data and is scheduled for completion in 2008. Digital climatological data for use in GIS applications are also available for such metrics as temperature, precipitation, humidity, and radiation. Two examples of accessible databases are the Spatial Climate Analysis Service (<http://www.ocs.orst.edu/prism/>) and the Daymet U.S. Database (<http://www.daymet.org/>).

APPROACHES TO LINKING BIRD DISTRIBUTION AND ABUNDANCE WITH HABITAT ASSESSMENTS AT THE BCR SCALE

Database Models

Database models are decision support tools employing decision rules for characterizing species-habitat associations (Thogmartin et al. 2006*a*). The Playa Lakes Joint Venture (PLJV) employed a database decision support tool for modeling species-habitat associations in the Shortgrass and the Central Mixed-grass Prairie Bird Conservation Regions (BCRs 18 and 19, respectively). These two regions include portions of six states: Colorado, Kansas, Nebraska, New Mexico, Oklahoma, and Texas. The PLJV has regional responsibility for priority species associated with both wetland and terrestrial systems. The JV needed a tool that would enable planning for all 53 priority species within their landscape and allow users to view habitat implementation implications for all species simultaneously and quickly. In response to that need, JV staff and partners developed the Hierarchical All-Bird Strategy (HABS), a system for maintaining and manipulating bird information within a relational Access-based database linked to a GIS (Fig. 22-2; Dobbs 2006). The best available land-cover data were acquired from a variety of sources and integrated into a seamless layer encompassing the entire JV planning unit. Land-cover types were cross-walked and renamed to reflect commonalities (e.g., for New Mexico, "Chihuahuan Mesquite Upland Scrub" and "Western Great Plains Mesquite Woodland and Shrubland" were grouped together as "Mesquite"). Spatial data depicting the location of roads, wetlands, soils, hydrography, and other information pertinent to conservation planning were also brought into the GIS.

The primary geographical planning polygon (within the PLJV boundaries) in the HABS database hierarchy is the portion of a state within a BCR, termed an "area." One of the goals of the PLJV is the development of Area Implementation Plans based on the HABS database. Within each area, general habitat types are defined, and within each habitat type, various habitat conditions are quantified. For example, where the acreage of riparian forest is given as a habitat type, the amount of that acreage that is composed of late successional cottonwood forest with understory vegetation is reported. Because habitat conditions are not well classified by large-scale land-cover data sets, and they vary annually and seasonally, various sources of information and expert opinion were used to approximate percentages of condition within each habitat, assuming average climatic conditions. The amounts of habitat in all condition categories were ultimately expressed in number of acres.

Data on the densities of priority species were compiled from both published and unpublished literature and were assigned to each habitat condition; bird density data were standardized to acres. Where density data were not available

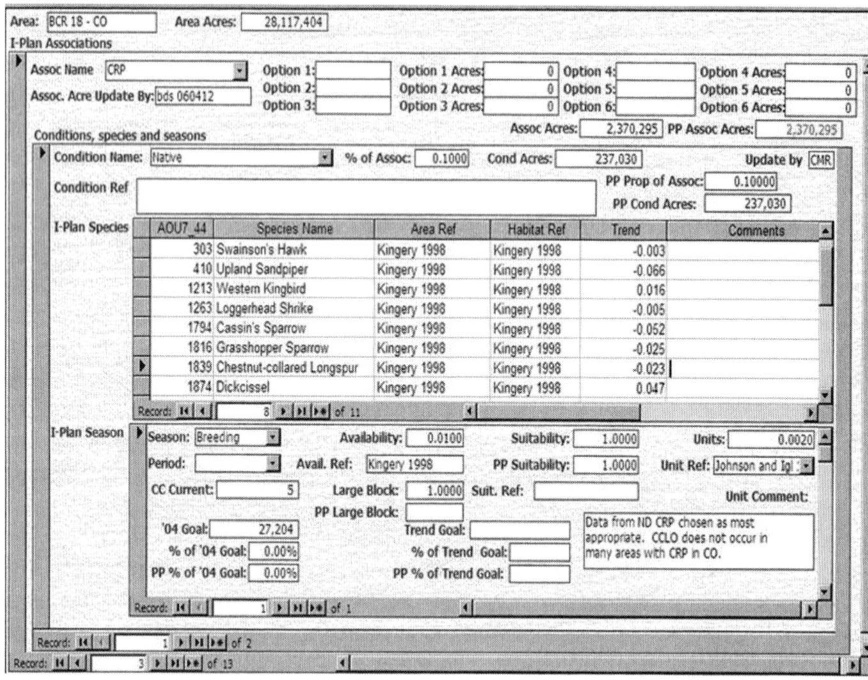

FIG. 22-2

A screen shot of the HABS Database details the components of the model for chestnut-collared longspur (*Calcarius ornatus*) in eastern Colorado in Conservation Reserve Program (CRP) grasslands (Dobbs 2006). "Assoc." = habitat name.

for an area, density values that were most similar in location and habitat condition were assigned, often adjusted using BBS relative abundance maps (Fig. 22-3; Sauer et al. 2007). Data comparability is an issue when dealing with data from multiple sources. In the case of the PLJV HABS database, the data from most of BCR 18 was from one source (Rocky Mountain Bird Observatory); this was not the case for BCR 19.

Additional correction factors were applied, if needed, to the species-habitat models to account for suitability or availability of habitat. Following are three examples of how particular data or needs were addressed:

1. Prairie-dog (*Cynomus* spp.) colonies host high densities of burrowing owls (*Athene cunicularia*), but Butts (1973) noted that in Oklahoma only 40% of all colonies across the landscape were utilized by owls. This correction factor was applied to the total acres of prairie-dog colonies in Oklahoma to modify estimates of number of owls.

2. Many grassland species in the region require a minimum patch size (Herkert 1994, Winter 1998, Johnson and Igl 2001). The standard management

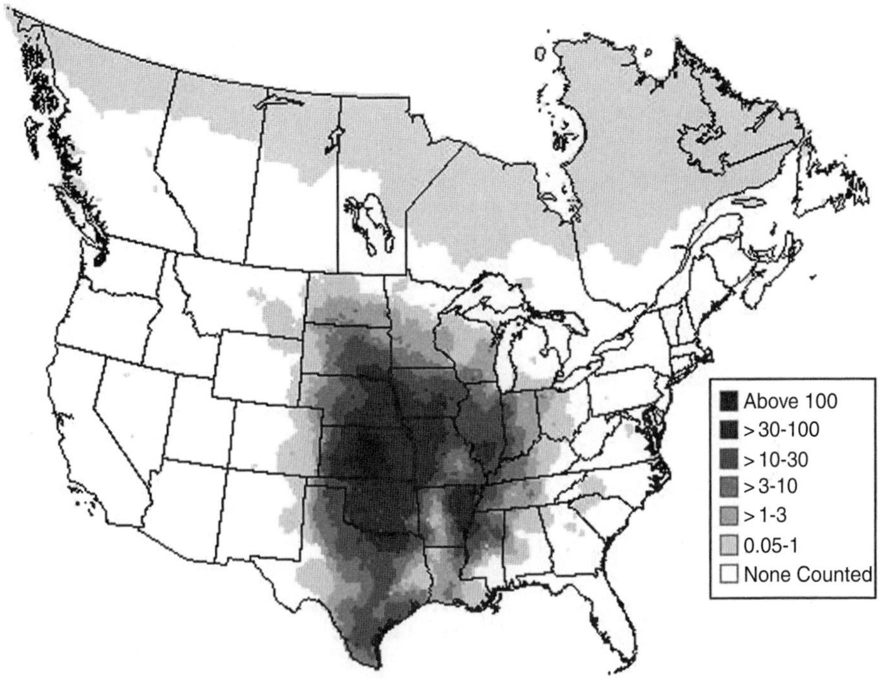

FIG. 22-3

North American Breeding Bird Survey relative abundance maps for dickcissel (*Spiza americana*) 1994–2003 (Sauer et al. 2007). Based on these data, the density for dickcissel from central Kansas was divided by 10 to determine an appropriate density for the eastern panhandle of Texas in the HABS database.

unit in the region is 160 acres. Because an initial analysis showed that less than 0.1% of all grassland areas were patches less than 40 acres, it was decided that accounting for minimum patch size requirements less than 40 acres was not necessary.

3. For species requiring more than 160 acres, more complex GIS models, similar to the Habitat Suitability Index (HSI) models described later in this chapter, were developed to identify blocks that are large enough to meet the needs of these species. For example, lesser prairie-chicken (*Tympanuchus pallidicinctus*) may require 5,000 acres or more of appropriate habitat (R. Rogers, Kansas Department of Wildlife and Parks, personal communication), and they are also affected by proximity to or amount of surrounding, potentially hostile habitat (Crawford and Bolen 1976). One of the lesser prairie-chicken model products that was included in HABS is a "large block factor" that describes the percentage of a particular habitat within a polygon that is actually suitable for this species (Fig. 22-4).

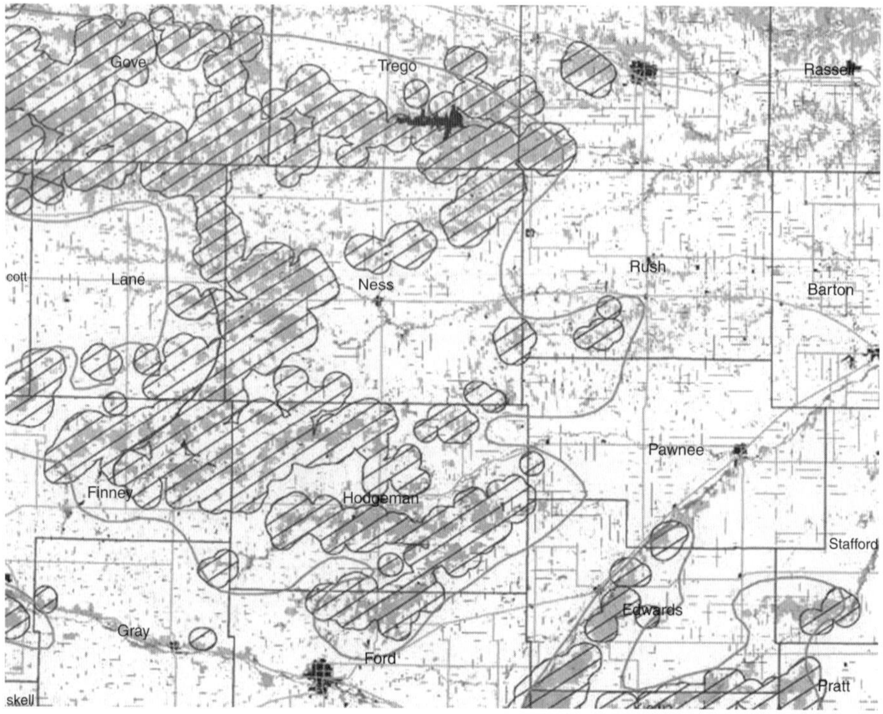

FIG. 22-4

Lesser prairie-chicken (*Tympanuchus pallidicinctus*) habitat in west-central Kansas. Occupied prairie-chicken areas are within gray lines (Kansas Department of Wildlife and Parks data) and PLJV-modeled habitat in crosshatch.

Bird densities are then multiplied by habitat condition acres and correction factors across all area habitats within HABS to arrive at a current estimated carrying capacity (a population estimate) for each area. HABS compares that figure to population goals and calculates the percentage of the goal already achieved. Habitat objectives, if needed to fully meet the goal, are determined by plugging varying acreage or percent of condition options into the HABS database to assess various means of reaching 100% of the population goal. Various habitat management options are typically available to meet a species' population goals, depending on the number of habitat types utilized or differing densities by condition. Once a habitat objective (needed number of acres of each habitat type) has been determined for one species, HABS can quickly project resulting declines or increases in other species' populations to be expected as a result of the targeted habitat changes for the first species. This provides planners with the ability to evaluate the differing effects of management on multiple species within the same area. Once the desired number of acres in each habitat condition in each area is determined, these recommendations are included in Area

Implementation Plans and made available to conservation practitioners. The HABS database and processes described here provide a valuable tool to managers for assessing and predicting impacts of habitat manipulations.

GIS-Based Habitat Suitability Models

Habitat Suitability Index models have been used to evaluate wildlife habitat and the effects of management activities and development since the early 1980s. Habitat Suitability Index models estimate habitat suitability on a scale of 0 (not suitable habitat) to 1 (highly suitable habitat) based on an assessment of resource attributes considered important to a species' abundance, survival, or reproduction (U.S. Fish and Wildlife Service 1980, 1981). Important habitat attributes (e.g., herbaceous ground cover, tree canopy cover, stem densities) are individually modeled based on a mathematical or graphical relationship, resulting in individual suitability indices (SIs) for each attribute. Overall habitat suitability, or the HSI, is calculated as some mathematical combination of the individual SIs. The HSI is typically calculated as the geometric mean of the individual SIs, although more complex formulas can be used, depending on how the SIs are thought to interact. Habitat Suitability Index models are developed from existing knowledge; this knowledge can be in the form of published studies, relationships derived from existing data or expert opinion, or hypothesized responses to habitat and other environmental correlates. Validation of HSI models is an important component of the modeling process because it tests how successfully the model has described the species-habitat relationship. Until models are validated, they represent hypotheses about these habitat relationships. However, even without final validation, HSI models may be useful for improved decision making and increased understanding of habitat relationships.

Traditionally HSI models were applied to an area or landscape habitat attributes were measured at a sample of locations within mapped land cover types or vegetation types and HSI values calculated. Habitat quality for the area was then summarized in terms of habitat units, which represent the product of the mean HSI score in each vegetation type and the area of land in that vegetation type, summed across the study area. Recent developments in HSI modeling have resulted in models that can be applied to large landscapes through the utilization of GIS. As with the database model previously described, these models typically rely on data layers derived from remote sensing and other existing spatial databases or large-scale inventories. Habitat Suitability Index values are calculated for each pixel in the landscape (Fig. 22-5A), and the distribution of HSI values for all the pixels in a landscape can be summarized in many different ways (Fig. 22-6; and see Dijak and Rittenhouse, this volume). Because of the focus on broad spatial extents and their use of GIS technology, these "next generation" HSI models can better address ecological and landscape effects on wildlife such as area sensitivity, edge effects, interspersion, landscape composition,

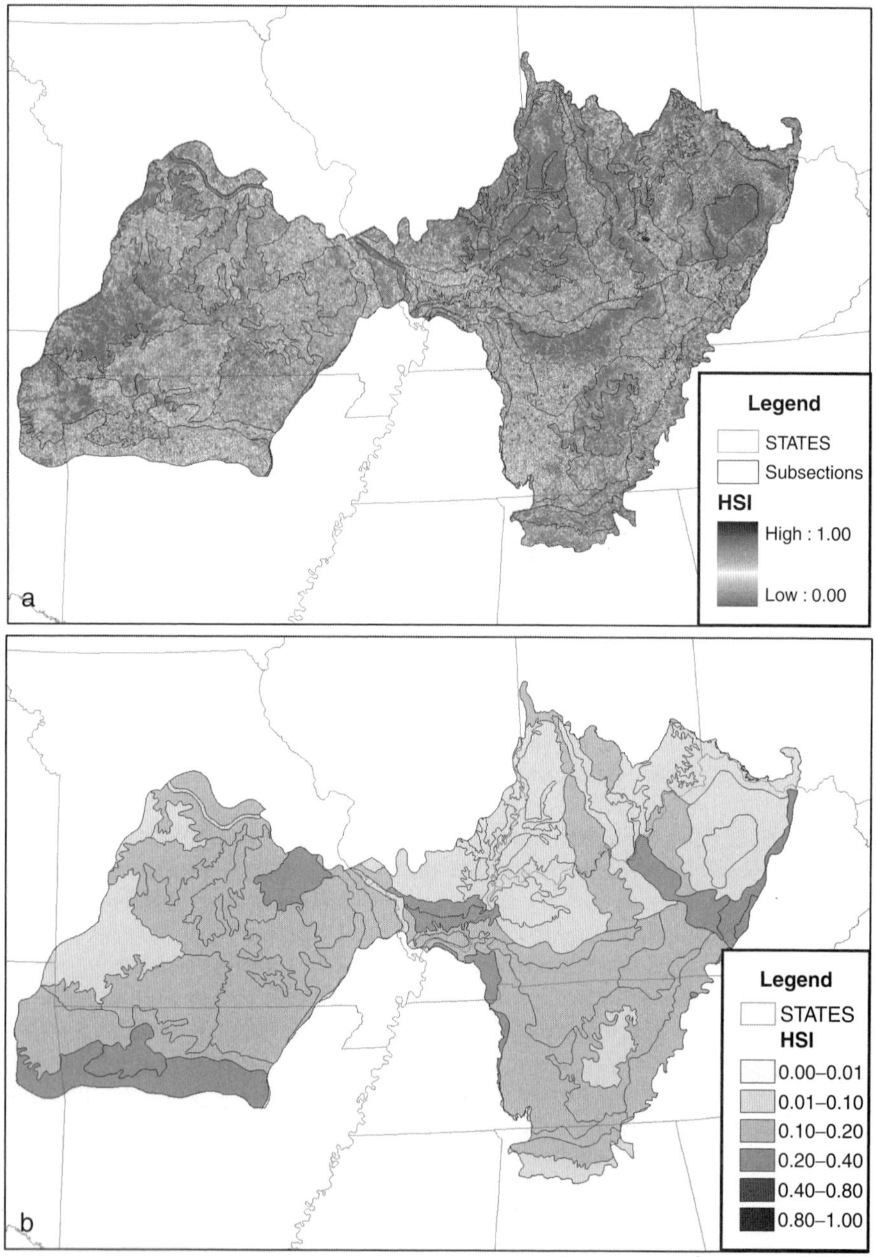

FIG. 22-5 (See color insert.)

Application of a habitat suitability model for Acadian flycatcher (*Empidonax virescens*) to the Central Hardwoods Bird Conservation Region in the Midwestern United States. Habitat suitability values are plotted at the 30 m pixel level (A) and mean habitat suitability values for ecological subsections (B). Because the models include data layers that were spatially interpolated from point data, the values are not spatially accurate at the pixel level but should be representative at the subsection level.

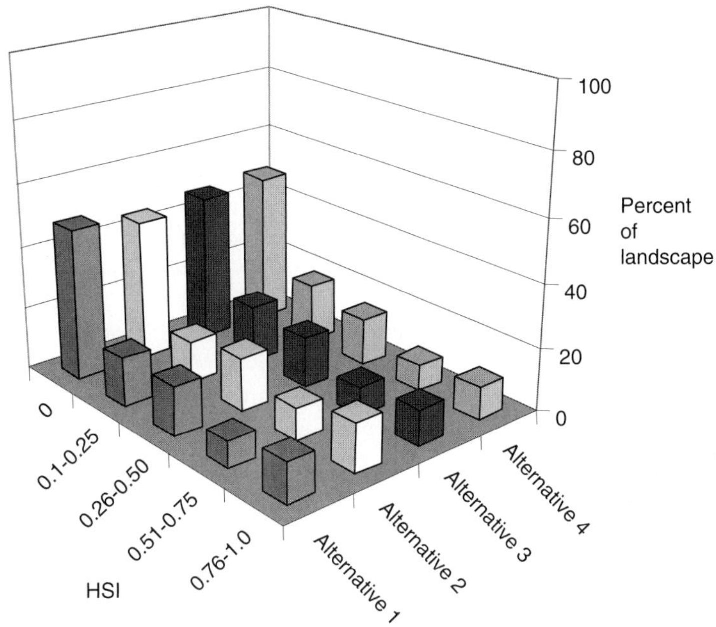

FIG. 22-6

Comparison of the distribution of HSI values for cerulean warblers in five classes of habitat suitability for four management alternatives on the Hoosier National Forest in Indiana, USA.

and juxtaposition of resources (Gustafson et al. 2001; Larson et al. 2003, 2004; Rittenhouse et al. 2007; Dijak and Rittenhouse, this volume).

Habitat Suitability Index models were recently developed for 40 bird species for application to bird conservation planning in the West Gulf Coastal Plain and the Central Hardwoods Bird Conservation Regions (Tirpak et al., in press). The goals of this approach are to summarize available habitat for high concern species in a region; to produce habitat-based estimates of bird numbers; and to link with other models in order to demonstrate how succession, disturbance, and management may affect the amount of habitat over time. A major challenge to applying models at this scale is the data needs. Any variables used in the SIs have to be mapped at a relevant pixel size across the entire region of interest. Landcover and landform features are generally available in GIS coverages spanning states, countries, or continents, but features such as vegetation structure may only be mapped as part of inventories on some managed lands such as state or national forests.

One approach to addressing this need in the United States is to spatially model vegetation structure based on plot data from the U.S. Forest Service Forest Inventory and Analysis (FIA) program (<http://fia.fs.fed.us>) and the GIS coverages mentioned previously. The FIA program measures vegetation features on plots distributed across forest land in the United States at a density of one plot per 6,000 acres. Tirpak et al. (2008) defined 36 potential strata in each ecological subsection within two BCRs by intersecting six possible NLCD forest classes with six landforms. For each forest patch defined by the intersection of NLCD class and landform class, they randomly selected an FIA plot from the pool in that stratum and applied its attributes to that patch. The result is a spatial map of any of the forest attributes measured by FIA. While at the subsection level, the overall composition and pattern are representative of forest conditions; at the pixel or patch level, they are not spatially accurate.

We plotted HSI values for 30 m pixels (Fig. 22-5A) and mean HSI values for the ecological subsection level (Fig. 22-5B) for the Acadian flycatcher (*Empidonax virescens*) in the Central Hardwood Bird Conservation Region. The model is composed of five SI functions that incorporate the following variables: landform, land cover, forest age class, distance to water, canopy cover, forest patch size, and percent forest. Coverages in GIS for forest age class and canopy cover were derived from FIA data as described previously. Because these methods produce maps of HSI values at a 30 m pixel size, there might be the temptation to interpret them at a finer scale (Fig. 22-5A). However, this would not be appropriate because suitability values derived from the spatially interpolated FIA data are not spatially accurate at the 30 m pixel scale but should be representative at larger scales such as ecological subsections (Fig. 22-5) (Tirpak et al. 2008).

These models can be used in bird conservation to identify subsections within BCRs with the highest habitat suitability to help focus conservation efforts. Because FIA data and national land cover data are periodically updated, the models can also be used to show changes in habitat suitability over time at the ecological subsection level, or to evaluate changes in suitability under simulated or hypothetical changes in landscapes. As part of model validation efforts, mean predicted HSI values were regressed on mean breeding bird survey counts at the ecological subsection scale. As expected, for most species, HSI values were positively related to the count data, demonstrating a link between predicted habitat suitability and population levels (T. Jones-Farrand, University of Missouri; J. Tirpak, U.S. Fish and Wildlife Service, personal communications). Furthermore, these regressions can be used to predict habitat-based estimates of population size for ecological subsections under similar assumptions used by Rosenberg and Blancher (2005) to estimate continental populations of birds from Breeding Bird Survey data. Future work by these investigators will determine if the addition of the spatially interpolated forest structure data from FIA substantially improved the models compared to those only based on existing spatial data such as land cover and land use, land form, and forest type. If spatially accurate input data are available, the resulting HSI maps will be spatially accurate at the level of resolution of the input data. This level of accuracy is more likely for project- or ownership-level planning than for regional-level planning.

Statistical Models

Various statistical techniques are useful for developing species-habitat relationship models (e.g., Pearce and Boyce 2006, Austin 2007); Scott et al. (2002) provided a good treatise on the subject. The majority of these habitat relationship models employ some form of regression model to characterize the relationship between species and their habitats. Such models can be developed for prediction or for elucidating ecological processes. Traditionally, the most commonly used statistical technique has been generalized linear modeling (Morrison et al. 1992, Trexler and Travis 1993, Jones et al. 2002). Continual improvements in the power and sophistication of personal computers and statistical software have given ecologists greater access to more sophisticated regression techniques and alternative modeling approaches. Some of these approaches are derived from classical statistical theory (e.g., hierarchical models), whereas others trace their origins to machine learning and data mining (e.g., classification and regression trees).

The most notable modern regression techniques are those based on generalized linear models (GLMs) that favor the logistic, Poisson, and negative binomial distributions over the normal (Gaussian) distribution (Hosmer and Lemeshow 1989, Agresti 1990, Menard 1995, Hastie and Pregibon 1997, Long 1997, Venables and Ripley 1997). All three of these regression approaches are parametric because they make the assumption that the data conform to a particular frequency distribution. Hierarchically based modeling techniques represent a more recent development (Bryk and Raudenbush 1992, Snijders and Bosker 1999, Thogmartin et al. 2004*b*). Hierarchical modeling is a generalization of linear modeling in which regression coefficients are themselves given a model whose parameters are also estimated from data. Hierarchical models, also called multilevel or random-coefficient models, are employed when correlated behavior occurs in the explanatory variables. Such correlated behavior often results from complex survey designs such as a clustered or multistage sample design. For instance, in any wildlife survey, observers may differ in how they count a species of interest (e.g., some observers may tend to overcount, whereas others may tend to undercount). This observer-related correlation is a nuisance, and failure to accommodate such nuisance behavior leads to undue bias in the parameter estimates for the remaining explanatory variables in the model.

As an alternative to using regression methods, it is possible to draw inferences about species-habitat relationships using approaches that are derived in other fields from pattern recognition and artificial intelligence (Ripley 1996). Of these techniques, the most commonly used approaches in ecology are classification and regression trees (CARTs; Breiman et al. 1984) and neural networks (Ripley 1996). Tree-based methods have the ability to detect structure in large, complex data sets in ways that might not be suspected *a priori*. Tree models are fit by a recursive binary splitting of the data set to create homogeneous groups (Clark and Pregibon 1992). The algorithms used in these analyses attempt to produce the most homogeneous groupings (nodes) of the response variable, thereby reducing the within-group measure of dispersion (i.e., variance or mean square deviance). Response variables can be either continuous (regression tree) or categorical (classification tree). Explanatory variables in either type of model can be continuous, categorical, or a combination of the two. Classification and regression trees have been widely used in developing habitat and landscape relationship models in ecology (Michaelsen et al. 1987, Moore et al. 1991, O'Connor et al. 1996, O'Connor and Jones 1997, Fertig and Reiners 2002). More recently developed techniques such as multivariate adaptive regression splines (MARS; Friedman 1991) have yet to see widespread use in addressing ecological problems. Unfortunately, these procedures require modestly sized data sets ($n \geq 150$; T. Jones, U.S. Fish and Wildlife Service, personal communication). We do not discuss these CART methods further in this chapter. Austin (2007) described these and other methods, which may be of interest to those developing species-habitat models.

With such a rich set of tools available to model species-habitat relationships, it can be difficult to determine which statistical method is best in any given situation. Survey data collected by biologists are often in the form of counts (e.g., birds counted along transects, seals counted at haul outs, bird and bat carcasses found at radio towers). When the sample size is sufficiently large and nonzero values are observed in most of the sampled units, the outcome may be considered continuous, and statistical methods that assume the data are normally distributed can be applied. As the sample size becomes smaller, however, at least three things can be expected. First, the number of counts observed in each survey decreases, and the distribution of counts becomes highly skewed (Fig. 22-7). Second, the proportion of survey units with zero values increases, thereby inflating the distribution of the outcome at zero. Third, differences in the number of counts that could have been observed in sample surveys, simply because of differences in the populations at risk of experiencing the event, become more pronounced, thus violating the underlying assumption of normally distributed data.

In these cases, it is usually more appropriate to employ Poisson-based regression models (Long 1997, Jones et al. 2002). The Poisson regression model assumes an underlying Poisson distribution, which is defined as

$$P(X) = (e^{-\mu}\mu^{X})/X!$$

where $P(X)$ is the probability of X occurrences and X is the count of events. As the mean (μ) of X increases, the Poisson distribution approximates a normal or Gaussian distribution (Long 1997), but the Poisson-based regression model is still often preferred because it is bounded by zero at its minima. Use of a linear regression model with count data results in the possibility of predicting a negative abundance estimate, a result that is not biologically sensible. There are two important assumptions of Poisson regression models. The first is that the data follow a Poisson distribution. In a Poisson, the variance is assumed to equal

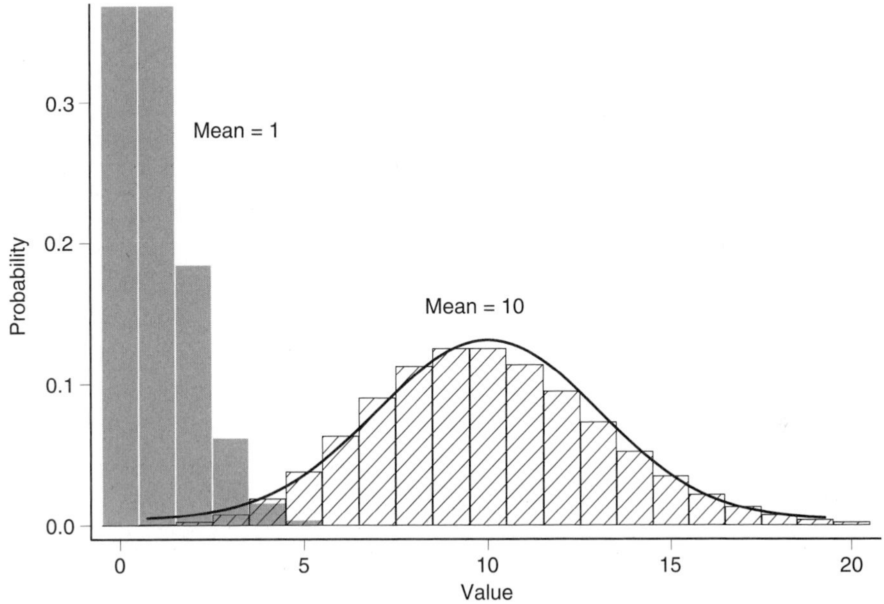

FIG. 22-7

Count data used in bird-habitat relationship modeling are often skewed, as in the case when the mean count for a survey (e.g., point count, Breeding Bird Survey route count) is 1. When the mean count of a survey reaches 10, the distribution is often roughly normally distributed (thick line).

the mean. The second assumption is that the data are independent. This latter assumption is typical of all generalized linear models but can be relaxed to accommodate various correlated data.

We present an example using Poisson regression for modeling bird abundance over large areas by modeling rare warbler abundance in the Appalachians of the United States with a hierarchical spatial count model (Thogmartin et al. 2004*b*, 2006*c*, 2007). These models were developed to aid in directing scarce conservation resources to those areas in which the resources would be most effective as opposed to broadly, but diffusely, applying the conservation resources over the region.

The relative abundance of worm-eating warbler (*Helmitheros vermivorus*) and Kentucky warbler (*Oporornis formosus*) in the Appalachians (Bird Conservation Region 28) was modeled as a function of explanatory variables. The response variable in these models was annual BBS counts collected between 1981 and 2001 (Fig. 22-8). Environmental explanatory variables included those associated with land-cover composition and configuration, topographical position, climate, brown-headed cowbird (*Molothrus ater*; a common nest parasite) abundance, deer forage, and annual acid rain deposition. This latter variable was

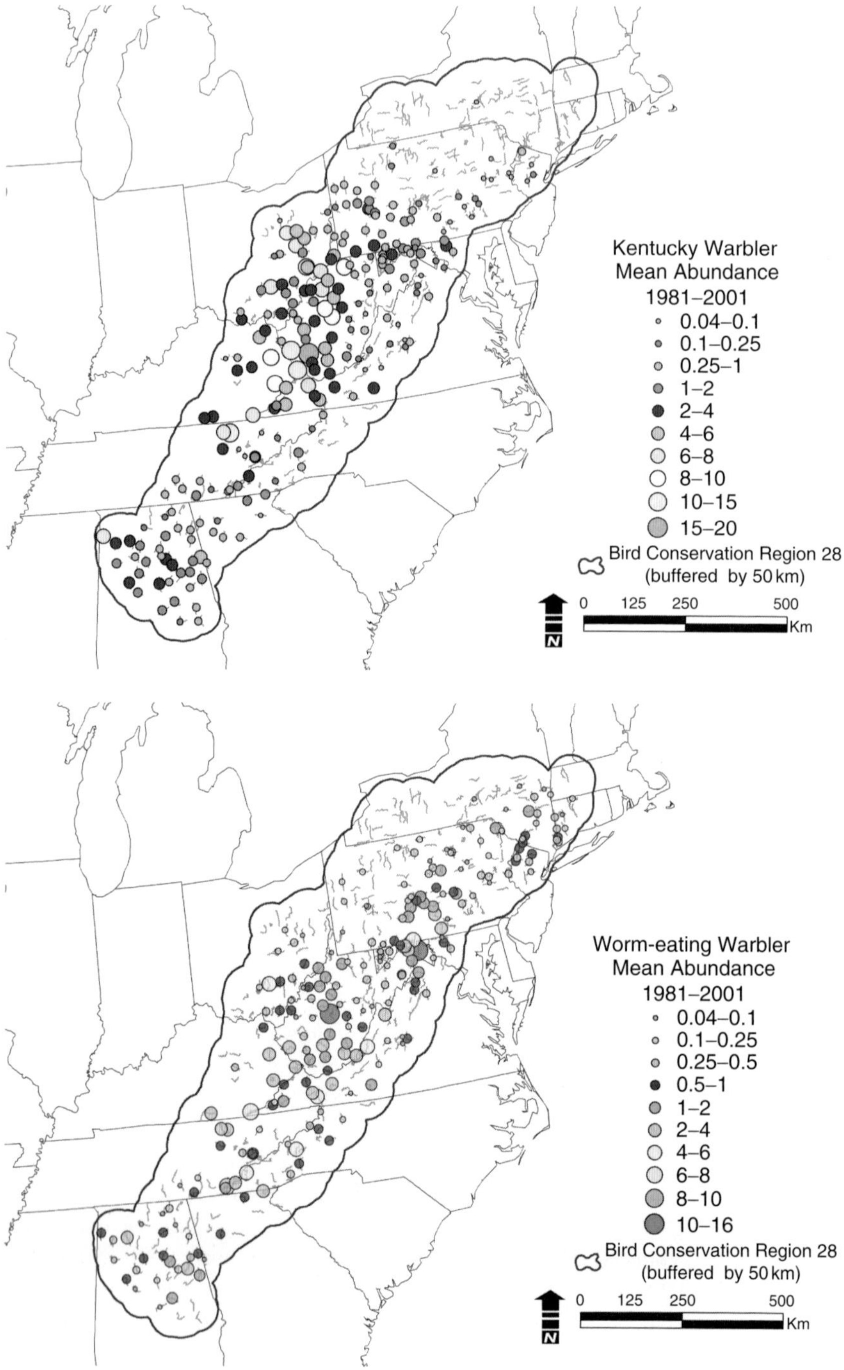

FIG. 22-8 (See color insert.)

(continued)

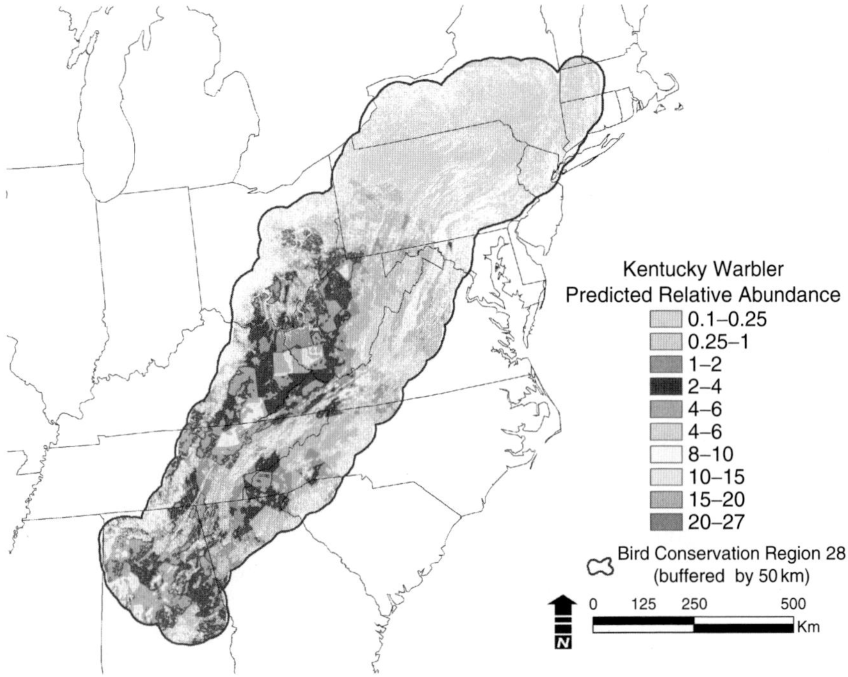

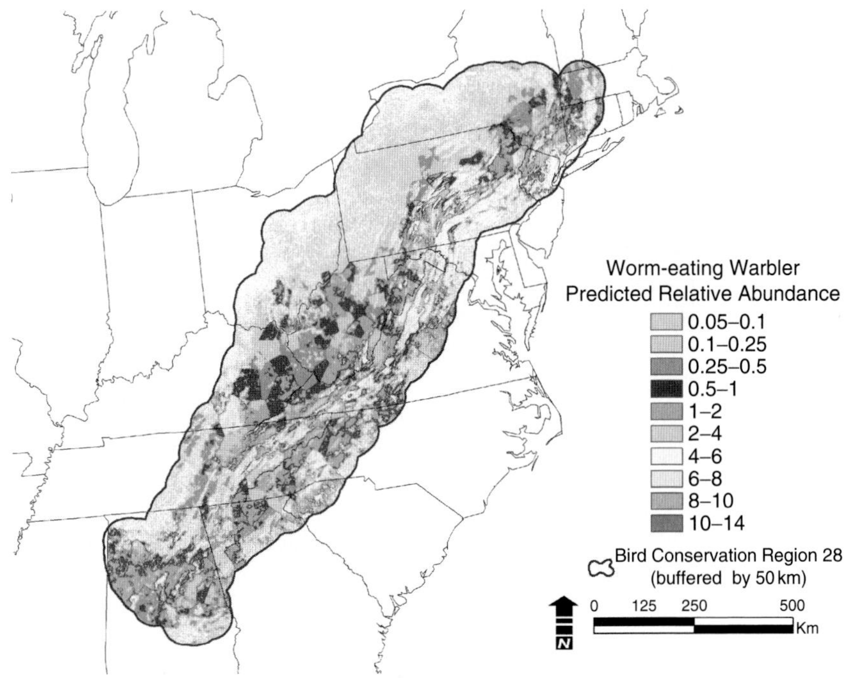

FIG. 22-8 cont'd

(continued)

FIG. 22-8 cont'd

(First Panel) Bubble plots indicating location and magnitude of mean Kentucky and worm-eating warbler relative abundance in the Appalachians, 1981–2001, as determined by the North American Breeding Bird Survey. (Second Panel) Predicted relative abundance circa 1995 for the Kentucky and worm-eating warbler in the Appalachians as determined by a hierarchical spatial count model.

considered in conjunction with soil pH to assess whether acid rain may be affecting regional warbler abundance through eggshell thinning and subsequent nest failure (Hames et al. 2002). The hierarchical aspect of these models included random effects associated with observer differences, year effects, and potential spatial autocorrelation in route counts (Thogmartin et al. 2004*b*, 2006*c*, 2007).

The model results for these two warblers were decidedly dissimilar (Table 22-1). Neither species appeared to be influenced by acid deposition, although there was a trend for higher abundance of both species in areas in which acid deposition was buffered by basic soils (W. E. Thogmartin, unpublished information). The commonality between the worm-eating and Kentucky warblers was in the effect of deciduous forest composition. As was expected, both species increased in abundance as deciduous forest increased in the landscape. The worm-eating warbler also increased in abundance as white-tailed deer (*Odocoileus virginianus*) forage increased and as precipitation decreased. The Kentucky warbler was more abundant in moister landscapes, and paradoxically in areas in which brown-headed cowbirds were most abundant.

Mapping these models was instructive in identifying spatial patterns in predicted abundance (Fig. 22-9) and therefore helping planners identify target areas for conservation actions. Both species were largely absent from the northern portion of the BCR. Kentucky and worm-eating warblers were more abundant west and east of the Appalachian divide, respectively. Peaks of predicted abundance for the Kentucky and worm-eating warblers occurred in southeastern Kentucky and western North Carolina, respectively.

One benefit of mapping predicted relative abundance is in locating gaps in our ability to manipulate or control conservation action (W. E. Thogmartin and J. J. Rohweder, U.S. Geological Survey, unpublished information) (Fig. 22-9). Fig. 22-9 illustrates that the location of the predicted peak of worm-eating warbler abundance is largely outside direct governmental stewardship (i.e., it occurs on land over which government or conservation agencies have little or no direct control). Much of the peak predicted abundance lies on private land to the north and east of the Green River Game Lands (North Carolina

Table 22-1 Parameter Estimates (with 2.5% and 97.5% Credibility Limits) from Spatial Hierarchical Count Models Describing Predicted Relative Abundance in the Appalachians, Circa 1981–2001, for the Kentucky and Worm-Eating Warblers. Estimates in Bold are those that Differ Credibly from Zero

	Kentucky Warbler			**Worm-eating Warbler**		
Variable	**LCL**	**Median**	**UCL**	**LCL**	**Median**	**UCL**
Slope of the temporal trend	−0.041	**−0.029**	−0.016	−0.012	0.004	0.021
Forest (%)	0.241	**0.483**	0.713	0.587	**0.990**	1.395
Deer Forage (%)	−0.112	0.012	0.138	0.162	**0.335**	0.499
Brown-headed Cowbird Relative Abundance	0.076	**0.272**	0.484	−0.193	0.107	0.403
Forest Edge Density (km/km^2)	−0.112	0.067	0.251	−0.150	0.069	0.293
Oak/Elm (%)	−0.058	0.117	0.284	−0.155	0.066	0.275
Mean Wetness Potential[a]	−0.500	**−0.287**	−0.078	−0.364	−0.076	0.212
Area-weighted Mean Patch Size of Forest	−0.148	−0.010	0.139	−0.563	−0.232	0.086
Wooded Wetland (%)	−0.187	0.086	0.347	−0.551	−0.272	0.016
Mean Precipitation	−0.487	−0.188	0.089	−0.801	**−0.441**	−0.089
Acid Deposition	−0.037	0.025	0.089	−0.055	0.032	0.120
Soil pH	−0.190	0.017	0.239	−0.279	−0.028	0.225
Acid Deposition × Soil pH	−0.048	0.010	0.067	−0.020	0.074	0.169
Intercept	−4.727	−0.642	2.210	−3.449	−0.809	0.506
Observer Effect	−4.070	−1.190	2.896	−2.405	−1.145	1.504

[a] *As determined by the topographic convergence index, ln([Catchment Area (m^2/m)]/tan(Slope(degrees)])).*

Wildlife Resources Commission) and Hickory Nut Gorge (The Nature Conservancy), respectively. Future efforts to conserve this species would benefit most by focusing conservation efforts in those areas where the species is predicted to be highly abundant.

A benefit of working with models of abundance, as opposed to those predicting occurrence (presence-absence), is that there is the potential to estimate population size. There is a need, in such an endeavor, to translate from a metric of relative abundance to true population size. The current impediments to direct estimation of population size from BBS data are too many to recount here (see Thogmartin et al. 2006*b*), but Rosenberg and Blancher (2005) have devised one approach that we employ here as a means of initiating discussions in this area. Using the methods employed by Rosenberg and

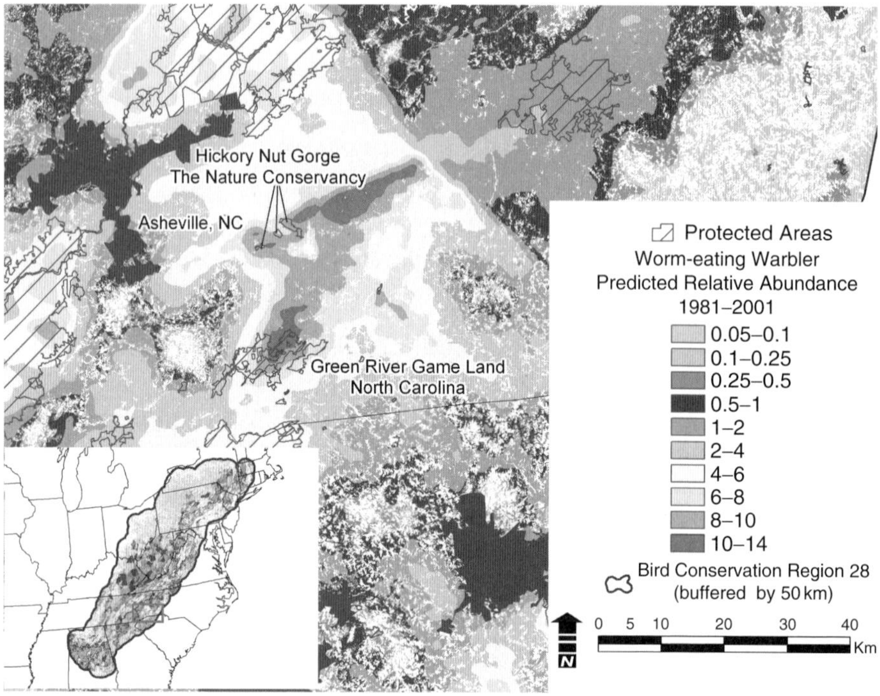

FIG. 22-9 (See color insert.)

Mapping the conservation estate relative to maps of predicted occurrence and abundance can aid in identifying gaps in stewardship. This example is conservation areas overlying worm-eating warbler relative abundance as determined from a spatial hierarchical count model.

Blancher (2005), we adjusted our previously described model estimates by factors accounting for the facts that (1) it is males of the species that are principally counted by the BBS; (2) these counts vary over the course of the survey day (i.e., typically highest nearest dawn); and (3) these species are generally heard at distances less than presumed by the standard survey methodology. Thus, the translation of relative abundance to population size occurs as 11,665 (relative estimate of the number of worm-eating warblers) × 2 (pair adjustment) × 1.29 (time-of-day adjustment) × 10.24 (detectability adjustment) = 308,180 worm-eating warblers in the Appalachian Mountains, circa 1995. Similarly, 21,181 × 2 × 1.11 × 4 = 188,087 Kentucky warblers circa 1995. These numbers are approximately 20% lower than those estimated for this region based on the global population estimates in the PIF North American Landbird Conservation Plan (Rich et al. 2004) (i.e., 389,000 and 243,600 birds, respectively), possibly because BBS sites are inequitably distributed in the Appalachians.

DISCUSSION

Not all the components of PIF's Five Elements of Conservation Design (Will et al. 2005) have been fully integrated into bird conservation planning at BCR scales in the United States. Most efforts to date have been directed toward (1) landscape characterization and assessment and (2) "population response" modeling focused on developing the ability to link data depicting the distribution and abundance of a bird species with habitat variables quantifiable across relatively large spatial scales. We describe here three conceptually different approaches that have been applied toward those ends. While we believe the products from those efforts are immediately useful and should help to establish a foundation for the next steps in the Five Elements process, each methodology has both similar and distinct sets of advantages and disadvantages. Planners and other end users should employ those methods most suited to their specific need and capacity. Some factors to consider are described next.

Complex Species-Habitat Relationships

Most species-habitat relationships are complex, involving many variables and interactions. Although the HABS database approach is perhaps the least able to incorporate such complex relationships, it is able to bring species-habitat relationship data from existing models into the database as a means of incorporating this information into the development of habitat objectives. The HSI and statistical models are better able to incorporate complex functions characterizing a species' relationship with its environment. These characterizations may occur over a range of spatial resolutions and extents.

Both the HABS database tools and the HSI models characterize the environmental requirements within which the species *may* occur (e.g., habitat suitability). Statistical models based on abundance surveys describe associations between the observed abundance of species and their habitats. However, if an important variable is missing from the design of any of these tools or models, any approach will likely misrepresent species-habitat relationships. As an example, many of the large-scale data sets used in model building may not contain habitat-specific or microhabitat variables that are known to be important components of species-habitat relationships (e.g., measures of bare ground, litter depth, vegetation density, species composition, etc.). Other important variables that often are not included are associated with nonbiological habitat factors or nonhabitat ecological factors such as competition, predation, or disease.

Data Constraints and Limitations

One of the biggest flaws in all these approaches is that the response data upon which they are based is often seriously constrained. Such constraints include temporal and spatial correlation and effects associated with the observation process (e.g., observer differences, species detectability). As an example, in the hierarchical spatial count models described previously, the most obvious limitation in the survey data is that they come from a roadside survey that does not account for imperfect detection (Thogmartin et al. 2006*b*,*c*). In addition, there are substantial gaps in the availability of landscape-scale bird data to be used in databases and models. As a result, conclusions must often be extrapolated from more localized data and relationships. Database and HSI models can be more conceptual and based on hypothesized relationships formulated from literature review, data, or expert opinion.

Assumptions

Because our knowledge of species-habitat relationships typically is limited and imperfect, it is important to identify the assumptions that are made in developing these tools or models. Habitat conditions are often difficult to quantify accurately, and deriving bird density estimates from different sources using different methodologies may not produce comparable and valid results. As a result, tools or models often must be based on limited data and conceptual knowledge of the factors that influence species' abundance, distribution, and vital rates. One of the concerns with the HABS database approach, for example, is that a large number of assumptions, many of them untested, enter into the basic models for many species. Uncertainty remains about whether species that are patchily distributed, or that occupy habitats not well represented within a GIS framework, are appropriately characterized by this approach. The key assumption associated with statistical models is that the final model that is chosen correctly characterizes the relationship between the response and the explanatory variables. Numerous model diagnostics and validation procedures are needed to assess the worthiness of a statistical model (Shifley et al., this volume), but too often this aspect of statistical model building is given short shrift. After a decision support tool or a model is developed, it is important that targeted research be conducted to test underlying assumptions in order to improve the accuracy of the estimates and models in the future.

Uncertainty

One of the weaknesses of HABS database tools and HSI models, as currently implemented, is that they do little to represent the uncertainty associated with the various assumptions in the models (see Millspaugh et al., this volume). This uncertainty arises from a number of sources including stochastic effects on species distributions, ambiguities in the presumed species-habitat relation, and inadequacies in data. However, Monte Carlo simulation can be used to calculate confidence intervals for HSI scores from uncertainty in input variables (Bender et al. 1996), and fuzzy math (Ferson et al. 1998) can be used to calculate reliability bounds on HSI scores from both statistical and structural uncertainty in the model (Burgman et al. 2001). Predictions from statistical models are usually accompanied by measures of uncertainty like standard errors or confidence intervals. These should be interpreted cautiously, however, because statistical models are usually built from data limited in their geographic scope, and statistical inference is appropriate only to the population sampled and sometimes only the sample. In these cases application of the model to a broader geographic area is a subjective inference, the observer assumes the data were representative of the broader area, and the original standard errors and confidence intervals are likely underestimates.

Model Fit

Because HABS database tools and HSI models are usually not evaluated using empirical data, it is unclear how well they capture the patterns in species occurrence and abundance, or in species-habitat relationships. Statistical models can be assessed by an array of goodness of fit procedures, measures of explained variability, measures of model parsimony relative to model fit, etc. Goodness of fit measures, however, evaluate how models fit the data they were built from. Usually, there is need to apply any of these types of models to a broader geographic scope than the original data. Therefore, validation with independent data is important for all these approaches. Some efforts are currently underway to validate HSI models developed for bird conservation planning in the Central Hardwoods and West-Gulf Coastal Plain BCRs (T. Jones-Farrand, University of Missouri; J. Tirpak, U.S. Fish and Wildlife Service; F. Thompson, U.S. Forest Service; D. Twedt, U.S. Geological Survey; personal communications).

Flexibility and Adaptability

The value of HABS database tools and HSI models, especially in an adaptive management context, is their flexibility and ability to be refined in the face of new data (e.g., new species, habitats, or habitat conditions). Unfortunately, this is rarely the case with most applications of statistical models, which upon their completion are often never revised. Statistical models are rarely updated in the face of new information, principally because of their "costly time to production" (although see below).

Spatial Scalability

Spatial scale is the integration of resolution and extent, and the ability of tools and models to be spatially scalable is important in conservation planning, which must incorporate decisions at multiple spatial scales. All the approaches mentioned here can scale to virtually any spatial extent. Where they differ is in the resolution to which they most appropriately apply. An asset of database decision support tools and the HSI models is that they are readily scalable. The HABS tool was designed to describe the area encompassed by the Playa Lakes Joint Venture and to operate at the spatial extent of a Bird Conservation Region × State intersection. It is possible to scale the applications down to the county level, but scaling below that level would not be appropriate. The relationships for the birds in the tool are fitted to conform to *a priori* notions for different areas at that scale. In HSI models, unless there is a specific recognition of the scales to which the relationships apply, the results of the models can be applied at any scale convenient to the user. The finest spatial resolution for the hierarchical spatial count model is ostensibly the finest resolution of the response (i.e., the BBS count). In the application described previously, the finest resolution is approximately 25 km^2 because the models are built from a route count, which is an aggregate of counts from the 50 survey stops on a BBS route. It is possible to map the model results at a finer resolution; Thogmartin et al. (2004*b*, 2006*c*, 2007) mapped at a finer resolution (1 ha), but there is some question as to the validity of interpolating to a finer resolution the results of a model derived from coarsely resolved data (McPherson et al. 2006). There are some efforts to statistically model species response using data from the individual stops on a BBS route (e.g., Hepinstall et al. 2002, Thogmartin 2002), which would then lower the "floor" of the spatial resolution to an area surveyed at a stop (i.e., 2–200 ha; Thogmartin et al. 2006*b*).

Future Projections

An important need in bird conservation is the ability to project real or hypothetical changes in landscapes and birds that may result from management decisions or environmental change. The HABS database method is capable of predicting these kinds of changes by plugging potential changes in habitat amounts and condition into the database to immediately project the effects on multiple bird species populations. If methods exist to update or project habitat and landscape conditions, both HSI and statistical models can be used to generate new predictions from these updated or predicted future conditions. For example, HSI models have been linked to outputs from LANDIS, a forest-landscape simulation model, to predict the consequences of forest management decisions, succession, and disturbance on wildlife habitat (Shifley et al. 2006). Similar approaches could be used to evaluate the simulated effects of urban and suburban development, fire, expansion of exotic plant species, and perhaps in the future, global climate change.

Time and Cost

The amount and kinds of resources required to produce the types of planning tools and models discussed here vary, and their component costs and time required are additional factors to consider when deciding which of the planning tools a user will develop. Comparing the products of each approach dollar for dollar or hour for hour is beyond the scope of this chapter; rather we will attempt to summarize the components and the relative amount of each component involved in each approach. First, each requires specialized software and computer hardware with above-average computing capacity; included in this component is the cost of maintaining these computer resources. Each approach also requires land-cover data; the end user must choose between low-cost and readily available data with lower resolution and higher quality data that might have to be purchased, reclassified in some way before it can be applied seamlessly across large landscapes, or newly created. Both the HABS database approach and HSI models require extensive literature searches. In addition, the database approach requires substantial time to populate the database (i.e., data entry). The kinds of technical expertise needed to work with sophisticated land-cover data sets, build complex biological models, and program computers will be a cost in each application, but likely is greater for statistical models than HSI models, and least for the database approach, the number of species being equal. In all cases, collaboration among this team of computer, mathematical/statistical, GIS, and biological experts is crucial to developing the tools or models that truly answer the questions being asked by conservation planners. Finally, all approaches become more time intensive as the number of species or habitat types to be addressed in the database or model increases.

Finally, each of these approaches has an inherent set of assumptions that should affect the user's confidence in, and use of, the products. They will undoubtedly misrepresent spatial patterns of bird population parameters at least somewhere on the landscape, even if just in response to changes in land cover and land use over time. The crucial importance of incorporating tool and model evaluation, assumption testing, and adaptive management concepts into conservation planning efforts is clear. However, the costs associated with these evaluations and refinements are not well understood.

SUMMARY

Partners in Flight (PIF), a public-private coalition for the conservation of land birds, has developed one of four international bird conservation plans recognized under the auspices of the North American Bird Conservation Initiative (NABCI). Partners in Flight prioritized species most in need of conservation attention and set range-wide population goals for 448 species of terrestrial birds. Partnerships are now tasked with developing spatially explicit estimates of the distribution and abundance of priority species across large ecoregions and identifying habitat acreages needed to support populations at prescribed levels. The PIF Five Elements process of conservation design identifies five steps needed to implement all bird conservation at the ecoregional scale. We reviewed the application of some newly developing geospatial techniques, tools, and models that are being used for (1) landscape characterization and assessment and (2) bird population response modeling, the first two elements in the Five Elements process. Habitat assessment and landscape characterization describe the current amounts of different habitat types and summarize patch characteristics and landscape configurations that define the ability of a landscape to sustain healthy bird populations and are a valuable first step to describing the planning area before pursuing more complex species-specific models. Spatially linked database models, landscape-scale habitat suitability models, and statistical models are viable alternatives (in order of increasing complexity and data needs) to predicting habitat suitability or bird abundance across large planning areas to help assess conservation opportunities, design landscapes to meet population objectives, and monitor change in habitat suitability or bird numbers over time. Decisions by conservation planners about what approach to use in a particular circumstance should be based on their specific needs and capability and should consider (1) complexity of species-habitat relationships; (2) data constraints; (3) model assumptions, uncertainty, fit, flexibility, scalability, and ability to make future projections; and (4) cost and time required.

ACKNOWLEDGMENTS

The planning tools discussed in this chapter were first described collectively at a Partners in Flight workshop held in St. Louis, Missouri, in April 2006. The slides from these and numerous other presentations related to landbird conservation planning, brief biographies of authors, and other pertinent information can be viewed at <www.partnersinflight.org/events/conserv_design_wkshp_0406/default.htm>. We thank all those who participated in the workshop for their insightful questions and comments, which improved our ability to prepare this manuscript. We also thank M. Suárez and an anonymous reviewer for their review of an early draft.

LITERATURE CITED

Agresti, A. 1990. *Categorical data analysis*. First edition. John Wiley and Sons, New York, New York, USA.

Andren, H., and P. Anglestam. 1988. Elevated predation rates as an edge effect in habitat islands: Experimental evidence. *Ecology* 69:544–547.

Austin, M. 2007. Species distribution models and ecological theory: A critical assessment and some possible new approaches. *Ecological Modelling* 200:1–19.

Bender, L. C., G. J. Roloff, and J. B. Haufler. 1996. Evaluating confidence intervals for habitat suitability models. *Wildlife Society Bulletin* 24:347–352.

Breiman, L., J. H. Freidman, R. A. Olshen, and C. J. Stone. 1984. *Classification and regression trees*. Wadsworth International Group, Belmont, California, USA.

Brown, S., C. Hickey, B. Harrington, and R. Gill, editors. 2001. *United States shorebird conservation plan.* Second edition. Manomet Center for Conservation Sciences, Manomet, Massachusetts, USA. <http://www.fws.gov/shorebirdplan/USShorebird/downloads/USShorebirdPlan2Ed.pdf>. Accessed 27 January 2007.

Bryk, A. S., and S. W. Raudenbush. 1992. *Hierarchical linear models*. Sage Publications, Newbury Park, California, USA.

Burgman, M. A., D. R. Breininger, B. W. Duncan, and S. Ferson. 2001. Setting reliability bounds on habitat suitability indices. *Ecological Applications* 11:70–78.

Butts, K. O. 1973. *Life history and habitat requirements of burrowing owls in western Oklahoma*. Thesis, Oklahoma State University, Stillwater, USA.

Clark, L. A., and D. Pregibon. 1992. Tree-based models. Pages 377–419 *in* S. J. M. Chamber and T. J. Hastie, editors. *Statistical models.* Wadsworth and Brooks/Cole, Pacific Grove, California, USA.

Cowardin, L. M., and D.H Johnson. 1979. Mathematics and mallard management. *Journal of Wildlife Management* 43:18–35.

Cowardin, L. M., T. L. Shaffer, and P. M. Arnold. 1995. *Evaluations of duck habitat and estimation of duck population sizes with a remote sensing-based system*. U.S. National Biological Service, Biological Science Report 2.

Crawford, J. A., and E. G. Bolen. 1976. Effects of land use on lesser prairie chickens in Texas. *Journal of Wildlife Management* 40:96–104.

DeZonia, B., and D. J. Mladenoff. 2004. *IAN 1.0.15*. Department of Forest Ecology and Management, University of Wisconsin, Madison, USA. <http://landscape.forest.wisc.edu/projects/ian/>. Accessed 26 January 2007.

Dobbs, R. 2006. *A review of distribution, habitat use, and population density data in the Hierarchical All Bird Strategy (HABS) Database*. Playa Lakes Joint Venture Report. Lafayette, Colorado, USA.

Ferson, S., W. Root, and R. Kuhn. 1998. *RAMAS/RiskCalc: Risk assessment with uncertain numbers*. Applied Biomathematics, Setauket, New York, USA.

Fertig, W. F., and W. A. Reiners. 2002. Predicting presence/absence of plant species for range mapping: A case study from Wyoming. Pages 483–489 *in* J. M. Scott, P. J. Heglund, M. L. Morrison, J. B. Haufler, M. G. Raphael, W. A. Hall, and F. B. Samson, editors. *Predicting species occurrences: Issues of scale and accuracy.* Island Press, Washington, D.C., USA.

Friedman, J. H. 1991. Multivariate adaptive regression splines. *Annals of Statistics* 19:1–141.

GRASS Development Team. 2006. *Geographic Resources Analysis Support System (GRASS) software*. ITC-irst, Trento, Italy. <http://grass.itc.it>. Accessed 26 January 2007.

Gustafson, E. J. 1998. Quantifying landscape spatial pattern: What is the state of the art? *Ecosystems* 1:143–156.

Gustafson, E. J., D. W. Murphy, and T. R. Crow. 2001. Using a GIS model to assess terrestrial salamander response to alternative forest management plans. *Environmental Management* 63:281–292.

Hames, R. S., K. V. Rosenberg, J. D. Lowe, S. E. Barker, and A. A. Dhondt. 2002. Adverse effects of acid rain on the distribution of the wood thrush *Hylocichla mustelina* in North America. *Proceedings of the National Academy of Sciences* 99:11235–11240.

Hartley, M. J., and M. L. Hunter, Jr. 1998. A meta-analysis of forest cover, edge effects, and artificial nest predation rates. *Conservation Biology* 12:465–469.

Hastie, T. J., and D. Pregibon. 1997. Generalized linear models. Pages 195–248 *in* M. J. Chambers and T. J. Hastie, editors. *Statistical models.* Chapman and Hall, London, United Kingdom.

Hepinstall, J. A., S. A. Sader, and W. B. Krohn. 2002. Effects of niche width on the performance and agreement of avian habitat models. Pages 593–606 *in* J. M. Scott, P. J. Heglund, M. L. Morrison, J. B. Haufler, M. G. Raphael, W. A. Wall, and F. B. Samson, editors. *Predicting species occurrences: Issues of scale and accuracy.* Island Press, Washington, D.C., USA.

Herkert, J. R. 1994. The effects of habitat fragmentation on Midwestern grassland bird communities. *Ecological Applications* 4:461–471.

Homer, C., J. Dewitz, J. Fry, M. Coan, N. Hossain, C. Larson, N. Herold, A. McKerrow, J. N. VanDriel, and J. Wickham. 2007. Completion of the 2001 National Land Cover Database for the conterminous United States. *Photogrammetric Engineering and Remote Sensing* 73:337-341.

Hosmer, D. W., and S. Lemeshow. 1989. *Applied logistic regression*. Wiley Series, New York, New York, USA.

Johnson, D. H., and L. D. Igl. 2001. Area requirements of grassland birds: A regional perspective. *Auk* 118:24-34.

Jones, M. T., G. J. Niemi, J. M. Hanowsk, and R. R. Regal. 2002. Poisson regression: A better approach to modeling abundance data. Pages 411-418 *in* J. M. Scott, P. J. Heglund, M. L. Morrison, J. B. Haufler, M. G. Raphael, W. A. Wall, and F. B. Samson, editors. *Predicting species occurrences: Issues of scale and accuracy.* Island Press, Washington, D.C., USA.

Kushlan, J. A., M. J. Steinkamp, K. C. Parsons, J. Capp, M. Acosta Cruz, M. Coulter, I. Davidson, L. Dickson, N. Edelson, R. Elliott, R. M. Erwin, S. Hatch, S. Kress, R. Milko, S. Miller, K. Mills, R. Paul, R. Phillips, J. E. Saliva, B. Sydeman, J. Trapp, J. Wheeler, and K. Wohl. 2002. *Waterbird conservation of the Americas: The North American waterbird conservation plan, Version 1.* Waterbird Conservation for the Americas, Washington, D.C., USA. <http://www.waterbirdconservation.org/pubs/ContinentalPlan.cfm>. Accessed 26 January 2007.

Larson, M. A., W. D. Dijak, F. R. Thompson, III, and J. J. Millspaugh. 2003. *Landscape-level habitat suitability models for twelve species in southern Missouri.* U.S. Forest Service, North Central Research Station, General Technical Report NC-233, St. Paul, Minnesota, USA.

Larson, M. A., F. R. Thompson, III, J. J. Millspaugh, W. D. Dijak, and S. R. Shifley. 2004. Linking population viability, habitat suitability, and landscape simulation models for conservation planning. *Ecological Modeling* 180:103-118.

Li, X. Z., H. S. He, R. C. Bu, Q. C. Wen, Y. Chang, Y. M. Hu, and Y. H. Li. 2005. The adequacy of different landscape metrics for various landscape patterns. *Pattern Recognition* 38:2626-2638.

Long, J. S. 1997. *Regression models for categorical and limited dependent variables.* Sage Publications, Thousand Oaks, California, USA.

McGarigal, K., S. A. Cushman, M. C. Neel, and E. Ene. 2002. *FRAGSTATS: Spatial Pattern Analysis Program for Categorical Maps.* Computer software program produced by the authors at the University of Massachusetts, Amherst, USA. <http://www.umass.edu/landeco/research/fragstats/fragstats.html>. Accessed 26 January 2007.

McPherson, J. M., W. Jetz, and D. J. Rogers. 2006. Using coarse-grained occurrence data to predict species distributions at finer spatial resolutions—Possibilities and limitations. *Ecological Modeling* 192:499-522.

Menard, S. 1995. *Applied logistic regression analysis.* Sage Publications, Thousand Oaks, California, USA.

Michaelsen, J., F. W. Davis, and M. Borchert. 1987. A non-parametric method for analyzing hierarchical relationships in ecological data. *Coenoses* 2:39-48.

Moore, D. M., B. G. Lees, and S. M. Davey. 1991. A new method for predicting vegetation distributions using decision tree analysis in a geographic information system. *Environmental Management* 15:59-71.

Morrison, M. L., B. G. Marcot, and R. W. Mannan. 1992. *Wildlife-habitat relationships: Concepts and applications.* University of Wisconsin Press, Madison, USA.

North American Waterfowl Management Plan Committee. 2004. *Strategic guidance: Strengthening the biological foundation.* Canadian Wildlife Service. U.S. Fish and Wildlife Service, Secretaria de Medio Ambiente y Recursos Naturales. <http://www.fws.gov/birdhabitat/NAWMP/files/NAWMP2004.pdf>. Accessed 26 January 2007.

O'Connor, R. J., and M. T. Jones. 1997. Using hierarchical models to index the ecological health of the nation. *Transactions of the North American Wildlife and Natural Resources Conference* 62:558-565.

O'Connor, R. J., M. T. Jones, D. White, C. Hunsaker, T. Loveland, B. Jones, and E. Preston. 1996. Spatial partitioning of environmental correlates of avian biodiversity in the conterminous United States. *Biodiversity Letters* 3:97-110.

Paine, D. P., and J. D. Kiser. 2003. *Aerial photography and image interpretation*. Second edition. John Wiley and Sons, Hoboken, New Jersey, USA.

Pearce, J. L., and M. S. Boyce. 2006. Modelling distribution and abundance with presence-only data. *Journal of Applied Ecology* 43:405–412.

Reynolds, R. E., D. R. Cohan, and M. A. Johnson. 1996. Using landscape information approaches to increase duck recruitment in the Prairie Pothole Region. *Transactions of the North American Wildlife and Natural Resource Conference* 61:86–93.

Rich, T. D., C. J. Beardmore, H. Berlanga, P. J. Blancher, M. S. W. Bradstreet, G. S. Butcher, D. W. Demarest, E. H. Dunn, W. C. Hunter, E. E. Iñigo-Elias, J. A. Kennedy, A. M. Martell, A. O. Panjabi, D. N. Pashley, K. V. Rosenberg, C. M. Rustay, J. S. Wendt, and T. C. Will. 2004. *Partners in Flight North American Landbird Conservation Plan*. Cornell Lab of Ornithology, Ithaca, New York, USA. <http://www.partnersinflight.org/cont_plan/>. Accessed 26 January 2007.

Ripley, B. D. 1996. *Pattern recognition and neural networks*. Cambridge University Press, Cambridge, United Kingdom.

Rittenhouse, C. D., W. D. Dijak, F. R. Thompson, III, and J. J. Millspaugh. 2007. *Development of landscape-level habitat suitability models for ten wildlife species in the central hardwoods region*. U.S. Forest Service, Northern Research Station, General Technical Report NRS-4, Newtown Square, Pennsylvania, USA.

Rollins, M. G., R. E. Keane, Z. Zhu, J. Menakis, W. Hann, and A. Shlisky. 2003. *LANDFIRE: A nationally consistent and locally relevant interagency fire, fuels, and risk assessment*. American Meteorological Society: Proceedings of the Second International Wildland Fire Ecology and Fire Management Congress and Fifth Symposium on Fire and Forest Meteorology, Orlando, Florida, USA.

Rosenberg, K. V., and P. J. Blancher. 2005. Setting numerical population objectives for priority landbird species. Pages 57–67 *in* C. J. Ralph and T. D. Rich, editors. *Bird Conservation Implementation and Integration in the Americas.* Proceedings of the Third International Partners in Flight Conference. U.S. Forest Service, Pacific Southwest Research Station, General Technical Report PSW-GTR-191, Albany, California, USA.

Sauer, J. R., J. E. Hines, and J. Fallon. 2007. *The North American Breeding Bird Survey, Results and Analysis 1966–2006*. Version 10.13.2007. USGS Patuxent Wildlife Research Center, Laurel, MD.

Scott, J. M., J. H. Heglund, M. L. Morrison, J. B. Haufler, M. G. Raphael, W. A. Wall, and F. B. Samson, editors. 2002. *Predicting species occurrences: Issues of accuracy and scale*. Island Press, Washington, D.C., USA.

Shifley, S. R., F. R. Thompson, III, W. D. Dijak, M. L. Larson, and J. J. Millspaugh. 2006. Simulated effects of forest management alternatives on landscape structure and habitat suitability in the Midwestern United States. *Forest Ecology and Management* 229:361–377.

Snijders, T. A. B., and R. J. Bosker. 1999. *Multilevel analysis: An introduction to basic and advanced multilevel modeling*. Sage Publications, Thousand Oaks, California, USA.

Stehman, S. V., J. D. Wickham, J. H. Smith, and L. Yang. 2003. Thematic accuracy of the 1992 National Land-Cover Data (NLCD) for the eastern United States: Statistical methodology and regional results. *Remote Sensing of Environment* 86:500–516.

Thogmartin, W. E. 2002. *Spatiotemporal dynamics of northern bobwhite (*Colinus virginianus*) in Illinois*. Dissertation, Southern Illinois University, Carbondale, USA.

Thogmartin, W. E., T. J. Fox, J. J. Rohweder, M. G. Knutson, and T. C. Will. 2006*a*. Emerging technologies: LINK: A land conservation decision support tool. *Bulletin of the Ecological Society of America* 87:229–236.

Thogmartin, W. E., A. L. Gallant, M. G. Knutson, T. J. Fox, and M. J. Suárez. 2004*a*. A cautionary tale regarding use of the 1992 National Land Cover Dataset. *Wildlife Society Bulletin* 32:960–968.

Thogmartin, W. E., F. P. Howe, F. C. James, D. H. Johnson, E. T. Reed, J. R. Sauer, and F. R. Thompson, III. 2006*b*. A review of the population estimation approach of the North American Landbird Conservation Plan. *Auk* 123:892–904.

Thogmartin, W. E., M. G. Knutson, and J. R. Sauer. 2006*c*. Predicting regional abundance of rare grassland birds with a hierarchical spatial count model. *Condor* 108:25–46.

Thogmartin, W. E., J. R. Sauer, and M. G. Knutson. 2004*b*. A hierarchical spatial model of avian abundance with application to cerulean warblers. *Ecological Applications* 14:1766–1779.

Thogmartin, W. E., J. R. Sauer, and M. G. Knutson. 2007. Modeling and mapping abundance of American woodcock across the Midwestern and Northeastern United States. *Journal of Wildlife Management* 71:376–382.

Tirpak, J. M., D. T. Jones-Farrand, F. R. Thompson, III, D. J. Twedt, M. D. Nelson, and W. B. Uihlein, III. 2008. Predicting bird habitat quality from a geospatial analysis of FIA data. *In* R. McRoberts, G. A. Reams, P. C. Van Deusen, and W. H. McWilliams, eds. *Proceedings of the 8th annual forest inventory and analysis symposium*, U.S. Forest Service General Technical Report WO-xx, Washington, D.C., USA.

Tirpak, J. M., D. T. Jones-Farrand, F. R. Thompson, III, D. J. Twedt, and W. B. Uihlein, III. *In press*. Multi-scale habitat suitability index models for priority landbirds in the central hardwoods and west gulf coastal plain bird conservation regions. U.S. Forest Service, Northern Research Station, General Technical Report NRS-xx, Newtown Square, Pennsylvania, USA.

Trexler, J. C., and J. Travis. 1993. Nontraditional regression analyses. *Ecology* 74:1629–1637.

U.S. Fish and Wildlife Service. 1980. *Habitat evaluation procedures (HEP)*. U.S. Fish and Wildlife Service, Division of Ecological Services, Manual 102, Washington, D.C., USA.

U.S. Fish and Wildlife Service. 1981. *Standards for the development of habitat suitability index models for use in the habitat evaluation procedure*. U.S. Fish and Wildlife Service, Division of Ecological Services, Manual 103, Washington, D.C., USA.

Venables, W. N., and B. D. Ripley. 1997. *Modern applied statistics with S-PLUS*. Second edition. Springer-Verlag, New York, New York, USA.

Vogelmann, J. E., S. M. Howard, L. Yang, C. R. Larson, B. K. Wylie, and N. Van Driel. 2001. Completion of the 1990s National Land Cover Data Set for the conterminous United States from Landsat Thematic Mapper Data and ancillary data sources. *Photogrammetric Engineering and Remote Sensing* 67:650–662.

Will, T.C, J. M. Ruth, K. V. Rosenberg, D. Krueper, D. Hahn, J. Fitzgerald, R. Dettmers, and C. J. Beardmore. 2005. *The five elements process: Designing optimal landscapes to meet bird conservation objectives*. Partners in Flight Technical Series No. 1. PIF website <http://www.partnersinflight.org/pubs/ts/01-FiveElements.pdf>. Accessed 2 June 2008.

Winter, M. 1998. *Effect of habitat fragmentation on grassland-nesting birds in southwestern Missouri*. Dissertation, University of Missouri, Columbia, USA.

CHAPTER

Modeling Bird Responses to Predicted Changes in Land Cover in an Urbanizing Region

23

Jeffrey A. Hepinstall, John M. Marzluff, and Marina Alberti

Humanity has transformed nearly half of Earth's land areas (Meyer and Turner 1992, Houghton 1994, Lambin et al. 2001). Most of this change has occurred in the last several thousand years as we developed into an agrarian, and then increasingly mechanized, society. A transformed Earth has provided material goods required by modern society and enabled spectacular growth of the human population, but at the cost of long-term provisioning of ecosystem services (Foley et al. 2005). Land transformation is also a major threat to biological diversity now, and increasingly so for temperate regions in the future (Sala et al. 2000).

Urbanization is a growing driver of worldwide change in land cover (Vitousek et al. 1997, Grimm et al. 2000, Alberti et al. 2003). In 1900, only 10% of humans lived in cities, but by 2000, nearly 50% did; by 2030 that number is expected to rise to 60% (Sadik 1999). Depending on economics, social preferences, and land-use policies, the growth of urban populations causes cities, and even more profoundly their suburbs, to spread across large expanses of former agricultural and natural lands (Matlack 1993, Ewing 1994). The worldwide extent of sprawling settlement is visible in the nighttime images of Earth from space (Elvidge et al. 1997). These images reveal that substantial portions of the north temperate zone are heavily settled, most ice-free coastlines are settled, our most fertile lands are quickly being developed, and overall about 3% of Earth's land area is urban (Lawrence et al. 2002, Imhoff et al. 2004).

The influence of rapid urbanization is particularly evident in the western United States (Hansen et al. 2005). People seeking to improve their quality of life are attracted to the west's scenic beauty and its strong service and light industrial job base. As a result of increased urban population growth and immigration, agricultural lands and undeveloped lands that hold great stores of biodiversity are being converted into developed land uses (Mörtberg et al. 2007).

Changed land cover, altered biogeochemical cycles and climate, pollutants, introduced invasive species, and other consequences of human activities directly and indirectly change the selective forces acting upon plants and animals causing some to go extinct and enabling others to thrive and expand (Marzluff 2001, Pickett et al. 2001, McKinney 2002, Kaye et al. 2006). The local difference in extinction and colonization determines standing diversity (Marzluff 2005).

While we are beginning to understand how local urbanization processes influence biodiversity, we know much less about how these altered processes of extinction and colonization will play out through time. To understand how biodiversity responds to land-cover changes requires large-scale modeling. Examples of land-cover and land-use change models abound and are diverse in their theoretical traditions, application, and geographic scope (e.g., Riitters et al. 1997, Pearson et al. 1999, Parker et al. 2003, Turner et al. 2003, Tang et al. 2005). Examples of linking models of land-cover or land-use change to changes in biodiversity are less numerous (Sala et al. 2000, Schumaker et al. 2004, Prato 2005, Mörtberg et al. 2007). While there is no generally accepted single method for predicting how biodiversity will change with landscape change (Doak and Mills 1994, Ruckelshaus et al. 1997), there is a rich literature of how species respond to changes (i.e., loss and fragmentation) of their habitat (e.g., Fahrig and Merriam 1994, McGarigal and McComb 1995, Villard et al. 1999).

Linking predictions from landscape change models to ecological models is generally accomplished by developing spatially explicit habitat models, either at coarse scales with coarse input and output such as habitat-association models (e.g., Scott et al. 1993) for a large number of species (White et al. 1997, Schumaker et al. 2004) or individually based finer resolution models for a single (or few) species (Dunning et al. 1995, Schumaker et al. 2004). In this chapter, we build on such approaches by using sophisticated models of urban development and land-cover change to model the influence of urbanization on biodiversity. We show how models of urban development can be translated into changes in land cover and how these changes can be projected to affect the abundance and diversity of birds in the rapidly urbanizing Central Puget Sound region of western Washington, United States. This area has experienced dramatic urban growth, especially during the past 30 years (Hansen et al. 2005, Robinson et al. 2005) and is projected to grow by 31% (an additional 1 million people from 2000) by 2025 (Office of Financial Management, State of Washington: http://www.ofm.wa.gov/pop/gma/projections.asp). Currently, remnant forests exist in a variety of sizes and settings, from small urban parks or undeveloped parcels to large blocks of contiguous forests (Donnelly and Marzluff 2004*b*, 2006). Songbird diversity peaks in landscapes with 50–60% forest cover, because such areas gain more synanthropic and early successional species than the native forest species they lose (Marzluff 2005). This dynamic response of birds to changing land cover allows us to demonstrate how one component of biological diversity might respond to urbanization. While birds have high public appeal that makes them attractive to policy makers and advocacy groups, their vagility also buffers them from extreme changes (Robinson et al. 1995, Tittler et al. 2006), which may lead us to underestimate the response of birds to local changes in land cover, especially in settings like Seattle, where nearby lands protected from conversion (national parks and forests) may act as source pools for urban populations with high turnover (Alberti and Marzluff 2004).

The objective of this chapter is to provide an example of how to link the output of spatially explicit land-use and land-cover change models to ecological process models. Our ecological response model is a two-scale approach to modeling avian diversity–species richness and relative abundance, but such an approach could have been used to link any spatially explicit ecological process model.

METHODS

Study Area

Our study area is the 3,200 km^2 area of temperate, moist forest around Seattle, Washington (Fig. 23-1). Forests were mostly coniferous, including western hemlock (*Tsuga heterophylla*), Douglas fir (*Pseudotsuga menziesii*), and western red cedar (*Thuja plicata*), with a few red alder (*Alnus rubra*), big-leaf maple (*Acer macrophyllum*), black cottonwood (*Populus trichocarpa*), and Oregon ash (*Fraxinus latifolia*) occurring near riparian and disturbed areas (Franklin and Dyrness 1988). Elevation varied from sea level to near 300 m on the lower slopes of the Cascade Range. Areas above 300 m, while predicted by our land-cover change model, were not considered in our avian biodiversity models, as these areas were not sampled by our field studies.

Study Sites

We chose 139 1-km^2 study sites within this study area (Fig. 23-1): 119 single-family residential (SFR) sites, 13 mixed use/commercial/industrial sites, and 7 forested ("control") sites with minimal development. We stratified 54 randomly selected SFR sites study area along three axes of urbanization: (1) percent of urban, forest, or grass in a landscape; (2) average patch size of urban land cover; and (3) the probability that two randomly chosen adjacent pixels belong to the same class (contagion) per km^2 using a 1999 land-cover map derived from Landsat Thematic Mapper imagery (Alberti et al. 2004). We reclassified the original land-cover types into urban (pixels containing >20% impervious area), forest (deciduous and mixed forest and coniferous forest), grass (grass and agriculture), and other (e.g., wetlands, shoreline, snow/rock/ice). These sites represented a gradient of those sites dominated by urban land cover to those dominated by remnant patches of forest (details in Donnelly and Marzluff

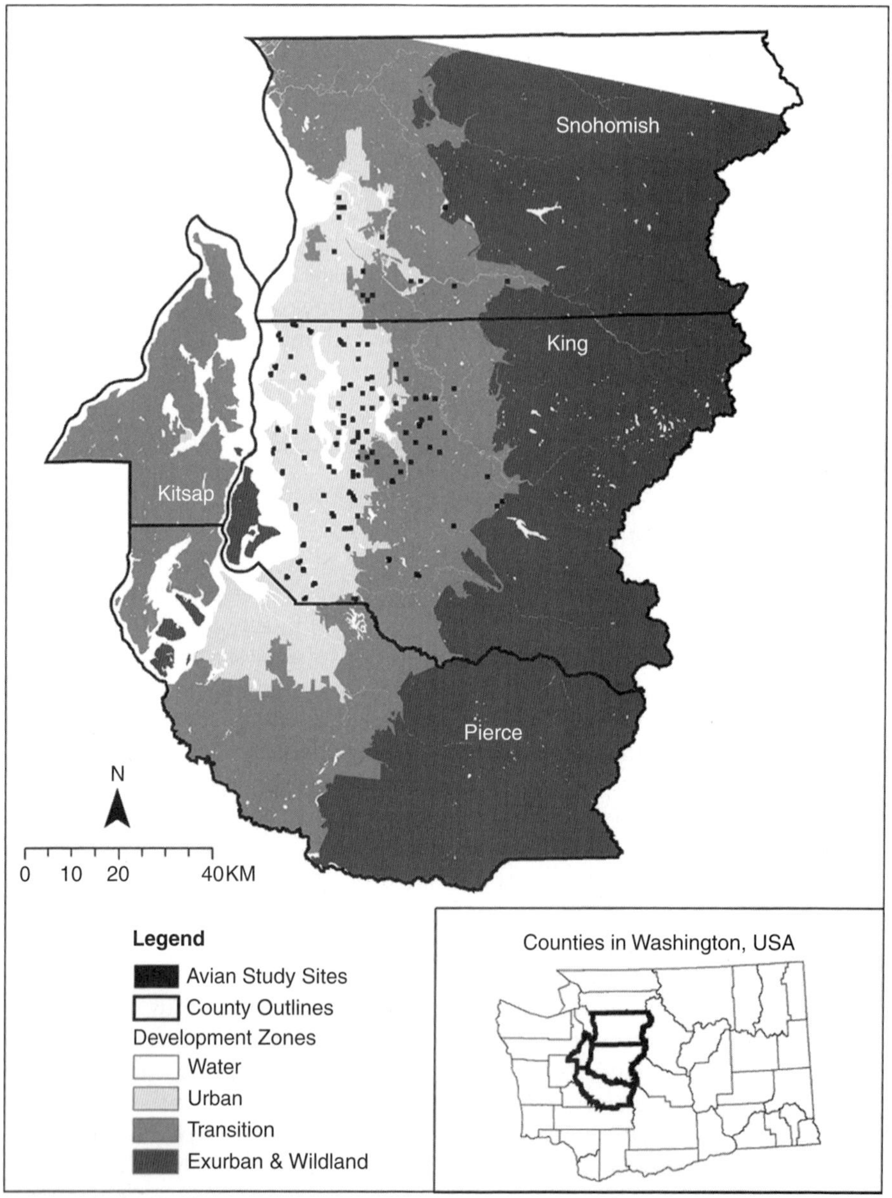

FIG. 23-1

Four-county study area in western Washington, USA, used to model bird response to changes in land cover and the locations of the 1 km^2 bird study sites (n = 139) and the three development zones (urban, transition, and exurban and wildland) defined by distance to urban center, population density, and elevation above sea level.

2004*a*, Blewett and Marzluff 2005). We selected an additional 65 SFR sites across the following gradients: (1) development age (i.e., housing age 5–15 years [young], 40–50 years [middle-age], and >70 years [old]) derived from 2002 parcel data for each county; (2) percent urban in the 1 km^2 landscape; and (3) settlement canopy composition (percent native versus exotic tree species as determined by vegetation surveys on each site [Donnelly and Marzluff 2004*a*]). We selected 13 highly developed sites using 2002 land-use maps derived from each county's parcel database (commercial, heavy commercial, industrial, heavy industrial, institutional/office, and multifamily residential) to improve our ability to understand bird communities in the more developed portions of the landscape. We selected our seven control sites in areas of large contiguous patches of forest with minimal (<5%) developed land within the 1 km^2 study site.

Avian Surveys

Trained observers conducted 6,437 fixed-radius (50 m) point count surveys of breeding birds at 992 locations within 139 study landscapes during the spring and summers of 1998 through 2005. Individual sites were sampled 1-7 years. We visited locations within sites 3–5 times per year (late March–late August). Observers arrived at point count locations between 30 minutes prior to and approximately six hours after sunrise, where all possible identifications by sight or sound were recorded during a 10-minute period at each point (details in Donnelly and Marzluff 2004*a*, *b*, 2006). For most sites, eight point counts were conducted in each site: six located in the developed areas and two in the remnant patches of forest. Greater effort was allocated to sampling birds in the developed areas than forests because a previous study of forest reserves in the same region indicated that birds and vegetation were more variable in developed areas (Donnelly and Marzluff 2006). All points were >100 m apart, with the exception of a small number (n = 83; mean distance between these points = 76 m) where the separation was maximized within the only forest fragment that existed within the study landscape. Our sampling effort resulted in a total of 2,866 counts at 302 locations in forested portions of sites and 3,571 surveys at 690 locations in the built portions of sites.

We categorized a subset of 57 common species into three development-sensitive guilds (Appendix A). The guild approach centers on different colonization and extinction probabilities of different birds. We grouped species into "native forest" birds, those typically found in intact, mature forest (n = 19), "synanthropic birds," those native or nonnative birds that thrive in human-dominated landscapes, making use of increased resources not available prior to human development (n = 9), and "early successional" birds, those that exploit the heterogeneous vegetation of fragmented landscapes following some type of physical disturbance (n = 29; Marzluff 2005). We calculated relative abundance for each site as the mean number of individuals detected per point, per survey. This method was preferable to others, such as maximum abundance per survey, because it prevented young of the year and migrating individuals from inflating abundances.

Land-Cover Change Modeling

The land-cover change model (LCCM) consists of a set of discrete choice equations of site-based land-cover transitions derived from observed land-cover change (Fig. 23-2) that are applied to geographic information system (GIS) layers to predict land-cover change at a 30 m resolution across four counties in western Washington, representing the central Puget Sound Region. A short description of the model follows; a complete description of the theoretical foundations of the model is available in Hepinstall et al. (2008).

The LCCM framework derives from the traditions of modeling landscape change as a dynamic interaction between socioeconomic and biophysical processes (Turner et al. 1996, Wear and Bolstad 1998, Wear et al. 1998). The LCCM is written in Python and is designed as a module within the larger Open Platform for Urban Simulation (OPUS) and UrbanSim modeling platforms (Waddell 2002, Waddell et al. 2003, <www.urbansim.org>). UrbanSim consists of a series of modules that have been developed to, among other things, model land-use change in response to changes in transportation networks, household and business location, property development and intensity, infrastructure changes, and policy choices. UrbanSim is designed to aid regional land-use planning.

Urban development models such as UrbanSim predict changes in land use (e.g., undeveloped, residential, commercial, mixed use, timberlands) and development intensity (number of residential units or square feet of commercial space), whereas avian communities respond to changes in vegetation type and structure. We must link models predicting change in land use to models of land-cover change, which then can be used to predict the effects of land development on avian communities. Our LCCM predicts future land cover in response to land-use change and biophysical constraints.

For our implementation of the LCCM, we simulated the potential change to one of eight land-cover classes: heavy urban (>80% impervious surfaces), medium urban (20-80% impervious surfaces), low urban (a mixed class with <20% impervious area and the remaining area in vegetation), grass, agriculture, deciduous and mixed forest, coniferous forest, and clearcut. Each land-cover class can transition from a variable number of other classes. We empirically estimated 26 transition equations as a function of observed land-cover and independent variables from two dates (Fig. 23-2; Turner et al. 1996). The focus of the LCCM is to model urban growth, which in the central Puget Sound is limited to the lower elevations that have little commercial forestland. We chose not to model forest regeneration and instead converted any predicted new clearcut into regenerating forest in the subsequent time step and retained all regenerating forest for the duration of the LCCM run (28 years in this application).

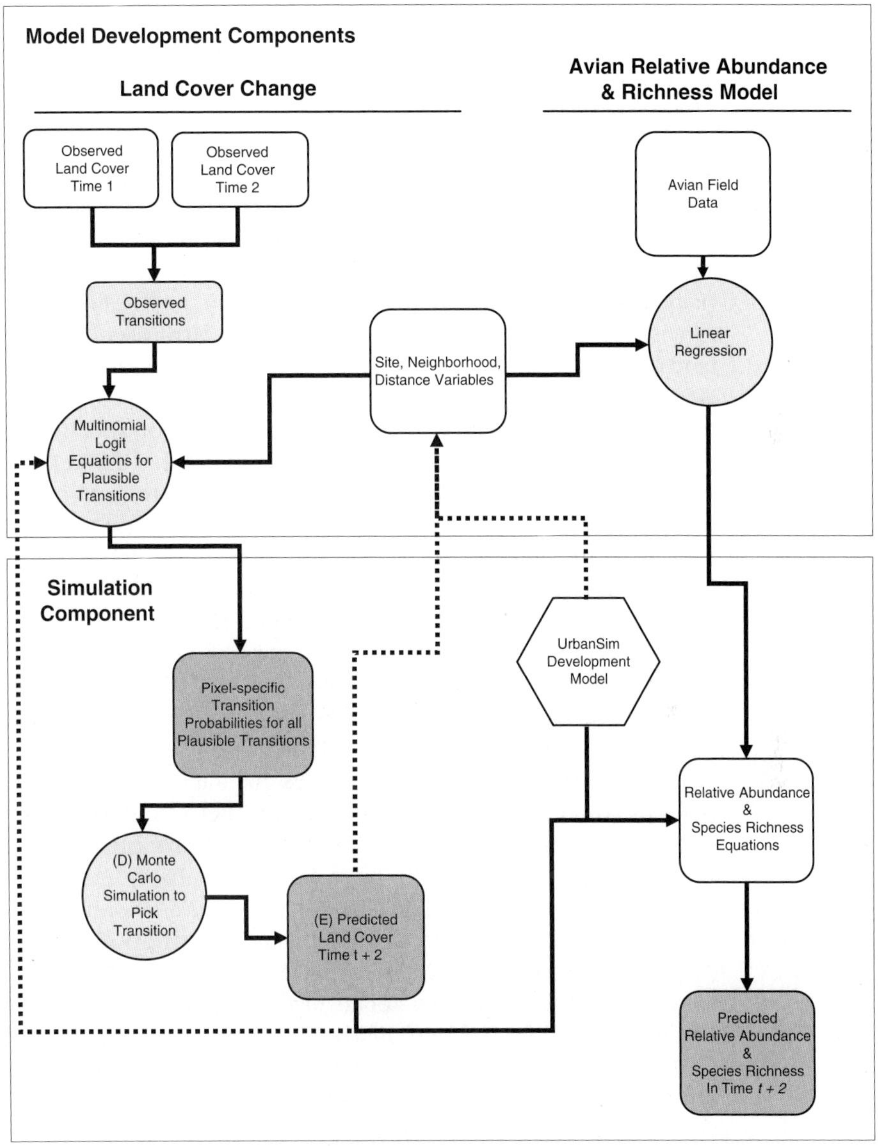

FIG. 23-2

Flow chart of steps performed as part of the land-cover change model and the avian richness and relative abundance models used to predict bird response to changes in land cover.

The central Puget Sound implementation of LCCM has multiple possible input dates of land cover to use for developing transition models including 1986, 1991, 1995, 1999, and 2002; we used equations developed from observed 1995–1999 transitions (Hepinstall et al., in press). We modeled land-cover change using discrete choice (multinomial logit) statistical models. Developing a discrete choice equation for each transition modeled is an iterative, semi-automated process that can be done directly within the LCCM code base, but still takes multiple days to complete. Transition probabilities for each 30 m pixel to change from one discrete land-cover class i to another cover class j is potentially influenced by many factors including (1) the predicted type and predicted intensity of a development event; (2) a set of attributes of the pixel; and (3) the land-cover composition and configuration of neighboring pixels (Fig. 23-2; Hepinstall et al., in press). In the Puget Sound implementation of the LCCM, 65 potential explanatory variables are available for specifying discrete choice equations. Land development, or the probability that a pixel will transition from an undeveloped to a developed state, is derived from UrbanSim development module output. UrbanSim output is also used to determine the type (residential, commercial/industrial, mixed use) and intensity (number of residential units or ft^2 commercial/industrial added) of development. The remaining variables were developed from spatial databases obtained from county, state, and federal GIS data repositories and required several months to compile and error check. Site attributes influence the ability to develop land through increasing the cost of development (e.g., steep slopes, unstable soils), limiting or prohibiting development (e.g., critical areas such as steep slopes, landslide hazard, riparian areas, etc.; proximity to endangered species habitat), or encouraging development (e.g., proximity to existing infrastructure). Because development events generally occur in patches that are greater than the size of an individual 30 m pixel (900 m^2), land-cover transitions in adjacent cells influence the probability of land-cover transitions in a focal cell. The LCCM, therefore, includes distance variables (e.g., distance to central business district) and variables measuring the spatial context of the target pixel, by calculating several measures (e.g., number of residential units added in the previous three years) within 150 m, 450 m, and 750 m moving windows.

The output of the discrete choice equations are probabilities that any given pixel will transition from its current class to one of the possible options for that class including the no-change option. For example, light urban can transition to medium urban, heavy urban, or remain as light urban. Parameter estimates from the discrete choice equations are applied to GIS layers to derive pixel-specific transition probabilities for each pixel to convert to a land-cover class (Fig. 23-2). Because only a small portion of the landscape changes to a new land-cover class over short time intervals (in our case four years), we used Monte Carlo simulations to pick what land-cover type each pixel will be in the next time step (Fig. 23-2). Specifically, transition probabilities for each land-cover class are normalized by the annualized observed transitions and scaled to sum to 1.0 for each possible transition from the starting class. Then predicted transitions are implemented by comparing the class-specific probabilities for each pixel to a random number chosen from a uniform distribution between 0 and 1. If the scaled transition probability to a new land-cover class matches the random value, the transition takes place; otherwise, the grid cell maintains its current land cover.

The LCCM is implemented in the Python language as a component of UrbanSim and can be downloaded and used as a template to develop a local implementation for any region with spatial data for at least two dates of land cover and drivers of land-cover change (i.e., biophysical and socioeconomic). While UrbanSim requires many socioeconomic data layers to fully implement, LCCM is independent of UrbanSim, is flexible, and can be implemented using output from any land-use change prediction.

Coupling Land-Cover Change and Avian Richness and Relative Abundance Models

As a proof of concept linking predicted landscape change to potential changes in the avian community, we developed preliminary statistical models to predict bird species richness and relative abundance as a function of land-use and land-cover composition and configuration. We used linear regression to develop separate models of species richness for all species and the three development-sensitive guilds using point count data from the 139 study landscapes. We also developed linear regression models to predict the relative abundance of three representative species, one from each habitat guild. Linear regression models were appropriate, since mean counts and species abundances were generally larger than 10. Poisson regression methods, while generally useful where average counts less than 10, were not further considered at this point.

We defined two *a priori* models based on previous studies (e.g., Donnelly and Marzluff 2004*a*, *b*, 2006), landscape measures relevant to urban planners, and variables available as output from UrbanSim and the LCCM to allow for predictions of future species richness and relative abundance. We used correlations between landscape variables derived from land cover and land use to eliminate highly correlated variables (Pearson correlation coefficient > 0.70) prior to developing our models, yielding 10 variables describing land-cover and land-use patterns. Within the 1 km^2 bird study sites, we used land-cover data from 2002 to calculate the percent forest, percent urban, the aggregation index (Fragstats 3.3; McGarigal et al. 2002) of forest, the number of patches of forest, and the number and mean patch size of urban patches. We used land use derived from 2002 county parcel data to calculate the percent, patch density, and aggregation index of residential parcels, and the mean age of development of parcels within each study site. Our *a priori* simple model (SM) included (1) percentage of forest; (2) aggregation of residential land use; and (3) mean age of development within a 1 km^2 window. Our *a priori* full model (FM) included these variables and the following: (1) percentage of grass and agriculture; (2) forest aggregation index; (3) the number of unique patches of forest land cover; (4) number of unique patches of urban land cover; (5) the mean patch size of unique patches of urban land cover; (6) the percent of residential land use; and (7) the patch density of patches of residential land use.

We then applied the parameter estimates from the regression models of species richness and relative abundance to the future landscapes generated by UrbanSim and LCCM to calculate total and guild-specific species richness and relative abundance for three representative species (Pacific-slope flycatcher [*Empidonax difficilis*], native forest guild; yellow-rumped warbler [*Dendroica coronata*], early successional guild; and American crow [*Corvus brachyrhynchos*], synanthropic guild). We calculated landscape variables required for our avian models for each 30 m pixel in our four-county study area with ArcGIS (Environmental Systems Research Institute, Redlands, California) and Fragstats 3.3 (for aggregation index) using a 1 km^2 moving window to match the study site design of the avian surveys, where field surveys were designed to characterize the bird community of 1 km^2 landscapes. We converted relative abundance estimates for 1 km^2 windows from predicted mean number of individuals per point per survey into mean number of individuals per 1 km^2. We smoothed species richness estimates to represent the mean predicted species richness within a 1 km^2 window. To evaluate trends over time in our predictions and because we could not compare each of the 11 million pixels in our study area directly, we randomly selected 100,000 pixels from our predictions of richness and relative abundance for our starting and ending dates (2003 and 2027). We also calculated the range of variability in our predictions using the upper and lower confidence bounds on the parameter estimates and grouped variability into three classes: low, medium, and high.

We compared avian model results for two zones of urbanization (urban and transition; Hepinstall et al., in press), corresponding to the zones where our avian study sites were located, allowing us to examine the changes in avian richness and relative abundance relative to proximity and proportion of development (Fig. 23-1). The urban zone is dominated by impervious areas (residential, commercial, industrial, institutional, and office) and contains all the major cities of the study area. The transition zone contains a more heterogeneous mix of impervious areas, agricultural lands, and remnant woodlands.

RESULTS

Predictions of Future Land Cover

Between 2003 and 2027 for the full four-county study area, we predict a decrease in mature forest types (deciduous, mixed, and coniferous) from 60% of the area to 38% and an increase in developed land (heavy, medium, and light

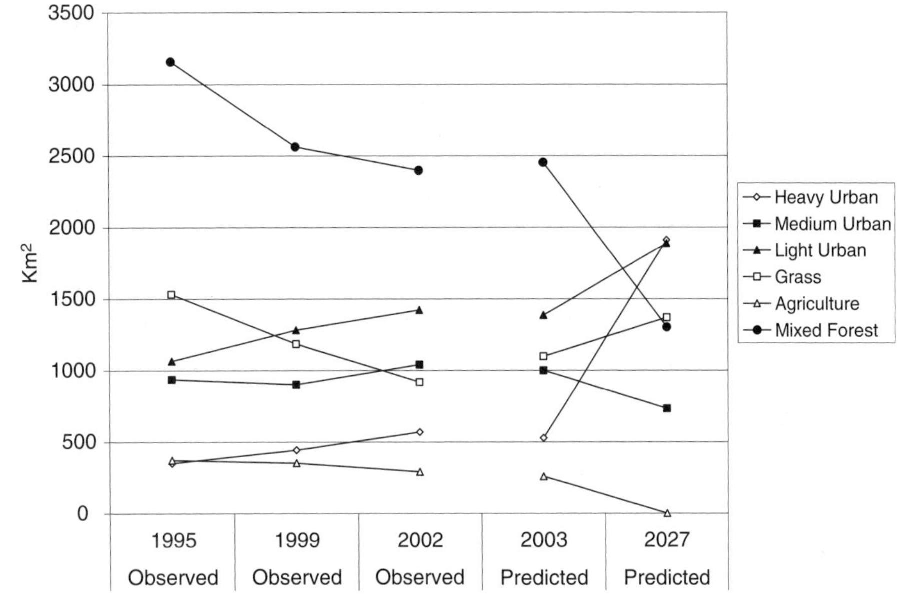

FIG. 23-3

Area (km^2) of observed and predicted land cover for four-county study area in western Washington, USA, within the urban and transition development zones derived from 1995–1999 observed transitions (classes include heavy urban, >80% impervious surfaces; medium urban, 50–80% impervious; light urban, 20–50% impervious; grass; agriculture; deciduous and mixed forest—100–25% deciduous species; coniferous forest – >75% conifers; clearcut—recent forest harvest with no regrowth yet visible on satellite imagery; and early regenerating forest—recent forestland harvest with regrowth visible).

urban classes) from 17% to 34% (Figs. 23-3 and 23-4). We predict a decrease in grass and agriculture from 14% to 10% of the area and an increase in clearcut and regenerating forest from 9% to 18%. We predict a decrease in percentage of forest (per 1 km^2) and a movement of primarily forested areas up the slopes of the Cascades (Fig. 23-4).

Avian Surveys

Across the 139 sites, we identified 114 species of birds. Each site contained an average of 31.5 ($\pm$1.0 SE) species. Average species composition per site was 11.2 $\pm$ 0.4 forest species, 11.3 $\pm$ 0.4 early successional species, and 5.2 $\pm$ 0.2 synanthropic species. We observed Pacific-slope flycatcher, yellow-rumped warbler, and American crow, on average, on 25.3 (yearly range 14–40, 80 total in 7 years of sampling), 9.5 (yearly range 3–12, 32 total), and 30.5 (yearly range

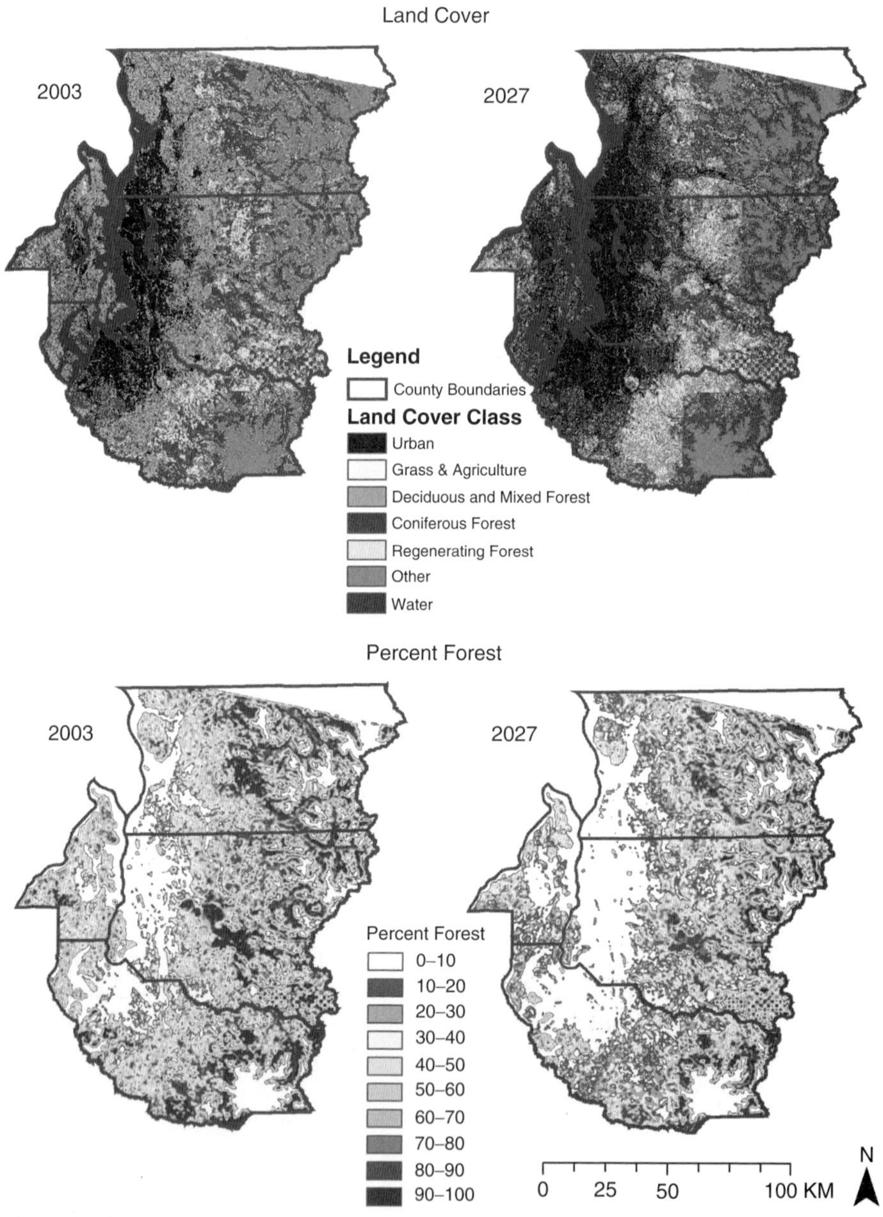

FIG. 23-4 (See color insert.)

Predicted land cover and percent forest within a 1 km^2 moving window in 2003 and 2027 for the four-county study area in western Washington, USA, based on the 1995–1999 observed land-cover transitions and the Land-Cover Change model.

13-60, 132 total) sites, respectively. We observed, on average, 7.8 ± 1.7 Pacific-slope flycatchers, 2.0 ± 0.6 yellow-rumped warblers, and 14.0 ± 8.3 American crows per site.

Models of Avian Species Richness and Relative Abundance

Approximately 18–20% of the variation in total species richness was accounted for in our simple and full models (Table 23-1). Richness increased with increasing percentage of forest and aggregation of residential development in the 1 km^2 study site, and decreased with age of residential development. These relationships were all significant in the simple model, but percent forest was not significant in the full model, possibly in response to including somewhat co-varying land-cover classes (percent grass) in that model.

Within habitat guilds, we explained more (44–51%) of the variation in native forest species richness than in the richness of early successional (12-15%) and synanthropic species (23–28%; Table 23-1). Native forest species richness mirrored total species richness and increased with increasing percentage of forest and aggregation of residential development in the 1 km^2 study site, and decreased with age of residential development (Table 23-1). Native forest species richness also was positively correlated with forest aggregation with our full model explaining 51% of the variation in observed native forest species richness. Early successional species richness was only significantly correlated with mean age of development (−) in both models and residential aggregation (+) in the full model, but even with the full model, only 15% of the observed variation in guild richness was explained by our models. Synanthropic species richness was significantly associated with percentage forest (−) and residential aggregation (+) in the simple model and year built (−) and percentage grass (+) in the full model.

In general, our full models of species richness explained little variation beyond that explained by the simple models (Table 23-1). The few variables that significantly contributed to the full models included those in the simple models and additional, biologically relevant variables. Specifically, percentage of grass was an important contributor to explaining variation in richness of those species utilizing developed landscapes, and forest configuration was important to native forest species.

Our *a priori* simple model for predicting relative abundance of three species had significant relationships with percent forest (Pacific-slope flycatcher [+], American crow [−]) and residential aggregation index and mean year built (yellow-rumped warbler [−]; Table 23-2). Percent forest was significant (−) for both yellow-rumped warbler and American crow for the full model. Only two other variables were significant predictors of relative abundance for these three species in the full models: number of patches urban (yellow-rumped warbler)

Table 23-1 Linear Regression Model Results for Simple Models of Species Richness (Species/point/survey) for Total Species Richness and for Three Guilds (Native Forest, Early Successional, Synanthropic) as a Function of Landscape Metrics for 139 Suburban Landscapes In Puget Sound, Washington, USA, 1998–2005, for (A) Simple and (B) Full Models. Standardized Coefficients (B) Lower 95% Confidence Interval (LCI) and Upper 95% Confidence Interval (UCI) Presented for Each Unstandardized Parameter Estimate. Simple Models had 3 Regression Degrees of Freedom and 135 Residual d.f. Full Models had 10 Regression Degrees of Freedom and 128 Residual d.f.

A) Simple Model							
Species Richness	**Adjusted R^2**	**P**		**Constant**	**Percent Forest**	**Residential Aggregation Index (AI)**	**Mean Age Development**
Total Species	0.178	0.000	B		0.201	0.165	–0.305
			LCI	18.910	0.014	0.006	–0.365
			UCI	38.534	0.159	0.188	–0.109
Native Forest Species	0.440	0.000	B		0.550	0.222	–0.221
			LCI	3.132	0.071	0.022	–0.111
			UCI	9.585	0.118	0.082	–0.026
Early Successional Species	0.120	0.031	B		0.033	–0.690	0.755
			LCI	7.675	–0.024	–0.003	–0.155
			UCI	15.740	0.035	0.072	–0.050
Synanthropic Species	0.225	0.000	B		–0.458	0.173	–0.095
			LCI	3.451	–0.050	0.003	–0.037
			UCI	7.028	–0.024	0.036	0.010

B) Full Model														
Species Richness	Adj–R^2	P		Constant	% Forest	Res. Agg. Index	Mean Age Dev.	% Grass	Forest Agg. Index	Number Patches Forest	Number Patches Urban	MPS Urban	% Resid.	Patch Density Resid.
Total	0.000	0.208	B		0.140	0.252	−0.348	0.185	0.148	0.055	−0.090	0.042	−0.203	0.068
Species			LCI	9.837	−0.072	0.024	−0.426	0.028	−0.044	−0.298	−0.616	−0.613	−0.194	−0.188
			UCI	35.265	0.192	0.273	−0.114	0.889	0.187	0.533	0.269	0.887	0.011	0.381
Native	0.000	0.508	B		0.347	0.199	−0.167	0.020	0.394	0.141	−0.072	−0.045	−0.049	0.090
Forest			LCI	−2.053	0.018	0.008	−0.101	−0.115	0.040	−0.010	−0.194	−0.293	−0.041	−0.038
			UCI	5.927	0.101	0.086	−0.003	0.155	0.112	0.251	0.084	0.178	0.023	0.140
Early	0.000	0.151	B		−0.018	0.225	−0.384	0.202	0.139	0.019	−0.076	0.086	−0.193	0.053
Succ.			LCI	3.891	−0.057	0.002	−0.183	0.022	−0.021	−0.155	−0.240	−0.198	−0.077	0.619
			UCI	14.345	0.051	0.104	−0.054	0.376	0.074	0.187	0.124	0.419	0.008	0.147
Syn.	0.000	0.281	B		−0.380	0.307	−0.205	0.245	−0.041	−0.159	−0.152	0.112	−0.139	0.026
			LCI	2.668	−0.054	0.012	−0.058	0.037	−0.024	−0.138	−0.134	−0.066	−0.030	−0.044
			UCI	7.214	−0.007	0.056	−0.002	0.191	0.017	0.011	0.024	0.202	0.007	0.058

Table 23-2 Linear Regression Model Results for Relative Abundance for One Representative Species from Each Habitat Guild as a Function of Landscape Metrics for 139 Survey Sites in Puget Sound, Washington, USA, 1998–2005, for (A) Simple and (B) Full Models. Standardized Coefficients (B) Lower 95% Confidence Interval (LCI) and Upper 95% Confidence Interval (UCI) Presented for Each Unstandardized Parameter Estimate. Simple Models had 3 Regression Degrees of Freedom and 135 Residual d.f. Full Models had 10 Regression Degrees of Freedom and 128 Residual d.f.

A) Simple Model Parameter Estimates

Guild/Species	R^2	P		Constant	Percent Forest	Residential Aggregation Index (AI)	Mean Age Dev.
Native Forest: Pacific-slope flycatcher	0.303	0.000	B		0.529	−0.079	−0.037
			LCI	−0.316	0.013	−0.011	−0.012
			UCI	1.162	0.024	0.003	0.007
Early Successional: yellow-rumped warbler	0.263	0.000	B		−0.068	−0.513	−0.174
			LCI	0.438	−0.002	−0.007	−0.005
			UCI	0.784	0.001	−0.004	0.000
Synanthropic American crow	0.327	0.000	B		−0.566	0.048	0.020
			LCI	1.474	−0.036	−0.006	−0.012
			UCI	3.553	−0.021	0.013	0.015

R^2	P		Constant	% Forest	Res. AI	Mean Age Dev.	% Grass	Forest AI	NP Forest	NP Urban	LN Urban MPS	% Res.	Res. PD
B) Full Model Parameter Estimates													
Native Forest: Pacific-slope flycatcher													
0.345	0.000	B		0.249	0.027	−0.014	−0.111	0.040	0.013	0.040	−0.362	−0.129	0.036
		LCI	0.356	−0.001	−0.008	−0.012	−0.055	−0.007	−0.029	−0.027	−0.151	−0.012	−0.017
		UCI	2.247	0.019	0.011	0.011	0.010	0.010	0.033	0.039	−0.040	0.003	0.025
Early Successional: yellow-rumped warbler													
0.269	0.000	B		−0.303	−0.605	0.101	0.043	0.169	−0.022	0.239	0.081	−0.046	−0.127
		LCI	0.345	−0.005	−0.009	−0.004	−0.010	−0.001	−0.008	0.001	−0.009	−0.002	−0.008
		UCI	0.800	0.000	−0.004	0.001	0.006	0.004	0.007	0.016	0.018	0.001	0.002
Synanthropic: American crow													
0.319	0.000	B		−0.596	0.097	0.080	−0.032	0.083	−0.045	−0.054	0.057	−0.082	−0.120
		LCI	0.703	−0.044	−0.007	−0.010	−0.056	−0.008	−0.056	−0.060	−0.060	−0.015	−0.051
		UCI	3.463	−0.016	0.020	0.024	0.037	0.017	0.034	0.036	0.103	0.007	0.011

and mean patch size urban (Pacific-slope flycatcher). Each of our *a priori* models explained approximately 26–35% of the observed variance in relative abundance for the three selected species.

Predictions of Future Bird Species Richness and Relative Abundance

We predicted species richness will decline slightly during the next 25 years as native forest species respond to the loss of forest cover and early successional species respond to the aging of current developments. Both the simple and full models project a decline in total species richness for the four counties from a mean of 34 or 37 in 2003 to 28 or 30 in 2027, respectively, in the transition zone (Fig. 23-5). The guild-specific results indicate that the loss of early successional (~1) and, more significantly, native forest (~2–4) species would be noticeable in both the urban and transition zones. Simple models predict the diversity of the synanthropic guild will increase slightly in the transition zone and remain relatively stable in the urban zone (Fig. 23-5). The full model predicts counterintuitive results for the synanthropic guild—a decrease in species richness in the transition zone (Fig. 23-5). The additional landscape variables in the full models consistently reduce mean predictions of species richness for the three habitat guilds by 1–3 species.

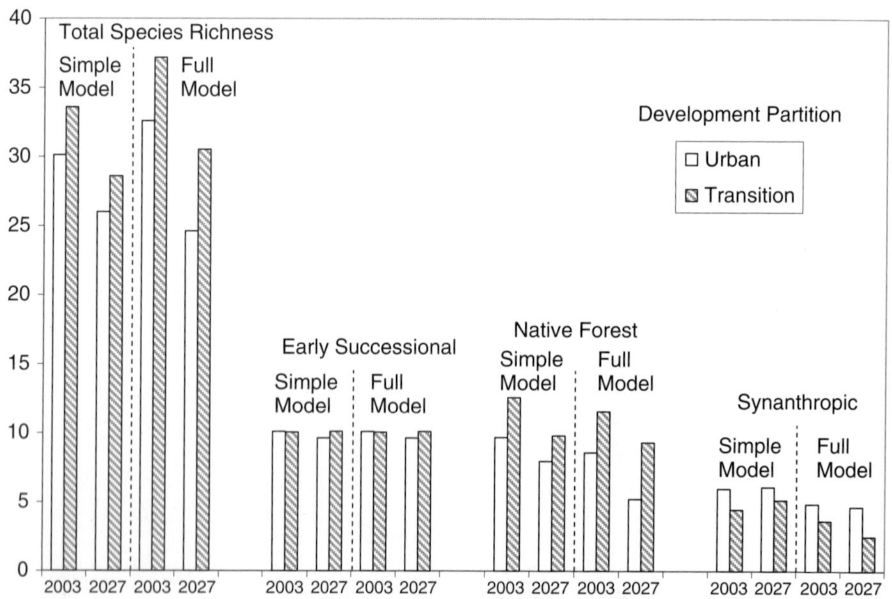

FIG. 23-5

Comparison of mean predicted bird species richness for the urban and transition zones (Fig 23-1.) for simple model and full models in western Washington, USA.

Maps generated from projecting the full and simple models show that changes in species richness are expected to be concentrated in those regions of the study area where land-cover change is most dramatic (Fig. 23-6). This is primarily in the transition development zone surrounding the present heavy

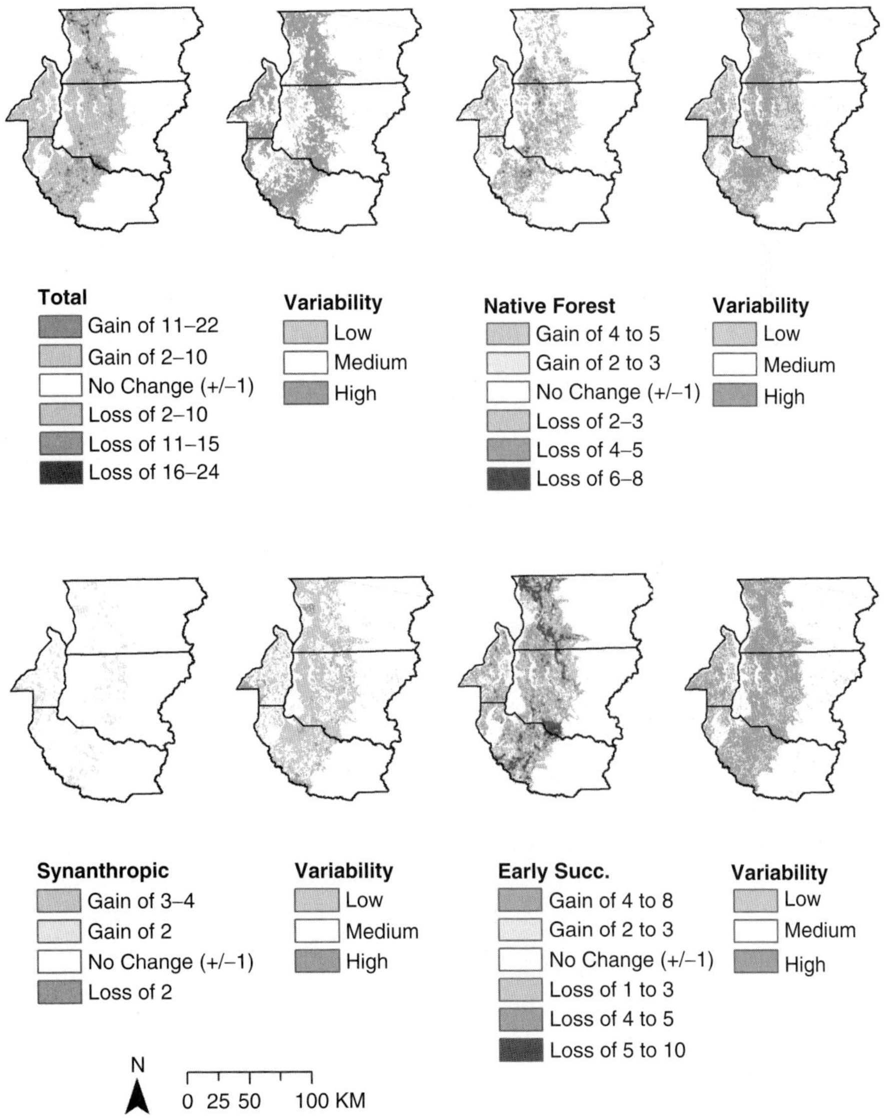

FIG. 23-6 (See color insert.)

Mapped predicted gain or loss and variability of model predictions between 2003 and 2027 for full (total, native forest, early successional) and simple (synanthropic) species models for urban and transition zones within the four-county study area in western Washington, USA.

urban core where forest loss and aging of developments cause the greatest changes in land cover (Figs. 23-1 and 23-2). Total species, native forest species, and early successional species richness show similar patterns of species loss with up to 24 species being predicted to be lost from the total species list at specific locations, and nine each from the native forest and early successional guilds (Fig. 23-6). The full model for the native forest species guild seems to overpredict the number of native forest species in the urban zone in 2003 and then predicted a large loss of species by 2027 (Fig. 23-6), creating a change map that shows a larger loss of species closer to the urban centers than the predicted change from the simple model output (Fig. 23-5). Because full model predictions for the synanthropic guild predicted a slight loss of species over time (Fig. 23-5) in the urban and transition zones, clearly an error in model formulation, we present the predicted distribution of synanthropic species as predicted by the simple model (Fig. 23-6). The majority of change for synanthropic species predicted by the simple model is a gain of 2–4 species in the transition zone (Fig. 23-6). The full model results for synanthropic species indicated that species in the group have a complex relationship (i.e., neither all positive nor all negative) to the variables we included in our *a priori* model. Further exploration of how this group responds to landscape patterns is clearly warranted. Model prediction variability varied by guild and location with total species predictions being most variable in the transition zone (Fig. 23-6).

The spatial pattern of forest loss has additional consequences for future avian communities. Currently, avian diversity is greatest in diverse land-cover characteristic of the transition zone (Fig. 23-5) or equally great in the transition and wildland zones (Fig. 23-5). However, in just a few years and increasingly through 2027, the diversity of the transition zone is expected to drop rapidly (Fig. 23-5). Future bird communities are predicted to increase gradually in richness with distance from development rather than peaking in diversity in the current intermediately settled transition zone. In fact, as the transition zone is transformed into dense development, the region is likely to end up with spatially partitioned bird communities dominated by either adaptable, synanthropic species (in dense developments) or resilient native forest birds (in the wildland zone). This is suggested by the projected distribution of forest cover (Fig. 23-4). Currently, forest cover is very low (<20%; red shades on Fig. 23-4) in the small area proximal to Seattle. Much of the transition zone east of Seattle is a 50:50 mixture of built and forested lands (blue-green in color; Fig. 23-4) with very high avian diversity, but this will not be the case by 2027.

The three species selected as examples of how to apply relative abundance equations to the future landscapes created by the LCCM show clear patterns of response to changes in land-use and land-cover amount and configuration. American crow abundance per km^2 was predicted to increase in both development zones (Fig. 23-7) with the greatest increases predicted in the transition zones by both simple (Figs. 23-7 and 23-8) and full models (Fig. 23-7). Pacific-slope flycatcher predictions from the full model were lower than from the

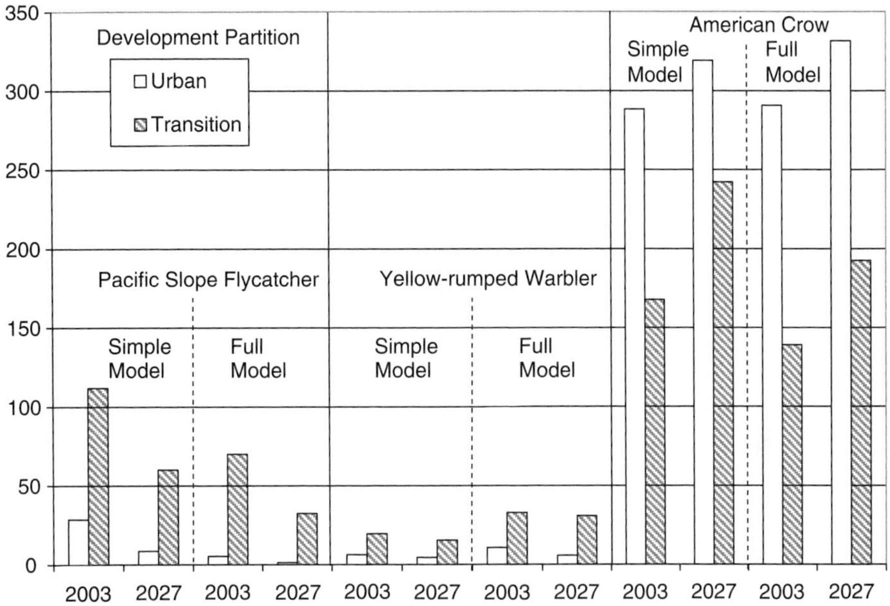

FIG. 23-7

Comparison of mean number of individuals per 1 km^2 for the urban and transition development zones (Fig 23-1.) for three individual species (Pacific-slope flycatcher, yellow-rumped warbler, and American crow) in western Washington, USA.

simple model (Fig. 23-7), with a 25% decline in relative abundance over time in the transition zone predicted by the full model. Model predictions for the yellow-rumped warbler were relatively stable over time with slight decreases in numbers for the transition zone using the full model (Fig. 23-7) and a thin strip of lower relative abundance along the zones of greatest new development as land-cover change reduces the amount of forest present in these areas.

DISCUSSION

We presented a method for combining land-use change models, land-cover change models, and avian biodiversity models as an example of how to couple sophisticated models developed from different modeling traditions. The output from the UrbanSim development model was combined with a discrete choice land-cover change model to predict land-use and land-cover change, which were then used as input to avian richness models to predict changes in avian species richness and individual species relative abundance 28 years into the future. Previous large-scale modeling efforts have used more generalized

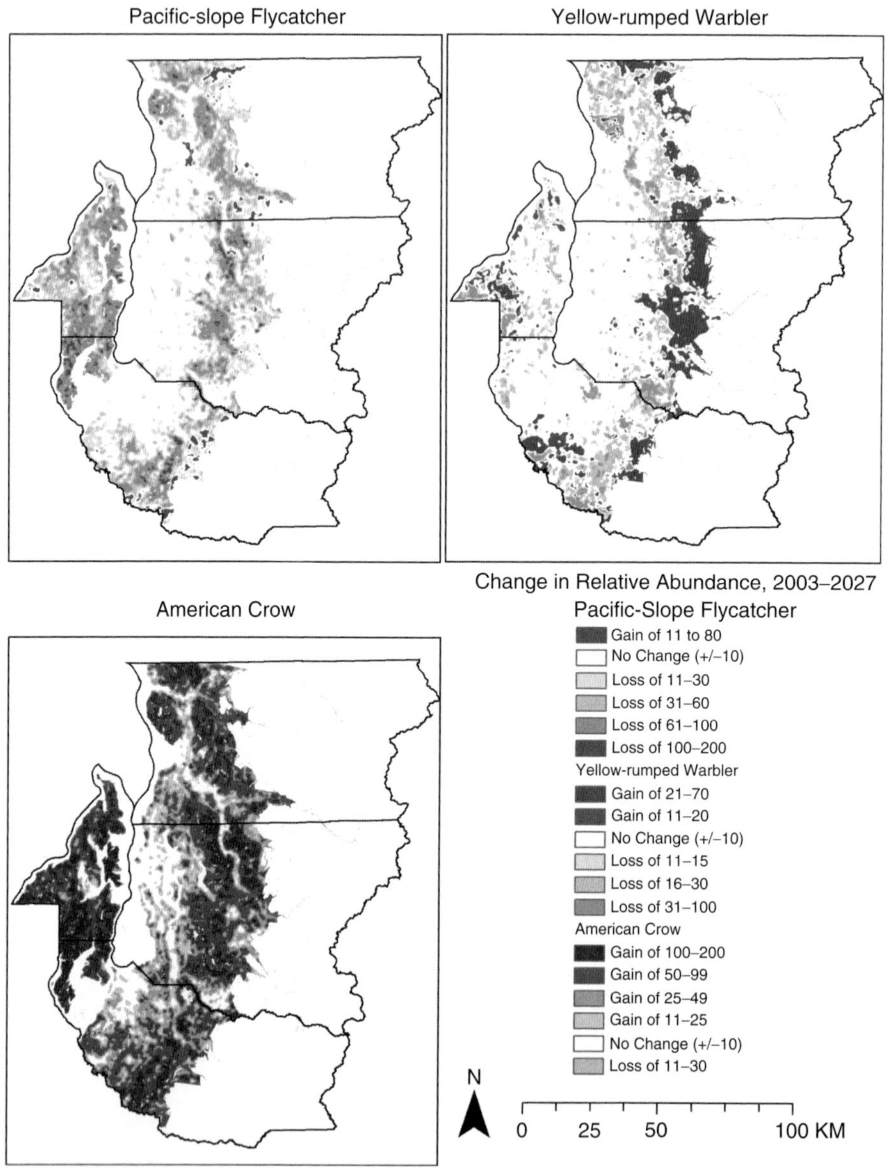

FIG. 23-8 (See color insert.)

Mapped differences in predicted relative abundance per km^2 between 2003 and 2027 for American crow (simple model), Pacific-slope flycatcher (full model), and yellow-rumped warbler (full model) within the urban and transition zones in the four-county study area in western Washington, USA.

land-use/land-cover change models such as Markov transition models (e.g., Turner et al. 1996, Wear and Bolstad 1998, Wear et al. 1998, Prato 2005) and/or more general habitat-association models (e.g., Scott et al. 1993, White et al. 1997, Schumaker et al. 2004) to accomplish similar tasks but at lower spatial resolutions and incorporating fewer drivers and constraints of landscape change. What is missing, generally, are links between models that predict how species respond to landscape change and models that predict the future extent and intensity of urban development (Pickett et al. 2001).

Studies that link future landscape changes and the potential effects on biodiversity often use alternative futures analysis (Steinitz 1990, Ahern 1999, Steinitz and McDowell 2001) where scenarios of future growth developed by planning agencies serve as "visions" of the future if different planning principles are followed, rather than explicitly modeling urban growth (e.g., Hulse et al. 2004, Mörtberg et al. 2007). In Oregon's Willamette River Basin, Hulse et al. (2004) worked with citizen groups to define three value-based future scenarios of development policy, including a continuation of current policies. Schumaker et al. (2004) took these scenarios and evaluated how change in land cover would affect species using simple habitat-association models for 279 vertebrate species and a life history simulator to simulate the potential population effects on a small subset of these species. Mörtberg et al. (2007) used three urbanization scenarios for the Stockholm, Sweden region and logistic regression models for three focal avian species (two requiring large tracts of forest and one requiring smaller patches of forest and potentially able to persist in remnant forests in urban and suburban areas). Their avian models included measures of landscape composition and configuration to predict the probability of species occurrence on the landscape.

We combined ideas derived from landscape ecology (i.e., that landscapes change in response to human and natural disturbances, which change the composition and configuration of resources, which then affect species diversity), with sophisticated micro-simulation economic models (Waddell 2002, Waddell et al. 2003) and preliminary avian community models developed from a large and long-term field data set designed to investigate community dynamics in the face of urban development. The central Puget Sound region is experiencing rapid human population growth and concomitant landscape change to accommodate the new individuals and businesses moving into the area. Because the region is bounded by water (e.g., Puget Sound and Lake Washington) and mountain ranges (Cascades and Olympics), development is concentrated in the lower elevations, increasing the pressure in these areas. Our LCCM model has clearly shown the potential impacts of landscape change on land-cover composition and configuration in the future. Development, however, will not occur at the same rate or intensity everywhere. Our method provided a first look at how areas that are likely to develop in the future are spread across the landscape and how these new patterns of land cover and land use may affect the regional avian communities. Spatially explicit predictions of change are important to inform large-scale conservation planning in urban and urbanizing regions (Mörtberg et al. 2007, Schumaker et al. 2004). The ability to change key assumptions, inputs, constraints, and even statistical approaches through the use of open-source programming modules makes our approach flexible and dynamic (e.g., Waddell 2002, Noth et al. 2003).

Responses of Birds to Urbanization

The substantial reduction in forest cover and increase in developed land that our models project for the Seattle area in the coming few decades (Fig. 23-1) are expected to challenge the region's avifauna. While the overall diversity of birds is expected only to decline by an average of 3–5 species (Fig. 23-4), the region will be significantly more vulnerable to further loss of forest. Avian diversity in our forested landscape exhibits a slight peak between 40% and 60% forest cover, but is approximately equal at both 40% and 60% forest (Marzluff 2005). Avian diversity declines rapidly and substantially as forest cover is reduced below 40% and bird communities hold fewer and fewer native forest and early successional species (Donnelly and Marzluff 2006). In contrast, avian diversity declines only slightly as forest cover exceeds 60% and bird communities come to be composed of mainly native forest species (Marzluff 2005). Thus, the reduction in overall forest cover from 60% in 2003 to 38% in 2027 that we project is not expected to substantially lower regional avian diversity, but it is expected to threaten the remaining bird communities with rapid and substantial reduction in diversity if loss of forest cover continues beyond 2027.

Forest species increase in landscapes with increasing amounts of forest, recent development, and aggregated (clumped) housing developments because in such landscapes substantial contiguous forests remain. Synanthropic species also increase in landscapes of aggregated development, but where built lands, not forested ones, are extensive. Overall, synanthropic species increase in young developments, but the response of individual species to development age is mixed (e.g., American crow density increases in older developments; Table 23-2).

Farther east, the elevation increases and large patches of contiguous forest likely contain viable populations of native forest birds. By 2027, our projection suggests that the rich 50:50 landscape of the transition zone will be mostly lost and replaced with dense development and low bird diversity. This will then contrast dichotomously with expected diverse bird communities farther east in the forested Cascade Mountain foothills.

In summary, while some of our conclusions may depend on our ability to model native forest and synanthropic species better than early successional species, we expect future bird communities to be slightly less diverse and more vulnerable to future losses than they are at present. We expect native forest birds to become increasingly reliant on higher elevation forests because most low elevation forests will be converted to development too dense to support viable populations. High elevation bird populations may be less sustainable due to harsher winters and shorter growing seasons that may limit survival and reproduction.

Policy and Management Implications of Approach

Planning agencies are increasingly challenged by the need to provide evidence of the social and economic benefits of conservation strategies. The assessment of these benefits against societal costs requires the ability to predict future conditions without conservation and under alternative strategies. The coupling of the three models presented here allows policy makers to explore policy options across the landscape and more effectively assess the implications of different policy choices on land development, land cover, and biodiversity. Agencies charged with regional metropolitan planning (Puget Sound Regional Council [PSRC] in the Seattle, Washington metropolitan area) require such integrated modeling systems to develop scenarios, assess alternative strategies, and make effective planning decisions and investment choices. Natural resource and wildlife conservation agencies (e.g., Washington Department of Natural Resources) and nongovernmental organizations (e.g., The Nature Conservancy) also require tools to assist with planning for future landscape changes. Currently, PSRC is using UrbanSim (P. Waddell, personal communication) to inform land use and transportation planning. In addition, we have been working with the Army Corps of Engineers to explore applications of the LCCM model to simulate what will happen to coastal areas if no large-scale projects to restore the nearshore ecosystem are undertaken. In addition, the outputs of the integrated models can potentially be used to inform a variety of new conservation initiatives. For example, they could effectively serve to assess and prioritize conservation strategies as part of the action plan to restore the Puget Sound ecosystem initiated by the new Washington State Puget Sound Partnership Agency.

While urban development is linked to loss of biodiversity, its impacts are by no means homogeneous across the landscape. Thus, the trajectory of urban development that we choose or encourage will influence biodiversity differently. This study clearly indicates that both the landscape composition and configuration resulting from urbanization has variable effects on the region's biodiversity.

Planning agencies will increasingly need spatially explicit models of coupled human-natural systems to realistically assess the effectiveness of conservation strategies. It is critical to better understand and represent mechanisms of interactions among human stressors, land-cover patterns, and ecosystem functioning before we can devise management strategies and target conservation actions. Such models can become extremely complex and data intensive, requiring a balance between simplicity and realism to make models usable. The simplicity of even our full models can be highly useful to urban planners; however, important aspects of biological reality may not be included in our models. For example, we do not model the response of birds to specific sorts of vegetation, important habitat elements like snags (Blewett and Marzluff 2005), or important aspects of the understory (Donnelly and Marzluff 2006). While these would likely increase realism, and perhaps predictability of avian diversity, they would not necessarily make such models more applicable to city and regional planning. More importantly, by showing how birds generally respond to the amount (% forest, % grass), pattern (housing aggregation), and age of development, we can provide planners with relevant tools to better understand how their decisions concerning zoning, housing density, and designation of conservation areas affect bird communities. Some species and guilds may require measuring more aspects of the landscape than others, but with the increasing ubiquity of spatial data and GIS skills, even local planning offices will increasingly be able to develop such measures.

Avian diversity, while generally less affected by the pattern than the amount of development (Alberti and Marzluff 2004), was sensitive to aspects of both in our current modeling effort (Table 23-1). The importance of residential aggregation to total diversity and the diversity of each guild suggests that a variety of birds in our region will benefit from future development that is aggregated rather than dispersed. Aggregating future residential development may be the only way to provide a variety of forested reserves and variously built areas that together maximize the region's bird diversity (Donnelly and Marzluff 2004*a*, 2006; Blewett and Marzluff 2005). However, planners should be especially concerned with our projected losses in the transition zone with its diverse land cover and rich bird communities. From an avian perspective, aggregating development in this zone and especially as it extends further east should be done with an eye toward increasing local diversity of land cover rather than simply using this area to accommodate dense development and limit the eastern extent of sprawl. The creation of locally diverse landscapes, while not beneficial to some native forest species and indicative of low density exurban sprawl, is beneficial to many early successional bird species, and therefore regional avian diversity (Marzluff 2005).

Maintaining a diverse bird community in the face of our projected wave of forest conversion will require more than regional and city planning. Individual landowners also need to be engaged in reducing the loss of native forest and early successional species on their properties (Marzluff and Ewing 2001; Donnelly and Marzluff 2004*a*, 2006; Blewett and Marzluff 2005). Projections of future conditions, such as we have developed, can catalyze the action of multiple stakeholders. Landowners alarmed by our projections can help conserve native birds in developed areas by maintaining native forest conditions (secluded forest tracts with native understories), large snags, and a variety of forest ages and types on their properties. Communities can reduce the loss of native birds by controlling their supplements (e.g., food waste eaten by generalist predators such as opossums and raccoons), keeping pets restrained when in forested areas, and fostering a functional ecosystem. Fostering ecosystem functionality can be messy and unappealing in urbanizing regions, but it is an important complement to land-use and land-cover planning efforts. This requires less manicured yards with minimal grass, letting trees die and rot, enabling native predators like coyotes to live in our neighborhoods, and seeing the good that comes from natural disturbances like wind (forest openings), floods (soil regeneration), and insect outbreaks (creating snags and feeding opportunities for insectivores). Tending to our lifestyle as well as to land cover will be increasingly important to future bird diversity if we intend to minimize the impacts we expect in the next 25 years.

Limitations of Approach

Our approach suffers from many potential drawbacks. First, the number of data layers required for each model is large and varied and, in the case of avian models, requires substantial field work. Data development was a large portion of the activities of each modeling team participating in this project. Ideally, input from the community of developers, land owners, and planners to identify those forces driving landscape change will improve both the models' performance and their potential future use. In addition, knowledge of the availability and reliability of spatial data will help determine what variables can easily be incorporated in any modeling exercise.

Because the LCCM does not model forest regrowth into mature forest, predictions of total and native forest species may have underestimated species richness in the commercial forestlands east of Seattle. The LCCM was designed primarily as an urban growth model, and as such does not capture the production forest landscape change as accurately as a model designed specifically for such lands. In addition, our avian field data are derived from the lower elevations, so we cannot be positive how these species will adapt to higher elevation habitat that may already have a full compliment of species occupying the habitat.

We have the greatest confidence in our predictions concerning native forest and synanthropic species. Our models account for substantial proportions of the variation in the diversity (Table 23-1) and abundance (Table 23-2) of these guilds. Their response to the processes of forest conversion is obvious to even the casual observer and captured in our regression equations by considering the amount of forest along with the pattern and age of development. Differences between predictions for the synanthropic guild for simple and full models indicates the importance of including only those landscape variables where we have an understanding of the mechanisms that relate landscape configuration to species responses. Our synanthropic guild predictions using the full model clearly contained errors with species richness predicted to decline in the transition zone where developed land cover was predicted to increase. The full models included more variables that measure landscape configuration to which species may have individualistic responses that are masked by guild-level models

(Taper et al. 1995). In addition, there may be errors in the form of the regression equation with respect to the functional response of the species making up the guild. To address these errors, an exploratory analysis of the exact relationship between landscape metrics and guild responses should be conducted. Poisson regression could be used instead of linear regression to account for the small number of counts for this guild and preclude prediction of negative species richness or relative abundance (e.g., Thogmartin et al. 2007).

Our predictions concerning early successional species diversity may be problematic. Understanding this group is vital to understanding the overall response of birds to urbanization in our region (Marzluff 2005) because half of the total bird diversity is included in this group (Appendix A). We may improve these models by relating diversity directly to light urban development (rather than grouping all urban classes) or by better accounting for the dynamics of regenerating forests in our land-cover change models. However, it is likely that the diversity of bird species in this guild will continue to defy unified modeling efforts. Rather, we should look for common responses to land-cover change by subsets of the early successional guild as models of individual species appear more robust (Table 23-2) than models of the full guild (Table 23-1). Improving our ability to model early successional species is an important future direction.

FUTURE DIRECTIONS

We have presented here a first step in linking landscape models of urban development, landscape change, and ecological responses. Work continues on increasing the accuracy, flexibility, and applicability of each model. For instance, the LCC accuracy for wildland areas can be increased by using readily available forest growth models and modeling systems (e.g., McCarter et al. 1998, Marzluff et al. 2002). Extrapolation of the avian biodiversity models to these higher elevations would require substantial additional field data to understand how species in these areas respond to changes in the landscape. Also, coupling the land-cover change model with biophysical process models (e.g., climate) is critical when aiming at predicting land-cover change and related avian diversity on a longer time scale.

Modeling total species richness as a single response to changing landscape conditions masks individual species' responses. Ideally, the summary of predictions for individual species abundance would equal the predicted total species richness; however, this is rarely the case due to confounding factors such as differential species detectability (Dorazio et al. 2006). In general, our models of single species relative abundance are better at capturing the diversity of observed responses than guild predictions of species richness (Table 23-1 versus Table 23-2). More realistic estimates of total and guild species richness could be obtained through developing mechanistic relationships of relative abundance for each species and using individual species predictions to build community-level responses to landscape change. Combined individual species predictions could then be compared against single guild or total species richness predictions to evaluate the validity of each approach. Avian models of relative abundance in their current form have not been corrected for the effects of species detectability, which is something that would be necessary to develop valid summed predictions of individual species richness to predict changes in community richness (A. Royle, personal communication). Additional exploration into and inclusion of the mechanisms (e.g., dispersal, territory size, source-sink dynamics) behind observed bird population responses to landscape change would improve individual and community predictions. In addition, validating avian models has not yet occurred. Data from two additional field seasons will soon be available to use as comparisons with our current and future model's predictions.

Such complex modeling steps we have done for our avian model would benefit from the development of an automated modeling system, such as UrbanSim and LCCM, where the entire process of equation development to prediction of future landscapes and community richness would be present in a single model development environment. Continued development of an open-source software development environment to allow seamless simulation of urban development, land-cover change, and ecological responses will increase the applicability of such integrated models. Research is needed to understand how these different systems influence each other and how to incorporate such feedback into a spatially explicit modeling system. Moreover, individual- or agent-based models will be critical to realistically represent human-nature feedbacks. Furthermore, due to the inherent limitations in predicting the behavior of coupled human-natural systems, future research will need to effectively link predictive models with scenario planning (Millennium Ecosystem Assessment 2003, Shearer 2005). The more flexible, adaptable, and transparent the model systems, the more likely they will be useful to planners and policy makers.

SUMMARY

We presented a unified modeling approach to predict urban development, land-cover change, and ecosystem response to landscape change. We focused on predicting the effects of future landscape change on avian communities as a case example of models that produce results useful to conservation planning across large landscapes. The Central Puget Sound of western Washington State, USA, is a 3,200 km^2 area undergoing significant urban development and resulting landscape change. We used a microeconomic development model of human behavior, UrbanSim, to predict land-use change. The land-cover change model incorporates output from UrbanSim, existing land cover, and biophysical attributes to predict land-cover change every four years, 28 years into the future. Land-cover and land-use predictions are input into models of avian species richness and relative abundance developed from five years of field studies across an urban to wildland gradient. We found that avian diversity was sensitive to both the amount and pattern of land cover. The amount of forest was a key determinant of species richness and abundance of native forest birds. Additionally, aggregation of residential development was important for total species diversity and the diversity of three habitat guilds modeled, suggesting that a variety of birds in our region will benefit from aggregating future development. The richness of future bird communities will increase gradually with distance from development. As the transition zone between landscapes dominated by human development and wildland areas is transformed into dense development, the region is likely to be composed of spatially partitioned bird communities dominated by either adaptable, synanthropic species (in dense developments) or resilient native forest birds (in the wildland zone). We expect native forest birds to become increasingly reliant on higher elevation forests because most low elevation forests will be converted to development too dense to support viable populations. With increased development, the location of new development in regards to existing and proposed conservation networks will need to be considered. Conservation and planning agencies can use our models to evaluate proposed policies and conservation strategies.

ACKNOWLEDGMENTS

We thank Roarke Donnelly, Tina Blewett, Cara Ianni, Kara Whittaker, Jack DeLap, Stan Rullman, Thomas Unfried, and Dave Oleyar for their help collecting data. We thank Stefan Coe for development of the land-cover maps used in the LCC and the staff at the Center for Urban Simulation and Policy Analysis at The University of Washington. Many private landowners graciously gave us their permission to conduct research on their property. This research was supported by the University of Washington (Tools for Transformation Fund), the National Science Foundation (DEB-9875041, BCS 0120024, IGERT-0114351), and the University of Washington's College of Forest Resources, particularly its Rachel Wood's Endowed Graduate Program and the Denman Sustainable Resource Sciences Professorship.

LITERATURE CITED

Ahern, J. 1999. Spatial concepts, planning strategies, and future scenarios: A framework method for integrating landscape ecology and landscape planning. Pages 175–201 *in* J. M. Klopatek and R. H. Gardner, editors. *Landscape ecological analysis: Issues and applications.* Springer-Verlag, New York, New York, USA.

Alberti, M., and J. M. Marzluff. 2004. Ecological resilience in urban ecosystems: Linking urban patterns to human and ecological functions. *Urban Ecosystems* 7:241–265.

Alberti, M., J. M. Marzluff, E. Shulenberger, G. Bradley, C. Ryan, and C. ZumBrunnen. 2003. Integrating humans into ecology: Opportunities and challenges for studying urban ecosystems. *BioScience* 53:1169–1179.

Alberti, M., R. Weeks, and S. Coe. 2004. Urban land-cover change analysis in central Puget Sound: 1991–1999. *Photogrammetric Engineering and Remote Sensing* 70:1043–1052.

Blewett, C. M., and J. M. Marzluff. 2005. Effects of urban sprawl on snags and the abundance and productivity of cavity-nesting birds. *Condor* 107:677–693.

Doak, D. F., and L. S. Mills. 1994. A useful role for theory in conservation. *Ecology* 75:615–626.

Donnelly, R., and J. M. Marzluff. 2004*a*. Importance of reserve size and landscape context to urban bird conservation. *Conservation Biology* 18:733–745.

Donnelly, R., and J. M. Marzluff. 2004*b*. Designing research to advance the management of birds in urbanizing areas. Pages 114–122 *in* W. W. Shaw, L. K. Harris, and L. Vandruff, editors. *Proceedings of the 4th International Symposium on Urban Wildlife Conservation.* University of Arizona Press, Tucson, USA.

Donnelly, R., and J. M. Marzluff. 2006. Relative importance of habitat quantity, structure, and spatial pattern to birds in urbanizing environments. *Urban Ecosystems* 9:99–117.

Dorazio, R. M., J. A. Royle, B. Söderström, A. Glimskär. 2006. Estimating species richness and accumulation by modeling species occurrence and detectability. *Ecology* 87:842–854.

Dunning, J. B., Jr., D. J. Stewart, B. J. Danielson, B. R. Noon, T. L. Root, R. H. Lamberson, and E. E. Stevens. 1995. Spatially explicit population models: Current forms and future uses. *Ecological Applications* 5:3–11.

Elvidge, C. D., K. E. Baugh, E. A. Kihn, H. W. Kroehl, and E. R. Davis. 1997. Mapping of city lights using DMSP Operational Linescan System. *Photogrammetric Engineering and Remote Sensing* 63:727–734.

Ewing, R. H. 1994. Characteristics, causes, and effects of sprawl: A literature review. *Environmental and Urban Studies* 21:1–17.

Fahrig, L., and G. Merriam. 1994. Conservation of fragmented populations. *Conservation Biology* 8:50–59.

Foley, J. A., R. DeFries, G. P. Asner, C. Barford, G. Bonan, S. R. Carpenter, F. S. Chapin, M. T. Coe, G. C. Daily, H. K. Gibbs, J. H. Helkowski, T. Holloway, E. A. Howard, C. J. Kucharik, C. Monfreda, J. A. Patz, I. C. Prentice, N. Ramankutty, and P. K. Snyder. 2005. Global consequences of land use. *Science* 309:570–574.

Franklin, J. F., and C. T. Dyrness. 1988. *Natural vegetation of Oregon and Washington.* Oregon State University Press, Corvallis, USA.

Grimm, N. B., J. M. Grove, S. T. A. Pickett, and C. L. Redman. 2000. Integrated approaches to long-term studies of urban ecological systems. *BioScience* 50:571–584.

Hansen, A. J., R. L. Knight, J. M. Marzluff, S. Powell, K. Brown, P. H. Gude, and K. Jones. 2005. Effects of exurban development on biodiversity: Patterns, mechanisms, and research needs. *Ecological Applications* 15:1893–1905.

Hepinstall, J. A., M. Alberti, and J. M. Marzluff. Predicting land cover change and avian community responses in rapidly urbanizing environments. *Landscape Ecology: In press.*

Houghton, R. A. 1994. The worldwide extend of land-use change. *BioScience* 44:305–313.

Hulse, D. W., A. Branscomb, and S. G. Payne. 2004. Envisioning alternatives: Using citizen guidance to map future land and water use. *Ecological Applications* 14:325–341.

Imhoff, M. L., L. Bounoua, R. DeFries, W. T. Lawrence, D. Stutzer, C. J. Tucker, and T. Ricketts. 2004. The consequences of urban land transformation on net primary productivity in the United States. *Remote Sensing of Environment* 89:434–443.

Kaye, J. P., P. M. Groffman, N. B. Grimm, L. A. Baker, and R. V. Pouyat. 2006. A distinct urban biogeochemistry? *Trends in Ecology and Evolution* 21:192–199.

Lambin, E. F., B. L. Turner, H. J. Geist, S. B. Agbola, A. Angelsen, J. W. Bruce, O. T. Coomes, R. Dirzo, G. Fischer, C. Folke, P. S. George, K. Homewood, J. Imbernon, R. Leemans, X. Li, E. F. Moran, M. Mortimore, P. S. Ramakrishnan, J. F. Richards, H. Skanes, W. Steffen, G. D. Stone, U. Svedin,

T. A. Veldkamp, C. Vogel, and J. Xu. 2001. The causes of land-use and land-cover change: Moving beyond the myths. *Global Environmental Change* 11:261–269.

Lawrence, W. T., M. L. Imhoff, N. Kerle, and D. Stutzer. 2002. Quantifying urban land use and impact on soils in Egypt using diurnal satellite imagery of the Earth surface. *International Journal of Remote Sensing* 23:3921–3937.

Marzluff, J. M. 2001. Worldwide urbanization and its affects on birds. Pages 19–47 *in* J. M. Marzluff, R. Bowman, and R. Donnelly, editors. *Avian ecology and conservation in an urbanizing world.* Kluwer Academic, Norwell, Massachusetts, USA.

Marzluff, J. M. 2005. Island biogeography for an urbanizing world: How extinction and colonization may determine biological diversity in human-dominated landscapes. *Urban Ecosystems* 8:157–177.

Marzluff, J. M., and K. Ewing. 2001. Restoration of fragmented landscapes for the conservation of birds: A general framework and specific recommendations for urbanizing landscapes. *Restoration Ecology* 9:280–292.

Marzluff, J. M., J. J. Millspaugh, K. R. Ceder, C. D. Oliver, J. Withey, J. B. McCarter, C. L. Mason, and J. Comnick. 2002. Modeling changes in wildlife habitat and timber revenues in response to forest management. *Forest Science* 48:191–202.

Matlack, G. R. 1993. Sociological edge effects: Spatial distribution of human impact in suburban forest fragments. *Environmental Management* 17:829–835.

McCarter, J. B., J. S. Wilson, P. J. Baker, J. L. Moffett, and C. D. Oliver. 1998. Landscape management through integration of existing tools and emerging technologies. *Journal of Forestry* 96:17–23.

McGarigal, K., S. A. Cushman, M. C. Neel, and E. Ene. 2002. *FRAGSTATS: Spatial pattern analysis program for categorical maps*. University of Massachusetts, Amherst, USA.

McGarigal, K., and W. C. McComb. 1995. Relationships between landscape structure and breeding birds in the Oregon Coast Range. *Ecological Monographs* 65:235–260.

McKinney, M. L. 2002. Urbanization, biodiversity, and conservation. *BioScience* 52:883–890.

Meyer, W. B., and B. L. Turner II. 1992. Human population growth and global land-use/cover change. *Annual Review of Ecology and Systematics* 23:39–61.

Millennium Ecosystem Assessment. 2003. *Ecosystems and human well-being*. Island Press, Washington D.C., USA.

Mörtberg, U. M., B. Balfors, and W. C. Knol. 2007. Landscape ecological assessment: A tool for integrating biodiversity issues in strategic environmental assessment and planning. *Environmental Management* 82:457–470.

Noth, M., A. Borning, and P. Waddell. 2003. An extensible, modular architecture for simulating urban development, transportation, and environmental impacts. *Computers, Environment and Urban Systems* 27:181–203.

Parker, D. C., S. M. Manson, M. A. Janssen, M. J. Hoffmann, and P. Deadman. 2003. Multi-agent systems for the simulation of land-use and land-cover change: A review. *Annals of the Association of American Geographers* 92:314–337.

Pearson, S. M., M. G. Turner, and J. B. Drake. 1999. Landscape change and habitat availability in the Southern Appalachian Highlands and Olympic Peninsula. *Ecological Applications* 9:1288–1304.

Pickett, S. T. A., M. L. Cadenasso, J. M. Grove, C. H. Nilon, R. V. Pouyat, W. C. Zipperer, and R. Costanza. 2001. Urban ecological systems: Linking terrestrial ecological, physical, and socioeconomic components of metropolitan areas. *Annual Review of Ecology and Systematics* 32:127–157.

Prato, T. 2005. Modeling ecological impacts of landscape change. *Environmental Modeling and Software* 20:1359–1363.

Riitters, K. H., R. V. Oneill, and K. B. Jones. 1997. Assessing habitat suitability at multiple scales: A landscape-level approach. *Biological Conservation* 81:191–202.

Robinson, L., J. P. Newell, and J. M. Marzluff. 2005. Twenty-five years of sprawl in the Seattle region: Growth management responses and implications for conservation. *Landscape and Urban Planning* 71:51–72.

Robinson, S. K., F. R. Thompson, III, T. M. Donovan, D. R. Whitehead, and J. Faaborg. 1995. Regional forest fragmentation and the nesting success of migratory birds. *Science* 267:1987–1990.

Ruckelshaus, M., C. Hartway, and P. Kareiva. 1997. Assessing the data requirements of spatially explicit dispersal models. *Conservation Biology* 11:1298–1306.

Sadik, N. 1999. *The state of world population 1999-6 billion: A time for choices*. United Nations Population Fund, New York, USA.

Sala, O. E., F. S. Chapin, III, J. J. Armesto, E. Berlow, J. Bloomfield, R. Dirzo, E. Huber-Sanwald, L. F. Huenneke, R. B. Jackson, A. Kinzig, R. Leemans, D. M. Lodge, H. A. Mooney, M. Oesterheld, N. L. Poff, M. T. Sykes, B. H. Walker, M. Walker, and D. H. Hall. 2000. Global biodiversity scenarios for the year 2100. *Science* 287:1770–1774.

Schumaker, N. H., T. Ernst, D. White, J. Baker, and P. Haggerty. 2004. Projecting wildlife responses to alternative future landscapes in Oregon's Willamette Basin. *Ecological Applications* 14:381–400.

Scott, J. M., F. Davis, B. Csuti, R. Noss, B. Butterfield, C. Groves, H. Anderson, S. Caicco, F. D'Erchia, T. C. Edwards, Jr., J. Ulliman, and R. G. Wright. 1993. Gap analysis: A geographic approach to protection of biological diversity. *Wildlife Monographs* 123:1–41.

Shearer, A. W. 2005. Approaching scenario-based studies: Three perceptions about the future and considerations for landscape planning. *Environment and Planning B* 32:67–87.

Steinitz, C. 1990. A framework for theory applicable to the education of landscape architects (and other environmental design professionals). *Landscape Journal* 9:136–143.

Steinitz, C., and S. McDowell. 2001. Alternative futures for Monroe County, Pennsylvania: A case study in applying ecological principles. Pages 165–193 *in* V. H. Dale and R. Haeuber, editors. *Applying ecological principles to land management.* Springer-Verlag, New York, New York, USA.

Tang, Z., B. A. Engel, B. C. Pijanowski, and K. J. Lim. 2005. Forecasting land use change and its environmental impact at a watershed scale. *Journal of Environmental Management* 76:35–45.

Taper, M. L., K. Böhning-Gaese, and J. H. Brown. 1995. Individualistic responses of bird species to environmental change. *Oecologia* 101:478–486.

Thogmartin, W. E., J. R. Sauer, and M. G. Knutson. 2007. Modeling and mapping abundance of American woodcock across the Midwestern and Northeastern United States. *Journal of Wildlife Management* 71:376–382.

Tittler, R., L. Fahrig, and M.-A. Villard. 2006. Evidence of large-scale source-sink dynamics and long-distance dispersal among wood thrush populations. *Ecology* 87:3029–3036.

Turner, M. G., S. M. Pearson, P. Bolstad, and D. N. Wear. 2003. Effects of land-cover change on spatial pattern of forest communities in the Southern Appalachian Mountains (USA). *Landscape Ecology* 18:449–464.

Turner, M. G., D. N. Wear, and R. O. Flamm. 1996. Land ownership and land-cover change in the southern Appalachian highlands and the Olympic peninsula. *Ecological Applications* 6:1150–1172.

Villard, M.-A., M. K. Trzcinski, and G. Merriam. 1999. Fragmentation effects on forest birds: Relative influence of woodland cover and configuration on landscape occupancy. *Conservation Biology* 13:774–783.

Vitousek, P. M., H. A. Mooney, J. Lubchenco, and J. M. Melillo. 1997. Human domination of Earth's ecosystems. *Science* 277:494–499.

Waddell, P. 2002. UrbanSim: Modeling urban development for land use, transportation and environmental planning. *Journal of the American Planning Association* 68:297–314.

Waddell, P., A. Borning, M. Noth, N. Freier, M. Becke, and G. Ulfarsson. 2003. UrbanSim: A simulation system for land use and transportation. *Networks and Spatial Economics* 3:43–67.

Wear, D. N., and P. Bolstad. 1998. Land-use changes in southern Appalachian landscapes: Spatial analysis and forecast evaluation. *Ecosystems* 1:575–594.

Wear, D. N., M. G. Turner, and R. J. Naiman. 1998. Land cover along an urban-rural gradient: Implications for water quality. *Ecological Applications* 8:619–630.

White, D., P. G. Minotti, M. J. Barczak, J. C. Sifneos, K. E. Freemark, M. V. Santelmann, C. F. Steinitz, A. R. Kiester, and E. M. Preston. 1997. Assessing risks to biodiversity from future landscape change. *Conservation Biology* 11:349–360.

Appendix A List of Common Bird Species by Guild Membership, as Detected in Point Count Surveys of 139 Study Sites in Puget Sound, Washington, 1998–2005

Common name	Genus species	Common name	Genus species
Native Forest		Early Successional	
American robin	*Turdus migratorius*	American goldfinch	*Carduelis tristis*
Black-throated gray warbler	*Dendroica nigrescens*	Band-tailed pigeon	*Columba fasciata*
Brown creeper	*Certhia americana*	Bewick's wren	*Thryomanes bewickii*
Chestnut-backed chickadee	*Poecile rufescens*	Black-capped chickadee	*Poecile atricapillus*
Dark-eyed junco	*Junco hyemalis*	Black-headed grosbeak	*Pheucticus melanocephalus*
Downy woodpecker	*Picoides pubescens*	Bushtit	*Psaltriparus minimus*
Golden-crowned kinglet	*Regulus satrapa*	Cassin's vireo	*Vireo cassinii*
Hairy woodpecker	*Picoides villosus*	Cedar waxwing	*Bombycilla cedrorum*
Hammond's flycatcher	*Empidonax hammondii*	Common yellowthroat	*Geothlypis trichas*
Hermit thrush	*Catharus guttatus*	Killdeer	*Charadrius vociferus*
Hutton's vireo	*Vireo huttoni*	MacGillivray's warbler	*Oporornis tolmiei*
Pacific-slope flycatcher	*Empidonax difficilis*	Northern flicker	*Colaptes auratus*
Red-breasted nuthatch	*Sitta canadensis*	Northern Rough-winged swallow	*Stelgidopteryx serripennis*
Spotted towhee	*Pipilo maculatus*	Olive-sided flycatcher	*Contopus cooperi*
Steller's jay	*Cyanocitta stelleri*	Orange-crowned warbler	*Vermivora celata*
Swainson's thrush	*Catharus ustulatus*	Pine siskin	*Carduelis pinus*

continues

Appendix A List of Common Bird Species by Guild Membership, as Detected in Point Count Surveys of 139 Study Sites in Puget Sound, Washington, 1998–2005 *cont...*

Common name	Genus species	Common name	Genus species
Western tanager	*Piranga ludoviciana*	Purple finch	*Carpodacus purpureus*
Wilson's warbler	*Wilsonia pusilla*	Red crossbill	*Loxia curvirostra*
Winter wren	*Troglodytes troglodytes*	Red-winged blackbird	*Agelaius phoeniceus*
Synanthropic		Rufous hummingbird	*Selasphorus rufus*
American crow	*Corvus brachyrhynchos*	Savannah sparrow	*Passerculus sandwichensis*
Anna's hummingbird	*Calypte anna*	Song sparrow	*Melospize melodia*
Barn swallow	*Hirundo rustica*	Tree swallow	*Tachycineta bicolor*
Brewer's blackbird	*Euphagus cyanocephalus*	Violet-green swallow	*Tachycineta thalassina*
Brown-headed cowbird	*Molothrus ater*	Warbling vireo	*Vireo gilvus*
European starling	*Sturnus vulgaris*	Western wood pewee	*Contopus sordidulus*
House finch	*Carpodacus mexicanus*	White-crowned sparrow	*Zonotrichia leucophrys*
House sparrow	*Passer domesticus*	Willow flycatcher	*Empidonax traillii*
Rock pigeon	*Columba livia*	Yellow-rumped warbler	*Dendroica coronata*

CHAPTER

A Decision Framework for Choosing Models in Large-Scale Wildlife Conservation Planning

24

Frank R. Thompson, III and Joshua J. Millspaugh

This volume presents principles, concepts, methods, and examples of several modeling approaches suitable for planning wildlife conservation in large landscapes. Many approaches are rooted in ecological theory (e.g., Rowland and Wisdom, this volume), with different data needs, assumptions, and methodologies, and have been applied successfully in a diversity of environments. Approaches range from fine filter, single species approaches to multiple species or coarse filter, ecosystem approaches. The diversity of approaches may seem overwhelming to land managers or planners who must select among these options for their project. In this chapter, we present a decision framework (Fig. 24-1) that represents the range of options for planning wildlife conservation considered in this book. The context for this framework, and for the book, is that the reader is interested in the conservation of wildlife species; we do not consider coarse filter approaches focused solely on higher ecological levels without consideration of species. As with any science-based planning effort, careful upfront consideration of objectives and the level of investment you can make will make subsequent methodological questions much easier to address. The topics here are generally arranged from those related to the goals of the project and refining objectives, to model development and application, to model or project evaluation, and conclude with some miscellaneous issues we believe deserve consideration. We use the framework to step through two case studies (Fig. 24-1).

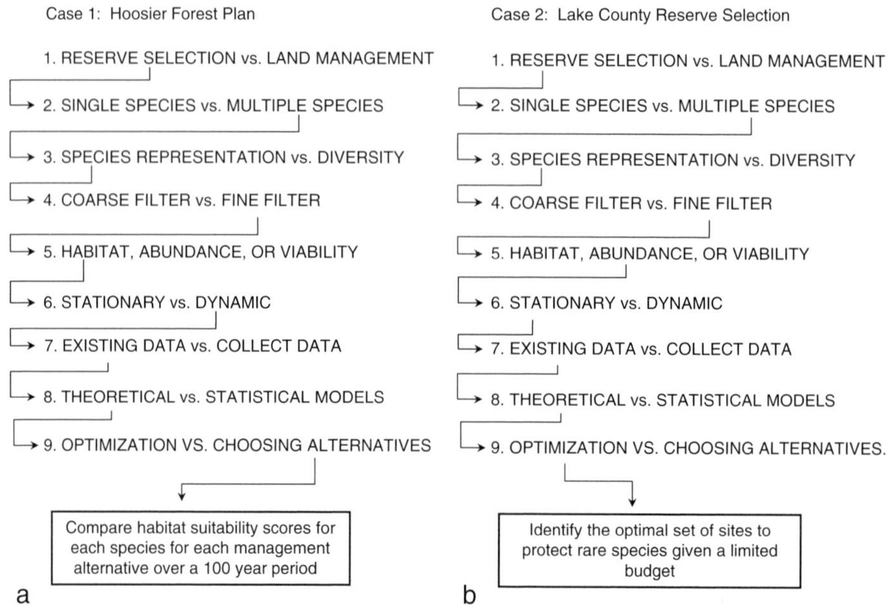

FIG. 24-1

Examples of decision pathways for two different conservation planning problems. Case 1 (A) is part of the Hoosier National Forest Plan that considered habitat suitability for focal species under five plan alternatives. Case 2 (B) is a reserve selection problem in which Lake County Illinois planners identified the optimal set of sites to protect rare plants and animals given a fixed budget. See text for details.

RESERVE SELECTION VERSUS LAND MANAGEMENT PLANNING

The goal of reserve selection is to select a geographic area that best addresses conservation objectives, while the goal of land management planning is typically to select land management practices that best meet conservation objectives for a defined geographic area. The boundary between these methods becomes blurred when the definition of reserves is loosened to include multiple-use lands for which conservation is one of many objectives, and reserve selection methods are used to select lands, perhaps already within public ownership, for different levels of protection. Both reserve selection and land management planning can address single or multiple species. Multiple species problems typically consider habitat needs of as many species as possible, or a subset of high-concern species (Flather et al.; Noon et al., this volume). Single species problems are more likely to focus on abundance and can use gap approaches to see if important areas (i.e., areas of high abundance) are covered (i.e., Fig. 22-6, Fitzgerald et al., this volume). Mathematical optimization procedures can be used to maximize benefits across several objectives (Haight and Gobster, this volume), or simple graphical or map-based assessments can be made, for example, by comparing maps of current protected areas to maps of species abundance.

SINGLE SPECIES VERSUS MULTIPLE SPECIES

The decision regarding single species versus multiple species is primarily a function of project objectives. Some projects may be focused on a single species, such as the recovery plan for an endangered species. Most land management decisions, however, involve multiple species. Reserve selection problems are generally multispecies approaches that focus on species diversity or maximizing coverage of a set of high-priority species. In land management planning, however, multispecies approaches usually involve reapplying single species models. For example, Habitat Suitability Index (HSI) models have been used to evaluate the impact of management alternatives on 10 wildlife species for the Hoosier National Forest (Dijak and Rittenhouse, this volume) or to develop bird conservation plans covering 40 species for multistate bird conservation regions (Fitzgerald et al., this volume). We are not aware of any land management planning exercise that evaluated habitat quality or animal abundance for multiple species in a single, multivariate model. The reason is that predictive statistical models would likely be too complicated, or be unrealistically simplistic, to address the diverse responses of multiple species simultaneously. Some software automates predictions for multiple species by implementing a number of single species models (Beck and Suring, this volume). An alternative to multiple single species models is a coarse filter approach or a focus on species diversity, both of which require additional assumptions and issues that we discuss in the following sections.

COARSE FILTER VERSUS FINE FILTER

Many conservation needs can be met through coarse filter approaches that focus on ecosystem representation and diversity. Coarse filters can be used where ecosystems or other high-level ecological units are the focus of conservation efforts, or where the focus is on species conservation and ecosystems are surrogates that address the needs of single or multiple species (Haufler and

Kernohan, this volume; Hicks et al., this volume; Noon et al., this volume), which is our approach here. For a single species, planners would select ecosystems or habitats as surrogates for a species abundance or viability. This approach is extended to multiple species by repeating the process for groups of species. Most projects, however, will involve the use of fine filters (i.e., species-level modeling), at some stage, whether as the primary planning approach (Fitzgerald et al., this volume), or as a component for selected high-profile species (Hicks et al., this volume), or to validate assumptions about the adequacy of coarse filters to meet species viability needs. At a superficial level, coarse filter approaches may appear like a less complex planning approach, but they rely on many untested assumptions such as the adequacy of vegetation type and structure to represent ecosystems or other ecological units (Noon et al., this volume). Multispecies approaches using a fine filter approach require models for all species of interest or a set focal species.

An alternative to the coarse filter/fine filter dichotomy is a multiscale approach that takes a stepwise approximation approach working from broad scales to successively finer scales (Probst and Gustafson, this volume). This species-based approach begins at the broadest scale in the problem with simple distribution data and then increases resolution by examining habitat gradients and occurrence information, then productivity and survival. Information is synthesized across scales for the problem and key assumptions addressed through monitoring (Probst and Gustafson, this volume).

So whether through a combination of coarse and fine filters, a stepwise multiscale approach, or a fine filter species-level approach, at some point species-level models will be required either as the primary approach or to validate assumptions made in coarser-scale analyses. Hence, much of our focus in this volume and the remaining discussion in this chapter relates to species-level approaches.

SPECIES REPRESENTATION VERSUS DIVERSITY

A focus on species diversity is one potential solution to the trade-offs involved between fine filter/species approaches versus coarse filter/ecosystem approaches. Perhaps the most popular measure of species diversity is species richness, a simple count of the number of species. However, species richness is generally not a suitable metric for conservation planning because it is generally driven by common, widespread species, which are not the species most conservation planning efforts are intended to address (Brooks et al. 2006). Furthermore, maximizing species richness at local scales can reduce species richness at larger scales (Noss 1987). So, simple diversity objectives are generally abandoned in favor of those that address species representation; we generally want all species (or other components of biodiversity) adequately protected in conservation plans (Sarkar et al. 2006). One way to address species representation in reserve design is to ensure that some target set of species pool members, typically species considered at risk or a high conservation concern, are adequately represented in the conservation plan (Flather et al., this volume). Additionally, reserves can consider those that complement, as opposed to those redundant with, species covered in existing reserves or other areas being considered (Flather et al., this volume). In land management planning, species representation is often addressed by focusing on high-concern species or selecting indicator species that serve as surrogates for a larger group of species. For example, bird conservation under the Partner's in Flight plan in North America focuses on a set of priority species (Fitzgerald et al., this volume). Land management planning on national forests in the United States has focused on threatened, endangered, and sensitive species and management indicator species, but is shifting to greater use of coarse filter approaches (Noon et al., this volume). Species representation problems can address presence/absence, abundance, or viability.

HABITAT, ABUNDANCE, OR VIABILITY

We and others in this volume have discussed the need to focus on viability (Akçakaya and Brooks, this volume; Beissinger et al., this volume; Bekessy et al., this volume; Millspaugh et al., this volume) or setting population goals to meet desired levels of viability (Fitzgerald et al., this volume; Johnson et al., this volume). Indeed, persistence is one of the key tenets to biodiversity conservation in addition to representation (Sarkar et al. 2006). However, there will be many applications where habitat will serve as a surrogate for populations and the amount of habitat or population size as a surrogate for viability. Projects dependent on existing knowledge such as the scientific literature and expert opinion will be more amenable to simpler models (i.e., habitat matrix or habitat suitability models) that predict habitat quality and not abundance. However, because of their simplicity and availability of data, habit models can be used for more species than population or viability models. Predicting relative or absolute abundance usually involves fitting a statistical model to a suitable data set and using it to predict the response variable of interest, often in the form of continuous surface maps (Fitzgerald et al., this volume; Niemuth et al., this volume). These approaches require all the usual assumptions of statistical models, and though often ignored, should consider issues of detection probabilities and potential biases if ignored. See related discussion later on conceptual versus statistical models.

The majority of projects have an ultimate goal of ensuring species viability, but default to modeling approaches that consider only habitat or abundance because of real or perceived project limitations. A lack of data about population vital rates and the impact of environmental factors on those rates often precipitates the use of habitat or abundance models. Most wildlife research and monitoring activities have focused on habitat use and abundance, in part because data to estimate productivity and survival can be more difficult to collect. We suggest projects that have an ultimate goal of ensuring species viability, but that take a habitat or abundance approach because of the preceding constraints, should attempt to develop population viability models based on their best understanding of population parameters to validate their approach for select species. Available software greatly facilitates the development of such models (Akçakaya and Brook, this volume; He, this volume; Roloff et al., this volume) and the experience of parameterizing a model and conducting simple sensitivity analyses can identify important knowledge gaps and areas of uncertainty related to the assumptions and data used in habitat or abundance-based models.

PROCEED WITH EXISTING DATA OR COLLECT NEW DATA

Often, it is tempting to postpone modeling until more and better data are available. However, we encourage moving forward with modeling efforts despite a lack of complete information (Millspaugh et al., this volume). Simple models without a full complement of data can be developed, which still allow for important evaluations, such as sensitivity analyses that can be used to guide future data collection. When data collection and needs are considered within an adaptive management framework, alternative models and key assumptions can be evaluated.

For many species, existing literature about habitat relationships can be derived from the literature. Such information can form the basis for preliminary model development and evaluation. Thus, one can often proceed with model development and application despite a lack of site-specific data. Subsequent investigation can evaluate the validity of those existing data and assumed relationships. At the very least, literature-derived estimates and relationships offer some insight into possible factors of importance and can help guide study objectives and experimental design (e.g., determination of sample size requirements). Without model validation, though, literature-derived models should be used with some suspicion.

Whereas general habitat requirements might be derived from the literature, models requiring vital rates might be more difficult to parameterize. Because habitat studies are more common and generally applicable across a species range, there might be less danger in applying simple habitat models versus viability models, which are more data hungry. Also, vital rates are often more likely to differ across a species range when one considers population processes that are site and population specific (e.g., density dependence, density-independent factors such as weather, predation, habitat fragmentation). Regardless of the approach, we believe it is prudent to make full use of existing data while acknowledging limitations and uncertainty and identifying ways of reducing both. One will never have all the data he wants or desires; however, it is necessary to move forward and make management decisions in a timely manner.

QUALITATIVE OR CONCEPTUAL VERSUS STATISTICAL MODELS

Most planners, managers, and scientists would rather have a model based on good empirical data (i.e., a statistically fit model) than a more qualitative or conceptual model based on existing knowledge in the form of literature and expert opinion. The reality is many more of the latter type models exist, and will be built, than the former. Qualitative or conceptual-based models can be based on a variety of data types (expert opinion, literature, and empirical) and built with whatever knowledge currently exists, which is a strength. So, for example, habitat matrix models have been built for literally thousands of vertebrate species across the northeastern (DeGraaf and Rudis 1986), southern (Hamel 1992), and western (Airola 1988, Fitzgerald et al., this volume; Hepinstall et al., this volume; Hicks et al., this volume) United States. GIS-based habitat suitability models, which add important landscape components, making them spatially explicit, are now available for many species (Larson et al. 2003; Rittenhouse et al. 2007; Tirpak et al. 2008).

The shortcoming of qualitative or conceptual models is that without some type of data-based validation, there is no way to assess model validity. With a statistical model we should at least know how well the model describes the data on which it was built. This assessment of the fit of a statistical model, however, can create a false sense of security and lead to the application of the model outside its true scope of inference, which could result in large prediction errors. Thus, appropriate assessment of a statistical model should consider standard metrics of model fit (e.g., deviance), but also whether the model is appropriate for the site.

Good modeling practices (see Millspaugh et al., this volume) should be used regardless of which approach is taken. We encourage hypothesis-based or mechanistic models because most models in landscape planning will be used for prediction with new data and simple correlative relationships from a single data set may not work well for this purpose. We see convergence between conceptual versus statistical models in current information-theoretic frameworks. For example, suitability indices in HSI models can serve as the basis for candidate models in an information theoretic framework evaluating model support when empirical data become available (Rittenhouse 2008). Also, this approach can guide data collection when applied in an adaptive management framework that includes model evaluation.

STATIONARY VERSUS DYNAMIC

Stationary approaches typically are either focused on current conditions; assume habitats, landscapes, or populations are not going to change significantly over the planning horizon; or assume current landscape conditions are representative of future conditions, just not in a spatially exact way. Dynamic approaches directly address landscape or population change over time. In a statistical model, this usually means fitting a model to a time series and making assumptions about the applicability of current trends to forecasting the future. In simulation modeling it involves parameterizing vital rates of populations or landscape processes to project current conditions into the future.

Others in this volume (Akçakaya and Brook, Bekessy et al., He, Hepinstall et al., Oliver et al., McKenzie et al.) and elsewhere (Akçakaya et al. 2004, Wintle et al. 2005, Pichancourt et al. 2006, Shifley et al. 2006) have demonstrated the benefits of modeling landscape change as part of wildlife conservation planning or viability assessments. A failure to account for succession in understory and overstory, natural disturbances, changes in land use, climate change, or planned management activities can result in inaccurate or biased estimates of habitat suitability, abundance, or viability. We believe dynamic approaches have great utility and will increase in use; however, we suggest users consider the following issues in their application. Dynamic modeling approaches can be a large undertaking and may not fit the time frame of the project or the objectives of the planning process. While some simple approaches undoubtedly exist, problems that consider multiple management alternatives in large landscapes can take years to assemble data, parameterize models, run the models, and compile and interpret output (Shifley et al. 2006). Second, consider how the time series generated by dynamic approaches will be used. The addition of time as an axis to analyses that already consider multiple species and multiple management alternatives may provide too much information. An alternative is to consider a single point in time, for example, at the end of the planning horizon. Dynamic landscape modeling approaches are constrained in spatial extent or require trade-offs between resolution and extent. For very large-scale planning efforts, the coarser resolution required to address large spatial extents may be too coarse to capture important spatial processes affecting the species of interest. These problems require rethinking the important dynamic process at larger spatial scales because given current computing limitations, small-scale processes simply cannot be replicated over larger spatial extents, although features such as understory vegetation growth might be important (McKenzie et al., this volume). Lastly, approaches that directly incorporate dynamic landscape modeling will almost certainly involve heuristic approaches to optimization or more likely "choosing from alternatives" approaches to decision making simply because they are too complex for true optimization approaches to landscape design.

One dynamic process that we have largely ignored in this volume is global change. The reason for this omission is that most operational land-management planning at landscape scales ignores global change, operating under the assumption that impacts resulting from land-use change and resource management practices are of greater concern under typical planning horizons. This assumption, however, may be challenged as new information on the magnitude or time frame of global change is discovered. Global change modeling generally focuses on larger scales than land management planning. Traditionally, global change modeling addressed vegetation and wildlife through modeling the impact of climate change on species distribution through characterizations of species bioclimatic envelopes. In response to critiques that factors other than climate change affect species distribution, however, models have become more complex and include factors such as biotic interactions, dispersal, and disturbance (Pearson et al. 2003, Beaumont et al. 2007). For example, a frame-based spatially explicit model (ALFRESCO) was developed to simulate landscape-level response of vegetation to interactions between fire, climate, and vegetation in the boreal forest of interior Alaska (Rupp et al. 2000, 2002). As climate change models become more realistic and step down to the landscape level, and concerns for global change impacts grow, we expect to see global change addressed by more large-scale wildlife conservation and land management planning. For example, it is likely that the incorporation of such spatio-temporal changes will become routine in future PVA modeling (Akçakaya and Brook, this volume).

OPTIMIZATION VERSUS CHOOSING FROM ALTERNATIVES

All but the simplest projects will ultimately have to address how to maximize benefits from competing objectives. Projects focused exclusively on wildlife conservation will need to address conflicts among species needs and maximizing viability given financial constraints (Haight and Gobster, this volume). Multiple-use projects will need to maximize benefits associated with species viability and other benefits such as recreation or wood or mineral production. Selecting a course of action to meet objectives can involve an optimization approach or choosing among defined alternatives. Optimization methods are generally empirical, and methods to choose among alternatives range from qualitative to empirical (Haight and Gobster, this volume).

Choosing among defined alternatives is a common approach used for national forest land management plans in the United States. A number of management alternatives are considered that span the interests of stakeholders and balance competing objectives in different ways. Information is gathered on the effects or outcomes of each alternative and can range from expert opinion to predictions from empirical models (Fig. 24-2; Dijak and Rittenhouse, this volume). A planning team considers which alternative best meets the public interests and legal mandates based on information gathered from the stakeholders and resources specialists. If objectives can be clearly defined in some measurable form, and some form of decision framework or weighting agreed upon for competing objectives, a model could be developed to choose the optimal alternative. We are not aware of any examples of this approach; one of the benefits of choosing from alternative models is the simplicity of relying on a consensus after reviewing the evidence for the alternatives. A potential important shortcoming of this approach is that the optimal solution is not likely among the

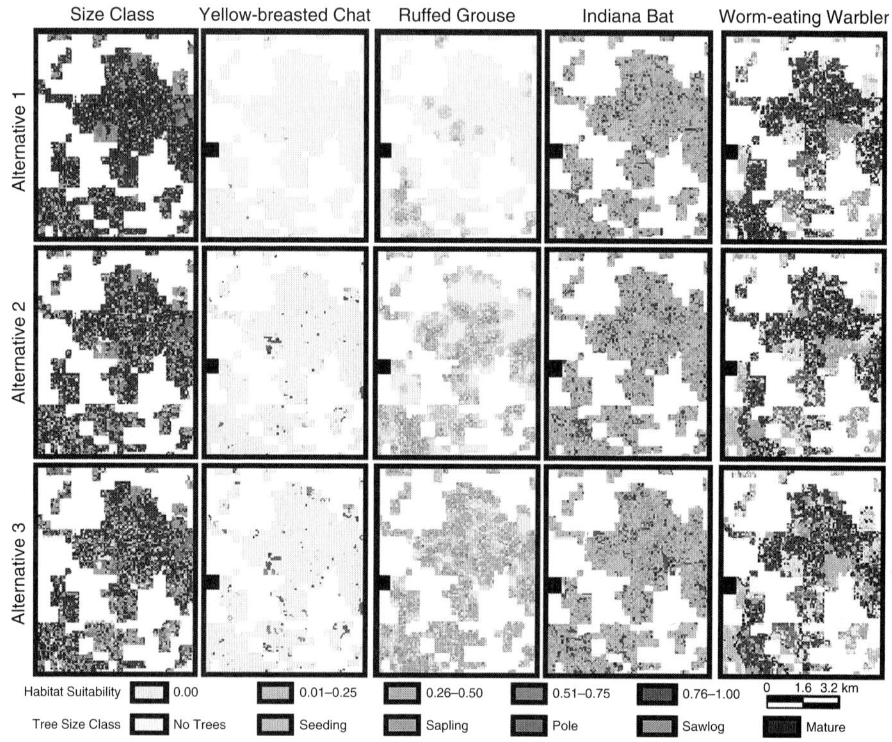

FIG. 24-2 (See color insert.)

An example of the "choosing from alternatives" approach. Maps of predicted changes in forest size-class and habitat suitability for four wildlife species under three management alternatives considered in the Hoosier National Forest Land Management Plan. Maps and graphs such as these can be used to inform selection of a preferred alternative.

few alternatives typically considered; however, we assume that one of the alternatives is close enough.

Optimization approaches seek the optimal solution based on well-defined objectives or evaluation criteria (Flather et al., this volume; Haight and Gobster, this volume). The use of optimization analysis in conservation is not common but is increasing (Rodrigues and Gaston 2002). There are two broad classes of optimization approaches: those that seek exact optimal solutions (Hof and Flather 2007) and more heuristic approaches that use iterative or stepwise algorithms that approximate an efficient design (Cabeza and Moilanen 2001). Some conservation problems can be reduced to a few important variables and solved through an exact optimal solution (i.e., Haight and Gobster, this volume). However, many conservation problems are too complex and intractable in a closed-form exact optimization model; in these cases, more heuristic approaches offer a reasonable alternative (Pressey et al. 1997). Often, more species in the analysis results in consideration of more management options and objectives, which makes the problem more difficult to solve. Somewhat ironically, these more difficult cases become too complex for any reasonable effort using mathematical optimization, and planners often fall back to simpler approaches such as choosing among a defined set of alternatives that have been characterized by an evaluation of their impact on a limited set of resources and species; models may or may not be used. However, in some cases spatial optimization has been used to maximize persistence or habitat suitability for multiple species (Noon et al., this volume).

Careful articulation of objectives is important in any decision, but especially so for empirical-based decisions. For example, maximizing (1) total habitat suitability across all species, (2) average total habitat suitability per species, or (3) average total habitat suitability per species, with the constraint that total habitat suitability must be greater than some minimum for each species, will likely result in different solutions.

CASE STUDIES

We present two case studies based on examples presented in this volume to illustrate the use of this decision framework. For illustrative purposes we present plausible reasons for the decisions made in these examples; these may or may not represent the thought process of the parties involved.

Case 1: Hoosier National Forest Plan

In Case 1 we consider one element of the Hoosier National Forest Plan: planning for the viability of a set of focal wildlife species (Rittenhouse 2008). This case study involves a land management problem where a plan will be developed to manage lands encompassing the Hoosier National Forest to meet stakeholder interests and legal mandates (Fig. 24-1A, step 1). By legal mandate, the agency must consider viability of native species, so it is a multiple species problem that addresses species representation (Fig. 24-1A, step 2 and 3). The planning team chose a fine filter approach by considering the habitat needs of a set of focal species representing a mix of management indicator species and species of concern (Fig. 24-1A, step 4). The plan considered the amount and quality of habitat, as opposed to species abundance or viability directly, because habitat could most easily be linked to the forecasted changes in forest composition and structure under the plan alternatives (Fig. 24-1A, step 5). The plan used a dynamic approach so it could address short- and long-term affects of management by estimating habitat suitability from the outputs of a dynamic landscape change model (Fig. 24-1A, step 6; Dijak and Rittenhouse, this volume). The plan had to be completed within a short time frame, so it relied on existing data and utilized habitat suitability models that could be developed from expert opinion and published studies (Fig. 24-1A, step 7 and 8). The plan considered management alternatives (Fig. 24-2) and relied on the development of a consensus based on input from resource specialists, stakeholder meetings, and deliberations of the planning team (Fig. 24-1A, step 9).

Case 2: Lake County Reserve Selection

In Case 2 we use the example presented by Haight and Gobster (this volume) where Lake County Illinois planners wanted to identify a cost-effective set of sites to be acquired to protect rare plants and animals. This is a reserve selection problem that addresses multiple species (Fig. 24-1B, step 1 and 2). The planners wanted to select sites for protection that optimized the number of species protected within a defined budget. While the quantity being optimized is a count of species, they considered only a pool of rare plants, so this is a species representation problem (Fig. 24-1A, step 3). They took a fine filter approach because they wanted to explicitly account for the occurrence of rare plants and animals (Fig. 24-1A, step 4). The approach was based on occurrence, the simplest metric of abundance, presumably because these data were available and they did not see enough additional benefit for the additional effort required to collect or analyze additional data on density or viability (Fig. 24-1A, step 5 and 7). This represents a stationary approach because the planners did not model how sites or populations might change over time (Fig. 24-1A, step 6). Instead, they relied on existing data about occurrence, which is essentially a qualitative model of occurrence, because they did not model occurrence from survey data (Fig. 24-1A, step 8). They used a true closed form optimization model to select the optimal set of sites given their budget (Haight and Gobster, this volume).

OTHER ISSUES

Many of the procedures discussed in this book require combined skills of GIS applications, vegetation modeling, wildlife-habitat modeling, and social and economic considerations. Given the data and technical expertise required, the availability of large spatial data sets and concepts being addressed, model sophistication is quickly outpacing the ability of agencies to apply them. Model and data complexity require teams of scientists, planners, and managers to work collaboratively to address planning requirements. Furthermore, the availability of high-quality data is not keeping pace with sophistication of analytical methods. As pressures increase for large-scale conservation planning, agencies will need to continually retool to meet mandates and planning needs.

SUMMARY

We outlined a decision framework for choosing among the many modeling approaches presented in this book. Elements in the framework include choosing models that address reserve selection versus land management; single species versus multiple species; species representation versus diversity; coarse filter versus fine filter; habitat, abundance, or viability; stationary versus dynamic; existing data versus new data; theoretical versus statistical; and optimization versus choosing from alternatives. After reviewing elements in the decision framework discussed in detail in other parts of the book, we worked through the decision process of two case studies. In doing so, we identified several pitfalls, such as lack of data, and offered guidance about the modeling process. Careful consideration of objectives is necessary to select the appropriate procedures, metrics, and tools in any conservation planning activity. Although reminding one to revisit objectives might seem like an overly generic recommendation, it is a fundamental consideration that drives all subsequent decisions. Without clearly articulated and followed objectives, one could easily become lost in the myriad of methods, available data, future data needs, and other decisions within the framework we have provided. One must avoid the temptation to use metrics that are simply easy to measure or readily available. Instead, it is far better to move forward with a lack of complete knowledge using the most appropriate metrics and concepts, while acknowledging the need to collect additional data to test assumptions and reduce uncertainty.

LITERATURE CITED

Airola, D. A. 1988. *A guide to the California Wildlife-Habitat Relationship system.* California Department of Fish and Game, Rancho Cordova, CA. 74p.

Akçakaya, H. R., V. C. Radeloff, D. J. Mladenoff, and H. S. He. 2004. Integrating landscape and metapopulation modeling approaches: Viability of the sharp-tailed grouse in a dynamic landscape. *Conservation Biology* 18:526–537.

Beaumont, L. J., A. J. Pitman, M. Poulsen, and L. Hughes. 2007. Where will species go? Incorporating new advances in climate modelling into projections of species distributions. *Global Change Biology* 13:1368–1385.

Brooks, J. M., R. A. Mittermeier, G. A. B. da Faonseca, J. Gerlach, M. Hoffman, J. F. Lamoreux, C. G. M. Mittermeier, J. D. Pilgrim, and A. S. L. Rodrigues. 2006. Global biodiversity conservation priorities. *Science* 313:58–61.

Cabeza, M., and A. Moilanen. 2001. Design of reserve networks and the persistence of biodiversity. *Trends in Ecology and Evolution* 16:242–248.

DeGraaf, R. M., and D. D. Rudis. 1986. New England wildlife: Habitat, natural history, and distribution. U.S. Forest Service, Northeastern Forest Experimental Station, General Technical Report NE-108. Broomall, Pennsylvania, USA.

Hamel, P. B. 1992. *Land manager's guide to the birds of the South.* The Nature Conservancy, Southeastern Region, Chapel Hill, North Carolina, USA.

Hof, J., and C. H. Flather. 2007. Optimization of landscape pattern. Pages 143–160 *in* J. Wu, and R. Hobbs, editors. *Key topics and perspectives in landscape ecology.* Cambridge University Press, Cambridge, United Kingdom.

Larson, M. A., W. D. Dijak, F. R. Thompson, III, and J. J. Millspaugh. 2003. Landscape-level habitat suitability models for 12 wildlife species in southern Missouri. U.S. Forest Service, North Central Research Station, General Technical Report NC-233, St. Paul, Minnesota, USA.

Noss, R. F. 1987. Do we really want diversity? *Whole Earth Review* 1987 (Summer):126–128.

Pearson, R. G., and T. P. Dawson. 2003. Predicting the impacts of climate change on the distribution of species: Are bioclimate envelope models useful? *Global Ecology and Biogeography* 12:361–371.

Pichancourt, J.-B., F. Burel, and P. Auger. 2006. Assessing the effect of habitat fragmentation on population dynamics: An implicit modelling approach. *Ecological Modelling* 192:543–556.

Pressey, R. L., H. P. Possingham, and J. R. Day. 1997. Effectiveness of alternative heuristic algorithms for identifying indicative minimum requirements for conservation reserves. *Biological Conservation* 80:207–219.

Rittenhouse, C. D. 2008. Wildlife response to spatial and tempered changes in forest habitat. PhD Dissertation. University of Missouri, Columbia, Missouri, USA.

Rittenhouse, C. D., W. D. Dijak, F. R. Thompson, III, and J. J. Millspaugh. 2007. Development of landscape-level habitat suitability models for ten wildlife species in the Central Hardwoods Region. U.S. Forest Service, Northern Research Station, General Technical Report NRS-4, Newtown Square, Pennsylvania, USA.

Rodrigues, A. S. L., and K. J. Gaston. 2002. Optimisation in reserve selection procedures—Why not? *Biological Conservation* 107:123–129.

Rupp, T. S., A. M. Starfield, and F. S. Chapin, III. 2000. A frame-based spatially explicit model of subarctic vegetation response to climatic change: Comparison with a point model. *Landscape Ecology* 15:383–400.

Rupp, T. S., A. M. Starfield, F. S. Chapin, III, and P. Duffy. 2002. Modeling the impact of black spruce on the fire regime of Alaskan boreal forest. *Climatic Change* 55:213–233.

Sarkar, S., R. L. Pressey, D. P. Faith, C. R. Margules, T. Fuller, D. M. Stoms, A. Moffett, K. A. Wilson, K. J. Williams, P. H. Williams, and S. Andelman. 2006. Biodiversity conservation planning tools: Present status and challenges for the future. *Annual Review of Environment and Resources* 31:123–159.

Shifley, S. R., F. R. Thompson, III, W. D. Dijak, M. A. Larson, and J. J. Millspaugh. 2006. Simulated effects of forest management alternatives on landscape structure and habitat suitability in the Midwestern United States. *Forest Ecology and Management* 229:361–377.

Tirpak, J. M., D. T. Jones-Farrand, F. R. Thompson, III, D. J. Twedt, M. D. Nelson, and W. B. Uihlein, III. 2008. Predicting bird habitat quality from a geospatial analysis of FIA data. *In* R. E., McRoberts, G. A. Reams, P. C. Van Duesen, W. H. McWilliams, eds. Proceedings of the 8th annual forest inventory and analysis symposium. U.S. Forest Service, Northern Research Station, General Technical Report WO-xx, Newtown Square, Pennsylvania, USA. *In press*.

Wintle, B. A., S. A. Bekessy, L. A. Venier, J. L. Pearce, and R. A. Chisholm. 2005. Utility of dynamic-landscape metapopulation models for sustainable forest management. *Conservation Biology* 19:1930–1943.

Index

A

Acadian flycatcher (*Empidonax virescens*), 604, 606
Adaptive Resource Management (ARM), models, 11-12
ALCES, *see* A Landscape Cumulative Effects Simulator
Allometric scaling relationships, multispecies conservation planning, 53
ALMaSS, *see* Animal, Landscape, and Man Simulation System
American bittern (*Botaurus lentiginosus*), 534
American coot (*Fulica americana*), 534
American crow (*Corvus brachyrhynchos*), 634, 640-641, 659
American goldfinch (*Curduelis tristis*), 658
American robin (*Turdus migratorius*), 658
American white pelican (*Pelecanus erythrorhynchos*), 534
American woodcock (*Scolopax minor*), 215, 380
Animal, Landscape, and Man Simulation System (ALMaSS), habitat-relationships modeling, 257, 265, 269
Anna's hummingbird (*Calypte anna*), 659
Arc Habitat Suitability Index (ArcHSI), habitat-relationships modeling, 257, 268
Arc-Habcap, habitat-relationships modeling, 257, 267
ArcHSI, *see* Arc Habitat Suitability Index
ARM, *see* Adaptive Resource Management
Artificial neural network (ANN)
  habitat-relationships modeling, 257, 267
  multispecies habitat modeling, 237

B

Bachma's sparrow (*Aimophila aestivalis*), 236
Bald eagle (*Haliaeetus leucocephalus*), 563
Balsam fir (*Abies balsamifera*), 485
Band-tailed pigeon (*Columba fasciata*), 658
Barn swallow (*Hirundo rustica*), 659
Barred owl (*Strix varia*), 580
Bayesian belief network (BBN)
  habitat-relationships modeling, 258, 269
  population viability evaluation, 234-235
BBN, *see* Bayesian belief network
Beech, 370
Bell's sage sparrow (*Amphispiza belli*), 240
BEST, *see* Biodiversity Expert System Tool
Bewick's wren (*Thyromanes beweickii*), 658
Bi-criteria species covering problem, reserve selection modeling, 135-139
BIOCLIM program, 367
Biodiversity conservation
  error effects
    analysis, 103-105
    implications, 105-107
    overview, 94-95, 101-103
  geographic approaches
    context setting, 87-88
    data types, 91-92
    overview, 92-94
    species criteria, 88-91
  overview, 85-86
  research needs
    data availability and quality, 107
    ecological process incorporation, 111
    extended geographic approaches, 108-110
    inferential basis for conservation planing, 108
    practitioner accessibility, 111-113
  scale effects
    analysis, 96-100
    extent, 95
    grain, 95
    implications, 100-101
    overview, 94-96
Biodiversity Expert System Tool (BEST), habitat-relationships modeling, 258, 269
BIOMAPPER program, 367
Bird conservation, *see also individual species*
  Five Elements process
    assumptions, 616
    complex species-habitat relationships, 615
    data constraints and limitations, 615-616
    flexibility and adaptability, 617
    model fit, 617
    projections, 618
    spatial scalability, 617-618
    time and cost, 617-619
    uncertainty, 616-617
  geographic information system

Bird conservation (*Continued*)
habitat assessment and landscape characterization, 596–598
Habitat Suitability Index models, 603–606
initiatives in United states, 593–596
linking bird distribution and abundance with habitat assessments in Bird Conservation Regions
database models, 599
geographic information system-based habitat suitability models, 603–606
statistical models, 607–614
urbanization response modeling
avian species richness coupling with relative abundance models, 633–634, 637–645
avian surveys, 629–630, 635–637
findings, 648–649
future land cover prediction, 634–635
land-cover change modeling, 630–633
limitations, 651–653
overview, 625–627
policy and management implications, 649–651
prospects, 653–654
study area and sites, 627–629
Urbansim, 655
BIRDHAB, habitat-relationships modeling, 258, 267, 278
Bison (*Bos bison*), 127
Black bear (*Ursus americana*), 65, 231, 380
Black-capped chickadee (*Poecile atricapillus*), 658
Black-capped vireo (*Vireo atricapillus*), 191
Black-footed ferret (*Mustela nigripes*), 167
Black-headed grosbeak (*Pheucticus melanocephalus*), 658
Black spruce (*Picea mariana*), 484
Black tern (*Chlidonia niger*), 534
Black-throated gray warbler (*Dendroica nigrescens*), 658
Black-throated sparrow (*Amphispiza bilineata*), 517
Blue-winged teal (*Anas discors*), 540
Bobcat (*Lynx rufus*), 380
Bobolink (*Dolichonyx orzivorus*), 211
Bobwhite (*Colinus virginianus*), 373, 380
BOREAL, habitat-relationships modeling, 258, 267
Brewer's blackbird (*Euphagus cyanocephalus*), 517, 659
Brewer's sparrow (*Spizella breweri*), 516
Brown creeper (*Certhia americana*), 298, 332, 485, 658
Brown trout (*Salmo trutta*), 69
Brown-headed cowbird (*Molothrus ater*), 609, 659
Buffalo (*Bubalus bubalis*), 458–459
Bull trout (*Salvenus confluentus*), 563
Burrowing owl (*Athene cunicularia*), 167, 600
Bushtit (*Psaltriparus minimus*), 658

C

California gnatcatcher (*Polioptila c. californica*), 461
California gull (*Larus californicus*), 534
California Wildlife Habitat Relationships (CWHR), habitat-relationships modeling, 258, 267, 278
Canada lyx (*Lynx canadensis*), 289, 563
Capture/mark/recapture (CMR), 68
CART, *see* Classification and regression tree
Cascades Habitat Conservation Plan, *see* Plum Creek Central Cascades Habitat Conservation Plan
Cassin's vireo (*Vireo cassinii*), 658
Cedar, 370
Cedar waxwing (*Bombycilla cedrorum*), 658
Cerulean warbler (*Dendroica cerulea*), 215, 371, 380
Cheatgrass (*Bromus tectorum*), 170, 403, 514, 526
Chestnut-backed chickadee (*Poecile ruescens*), 658
Chestnut-collared longspur (*Calcarius ornatus*), 161
Chisel-toothed kangaroo rat (*Dipodomys microps*), 516
Classification and regression tree (CART), 607–608
CMR, *see* Capture/mark/recapture
Coarse filter
versus fine filter approach, 663–664
mixed coarse and fine filter approach, 61–62
multispecies conservation planning, 56–58
Coho salmon (*Oncorhynchus kisutch*), 236
Common sagebrush lizard (*Sceloporus graciosus*), 517
Common yellowthroat (*Geothlypis trichas*), 658
Complementarity, species representation, 89
Complex models, utility, 6–7
CompPATS, *see* Computerized Project Analysis and Tracking System

Computerized Project Analysis and Tracking System (CompPATS), habitat-relationships modeling, 258, 265-266, 269
Conservation efficiency, 203
Cowbird (*Molothrus ater*), 159
CWHR, *see* California Wildlife Habitat Relationships

D

Dark kangaroo mouse (*Microdipodops megacephalus*), 517
Dark-eyed junco (*Junco hyemalis*), 658
Decision-support model, 2
Desert horned lizard (*Phrynosoma platyrhinos*), 516
Desert spiny lizard (*Sceloporus magister*), 516
DISPATCH, 323
DLMP models, *see* Dynamic landscape metapopulation models
DOMAIN program, 367
Douglas fir (*Pseudotoga menziesii*), 567
Downy woodpecker (*Picoides pubescens*), 658
Dynamic landscape metapopulation (DLMP) models
  case studies
    key findings, 487-490
    modeling approach and scenarios, 485-487
    Ontario White River management area, 484-485
    species, 485
  development steps, 480-481
  feasibility, 490-491
  habitat supply model comparison, 491
  indicator species, 477-479
  interpretation, 482-484
  limitations, 492-494
  overview, 476-477
  prospects, 494
  software, 481
Dynamic species covering problem, reserve selection modeling, 140

E

EAM, *see* Effective Area Model
Earth Systems Research Institute (ESRI), 290
Eastern white cedar (*Thuja occidentalis*), 485
Ecological engineer species, multispecies conservation planning, 61
Ecosystem Management Decision Support (EMDS), habitat-relationships modeling, 259, 267
Ecosystem Management Research Institute (EMRI), 172
Effective Area Model (EAM), habitat-relationships modeling, 258, 268, 278
Effective population size, population viability analysis, 37-38
Elk (*Cervus elaphus*), 127, 167, 239, 416, 437-438, 576
EMBYR, 324
EMDS, *see* Ecosystem Management Decision Support
EMP, *see* Expected minimum population size
Empirical observation, species occupancy, 91
EMRI, *see* Ecosystem Management Research Institute
Endangered Species Act (ESA), 51, 153, 155, 163, 561
Engelman spruce (*Picea engelmanni*), 568
Environmental Quality Incentives Program (EQIP), 173
EQIP, *see* Environmental Quality Incentives Program
ESA, *see* Endangered Species Act
ESRI, *see* Earth Systems Research Institute
European starling (*Stumus vulgaris*), 659
Evaluation of models, *see* Validation, landscape models
Expected minimum population size (EMP), dynamic landscape metapopulation modeling, 482
Expert judgment, biodiversity conservation, 91
Expert systems, habitat-relationships modeling, 259, 268
Explanatory model, 2

F

Federal lands
  legislation, 51
  multispecies conservation planning, 73-76
Ferruginous hawk (*Buteo regalis*), 167, 517
FIA, *see* Forest Inventory and Analysis
Fine filter, multispecies conservation planning
  versus coarse filter, 663-664
  ecological engineer species, 61
  focal species, 59-60
  keystone species, 60
  link species, 60
  mixed coarse and fine filter approach, 61-62
  strong interactor species, 60
  threatened species, 58-59
  umbrella species, 60
Fir, understory vegetation modeling, 403-405

FIRESCAPE, 323
Flammulated owl (*Otus flammeolus*), 574
Focal species
classification, 59-60
fine filter approach for multispecies conservation planning, 59-60
identification, 71-73
strategic habitat conservation, 205-206, 223
Forest Inventory and Analysis (FIA), 74, 289, 375, 377, 605-606
Forest landscape models, *see also* LANDIS; Landscape Management System
landscape process models, 323-324
overview, 321-322
program comparison, 359-360
succession pathway models, 324
types, 424-426
validation, *see* Validation, landscape models
vital attribute models, 325
Forest Vegetation Simulator (FVS), 225, 398-399, 425
FORHAB, habitat-relationships modeling, 259, 269
FRAGSTATS program, 225, 598
Franklin's gull (*Larus pipixcan*), 534
FREIGHTS model, 239
Fuzzy logic, habitat quality indexing, 231
FVS, *see* Forest Vegetation Simulator

G

Gadwall (*Anas strepera*), 540
Gap Analysis Program (GAP), 309, 596
Gap models, ecosystem dynamics models, 397-398
GAP, *see* Gap Analysis Program
General Land Office (GLO), vegetation maps, 292-294, 296
Generalized linear model (GLM), 607
Genetic monitoring, multispecies conservation planning, 68-69
Geographic information system (GIS)
bird conservation
habitat assessment and landscape characterization, 596-598
Habitat Suitability Index models, 603-606
habitat modeling
analysis structure, 288-290
classification systems
attribute variability, 300-301
language, 296-300
data abundance, 290-292
large landscape habitat studies, 225, 227, 230, 240-241
Geographic information system (GIS) (*Continued*)
spatial scale considerations, 302-305
technological considerations, 308-312
temporal scale considerations, 305-308
vegetation change analysis
boundary inaccuracy limitation, 294-295
scale-related limitations, 292-294
underrepresentation of vegetation limitation, 296
private land conservation planning, 170
Glanville fritillary butterfly (*Melitaea cixia*), 457
GLM, *see* Generalized linear model
GLO, *see* General Land Office
GNN, *see* Gradient-nearest-neighbor
Golden-crowned kinglet (*Regulus satrapa*), 658
Gradient-nearest-neighbor (GNN), understory vegetation modeling, 401-402
Grand fir (*Abies grandis*), 568
GRASS, 598
Gray flycatcher (*Empidonax wrightii*), 517
Gray squrrel (*Sciurus carolinensis*), 380
Gray wolf (*Canis lupus*), 127, 563
Great Basin collard lizard (*Crotaphytus bicinctores*), 516
Great Basin Ecoregion, 514-521
Great Basin spadefoot (*Spea intermontana*), 517
Green-tailed towhee (Pipilo chlorurus), 517
Grizzly bear (*Ursos arctos*), 126, 153, 563
Groundsnake (*Sonora semiannulata*), 516

H

HABIT@, habitat-relationships modeling, 259, 269
Habitat-based species viability (HBSV), habitat-relationships modeling, 259, 267
Habitat Capacity Index (HCI), habitat-relationships modeling, 260, 267
Habitat Conservation Plan, *see* Plum Creek Central Cascades Habitat Conservation Plan
Habitat Effectiveness Index (HEI), habitat-relationships modeling, 260, 267
Habitat Evaluation Procedure (HEP), 229, 368
Habitat networks
benefits, 502
case studies
Great Basin Ecoregion, 514-521
Interior Columbia Basin Ecosystem Project, 509-513
overview, 508-509
conceptual basis, 502-504
conservation objectives, 506-507
definition, 501
development steps, 504-506

Habitat networks (*Continued*)
management
direct links, 522-523
integration within and among species groups, 521-522
prospects
implementation through adaptive management, 525-526
knowledge gaps, 524-525
reconciliation of network strategies with other objectives, 523-524
spatial and temporal basis, 507-508
HABITAT program, 367
Habitat Quality (HQ), habitat-relationships modeling, 260, 268
Habitat quality indexing (HQI), 229-231
Habitat-relationships modeling
applications
cluster 1 frameworks, 273-274
cluster 2 frameworks, 274-275
cluster 3 frameworks, 275
cluster 4 frameworks, 276
cluster 5 frameworks, 276
cluster 6 frameworks, 276
cluster 7 frameworks, 276
frameworks
description, 255-256, 266-272
identification and rating, 252-266
overview, 251-252
prospects, 277-278
Habitat Suitability Index (HSI)
equations, 379-380
geographic information system-based bird conservation, 603-606
habitat-relationships modeling, 260
Hoosier National Forest case study, 382-385
input data layers, 375-379
Landscape HSImodels software, 380-382
large landscape habitat quantification, 229, 230, 234, 237-238
model development
data sources, 371-372
geographic extent, 370
spatial grain and extent, 370-371
suitability functions
area sensitivity, 372
distance, 372-373
edge effects, 373
habitat type and structure, 372
landscape composition, 371-373
overview of models, 368-369
validation of models, 432
HABS, *see* Hierarchical All-Bird Strategy
HABSCAPES, habitat-relationships modeling, 259, 266, 269
HABSIM, habitat-relationships modeling, 259, 265, 268
Hairy woodpecker (*Picoides villosus*), 658
Hammond's flycatcher (*Empidonax hammondii*), 658
HARVEST, 323, 426-427
HBSV, *see* Habitat-based species viability
HCI, *see* Habitat Capacity Index
HEI, *see* Habitat Effectiveness Index
Henslow's sparrow (*Ammodramus henslowii*), 380
HEP, *see* Habitat Evaluation Procedure
Hermit thrush (*Catharus guttatus*), 658
Hickory, 370
Hierarchical All-Bird Strategy (HABS), 599-602, 615-617
HNF, *see* Hoosier National Forest
Hooded warbler (*Wilsonia citrina*), 333, 380
Hoosier National Forest (HNF)
decision framework in model selection, 670-672
Habitat Suitability Index case study, 382-385, 605
Hotspot analysis, biodiversity conservation, 93
House finch (*Carpodacus mexicanus*), 659
House sparrow (*Passer domesticus*), 659
HQ, *see* Habitat Quality
HQI, *see* Habitat quality indexing
HQI program, habitat-relationships modeling, 260, 268
HQI+ program, habitat-relationships modeling, 260
HSI, *see* Habitat Suitability Index
Human-wildlife interactions, *see* People-wildlife interactions
Hunting, 125-126
Hutton's vireo (*Vireo huttoni*), 659

I

IAN, 598
IBMs, *see* Individual-based models
IBR model, *see* Isolation-by-resistance model
ICBEMP, *see* Interior Columbia Basin Ecosystem Project
IFM, *see* Incidence function model
IFOV, *see* Instantaneous Field of View
Incentives, private land conservation, 163-165
Incidence function model (IFM), 457
Indiana bat (*Myotis sodalis*), 380

Individual-based models (IBMs), large-scale population viability analysis, 463
Instantaneous Field of View (IFOV), 308–309
Interior Columbia Basin Ecosystem Project (ICBEMP), 509–513
Irreplaceability, land, 90
Isolation-by-resistance (IBR) model, 65

J

Jack pine (*Pinus banksiana*), 127, 191, 331, 484
Joint ventures, bird conservation, 584–595

K

K-nearest-neighbor (KNN), understory vegetation modeling, 401
Kentucky warbler (*Oporornis formosus*), 609–613
Keystone species
  dynamic landscape metapopulation models, 477
  multispecies conservation planning, 60
Kildeer (*Charadrius vociferus*), 658
King rail (*Rallus elegans*), 207
Kirtland's warbler (*Dendroica kirtlandii*), 127, 191, 193
Kit fox (*Vulpes macrotis*), 517
KNN, *see* K-nearest-neighbor

L

LAFS, 324
Lake sturgeon (*Acipenser fulvescens*), 207
Land Use Evolution and Impact Assessment Model (LEAM), habitat-relationships modeling, 261, 267
Land-cover change (LCC) modeling, 630–633, 647
LANDFIRE, *see* Landscape Fire and Resource Management Planning Tools Project
LANDIS
  dynamic landscape metapopulation modeling, 481
  Landscape Management System comparison, 359
  natural disturbance, 327–328
  operational design, 328–329
  overall design and structure, 325–326
  succession, 326–327
  timber harvest and fuel treatment, 328
  versions, 334
  wildlife model integration
    advantages and limitations, 333–335
    approaches, 329–333
Landscape connectivity, addressing for multispecies conservation planning, 54, 64–65
A Landscape Cumulative Effects Simulator (ALCES), habitat-relationships modeling, 257, 267
Landscape Evaluation Effects of Management Effects on Timber and Habitat (LEEMATH), habitat-relationships modeling, 239, 261, 265, 267
Landscape Fire and Resource Management Planning Tools Project (LANDFIRE), 597
Landscape HSI, habitat-relationships modeling, 261, 268
Landscape Management System (LMS)
  comparison with other programs
    LANDIS, 359
    Spatial Woodstock, 360
  decision analysis
    alternate landscape plan development, 355–358
    grouping stands, 355
    overview, 354
    stand-specific prescriptions from groups, 357, 359
  geographic area of study and application, 344–345
  habitat-relationships modeling, 261, 267
  management application, 351–354
  modules, 341–343
  output
    criteria for sustainable forestry, 349–350
    inventory information, 348
    summarized information, 348
    visualization, 348–349
  portfolio creation, 345–346
  prospects, 360–362
  scientific basis
    complexity and systems theory, 341–342
    decision analysis, 342
    hierarchies and top-down planning, 342
    stand and landscape dynamics, 340–341
    trade-offs, 341
  stand treating and projecting, 346–348
  Toggle program, 356–357
Landscape models, *see also* Forest landscape models
  Adaptive Resource Management, 11–12
  decision framework in model selection
    case studies, 671–672
    coarse versus fine filter, 663–664
    conceptual versus statistical models, 667

Landscape models (*Continued*)
existing versus new data collection, 666
habitat, abundance, or viability, 665-666
optimization versus choosing from alternatives, 669-671
overview, 661
reserve selection versus land management planning, 662-663
single species versus multiple species, 663
species representation versus diversity, 664-665
stationary versus dynamic approaches, 667-669
overview, 1-2
philosophy of modeling, 3-10
uncertainty, 12-16
unreliable model avoidance, 10-11
uses, 2
validation, *see* Validation, landscape models
viability a general management metric, 16-20
visualization, 20-26
LANDSIM, 325
LANDSUM, 324, 426-427
Large landscape habitat quantification
comparison of landscapes
spatial comparisons, 238-239
temporal comparisons, 239-240
complexity gradient of models, 226-229
geographic information system
analysis structure, 288-290
classification systems
attribute variability, 300-301
language, 296-300
data abundance, 290-292
examples, 225, 227, 230, 240-241
spatial scale considerations, 302-305
technological considerations, 308-312
temporal scale considerations, 305-308
vegetation change analysis
boundary inaccuracy limitation, 294-295
scale-related limitations, 292-294
underrepresentation of vegetation limitation, 296
Habitat Suitability Index modeling, *see* Habitat Suitability Index
model evaluation, 240-241
multispecies approaches, 236-238
occurrence probability modeling, 231-233
overview, 225-226
population density estimation, 233-234
population viability evaluation, 234-236
prospects, 242
Large landscape habitat quantification (*Continued*)
quality indexing, 229-231
Lark bunting (*Calamospiza melanocorys*), 161
Lark sparrow (*Chodestes grammacus*), 517
Lattice models, large-scale population viability analysis, 458-460
LCC modeling, *see* Land-cover change modeling
LEAM, *see* Land Use Evolution and Impact Assessment Model
LEEMATH, *see* Landscape Evaluation Effects of Management Effects on Timber and Habitat
Lesser prairie-chicken (*Tympanuchus pallidicinctus*), 601-602
LiDAR, *see* Light Detection and Ranging
Light Detection and Ranging (LiDAR), Habitat Suitability Index input, 378-379
Link species, multispecies conservation planning, 60
Little pocket mouse (*Perognathus longimembris*), 517
LMS, *see* Landscape Management System
Lodgepole pine (*Pinus contorta*), 568
Loggerhead shrike (*Lanus ludovicianus*), 167, 517
Long-billed curlew (*Numenius americanus*), 161, 167
Long-nosed leopard lizard (*Gambelia wislizenii*), 516
Long-nosed snake (*Rhinocheilus lecontei*), 516

M

MacGiillivray's warbler (*Oporornis tolmei*), 659
Magpie goose (*Anseranas semipalmata*), 460
Mallard (*Anas platyrhynchos*), 540
Management-oriented model, 9-10
Maple, 370
Marbled godwit (*Limosa fedoa*), 210, 533, 545-546, 549-551
Marbled murrelet (*Brachyramphus mamoratus*), 563
MARS, *see* Multivariate adaptive regression splines
Marten (*Martes americana*), 239
MARXAN program, 97
Matrix models, population viability analysis, 35
MAXENT program, 64
Maximum expected species covering problem, reserve selection modeling, 139
Maximum species covering problem, reserve selection modeling, 131-135
MBCF, *see* Migratory Bird Conservation Fund
McCown's longspur (*Calcarius mccownii*), 161

MDNR, *see* Michigan Department of Natural Resources
Measurement scale, geographic information system, 303–305
Merriam's kangaroo rat (*Dipodomys merriami*), 516
Merriam's shrew (*Sorex merriami*), 517
Mesofilter, multispecies conservation planning, 58
Metadata, geographic information system, 290
Metapopulation models
  dynamic landscape metapopulation models, *see* Dynamic landscape metapopulation models
  large-scale population viability analysis, 460–462, 475
  population viability analysis, 36
Metapopulation size problem, reserve design modeling, 143
Michigan Department of Natural Resources (MDNR)
  map boundary limitations, 294–295
  vegetation classification, 297–298
Migratory Bird Conservation Fund (MBCF), 535
Missouri Forest Ecosystem Project (MOFEP), 441
MOFEP, *see* Missouri Forest Ecosystem Project
Moose (*Alces alces*), 207
Most-similar-neighbor (MSN), understory vegetation modeling, 401
Mountain hemlock (*Tsuga mertensiana*), 567
Mountain plover (*Charadrius montanus*), 161, 167
Movement, landscape connectivity in multispecies conservation planning, 54
MSN, *see* Most-similar-neighbor
Mule deer (*Odocoileus hemionus*), 167, 416
Multispecies conservation planning
  allometric scaling relationships, 53
  demography, 54
  federal public lands, 73–76
  filtering approaches
    coarse filter, 56–58
    fine filter
      threatened species, 58–59
      focal species, 59–60
      umbrella species, 60
      keystone species, 60
      strong interactor species, 60
      link species, 60
      ecological engineer species, 61
    mesofilter, 58
    mixed coarse and fine filter approach, 61–62
    overview, 55–56
Multispecies conservation planning (*Continued*)
  focal species identification, 71–73
  genetic monitoring, 68–69
  landscape connectivity addressing for multiple species, 64–65
  movement and landscape connectivity, 54
  The Nature Conservancy strategy, 69–70
  niche theory, 52–53
  persistence likelihood, maximization for multiple species, 67–68
  probability surfaces for species occurrences, 62–64
  reserve selection algorithms, 65–67
  versus single species, 663
  trophic structure, 53
Multivariate adaptive regression splines (MARS), 608

N

NABCI, *see* North American Bird Conservation Initiative
National Elevation Dataset (NED), 598
National Forest Management Act (NFMA), 51, 59, 73
National Hydrography Dataset (NHD), 598
National Land Cover Dataset (NLCD), 596–597
National Vegetation Classification, 299
National Wetlands Inventory (NWI), 294–295, 539–540
The Nature Conservancy (TNC), multispecies conservation planning strategy, 69–70
NAWCA, *see* North American Wetlands Conservation Act
NAWMP, *see* North American Waterfowl Management Plan
NED, *see* National Elevation Dataset
NFMA, *see* National Forest Management Act
NHD, *see* National Hydrography Dataset
Niche theory, multispecies conservation planning, 52–53
Nightsnake (Hypsiglena torquata), 517
NLCD, *see* National Land Cover Dataset
Noble fir (*Abies procera*), 567
North American Bird Conservation Initiative (NABCI), 594
North American Waterfowl Management Plan (NAWMP), 594
North American Wetlands Conservation Act (NAWCA), 535
Norther rough-winged swallow (*Stelgidopteryx serripennis*), 659
Northern flicker (*Colaptes auratus*), 659

Northern goshawk (*Acciper gentilis*), 191
Northern grasshopper mouse (*Onychomys leucogaster*), 517
Northern harrier (*Circus cyaneus*), 517
Northern long-eared bat (*Myotis sepentrionalis*), 380
Northern pintail (*Anas acuta*), 540
Northern shoveler (*Anas clypeata*), 540
Northern spotted owl (*Strix occidentalis*), 59, 124, 191, 235, 506, 561-588
Nuthatch (*Sitta europaea*), 233
NWI, *see* National Wetlands Inventory

O

Oak, 370, 429-430
Object-oriented programming (OOP), habitat quality indexing, 231
Observational scale, geographic information system, 302-303
Occam's razor, 6
Occupancy models, large-scale population viability analysis, 455, 457-458
Olive-sided flycatcher (*Contopus cooperi*), 659
ONFIRE, 323
OOP, *see* Object-oriented programming
Optimization analysis, biodiversity conservation, 94
OPTIONS program, futuring of wildlife habitat, 574-576
Orange-crowed warbler (*Vermivora celata*), 659
Ord's kangaroo rat (*Dipodomys ordii*), 517
Ornate box tutle (*Terrapene ornata*), 212
Ovenbird (*Seiurus aurocapillus*), 240, 332, 380

P

Pacific fisher (*Martes pennanti*), 574
Pacific silver fir (*Abies amabilis*), 567
Pacific-slope flycatcher (*Empidonax difficilis*), 634, 640-641, 646, 659
Parameters, model
  estimation, 10, 14-15
  dependence on too many, 10
Partners in Flight (PIF), 594-595, 614-615
Patch models, population viability analysis, 36
PATCH, *see* Program to Assist in Tracking Critical Habitat
PATREC, *see* Pattern Recognition
Pattern Recognition (PATREC)
  habitat-relationships modeling, 262, 267
  population density estimation, 234
People-wildlife interactions
People-wildlife interactions (*Continued*)
  amenity migration, 127-129
  recreation, 124-126
  restoration, 126-127, 148
Peregrine falcon (*Falco peregrinus*), 563
Persistence likelihood, maximization for multiple species, 67-68
Pied-billed grebe (*Podilymbus podicep*), 534
PIF, *see* Partners in Flight
Pig, (*sus scrofa*), 458
Pileated woodlepcker (*Dryocopus pileatus*), 574
Pine siskin (*Carduelis pinus*), 659
Pine warbler (*Dendroica pinus*), 333, 380
Pinyon-juniper woodlands, 514
Playa Lakes Joint Venture (PLJV), 599-600, 602
PLJV, *see* Playa Lakes Joint Venture
Plum Creek Central Cascades Habitat Conservation Plan
  adaptive management and monitoring, 582-584
  forest stand classification, 570-571
  futuring of wildlife habitat, 574-577
  Habitat Conservation Plan process, 563-566
  land ownership and area, 566-568
  lifeforms and habitat analysis, 571-574
  multispecies approach, 568-570
  northern spotted owl
    centers, 562
    resource selection function analysis, 578-579
    resource selection probability function, 580-581
  policy and management implications, 584-587
  prospects, 587-588
  threatened species, 563
Point Specific Estimator (PSE), habitat-relationships modeling, 262, 268
Ponderosa pine (*Pinus ponderosa*), 568
Population viability analysis (PVA), *see also* Viability
  decision theory, 40
  demographic models
    matrix models, 35
    metapopulation models, 36
    spatially explicit population models, 36
    stochastic population viability analysis models, 35-36
  effective population size, 37-38
  large landscape applications
    connectivity, 453
    habitat quantification, 234-236
    model selection and parameterization, 464
    model types and applications
      individual-based models, 463

Population viability analysis (PVA) (*Continued*)
lattice models, 458-460
metapopulation models, 460-462
occupancy models, 455, 457-458
overview, 456
prospects, 465-466
rationale, 449-450
spatial extent, 451-452
spatial heterogeneity, 453-454
spatial resolution, 452-453
temporal dynamics, 454-455
model selection factors, 38-39
optimization framework applications, 41-42
relationship to other methods, 450-451
scope, 33
stochastic patch occupancy models, 37
trend analysis, 37
uncertainty, 42-43
PPR, *see* Prairie Pothole Region
Prairie dog (*Cynomys*), 143, 159, 167, 600
Prairie falcon (*Falco mexicanus*), 517
Prairie Pothole Region (PPR)
conservation efforts, 534-535
conservation planning approach
data availability and quality, 539-540
ducks, 540
model development and integration, 536-539
realities, 535-536
waterbirds and shorebirds, 540-542
overview, 533-534
policy and management implications of conservation, 554-555
prospects for conservation, 555-556
species abundance and distribution models
conservation applications, 551-554
marbled godwit
conceptual model, 546
outcomes, 549-551
statistical models, 545-546
sora, 543-545, 548-549
waterfowl, 542-543, 547-548
Prairie warbler (*Dendroica discolor*), 333, 380
Predicted occurrence, species in biodiversity conservation, 92
Predictive model, 2, 5
Primary viability, 18
Private lands
abundance, 153
conservation planning challenges
funding, 155-156
landowner engagement, 156-157
mixed ownership, 159-160
Private lands (*Continued*)
overview, 154
property rights, 154-155
proprietary information, 157-158
solutions, 160-165
species management focus, 158-159
ecosystem diversity approach for conservation planning, 160-163
incentives for conservation planning, 163-165
modeling
forestlands, 166-167
Thunder Basin Grassland Prairie Ecosystem Association, 167-173
PROC CLUSTER, habitat-relationships modeling, 256
PROC DISTANCE, habitat-relationships modeling, 256
Program to Assist in Tracking Critical Habitat (PATCH), habitat-relationships modeling, 261, 267
Pronghorn (*Antilocapra americana*), 517
Pronghorn antelope (*Antilocapra americana*), 167
PSE, *see* Point Specific Estimator
Public lands, *see* Federal lands
Purple finch (*Carpodacus purpureus*), 659
Purpose, defining for model, 4-5
PVA, *see* Population viability analysis
Pygmy rabbit (*Brachylagus idahoensis*), 516

R

RAMAS LANDIS, 331
RAMAS Landscape
applications, 239-240
habitat-relationships modeling, 262, 268
limitations, 492-494
RAMAS GIS module
dynamic landscape metapopulation modeling, 481, 485, 490
LANDIS integration, 332
Real estate investment trust (REIT), 166
Recreation, 124-126
Red-backed salamander (*Plethodon cinereus*), 485
Red-backed vole (*Clethrionomys gapperi*), 485
Red bat (*Lasiurus borealis*), 380
Red-breasted nuthatch (*Sitta canadensis*), 659
Red-cockaded woodpecker (*Picoides borealis*), 157
Red crossbill (*Loxia curvirostra*), 659
Red pine (*Pinus resinosa*), 331
Red-winged blackbird (*Agelaius phoeniceus*), 659
Redundancy, species incidence, 90

Region specificity, models, 8-9
REIT, *see* Real estate investment trust
Reserve-based modeling
  advantages and limitations, 146-147
  overview, 129-131
  prospects, 147-148
  reserve design models
    population dynamics incorporation
      metapopulation size problem, 143
      overview, 142-143
      safe-minimum-standard problem, 144-145
      surviving populations problem, 145-146
    reserve compactness problem, 142
    reserve connectivity problem, 141-142
    reserve proximity problem, 141
  reserve selection models
    bi-criteria species covering problem, 135-139
    decision framework in model selection, 662-663, 672
    dynamic species covering problem, 140
    maximum expected species covering problem, 139
    maximum species covering problem, 131-135
    multispecies conservation planning, 65-67
Resource selection function (RSF), 578-579
Resource selection probability function (RSPF), 580-581
Restoration, habitats, 126-127, 148
Robustness, biodiversity conservation, 90
Rock pigeon (*Columba livia*), 659
Rock squirrel (Spermophilus variegatus), 518
Rock wren (*Salpinctes obsoletus*), 518
RPF, *see* Resource selection probability function
RSF, *see* Resource selection function
Ruffled grouse (*Bonasa umbellus*), 380
Rufous hummingbird (*Selasphorus rufus*), 659

S

Safe-minimum-standard problem, reserve design modeling, 144-145
Sage grouse (*Centrocercus urophasianus*), 156, 236, 403, 516
Sage sparrow (*Amphispiza belli*), 516
Sage thrasher (*Oreoscoptes montanus*), 516
Sagebrush (*Artemisia*), 157, 518
Sagebrush vole (*Lemmiscus curtatus*), 517
Savannah sparrow (*Passerculus sandwichensis*), 659
Secondary viability, 18
SEM, *see* Spatially explicit model
SEPMs, *see* Spatially explicit population models
SESI, *see* Spatially Explicit Species Index
Sharp-tailed grouse (*Tympanuchus phastanellus*), 240, 310, 331, 462
SHC, *see* Strategic habitat conservation
SHM, *see* Species-habitat matrices
Short-eared owl (*Asio flammeus*), 161, 517
SIMFOR, habitat-relationships modeling, 262, 268
SIMPPLLE, 324, 426-427
SNBN, *see* Spatially Neutral Bayesian Model
Snowshoe hare (*Lepus americanus*), 301
Song sparrow (*Melospize melodia*), 659
Sora (*Porzana carolina*), 210, 533-534, 543-545, 548-549
Southern redback salamander (*Plethedon serratus*), 280
Spatial optimization, habitat-relationships modeling, 263, 268
Spatial Woodstock, Landscape Management System comparison, 360
Spatially Explicit Species Index (SESI), habitat-relationships modeling, 262, 268
Spatially explicit model (SEM), strategic habitat conservation, 212
Spatially explicit population models (SEPMs), 36
Spatially Neutral Bayesian Model (SNBN), habitat-relationships modeling, 263, 268
Species-area relationship (SPPAREA), habitat-relationships modeling, 263, 268
Species count, biodiversity conservation, 88.89
Species-habitat matrices (SHM), habitat-relationships modeling, 263, 267, 278
Species representation, biodiversity conservation, 89
Species richness, *see* Species count
Species sorting algorithm (SSA), habitat-relationships modeling, 263, 269
Spectral-distance error, geographic information system, 309-310
SPOMs, *see* Stochastic patch occupancy models
Spotted owl, *see* Northern Spotted owl
Spotted towhee (*Pipilo maculatus*), 659
SPPAREA, *see* Species-area relationship
SSA, *see* Species sorting algorithm
Steller's jay (*Cyanocitta stelleri*), 659
Stepwise approximation approach, conservation assessment and planning
  broad-scale multispecies ranges, 181
  examples
    multiscale bird distribution assessment, 182-184
    openland bird assessment, 189-191

Stepwise approximation approach, conservation assessment and planning (*Continued*)
habitat-quality evaluation
abundance estimates and temporal change, 187–188
refinement with productivity and survivorship estimates, 188–189
implementation guidelines, 194–195
maps and summary models, 191
overview, 179–181
rationale, 177–179
resolution levels
level 1, 181–184
level 2, 185–186
level 3, 186–187
level 4, 187–188
level 5, 188–189
level 6, 191–192
level 7, 192–194
species distribution refinement via habitat-gradient relationships, 185–186
testing and integrating monitoring into science, 192–194
test species habitat or gradient relationships across landscapes, 186–187
Stochastic patch occupancy models (SPOMs), population viability analysis, 37
Stochastic population viability analysis models, 35–36
Strategic habitat conservation (SHC)
advantages, 220–221
biological planning
focal species selection, 205–206, 223
limiting factor and conservation treatment identification, 208
model development and application, 208–211
population objective setting, 206–207
conservation design
biodiversity and species richness maps, 214–216
ecosystem assessment, 212
habitat objectives, 213
priority area designation, 213–214
spatially explicit model development, 212
evaluation
assumption-driven research, 216–218
conservation action effects, 218–219
outcome-based monitoring, 218
population objective progress, 220
program and agency accomplishments, 219–220
Strategic habitat conservation (SHC) (*Continued*)
forest type matrix, 223
overview, 202–204
rationale, 201–202
Striped whipsnake (*Masticophis taeniatus*), 517
Strong interactor species, multispecies conservation planning, 60
Subalpine fir (*Abies lasiocarpa*), 568
Surrogates, biodiversity conservation, 92
Surviving populations problem, reserve design modeling, 145–146
Swainson's hawk (*Buteo swainsoni*), 517
Swainson's thrush (*Catharus ustulatus*), 659

T

Tamarack (*Larix larcina*), 485
TBGPEA, *see* Thunder Basin Grassland Prairie Ecosystem Association
Tertiary viability, 18–19
Thunder Basin Grassland Prairie Ecosystem Association (TBGPEA), private land conservation modeling, 167–173
Tiger salamander (*Ambystoma tigrinum*), 522
Timber rattlsnake (*Crotalus horridus*), 380
Timber wolf (*Canis lupus lycaon*), 126
Timberland investment management organization (TIMO), 166
TIMO, *see* Timberland investment management organization
TNC, *see* The Nature Conservancy
Toggle program, *see* Landscape Management System
Transparency, models, 7–8
Tree swallow (*Tachycineta bicolor*), 659
Trembling aspen (*Populus tremuloides*), 484–485
Trend analysis, population viability analysis, 37
Turkey (*Melagris gallopavo*), 230, 380

U

Umbrella species, multispecies conservation planning, 60
Uncertainty, models
hypotheses and model structure, 13–14
model user bias, 16
parameter estimates, 14–15
population viability analysis, 42–43
verification, 15–16
Understory vegetation modeling
ecosystem dynamics models, 397–399
empirical models
abundance data, 395–396

Understory vegetation modeling (*Continued*)
binary data, 394-395
data requirements and limitations, 397
definition, 391-392
multivariate data, 397
techniques, 393-394
fire response, 403-405
ideal model, 400
importance, 391
large landscape predictions, 400-402
paradigms, 392-393
process-based models, 392
prospects, 405-407
qualitative models, 392, 399-400
Unreliable models, avoiding, 10-11
Upland sandpiper (*Bartramia longicauda*), 161, 167
Urbanization response modeling, *see* Bird conservation

V

Validation, landscape models
animal location data, 436-440
autocorrelation, 435-436
comparison of models, 418
data, 431-432, 434
definitions
calibration, 419
evaluation, 419
validation, 419
verification, 419
evaluation elements
application environment, 420, 422
conceptual design, 422-423
overview, 421
quantitative performance, 423-424
forest vegetation models
forest structure change validation, 427
modeled disturbance evaluation, 42-428
species change evaluation, 428-430
types, 424-426
metrics, 418-419, 430
model uncertainty, 433-434
objectives, 418
overview, 415-418
prospects, 440-442
spatial scale and validation, 434-435
VDDT/TELSA, 426-427
Verification, models, 15-16
Vesper sparrow (*Pooecetes gramineus*), 516
Viability, *see also* Population viability analysis
definition, 17
modeling in large landscapes, 19-20
primary viability, 18
secondary viability, 18
tertiary viability, 18-19
Violet-green swallow (*Tachycineta thalassina*), 659
Visualization, models
context, 23-24
human perception issues, 24
misrepresentation, 24-26
motion utilization, 22-23
overview, 20-22
Vulnerability, biodiversity conservation, 90

W

Warbling vireo (*Vireo gilvus*), 659
Weighted linear combination (WLC), habitat quality indexing, 230-231
Western burrowing owl (*Athene cubicaria hypugaea*), 517
Western hemlock (*Tsuga heterophylla*), 567
Western larch (*Larix occidentalis*), 568
Western spruce budworm (*Choristoneura occidentalis*), 577
Western tanager (*Piranga ludoviciana*), 659
Western white pine (*Pinus monticola*), 568
Western wood pewee (*Contopus sordidulus*), 659
White birch (*Betula papyrifera*), 485
White-crowned sparrow (*Zonotrichia leucophrys*), 659
White-headed woodpecker (*Picoides albolarvatus*), 522
White spruce (*Picea glauca*), 485
White-tailed deer (*Odocoileus virginianus*), 167, 234, 416, 612
White-tailed jackrabbit (*Lepus townsedii*), 517
WHQ, *see* Wildlife habitat quality
Wildlife habitat quality (WHQ), habitat-relationships modeling, 264, 268
Willow flycatcher (*Empidonax traillii*), 659
Wilson's warbler (*Wilsonia pusilla*), 659
Winter wren (*Troglodytes troglodytes*), 659
WLC, *see* Weighted linear combination

Wolverine (*Gulo gulo*), 153
Wood thrush (*Hyocichla mustelina*), 370
Woodland caribou (*Rangifer tarandus*), 239
Worm-eating warbler (*Helmitheros vermivorous*), 609–613
Wyoming ground squirrel (*Spermophilus elegan nevadensis*), 516

Y

Yellow-breasted chat (*Icteris virens*), 370
Yellow-rumped warbler (*Dendroica coronata*), 634, 640–641, 646, 659

Z

ZONATION program, 63, 65

COLOR PLATE 1-7

The image is from a New England hardwood stand generated using NE-twigs for the stand conditions and Visual Nature Studio 2 (<http://3dnature.com/>) for the image. This image is the work of Anna Lester (DeGraaf et al. 2005).

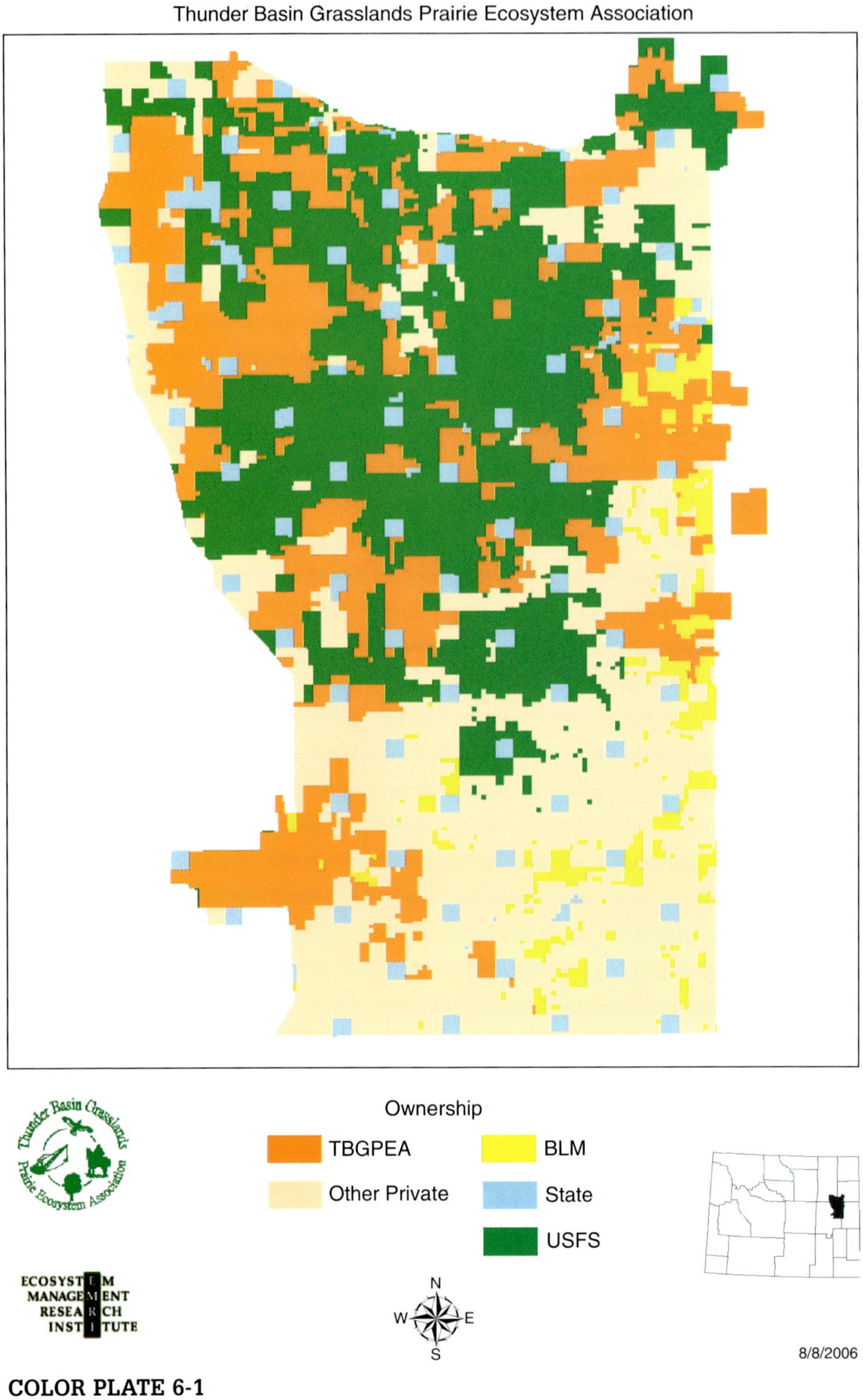

COLOR PLATE 6-1

The Thunder Basin Grasslands Prairie Ecosystem Association (TBGPEA) manages over 300,000 acres within a delineated 945,000-acre landscape in Converse, Campbell, Weston, and Niobrara Counties in eastern Wyoming, USA.

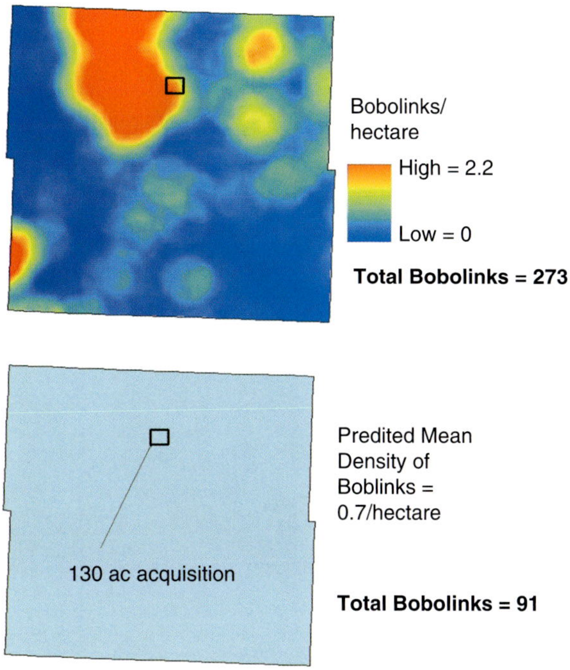

COLOR PLATE 8-4

The predicted relative abundance of bobolinks (*Dolichonyx oryzivorus*) in Grant County, Minnesota, assessed for 1 ha pixels (top) and at a county scale (bottom) and a hypothetical 52 ha grassland acquisition. County or large watershed-scale assessments make the unreasonable assumption that the entire area has homogeneous conservation potential.

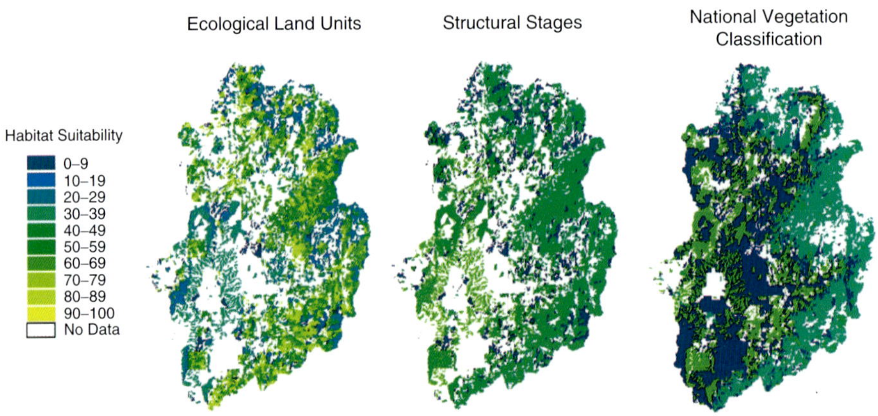

COLOR PLATE 11-3

Brown creeper habitat suitability scores on a scale of 0 (nonhabitat) to 100 (optimal) for three ecological classification systems: ecological land units (Haufler 1996), structural stages (Johnson et al. 1994) with Daubenmire plant series (sensu Daubenmire 1952), and the National Vegetation Classification (Grossman et al. 1998).

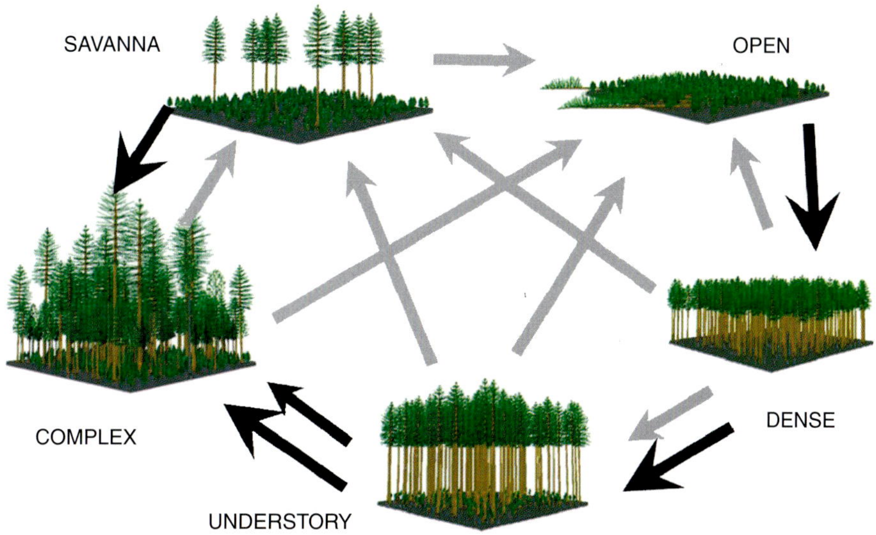

COLOR PLATE 13-1

Forests exist in many structures, such as the five-structure classification system shown here (Oliver and Larson 1996). The stands change constantly among structures through growth and disturbances. Each structure provides habitat for some species. *(Copyright, C. Oliver, Yale University, 2006).*

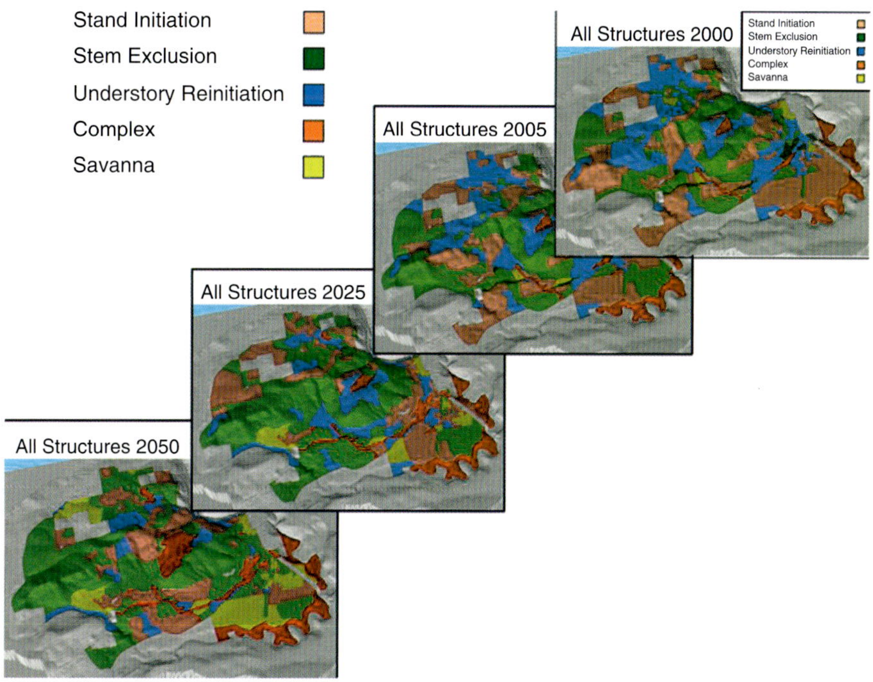

COLOR PLATE 13-5

The anticipated results of different attributes can be visually displayed across the landscape over time, as shown here for stand structures (Fig. 13-1) in the Pacific Northwest Pack Forest Plan, USA. *(Copyright, C. Oliver, Yale University, 2006).*

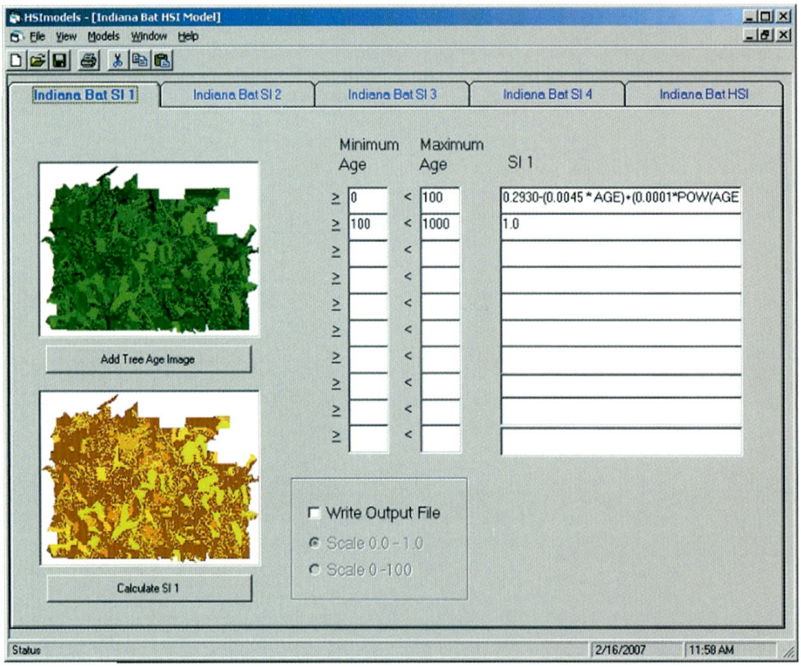

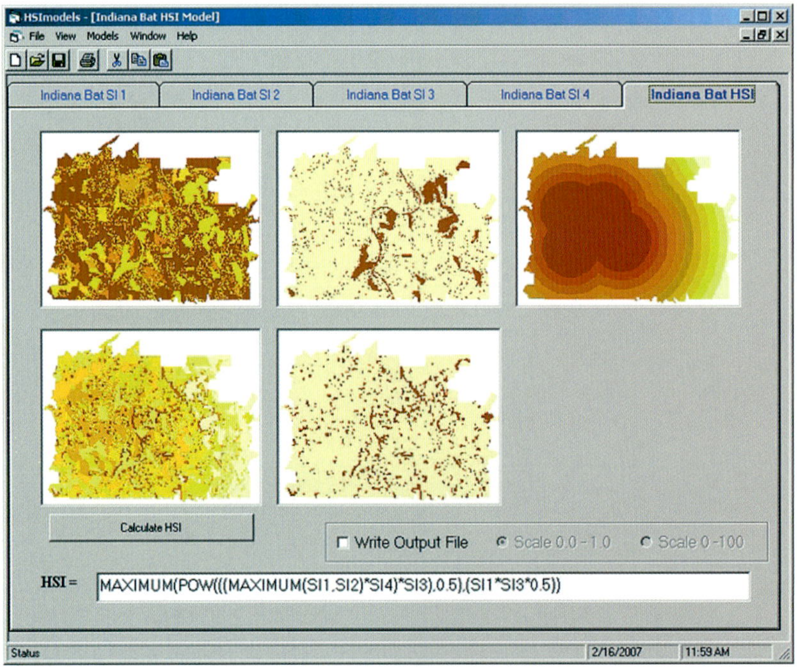

COLOR PLATE 14-3

Examples of the Indiana bat model in Landscape HSImodels software. The top window is a map of tree age (light green for young forest and dark green for older forest) and the resulting suitability index values ranging from light yellow (SI 1 = 0.0) to dark red (SI 1 = 1.0). The lower window is a map of the overall Habitat Suitability Index (lower, left map) and four suitability index maps.

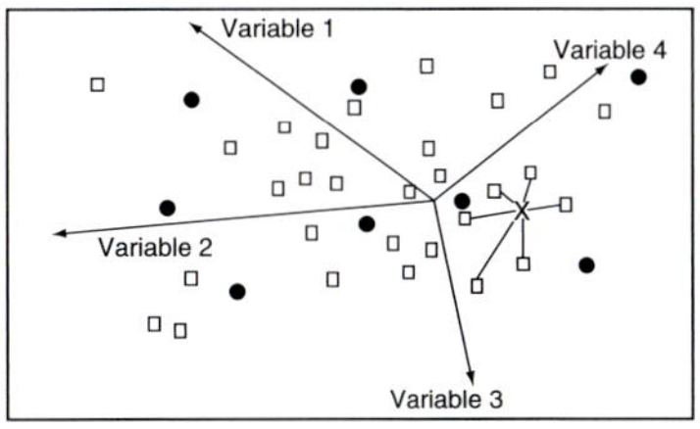

(1) Independent variables at a range of spatial scales are used for a nonlinear gradient model of niche structure for each species. This example uses CCA. Axes are orthogonal dimensions of ecological space. Species optima locate the center of their environmental niche with respect to the measured predictor variables.

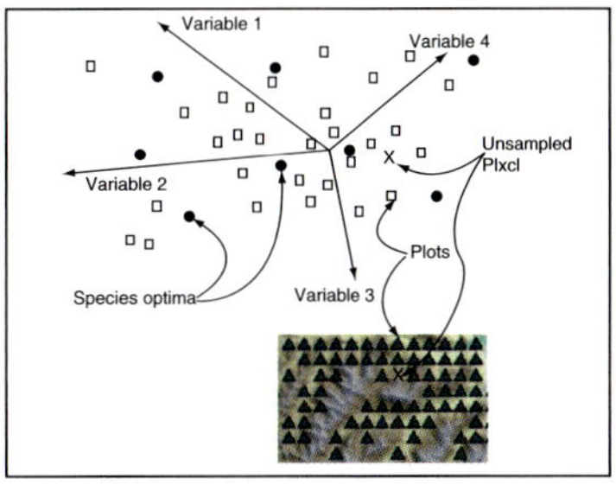

(2) The niche model uses ecological space, but we can make inferences about suitability in geographic space using *GNN*. Each unsampled location (cell) is projected iteratively into the environmental space defined by the niche model. By imputing the value of the species at the nearest sampled plot (or a weighted average of nearby plots) we estimate the suitability of the new location for that species.

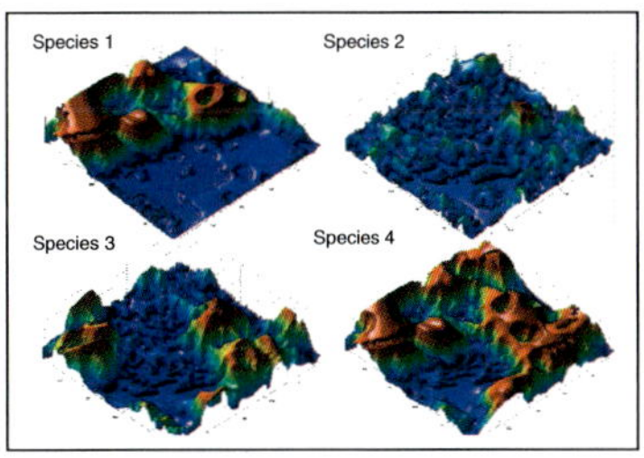

(3) Gradient imputation allows the explicit translation from quantitative models in environmental space to suitability maps in geographic space. Normalized values for multiple species at each location provide equilibrium estimates of relative competitive ability, useful for defining parameters in ecosystem dynamics (gap) models.

COLOR PLATE 15-4

Gradient imputation example using canonical correspondence analysis (ter Braak 1986) and gradient-nearest-neighbor (Ohmann and Gregory 2002). Adapted from Cushman et al. (2007).

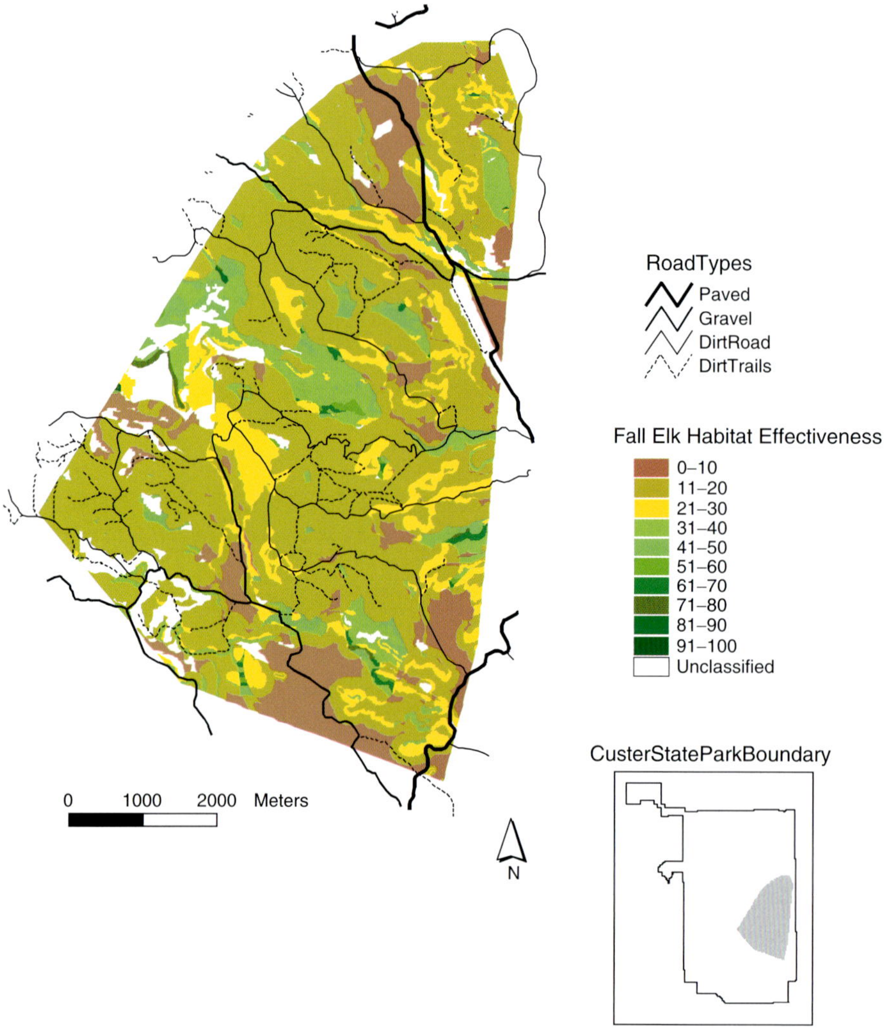

COLOR PLATE 16-5

Fall habitat quality scores for an elk subherd in Custer State Park, South Dakota, USA (from Roloff et al. 2001).

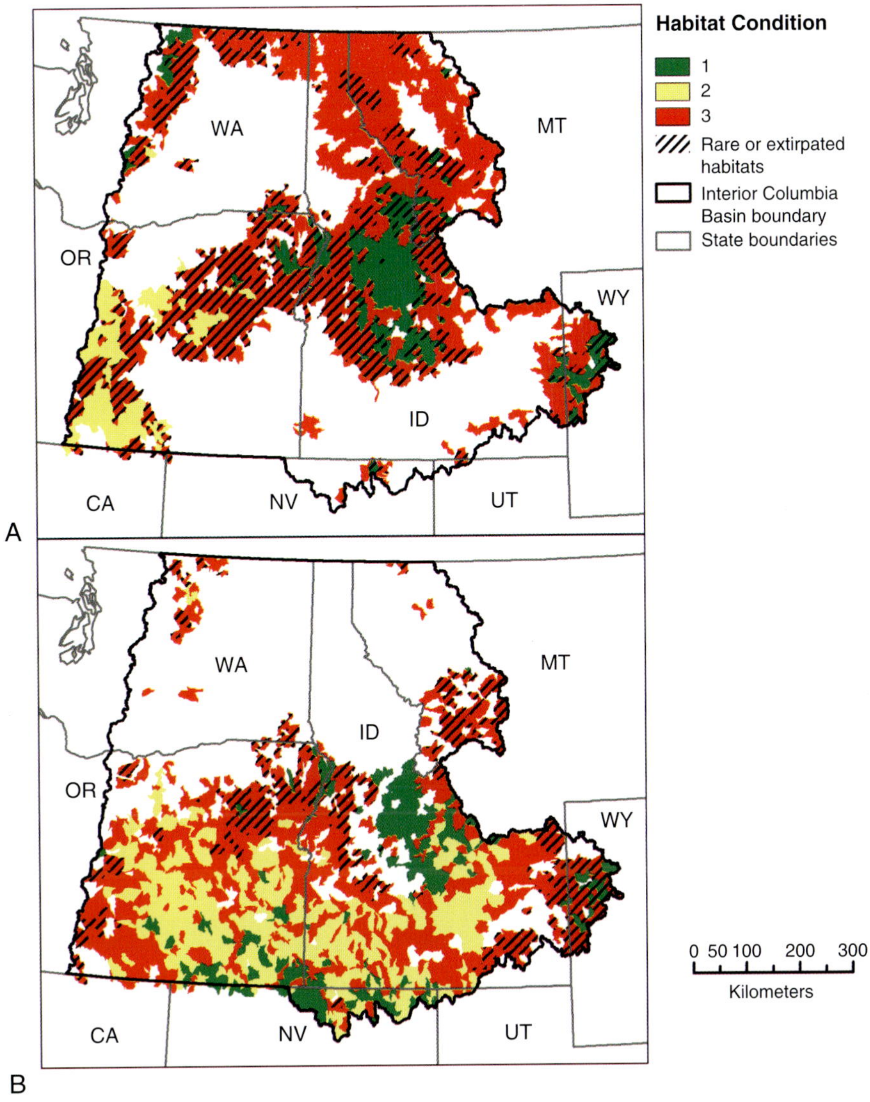

COLOR PLATE 19-2

Habitat network for Family 1 (old forest, low elevation; A) and Family 11 (sagebrush; B) in the Interior Columbia Basin, USA. See text for explanation of habitat condition classes and rare or extirpated habitats. Blank areas are watersheds that contain no public lands or are outside the range of the family. Adapted from Wisdom et al. (2002).

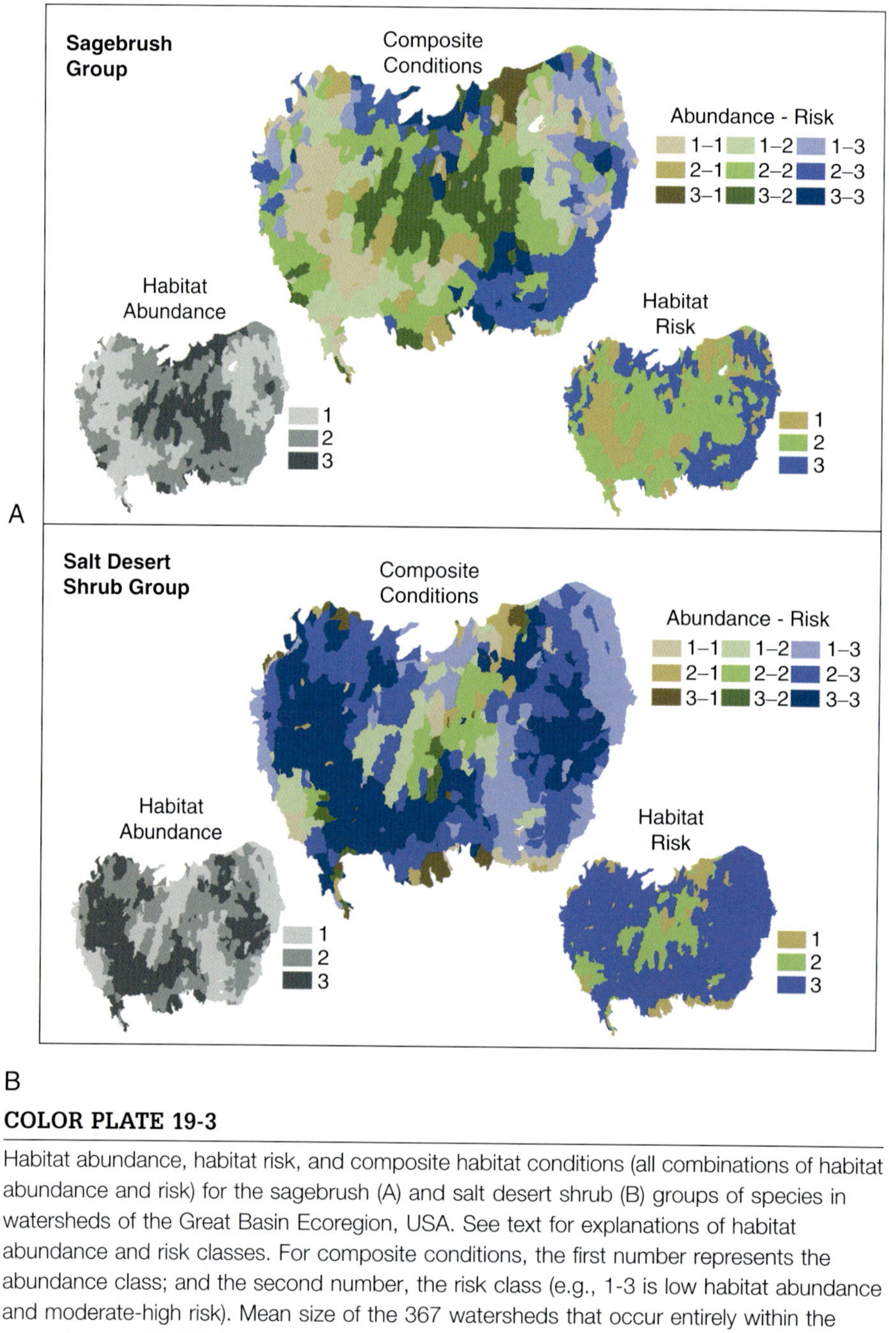

COLOR PLATE 19-3

Habitat abundance, habitat risk, and composite habitat conditions (all combinations of habitat abundance and risk) for the sagebrush (A) and salt desert shrub (B) groups of species in watersheds of the Great Basin Ecoregion, USA. See text for explanations of habitat abundance and risk classes. For composite conditions, the first number represents the abundance class; and the second number, the risk class (e.g., 1-3 is low habitat abundance and moderate-high risk). Mean size of the 367 watersheds that occur entirely within the ecoregion was 66,000 ha; n = 521 for all watersheds intersecting the ecoregion. Adapted from Wisdom et al. (2005*b*).

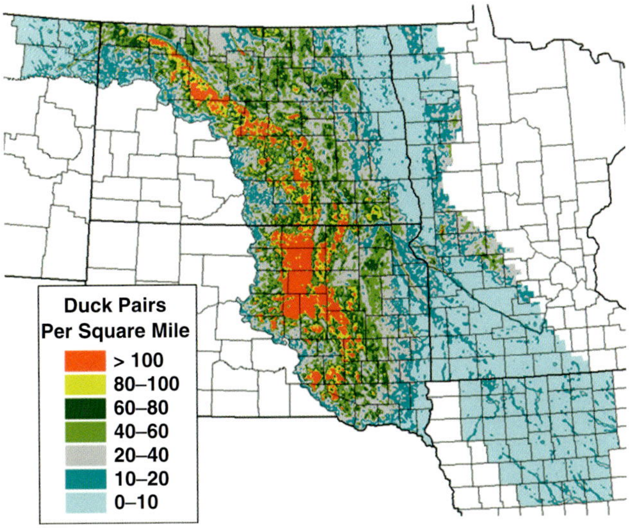

COLOR PLATE 20-4

Results of applying duck pair-wetland regression models to wetland basins, pair ratio models to riverine wetlands, and accessibility models to 390 × 390 m (15.2 ha) land units to the primary working area within the Prairie Pothole Region of the United States. This figure (referred to as the "Thunderstorm Map") shows which land units would be accessible to different densities of nesting hens and, thus, where grassland conservation efforts would provide the greatest benefits. Results are presented as pairs per km^2 for proximity zones around each land unit, where the area of proximity for each of five upland-nesting duck species was the approximate distance hens have been known to travel from core wetlands to nesting cover (derived from Reynolds et al. 2006; and R. R. Johnson, unpublished data).

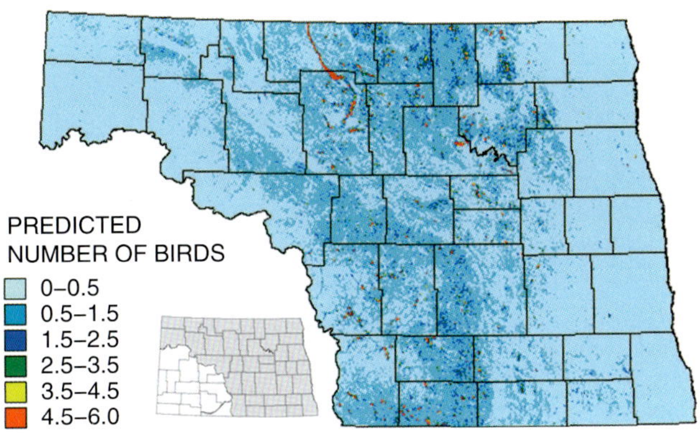

COLOR PLATE 20-6

Number of soras predicted to be present per 5 ha unit in the Prairie Potholes Bird Conservation Region portion of North Dakota, USA, in 1995 as a function of landscape-level spatial model. Low numbers relative to waterfowl model (Fig. 20-4) are due in part to different spatial scales used in model output.

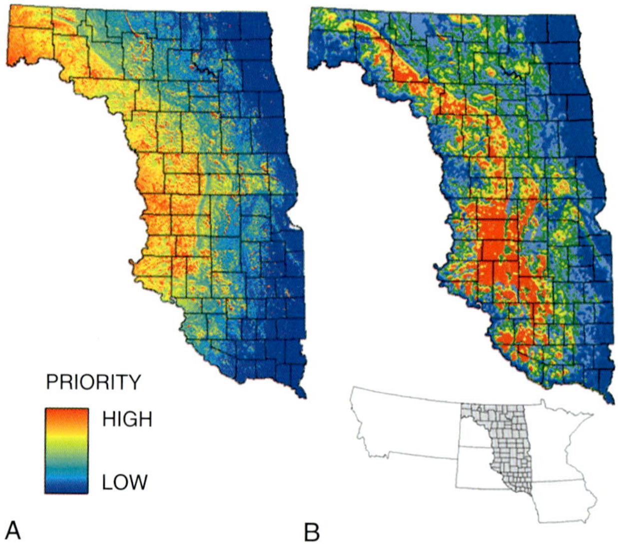

COLOR PLATE 20-7

(A) Predicted distribution of marbled godwit in the Prairie Pothole Region of North Dakota and South Dakota, USA, in 2004, where landscapes with high probability of detecting marbled godwits are given high priority. (B) Breeding duck "Thunderstorm Map" from Fig. 20-5 for the same geographic region and using same color scheme as (A); priority in this example was defined by number of five species of waterfowl predicted to have access to 390 × 390 m land units. Southern South Dakota is outside the range of marbled godwit; hence, predicted marbled godwit presence is low even though predicted duck numbers are high.

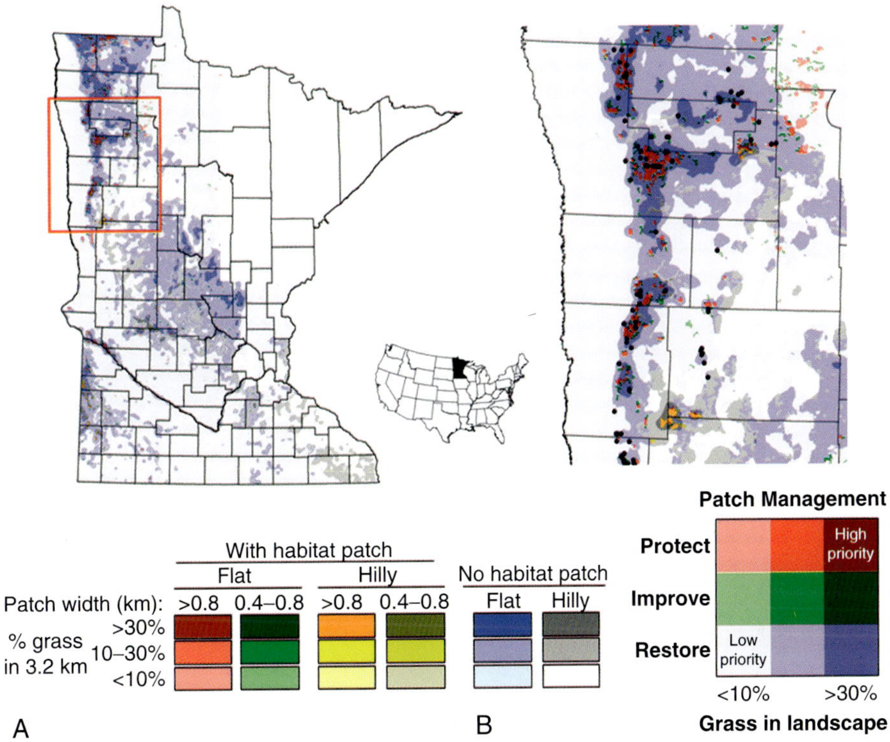

COLOR PLATE 20-8

(A) Conceptual model for marbled godwit habitat quality in Minnesota, USA, driven by expert knowledge. Dark red areas indicate the full complement of requirements are met and should be protected; purple areas would require restorations to meet species needs (i.e., patch development: increase grass patch size, wetland restoration within patch, etc.). Yellow and colorless areas should not be targeted for godwit management because of the godwit's preference for low-relief areas. (B) Portion of conceptual marbled godwit model in northwestern Minnesota showing location (●) of marbled godwits observed along *ad hoc* roadside survey.

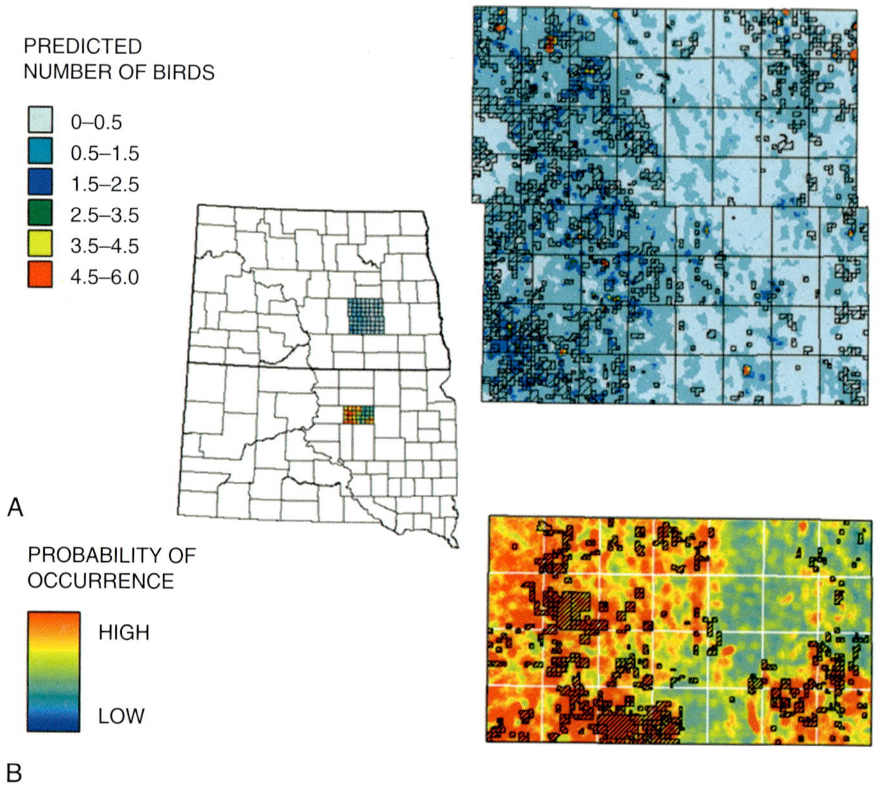

COLOR PLATE 20-9

(A) Waterfowl Production Areas and wetland easements (black hatching) overlaid on map of predicted number of soras in Stutsman County, North Dakota. (B) Waterfowl Production Areas and grassland easements (black hatching) overlaid on map of predicted probability of occurrence of marbled godwit in Faulk County, South Dakota. Gridlines in both maps indicate civil townships at 10 km intervals.

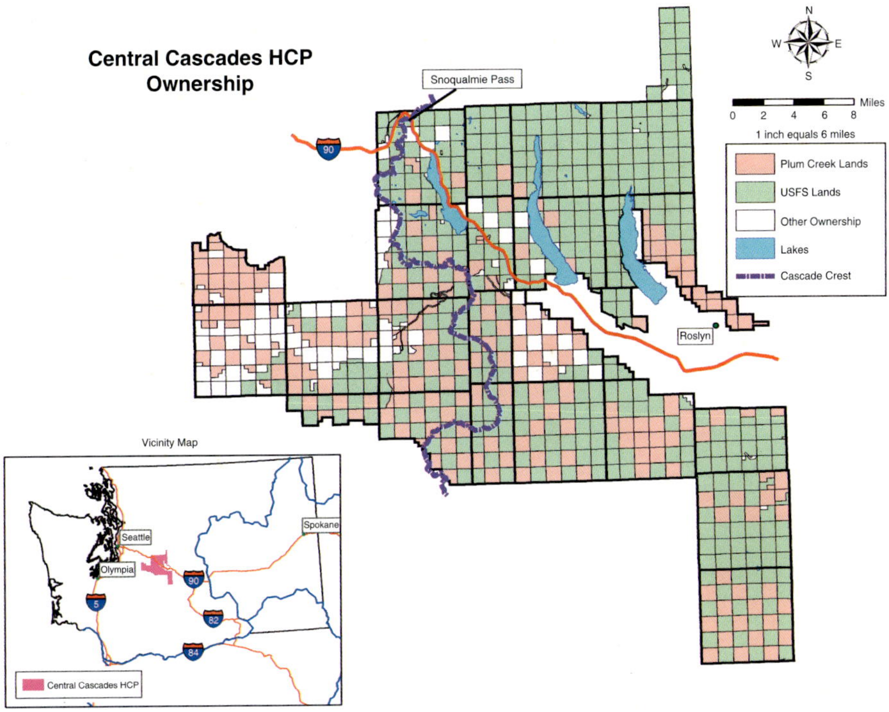

COLOR PLATE 21-2

Land ownership pattern in the 170,000 ha Cascades Habitat Conservation Plan Area, central Cascades Mountain Range, Washington, USA.

Spotted Owl Pair Nest Site Probability

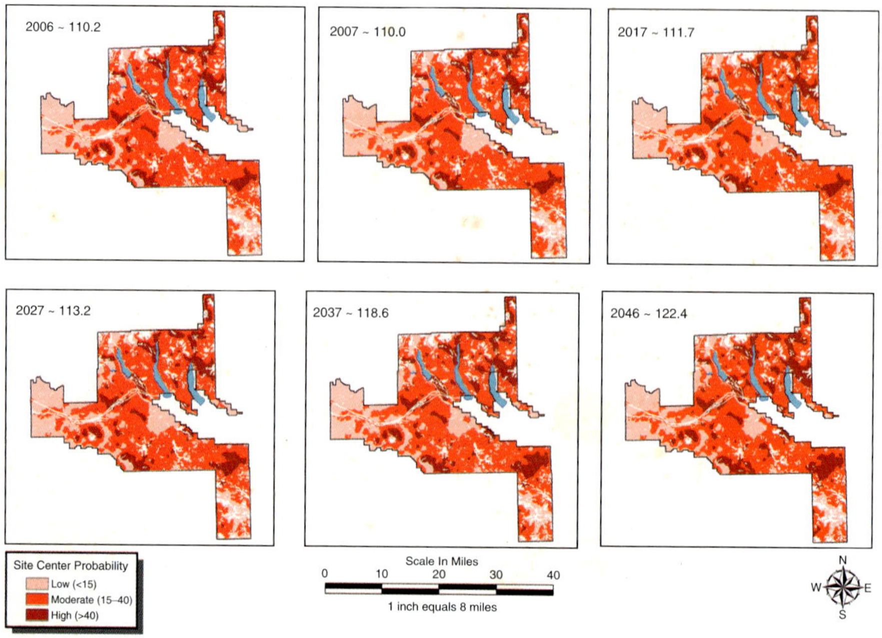

COLOR PLATE 21-5

Map of estimated values from the resource selection probability function (RSPF) for the Central Cascades Habitat Conservation Plan area, Washington, USA, for the present (2006) and as projected for 2045. The value of the RSPF for every pixel in the geographic information system was evaluated and plotted using a gradient from nonforested areas (white) to areas with the highest probabilities of spotted owl use (maroon). Map headings are the year, followed by the estimated carrying capacity of owl pairs in that year.